Semiconductors for Room Temperature Nuclear Detector Applications

SEMICONDUCTORS
AND SEMIMETALS
Volume 43

Semiconductors and Semimetals

A Treatise

Edited by R. K. Willardson
CONSULTING PHYSICIST
SPOKANE, WASHINGTON

Albert C. Beer
CONSULTING PHYSICIST
COLUMBUS, OHIO

Eicke R. Weber
DEPARTMENT OF MATERIALS SCIENCE
AND MINERAL ENGINEERING
UNIVERSITY OF CALIFORNIA AT
BERKELEY

Semiconductors for Room Temperature Nuclear Detector Applications

SEMICONDUCTORS
AND SEMIMETALS

Volume 43

Volume Editors

T. E. SCHLESINGER

DEPARTMENT OF ELECTRICAL AND COMPUTER ENGINEERING
CARNEGIE MELLON UNIVERSITY
PITTSBURGH, PENNSYLVANIA

RALPH B. JAMES

ADVANCED MATERIALS RESEARCH DEPARTMENT
SANDIA NATIONAL LABORATORIES
LIVERMORE, CALIFORNIA

ACADEMIC PRESS
San Diego   Boston   New York
London   Sydney   Tokyo   Toronto

This book is printed on acid-free paper. ∞

Copyright © 1995 by ACADEMIC PRESS, INC.

All Rights Reserved.
No part of this publication may be reproduced or transmitted in any form or by any means, electronic or mechanical, including photocopy, recording, or any information storage and retrieval system, without permission in writing from the publisher.

Academic Press, Inc.
A Division of Harcourt Brace & Company
525 B Street, Suite 1900, San Diego, California 92101-4495

United Kingdom Edition published by
Academic Press Limited
24-28 Oval Road, London NW1 7DX

International Standard Serial Number: 0080-8784

International Standard Book Number: 0-12-752143-7

PRINTED IN THE UNITED STATES OF AMERICA
95 96 97 98 99 00 BC 9 8 7 6 5 4 3 2 1

Contents

LIST OF CONTRIBUTORS xiii
PREFACE xv

Chapter 1 Introduction and Overview
T. E. Schlesinger and R. B. James

I. Introduction 1
II. Semiconductor Nuclear Detectors 2
    1. Types of Radiation 2
    2. Detector Operation 3
    3. Practical Detectors 10
III. Applications 11
    1. National Security 11
    2. Commercial 12
    3. Medical 14
    4. Environmental Remediation and Safety 16
    5. Space Applications and Basic Science 18
IV. Outline of Text 19
    References 20

Chapter 2 High-Purity Germanium Detectors
Larry S. Darken and Christopher E. Cox

I. Introduction 23
II. Crystal Growth 25
    1. Crystal Growth Technique 26
    2. Purification at the Liquid–Solid Interface 27
    3. Purification by Oxidation 28
III. Characterization 31
    1. Carrier Transport Techniques 32
    2. Capacitance Transient Techniques 35
    3. Crystallographic Techniques 38
    4. Neutral Defects 38
IV. Large Volume Detectors 38
    1. Introduction 38
    2. Geometry and Size 40
    3. Closed End Coaxial Detector Fabrication 41
    4. Closed End Coaxial Detector Performance 47

V.	Charge Collection	51
	1. Geometric Effects	52
	2. Trapping by Extended Defects	54
	3. Trapping by Point Defects	55
	4. Modeling Charge Collection in Spectrometers	60
	5. Radiation Damage	64
VI.	Germanium X-Ray Detectors	69
	1. Introduction	69
	2. Detector Fabrication	70
	3. Detector Housing	71
	4. Development of the Detector Entrance Window	72
	5. Performance at Low Energies, down to 200 eV	73
	6. Performance at High Energies, up to 100 keV	76
	7. Comparison with Si (Li) Detectors	76
VII.	Summary	79
	References	80

Chapter 3 Growth of Mercuric Iodide

A. Burger, D. Nason, L. van den Berg, and M. Schieber

I.	The Crystal Structure and Phases of Mercuric Iodide	86
	1. Crystal Structure	86
	2. Phase Diagram	87
II.	Physical Properties Relevant to Crystal Growth	87
	1. Vapor Pressure	88
	2. Stoichiometry	88
	3. Thermal Properties	89
III.	Growth of High Purity Mercuric Iodide Crystals	91
	1. Purification of Starting Materials	91
	2. Physical Vapor Transport Growth	93
	3. Solution Growth	98
IV.	Crystal Perfection	98
	1. Microscopy	98
	2. Gamma Ray and X-Ray Diffraction Studies	99
V.	Recent Developments	102
	1. Seeded Growth	102
	2. *In-Situ* Process Monitoring	103
	3. Fundamental Studies	104
VI.	Challenges in Crystal Growth	106
	1. Impurity Analysis	106
	2. Stoichiometry	107
	3. Bulk Growth	107
	4. Film Growth	107
	References	108

Chapter 4 Electrical Properties of Mercuric Iodide

X. J. Bao, T. E. Schlesinger, and R. B. James

	LIST OF SYMBOLS	111
I.	Introduction	112
II.	Carrier Transport	113

	1. Dark Resistivity and I–V Characteristics	113
	2. Mobility and Lifetime	119
	3. Effective Mass	131
III.	Deep Levels	132
	1. Thermally Stimulated Current Technique	133
	2. Other Techniques	144
IV.	Photoconductivity	148
	1. Behavior of Photocurrent	149
	2. Photocurrent Quenching	153
V.	Surface Effects	154
	1. Electrical Contacts	154
	2. Surface Recombination	156
VI.	Detector Performance	158
	1. Peak-to-Background Ratio	158
	2. Polarization Effect	160
VII.	Conclusions	165
	References	165

Chapter 5 Optical Properties of Red Mercuric Iodide

X. J. Bao, R. B. James, and T. E. Schlesinger

I.	Introduction	169
II.	Band Structure	170
III.	Experimental Techniques and Measured Values for Optical Constants	177
	1. Measurement of Bandgap and Shift with Temperature	177
	2. Absorption	183
	3. Ellipsometry Measurements of Optical Constants	185
	4. Optical Properties near and below the Bandgap	187
	5. Phonon Structure of Red Mercuric Iodide	192
	6. Radiative Recombination of Nonequilibrium Electron–Hole Pairs	195
	7. Phonon-Assisted Electron–Hole Pair Photoluminescence	202
	8. Effects of Geometrical Configuration on Photoluminescence Spectra	202
	9. Temperature Dependence of Photoluminescence Spectrum	204
IV.	Study of Processing by Photoluminescence Spectroscopy	205
	1. Purification and Stoichiometry	205
	2. Etching and Vacuum Exposure	207
	3. Contacts	210
	4. Detector Performance	214
V.	Conclusions	216
	References	216

Chapter 6 Growth Methods of CdTe Nuclear Detector Materials

Makram Hage-Ali and Paul Siffert

I.	Introduction	219
II.	Phase Diagram	220
	1. $T(x)$ Projection	220
	2. $P(T)$ Projection	221
	3. Field of Existence	225
III.	Synthesis and Purification	229
	1. Purification of Elements	229

		2. Tube Graphitization	231
		3. Synthesis	231
IV.	Growth of Bulk CdTe		232
	A.	Stoichiometric Growth Methods: Congruent or Near-Congruent Melts	233
		1. Zone Melting	233
		2. Bridgman Methods	234
	B.	Solvent Growth Methods	239
		1. Te-Rich Solution Bridgman	239
		2. Traveling Heater Method	240
		3. Direct THM Synthesis and Growth	244
V.	High Resistivity Materials		245
		1. High Purity Materials	245
		2. Compensation	246
VI.	Experimental Results and Conclusion		254
	References		255

Chapter 7 Characterization of CdTe Nuclear Detector Materials

Makram Hage-Ali and Paul Siffert

I.	Introduction	259
II.	Impurities Analysis	260
	1. Spark Mass Spectrography Analysis	260
	2. Atomic Absorption	261
	3. Nuclear Activation	261
	4. X-Ray Fluorescence	263
	5. Secondary Ion Mass Spectrometry	265
	6. Ion Chromatography–IR Absorption	265
III.	Surface Analysis	266
	1. Lapped and Polished Surfaces	267
	2. Chemically Etched Surfaces	268
	3. Oxidized Surfaces	270
	4. Interfaces and Contact Analysis	270
IV.	Electrical and Optical Characterization	277
	1. Resistivity	277
	2. Mobility, Lifetime, and $\mu\tau$ Product	279
	3. Photoluminescence Analysis	281
	4. ODMR and Related Methods	282
	5. TSC and PICTS Measurements	283
V.	Discussion and Conclusions	286
	References	287

Chapter 8 CdTe Nuclear Detectors and Applications

Makram Hage-Ali and Paul Siffert

I.	Introduction	291
II.	Detection Parameters	292
III.	CdTe Detectors	293
	1. Starting Materials	293
	2. Detector Devices	294
	3. Electrical Characteristics	296
	4. Main Detector Properties	305

IV. Improvement of Detector Quality 316
  1. New Growth Process Materials 316
  2. Single Type Carrier Collection 317
  3. (n-i-p), (M-π-n) Structures and Cooling 320
  4. Electronic Treatment 322
V. Applications of CdTe Detectors 326
  1. Spectrometers 327
  2. Safeguard Systems 327
  3. Dosimetry 328
  4. Space and Astrophysics 328
  5. Medical Applications 328
  6. Industrial Applications 330
  References 331

Chapter 9 $Cd_{1-x}Zn_xTe$ Spectrometers for Gamma and X-Ray Applications

R. B. James, T. E. Schlesinger, J. C. Lund, and Michael Schieber

I. Introduction 336
II. Growth of $Cd_{1-x}Zn_xTe$ Crystals 337
III. Material Properties of $Cd_{1-x}Zn_xTe$ 339
  1. Resistivity 340
  2. Alloy Composition 341
  3. Photoluminescence Spectrum 342
  4. Charge Transport 342
  5. Absorption Coefficient for X-Rays and Gamma Rays 345
IV. Defect Characterization and Effects on Device Response 346
  1. Etch Pit Densities 346
  2. X-Ray Rocking Curves 346
  3. Precipitates 348
  4. Impurities 349
V. Detector Characterization 350
  1. Detector Fabrication 350
  2. Nuclear Spectroscopic Data at Room Temperature 350
  3. Detector Current–Voltage Characteristics 353
  4. Large-Volume Gamma Ray Detectors 355
  5. p-i-n Gamma Ray Detectors 362
  6. X-Ray Detector Response 364
  7. Detector Polarization 368
  8. Temperature Dependence 369
  9. Pulse Risetime Discrimination and Compensation 372
VI. Imaging Applications 375
VII. Future Work 378
  References 378

Chapter 10 Gallium Arsenide Radiation Detectors and Spectrometers

D. S. McGregor and J. E. Kammeraad

LIST OF SYMBOLS 384
I. Introduction 386
II. Basic Properties of GaAs 391

	1. Band Structure, Effective Mass, Density of States, and Intrinsic Carrier Concentration	391
	2. Mobility and Velocity	394
	3. Charge Carrier Lifetimes	395
	4. Ionization Energy and the Fano Factor	399
	5. Techniques of Material Growth	400
	6. Compensation in Bulk GaAs	401
III.	General Detector Operation	403
IV.	Epitaxial GaAs Detectors	407
	1. Detector Configurations	408
	2. Detector Performance	409
	3. Discussion	413
V.	Bulk GaAs Detectors Operated in Quantum Pulse Mode	414
	1. Detector Configurations	414
	2. I–V Characteristics	416
	3. C–V Characteristics	418
	4. Active Region Measurements	420
	5. Radiation Measurements	425
	6. Proposed Models for Observed Behavior	427
	7. Discussion	431
VI.	Bulk GaAs Photoconductive Detectors Operated in Current Mode	432
VII.	Summary	437
	References	437

Chapter 11   Lead Iodide Crystals and Detectors

J. C. Lund, F. Olschner, and A. Burger

I.	Introduction	444
II.	Physical Properties	444
	1. Crystal Structure and Lattice Properties	444
	2. Semiconducting Properties	445
III.	Preparation of Lead Iodide Crystals	445
	1. Phase Behavior	446
	2. Purification	446
	3. Crystal Growth	448
	4. Summary of Crystal Growth Preparation	451
IV.	Radiation Detector Fabrication and Implementation	451
	1. Detector Fabrication	451
	2. Electronic Readout	453
	3. Radiation Testing	456
V.	Potential Applications of Lead Iodide	459
	1. X-Ray Spectrometers	459
	2. Gamma Ray Detectors	459
	3. Flux Detectors	462
VI.	Conclusion	463
	1. Summary	463
	2. Future Research Directions	463
	References	463

Chapter 12 Other Materials: Status and Prospects
Michael R. Squillante and Kanai S. Shah

I. Introduction	465
II. Detector Materials	467
1. Overview	467
2. Fundamentals of Crystal and Device Preparation	468
3. III–V Materials	469
4. II–VI Semiconductors	473
5. Thallium Bromide	475
6. Amorphous Silicon	477
7. Ternary Materials	478
8. Other, Less Studied, Crystalline Materials	481
9. Other, Less Studied, Thin Film Materials	482
III. Current Status and Prospects	484
1. Comparison of Material Properties	484
2. Future Directions	486
3. Summary and Conclusions	487
References	487

Chapter 13 Characterization and Quantification of Detector Performance
Vernon M. Gerrish

I. Introduction	493
II. X-Ray and Gamma Ray Spectroscopy	496
1. Interaction of X-Rays and Gamma Rays with Matter	496
2. Detector Response Function	498
3. Charge Collection	499
4. Electronics	502
5. Detector Performance	503
6. Polarization	510
7. Radiation Damage Resistance	513
III. Electronic Characterization	513
1. Bulk Measurements	513
2. Contacts	520
IV. Correlation of Material Properties with Detector Performance	524
V. Concluding Remarks	527
References	527

Chapter 14 Electronics for X-Ray and Gamma Ray Spectrometers
Jan S. Iwanczyk and Bradley E. Patt

I. Introduction	531
II. Electronic Noise Limited Systems	534
1. The Charge-Sensitive Preamplifier and Sources of Electronic Noise	536
2. Configurations for Low Noise Preamplifiers	543
3. Low Noise FET Structures	547
III. Statistical Noise Limited Systems	548

IV.	Trapping Noise Limited Systems	549
	1. Specialized Electronics Accomodating Long Shaping Times	552
	2. Single Carrier Techniques	555
	3. Charge Deficit Correction	557
V.	Miniaturized Electronics and Multielement Systems	558
	References	559

Chapter 15 Summary and Remaining Issues for Room Temperature Radiation Spectrometers

Michael Schieber, R. B. James, and T. E. Schlesinger

I.	Introduction	561
II.	Materials Requirements	562
III.	Issues in HgI_2 Detector Technology	563
	1. Precursors and Starting Materials	564
	2. Purification and Crystal Growth	566
	3. Device Fabrication	569
	4. Nuclear Spectroscopic Results	574
IV.	Materials Issues in CdTe and CdZnTe	575
	1. Purification, Precursors, and Growth of CdTe and CdZnTe	575
	2. Device Fabrication	578
	3. Nuclear Spectroscopic Data	579
V.	Unresolved Problems and Conclusions	580
	References	581
Index		585
Contents of Volumes in This Series		595

Contributors

Numbers in parentheses indicate the pages on which the authors' contributions begin.

X. J. BAO (111, 169), *Department of Analytical Instruments, TN Technologies Inc., Round Rock, Texas 78664*

A. BURGER (85, 443), *NASA Center for Photonic Materials and Devices, Department of Physics, Fisk University, Nashville, Tennessee 37208*

CHRISTOPHER E. COX (23) *Nuclear Measurements Group, Oxford Instruments, Oak Ridge, Tennessee 37831*

LARRY S. DARKEN (23) *Nuclear Measurements Group, Oxford Instruments, Oak Ridge, Tennessee 37831*

VERNON M. GERRISH[1] (493) *EG&G Energy Measurements Inc., Santa Barbara Operations, Goleta, California 93117*

MAKRAM HAGE-ALI (219, 259, 291) *Centre de Recherches Nucléaires, Laboratoire PHASE, 67037 Strasbourg, France*

JAN S. IWANCZYK (531) *Xsirius, Inc., Camarillo, California 93012*

R. B. JAMES (1, 111, 169, 335, 561) *Advanced Materials Research Department, Sandia National Laboratories, Livermore, California 94550*

J. E. KAMMERAAD (383) *Lawrence Livermore National Laboratory, Livermore, California 94551*

J. C. LUND (335, 443) *Advanced Materials Research Department, Sandia National Laboratories, Livermore, California 94550*

D. S. MCGREGOR (383) *Advanced Materials Research Department, Sandia National Laboratories, Livermore, California 94550*

D. NASON (85) *EG&G Energy Measurements Inc., Santa Barbara Operations, Santa Barbara, California 93111*

F. OLSCHNER (443) *Radiation Monitoring Devices, Inc., Watertown, Massachusetts 02172*

BRADLEY E. PATT (531) *Xsirius, Inc., Camarillo, California 93012*

MICHAEL SCHIEBER (85, 335, 561) *School of Applied Science and Technology, Hebrew University, Jerusalem 91904, Israel*

T. E. SCHLESINGER (1, 111, 169, 335, 561) *Department of Electrical and Computer Engineering, Carnegie Mellon University, Pittsburgh, Pennsylvania 15213*

[1]Present address: Constellation Technology Corporation, St. Petersburg, Florida 33702.

KANAI S. SHAH (465) *Radiation Monitoring Devices, Inc., Watertown, Massachusetts 02154*

PAUL SIFFERT (219, 259, 291) *Centre de Recherches Nucléaires, Laboratoire PHASE, 67037 Strasbourg, France*

MICHAEL R. SQUILLANTE (465) *Radiation Monitoring Devices, Inc., Watertown, Massachusetts 02154*

L. VAN DEN BERG (85) *EG&G Instruments Inc., Oak Ridge, Tennessee 37830*

Preface

The ability to detect and perform energy-dispersive spectroscopy of X-rays and gamma rays is of great importance since it makes possible a wide variety of analysis and imaging techniques. X-ray fluorescence spectroscopy, for example, is a powerful tool often employed for atomic analysis of materials and is very familiar to scientists and engineers working in many fields. However, the use of this technique has generally been limited to the laboratory since, in the past, the detectors and spectrometers that offered high resolution spectroscopic capabilities were limited to operation at cyrogenic temperatures (77K). These devices included lithium-drifted silicon or germanium and more recently high purity germanium. Thus, the widespread application and use of these and other spectroscopic techniques have been limited by the need for a cumbersome cooling apparatus and the constant attention these systems require. Size alone often impedes the use of these systems in some applications.

In recent years, however, the technology of X-ray and gamma ray detectors and spectrometers operating at room temperature has matured to the point that it is poised to impact areas far beyond those associated with basic scientific applications. The reason for this is that the ability to grow a number of semiconductor materials with the properties required for high performance spectrometers has been developed. While some of the requirements demanded of the semiconductors are similar to those necessary for other types of electronic or optical devices, there are also significant differences in the material properties necessary to fabricate high resolution spectrometers. Thus a complementary effort has been undertaken, over the years, to produce the materials upon which this technology is based. The requirements include large volumes of high atomic number, high resistivity, highly uniform, and stoichiometric material with a minimum of structural and chemical defects. At the same time, the development of pulse detection and pulse processing electronics has also been advanced to the point that many materials limitations can be minimized. These advances have allowed for the production of systems operating at room temperature that, in many cases, can rival the performance of cryogenically cooled detectors in terms of energy resolution, signal-to-noise ratio, collection efficiency, and sensitivity. With the need for expensive and cumbersome cooling apparatus eliminated, these systems can be compact, even hand-held, operating unattended for long periods of time, and may be produced in large quantities at reasonable cost. General purpose commercial systems capable of performing X-ray fluorescence spectroscopy over a wide range of elements are already available, as are less expensive systems for the identification of particular elements such as lead. These systems are portable and can be used in the field in a variety of environments. Spectrometers that can aid in the detection and identification of nuclear materials and enhance personal and environmental safety or aid in international treaty verification are already being employed. Practical imaging systems operating at X-ray or gamma ray energies are a few years away from realization with prototype systems already having been demonstrated. These systems offer the potential of eliminating the use of medical X-ray film or may offer high resolution images of scenes at a variety of X-ray or gamma ray photon energies.

The importance of this technology for fields as diverse as medicine, national security and treaty verification, industrial process monitoring, environmental remediation, and safety, as well as the traditional applications in basic science and space exploration, should not be underestimated. As this technology finds its way increasingly into these fields, it is clear that a text which brings together and summarizes some of the most important aspects of this technology will be of use to a wide audience. Thus, we have attempted to bring together a comprehensive overview of the current state-of-the-art in room temperature X-ray and gamma ray spectrometers with an emphasis on the materials aspects of this technology. Chapter 1 offers a brief explanation of the principles of operation of semiconductor spectrometers along with a discussion of some of the applications of these devices in the areas mentioned above. The text then provides detailed discussions of the growth, characterization, and device performance associated with the most important materials currently being employed, namely, mercuric iodide and cadmium telluride. Significant discussions of other important materials including cadmium zinc telluride, lead iodide, and gallium arsenide are also provided, and we have also included discussions of other promising, though less well studied, materials. The chapter on high purity germanium is intended to provide an overview of the state-of-the-art in cooled spectrometer technology and thus establish a point of reference for the room temperature devices. Finally, while the focus of this text is on the materials issues, the discussions of detector testing and electronics is meant to provide an introduction to these aspects of this technology. We hope that scientists and engineers who are already expert in this field will find this text to be a useful reference, while those just becoming familiar with the field will be able to use this book as a starting point for their reading of the literature.

The success in the development of X-ray and gamma ray spectrometers operating at room temperature is based on many years of effort on the part of large numbers of workers around the world. These individuals have contributed to the understanding of the fundamental materials issues associated with the growth of semiconductors for this application, the development of device fabrication and processing technology, and advances in low noise electronics and pulse processing. Progress in this field continues at an accelerated pace, as is evidenced by the improvements in detector performance and by the growing number of commercial products. The authors of the various chapters are individuals who have been and continue to be active in these areas and without their contributions this text would not have been possible. Their time and effort in writing their respective chapters are greatly appreciated by the editors of this volume, and we thank them for their tremendous help in putting together this book.

T. E. Schlesinger
R. B. James

CHAPTER 1

Introduction and Overview

T. E. Schlesinger

DEPARTMENT OF ELECTRICAL AND COMPUTER ENGINEERING
CARNEGIE MELLON UNIVERSITY
PITTSBURGH, PENNSYLVANIA

R. B. James

ADVANCED MATERIALS RESEARCH DEPARTMENT
SANDIA NATIONAL LABORATORIES
LIVERMORE, CALIFORNIA

I. INTRODUCTION	1
II. SEMICONDUCTOR NUCLEAR DETECTORS	2
1. *Types of Radiation*	2
2. *Detector Operation*	3
3. *Practical Detectors*	10
III. APPLICATIONS	11
1. *National Security*	11
2. *Commercial*	12
3. *Medical*	14
4. *Environmental Remediation and Safety*	16
5. *Space Applications and Basic Science*	18
IV. OUTLINE OF TEXT	19
References	20

I. Introduction

This book discusses the properties of semiconductors that are being employed for the fabrication of room temperature nuclear detectors and spectrometers. Nuclear radiation detection and spectroscopy, and particularly semiconductor detector systems that can be operated at room temperature, are finding increasing applications in fields as diverse as national security, medicine, industrial process monitoring, astronomy, high energy physics, radioactive waste management, environmental remediation, and elemental analysis of materials. Today, the best performance in terms of efficiency and energy resolution is achieved by radiation spectrometers fabricated from semiconducting materials. The leading candidate materials for the fabrication of these room temperature devices are mercuric iodide (HgI_2), cadmium telluride (CdTe), and cadmium zinc telluride (CZT). However, a variety of other materials also show promise for the fabrication of room temperature detectors, and they too are discussed. To fully appreciate the issues that drive the research in this field it is important that the reader have some understanding of the nature of the interaction of nuclear radiation with semiconducting materials and the material properties required to detect radiation with high sensitivity and energy resolution. It is also useful to understand the operation of more traditional solid state detectors such as lithium drifted silicon or germanium, Si(Li) or Ge(Li), and high purity germanium as well as some of the considerations that go into the design of the pulse detection electronics. These topics are also given consideration within this text. Some familiarity with the present and potential applications for these detectors is important since this will allow the reader to understand the deficiencies of Si and Ge at room temperature and more fully appreciate the advantages of detectors that are operable at higher temperatures.

This introductory chapter will serve as a brief and general overview of the types of radiation encountered in practice and some of the applications of radiation detectors. The discussion presented here is by no means comprehensive and the reader is referred to a more detailed review of radiation sources and detection which can be found in the text by Knoll (1989). The introductory comments do not focus on a particular detector material, thus specific issues related to particular semiconductor materials are reviewed in the following chapters of this text. Following the general comments on radiation detectors and applications we describe briefly the subsequent chapters in hopes that this will serve as a guide for the reader who may not wish to cover the text in its entirety.

II. Semiconductor Nuclear Detectors

1. TYPES OF RADIATION

Radiation may be broadly divided into two categories: uncharged radiations, which include x-rays, gamma rays, and neutrons; and charged radiations, including beta particles (electrons and positrons), protons, alpha particles, and fission fragments. There are numerous processes by which these radiations can be produced. Electrons, for example, may be produced when a radioisotope decays via the emission of a beta particle, when an excited nuclear state relaxes to its ground state via the ionization of an orbital electron (conversion electron), or when an excited atomic state relaxes through the emission of an orbital electron (Auger electron). The spontaneous decay of a heavy nucleus may produce alpha particles or fission fragments. Such a decay process can also produce neutrons. Neutrons may also be produced through nuclear reactions with either gamma ray photons, alpha particles, or fusion events. Gamma rays may be emitted by an excited nucleus when it relaxes to a lower energy state or from the annihilation of electrons and positrons. Both these processes typically produce gamma rays of discrete energies. X-rays may be produced when fast electrons transfer their kinetic energy to photons (bremsstrahlung) through their interaction with the medium in which they are moving. This produces a broad continuous energy spectrum. On the other hand, x-rays produced when an excited orbital electron of an atom relaxes to a lower energy state are of discrete energies (characteristic of the atom on which the transition is taking place). Energies of x-ray photons are typically in the range of 1 keV to about 100 keV, while gamma photon energies will range from about 100 keV to 10 MeV. Alpha particles are limited in energy from about 3 to 7 MeV. Neutron energies that average around 0.025 eV are labeled as "slow" or "thermal" neutrons while those with energies of 10 keV to as high as 15 MeV are considered to be "fast" neutrons. Electron energies may also range from a few keV to as high as 100 MeV.

2. Detector Operation

In general terms all semiconductor detectors of nuclear radiation operate by exploiting the fact that an incident radiation will, through some interaction in the detector volume, create a charge pulse that can be detected. This charge pulse consists of electrons and holes, which are then separated under the influence of an applied electric field, and the current is detected by an external circuit. Therefore, in considering detector operation we must be concerned with the nature of the interaction between the incident radiation and the volume of the detector material where the charge is created, the efficiency of the excitation process, the efficiency of the charge collection process, the external circuit that detects the charge pulse that has been created, and finally the background noise of the device. The detector noise as well as the nature and efficiency of the interaction between the particular incident radiation and the detector volume and the charge collection process will determine what materials may be employed for the fabrication of nuclear detectors that can operate at room temperature.

We consider first the interaction of x-rays and gamma rays with a solid state detector. Electromagnetic radiation can interact with a material via four mechanisms (Burcham, 1963):

1. elastic scattering,
2. photo-electric absorption,
3. Compton scattering,
4. pair production.

In the case of elastic scattering, the energy of the incident photon is not changed; rather the photon is merely deflected out of the incident beam. Thus, this process does not deposit any energy in the detector and it is important only for this discussion in that it may contribute to the detector efficiency. The last three processes each involve the deposition of all or part of the energy of the incident photon within the detector volume.

Photoelectric absorption is, in most cases, the ideal process for detector operation. All of the energy of an incident photon is absorbed by one of the orbital electrons of the atoms within the detector material. This photoelectron will have a kinetic energy equal to that of the incident photon minus the atomic binding energy of the electron. For typical gamma photon energies, this electron will most often originate in the K shell of the atom. In a semiconductor the photoelectron will then lose its kinetic energy as it interacts via coulomb interactions with the semiconductor lattice, creating many electron–hole pairs. The number of electron–hole pairs created will be proportional to the energy of the incident photon.

Compton scattering may be regarded as a collision between an incident photon and an orbital electron. The photon's direction as well as energy is altered, and some of its energy will be lost to the electron with which it collided. This electron will then lose its energy through the creation of electron–hole pairs. A photon does not transfer all of its energy to an electron in a Compton scattering event, and the number of electron–hole pairs produced in the detector varies significantly between different Compton events.

If the energy of the incident photon is above 1.02 MeV then an electron–positron pair may be created by the incident photon. Any energy in excess of this amount can go into the kinetic energy of the electron or positron. The positron has a very short lifetime in the material and will subsequently annihilate with an electron in the material producing two annihilation photons of energy 0.511 MeV. These photons can then undergo interactions of the sort described previously. Of course, any combination of the interactions described may occur as the photon gives up all or part of its energy to the detector volume.

Charged particle radiations such as electrons, protons, or alpha particles may also be detected with semiconductor detectors. These particles may interact directly with the electrons of the detector material through coulomb scattering or may be involved in nuclear reactions. Uncharged slow neutrons are detected through the use of nuclear reactions in which an incident neutron initiates a nuclear reaction that produces a charged particle or gamma ray. Thus neutron detectors must include a target material chosen to produce the desired nuclear reaction. The charged particle is then detected through the same processes discussed earlier. Fast neutron detection is based on the detection of a charged recoil nucleus, which is produced in a scattering event between the neutron and the nucleus of the material making up the detector. Alternatively a moderator material may be employed to take advantage of the higher cross sections for slow neutrons. The moderator first slows the fast neutrons, so that they can be detected by a slow neutron detector. Semiconductor detectors have been demonstrated to operate as charged particle (see, for example, Becchetti *et al.*, 1983) and neutron detectors (James *et al.*, 1990; Beyerle and Hull, 1987) as well as x-ray and gamma ray detectors.

The probability of a gamma– or x-ray–photon interaction with a detector material of atomic number Z is proportional to Z^n ($4 < n < 5$) for photoelectric interactions, Z for Compton scattering, and Z^2 for pair production (Malm, 1972). (Table I lists the symbols used in this chapter.) For binary materials some weighted average of the atomic mass of the components would be used where the weight depends on the strength of each scattering mechanism. It can therefore be seen that, in general, materials with high atomic mass will have significantly higher sensitivity to gamma and x-ray photons than low atomic mass materials. This is particularly true in energy ranges where the photoelectric interactions dominate.

TABLE I

LIST OF SYMBOLS

A detector area	t_r transit time of electrons
E applied electric field	v_d electron drift velocity
F generation rate of free carriers	Z atomic number
I current	α attenuation coefficient
L distance between detector electrodes	μ mobility of electrons
e electron charge	τ lifetime of electrons
n number of electrons per unit volume	

The attenuation coefficients for photoelectric absorption, Compton scattering, and pair production for silicon, germanium, cadmium telluride, and mercuric iodide are shown in Fig. 1. It can clearly be seen that an issue driving the choice of detector material is the desirability that the atomic number be as high as possible.

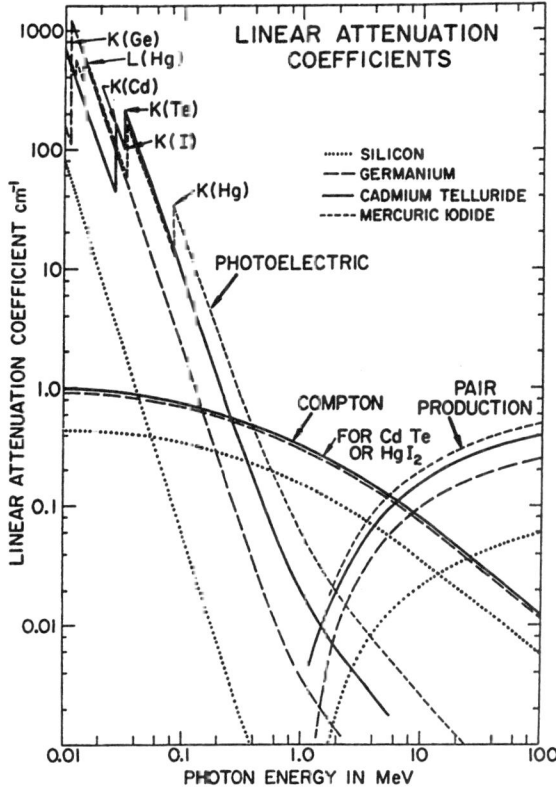

FIG. 1. Linear attenuation coefficient for photoelectric absorption, Compton scattering, and pair production in silicon, germanium, cadmium telluride, and mercuric iodide. (Reprinted with permission from Malm, 1972, © 1972 IEEE).

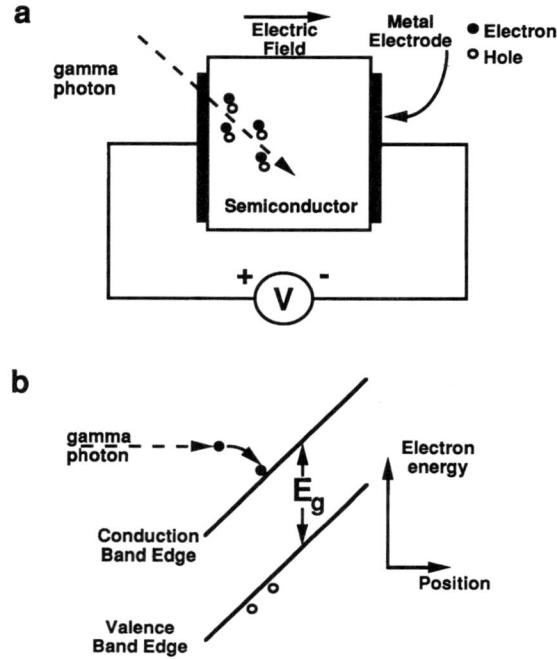

FIG. 2. A simple detector geometry and associate schematic of the process by which a charge pulse is created within the semiconductor material.

An important figure of merit for a detector is its intrinsic efficiency, that is, the ratio of the number of pulses recorded to the number of radiation particles incident on the detector.

The simplest geometry employed in the fabrication of a semiconductor detector is shown in Fig. 2. In the case of Si(Li) or Ge(Li) detectors, lithium is introduced into the semiconductor to precisely compensate the material so that it behaves nearly as an intrinsic semiconductor. In more complex structures a diffused or implanted pn junction may be employed to provide both an internal electric field and a charge depleted region within the material. High purity Ge has to a great extent replaced Ge(Li) detectors, but Si(Li) detectors are still widely employed for x-ray detection. In the case of HgI_2 and CdTe, though, detectors are fabricated from material that is very high resistivity and the configuration employed is as shown in Fig. 2. In this case a semiconductor slice a few millimeters to a centimeter thick is contacted by two metal electrodes. The thickness of material employed for the fabrication of a detector will be determined by the range of the incident radiation to be detected and the efficiency of the charge collection process. This, in turn, will depend on the type of interaction between the semiconductor and the incident radiation, the density of the detector material, the carrier lifetimes, and the dynamics of charge transport. A typical absorption depth for photon radiations of various energies can be obtained from Fig. 1 for Ge, Si, CdTe, and HgI_2. For example, for x-rays in mercuric iodide having energies of tens of keV, a typical absorption depth is several tens of microns. A competing consideration in determining the thickness of the semiconductor slice employed, the carrier drift length, is also important for x-ray and gamma ray spectrometers. Generally, one would prefer that the drift length of carriers be several times greater than the distance that the carriers must traverse to reach the electrode, with a typical rule of thumb being that the thickness of the detector be one tenth the drift length in the material. These two competing considerations result in typical detector thicknesses of about 1 mm for mercuric iodide and cadmium telluride x-ray spectrometers and on the order of 1 cm for gamma spectrometers (Szymczyk et al., 1983). Large volume gamma-ray detectors based on CdZnTe materials are under current development (C. Johnson and R. James, private communication, 1994).

In a semiconductor material the incident radiation can create a very large number of electron–hole pairs since the energy required to produce one electron–hole pair is about 3 to 6 eV, depending on the bandgap and other properties of the material being employed. The electron–hole pairs are created either directly, as might be the case if the incident radiation is a charged particle such as an electron, or indirectly, where the incident radiation undergoes any number of the processes already discussed and the secondary particles produced lose their energy via the production of electron–hole pairs. Since many electron–hole pairs typically are produced, semiconductor detectors benefit from at least two advantages compared to gas filled detectors, where the ionization of an atom is required and ionization energies are on the order of 30 eV. The larger quantity of charge makes the detection scheme simpler and allows for detection of smaller energies. Also resolution may be improved since the fluctuation in pulse height is proportional to the square root of the number of carriers in the pulse, and hence fluctuations in the pulse height originating from purely statistical sources are reduced. The relatively small energy required to produce electron–hole pairs in semiconductors and the high quantum efficiencies are two major advantages semiconductor nuclear detectors enjoy. Another is that their density is far higher than that of gas filled detectors and the probability of an interaction taking place between the detector and incident radiation is thus much higher.

The metal electrodes are used to impose an electric field across the semiconductor and collect the charge created by the incident radiation. In principle, for high resistivity materials, one would prefer that the metal–semiconductor contact be ohmic, that is, have a linear current versus voltage relationship, and of low resistance. In the case of silicon or germanium detectors this is impractical because the currents generated due to the application of an external bias would often be larger than those created by the incident radiation. Thus, blocking (or rectifying) contacts are typically employed. Even so, in the case of Ge or Si, their relatively small bandgaps require that the detectors be operated at low temperature to further reduce the leakage currents (and also maintain the Li profile in Si(Li) and Ge(Li) detectors). In the case of cadmium telluride and even more so with mercuric iodide, the larger bandgap means that sufficiently high resistivities can be achieved for intrinsic material so that the detectors may be operated at room temperature. Larger bandgaps and the ability to produce intrinsic (very high resistivity) material is, therefore, additional criteria that must be considered when choosing a material for room temperature detector operation.

In the case of larger bandgap materials, such as mercuric iodide and cadmium telluride, ohmic contacts might be acceptable in that the leakage currents can be kept acceptably small in this case. However, deposited metal electrodes appear to produce Schottky barrier-like contacts that can limit charge collection efficiency for at least one type of carrier. Thus contact technology continues to be an area of active research in the fabrication of room temperature nuclear detectors.

The externally applied electric field separates the charge carriers and the motion of the carriers constitutes the current pulse, whose size is proportional to the energy of the incident radiation. Thus a typical energy dispersive detection scheme consists of detecting pulses and placing them in a particular channel of a multichannel analyzer where the channel selected is based on the pulse amplitude. Here, the channel number corresponds to the energy of the detected radiation. There are other detection schemes, but this pulse mode operation is the one usually employed when spectral information is required. A simple scheme for analyzing the photoconductive process has been outlined by Rose (1963) and follows. For simplicity we consider only one carrier type (electrons).

Consider a uniform volume excitation that generates free electrons at a rate of F per second. Then

$$n = F\tau \tag{1}$$

where τ is the lifetime of the electrons. The photocurrent will be

$$I = \frac{ne}{t_r} \tag{2}$$

where t_r is the transit time of an electron from cathode to anode and is given by

$$t_r = \frac{L}{v_d} = \frac{L}{E\mu} \tag{3}$$

Here, L is the distance between the contacts, v_d is the drift velocity, E is the applied electric field, and μ is the mobility of the electrons. Combining equations (2) and (3), we obtain

$$I = (eF/L)(\mu\tau E) \tag{4}$$

Based on this simple model, it can be seen that, for a given size of detector and a fixed applied electric field, the size of the current pulse is determined by the $\mu\tau E$

product, with a larger $\mu\tau E$ product being more desirable. Thus this product, which is determined by material parameters μ and τ, will give an indication of how well a material will perform for spectrometer applications. In many cases, only the $\mu\tau$ product will be quoted for assessing the potential of a particular material. Moreover, it is clear that, since the mobility of both holes and electrons are important and the hole mobility is generally lower than the electron mobility, the mobility–lifetime product of the holes is usually of greatest concern when considering limitations of the detector thickness and overall device performance.

Both mobility and lifetime are material parameters that depend very much on the density of carrier traps and recombination centers within the semiconductor. In general for the highest mobilities one wishes to have the lowest concentrations of charged and neutral defect levels possible. Recombination centers will reduce carrier lifetimes, and while carrier traps can, in some cases, increase lifetimes, they will generally act as charge scattering centers and lead to a reduction in the mobility. Thus when considering a semiconductor for the fabrication of nuclear detectors, one must determine what ultimate intrinsic $\mu\tau$ product can be achieved in principle, which defects are present and in what quantities, how effective are the defects as traps and recombination centers, and to what degree can they be removed from the semiconductor by employing well-controlled growth and processing procedures.

Another issue that faces many semiconductor materials employed for the fabrication of nuclear detectors is the so-called polarization effect. The term "polarization effect" has come to mean any long term change in the performance of the detector that results from the application of the bias field (Holzer and Schieberg, 1980). These polarization effects can cause a change in the electric field within the device. Often this change tends to decrease the effect of the applied external field as internal fields are developed. These internal fields lead to a decrease in the charge collection efficiency of the detector. Such a process may occur, for example, as carriers are trapped on defect centers within the bulk or near surface region of the material or if mobile ions migrate under the influence of the applied field. Whatever the origin might be, polarization behavior is detrimental to the long term stability and performance of detectors, and the origin and elimination of these effects are of much concern in the area of semiconductor nuclear detectors.

In summary, the fabrication of room temperature nuclear radiation detectors requires a semiconductor material that has a number of physical properties. These properties and some of the central issues follow:

1. High atomic number (Z) for efficient radiation–atomic interactions.
2. Large enough bandgap for high resistivity and low leakage currents.
3. Small enough bandgap that the electron–hole ionization energy is small (< 5 eV).
4. High intrinsic $\mu\tau$ product.
5. High purity, homogeneous, defect-free material with acceptable cross-sectional area and thickness

6. Electrodes that produce no defects, impurities, or barriers to the charge collection process.

3. PRACTICAL DETECTORS

In this text a large number of material systems and working detectors are discussed and compared. It is important, however, that the reader keep in mind a number of issues and parameters when comparing the performance of various detector systems. One may be misled if the energy resolution presented in a single spectrum is used as the only measure of the performance of a spectrometer or the success of the detector in field applications. Of equal or greater importance are a number of other parameters, such as the detector volume and thickness, the energy of the photons being detected, the intensity of the photopeak relative to any x-ray escape peaks, the efficiency of the photopeaks, the applied bias, the dark current, the charge collection time or transit time for carriers, the fabrication yield of devices in this material system, and the stability of the detector to environmental conditions and aging.

For a particular photon energy and an attenuation coefficient, α, at that energy one generally would prefer a detector thickness, L, such that $\alpha L \sim 3$. This ensures that nearly all the photons are stopped in the detector volume, and thus one may expect a reasonable efficiency and large photopeak. Larger areas are also desirable so that for a given flux, depending on the source to be analyzed, an acceptable spectrum can be obtained in a few seconds or minutes. Thus, in comparing two detectors one must note both the area and thickness (volume) of the devices. One must also note the energy of the photons that are detected. It is generally more difficult to produce spectrometers that operate at higher gamma ray energies because higher energy gamma photons penetrate much more deeply than lower energy ones. Thus, if the semiconductor material is not stoichiometric, defect free, and homogeneous over large dimensions one cannot fabricate detectors having both sufficient thickness to stop a large fraction of the incident photons and good charge collection efficiency. Consequently, in comparing two detectors, the ability to detect higher energy gamma rays with good resolution and efficiency typically indicates a higher quality material system.

The dark current in the detectors should ideally contribute less unwanted noise than noise contributions from the system electronics. These issues are discussed in detail in a later chapter of this text. The charge transit time should also be as short as possible to ensure complete (or nearly complete) charge collection. For portable systems the ability to operate the spectrometer with as low an applied bias as possible is desirable. Typically the ability to operate the detector at lower bias allows for lighter power supplies and a more compact portable system. The operating temperature is also critical particularly if the detector is incorporated into a portable system, used in an unattended mode, or desired for measurements requiring an "anywhere anytime" mode of operation. While liquid nitrogen or

1. INTRODUCTION AND OVERVIEW   11

peltier coolers may improve detector performance, operation at and above room temperature may be the driving consideration in particular applications where the use of cooled detectors is not feasible.

One must consider fabrication yields in comparing different material systems. As a practical matter it is not sufficient to produce a single working detector from a candidate material. Fabrication yields directly affect manufacturing costs and hence the likelihood of success for a detector in any spectroscopic system. Finally, issues such as stability with respect to temperature, ruggedness, exposure to radiation, exposure to gases and other materials, and aging must be considered when comparing materials and spectrometers. All these issues must be kept in mind and temper any comparison of nuclear spectrometers discussed in this text or in the literature in general.

There is a great need for high resolution, large volume, gamma ray spectrometers that operate at ambient temperatures, require little or no maintenance, are easy to use, are rugged and stable over time, and can be manufactured at an acceptable cost. Some of these applications are discussed in the sections that follow.

III. Applications

The discussion that follows is focused primarily on applications of room temperature detectors fabricated from CdTe, HgI_2, and GaAs. However, the use of silicon based detectors for room temperature operation is also an area of active investigation. Examples of applications of room temperature Si detectors include the use of commercially available CCD arrays along with image processing systems to detect beta particles and low energy x-rays (Ellila and Pollari, 1990; Mayer, 1991); a two-dimensional array of Si(Li) detectors that have been employed as an auroral x-ray imager (Hirasima et al., 1987) by cooling the detectors to ambient temperature at high altitudes and for their use in alpha spectroscopy for isotopic ratio measurements or their use in the current mode to monitor high intensity radiation fields (Burger and Beroud, 1984). Silicon based detector technology may also be applicable to some of the applications discussed in the context of other materials.

1. NATIONAL SECURITY

National security applications of room temperature nuclear detectors are associated primarily with locating, monitoring, or identifying radioactive materials. Gamma ray spectrometers and neutron detectors are imperative for monitoring controlled nuclear materials such as plutonium. Room temperature, high resolution detectors are particularly needed because they are compact, low maintenance, and operable in an unattended mode. This capability is also important in arms control, treaty verification, and International Atomic Energy Agency (IAEA) operations. Thus, portable lightweight radiation detectors and spectrometers that can be used in the field without cooling and offer energy resolutions greater than scintillators find immediate uses in national security applications.

In treaty verification applications a portable spectrometer system is required. This capability allows for monitoring not only the presence of nuclear material but also determining information on its composition, while at the same time not revealing sensitive weapons design information. This capability was demonstrated by Fetter *et al.* (1990). The International Atomic Energy Agency requires detector systems that are compact, portable, robust, simple to troubleshoot, and easy to calibrate for its safeguards verification activities. Passive techniques based on gamma signature measurements are particularly well suited to the agency's work. CdTe based systems have proven to be well suited for verification measurements of the U enrichment of inner rods of a complete fuel assembly, since the small detector size makes it possible to insert this detector between rows of rods (De Carolis, Dragnev, and Waligura, 1976; Arlt *et al.*, 1993).

In the field of national security (and environmental remediation), one does not always have to perform spectral analysis of the radiation source. In some instances merely detecting the presence of radioactive sources above the natural background is sufficient, especially if the detector is small, lightweight, has a high sensitivity, requires low power, and is suitable for field uses. Thick mercuric iodide slices have been fabricated into gamma ray counters (Warren, 1983) that can provide this function.

In a related field of application, a compact handheld system has been developed to detect a variety of substances including narcotics and explosives. The detection of these materials is accomplished through the detection of the gamma photon backscatter from a ^{57}Co source using a CdTe spectrometer (Entine *et al.*, 1989). The large scale deployment of room temperature gamma ray spectrometers based on semiconductors is still limited by problems with detector efficiency and energy resolution.

2. COMMERCIAL

Elemental analysis of materials through the analysis of the characteristic x-ray fluorescence of compounds is an important area of application for room temperature x-ray spectrometers. In this application the elimination of the need for cryogenic cooling and the associated cryostat allows for a significant reduction in the size and weight of x-ray fluorescence spectrometers (Dabrowski *et al.*, 1983). In turn the increased operational flexibility significantly expands their range of applications into areas such as field studies of geological samples, analysis of art and archeological artifacts, bore hole logging, mineral exploration, pollution monitoring, and medical diagnostics. Indeed commercial instruments of this type are already available, and one such instrument is shown in Fig. 3. This instrument can

1. INTRODUCTION AND OVERVIEW   13

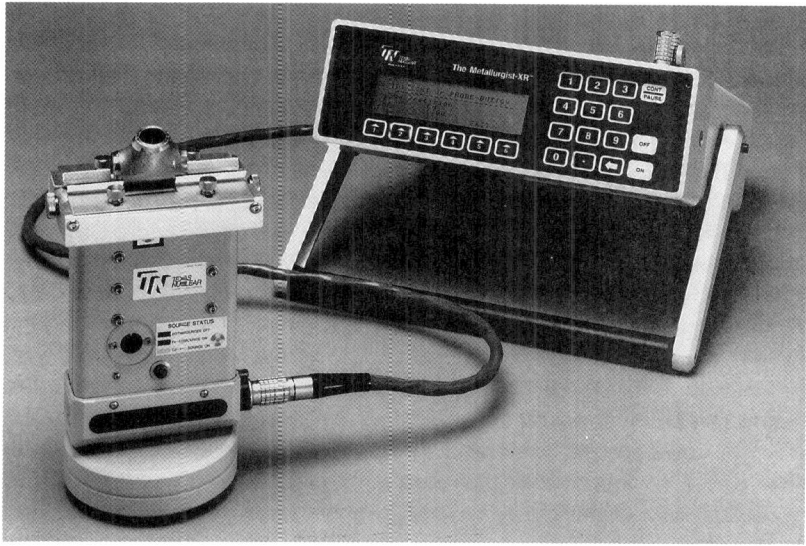

FIG. 3. Portable x-ray fluorescence spectrometer processing unit and probe (courtesy of TN Technologies, Inc.).

be employed for alloy analysis and material sorting and uses ^{55}Fe or ^{109}Cd as the x-ray excitation source. Measurement times are on the order of half a minute or less, depending on the particular application. This type of instrument is employed in a number of industries, including chemical–refining, detection of toxics, scrap, pump and valve, and shipyards. It provides both high sensitivity and high resolution in a system that is both compact and lightweight. The unit pictured in Fig. 3 includes a processing unit weighing 6.7 kg and a probe weighing 1.1 kg.

A major problem facing coal mining operations is the buildup of methane gas and the resulting risk of explosion. Many mining operations, therefore, are employing techniques to remove trapped methane gas within coal seams by drilling bore holes that allow the gas to be removed. By detecting the level of naturally occurring radioactivity originating from the shale that often surrounds coal seams, one can determine the position of the bore drilling bit within the coal seam and thus ensure that the bit moves along the appropriate path. For this application semiconductor nuclear detectors operating at room temperature and above are well suited not only because of their small size, elimination of the need for cryogenic cooling, and high sensitivity to x-rays and gamma photons, but also because the ability of these detectors to operate with low power requirements reduces the risk of sparks that may ignite the methane gas (Entine *et al.,* 1989; Thakur and Dahl, 1982). A second application is the control of drill bits so that only the low sulphur and low ash content coal, which tends to be near the center of the coal seam, is removed. In a related application, underground miners face the risks of lung cancer from the inhalation of radon daughter products, especially in uranium ore mines. A common method of measuring working level exposures is to collect a sample on a filter and measure the alpha or beta activity; however, this approach is time consuming and not always satisfactory. A continuous working level (CWL) detector would be preferred to provide an immediate readout of current radiation levels. Thus, the Bureau of Mines has investigated the development of systems that continuously monitor air samples for mines and dwellings (Droullard and Holub, 1985). The power requirements of these systems could be significantly reduced by the incorporation of semiconductor detectors designed for beta and alpha particle detection, as opposed to Geiger–Mueller tubes typically used today.

Monolithic and array detectors with large numbers of elements are of interest for a number of imaging applications. Arrays may be used for imaging of radiation sources or to provide higher effective count rates and larger solid angle collection efficiencies. Detector arrays and the associated electronics have been developed by a number of workers (Iwanczyck et al., 1990; Ortale, Padgett, and Schnepple, 1983) for these applications including a 90 element array fabricated in CdTe (Iwase, Takamura, and Ohmori, 1991). A linear array composed of several hundred miniature CdTe detectors has been demonstrated as an imaging system for luggage scanning. When operated in an energy dispersive mode it is possible to highlight particular atomic elements in the luggage (Eisen and Polak, 1993). An x-ray and gamma ray imaging camera has been developed that employs a crossed grid of electrodes on the top and bottom of a HgI_2 semiconductor to form a 32 × 32 element array. This camera has been demonstrated to have a spatial resolution of 1–2 mm at 59.6 keV and 5 or 6 mm at 662 keV (Patt et al., 1986; Patt, 1993). Such instruments can be used in on-line industrial process monitoring, for example, to detect flaws in metal parts or to inspect welds and in a variety of medical applications.

3. MEDICAL

Medical applications of radiation include imaging techniques such as x-ray radiography, x-ray or gamma ray tomography, positron emission tomography, in vivo x-ray fluorescence analysis, the use of radioactive tracers and markers, and therapeutic applications of radioactive isotopes (IAEA/WHO Report 1972). In imaging applications good spatial resolution with little noise due to scattered counts from multiple scattering of photons must be obtained. This is achieved by employing arrays of very small detectors, limiting the collimation angle of the detector, or using post detection data processing techniques that selectively reject counts at low energies, below the photopeak (Atkins et al., 1977). These techniques, however, will decrease the system efficiency. Hence, small but highly efficient semiconductor detectors can have significant impact in this field. Small single detector probes that can operate at body temperature and provide high efficiency and low noise have potential use in a variety of nuclear medical applications including their use in esophageal probes, eye probes, or probes introduced by catheterization or employed during surgery. These compact detectors offer a number of advantages over traditional detectors such as a scintillator–photomultiplier combination, including their portability, the elimination of the need to employ high voltage and precise bias voltage regulation, and the applicability of employing simpler data analysis methods.

A small mercuric iodide probe has been designed and tested for the location of Pu contamination in wounds (Friant et al., 1989). A handheld probe that relies on a CdTe detector and includes battery powered electronics has been developed for use as a surgical tool to locate radio-isotope labeled tumors (see Fig. 4) (Entine et al., 1989). While the types of tumors that can be labeled and hence located in this manner is limited, there remains a high level of interest in this technique, especially as radio-isotope labeling techniques improve using ^{111}In and ^{125}I. A comparison of the relative merits in terms of counting efficiency and energy resolution of three different surgical probes containing a NaI(Tl) scintillator with a flexible fiber-optic light guide, a CdTe detector, and a HgI$_2$ detector has been made by Barber and co-workers (1991). This comparison concluded that above 120 keV the counting efficiency of the Na(Tl) scintillator probe was an order of magnitude better than either semiconductor detector but that the energy resolution of the scintillator was five times worse. The HgI$_2$ detector was shown to have a slightly better efficiency and resolution than the CdTe device. Needle probes that

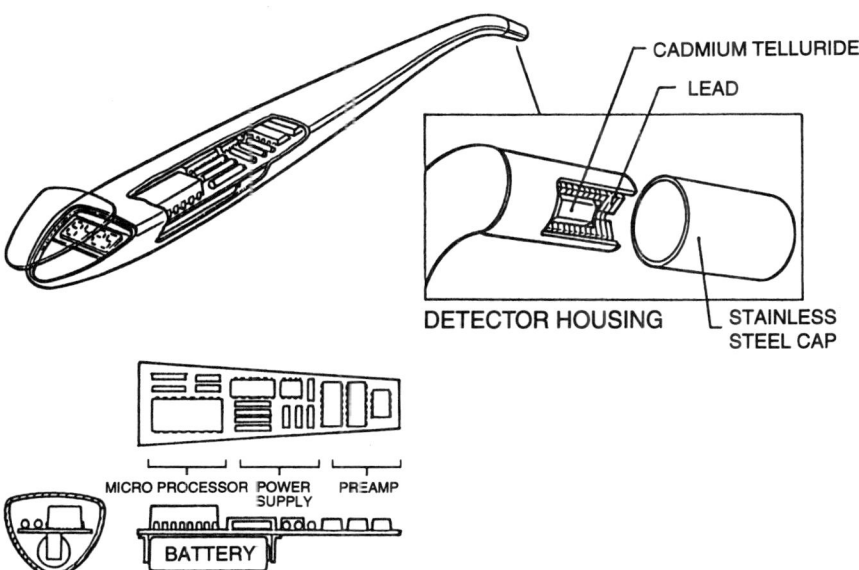

FIG. 4. Self-contained, handheld surgical probe for use in the operating room during surgery. The probe is battery powered The enlarged sketch shows the CdTe sensor and collimator. (Reproduced with permission from Entine et al., 1989 and Elsevier Science Publishers.)

fit into a syringe of 0.9 mm diameter have been fabricated (Caine et al., 1978) for in vivo dosimetry. ^{125}I can be used as a therapeutic tracer in the treatment of Graves' disease (Weidinger, Johnson, and Werner, 1974), and ^{133}Xe can be employed to monitor regional cerebral blood flow (rCBF) by observing the rate at which the body removes this isotope from the patient. Both HgI$_2$ and CdTe detectors have been developed for this purpose (Levi et al., 1982; Correia et al., 1981). A more advanced and very portable system has been developed and tested (Entine et al., 1989). This system employs 25 CdTe detectors and a microcomputer based data acquisition system placed on a mobile instrument cart. In applications such as rCBF the necessity of using more than one detector makes the availability of small compact systems a particular advantage. An analogous technique employs ^{133}Xe to measure blood flow in the myocardium immediately after a coronary bypass operation but before closing the chest. It also requires the use of miniature detectors since these are placed directly in the heart and are intended to provide an immediate indication of blood flow to the myocardial tissue fed by the bypass (Entine et al., 1989). GaAs probes have also been tested for applications such as rCBF and the location of malignant tumors (Kobayashi et al., 1973). ^{125}I can also be employed to label blood clots, and a CdTe probe has been shown to be useful in measuring ^{125}I accumulation in selected sites. The small size of the probe compared to standard NaI probes makes the instrument more accurate and reproducible in placement over a site, especially when attempting to locate a clot. Similar probes have been tested for their applicability to localizing dental infections or measuring lung density (Entine, Garcia, and Tow, 1977). The determination of the mineral content of bone is important in the diagnosis of osteoporosis and the determination of the effectiveness of therapies. Isotopic radiation sources such as ^{125}I, ^{241}Am, and ^{153}Gd are employed to determine bone density through transmission measurements (Vogel et al., 1979), and many commercial systems used to make such measurements employ CdTe detectors.

4. ENVIRONMENTAL REMEDIATION AND SAFETY

Both naturally occurring sources of radiation, such as radon in mines and dwellings, and manufactured sources of radiation, such as nuclear power plants, nuclear waste handling facilities, research facilities, and medical equipment, often require the monitoring, identification, or removal of radioactive materials. The portability and relatively simple data processing systems associated with semiconductor detectors make them well suited for many field applications.

Compact detector systems can be easily shielded from several types of undesired radiation sources or from high levels of radiation. Thus, these detectors are well suited for operation in nuclear power facilities and nuclear waste handling plants. A CdTe detector system has been developed for use at sites such as the Savannah River Plant (Aiken, South Carolina) to monitor the separation of high level radioactive sludge from a lower level radioactive liquid stream. The central design feature of this system was the necessity to shield the detectors from the high level radiation of the sludge while still remaining sensitive to small radiation changes in the liquid (Entine et al., 1989). The CdTe system is expected to provide the sensitivity, stability, and reliability required for this application. Portable systems can also be employed in power plants for pipe inspection as well.

Another application of these compact solid state detector systems is in radiation monitoring systems. A detector system that has been employed in a number of nuclear power plants is used to monitor radioactive levels in gas streams that would be released into the atmosphere in the event of an accident. These detectors would provide information on the amount of radioactive gas released, while remaining shielded from the high radiation levels expected to be incident on the detector from within the plant in the event of such an accident. A system which includes three detectors, including two CdTe detectors for the upper energy ranges and encased in several hundreds of kilograms of lead has been tested. The long term stability expected of CdTe detectors and the reduced shielding required make them ideal for this application (Entine et al., 1989).

The use of gamma or x-ray spectroscopy to detect contamination of soil is of interest. This is preferred over alpha particle detection because the very short range of alpha particles in dry air (3.7 cm) and water (40 μm) makes alpha particle detection both difficult and hazardous. A mercuric iodide array has been tested to monitor contamination of soils by gamma emitting radioisotopes (Friant et al., 1989). Many environmental remediation applications may involve working in high radiation areas. In these circumstances it would be highly desirable for robotic and automated systems to survey the area, identify sources of radiation, and perform the operations necessary for the removal of the source. Toward this goal the use of compact semiconductor detectors mounted on robotic manipulators and that obtain both intensity and spectroscopic information is being investigated (Kume and Khosla, 1993). This effort is aimed at not only identifying the sources of radiation in an environment and creating maps of radiation levels but also at integrating radiation sensor information with data from other instruments such as vision and range sensors. This type of multidimensional mapping would provide the type of information required for an automated system to perform unstaffed site characterization and make intelligent decisions about the strategies to be employed in a variety of cleanup and remediation operations.

Semiconductor nuclear detectors have also found application as personal safety systems in facilities where individuals process or use large amounts of radioactive material. Rather than carrying a radiation survey instrument, a small CdTe detector along with the necessary low power electronics has been developed as a personal radiation "chirper" (a device that emits an audible chirp or beep at a rate proportional to the photon exposure rate [Wolf, Umbarger, and Entine, 1979]). This device, unlike its predecessors whose operation relied on a gas filled Geiger-Muller tube, is more efficient, smaller, and less complex in design. These devices can also serve as simple counters for use by police and custom officers to monitor radiation.

5. SPACE APPLICATIONS AND BASIC SCIENCE

X-ray astronomy is a field that has advanced significantly since its beginnings in the early 1960s. To make measurements of the x-ray spectra of celestial objects, x-ray detectors must be placed above the atmosphere in balloons, rockets, or satellites. The detectors employed for this application must be extremely lightweight, with high efficiency, over large areas, operable at ambient temperatures, and extremely reliable for long term missions. While a detector that meets all these criteria is still not available, it is clear that a detector operating at ambient temperature would be of great advantage if it were capable of providing the same performance as a cooled detector. Advantages lie in the reduced size and lower weight as well as the potential for operating for much longer periods of time, since there would be no maintenance limitation based on cryogen depletion. Development of mercuric iodide detectors for this application have been carried out as well as testing of these detectors on balloon flights (Vallerga, Vanderspek, and Ricker, 1983; Ricker, Vallerga, and Wood, 1983). Detectors fabricated from mercuric iodide have also been shown to be extremely reliable for periods of up to four years when encapsulated in parylene-C and placed under high vacuum, controlled temperatures, and continuous bias. HgI_2 detectors have also been designed for use in a scanning electron microscope and particle analyzer instrument. This instrument was used to perform in-flight morphological and elemental analysis of comet dust, but its application can easily be extended to a planetary or asteroid lander. These detectors have achieved an energy resolution of 198 eV (FWHM) for the 5.9 keV Mn kα line (Hart et al., 1981; Iwanczyk et al., 1989; Bradley et al., 1989; Iwanczyk, 1993). The determination of the chemical composition of materials can also be accomplished with an alpha particle instrument in which the material to be analyzed is exposed to an alpha emitting radioactive source and energy spectra are acquired of the backscattered alpha particles, protons, and x-rays. Such an instrument employing a mercuric iodide charged particle detector has been considered (Economou and Iwanczyk, 1989). GaAs detectors also have a long history in x-ray and gamma ray detectors for astronomical applications and have been examined again for their potential application in this field, as well as in high energy physics and particle astrophysics (Sumner et al., 1991).

Scintillation detectors are employed often to detect alpha and beta particles, gamma photons, neutrons, protons, and even x-rays. In scintillation detectors one must employ a photodetector of some type, usually a photomultiplier tube, to detect the light pulses produced by a scintillation material such as CsI(Tl) or NaI(Tl). Mercuric iodide has been employed as a photocell in these applications (Iwanczyk et al., 1983; Markakis et al., 1985) since its photoresponse spectrum tends to peak near the wavelength range of some scintillator materials: 400 nm to 500 nm (Knoll, 1989). One particular advantage is that the very high resistivity of mercuric iodide ensures an extremely low dark current even when employed in a simple photoconductive mode. This device has the same configuration as the one shown in Fig. 2, except that transparent electrodes are used. Additional advantages of employing this type of photodetector lies in its small size compared with a PMT, potential for 100% efficiency in detecting optical pulses, no need for magnetic shielding, and simplicity of fabrication. A disadvantage is that the maximum size of HgI_2 photocells is about 4×4 cm. An energy resolution of 19% for the photopeak from annihilation gamma rays was obtained with CsI(Tl), and 24% with a Bismuth Germanate (BGO) crystal (Iwanczyk et al., 1983). Energy resolutions of better than 6% have recently been achieved for 662 keV gamma-rays using CsI(Tl) scintillators and HgI_2 photocells (Iwanczyk, private communication).

In high energy physics investigations a variety of particles must be detected. Often, however, the particles of interest are of very high energy and can travel with very little energy loss through a thin semiconductor detector. This can make their detection quite a challenge, particularly for high energy gamma photons and neutrons. Nonetheless, preliminary tests of a GaAs detector designed with this application in mind have been conducted for the detection of beta particles, alpha particles, gamma photons, and hadrons (Bertin et al., 1990; Buttar et al., 1991). A bulk GaAs ionization detector operating at 4 K has also been investigated for use at energies below 200 keV in the search for weakly interacting massive particles (Spooner et al., 1991). Applications for research instrumentation continue to expand as the performance of semiconductor radiation detectors improves. The ability to produce large one dimensional and two dimensional arrays of semiconductor detectors and to fabricate high quality thick semiconductor spectrometers with high stopping power also expands the applicability of these devices into other fields, such as high energy particle detection, x-ray microscopy, and synchrotron research.

IV. Outline of Text

Following this chapter is a detailed discussion of the most well-established semiconductor nuclear detector material; high purity Ge (HPGe). The HPGe chapter is followed by six chapters devoted to a discussion of mercuric iodide and cadmium telluride. Included are detailed treatments of the growth, optical, and electrical properties of these materials. Next, one chapter each is devoted to gallium arsenide, lead iodide, and cadmium zinc telluride all of which are materials that offer alternatives to CdTe and HgI_2 and have already been demonstrated to operate as x-ray or gamma ray detectors. A chapter discussing other promising materials follows these and the relative merits and reasons these semiconductors may yet play a role in nuclear radiation detection technology is considered. The electrical characteristics of charge detection electronics and the very stringent low noise requirements necessary of high resolution spectroscopy is detailed in the next chapter. Finally, the methods employed to test and quantify detector performance are presented. The concluding chapter summarizes the state of the art in this field and the problems and challenges that remain before semiconductor room temperature detectors reach their full potential in nuclear radiation detection.

REFERENCES

Atkins, F. B., Beck, R. N., Hoffer, P. B., and Palmer, D. (1977). *Proceedings of the 1976 Symposium on Medical Radionuclide Imaging* **1**, 101.
Arlt, R., Rundquist, D. E., Bot, D., Siffer, P., Richter, M., Khusainov, A., Ivanov, V., Chrunov, A., Petuchov, Y., Levai, F., Desi, S., Tarvainen, M., and Ahmed, I. (1993). *Materials Research Society Proceedings* **302**, 19.
Barber, H. B., Barrett, H. H., Hickernell, T. S., Kwo, D. P., Woolfenden, J. M., Entine, G., and Ortale-Baccash, C. (1991). *Med. Phys.* **18**, 373.
Becchetti, F. D., Raymond, R. S., Ristinen, R. A., Schnepple, W. F., and Otale, C. (1983). *Nucl. Instr. Meth.* **213**, 127.
Bertin, R., D'Auria, S., Del Papa, C., Fiori, F., Lisowski, B., O'Shea, V., Pelfer, P. G., Smith, K., and Zichichi, A. (1990). *Nucl. Inst. Meth. Phys. Res.* **A294**, 211.
Beyerle, A. G., and Hull, K. L. (1987). *Nucl. Inst. Meth. Phys. Res.* **A256**, 377.
Bradley, J. G., Conley, J. M., Albee, A. L., Iwanczyk, J. S., Dabrowski, A. J., and Warburton, W. K. (1989). *Nucl. Inst. and Meth.* **A283**, 348.
Burcham, W. E. (1963). *Nuclear Physics: An Introduction.* Longmans, London.
Burger, P., and Beroud, Y. (1984). *Nucl. Inst. Meth. Phys. Res.* **226**, 45.
Buttar, C. M., Combley, F. H., Dawson, I., Dogru, M., Harrison, M., Hill, G., Hou, Y., and Houston, P. (1991). *Nucl. Inst. Meth. Phys. Res.* **A310**, 208.
Caine, S., Holzer, A., Beinglass, I., and Schieber, M. (1978). *IEEE Trans. Nucl. Sci.* **NS-25**, 649.
Correia, J. A., Ackerman, R. H., Buonanno, F., Kaufman, D., Skiver, J., Alpert, N., Taveras, J., and Entine, G. (1981). *IEEE Trans. Nucl. Sci.* **NS-28**, 50.
Dabrowski, A. J., Szymczyk, W. M., Iwanczyk, J. S., Kusmiss, J. H., Drummond, W., and Ames, L. (1983). *Nucl. Inst. and Meth.* **213**, 89.
De Carolis, M., Dragnev, T., and Waligura, A. (1976). *IEEE Trans. Nucl. Sci.* **NS-23**, 70.
Droullard, R. F., and Holub, R. F. (1985). *Bureau of Mines Information Circular* **IC9029**.
Economou, T., and Iwanczyk, J. (1989). *Nucl. Intr. Meth. Phys. Res.* **A283**, 352.
Eisen, Y., and Polak, E. (1993). *Materials Research Society Proceedings* **302**, 487.
Ellila, M., and Pollari, K. (1990). *Nucl. Inst. Meth. Phys. Res.* **A288**, 267.
Entine, G., Garcia, D. A., and Tow, D. E. (1977). *Rev. Phys. Appl.* **12**, 355.
Entine, G., Waer, P., Tiernan, T., and Squillante, M. R. (1989). *Nucl. Inst. and Meth. Phys. Res.* **A283**, 282.
Fetter, S., Cochran, T. B., Grodzins, L., Lynch, H. L., and Zucker, M. (1990). *Science* **248**, 828.
Friant, A., Mellet, J., Barrandon, G., and Csakvary, E. (1989). *Nucl. Inst. and Meth.* **A283**, 227.
Hart, R. K., Albee, A. L., Finnerty, A. A., and Frazer, R. (1981). *Scanning Electron Microscopy* **96**, 97.
Hirasima, Y., Nakamoto, A., Murakami, H., Okudaira, K., and Yamagami, T. (1987). *Nucl. Inst. Meth. Phys. Res.* **A262**, 503.
Holzer, A., and Schieber, M. (1980). *IEEE Trans. Nucl. Sci.* **NS-27**, 266.
IAEA/WHO Expert Committee Report. (1972). World Health Organization, Geneva.
Iwanczyk, J. S. (1993). *Materials Research Society Proceedings* **302**, 79.
Iwanczyk, J. S., Barton, J. B., Dabrowski, A. J., Kusmiss, J. H., Szymczyk, W. M., Huth, G. C., Markakis, J., Schnepple, W. F., and Lynn, R. (1983). *Nucl. Inst. and Meth.* **213**, 123.
Iwanczyk, J. S., Wang, Y. J., Bradley, J. G., Conley, J. M., Albee, A. L., and Economou, T. E. (1989). *IEEE Trans. Nucl. Sci.* **NS-36**, 841.
Iwanczyk, J. S., Dorri, N., Wang, M., Szawlowski, M., Patt, B. E., Warburton, W. K., Hedman, B., and Hodgson, K. O. (1990). *IEEE Trans. Nucl. Sci.* **NS-37**, 198.

Iwase, Y., Takamura, H., and Ohmori, M. (1991). *International Conference on Solid-State Sensors and Actuators,* 840.

James, R. B., Lathrop, J. F., Haney, S. J., Barry, D. J., Clark, D., Harrison, T. R., Harris, J., and Stulen, R. H. (1990). *Nucl. Inst. Meth. Phys. Res.* **A294,** 229.

Knoll, Glen F. (1989). *Radiation Detection and Measurement.* John Wiley & Sons, New York.

Kobayashi, T., Sugita, T., Takayanagi, S., Iio, M., and Sasaki, Y. (1973). *IEEE Trans. Nucl. Sci.* **NS-20,** 310.

Kume, M., and Khosla, P. K. (1993). *Materials Research Society Proceedings* **302,** 55.

Levi, A., Roth, M., Schieber, M., Lavy, S., and Cooper, G. (1982). *IEEE Trans. Nucl. Sci.* **NS-29,** 457.

Malm, H. L. (1972). *IEEE Trans. Nucl. Sci.* **NS-19,** 263.

Markakis, J., Ortale, C., Schnepple, W., Iwanczyk, J., and Dabrowski, A. (1985). *IEEE Trans. Nucl. Sci.* **NS-32,** 559.

Mayer, R. (1991). *Rev. Sci. Instrum.* **62,** 360.

Ortale, C., Padgett, L., and Schnepple, W. F. (1983). *Nucl. Inst. and Meth.* **213,** 95.

Patt, B. E. (1993). *Materials Research Society Proceedings* **302,** 43.

Patt, B. E., Del Duca, A., Dolin, R., and Otale, C. (1986). *IEEE Trans. on Nucl. Sci.* **NS-33,** 523.

Ricker, G. R., Vallerga, J. V., and Wood, D. R. (1983). *Nucl. Inst. and Meth.* **213,** 133.

Rose, Albert. (1963). *Concepts in Photoconductivity and Allied Problems.* John Wiley & Sons, New York.

Spooner, N. J. C., Bewick, A., Holmes, S. N., Phillips, C. C., Quenby, J. J., Stradling, R. A., Sumner, T. J., Thomas, R. H., and Wang, P. D. (1991). *Nucl. Inst. Meth. Phys. Res.* **A310,** 227.

Sumner, T. J., Grant, S. M., Bewick, A., Li, J. P., Spooner, N. J. C., Smith, K., and Beaumont, S. P. (1991). *Proceedings of the SPIE Meeting on EUV, X-Ray, and Gamma-Ray Instrumentation for Astronomy II* **1549,** 256.

Szymczyk, W. M., Dabrowski, A. J., Iwanczyk, J. S., Kusmiss, J. H., Huth, G. C., Hull, K., Beyerle, A., and Markakis, J. (1983). *Nucl. Inst. and Meth.* **213,** 115.

Thakur, P. C., and Dahl, H. D. (1982). *Mining Engineering* (March), 301.

Vallerga, J. V., Vanderspek, R. K., and Ricker, G. R. (1983). *Nucl. Inst. and Meth.* **213,** 145.

Vogel, J. M., Cline, J. W., Harrison, J. F., Ulloa, G. A., and McDonald, R. J. (1979). *IEEE Trans. Nucl. Sci.* **NS-26,** 576.

Warren, J. L. (1983). *Nucl. Inst. and Meth.* **213,** 103.

Weidinger, P., Johnson, P. M., and Werner, S. C. (1974). *Lancet* **2,** 74.

Wolf, M. A., Umbarger, J. C., and Entine, G. (1979). *IEEE Trans. Nucl. Sci.* **NS-26,** 777.

CHAPTER 2

High-Purity Germanium Detectors

Larry S. Darken and Christopher E. Cox

NUCLEAR MEASUREMENTS GROUP
OXFORD INSTRUMENTS
OAK RIDGE, TENNESSEE

I. INTRODUCTION	23
II. CRYSTAL GROWTH	25
1. *Crystal Growth Technique*	26
2. *Purification at the Liquid–Solid Interface*	27
3. *Purification by Oxidation*	28
III. CHARACTERIZATION	31
1. *Carrier Transport Techniques*	32
2. *Capacitance Transient Techniques*	35
3. *Crystallographic Techniques*	38
4. *Neutral Defects*	38
IV. LARGE VOLUME DETECTORS	38
1. *Introduction*	38
2. *Geometry and Size*	40
3. *Closed End Coaxial Detector Fabrication*	41
4. *Closed End Coaxial Detector Performance*	47
V. CHARGE COLLECTION	51
1. *Geometric Effects*	52
2. *Trapping by Extended Defects*	54
3. *Trapping by Point Defects*	55
4. *Modeling Charge Collection in Spectrometers*	60
5. *Radiation Damage*	64
VI. GERMANIUM X-RAY DETECTORS	69
1. *Introduction*	69
2. *Detector Fabrication*	70
3. *Detector Housing*	71
4. *Development of the Detector Entrance Window*	72
5. *Performance at Low Energies, down to 200 eV*	73
6. *Performance at High Energies, up to 100 keV*	76
7. *Comparison with Si(Li) Detectors*	76
VII. SUMMARY	79
References	80

I. Introduction

High-purity germanium material fabricated into semiconducting diodes has been used for both charged particle and photon detection employing a range of crystal shapes and geometries. The major use of high-purity germanium is in gamma ray spectroscopy, for which it was specifically developed. This chapter reviews the crystal growth and characterization for this application and the current state of the art in detector fabrication. Germanium used for low energy X-ray detection is also discussed.

A fundamental consideration for detectors of ionizing nuclear radiation is the efficiency with which the radiation is absorbed and detected. Gamma rays are highly penetrating, and therefore detector thickness and composition are critical issues. In a semiconductor diode detector the active thickness is the depletion depth. For example, in a germanium planar diode the reverse bias V_R required for a depletion depth d is given by

$$V_R = 565 \text{ V} \frac{|N_A - N_D|}{10^{10} \text{ cm}^{-3}} \frac{d^2}{\text{cm}^2} \qquad (1)$$

where $|N_A - N_D|$ is the absolute value of the net electrically active impurity concentration. Thus for example to deplete 2 cm with $|N_A - N_D| = 10^{10}$ cm^{-3}, a reverse bias of $2260 V$ is required. Two centimeters of germanium would attenuate about 46% of incident 1 MeV gamma rays. All three mechanisms for gamma ray attenuation—photoelectric absorption, Compton scattering, and pair production—are more efficient as the atomic number (Z) increases. Therefore germanium is a significantly more efficient absorber than silicon but is less efficient than some high-Z scintillators and compound semiconductors.

However, for energy resolution of absorbed gamma rays, germanium detectors are unequalled among practical devices. Resolutions less than 2 keV full-width at half-maximum (FWHM) are typical for 1.33 MeV gamma rays. When energy resolution is the most important consideration, the only issue is the size and electrode configuration of the germanium detector to be used. Obtaining a low noise contribution to germanium detector resolution depends on cooling to near the liquid nitrogen boiling point (77 K) in order to reduce leakage current. This is a significant limitation in some applications.

In addition to its unique combination of stopping power and resolution, germanium detectors have several other advantages (which they share with other semiconductor diode-type detectors): fast response time, small size, and relative insensitivity to magnetic fields.

Germanium has disadvantages in addition to the need for cooling: energy resolution is sensitive to radiation damage (a feature common to all semiconductor diode-type detectors) and efficiency is low for higher energy gamma rays. The size and, to a lesser degree, the performance of germanium detectors have always been limited by the material technology.

Practical germanium gamma ray detector material was first produced by the lithium drifting process (Pell, 1960). In this process germanium with $N_A - N_D \approx 10^{14}$ cm^{-3} was closely compensated after crystal growth by interstitial lithium donors that were drifted into the bulk somewhat above room temperature. It was later proposed (Hall, 1966) that germanium could be purified and directly grown to the high purity standards required without the need for subsequent lithium drifting. Although problems associated with crystal growth are significantly greater for high-purity germanium (HPGe) than for lithium drifted germanium [Ge(Li)], HPGe technology had several advantages:

1. Ge(Li) detectors had to be kept near 77 K or the close compensation would be lost and the detector would have to be returned to the manufacturers for repair.
2. HPGe detector fabrication was simplified not only by skipping the lithium-drifting process (which could take several weeks) but by the ability to store at room temperature devices that had demonstrated good I–V characteristics at 77 K. This facilitated production in general, but was particularly important for the transfer of crystals between cryostats.
3. The reverse electrode configuration, crucial for minimizing the effect of radiation damage on resolution, is practical only for n-type HPGe detectors.

The market shifted abruptly to HPGe detectors around 1980 as soon as HPGe crystal production could reliably support the transition. The primary trend after this has been toward larger detectors. The maximum diameter crystals available has increased from 50 mm in 1980 to 90 mm today.

There has also been an evolution in the use of germanium detectors from single detector applications to multiple detector arrays to either increase absolute efficiency or to obtain information on the position of the absorption. Most notable is the interest of the nuclear physics community in assembling large arrays to study super deformed nuclei (major to minor axis 2 to 1). Eurogam is a France–United Kingdom collaboration, which will initially employ 72 n-type HPGe detector systems ($\sim$1400 g/each) shaped to fit in clusters among 45 bismuth germanate (BGO) scintillation detectors. The detector element will be either a single coaxial detector ($\sim$1400 g/each) or a clover array of four smaller coaxial detectors shaped for close packing (2300 g/clover). Gammasphere is funded by the Department of Energy (United States) and current plans include 110 coaxial n-type HPGe detector systems.

The three sections that immediately follow are entitled "Crystal Growth," "Characterization," and "Large Volume Detectors." We shall try to convey the current status of this production sequence. This is followed by a separate section on charge collection Charge collection problems in HPGe have generated considerable study, and the results may have interest to a broader community. We also include a section on germanium for X-ray detection and discuss recent developments demonstrating resolution at low energies comparable to that from a lithium-drifted silicon detector.

II. Crystal Growth

The most troublesome aspects of HPGe crystal growth are related to the specific requirements for HPGe detectors. Crystal growth of germanium itself is relatively straightforward compared to that in many other semiconductors. It is an elemental semiconductor with a congruent melting point (938°C), and the vapor pressure of germanium at its melting point is low (3×10^{-9} atm). Growth of single crystal germanium by Teal and Little (1950) using the Czochralski technique preceded silicon single crystal growth by this same technique. HPGe crystal growth, on the other hand, is much more demanding. In it, $|N_A - N_D|$ must be reduced to the order of 10^{10} cm^{-3}, and restrictions on centers that can be effective trapping centers are even more severe. Achieving and maintaining procedures for HPGe crystal growth has proven difficult.

1. CRYSTAL GROWTH TECHNIQUE

R. N. Hall was the pioneer of high purity germanium growth techniques (for a summary, see Hall, 1974a). Much seminal work was also done by W. L. Hansen, E. E. Haller, and co-workers at Lawrence Berkeley Laboratory (summarized in Hansen and Haller, 1983). Through the efforts of these workers the basic techniques for HPGe crystal growth were developed. Figure 1 shows HPGe being grown from a SiO_2 (synthetic) crucible under a hydrogen ambient. Growth under more oxidizing conditions (Hall, 1972) or vacuum (Hansen and Haller, 1983) results in severe charge collection problems. Graphite crucibles were tried early and fair results were obtained, but they are rarely used today. In addition to the excellent purity of synthetic SiO_2, it also getters aluminum.

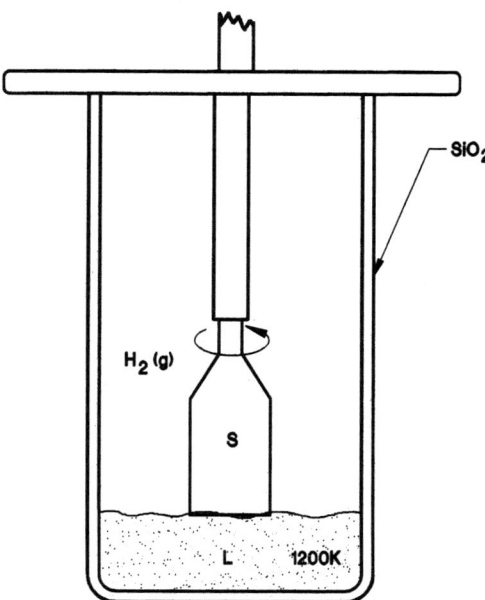

FIG. 1. Czochralski growth of an HPGe crystal.

A significant feature of HPGe crystal growth is the necessity for a broadly controlled dislocation density ($10^2 - 10^4$ cm^{-2}). For dislocation densities less than 10^2 cm^{-2}, excess vacancies cannot be annihilated during normal cooling after solidification, and they form hole trapping centers (Hansen and Haller, 1972; Hall and Soltys, 1971). Consequently, dislocation-free germanium is not suitable for detector fabrication. At dislocation densities near 10^4 cm^{-2}, charge trapping by the dislocations themselves is an issue (Glasow and Haller, 1976; van Sande et al., 1986). Furnace design and crystal growth procedures must be compatible with the major portion of HPGe crystals meeting these criteria.

The art of crystal growing has been to balance conflicting requirements. Thermal gradients are required to remove the heat of crystallization. However, stress due to thermal gradients can initiate and multiply dislocations. Present technology has been achieved through general insight and trial and error rather than through any attempt at realistic modeling of the thermal conditions.

2. Purification at the Liquid–Solid Interface

A typical process flow chart for HPGe growth and characterization is shown in Fig. 2. Polycrystalline electronic-grade germanium with $|N_A - N_D| \approx 10^{13}$ cm^{-3} must be purified three orders of magnitude, grown into large (50–90 mm diame-

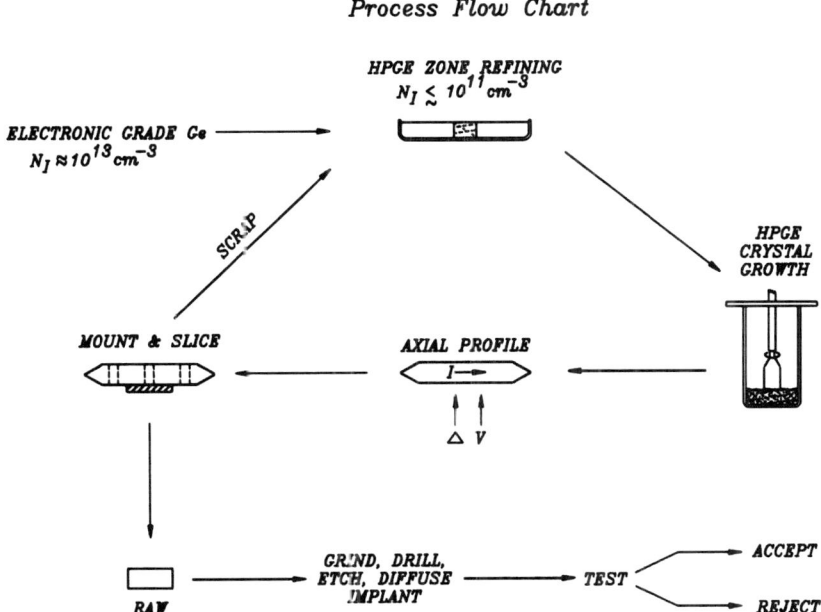

FIG. 2. Flow chart for HPGe purification, crystal growth, characterization, and detector fabrication.

ter) single crystals, and characterized to identify sections suitable for coaxial detectors.

Purification is achieved primarily through rejection of impurities into the liquid at the crystallization front. This process is characterized by an equilibrium distribution coefficient, k, for each impurity, x:

$$k_x = N_x(s)/N_x(l) \qquad (2)$$

where $N_x(s)$ is the atom fraction concentration of impurity x in the solid and $N_x(l)$ is the atom fraction concentration in the liquid. The common residual impurities in HPGe crystals are boron ($k = 20$), aluminum ($k = 0.1$), gallium, ($k = 0.1$), phosphorus ($k = 0.12$), and lithium ($k = 10^{-3}$). Many metallic impurities in germanium have values of k on the order of 10^{-5}, reflecting relatively low solubilities in crystalline germanium. Germanium is singularly well suited for zone refining, since no impurity has a distribution coefficient near unity. Considerable purification is also achieved in the first portion of the crystal to solidify in Czochralski crystal growth. Currently, however, laboratories that grow HPGe rely on zone refining for initial purification. An early technique of Hall's (Hall and Soltys, 1971) was to achieve purification by combining the top halves of two crystals for a subsequent crystal growth. While in principle purification by zone refining or repeated crystal growth could proceed indefinitely, in practice a level is reached where purification is matched by contamination. Possible contamination sources include chemicals, solvents, gases, and materials of constructions. In particular, the boat material in zone refining and the crucible material in crystal growth are liable to be sources of contamination.

3. Purification by Oxidation

Purification can also be achieved through oxidation. The thermochemical system under consideration consists of the liquid germanium Ge(l) and the phases in contact with it: solid germanium Ge(s), the glassy quartz crucible SiO_2 (gl), and the H_2 (g) ambient. For simplicity, all reactions considered are assumed to proceed at 1200 K. Equilibrium of the chemical potential of oxygen among three components of this system (Ge(s), Ge(l), and the gas phase) has been demonstrated by consideration of the following reaction (Darken, 1979a, 1982a):

$$H_2O\ (g) = H_2(g) + O \quad \text{(in solid solution)} \qquad (3)$$

where the trace water vapor dissociates at the gas–liquid interface and the oxygen proceeds through the liquid germanium to be incorporated in the solid germanium. From published thermochemical data on $GeO_2(s)$ reactions and oxygen in solubility in germanium, a relationship between (O(s)), the concentration of oxygen in the solid phase of germanium to the ratio of P_{H_2O}, the partial pressure of water vapor, to P_{H_2}, the pressure of hydrogen, may be calculated

$$\log [O(s)] = 18.45 + \log (P_{H_2O}/P_{H_2}) \qquad (4)$$

where [ ] indicates concentration in atoms per cm^3.

Now in HPGe the concentration of oxygen incorporated in the solid can be determined by the lithium precipitation technique, and the partial pressure of water vapor of the furnace effluent may be measured during crystal growth. Crystals were grown under varying P_{H_2O}/P_{H_2} ratios and the oxygen contents measured. The experimental results are shown in Fig. 3, together with Eq. (3). The close fit of the experimental data to the theoretical line proves that chemical equilibrium of oxygen exists among Ge(s), Ge(l), and the gas phase. The significance of this result is that the stability of solid oxide phases of impurities in the melt and the volatility of oxides or hydroxides of impurities may be quantitatively assessed.

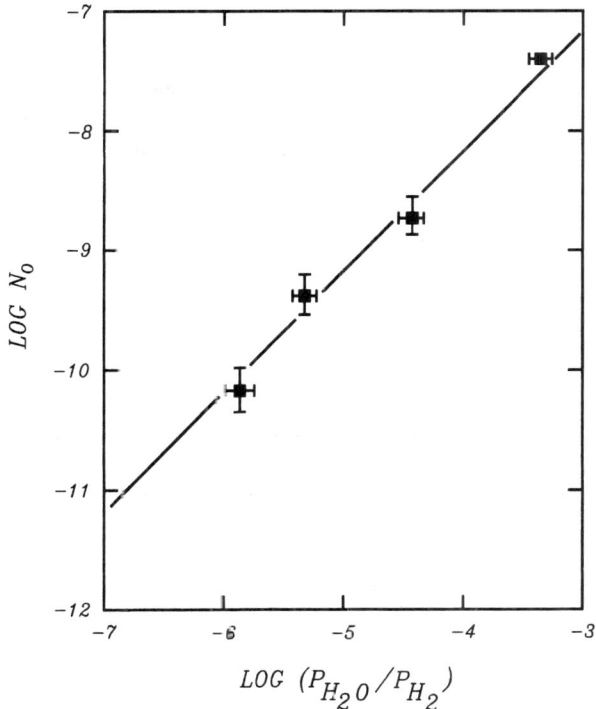

FIG. 3. Atom fraction of oxygen in germanium crystals for different P_{H_2O}/P_{H_2} ratios during growth (Darken, 1982a). The line is not the best fit to the data but represents a thermodynamic calculation, Eq. (4) (Darken, 1982a; reprinted by permission of the publisher, The Electrochemical Society, Inc.).

For example, consider the possibility of volatilizing lithium from the melt as LiOH(g) *by the reaction*

$$(\underline{\text{Li}}(l) + \underline{\text{O}}(l) + 1/2\text{H}_2(g) = \text{LiOH}(g) \tag{5}$$

where the underscore means that the species is in solution. The partial pressure of LiOH(g) could not be directly calculated from reaction (4) because thermochemical data on all the reacting species are not available. However, by use of reaction (3) and

$$\text{O (in solid solution)} = \text{O (in liquid solution)} \tag{6}$$

proceeding at equilibrium, reaction (5) may be reformulated as

$$\underline{\text{Li}}(l) + \text{H}_2\text{O}(g) = \text{LiOH}(g) + 1/2\text{H}_2(g) \tag{7}$$

The thermochemical functions for the gaseous species in reaction (7) are available in the JANAF tables. We take the standard state of lithium at 1200 K to be the pure liquid. The calculated volatility of LiOH(g) as a function of $P_{\text{H}_2\text{O}}/P_{\text{H}_2}$ is shown in Fig. 4. The partial pressure is normalized by $a_{\text{Li}(l)}$, the activity of lithium in the liquid. In an ideal solution this would be the atom fraction of lithium in the melt. The volatility of Li(g) and LiH(g) are also plotted for comparison (all assuming $\text{H}_2 = 1$ atom). Figure 4 indicates that additional volatilization of lithium through oxidation will be significant for $P_{\text{H}_2\text{O}}/P_{\text{H}_2} > 10^{-4}$.

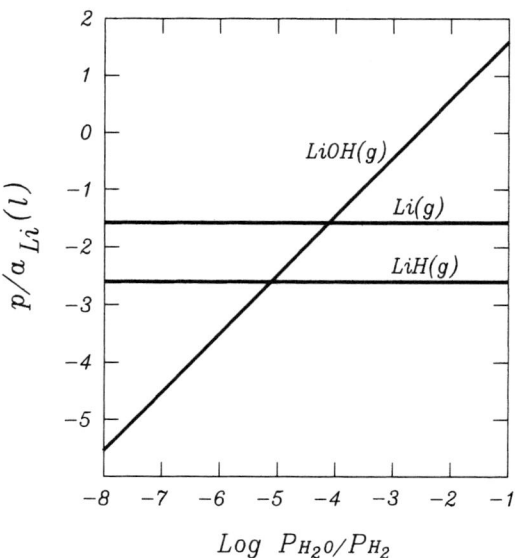

FIG. 4. Effect of oxygen chemical potential on the vapor pressure of lithium containing species (Darken, 1979a; reprinted by permission of the publisher, The Electrochemical Society, Inc.).

The most important solid oxide phase in HPGe growth is SiO_2. Thermochemical analysis indicates that the dissolution of the crucible

$$SiO_2(gl) = \underline{Si}(l) + 2\underline{O}(l) \tag{8}$$

is not proceeding at equilibrium under conditions normal for HPGe crystal growth $P_{H_2O}/P_{H_2} \approx 10^{-6} - 10^{-5}$ (Darken, 1979a). However, when HPGe is grown under an inert atmosphere or when the $H_2(g)$ is passed over ice at 0°C $P_{H_2O}/P_{H_2} \approx 10^{-3}$ (Hall, 1972), reaction (8) proceeds to the left due to residual silicon and a dispersion of SiO_2 particulates formed in the melt is included in the grown crystal. These inclusions produce smooth pits on a slice given a dislocation etch and are correlated with strong hole trapping (Hall, 1972).

Analogous to Eq. (2), a distribution coefficient can be defined for an impurity between the germanium liquid phase and the glassy oxide phase of the crucible, SiO_2. Consider the silicon–aluminum exchange reaction:

$$3/4 SiO_2(gl) + \underline{Al}(l) = \underline{AlO_{3/2}}(gl) + 3/4 Si(l) \tag{9}$$

where $\underline{AlO_{3/2}}(gl)$ represents oxidized aluminum dissolved in glassy SiO_2. Reaction (9) is written to preserve charge neutrality assuming that the cations are in their normal valence states in the oxide. The distribution coefficient is defined as

$$k_{\underline{Al}}^{SiO_2} = N_{\underline{AlO_{3/2}}}/N_{\underline{Al}}(l) \tag{10}$$

where $N_{\underline{AlO_{3/2}}}(gl)$ is the atom fraction of aluminum among the cations in the oxide. This distribution coefficient depends on the silicon content of the melt. For $N_{Si}(l) = 10^{-8}$ (typical for HPGe) it is estimated that $k_{\underline{Al}}^{SiO_2} \approx 10^8$ (Darken, 1979a). This considerable enrichment of aluminum in the oxide phase with respect to the liquid reflects the thermochemical stability of $Al_2O_3(s)$. Thus even very limited diffusion of aluminum into SiO_2 may deplete the melt of aluminum. In fact, aluminum ($k = 0.1$) is observed not to segregate normally when grown from an SiO_2 crucible (Hall and Soltys, 1971; Haller and Hansen, 1974). The aluminum concentration is nearly constant in the crystal. Gettering of aluminum from the melt via reaction (9) would explain this phenomena. No element with a solid oxide more stable than aluminum has ever been observed as an isolated defect in HPGe. It is presumed that their absence is due in part to exchange reactions like reaction (9).

III. Characterization

Measurement techniques and material properties of HPGe crystals have been reviewed by Haller et al. (1981) and by Darken (1983). We will emphasize here techniques that have found general use in industry, as well as more recent developments relevant to determining the suitability of HPGe for fabrication into radiation detectors.

1. CARRIER TRANSPORT TECHNIQUES

The most important characteristic of HPGe is the net electrically active impurity concentration, $N_A - N_D$, in the depletion region of a reverse biased diode at the operating temperature of a radiation detector (80–100 K). The depletion voltage for any detector geometry depends linearly on $N_A - N_D$. Although in principle a crystal could be characterized by fabricating test diodes and measuring capacitance voltage curves, practice has been to determine $N_A - N_D$ from transport measurements on samples immersed in liquid nitrogen.

A quick and nondestructive initial characterization procedure is an axial conductivity measurement on the as-grown crystal. Eutectic mixtures of gallium–indium or indium–mercury can be used to make ohmic contacts to the ends of the crystal. A current of 100 mA is typical, and voltage drops are measured along the side of the crystal. If there is a p-n junction, conductivity is measured in the forward direction. The conductivity σ is proportional to the carrier concentration (n for electron concentration, p for hole concentration):

$$\sigma_n = ne\mu_n; \quad \sigma_p = pe\mu_p \tag{11}$$

where μ_n and μ_p are the carrier drift mobilities (μ_n = 42,000 cm^2/volt-sec, μ_p = 45,000 cm^2/volt-sec at 77 K). Typically, crystals are p-type at the seed end due to boron and aluminum and eventually convert to n-type toward the tail end due to the segregating donor phosphorus. While the technique gives a useful general indication of net impurity distribution, it is subject to error if there is a radial gradient in the net impurity concentration.

If the crystal passes the axial profile test, slices are cut for van der Pauw (1958) measurements to identify sections of the crystal suitable for radiation detectors. The measurement may be performed on the whole slice or on pieces cut or diced out. In a van der Pauw measurement, the measured quantities are the resistivity ρ and the Hall coefficient R_H. The value of $N_A - N_D$ may be determined from ρ using Eq. (11) ($|N_A - N_D| = n$ or p), or from R_H using

$$N_A - N_D = \frac{r}{eR_H} \tag{12}$$

where r is a factor near unity that depends on carrier type, temperature, magnetic field strength, and crystallographic orientation of the sample with respect to the contacts and the magnetic field (see Miyazawa and Maeda, 1960 for n-type and Beer and Willardson, 1958 for p-type values). The carrier type is determined from the sign of R_H. A drift mobility may be obtained from each van der Pauw measurement,

$$\mu_{n \text{ or } p} = |R_H|/r\rho \qquad (13)$$

The degree of agreement with bulk values indicates the reliability of the measurement. Values may be lower than accepted bulk values for several reasons: poor contacts, the effects of surface charge, macroscopic nonuniformity, or microscopic inhomogeneities. Microscopic inhomogeneities in $N_A - N_D$ are caused by the effect of crystal rotation or thermal fluctuations in the melt on the microscopic growth rate of the crystal. The effective distribution coefficient is a function of microscopic growth rate. This effect is most likely to cause measurement problems near p-n junctions where compensation is high. Poor mobility at low electric field strength (<1 V/cm) does not forecast low mobility at typical detector field strengths ($\sim 10^3$ V/cm).

Care should be taken in the preparation and handling of the sample to minimize the contribution of the surface states to the measured transport properties. It is standard to polish etch (for example, three parts nitric acid (70%) to one part hydrofluoric acid (49%)) the sample, quenching and rinsing with either deionized water or methanol. If quenched with deionized water, at least one methanol rinse is recommended or the surface may be strongly n-type. The surface may also be turned n-type by exposure to humidity in the ambient. Laboratory to laboratory reproducibility of $N_A - N_D$ for p-type samples is $\pm 5\%$, and for n-type samples $\pm 25\%$.

Photothermal ionization spectroscopy (PTIS) (Lifshits and Nad, 1965) may not be used for routine characterization of HPGe crystals, but this technique has contributed strongly to the present understanding of residual impurities. PTIS was used by Haller and Hansen (1974) and by Hall (1974a) to demonstrate that residual $N_A - N_D$ in HPGe was due primarily to shallow donors (P, Li) and acceptors (B, Al, Ga). In PTIS a bound electron (hole) is raised from the ground state to an excited state of the donor (acceptor) optically, and the event is photoconductively detected if the electron (hole) is then thermally promoted into the conduction (valence) band before cascading back to the ground state. The excellent energy resolution of these optical transitions is shown in Fig. 5 (Darken, 1989) for phosphorus. The transitions are in the far infrared range. The effect of simultaneously irradiating a p-type sample with bandedge light is shown in Fig. 6 (Darken and Hyder, 1983). The different acceptors B, Al, and Ga are clearly resolved and the compensating phosphorus (neutralized by the free electrons generated by the bandedge light) is detected by its negative photoconductivity. The distribution of acceptors as determined from PTIS in an HPGe crystal is shown in Fig. 7 (Darken, 1982b). The x-axis is chosen so that the concentration of normally segregating impurities will be linear. The anomalous distribution profile of aluminum discussed earlier is evident. If aluminum in the melt were nonreactive, it would be expected to have a distribution profile similar to gallium ($k = 0.1$). The long tail on the boron ($k > 1$) distribution indicates a continuing source of boron.

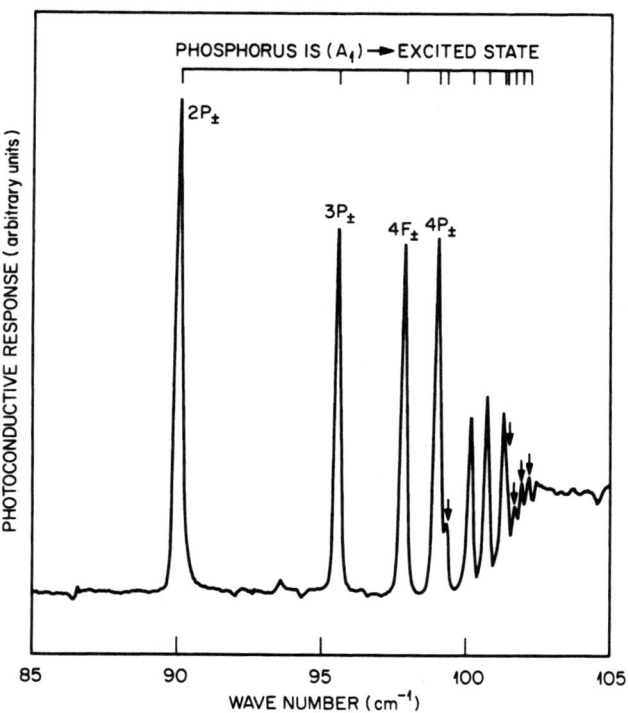

FIG. 5. Photothermal ionization spectrum showing 12 $p\pm$-like transitions from the IS (A_1) ground state of phosphorus. Arrows indicate previously unresolved transitions (Darken, 1989; reprinted with permission of the American Institute of Physics).

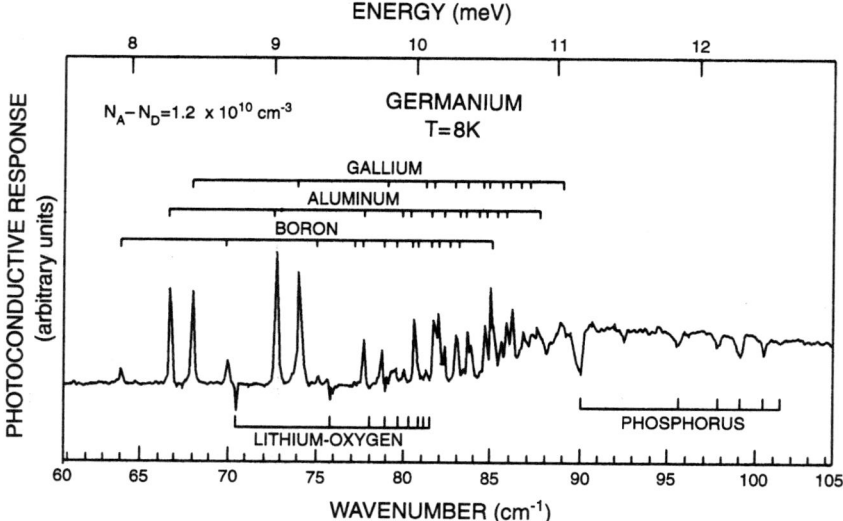

FIG. 6. Photothermal ionization spectrum of a p-type HPGe sample simultaneously illuminated with bandedge light. Ionization of neutralized compensating centers (P, Li-related) contributes negatively to the photoconductivity (Darken and Hyder, 1983; reprinted with permission of the American Institute of Physics).

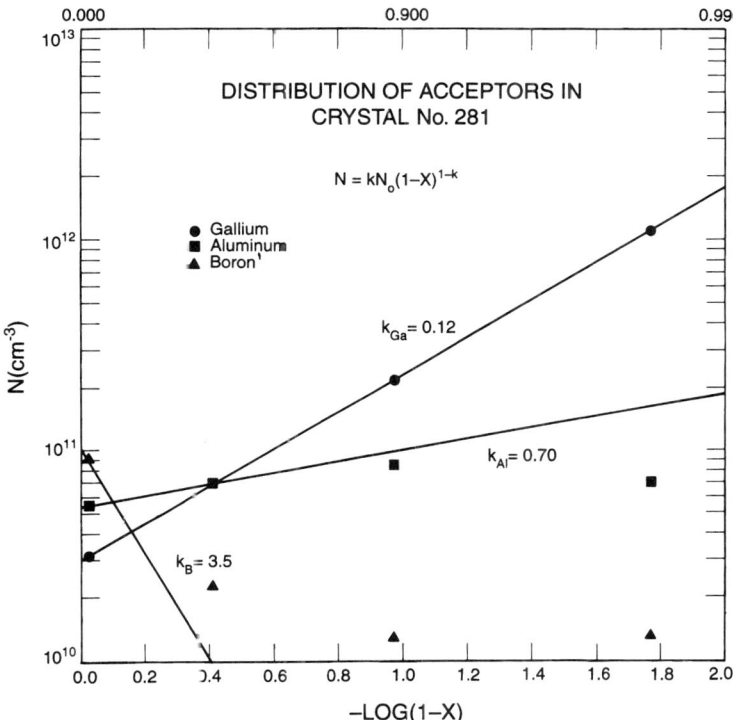

FIG. 7. Distribution of acceptors as determined by PTIS and semiconductor statistics in an HPGe crystal doped with a known amount of gallium (Darken, 1982b; reprinted with permission of the American Institute of Physics)

PTIS is very sensitive for shallow hydrogenic centers. However, this sensitivity diminishes quickly with increasing binding energy of the ground state due to the rapid decrease in the cross section for optical absorption. The deepest acceptor level seen in germanium by PTIS is the substitutional copper level at $E + 43$ meV (Butler and Fischer, 1974; Darken, 1982).

2. CAPACITANCE TRANSIENT TECHNIQUES

Defects with levels deeper than the hydrogenic levels are best detected and characterized by capacitance transient techniques. In particular, the deep level transient spectroscopy (DLTS) technique first implemented by Lang (1974) has been widely applied to HPGe. The procedure is schematically shown in Fig. 8. A semiconductor diode with V_R reverse bias is repetitively pulsed toward zero bias. During this pulse, the capacitance increases as the depletion width decreases, and deep levels in the region just exposed to majority carriers are filled. When V_R is reapplied, these deep levels emit the carriers with a temperature dependent time

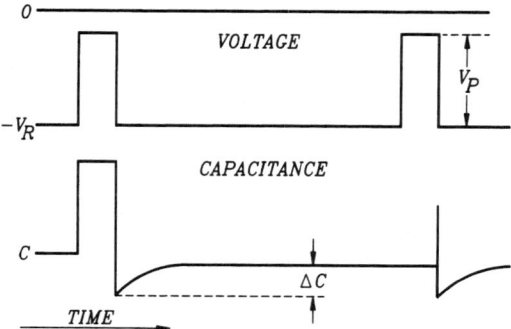

FIG. 8. DLTS voltage pulsing scheme.

constant. There is a corresponding capacitance transient with the same time constant. In DLTS the capacitance signal is processed to produce a DLTS signal that is a maximum when the time constant of a capacitance transient of fixed magnitude is some preset τ_{max}. The sample temperature is ramped through an appropriate temperature range (15–180 K for HPGe), and the DLTS signal is recorded. A typical spectrum is shown in Fig. 9. Peaks due to deeper levels are observed at higher temperatures. For small capacitance transients ($\Delta C/C \leq 0.1$), the DLTS peak height is proportional to DC. The first published results on HPGe were by Haller et al. (1979) and by Evwaraye et al. (1979). Other groups reporting on residual deep levels in HPGe focused their attention primarily on Cu-related acceptor levels (Simoen et al., 1982, 1986a, b; van Goethem et al., 1985; Darken and Jellison, 1989; Kamiura and Hashimoto, 1989). Some problems with using DLTS on high purity germanium have been discussed by Darken (1994).

Broader structures have also been observed and attributed to dislocations (Hubbard and Haller, 1980; Simoen, Clauws, and Vennik, 1985a; van Sande et al., 1986). Van Sande et al. (1986) combined DLTS with transmission electron mi-

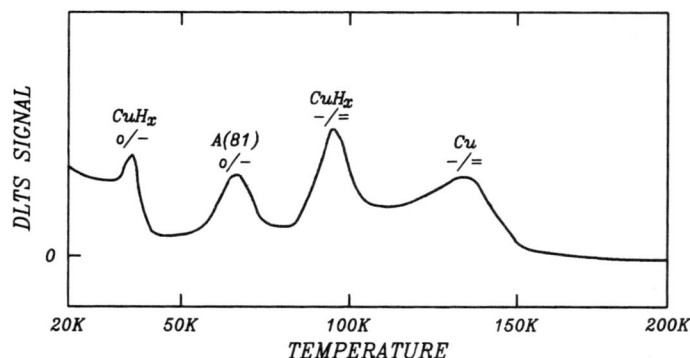

FIG. 9. Deep level transient spectrum of a p-type HPGe sample.

croscopy to identify the specific dislocation types responsible for the different DLTS structures observed. Due to carrier–carrier interactions, the kinetics of carrier capture and emission at extended defects are not as simple as at isolated defects. Capture at an extended defect is usually logarithmic; that is, the strength of DLTS signal depends linearly on the logarithm of the duration of the filling pulse over several orders of magnitude. The evidence for this effect has been summarized by Figielski (1990), but nothing has been published concerning HPGe. We believe this logarithmic filling has been verified but not formally reported by several groups for the dislocation related DLTS structures in HPGe just mentioned. Figielski (1990) has also discussed the origin of peak broadening in the case of the Read model (Read, 1954) for acceptorlike (neutral when empty) electron traps at dislocations.

The concentrations of isolated defects can be determined from DLTS measurements provided that relevant factors are taken into consideration. Simoen et al. (1985b) derived the following expression for r_T, the ratio of the concentration of a particular deep level to the net impurity ionized concentration when the level is filled (but all shallower levels are empty):

$$r_T = \frac{2\Delta C}{C} \left[\frac{1 - \Delta C/2C}{(1 - \Delta C/C)^2} \right] \frac{V_R + V_{bi}}{V_P} \quad (14a)$$

$$\frac{\Delta C}{C} = \frac{1 + D^2}{1 - D^2} \left(\frac{\Delta C}{C} \right)_{\text{EXPT}} \quad (14b)$$

The terms V_R, V_P, C, and ΔC are defined in Fig. 8. V_{bi} is the built-in potential (taken to be 0.5 V), and D is ωRC, the dissipation factor of the sample with series resistance R and ω being the angular velocity of the measurement. The second term in Eq. (14a) is a high trap concentration correction. The third term corrects for regions still depleted during the voltage pulse V_P. Equation (14b) is an equivalent circuit correction based on the assumption that the series resistance of the undepleted germanium is the important equivalent circuit effect while the capacitance meter is measuring equivalent parallel capacitance. In the temperature range of the measurement the series resistance of the undepleted germanium increases strongly with temperature. Thus the equivalent circuit correction (Eq. (14b)) is temperature dependent. This is a problem that must be addressed to quantitatively interpret a spectrum. Simoen et al. (1985b) suggest recording a capacitance–temperature plot simultaneously with the DLTS plot and inferring D as a function of temperature from it. Haller, Hansen, and Goulding (1981) suggest using a lower frequency bridge. Thinner samples would also be helpful. Darken (1994) has suggested a constant capacitance technique that simplifies analysis. In a few important cases (Cu, CuH) where the capture by the doubly ionized state degrades detector performance, the concentration of the center can be more conveniently determined by the DLTS peak due to capture by the shallower singly ionized state.

3. CRYSTALLOGRAPHIC TECHNIQUES

Crystallographic properties are routinely characterized using preferential etch techniques by which pits are formed where dislocations intersect a particular crystallographic surface. Preferential etches for germanium surfaces of various crystallographic orientations are compiled in the *Book of ASTM Standards,* Vol. 10.05 F389-88. Usually the etch pit pattern is noted and the density measured on the whole slice samples. Twin planes, grain boundaries, lineage, mosaic patterns, and smooth pits due to inclusions can also be observed. Detectors made from material containing twin planes or grain boundaries always have excessive leakage current. Severe lineage or mosaic structures are also related to excessive leakage current. Smooth pits have been correlated with severe hole trapping.

4. NEUTRAL DEFECTS

Purification and device fabrication are not usually affected by neutral impurities unless they form second phases. However, some care is required in thermal treatment due to the hydrogen content of HPGe crystals. The hydrogen content of HPGe crystals is about 10^{15} cm^{-3} (Hansen, Haller, and Luke, 1982). At room temperature the hydrogen is stable in some electronically inactive form, possibly as molecular H_2. However, if heated above about 300°C and cooled too rapidly, the atomic hydrogen generated by dissociation may form metastable electronically active complexes. Most frequently observed is the oxygen–hydrogen donor. These effects are discussed by Hall (1984) and the origin of the relevant defects by Haller, Pankove, and Johnson (1989).

Hydrogen passivates some defects in PHGe. Diffusing atomic hydrogen generated from a plasma at 300°C into germanium has been shown to reduce the electrical activity of copper (Pearton, 1982) (perhaps CuH_3 has no levels in the energy gap). The electrical activity of dislocations in p-type HPGe has also been reduced by exposure to a hydrogen plasma (Pearton and Kahn, 1983). However, practical application to HPGe for gamma spectroscopy use has not been possible due to the limited penetration of the passivation ($-100\ \mu$). These results may indicate, though, another advantage of growing HPGe under $H_2(g)$.

IV. Large Volume Detectors

1. INTRODUCTION

The following subsections discuss mostly large volume detectors for use in gamma ray spectroscopy, ranging from a few cubic centimeters to over 500 cc in volume. In x-ray and lower energy gamma ray applications 1 cm germanium or less may be required for complete absorption. A detector size and electrode configuration consistent with good detection efficiency may be selected, which optimizes energy resolutions by reducing device capacitance. However, for higher energy gamma rays complete absorption with HPGe detectors has not been achieved. A coaxial geometry detector oriented with its axis toward the gamma ray source can offer both higher efficiency and lower depletion voltages than a planar device from HPGe with the same N_A-N_D. Very large volume devices are known to offer advantages over the use of multiple detectors (Keyser, Twomey, and Wagner, 1990), particularly for low level environmental measurements. Spectra are presented from the largest high purity germanium detector so far reported.

Applications of x-ray and gamma ray spectroscopy with HPGe detectors include fundamental physics research (e.g., Pehl, 1982), nuclear well logging (Schweitzer, 1991), nuclear power technology, chemical analysis and nuclear medicine (Glasow, 1982), and position sensitive devices have also been developed (Varnell et al., 1984; Luke, Madden, and Goulding, 1985; Luke, 1988). In most applications the detector is operated with a pulse shaping and amplification module, an ADC and either a hardwired or computer based analyzer that generates an energy spectrum. The principles of operation of such systems have been described in the literature (Kandiah and White, 1981; Goulding and Landis, 1982; Britton et al., 1984) and are available from several vendors. Some applications, mostly in the high energy physics arena, involve timing spectroscopy of the detector pulses (Paulus et al., 1981; Quaranta et al., 1983); timing resolutions of about 5 ns are typical for a germanium detector at 1.33 MeV. The following subsections will be concerned mostly with energy spectroscopy performance.

The bandgap of germanium (0.7 eV) is such that operation at room temperature is impossible due to the large leakage current from thermally generated carriers. High leakage degrades the signal to noise ratio so that the device is not usable. The most common mode of operation involves cooling the detector to about 100 ± 20 K using an evacuated cryostat cooled with liquid nitrogen from a dewar. Studies have been made operating at higher (Pehl, Haller, and Cordi, 1973; Nakano, Simpson, and Imhof, 1977) and lower (Stuck et al., 1972) temperatures and also using mechanical cooling devices (Alberti, Clerici, and Zambra, 1979; Sakai, Murakamiand, and Nakatani, 1982; and Stone, Barkley, and Fleming, 1986), but nearly all applications at the present time use liquid nitrogen cooling. Operating leakage currents in the picoamp range eliminates the effect of this source of noise from the final energy spectrum.

Since the introduction of high purity germanium for gamma ray spectroscopy there has been a trend toward fabricating larger volume crystals as material has become available. The standard size crystal of a few years ago (370 g or 70 cc) is now considered a small device, and detectors up to 2.8 kg (526 cc) have recently been constructed by the authors. Another development has been the introduction of "low background" cryostats (Reeves et al., 1984; Brodzinski et al., 1985, 1991; Jagam et al., 1985, 1988) constructed from special materials to reduce unwanted background radiation.

An overview of many aspects of high purity germanium detector operation and performance is presented by Knoll (1989), which should be referred to for principles of radiation detection with a semiconductor diode. The present work emphasizes aspects of detector fabrication and performance that are linked to the germanium material properties.

2. Geometry and Size

The choice of geometry and size of a detector for optimum performance in a particular application usually depends on the required spectral resolution and data collection rate. Figure 10 shows schematic cross sections and electrostatic field distributions in a range of crystal geometries commonly employed in gamma ray spectroscopy.

The planar devices (Fig. 10a, b, c) offer high resolution for photoenergies from about 3 keV up to a few hundred eV but suffer from low collection efficiency at higher energies. Detectors with a small electrical contact (Fig. 10c) combine the benefits of the low capacitance and high resolution of a small planar with a relatively large surface area, but with some loss of field strength. The truncated coaxial detector (Fig. 10d) is ideal for collecting data at a few hundred keV in the presence of a high energy background, which is not well absorbed by the crystal. The "well" geometry (Fig. 10f) is used when high collection efficiency is required, the source being placed inside the central core of the detector and surrounded by nearly 4π solid angle of germanium.

However, the closed end coaxial geometry (Fig. 10e) is by far the most popular device for most applications, being easier to manufacture and with more active volume and lower depletion voltage than the equivalent true coaxial geometry, in

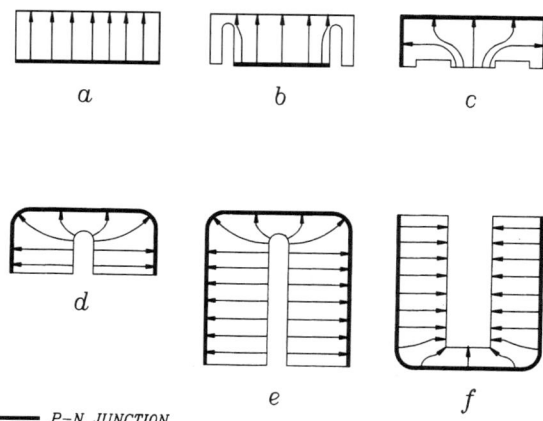

FIG. 10. Schematic cross sections and electrostatic field distributions in high purity germanium detectors. The dark line represents the p-n junction: (a) true planar, (b) grooved planar, (c) low capacity planar, (d) truncated coaxial, (e) closed end coaxial, (f) well geometry.

which the central hole is the length of the detector. The crystal is cut to a right circular cylinder with the ratio of the diameter to the length usually being 1:1 to within 20% and the corners rounded to improve the field strength at the edges of the crystal. The hole is typically 10 mm diameter and cut to within about 10 mm of the closed end. These dimensions offer a solution to the problem of generating field strengths in excess of 1000 V/cm for saturated drift velocity while maintaining a relatively low detector capacitance.

The sizes of coaxial detectors are often classified by the relative collection efficiency (%) of the 1.33 MeV photon from a ^{60}Co source at 25 cm from the detector, compared to that for a 3 in. × 3 in. sodium iodide detector (IEEE Std 325-1986). Current technology provides relative efficiencies in the range from less than 10% to over 150%.

3. CLOSED END COAXIAL DETECTOR FABRICATION

a. Material Considerations

The maximum voltage that can be applied to the reverse bias diode is limited to about 5 kV due to practical constraints in the fabrication of the detector and supporting housing; these include leakage current in the crystal and high voltage filter components, corona discharge, and arcing in the cryostat. Full depletion of very large devices can therefore be achieved only if the net impurity concentration does not exceed about 10^{10} per cc. Conversely, acceptable field strength cannot be accomplished throughout the crystal, even with overvoltage, if the net impurity level is too low. In this situation the field strength at the central core can be sufficiently high to increase the leakage current across the inner contact, while the field strength at the outside is still weak and the charge collection may be poor. There is therefore a minimum net concentration level of about 3×10^9 per cc for optimum detector performance.

Quaranta (1982) has demonstrated that the most uniform field strength in a closed end coaxial detector is accomplished using material with a linearly increasing impurity concentration from the open end to the closed end. Most detector grade crystals utilize this effect to improve the charge collection properties and the resultant energy resolution. Factors affecting charge collection are discussed in detail in Section V.

Closed end coaxial detectors can be constructed from either p-type or n-type material. The fabrication process is discussed in the following sections, but in either case the p-n junction is on the outside surface. This creates a high electric field near the outer diameter where most of the photon interactions take place. Another consequence of this configuration is that the p-type device has a thick (up to 1 mm) undepleted lithium diffused layer on the outside and is therefore not suitable for the detection of photons below about 40 keV. The n-type detector, however, usually has only a 0.3 μm surface dead layer (from the boron implant) and is useful down to 3 keV. Coaxial detectors (independent of polarity) are gen-

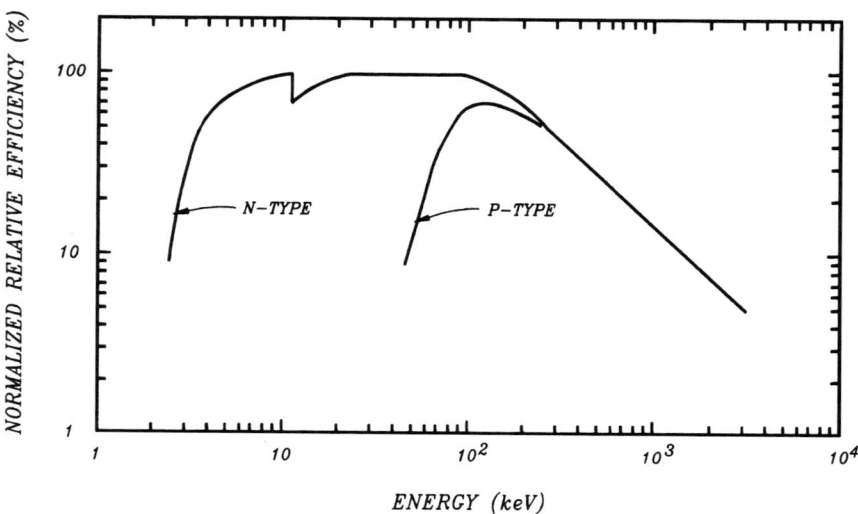

FIG. 11. Relative absorption efficiencies for typical n- and p-type detectors. (Darken and Cox, 1993; reprinted with permission of Oxford Instruments, Inc.).

erally considered to be useful for high energy collection up to about 10 MeV, above which the probability of total absorption and charge collection from a photon is low and therefore the peak to background ratio is poor. Figure 11 shows the relative absorption efficiencies as a function of energy for typical mid-sized n- and p-type coaxial detectors. The plateau in absolute efficiency has been normalized to 100% in each case. The dip in absorption just above 10 keV is due to the germanium K absorption edges.

The outside surface of the crystal can sometimes be damaged during the manufacturing process, for example, while grinding to the required dimensions or cutting the central hole. Precautions can be taken, such as using a blunt tool and abrasive paste to lap out the hole rather than cutting with a diamond tool. Other techniques include ultrasonic cutting and spark cutting. However, it is still necessary to remove up to 0.3 mm of the exposed surface by sanding, polishing, and chemical etching prior to the formation of the electrical contacts and surface passivation. A mechanically damaged surface causes excessive leakage current.

b. Contact Formation

A simple description of a semiconductor radiation detector is that for a reverse bias p-n junction diode with electrical contacts for the collection and measurement of charge. A junction is formed by a highly doped region on the outside surface of the lightly doped bulk material; under operating conditions the depleted volume is therefore almost equal to the physical dimensions of the device. A second "blocking" contact is made to the central core, which greatly reduces the passage of majority carriers and therefore decreases the bulk leakage to acceptable levels when cooled.

Early processes for contact formation on germanium included solution regrowth (Baertsch and Hall, 1970; Llacer, 1972) and Schottky barrier (Pehl et al., 1972) electrodes. Although the Schottky barrier contact has been used extensively and still finds favor it can suffer from lack of ruggedness and reliability particularly on the outer surface of n-type detectors and has now been superseded by some manufacturers. There is a great advantage during the detector fabrication in keeping the crystal temperature below 300°C to avoid the diffusion of copper into the germanium; copper forms deep level traps that severely degrade the charge collection properties. Two contact types that can be formed at low temperature are now commonly used; that is, lithium diffusion and boron implantation.

Lithium diffused at 270–330°C can be used to create a highly doped n-type region several hundred millimeters thick. In p-type material the lithium is on the outside of the crystal and forms the p-n junction (see Fig. 12a), leaving a 1 m layer of undepleted germanium that reduces the low energy detection sensitivity. In the n-type device, the lithium diffusion takes place in the central core and forms a rugged n^+ blocking contact. In a coaxial detector lithium will not diffuse significantly into the crystal should it be stored at room temperature for an extended period of time. A phosphorus implanted n-type electrode has been examined (Hubbard, Haller, and Hansen, 1977) as an alternative in applications where a thin contact is needed, but the process is involves a careful postimplant annealing schedule. The lithium diffusion method is presently used exclusively in coaxial detector fabrication. The boron implanted contact (Ponpon et al., 1972; Jones and Haller, 1987) is ideally suited for detector fabrication, since it has the unusual property of being electrically active in germanium after a room temperature im-

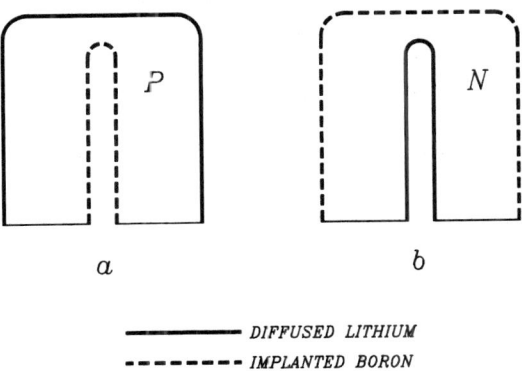

FIG. 12. Contacts on coaxial detectors: (a) p-type, (b) n-type.

plant without a high temperature anneal. An implant of typically 10^{14} to 10^{15} ions per cm^2 at 25 keV with no anneal provides the p-n junction on the outside of the n-type detector and the p^+ blocking contact in the central core of the p-type (Fig. 12). The n-type device has the advantage of a thin entrance window (0.3 μm) through the implant.

There can be some advantage in evaporating a thin metal film (e.g., a few hundred angstroms of gold or aluminum) over the lithium and boron contacts to reduce the electrical contact resistance to the metal mounting components in the cryostat. This is usually performed by a simple vacuum deposition from a hot filament. The improvement gained from this step is most pronounced for the metalization of the lithium core in the n-type crystal; reductions in electronic noise for other metalized surfaces are usually insignificant with coaxial detectors.

c. *Surface Passivation*

Figure 10e shows electrical field lines in an ideal detector orthoganol to the surface of the central core and parallel to the intrinsic surface of the crystal, which supports no electrical contacts. The intrinsic area provides isolation between the electrodes, and approximate electrical neutrality of the surface is important in attempting to establish the ideal field pattern and low reverse bias leakage current (Dinger, 1975). Control of the surface states is attempted by careful chemical treatment followed usually by the evaporation of an insulting or "passivation" layer on the order of 0.1 μm, which stabilizes the surface and protects it from the environment. In practice, process control is best maintained by fabricating devices with a slightly inverted surface.

Various techniques of surface passivation have been tried. Typically, these involve a chemical treatment and vacuum deposition of a silicon oxide or germanium, or sometimes no surface coating at all apart from the native oxide growth (Llacer, 1972). A technique using sputtered hydrogenated amorphous germanium has been well documented (Hansen, Haller, and Hubbard, 1980; Hansen and Haller, 1981). During this process amorphous germanium is sputtered onto the etched surface of the crystal with a gas mixture of argon and hydrogen. The proportion of hydrogen determines the electrical state of the coated surface and can be adjusted to optimize surface state conditions. In theory, therefore, neutrality of the surface can be achieved despite differences in the original electrical activity. The successful application of the surface treatment and passivated layer is usually the most difficult step in the manufacturing process. Contamination and subsequent degradation of the surface can take place even after passivation from solvent or acid fumes, and the finished device must be handled with great care to avoid damage.

d. *Detector Housing*

The completed detector requires cooling to about 100 K in an evacuated atmosphere, which provides thermal insulation and a clean environment for the sensitive passivated surface of the crystal. The most common method employs liquid nitrogen cooling from a copper cold finger dipstick surrounded by an evacuated tube, which connects to a crystal mounting cup. The cup supports the electronic components (field effect transistor, feedback capacitor and resistor) for the first stage of signal amplification, as well as the crystal itself. The vacuum space is enclosed by an endcap over the cup, with outside dimensions ranging from 2.75 in. diameter to 4 in. diameter, depending on the crystal cup size. A schematic of this configuration is shown in Fig. 13. The dipstick is typically immersed in a 30 liter dewar of liquid nitrogen.

This simple cryostat design incorporates a number of interesting features that are worth mentioning. Small vibrations caused by liquid nitrogen boiling at the bottom of the dipstick can potentially be transmitted through the copper cold finger into the crystal. The crystal supports a high voltage of a few thousand volts when operating, and any movement relative to the ground plane of the endcap will

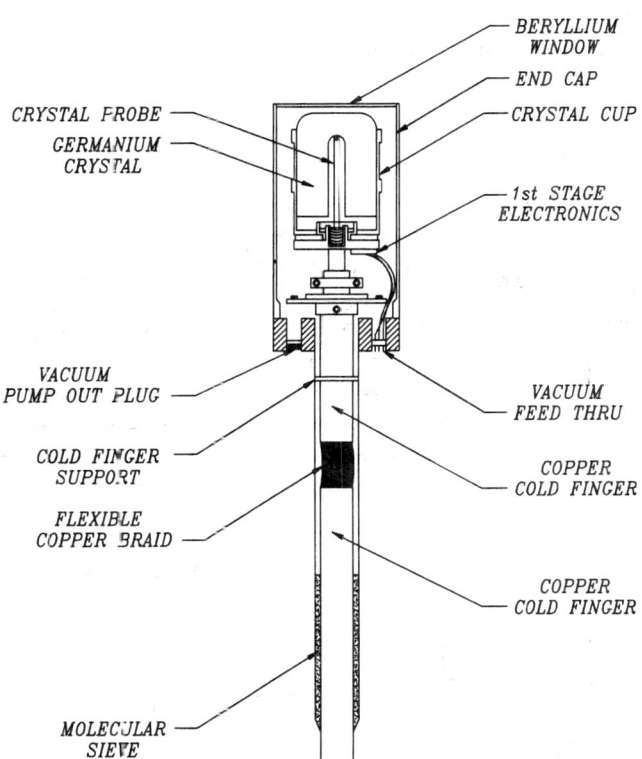

FIG. 13. Schematic cross section of a dipstick cryostat (Darken and Cox, 1993; reprinted with permission of Oxford Instruments, Inc.).

induce an additional noise signal into the crystal and the signal path, thus degrading the signal to noise ratio and therefore the energy resolution. This phenomenon is known as "microphonic pickup" and can be reduced significantly by careful cryostat design, including the use of a flexible section in the cold finger rod, as shown in Fig. 13.

It is important to maintain the vacuum inside the cryostat as high as possible during the operating life of the detector. This helps to protect the passivation layer from contamination and also ensures good thermal insulation and a predictable liquid nitrogen boil-off rate. A small amount of molecular sieve material is placed inside the cryostat near the bottom of the cold finger, which pumps the evacuated space when the detector is cold and absorbs outgassing from the walls of the cryostat for a period of several years. The detectors can even be thermally cycled to room temperature and cooled again without significant degradation of the operating vacuum and detector performance.

Many liquid nitrogen cooled cryostat configurations are available commercially in addition to the simple dipstick model. Figure 14 shows some common examples, some of which incorporate a dewar as part of the cryostat itself. The application of the spectroscopy system often dictates the cryostat model, which sometimes must be custom designed.

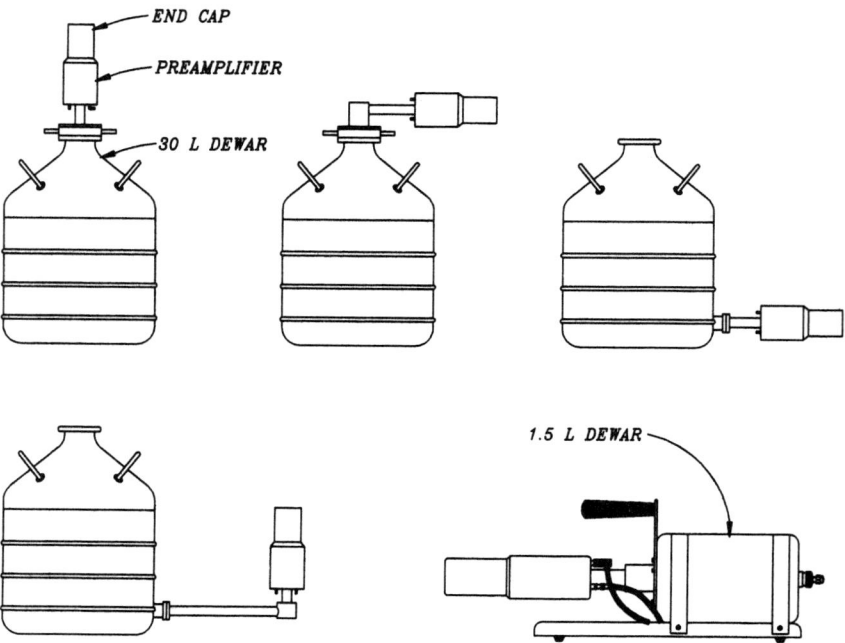

FIG. 14. Examples of germanium cryostat configurations.

4. Closed End Coaxial Detector Performance

a. Noise and Resolution

A typical pulse shaping and amplification system and the various components of the total system noise and spectral resolution are described by Goulding and Landis (1982). For an ideal detector, the spectral resolution L (full-width at half-maximum, or FWHM) is the sum of the electronic noise contribution and the statistical fluctuations in the amount of charge generated for a given energy of incident photon.

$$L^2 \text{ (eV FWHM)} = L_F^2 + L_N^2 \qquad (15)$$

where L_N is the electronic noise contribution in eV FWHM and L_F is the statistical broadening, given by

$$L_F^2 \text{ (eV FWHM)} = 5.52 \times F \times E \times W \qquad (16)$$

W is the energy (eV) required to produce one ion pair in the detector (2.98 eV for liquid nitrogen cooled germanium); E is the energy (eV) of the incident photon; and F is the Fano factor, which quantifies the division of the absorbed energy between phonon and ion pair production in the crystal. A value for F of 0.105 may be used.

The electronic noise contribution, L_N, is a function of the FET and feedback resistor noise characteristics, the total capacitance at the FET gate lead to ground including that of the detector, the leakage current, any series resistance, and dielectric noise components (Goulding and Landis, 1982). Some of these factors can vary with the size and geometry of the crystal, and all depend on the amplifier shaping time used. The optimum shaping time for minimum noise with a large volume germanium detector is typically 4 or 6 μsec, or equivalently, 8 or 12 μsec peaking time. Figure 15 shows the minimum noise contribution as a function of crystal volume for a sample of 50 detectors from 400 to 1900 g. Note the trend of increasing noise with volume.

The statistical broadening, L_F, is independent of the detector geometry from Eq. (16), but in practice the resolution can be worse than that calculated due to incomplete charge collection caused by deep level traps in the material. This is not a fundamental limit to the resolution, and as the crystal growth technology improves, even very large detectors will approach the theoretical resolution defined by Eqs. (15) and (16). Charge trapping effects are discussed in detail in section V. The peak shape can be characterized by the ratio of the full-width at tenth-maximum (FWTM) and full-width at fiftieth-maximum (FWFM) to the FWHM, the theoretical values for a true Gaussian being 1.82 (FWTM/FWHM) and 2.38 (FWFM/FWHM).

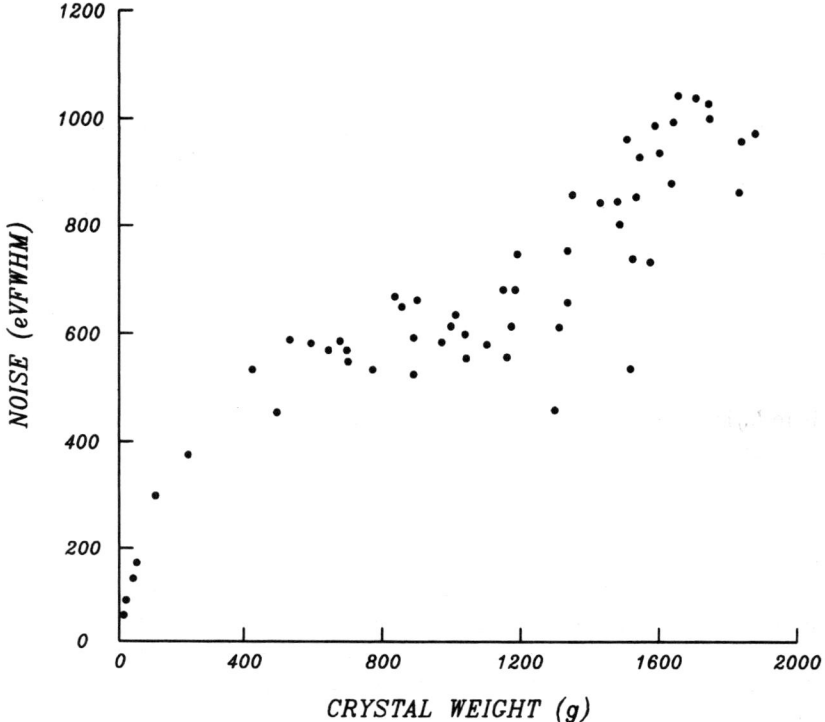

FIG. 15. The minimum noise contribution (eV FWHM) as a function of crystal weight (g) for a sample of 50 detectors. The six lowest weight data points are noise line widths from manufacturer's specifications for truncated coaxial and planar devices.

Figure 16 is a spectrum from a ^{60}Co source collected with a relatively small 600 g (22% relative efficiency) p-type detector. The resolution at 1332 keV is 1.64 keV FWHM, which is about the best achievable at this energy with germanium. The FWTM/FWHM is 1.86, and the FWFM/FWHM is 2.48, which are typical figures for a high performance germanium detector and close to the theoretical values for a Gaussian.

In contrast, Fig. 17 is part of a ^{60}Co spectrum collected with the largest germanium detector so far reported; that is, 2.8 kg or 157% relative efficiency p-type. The resolution at 1332 keV is 1.96 keV FWHM; the FWTM/FWHM is 1.94, and the FWFM/FWHM is 2.81. This detector represents the state of the art in large germanium crystal growth and detector fabrication. Figure 18 is the same ^{60}Co spectrum displayed over a large energy range and illustrates the typical features of a germanium detector spectrum. The peak to Compton ratio is a standard measurement of the detector performance with a ^{60}Co source, being a function of both the energy resolution and active volume. It is defined (ANSI/IEEE Std 325-1986)

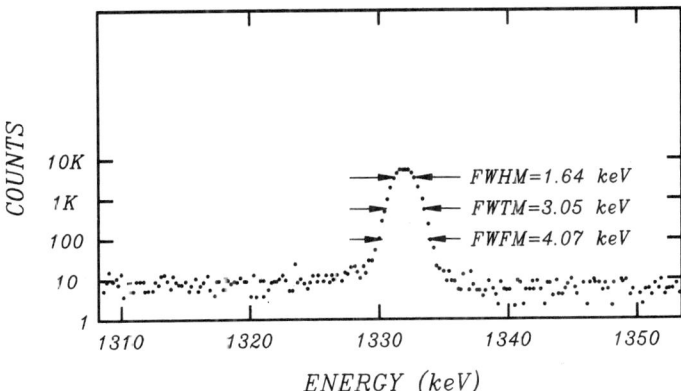

FIG. 16. A ^{60}Co spectrum collected with a 22% relative efficiency *p*-type detector (Darken and Cox, 1993; reprinted with permission of Oxford Instruments, Inc.).

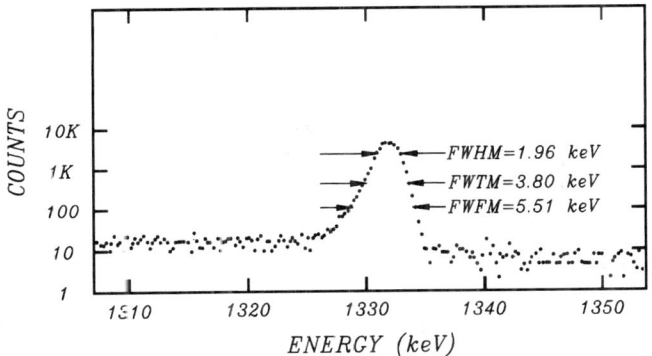

FIG. 17. A ^{60}Co spectrum collected with a 157% relative efficiency *p*-type detector (Darken and Cox, 1993; reprinted with permission of Oxford Instruments, Inc.).

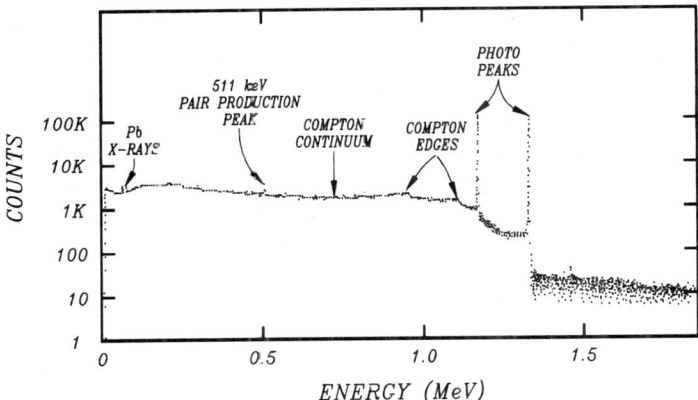

FIG. 18. A ^{60}Co spectrum collected with a 157% *p*-type detector showing typical features of a germanium detector spectrum.

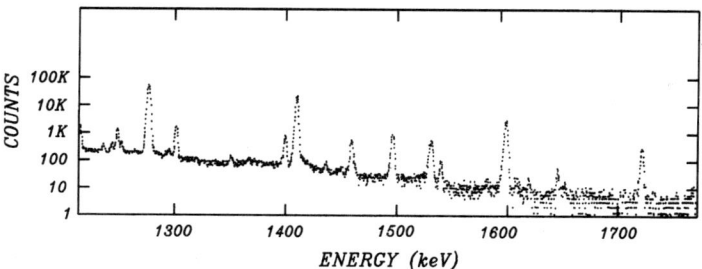

FIG. 19. A mixed isotope spectrum collected with a 157% p-type detector.

as the ratio of the number of counts in the 1.33 MeV peak channel to the average number of counts in the channels from 1.040 to 1.096 MeV. For this 157% device a measurement was performed with the detector inside a 2 in. lead shield to reduce the natural background from K-40. A 0.05 μCi ^{60}Co source was on the endcap, and a peak to Compton ratio of 94 was measured. Figure 19 is part of a spectrum collected with the same 157% detector using a mixed isotope source of ^{152}Eu, ^{154}Eu, and ^{155}Eu, and it illustrates good collection efficiency and peak shapes at energies approaching 2 MeV.

b. Performance at High Counting Rates

The energy resolution performance discussed so far has been concerned with data collected at relatively low counting rates of a few thousand counts per second or less. There are applications, for example, in the fields of nuclear physics research, the nuclear power industry, neutron activation analysis, and others, where it is necessary to analyze data at the fastest possible rate while maintaining the high resolution realizable with germanium detectors. Complete high counting rate systems and their design features have been described in the literature (Kandiah and White, 1981; Howes and Allsworth, 1986; Simpson *et al.*, 1991), and most involve replacing the continuous feedback resistor in the cryostat by either an optical coupled device (Kandiah and White, 1981) or an additional transistor (Landis *et al.*, 1982). This technique periodically removes the charge built up on the feedback capacitor without the degradation of signal to noise ratio associated with the Nyquist current noise of a relatively low value feedback resistor, which would be necessary for high rate counting.

High rate data collection can be achieved only by operating the pulse shaping amplifier at a peaking time shorter than that for the optimum signal to noise ratio condition. Under these circumstances the energy resolution is degraded not only by the increased electronic noise component but also by the so-called ballistic deficit effect in the detector itself. This is a result of the finite transit time of the charge carriers generated in the germanium after the photon interaction; full

2. HIGH-PURITY GERMANIUM DETECTORS

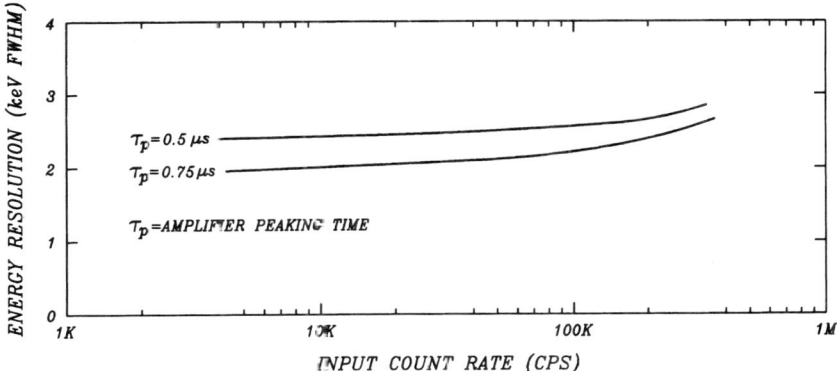

FIG. 20. Energy resolution (eV FWHM) at 1332 keV as a function of input count rate from a ^{60}Co source using an *n*-type detector with transistor reset feedback.

charge collection may take several hundred nanoseconds, and therefore at shorter peak shaping times there is a probability that not all of the deposited charge will be analyzed, thus broadening the spectrum peaks on the low energy side. Various electronic circuits have now been devised (Kandiah and White, 1981; Loo and Goulding, 1988; Hinshaw and Landis, 1990; Goulding, Landis, and Hinshaw, 1990; Simpson *et al.*, 1990) that significantly reduce this effect and allow good spectral resolution with data collection rates of up to 100,000 counts per second into the analyzer.

Figure 20 shows the resolution at 1332 keV as a function of input count rate from a ^{60}Co source using a 16% efficient *n*-type coaxial detector operating with transistor reset feedback (Nashashibi and White, 1990) and ballistic deficit correction in the amplifier (Hinshaw and Landis, 1990) at short peaking times (τp). Note that reasonable resolution is maintained even at the maximum data acquisition rate. Figure 21 shows the data output rate from the ADC into the analyzer as a function of input count rate for the same system.

V. Charge Collection

The dominant characteristic of germanium detectors is their excellent energy resolution in gamma ray spectroscopy. Nearly complete charge collection of both electrons and holes make this possible. Coaxial devices with volumes up to 500 cm^3 with greater than 99.9% collection for holes and up to 350 cm^3 with 99.9% collection for electrons have been fabricated. In this section charge collection deficits due to trapping are considered. Issues relevant to current production, as well as more recent published work, is reviewed.

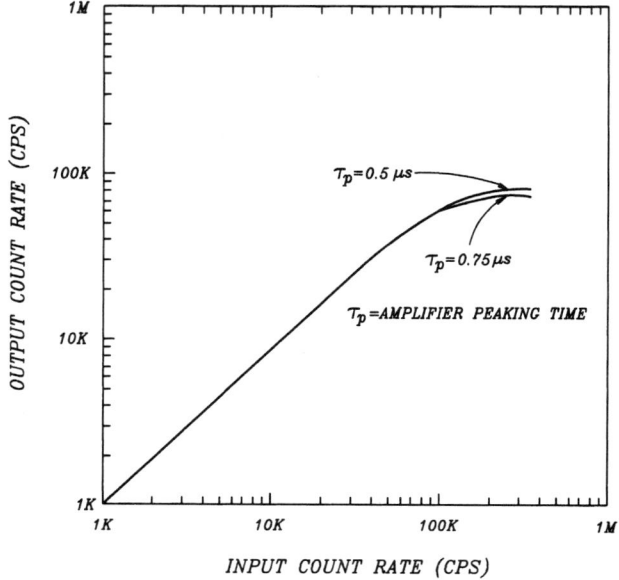

FIG. 21. Data output rate from the ADC as function of input count rate for the same system used in Fig. 20.

1. GEOMETRIC EFFECTS

The geometry and polarity of the contacts have a decisive effect on the relative sensitivity of energy resolution degradation to either electron trapping or hole trapping. In the closed end coaxial geometry with the standard electrode configuration (usually applied to p-type material), resolution degradation is more sensitive to hole trapping. In this configuration the outer contact is at a positive bias and the inner contact is near ground potential. In the reverse electrode configuration (usually applied to n-type material) resolution degradation is more sensitive to electron trapping. Here the outer contact is negatively biased and the inner contact is near ground potential. The difference in resolution sensitivity to hole trapping between these two configurations was dramatically demonstrated by irradiating a pair of detectors (one of each electrode configuration) with the same fluence of fast neutrons (Pehl et al., 1979). Fast neutrons introduce significant hole trapping, but no reported electron trapping. The performance of the detectors after about 10^{10} cm^{-2} neutrons is shown in Fig. 22. A concentration of hole traps sufficient to devastate the energy resolution obtained with the conventional electrode detector only moderately affected the resolution of the reverse electrode detector. There are two reasons for this uneven dependence of charge collection on the different carrier types. In the first place for a detector uniformly irradiated with gamma rays, the carrier type collected at the inner contact will on average have to traverse a longer distance than the carrier type collected at the outer contact.

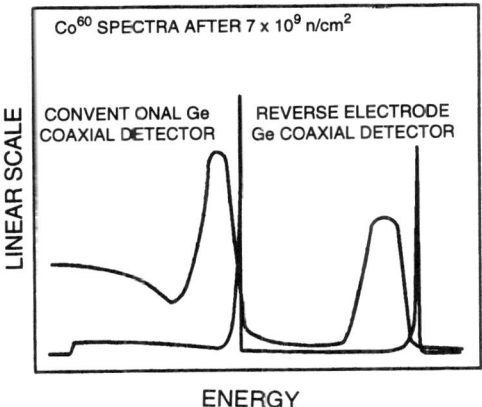

FIG. 22. ^{60}Co energy spectra obtained from both the conventional and reverse electrode configuration HPGe coaxial detectors after a neutron fluence of 7×10^9 n/cm^2 (reprinted with permission from Pehl et al., 1979. ©1979 IEEE).

The second reason has to do with the mechanism by which the signal pulse is generated in the external circuit. Charge is induced on the contacts concurrent with the movement of mobile carriers inside the detector. The amount of charge dq induced by the movement of charge q_o from r to $r + dr$ is given by

$$dq = q_o \frac{\hat{E}_1(r) \cdot dr}{V_a} \tag{17}$$

where V_a is the total electrostatic potential across the contacts, $\hat{E}_1(r)$ is the electric field at r due only to the electrostatic potential difference between the contacts. Cavelleri et al. (1963) were the first to point out that by Ramo's theorem this theoretical electric field, rather than the actual field including space charge, should be used in Eq. (17).

In a coaxial detector we refer to the carrier type that moves to the inner contact as the majority carrier and to the carrier type that moves to the outer contact as the minority carrier. For an ionizing event at radius r that generates a separation of charge q_o, the fraction of induced charge at the electrodes generated by the movement of the majority carrier to the inner contact is (for perfect charge collection)

$$\frac{q_{maj}}{q_o} = \frac{\ln r/r_1}{\ln r_2/r_1} \quad \text{(coaxial case)} \tag{18a}$$

$$\frac{q_{maj}}{q_o} = \frac{r_1^{-1} - r^{-1}}{r_1^{-1} - r_2^{-1}} \quad \text{(hemispherical case)} \tag{18b}$$

where r_1 is the inner radius and r_2 is the outer radius. The closed end of the coaxial

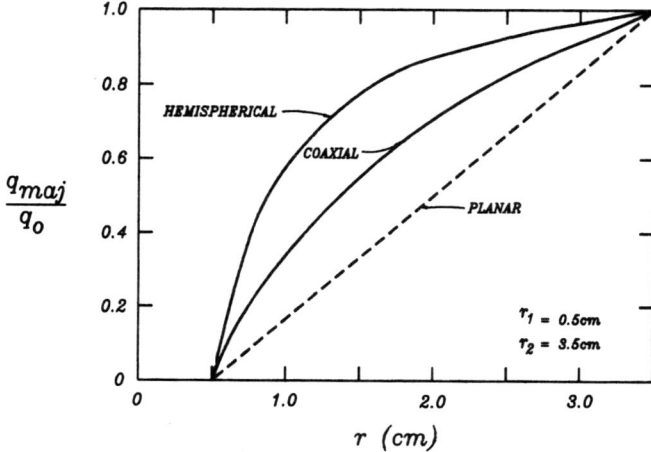

FIG. 23. Majority carrier contribution to charge collection as a function of the position of the ionizing event for hemispherical, coaxial, and planar electrode geometries.

detector may be better represented by concentric hemispheres than by coaxial cylinders. In Fig. 23 the fraction of charge collection due to majority carriers is plotted as function of r for a detector with a 70 mm outer diameter and a 10 mm inner diameter. The corresponding planar case has been added for comparison. The majority carriers, in addition to traveling on average a longer distance, also traverse the region of highest electric field $\hat{E}_1$ near the inner contact, and their relative importance in charge collection is further enhanced.

As a practical matter, charge collection in standard electrode devices is limited by residual hole traps and in reverse electrode devices by residual electron traps. From a detector user's point of view, if radiation damage (causing hole trapping) is anticipated, a reverse electrode detector is preferred.

2. TRAPPING BY EXTENDED DEFECTS

Extended defects usually have minimal effect on transport properties, may be difficult to detect or quantify via DLTS, but can have a devastating effect on charge collection. Oxide inclusions were the first extended defect whose effect on the charge collection of detectors was reported (Hall, 1972). Some of the worst charge collection ever observed in HPGe detectors has been due to hole trapping by these defects.

Dislocation related bands have been observed in DLTS of p-type material at about 25 K (Hubbard and Haller, 1980; Simoen *et al.*, 1985a; van Sande *et al.*, 1986) and in n-type material at 40 K and at 60 K (Simoen *et al.*, 1985; van Sande *et al.*, 1986). Van Sande *et al.* (1986) used electron microscopy to identify the dislocation types responsible. They correlated the 25 K p-type band and the 40 K n-type band with [112]-type dislocation (30° mixed and 90° edge) and the 60 K n-type band with [100]-type dislocations (90° edge). These authors also report a correlation between charge collection and EPD, recommending EPD <5,000 cm^{-2} in the reverse electrode configuration (n-type) and EPD <10,000 cm^{-2} in the standard electrode configuration (p-type). These limits are in general agreement with the resolution degradation observed by Glasow and Haller (1976) on planar detectors.

Extended defects can be generated in germanium detectors by exposure to energetic particles. Fast neutron damage is common in detectors operated around accelerators, and protons (10^5–10^6 eV) degrade detectors used in space. For example, a 1 MeV neutron can transfer up to 54 keV to a germanium atom in a "knock-on" collision. This energetic atom then creates a localized disordered region where electrical properties are determined by the locally high concentration of lattice point defects (Gossick, 1959; Crawford and Cleveland, 1959). The resolution degradation characteristics of the detectors described by Pehl et al. (1979) were interpreted by Darken et al. (1980, 1981) as originating from the hole kinetics of these disordered regions. Further discussion will be deferred to the section on radiation damage.

3. TRAPPING BY POINT DEFECTS

As discussed in subsection III.2, deep level point defects can be detected with great sensitivity ($\sim 10^7$ cm^{-3}) in HPGe by DLTS. Reports of point defect electron traps in n-type HPGe have been rare and controversial. Published reports on point defect hole traps have been more frequent, and there has been general agreement on the emission and capture kinetics of the levels observed. The levels due to simple substitution of a metal atom in the germanium lattice (e.g., Cu, Zn) are firmly identified, and some levels are known to be due to complexes. However, the composition and arrangement of the complexes are usually uncertain. Simoen et al. (1982a) and later Darken and Jellison (1993) related the energy resolution degradation of a set of p-type coaxial detectors to residual copper related defects observed by DLTS. The kinetics of hole capture and emission at defect sites and their effect on charge collection is discussed in this subsection.

The kinetics of hole emission and capture at acceptors are related to each other by the principle of detailed balance. According to this principle, at thermodynamic equilibrium all reaction paths are equally traveled in both directions. The following expression relates the emission rate e from an acceptor to the capture cross section σ of the reverse capture process (see for example Miller, Lang, and Kimerling, 1977):

$$e = \frac{\sigma <v> N_v}{g} \exp - \Delta G/kT \qquad (19)$$

where $<v>$ is the average thermal velocity of a hole, N_v is the effective density of states, ΔG is the Gibbs free energy for the ionization reaction (for example)

$$A^0 = A^- + e^+ \quad (20)$$

and g is the ratio of electronic degeneracy of the acceptor before ionization to its degeneracy after ionization. (This factor is normally contained in ΔG, but the form of Eq. (19) has gained wide currency.) Finally, σ is related to the mean capture time τ_c and the hole concentration p by

$$\sigma = (p<v>\tau_c)^{-1} \quad (21)$$

Using Eq. (21), Eq. (19) can be rewritten

$$e = \frac{N_v}{4p\tau_c} \exp(-\Delta G/kT) \quad (22)$$

where g has been taken to be 4 and the experimentally determined parameters p and τ_c are explicitly displayed in the prefactor of the right-hand side. This prefactor term may be thought of as an attempt-to-escape frequency of a bound hole. It has been shown that, for all the residual deep level point defect acceptors observed in p-type high purity germanium, the prefactor term can be expressed in the form (Darken and Jellison, 1989; Darken, Sangsingkeow, and Jellison, 1990)

$$\frac{N_v}{4p\tau_c} = \eta_i \exp(-E_i/kT) \, kT/h$$

$$0 < \eta_i \leq 1 \quad (23)$$

$$E_i \geq 0$$

where k is Boltzmann's constant, h is Planck's constant, and η_i and E_i are kinetic parameters of the center i. In Figs. 24 and 25, $N_v/4p\tau_c$ is plotted for capture at several singly and doubly bound ionized centers. In many important cases $\eta_i = 1$ and $E_i = 0$. Such levels possess "standard" kinetics, the maximum rate of ionization and capture allowable by Eq. (23). The attempt-to-escape frequency is that of a thermal phonon, and the temperature dependence of σ may be recast as a wave vector k dependence (Darken, 1992):

$$\sigma = \pi/k^2 \quad (24)$$

A cross section of $2\pi/k^2$ would represent capture of all incident s-wave holes

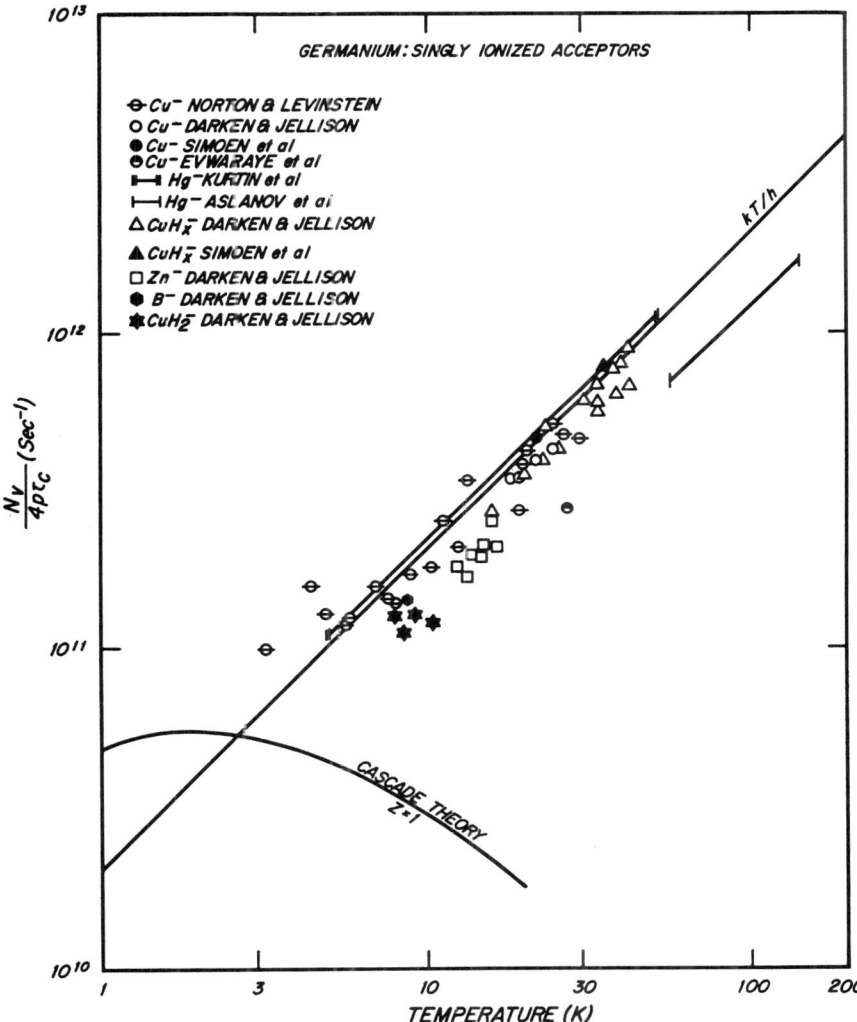

FIG. 24. Recombination at singly ionized acceptors in germanium (Darken, 1992; reprinted with permission of the American Physical Society).

that could make an angular momentum conserving transition to an s ground state. It would appear that Eq. (24) represents capture of such holes limited at most, for example, by the phase of a phonon mode. The electric field dependence of σ may be inferred from Eq. (24), assuming spherical bands

$$\sigma(T, \hat{E}) = \sigma(T)<v^2>_o/<v^2>_E \qquad (25)$$

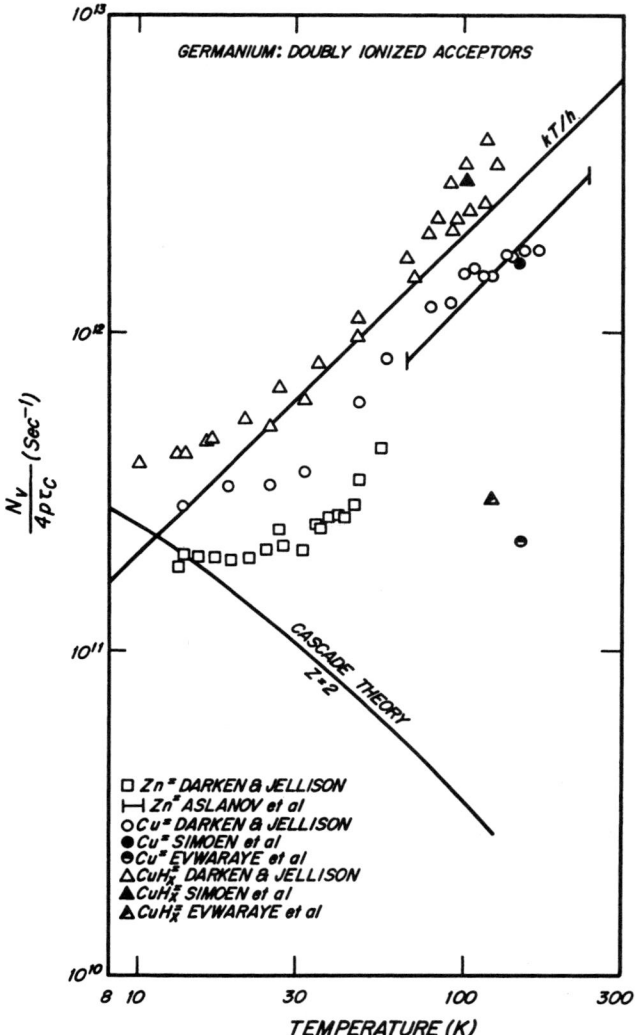

FIG. 25. Recombination at doubly ionized acceptors in germanium (Darken, 1992; reprinted with permission of the American Physical Society).

where $<v^2>_o$ is averaged over the states occupied at thermal equilibrium and the average $<v^2>_{\hat{E}}$ is over states occupied in the presence of a field $\hat{E}$. The results of a calculation for an acceptor level with standard kinetics at 80 K is shown in Fig. 26 (Darken and Jellison, 1993). Also plotted in Fig. 26 (for 80 K) is $v_d(\hat{E})/\sigma(\hat{E})<v>_{\hat{E}}$, where $v_d(\hat{E})$ is the drift velocity. This quantity is important (Raudorf and Pehl, 1987) in determining the mean drift length for holes $\lambda_h(r)$. In the pres-

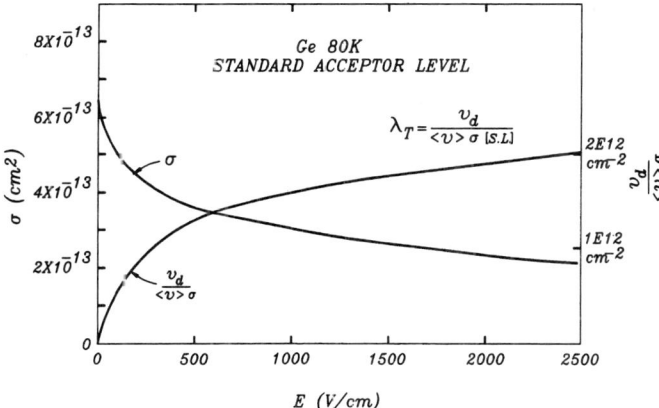

FIG. 26. Calculated values for σ and $v_d/\langle v \rangle \sigma$ vs. electric field strength for standard acceptor levels in germanium at 80 K (Darken and Jellison, 1993; reprinted with permission of American Institute of Physics).

ence of a concentration of standard levels [S.L.(r)], $\lambda_h(r)$] is given by

$$\lambda_h(r) = \frac{v_d}{\sigma \langle v \rangle_{\hat{E}} [\text{S.L.}(r)]} \tag{26}$$

The relationship between the emission time τ_e of the trapped carrier from the defect and conditions of the measurement determine the effectiveness of the trap. If the time constant for emission is comparable to or shorter than the amplifier time constant t_A, the trapped charge that is reemitted will contribute to the pulse height. Under simplifying assumptions the effect on η (charge collection efficiency) of reemission may be written (McMath and Sakai, 1972)

$$\eta = 1 - \delta \exp(-t_A/\tau_e) \tag{27}$$

where δ is the charge collection deficit from the defect in the absence of reemission.

On the other hand, if the trap is filled at thermal equilibrium, the trap might not activate over the duration of a particular experiment if the emission time is too long. Even if the trap is not filled at thermal equilibrium, hole current due to ionizing sources (even background radiation might be enough) may evolve the trap toward a filled condition. These considerations were applied quantitatively to isolated defects in germanium by Simoen et al. (1986). Figure 27 is a plot of τ_e vs. 1000/T for several commonly observed acceptors in germanium (Darken and Jellison, 1993). The lower limit for an effective trap is set at $\tau_e = 2.5$ μsec for which, by Eq. (27), 90% of the trapped charge will be recovered for $t_A = 6$ μsec. The upper limit is rather broadly put at $\tau_e = 10^4$ sec. Conditions during a particu-

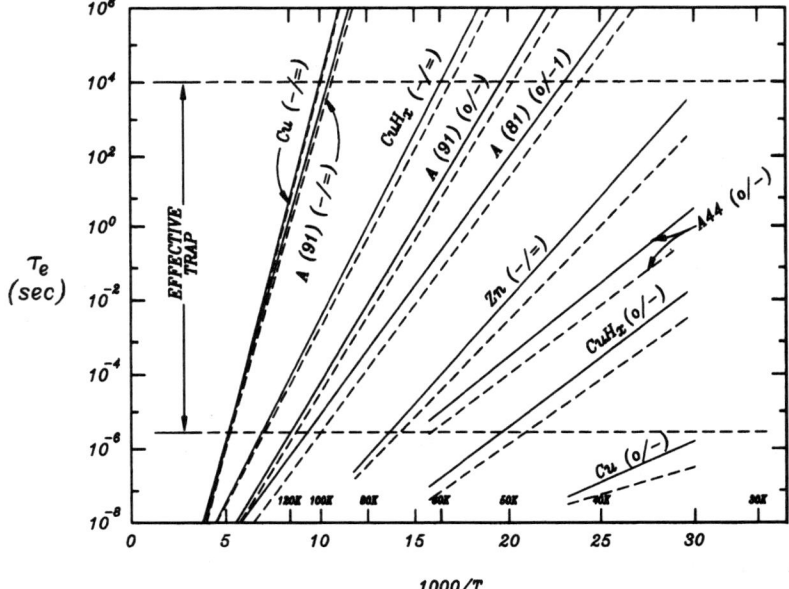

FIG. 27. Emission time τ_e vs. $1000/T(K^{-1})$ for commonly observed deep acceptor levels in HPGe. The full line is for $\hat{E} = 0$, and the dashed line is for $\hat{E} = 1500$ V/cm.

lar spectral run would have to be taken into consideration. Two observations made by Simoen et al. (1986c) are relevant. First, over a broad temperature range substitutional copper is not an effective trap: $Cu^0 \rightarrow Cu^- + e^+$ is too fast, and $Cu^- \rightarrow Cu^= + e^+$ is too slow. Second, $A(44)$ (acceptor defect at 44 K in DLTS), observed in high concentrations in dislocation free HPGe, is not an effective trap either because reemission is too fast at normal HPGe detector operating temperatures.

4. Modeling Charge Collection in Spectrometers

The modeling of gamma ray line shapes in germanium spectroscopy has followed the approach of Trammell and Walter (1969), in which each ionizing event is considered to produce a well-localized charge. Compton scattering and pair production are ignored. The basic equation for uniform irradiation of the detector with a gamma ray source of energy E is

$$\frac{dN(E)}{dE} = \frac{1}{V\sqrt{2\pi}} \int \frac{1}{\sigma(r)} \exp\left[-1/2 \frac{E - \eta(r)E_o}{L^2(r)}\right] dv \qquad (28)$$

where V is the active volume of the detector. Here, $dN(E)$ is the fraction of total events counted between E and $E + dE$, and $dN(E)/dE$ is proportional to the counts per multichannel analyzer channel at energy E. The spatial integration indicated in Eq. (28) would give the equivalent of one point in the energy spectrum of a particular peak. The spectrum would be generated by calculating many such points. Finally, $L^2(r)$ is the variance of the peak shape, and $\eta(r)$ is the induced charge collection efficiency due to both hole and electron collection for a photoelectric event recurring at r:

$$\eta(r) = \eta_h(r) + \eta_e(r) \tag{29}$$

The relative contribution of holes and electrons in the absence of trapping depends on r and the geometry of the electrodes as discussed in Subsection V.1. The effects of trapping are described by the local mean drift lengths (defined by Eq. (26) for holes with trapping by standard levels). The effect of different traps on $\lambda(r)$ is given by

$$\frac{1}{\lambda(r)} = \frac{1}{\lambda_1(r)} + \frac{1}{\lambda_2(r)} + \cdots \tag{30}$$

where $1, 2, \ldots$ are here different species of traps characterized by a λ determined by trap concentration and cross section using an analogous form of Eq. (26). A calculation of $\eta(r)$ will depend on λ_h along the collection path of the holes to the negative electrode and on λ_e along the collection path of electrons to the positive electrode. For simple geometries, analytical expressions are available.

The variance $L^2(r)$ in Eq. (28) is here considered due to three distinct mechanisms:

$$L^2(r) + L_F^2 + L_N^2 + L_T^2(r) \tag{31}$$

where L_F^2 is due to the statistics of charge production, L_N^2 is due to electronic noise, and $L_T^2(r)$ is due to trapping effects. Although it would seem that L_T^2 could be calculated knowing $\eta_h(r)$ and $\eta_e(r)$, in fact, practice has been to use an experimental relationship between $L_T(r)$ and $\eta(r)$ (Raudorf and Pehl, 1987).

Raudorf and Pehl (1987) modeled the resolution degradation of the neutron damaged detectors experimentally reported by Pehl et al. (1979). They assumed a coaxial geometry (ignoring the closed end) and could take the trap distribution as uniform. They found $\lambda_h = (2.5 \times 10^{11})/n_f$ (where n_f is the first neutron fluence) best fit the data. The observed and calculated resolution degradation FWHM at 1.33 MeV of the two detectors are shown in Fig. 28. The calculated resolution of a conventional electrode coaxial detector as a function of outer diameter is shown in Fig. 29 for various values of λ_h.

A 1.33 MeV gamma ray will on average lose energy through three Compton

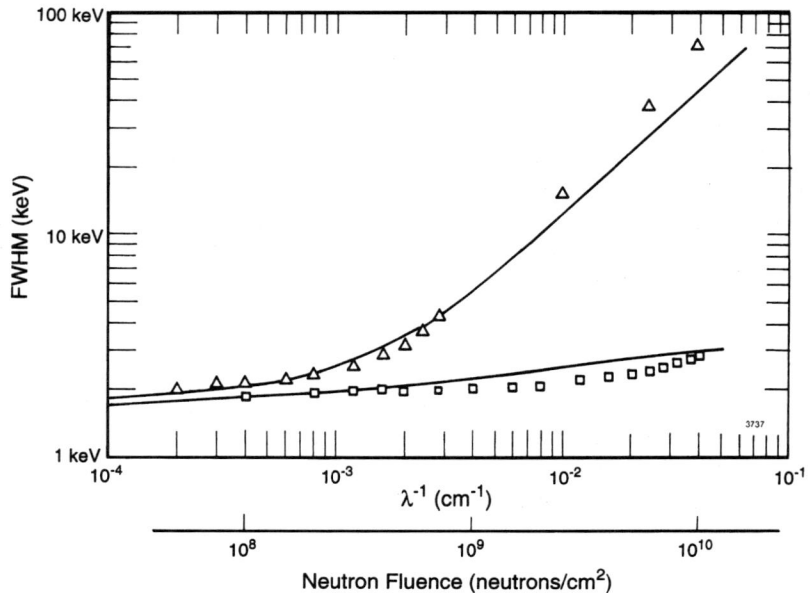

FIG. 28. Comparison of calculated FWHM measurements at 1332 keV to experimentally measured spectra (Pehl *et al.*, 1979) for conventional (triangles) and reverse electrode (squares) geometry HPGE coaxial detectors at various fast neutron fluences (Raudorf and Pehl, 1987; reprinted with permission of Elsevier Science Publishers).

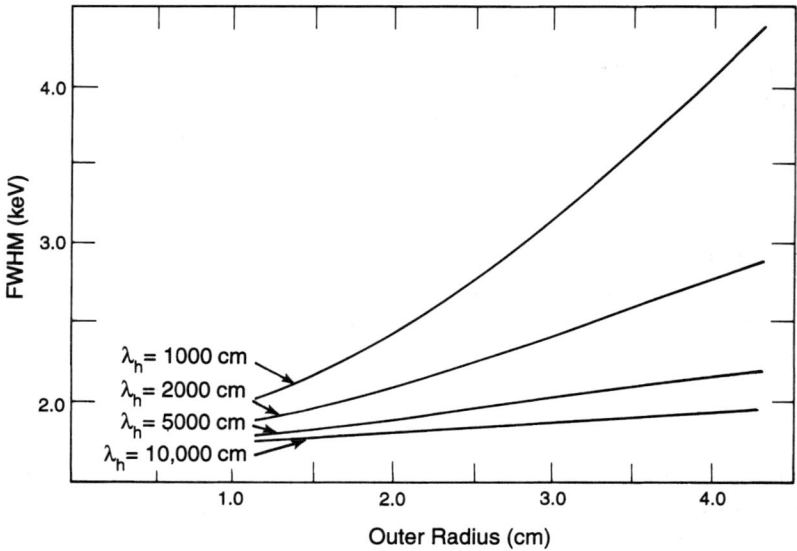

FIG. 29. Calculated effect on FWHM at 1332 keV when detector radius is varied for standard electrode HPGe coaxial detectors. L_N is 0.55 keV, $L_F = 1.62$ keV, and λ_h is treated as a parameter (Raudorf and Pehl, 1987; reprinted with permission of Elsevier Science Publishers).

events before a photoelectric absorption. Ignoring this distribution of the ionization should create more distortion when the trap concentration is nonuniform.

Modeling nonuniform trap concentrations has not been seriously attempted. This may be at least partly because the spatial distribution of trapping defects is not completely understood. Copper related defects may be strongly concentrated near the skin (Simoen et al., 1986). Simoen et al. (1986) plotted L_T for various detectors against λ_h where concentrations of traps and their capture cross sections were determined by DLTS. Despite scatter in the data (sample location was not specified), overall energy resolution less than 2 keV was found to be consistent with an upper limit of $4-5 \times 10^9$ cm^{-3} copper related defects.

Darken and Jellison (1993) studied a set of conventional electrode detectors known to have hole collection problems. Several DLTS samples were taken per detector and the average standard level defect concentration encountered by a hole traversing the detector from the outer diameter to the inner diameter was calculated. The result was expressed in terms of an equivalent concentration of standard levels [S.L.]. In most cases trapping was dominated by CuH$_x$ a defect with "standard" kinetics. In Fig. 30, L_T is plotted against [S.L.] for several detectors. The experimental scatter is reasonable considering the approximations being made. Some theoretical results from Fig. 29 (Raudorf and Pehl, 1987) are also included. These theoretical results were plotted by calculating a standard level concentration from λ_h using Eq. (26) and Fig. 25 taking $E = 1500$ V/cm (the average value for the detectors in Fig. 30). The agreement of the theoretical results with the experimental points is also good.

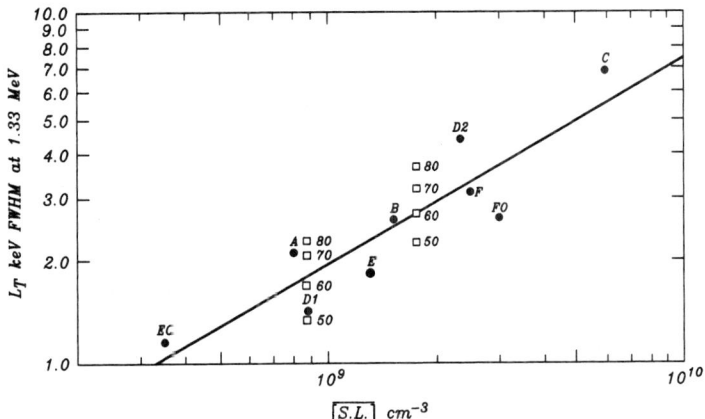

FIG. 30. Observed trapping contribution to detector resolution, L_T as a function of average standard level concentration [S.L.] for various coaxial detectors (solid circles). Suffix O indicates irradiation of open end only. Theoretical points (open squares) are labeled by the detector O.D. (mm). Data are for 1332 keV gamma rays. See text and reference for further details (Darken and Jellison, 1993; reprinted with permission of the American Institute of Physics).

5. RADIATION DAMAGE

HPGe gamma ray spectrometers are often used in experimental environments where there is a background of energetic particles. These particles, most commonly fast neutrons (10^6–10^7 eV) and energetic protons (10^6–10^9 eV), damage the germanium lattice, creating hole trapping centers. With continued exposure, the resolution degrades continuously. Reverse electrode coaxial detectors can withstand about 20 times the fluence incident on a standard electrode detector before equivalent resolution degradation is observed (Pehl et al., 1979). The original performance of the detector can be recovered by extended annealing (days) near 100°C.

Kraner (1980) has provided an overview and summary of the experimental situation to 1980 in a review of neutron damage to germanium detectors and to 1982 in a more general review of all kinds of radiation damage to semiconductor detectors (Kraner, 1982). More recent work has focused attention on the effect of operating temperature on the performance of radiation damaged HPGe detectors. Pehl and Friesel (1988) irradiated HPGe planar detectors with 150 MeV protons and observed that the spectral degradation increased markedly as the operating temperature of the detector was increased over 90 K. Brückner et al. (1991) studied reverse electrode coaxial HPGe detectors irradiated with 1.5 GeV protons and observed a similar effect. Peak shape at 1.33 MeV continuously worsened as the operating temperature was increased from 90 K to 130 K. Both these groups also conducted detailed annealing studies. Fourches et al. studied annealing stages in HPGe irradiated with 1.7 MeV neutrons, using capacitance techniques (1991a) and resolution degradation of planar detectors (1991b).

The effects of proton damage and neutron damage are qualitatively similar. All the phenomena first observed in neutron damaged germanium detectors—preponderance of hole trapping, further resolution degradation by annealing below room temperature, recovery after prolonged annealing near 100°C, and resolution transients after applying bias or removing a hot source—were later also observed in proton damaged germanium. Although protons can coulombically scatter off germanium nuclei creating damage while neutrons cannot, comparable resolution degradation is observed on detectors receiving equal fluences of either neutrons or protons, provided the energies are comparable and >20 MeV, the coulomb barrier of the germanium nucleus (Pehl, private communication, 1993). The effects of particle energy and dosage on resolution degradation generally appear to be quantitative rather than qualitative over the investigated ranges (i.e., there are no reports of new effects as damage increases). Thus there is a presumption that there is a common mechanism or mechanisms for resolution degradation in radiation damaged detectors and that observations on detectors irradiated with fast neutrons and with protons have relevance to each other.

In Fig. 31 the effect of the detector operating temperature on detector resolution after various fluences of 148–181 MeV protons is shown (Pehl, 1992). The data

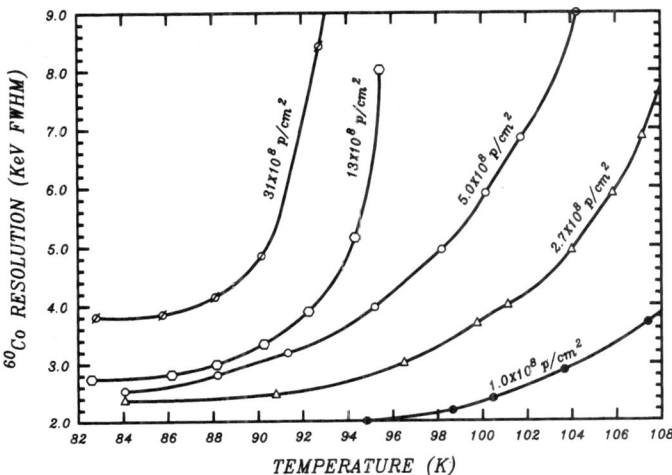

FIG. 31. ^{60}Co resolution as a function of operating temperature for various 1 cm thick planar HPGe detectors irradiated with energetic protons. Proton energies for indicated fluences are 198 MeV (1.0 × 10^8 p/cm^2), 181 MeV (2.7 × 10^8 p/cm^2), 163 MeV (5 × 10^8 p/cm^2), and 148 MeV (31 × 10^8 p/cm^2). (Reprinted with permission from Pehl, 1992. © 1992 IEEE.).

were obtained on planar detectors 1 cm thick. A strong temperature dependence is observed, particularly at higher fluences. This phenomena is closely related to transient detector resolutions observed when bias is applied to neutron damaged (Darken et al., 1980) and proton damaged (Brückner et al., 1991) detectors. At the detector operating temperature and zero bias voltage, the hole traps are relatively empty in n-type material and relatively full in p-type material. In each case, when bias is applied a steady state condition is reached in which hole capture and emission are balanced. The time required for equilibration is long enough (~100 min.) that resolution changes can be tracked. The temperature dependence of charge collection can also be understood qualitatively from this viewpoint. At higher temperatures the hole emission rate increases and the traps become relatively more empty; thus hole trapping increases. A model of the mechanism for radiation damage to germanium detectors should address this phenomena.

Two mechanisms have been proposed to explain resolution degradation in fast neutron irradiated HPGe detectors. They are also relevant to proton damaged detectors. The mechanism for initial damage to the lattice is clear. The mean free path of a fast neutron in germanium between scattering events is about 6 cm. The energy transfer of these "primary knock-on" collisions to a single germanium atom can be up to 5.5% of the kinetic energy of the incident neutron. This highly energetic atom then creates a localized disordered region as this energy is dissipated. It was proposed by Darken et al. (1980) that such structures could be much more efficient trapping centers than isolated defects. The physical cross section of the structures is $\sim 10^{-12}$ cm^2. However, if the net ionization of the disordered region is $Q = -ne$, where n is the number of ionized acceptors in the disordered region and e is the electron charge, a maximum cross section can be defined by the cross section of the electric field lines that terminates at the disordered region (Darken et al., 1980):

$$\sigma_{max} = Q/\epsilon \hat{E} \qquad (32)$$

where ϵ is the dielectric constant for germanium and $\hat{E}$ is the electric field that would exist if the disordered region were uncharged. Equation (32) describes the cross section for hole capture assuming the holes "stream" with the electric field lines and are retained at the disordered region for a period significantly longer than the electronic shaping time. For $Q = -100$ e and $\hat{E} = 1$ kV/cm, $\sigma_{max} = 1.1 \times 10^{-8}$ cm^2, this is four orders of magnitude larger than the physical cross section. In this limit the cross section per ionized acceptor is $\sim 10^{-10}$ cm^2 compared to $\sim 10^{-13}$ cm^2 for an isolated defect (see Fig. 26). In the disordered region model, capture is a two step process. The hole is first localized in the vicinity of the disordered region by coulombic attraction, then captured at a particular defect site there. Ionization would proceed by the reverse process. A hole emitted from an acceptor in the disordered region would also have to escape from the collective potential of the other ionized acceptors. Thus, as Q becomes more negative, the capture rate increases and the emission rate decreases. In a biased detector Q would reach a steady state value depending not only on characteristics of the disordered region but also on the local hole flux due to ionizing radiation (promoting capture) and temperature (promoting emission). The disordered region model

1. provides an explanation for the magnitude of the resolution degradation observed,
2. accounts naturally for transient effects related to applying bias and placing high activity sources near the detector and for temperature effects, as the parameter Q will be sensitive to local hole flux and temperature. The temperature dependence of capacitance transients observed in detectors could not be consistently interpreted in terms of isolated defects (Darken et al., 1981).

However, a case has also been made that isolated defects are largely responsible for the resolution degradation of fast-neutron irradiated HPGe detectors, particularly after any annealing cycles. Fourches, Walter, and Bougoin (1991a) conducted capacitance studies on HPGe samples irradiated with fast neutrons at 10 K and subsequently annealed. In DLTS, broad structures attributed to the disordered regions were observed in both p-type and n-type samples using the technique of minority carrier injection. Different annealing stages were identified. At 100 K in n-type HPGe (stage I) and at 200 K in p-type HPGe (stage II), the DLTS structures attributed to disordered region started to anneal away and the net acceptor concentration as determined by a capacitance–voltage measurement increased.

After stage II annealing, four deep acceptors appeared in the p-type DLTS spectra. In stage III annealing ($T \sim 125°C$), these secondary defects annealed away.

Thus, alerted to potential differences in annealing between n-type and p-type HPGe, Fourches, Huck, and Walter (1991b) conducted irradiation and annealing experiments on n-type and p-type liquid-nitrogen-cooled planar detectors. Unfortunately, the n-type detector could not be cooled below 100 K; irradiation occurred at $T = 107$ K. Thus, partial stage I annealing occurred concurrent with irradiation. Nevertheless, a striking difference between the n-type and the p-type detectors was observed on annealing. The resolution of the n-type planar degraded further immediately with annealing temperatures only slightly above the operating temperature (all spectra recorded at 107 K), and further degradation of the p-type detector did not commence until annealing temperatures exceeded 150 K (stage II). Fourches et al. (1991b) took this correlation to indicate that, at least after stage I and stage II annealing, isolated defects were the predominant cause of resolution degradation.

However, a different picture is presented in Fig. 32 (Darken, 1993). Here the

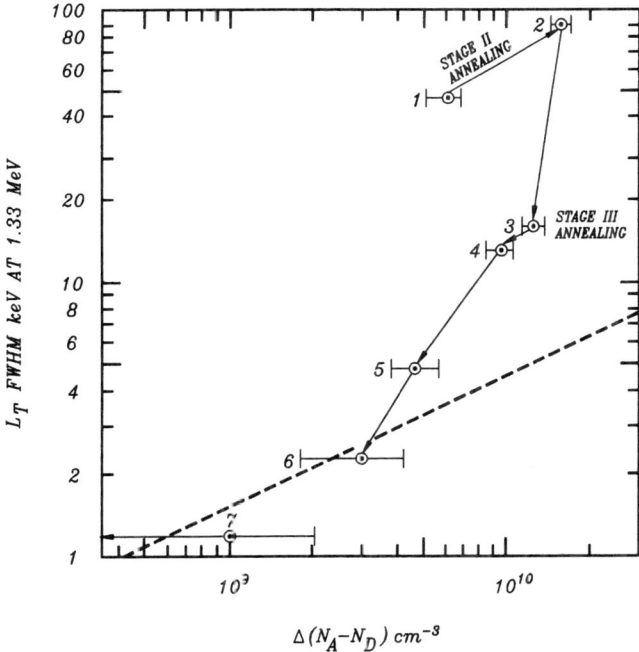

FIG. 32. Trapping contribution L_T to the resolution at 1332 keV of a fast-neutron damaged standard electrode coaxial detector after several annealing steps (enumerated in Table I). The x-axis is the increase in net acceptor concentration $\Delta(N_A - N_D)$ with respect to the pre-irradiation concentration. The line for resolution degradation due to standard levels is from Figure 30, drawn here assuming standard-level kinetics for $\Delta(N_A - N_D)$ The annealing steps are detailed in Table I (Darken, 1993; reprinted with permission of Elsevier Science Publishers).

TABLE I

ANNEALING OF NEUTRON DAMAGED P-TYPE COAXIAL HPGe DETECTOR[a]

Annealing Step	$\Delta(N_A - N_D)$ (10^{10} cm^{-3})	L_T (keV)
1. Initial[a]	0.62	48
2. 150 hr @ 22°C	1.58	90
3. 41 hr @ 100°C	1.29	16
4. 62 hr @ 120°C	0.98	13
5. 115 hr @ 120°C	0.49	4.8
6. 150 hr @ 120°C	0.25	2.0
7. 150 hr @ 120°C	0.13	0.9

[a] Detector performance after irradiation with a fluence of 10^{10} cm^{-2} fast neutrons from an unmodified 252 cf source and immediately before annealing. (Reprinted with permission from Pehl et al., 1979. © 1979 IEEE.)

evolution during annealing of the trapping component of the resolution of a neutron-damaged standard electrode configuration detector (reported by Darken et al., 1981) is plotted against the radiation induced change in net ionized acceptor concentration inferred from changes in the depletion voltage that occurred on annealing. The data points in Fig. 32 were taken after the indicated annealing steps listed in Table I and resolutions refer to steady state values FWHM at 1.33 MeV. The detector was in a standard liquid-nitrogen-cooled cryostat, and no effort was made to vary temperature. The portions of the curve corresponding to stage II and stage III annealing are labeled accordingly. The line is taken from Fig. 30, assuming the net ionized acceptor increase with respect to the preirradiation value is due to the ionization of standard levels and that these levels are effective traps. According to the data presented in Fig. 32, isolated defects cannot dominate resolution degradation until near the conclusion of stage III annealing. Apparently as the secondary (isolated) defects emerge from the disordered regions during stage I and stage II annealing, the reaction products left behind are more damaging to hole collection than the isolated defects themselves. Alternatively, the secondary defects might not migrate far enough away from the disordered region to be considered isolated from it and Eq. (32) may have some relevance. In this case nonexponential emission and capture should be observed from these levels in capacitance transient spectroscopy.

Thomas et al. (1993) studied the effects of temperature on the fast-neutron (1.3 MeV) degradation of an n-type coaxial HPGe detector. They found the same resolution degradation versus neutron flux curve for irradiation and operation at 95 K as for 83 K. In both cases the detector degrades from 1.93 KeV FWHM at 1.33 MeV to 2.35 keV FWHM with 2×10^9 neutrons/cm^2. However, in both cases they found additional and irreversible resolution degradation on annealing above

100 K. This is consistent with the already mentioned observations of Fourches *et al.* (1991b) on a neutron irradiated *n*-type planar HPGe detector. However, Thomas *et al.* made the additional observation that, if one looked at the resolution initially after applying bias (results quoted to this point are for steady state performance—*n*-type detector performance improves with time under bias as hole traps fill), no additional charge collection related degradation, reversible or irreversible, is observed on heating to 140 K. A similar observation was made by Brückner *et al.* (1991, see Fig. 10) on the annealing behavior of resolution transients in a proton damaged *n*-type coaxial HPGe detector. In the disordered region model these results imply that the initial value of Q for a disordered region in an *n*-type HPGe matrix is relatively insensitive to thermal history or temperature (80–140 K), but that the steady state value of Q depends markedly on both. Stage I annealing apparently affects the reemission rate of holes from disordered regions.

In summary, isolated defects cannot explain the magnitude of resolution degradation observed in radiation damaged HPGe detectors. Most of the features observed can be explained in terms of the disordered region model. A more detailed elaboration of the two step electronic kinetics of disordered regions during detector operation would be useful, particularly in elucidating the effect of disordered region size on kinetics.

VI. Germanium X-Ray Detectors

1. INTRODUCTION

For many years there has been an application for small volume, high resolution solid state detectors in the field of x-ray microanalysis. Such devices can be used to detect excited characteristic x-rays with energies as low as the boron $K\alpha$ at 110 eV. Until recently, lithium drifted silicon Si(Li) detectors have been used almost exclusively. In the pursuit of improved detector performance, high purity germanium detectors have been considered as an alternative to the Si(Li) (Llacer *et al.*, 1977; Barbi and Lister, 1981; Fink, 1981; Slapa *et al.*, 1982; Steel, 1986; Cox, Lowe, and Sereen, 1988; Rossington *et al.*, 1992; Darken and Cox, 1993). The advantages of germanium over silicon for this application are improved signal to noise ratio and energy resolution and the capability of detecting photons up to 100 keV with reasonable efficiency even with small volume devices.

The development of semiconductor crystals for low energy detection has concentrated on improvements in spectral resolution, energy linearity, charge collection, and detection sensitivity (Zullinger and Aitken, 1967, 1968, 1969; Pehl *et al.*, 1972; Llacer, Haller, and Cordi, 1977; Barbi and Lister, 1981; Fink, 1981; Slapa *et al.*, 1982; Rosner and Mingay, 1983; Craven, Adam, and Howe, 1985; Cox *et al.*, 1988; Rossington *et al.*, 1992). Various contacts for the front electrode have been tried, and recent developments (Cox *et al.*, 1988; Darken and Cox, 1993) have led to the dramatic improvement of HPGe detector performance at low energies. The Si(Li) detector is still employed for most microanalysis applications, but HPGe detectors are starting to be used as an alternative, particularly where the aforementioned advantages are of practical importance. The following sections consider the fabrication and performance of the HPGe x-ray detector and its comparison with the Si(Li) device.

2. DETECTOR FABRICATION

Figure 15 shows some typical electronic noise contributions for various HPGe detectors of different volumes. Using Eqs. (15) and (16) in Section IV.4.a, it can be calculated that to approach resolutions around 140 eV FWHM or better at 5.9 keV, which is a typical specification for a Si(Li) device, the detector volume must be less than 1 cc, so that the detector capacitance and other noise sources can be kept sufficiently low. The preferred geometry is that of a planar detector as shown in Fig. 33, which allows a relatively large surface area for a low energy x-ray collection. The grooved geometry improves the detector leakage current and ease of handling. Typical dimensions for the highest resolution devices are from 10 to 30 mm² active area and 3 mm thickness. An outline of one method of fabrication follows.

High purity p-type germanium material with a net impurity concentration between 5×10^9 and 1×10^{10} per cc is cut to a disk of a few centimeters diameter and 5 mm thickness. One side is lapped and etched and has lithium evaporated and diffused to a depth of 0.3 mm, forming a p-n junction. The lithium surface is metalized to reduce series resistance of the contact. Individual detectors are then cut from the disk to the required diameter with the groove geometry shown in Fig. 33. The groove area is carefully polished to remove mechanical damage from cutting. The front surface is lapped to remove the required thickness, and the active area is recessed to avoid damaging the sensitive front contact when mounting in the cryostat.

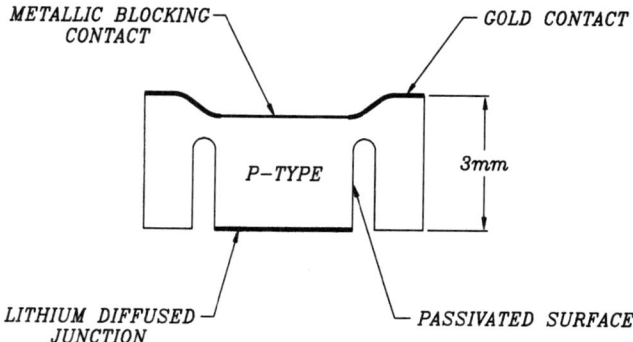

FIG. 33. Schematic cross section of an HPGe planar detector used in x-ray spectroscopy.

Best results have so far been achieved, using an evaporated metal blocking contact, by Cox *et al.* (1988), who applied a chemical treatment to reduce the thickness of the surface dead layer followed by the evaporation of a thin metal layer, for example, nickel. Finally, a thick gold ring is evaporated around the outside, but overlapping the thin electrode at the edges to provide a good ohmic contact to the high voltage bias (Fig. 33). The front electrode is discussed in some detail in Subsection 4.

The final process step is to etch the grooved area of the detector and deposit a thin insulating passivation layer (e.g., silicon monoxide) to neutralize the exposed surface states and so reduce the detector leakage current when operating in reverse bias.

3. DETECTOR HOUSING

The performance required from the finished detector assembly in terms of spectral resolution and sensitivity constrains the mechanical design of the detector housing, which is very similar to that for the Si(Li) device.

Most detector assemblies are designed to be mounted onto an electron microscope with the crystal positioned close to the electron beam target so as to offer a good solid angle for the collection of the emitted x-rays. Figure 34 is a schematic diagram of a crystal housing for such a detector. The crystal is normally cooled to at least 120 K to keep the reverse bias leakage current, and therefore the electronic noise contribution, sufficiently low. Figure 34 shows that the amplifying FET

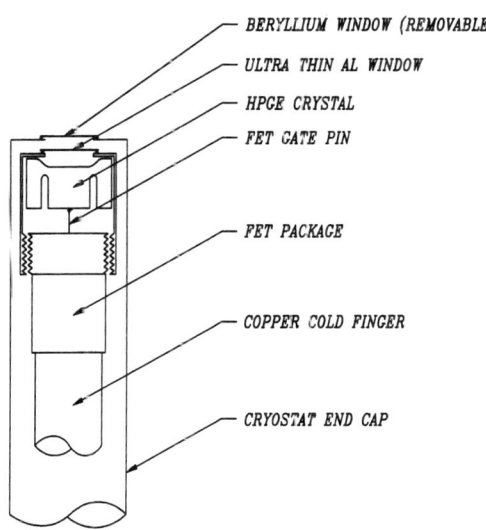

FIG. 34. Schematic diagram of the germanium x-ray detector crystal housing.

package is connected directly to the copper cold finger and provides the cooling path to the crystal. The FET employs a charge restoration circuit built into the chip (Nashashibi and White, 1990), and no feedback resistor or other charge restoration components are required. The copper cold finger and its metal housing lead out of the microscope through a vacuum seal and connect to a liquid nitrogen dewar. The crystal is mounted into an aluminum cup that connects to the FET package by an anodized screw thread. The spacing between the FET gate pin and the threaded ring forms the feedback capacitor, the ring thus forming one contact of the capacitor. A negative high voltage bias is applied to the front electrode via the aluminum cup, and the crystal depletes from the p-n junction formed by the lithium layer at the back contact.

The front of the crystal is screened from room temperature infrared radiation by an ultrathin aluminum reflector of about 0.1 μm thickness. The outer vacuum cover has a few micrometer thick beryllium window in front of the detector, which allows detection of x-rays down to 1 keV. For detection of energies below 1 keV the beryllium window must be removed and the system run "windowless" either by rotation of the cryostat outer cover to expose a hole or by some other means. Windowless operation can be initiated only once the detector is mounted inside the microscope vacuum.

4. DEVELOPMENT OF THE DETECTOR ENTRANCE WINDOW

The front electrode of the planar detector described previously is of critical importance to the charge collection properties of incident x-rays below a few keV in energy. Surface dead layers associated with this contact and the resulting degradation in peak shape and energy linearity have until recently limited the use of HPGe detectors to higher energies.

It has long been recognized (Zullinger and Aitken, 1968, 1969; Llacer et al., 1977), even before the advent of high purity germanium detectors, that one mechanism of charge loss at low energies was charge carriers diffusing into the front contact dead layer. The most significant charge loss for the detector geometry described previously is that for electrons. It is known that the surface band bending and the resultant dead layer is dependent on the surface preparation and type of contact used (Baertsch, 1971; Pehl et al., 1972; Malm, 1975; Llacer et al., 1977; Fink, 1981; Slapa et al., 1982; Cox et al., 1988; Rossington et al., 1991; Rossington et al., 1992; Darken and Cox, 1993). This has been the key in the development of improved contacts for low energy x-rays. It is beyond the scope of this work to present a detailed discussion of semiconductor surface behavior; such analysis can be found in the references quoted. However, some of the various methods employed in the front contact formation and the resulting detector performance will be considered.

An early work by Baertsch (1971) explored the technique of depositing a thin p^+ layer of boron on n-type germanium by gas discharge, but this method has not been favored. Pehl et al. (1972) worked with p-type planer HPGe detectors, initially using a gold evaporation for the p^+ entrance window contact, but found that chromium or platinum evaporated contacts supported a higher overvoltage. Also, palladium–germanium contact was used successfully to produce a resolution of 180 eV FWHM at 5.9 keV. No performance at lower energies was considered.

The Schottky barrier contact was discussed in detail by Malm (1975) who concluded that the barrier height, and therefore leakage current, depended on the metal work function and the surface state density, and that nickel or palladium were good candidates for blocking contacts. Llacer et al. (1977) were among the first to study the performance of HPGe detectors at energies below 5.9 keV and used a palladium Schottky barrier. Poor charge collection from surface effects were clearly shown at energies below 2 keV, which was attributed to the loss of nonthermalized electrons at the front electrode. Improvement in line shape was shown for x-rays with energies just below the Ge L absorption edges at about 1 keV, proving that the dead layer is in the germanium itself. Although the detector performance demonstrated was unsuitable for low energy x-ray spectroscopy, the effect of a Ge dead layer was quantified. A dead layer thickness of 0.4 μm was calculated from measurements of the spectrum peak tailing. A gallium ion implanted p^+ contact was used by Slapa et al. (1982), who demonstrated reduced tailing at energies down to 1.25 keV, but the peak shape was still inadequate for high resolution spectroscopy. Fink (1981) conducted a detailed review of germanium x-ray detector development until 1981 and reported surface dead layers of 0.3 μm, a figure similar to that from Pebara (1983), but performance comparable to Si(Li) detectors at low energies had still not been achieved.

Cox et al. (1988) later studied HPGe detectors made by processing the front crystal surface with a technique to modify the surface states before evaporation of a metal blocking contact. These devices demonstrated some aspects of performance surpassing that of the best Si(Li) detectors even at ultralow energies and achieved resolutions previously unattainable by semiconductor detectors. A dead layer thickness of 0.2 μm was measured, which is similar to that for a Si(Li) detector. Results obtained with these detectors are discussed in the following subsections.

5. Performance at Low Energies, down to 200 eV

There are a number of ways of quantifying detector performance at low energies, including peak position, peak shape, resolution, and peak to background ratios. The data presented here demonstrate the state of the art performance generated with a 10 mm^2 HPGe detector. A transistor reset FET (Nashashibi and White, 1990) and an active filter pulse processor have been used to minimize the electronic noise contribution of the detector system.

Figure 35 is a spectrum of electron beam excited Cu and Zn x-rays from a sample of brass measured with an HPGe detector. The peaks demonstrate excel-

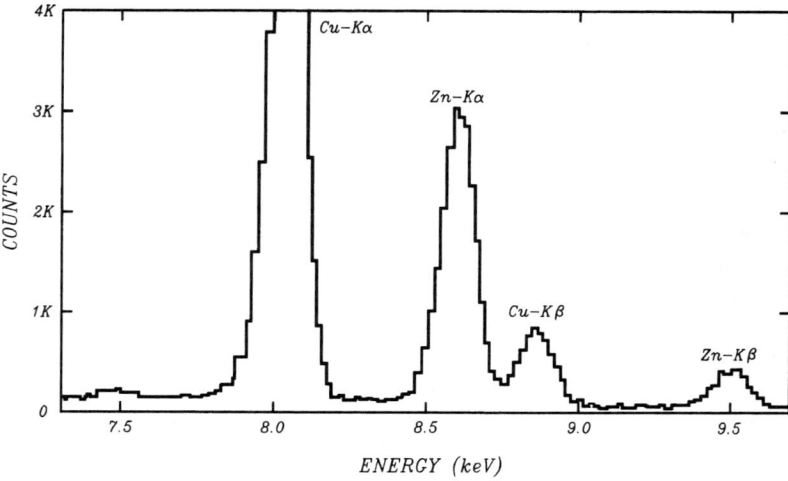

FIG. 35. Spectrum of electron beam excited zinc and copper x-rays from a sample of brass collected with an HPGe detector.

lent energy resolution and can easily be resolved. Figure 36 shows a boron Kα peak at 180 eV and a small carbon peak at 260 eV. The boron line is completely separated from the triggered noise peak at zero energy and is only 57 eV FWHM.

To further illustrate the performance at low energies, Fig. 37 shows a plot of the statistical line broadening or dispersion (FWHM) of the peaks (i.e., L_F in Eq. (16), Subsection IV.4.a) vs. the square root of energy. This is a measure of

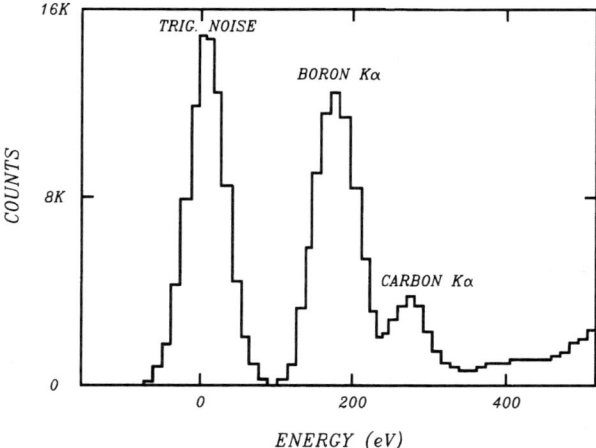

FIG. 36. Spectrum of boron and carbon K peaks collected with an HPGe detector (Darken and Cox, 1993; reprinted with permission of Oxford Instruments Inc.).

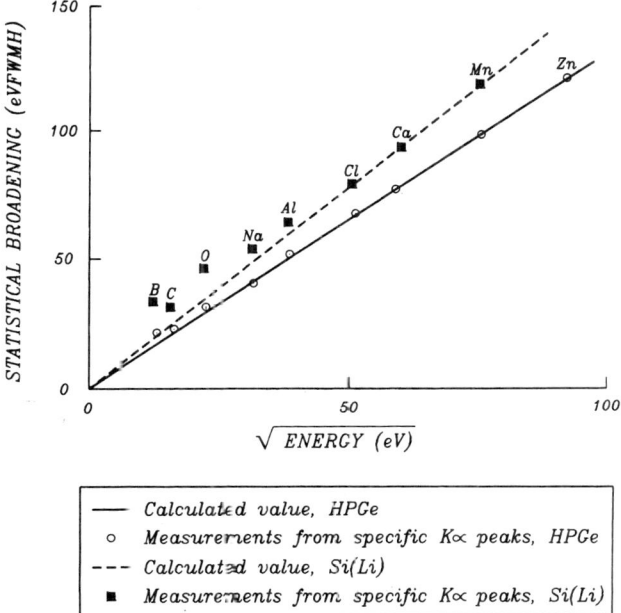

FIG. 37. Plot of the statistical line broadening (eV FWHM) vs. the square root of energy for x-ray peaks collected with HPGe and Si(Li) detectors (Darken and Cox, 1993; reprinted with permission of Oxford Instruments Inc.).

how the peak width is broadened by poor charge collection. The solid line is the theoretical value for an ideal germanium detector with no charge loss, and the circles are measured values for a real detector. It can be seen that there is almost no deviation from the ideal values.

An ^{55}Fe radioactive source generates manganese x-rays at about 5.9 keV and is commonly used for quantifying x-ray detector performance. Spectrum parameters sometimes quoted are the FWHM at 5.9 keV, the peak to background ratio between 5.9 keV and 1 keV, and the "tail factor," which measures the ratio of the peak height at 5.9 keV (Energy E_0, FWHM L eV), to counts in the energy region E_1 to E_2, where

$$E_1 = E_0 - 1740 \ eV + (L \times 1.58) \tag{33}$$

$$E_2 = E_0 - (L \times 1.58) \tag{34}$$

Tail factors are typically 0.1% and peak resolutions as low as 110 eV FWHM at 5.9 keV have been recorded by the author. Figure 38 shows a manganese spectrum from ^{55}Fe with 114 eV FWHM resolution at 5.9 keV. Peak to background ratios can be as high as 15000:1, but at this low level of background, scattering from the ^{55}Fe high energy gamma ray contributes to the count.

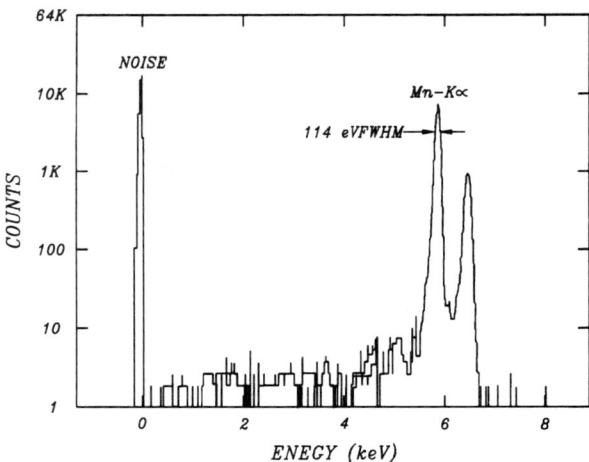

FIG. 38. Manganese x-ray spectrum from an ^{55}Fe source collected with an HPGe detector (Darken and Cox, 1993; reprinted with permission of Oxford Instruments Inc.).

6. PERFORMANCE AT HIGH ENERGIES, UP TO 100 KEV

In addition to the good performance at low energy, the higher atomic number of germanium with respect to silicon allows collection of high energy x-rays with smaller devices. Figure 39 shows a spectrum collected with a 10 mm^2 × 3 mm thick germanium detector using a transmission electron microscope (TEM) and shows indium, gadolinium, and platinum K x-ray peaks together with lower energy L x-ray peaks. The high energy lines would hardly be detectable with a Si(Li) detector of the same dimensions. However, it should be noted that when analyzing heavy elements by electron beam excitation the K ionization cross section decreases rapidly with increasing atomic number (Steel, 1986), resulting in relatively small peaks for high energy x-rays even with a germanium crystal.

7. COMPARISON WITH Si(Li) DETECTORS

A number of comparisons between small area HPGe and Si(Li) detectors have been mentioned in the previous sections. A more detailed study is now considered for x-ray applications.

a. Cryostat Design

The bulk and surface leakage currents for germanium detectors are generally higher than those for the equivalent silicon device at a given temperature, due to the smaller bandgap of the former. The mechanical mounting and cryostat design

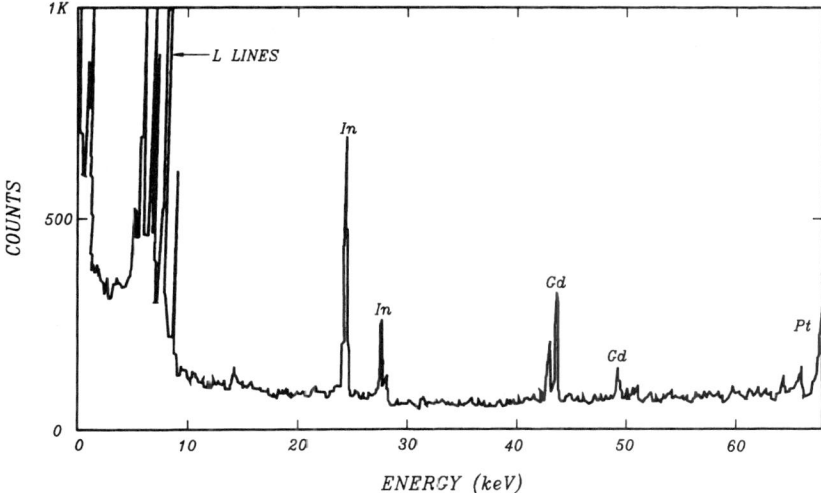

FIG. 39. Spectrum collected on a TEM with an HPGe detector, showing high energy x-ray peaks. (Reprinted with permission from Cox *et al.*, 1988. © IEEE 1988).

of both devices is essentially the same, but careful attention must be paid to the cooling path for germanium to achieve a temperature of around 120 K or better. Higher temperatures can lead to an unacceptably high leakage current and noise.

Germanium is sensitive to infrared absorption and therefore must be completely shielded from room temperature surfaces inside the cryostat to avoid increasing the leakage current. The 0.1 μm aluminum reflector described in Subsection VI.3 is not essential for silicon detectors.

b. Low-Energy X-Ray Detection

Entrance window effects quantified by the peak-to-background ratios and tail factors defined from the ^{55}Fe source manganese x-ray peaks (see Fig. 38) were discussed in Subsection VI.5 and are very similar for Si(Li) and germanium devices. The measured dead layers at the front surfaces (from 0.1 to 0.2 μm) are also comparable (Cox *et al.* 1988). The adherence to the theoretical statistical broadening at low energies, as shown by Fig. 37, is a measure of the effect of entrance window dead layers, and it can be seen in this figure that the HPGe detector can be superior in this respect, even to a premium Si(Li) device.

A reduction of the electronic noise contribution of between 5 and 10% has been reported (Cox *et al.*, 1988) when using a 10 mm^2 germanium detector compared to a Si(Li) of the same dimensions. The effect of the smaller bandgap of germanium is to increase the number of electron–hole pairs generated per photon and increase the signal with respect to the noise. This is offset somewhat by the higher dielectric constant and therefore capacitance of germanium, which increases the noise with respect to the signal, but the net effect is to increase the signal to noise ratio. This produces a theoretical improvement in the electronic recognition sensitivity at low energies, as well as the peak width.

The detection sensitivity of germanium to ultralow energy x-rays, in contrast to the energy linearity and resolution, is not well understood. The boron peak at 180 eV is easily detectable with excellent resolution, as illustrated in Fig. 36, but to the authors' knowledge the beryllium peak at 110 eV has never been satisfactorily identified. However, Si(Li) technology has been able to detect beryllium for a number of years (Statham, 1984), and the electronic processing effects that might influence the detection of beryllium are either identical or more favorable when using germanium.

c. Detection of High Energy X-Rays

The smaller bandgap in germanium with respect to silicon not only improves the signal to noise ratio but also decreases the statistical broadening (L_F in Eq. (16)). The dashed line in Fig. 37 shows the theoretical (best case) statistical broadening of a Si(Li) detector compared to germanium, and it can be seen that the difference between the two increases with energy. Therefore, although improvements in resolution with germanium are small only at very low energies, at higher energies they become significant. This is particularly helpful when trying to resolve overlapping spectrum peaks.

Germanium also has a higher absorption coefficient than silicon, particularly for x-rays above about 20 keV in a small germanium crystal. Figure 40 is a plot of the photon absorption in a 3 mm slice of germanium at energies between 10

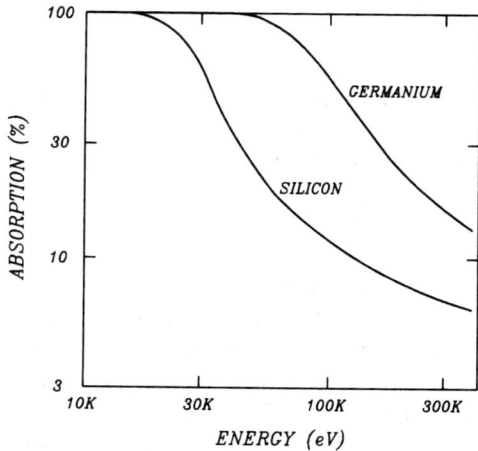

FIG. 40. Photon absorption as a function of energy in a 3 mm thick slice of germanium and silicon. (Reprinted with permission from Cox et al., 1988. © IEEE 1988.)

and 300 keV compared to that for silicon. Notice that the silicon absorption efficiency decreases rapidly above about 30 keV, whereas germanium is still adequate at 100 keV.

d. Escape Peaks

Spectra from semiconductor detectors can display escape peaks generated by the escape of a characteristic x-ray following the initial photoelectric absorption process. In Si(Li) spectra there is usually only one noticeable escape energy at 1.74 keV (Si K x-ray energy) below the main line.

The situation with germanium, however, can be more complicated. The L energies are about 1 keV and the associated escape peaks are usually undetected in the background. However, K escape peaks can be prominent for lines above the germanium K x-ray energies (about 11 keV). Thus for excited x-rays below 11 keV the spectra generated from germanium are less complicated than the equivalent Si(Li) spectra, due to the lack of discernible escape peaks. However, above 11 keV the K escape peaks can interfere with the detection of certain elements. The effects of the interference of escape peaks can be removed in the first order by suitable analytical software, and the degradation of quantitative analysis, if any, will depend on the sample composition and fluorescence conditions.

VII. Summary

The use of high-purity germanium for gamma ray and x-ray detection has been examined, starting with a discussion of the crystal growth and characterization techniques. The temperature and electric field dependence of hole trapping by point defects are discussed in terms of "standard levels." Improved quantification of peak shape degradation due to point defect traps has helped distinguish between the effects of point defect and extended defect trapping. This has been most notable in the case of radiation damaged detectors. Although published reports most frequently link excessive hole trapping to copper-related defects, the most severe hole collection problems have been observed in HPGe with an unusually high density of extended defects (oxide inclusions, disordered regions created by energetic neutrons or protons).

Detector fabrication methods were reviewed and results presented for photon detection from less than 200 eV to 2 MeV. Two developments in germanium detector technology were noted. First, the development of larger HPGe detectors has continued with no fundamental limit apparent. The 157% p-type detector discussed here demonstrated again that increases in efficiency are possible without undue sacrifices in resolution and peak shape. Second, the energy range of very small planar detectors has been extended to below 200 eV with good peak shape and resolution. These devices provide a replacement for the Si(Li) detector in x-ray microanalysis applications where energy resolution is critical.

Although germanium detectors suffer from the disadvantage of operating around liquid nitrogen temperature, the energy resolution demonstrated is superior to most other energy dispersive devices. Germanium therefore sets the standard to which other alternative materials should aspire.

References

Alberti, G., Clerici, R., and Zambra, A. (1979). *Nucl. Inst. Meth.* **158,** 425.
Baertsch, R. D. (1971). *IEEE Trans. Nucl. Sci.* **NS-18,** 166.
Baertsch, R. D., and Hall, R. N. (1970). *IEEE Trans. Nucl. Sci.* **NS-17**(3), 235.
Barbi, N. C., and Lister, D. B. (1981). NBS special publication no. 604, p. 35.
Beer, A. C., and Willardson, R. K. (1958). *Phys. Rev.* **110,** 1288.
Betler, N. R., and Fischer, P. (1974). *Bull. Am. Phys. Soc.,* series 11, **19,** 92.
Britton, C. L., Becker, T. H., Paulus, T. J., and Trammell, R. C. (1984). *IEEE Trans. Nucl. Sci.* **NS-31**(1), 455.
Brodzinski, R. L., Brown, D. P., Evans, J. C., Jr., Hensley, W. K., Reeves, J. H., Wogman, N. A., Avignone, F. T., III, and Miley, H. S. (1985). *Nucl. Instr. Meth.* **A239,** 207.
Brodzinski, R. L., Miley, H. S., Reeves, J. H., and Avignone, F. T., III (1991). *Proc. 2nd Int. Conf. Methods and Applications of Radioanalytical Chemistry,* Hawaii.
Brückner, J., Korfer, M., Wanke, H., Schroeder, A. N., Filges, D., Dragovitch, P., Englert, P. A. J., Starr, R., Trombka, J. I., Taylor, I., Drake, D. M., and Shunk, E. R. (1991). *IEEE Trans. Nucl. Sci.* **NS-38,** 209.
Butler, N. R., and Fisher, P. (1974). *Phys. Lett.* **47A,** 391.
Cavelleri, G., Fabri, G., Gatti, E., and Svelto, V. (1963). *Nucl. Instr. Meth.* **21,** 177.
Cox, C. E., Lowe, B. G., and Sereen, R. A. (1988). *IEEE Trans. Nucl. Sci.* **35**(1), 28.
Craven, A. J., Adam, P. F., and Howe, R. (1985). *Electron Microscopy and Analysis 1985.* Inst. Phys. Conf. Ser. **78,** 189.
Crawford, J. H., Jr., and Cleveland, J. W. (1959). *J. Appl. Phys.* **30,** 1204.
Darken, L. S. (1979a). *Electrochem. J. Soc.* **126,** 827.
Darken, L. S. (1979b). *IEEE Trans. Nucl. Sci.* **NS-26,** 324.
Darken, L. S. (1982a). *Electrochem. J. Soc.* **129,** 226.
Darken, L. S. (1982b). *J. Appl. Phys.* **53,** 3754.
Darken, L. S. (1983). *Matls. Res. Soc. Symp. Proc.* **16,** 47.
Darken, L. S. (1989). *J. Appl. Phys.* **65,** 1118.
Darken, L. S. (1992). *Phys. Rev. Lett.* **69,** 2839.
Darken, L. S. (1993). *Nucl. Instr. Meth. Phys. Res.* **B74,** 523.
Darken, L. S. (1994). *IEEE Trans. Nucl. Sci.* **41,** 343.
Darken, L. S., and Hyder, S. A. (1983). *Appl. Phys. Lett.* **42,** 731.
Darken, L. S., and Cox, C. E. (1993). *Semiconductors for Room-Temperature Detector Applications,* ed. R. B. James, P. Siffert, T. E. Schlesinger, and L. Franks. Materials Research Society Symposia Proceedings, **302,** p. 31.
Darken, L. S., and Jellison, G. E. (1989). *Appl. Phys. Lett.* **55,** 1424.
Darken, L. S., and Jellison, G. E. (1993). *J. Appl. Phys.* **74,** 4557.
Darken, L. S., Jr., Trammell, R. C., Randorf, T. W., Pehl, R. H., and Elliott, J. H. (1980). *Nucl. Instr. Meth.* **171,** 49.
Darken, L. S., Trammell, R. C., Randorf, T. W., and Pehl, R. H. (1981). *IEEE Trans. Nucl. Sci.* **NS-28,** 572.
Darken, L. S., Sangsingkeow, P., and Jellison, G. E. (1990). *J. of Electron. Mat.* **19,** 105.
Dinger, R. J. (1975). *IEEE Trans. Nucl. Sci.* **NS-22,** 135.
Evwaraye, A. O., Hall, R. N., and Soltys, T. J. (1979). *IEEE Trans. Nucl. Sci.* **NS-26**(1), 271.

Figielski, T. (1990). *Phys. Stat. Sol. (A)* **121**, 187.
Fink, R. W. (1981). NBS special publication no. 604, p. 5.
Fourches, N., Walter, G., and Bourgoin, J. C. (1991a). *J. Appl. Phys.* **69**, 2033.
Fourches, N., Huck, A., and Walter, G. (1991b). *IEEE Trans. Nucl. Sci.* **38**, 1728.
Fox, R. F. (1966). *IEEE Trans. Nucl. Sci.* **NS-13**(3), 367.
Glasow, P. A. (1982). *IEEE Trans. Nucl. Sci.* **NS-29**(3).
Glasow, P., and Haller, E. E. (1976). *IEEE Trans. Nucl. Sci.* **NS-23**, 92.
Gossik, B. R. (1959). *J. Appl. Phys.* **30**, 1214.
Goulding, F. S., and Landis, D. A. (1982). *IEEE Trans. Nucl. Sci.* **NS-29**(3), 1125.
Goulding, F. S., Landis, D. A., and Hinshaw, S. M. (1990). *IEEE Trans. Nucl. Sci.* **37**(2), 417.
Hall, R. N. (1966). In *Semiconductor Materials for γ-Ray Detectors—Proceedings of the Meeting,* ed. W. L. Brown and S. Wagner, p. 27.
Hall, R. N. (1972). *IEEE Trans. Nucl. Sci.* **NS-19**, 266.
Hall, R. N. (1974a). *IEEE Trans. Nucl. Sci.* **21**, 260.
Hall, R. N. (1974b). *Int. Conf. on Physics of Semiconductors.*
Hall, R. N. (1984). *IEEE Trans. Nucl. Sci.* **NS-31**, 320.
Hall, R. N., and Soltys, T. J. (1971). *IEEE Trans. Nucl. Sci.* **NS-18**, no. 1, 160.
Haller, E. E., and Hansen, W. L. (1974). *IEEE Trans. Nucl. Sci.* **NS-21**, 279.
Haller, E. E., Li, P. P., Hubbard, G. S., and Hansen, W. L. (1979). *IEEE Trans. Nucl. Sci.* **NS-26**(1), 265.
Haller, E. E. (1989). *Hydrogen in Semiconductors* (Pankove, J., and Johnson, N., Eds.), *Semiconductors and Semimetals,* Volume 34. Academic Press, Orlando, FL, p. 351.
Haller, E. E., Hansen, W. L., and Goulding, F. S. (1981). *Adv. in Phys.* **30**, 93.
Hansen, W. L., and Haller, E. E. (1972). *IEEE Trans. Nucl. Sci.* **NS-19**(1), 260.
Hansen, W. L., and Haller, E. E. (1981). *IEEE Trans. Nucl. Sci.* **NS-28**(1), 541.
Hansen, W. L., and Haller, E. E. (1983). *Nuclear Radiation Detector Materials,* ed. E. E. Haller and H. W. Kraner, Materials Research Society Symposia Proceedings, **16**, p. 1.
Hansen, W. L., Haller, E. E., and Hubbard, G. S. (1980). *IEEE Trans. Nucl. Sci.* **NS-27**(1).
Hansen, W. L., Haller, E. E., and Luke, P. N. (1982). *IEEE Trans. Nucl. Sci.* **NS-29**(1), 738.
Hinshaw, S. M., and Landis, D. A. (1990). *IEEE Trans. Nucl. Sci.* **NS-37**(2), 374.
Howes, J. H., and Allsworth, F. L. (1986). *IEEE Trans. Nucl. Sci.* **NS-33**(1), 283.
Hubbard, G. S., and Haller, E. E. (1980). *J. Electron Materials* **9**, 51.
Hubbard, G. S., Haller, E. E., and Hansen, W. H. (1977). *IEEE Trans. Nucl. Sci.* **NS-24**(1), 161.
Jagam, P., Simpson, J. J., Campbell, J. L., Robertson, B. C., and Malm, H. L. (1985). *Nucl. Instr. Meth.* **A239**, 214.
Jagam, P., Simpson, J. J., Robertson, B. C., Hahn, L. J., and Aardsma, G. E. (1988). *Nucl. Inst. Meth.* **A267**, 486.
Jones, K. S., and Haller, E. E. (1987). *J. Appl. Phys.* **67**(7), 2469.
Kamiura, Y., and Hashimoto, F. (1989). *Jap. J. Appl. Phys.* **28**, 763.
Kandiah, K., and White, G. (1981). *IEEE Trans. Nucl. Sci.* **NS-28**(1), 613.
Keyser, R. M., Twomey, T. R. and Wagner, S. E. (1990). *Radioactivity and Radiochemistry* **1**, 47.
Knoll, G. F. (1989). *Radiation Detection and Measurement,* 2nd ed. John Wiley & Sons, New York.
Kraner, H. W. (1980). *IEEE Trans. Nucl. Sci.* **NS-27**, 218.
Kraner, H. W. (1982). *IEEE Trans. Nucl. Sci.* **NS-29**, 1088.
Landis, D., Cork, C., Madden, N. W., and Goulding, F. S. (1982). *IEEE Trans. Nucl. Sci.* **NS-24**(1), 619.
Lang, D. V. (1974). *J. Appl. Phys.* **45**, 3022.
Lifshits, T. M., and Nad, F. Ya (1965). *Soviet Phys. Dokl.* **10**, 532.
Llacer, J. (1972). *Nucl. Instr. Meth.* **98**(2), 259.
Llacer, J., Haller, E. E., and Cordi, R. C. (1977). *IEEE Trans. Nucl. Sci.* **NS-24**(1), 53.
Loo, B. W., and Goulding, F. S. (1988). *IEEE Trans. Nucl. Sci.* **35**(1), 114.
Luke, P. N. (1988). *Nucl. Instr. Meth.* **A271**, 567.
Luke, P. N., Madden, N. W., and Goulding, F. S. (1985). *IEEE Trans. Nucl. Sci.* **NS-32**(1), 457.

Malm, H. L. (1975). *IEEE Trans. Nucl. Sci.* **NS-22**(1), 140.
McMath, T. A., and Sakai, E. (1972). *IEEE Trans. Nucl. Sci.* **NS-19**, 289.
Miller, G. L., Lang, D. V., and Kimerling, L. C. (1977). *Ann. Rev. Mat. Sci.* **7**, 377.
Miyazawa, H., and Maeda, H. (1960). *J. Phys. Soc. Japan* **15**, 1924.
Nakano, G. H., Simpson, D. A., and Imhof, W. L. (1977). *IEEE Trans. Nucl. Sci.* **NS-24**(1), 68.
Nashashibi, T., and White, G. (1990). *IEEE Trans. Nucl. Sci.* **37**(2), 452.
Paulus, T. J., Raudorf, T. W., Coyne, B., and Trammell, R. C. (1981). *IEEE Trans. Nucl. Sci.* **NS-28**(1), 5.
Pearton, S. J. (1982). *Appl. Phys. Lett.* **40**, 253.
Pearton, S. J., and Kahn, J. M. (1983). *Phys. Stat. Sol.* (a) **78**, K65.
Pebara, W. (1983). *Int. J. Appl. Radiat. Isot.* **34**(2), 519.
Pehl, R. H. (1982). *Physics Today,* November, p. 50.
Pehl, R. H. (1992). Progress report 2, NASA Grant NAG 5-1394.
Pehl, R. H., and Friesel, D. L. (1988). Progress Report 4, NASA grant NAG 5 721, June 30 (unpublished).
Pehl, R. H., Cordi, R. C., and Goulding, F. S. (1972). *IEEE Trans. Nucl. Sci.* **NS-19**, 265.
Pehl, R. H., Haller, E. E., and Cordi, R. C. (1973). *IEEE Trans. Nucl. Sci.* **NS-20**, 494.
Pehl, R. H., Madden, N. W., Elliott, J. H., Raudorf, T. W., Trammell, R. C., and Darken, L. S., Jr. (1979). *IEEE Trans. Nucl. Sci.* **NS-26**, 321.
Pell, E. M. (1960). *J. Appl. Phys.* **31**, 291.
Ponpon, J. P., Grob, J. J., Stuck, R., Burger, P., and Siffert, P. (1972). *2nd Int. Conf. Ion Implantation,* ed. P. Garmish. Springer Verlag, Berlin.
Quaranta, A. A., Andretta, M., and Zanarini (1982). *IEEE Trans. Nucl. Sci.* **NS-29**, 1370.
Quaranta, A. A., Catellani, A., Cuzzilla, M., and Zanarini, G. (1983). *IEEE Trans. Nucl. Sci.* **NS-20**(3), 1862.
Raudorf, T. W., and Pehl, R. H. (1987). *Nucl. Instr. Meth. Phys. Res.* **A255**, 538.
Raudorf, T. W., Trammell, R. C., Wagner, S., and Pehl, R. H. (1984). *IEEE Trans. Nucl. Sci.* **NS-31**, 253.
Read, W. T. (1954). *Phil. Mag.* **45**, 775.
Reeves, J. H., Hensley, W. K., Brodzinski, R. L., and Ryge, P. (1984). *IEEE Trans. Nucl. Sci.* **NS-31**, 697.
Rosner, B., and Mingay, D. W. (1983). *X-Ray Spect.* **12**(2), 82.
Rossington, C. S., Walton, J. T., and Jaklevic, J. M. (1991). *IEEE Trans. Nucl. Sci.* **38**, 239.
Rossington, C. S., Giauque, R. D., and Jaklevic, J. M. (1992). *IEEE Trans. Nucl. Sci.* **39**(4), 570.
Sakai, E., Murakamiand, Y., and Nakatani, H. (1982). *IEEE Trans. Nucl. Sci.* **NS-29**(1), 760.
Schoemaekers, W. K. H., Clauws, P., van der Steen, K., Broeckx, J., and Henck, R. (1979). *IEEE Trans. Nucl. Sci.* **NS-26**, 256.
Schweitzer, J. S. (1991). *IEEE Trans. Nucl. Sci.* **38**(2), 497.
Simoen, E., Clauws, P., Broeckx, J., Vennik, J., Van Sande, M., and De Laet, L. (1982a). *IEEE Trans. Nucl. Sci.* **NS-29**, 789.
Simoen, E., Clauws, P., and Vennik, J. (1985a). *Sol. St. Comm.* **54**(12), 1025.
Simoen, E., Clauws, P., and Vennik, J. (1985b). *Phys. D, Appl. Phys.* **18**, 2041.
Simoen, E., Clauws, P., and Vennik, J. (1986a). *Sol. St. Commun.* **54**, 1025.
Simoen, E., Clauws, P., Huylebroeck, G., Vennik, J., van Goethem, L., van Sande, M., De Laet, L., and Guislain, H. (1986b). *Nucl. Instr. Meth. Phys. Res.* **A251**, 519.
Simoen, E., Clauws, P., Huylebroeck, G., Vennik, J., Van Goethem, L., van Sande, M., De Laet, L., and Guislain, H. (1986c). *Nucl. Instr. Meth. Phys. Res.* **A251**, 519.
Simoen, E., Clauws, P., Lamon, M., and Vennik, J. (1986d). *Semicond. Sci. Technol.* **1**, 53.
Simpson, M. L., Raudorf, T. W., Paulus, T. J., and Trammell, R. C. (1990). *IEEE Trans. Nucl. Sci.* **37**(2), 444.
Simpson, M. L., Becker, T. H., Bingham, D. R., and Trammell, R. C. (1991). *IEEE Trans. Nucl. Sci.* **38**(2), 89.
Slapa, M., Chwaszczewska, J., Huth, G. C., and Jurkowski, J. (1982). *Nucl. Inst. Meth.* **196**, 575.

Statham, P. J. (1984). *J. Phys.* C2(2) **45,** 175.
Steel, E. B. (1986). *Microbeam Analysis 1986.* Proc. 21st Annual Conf. Microbeam Analysis Soc., p. 439.
Stone, R. E., Barkley, V. A., and Fleming, J. A. (1986). *IEEE Trans. Nucl. Sci.* **NS-31**(1), 299.
Stuck, R., Ponpon, J. P., Siffert, P., and Ricaud, C. (1972). *IEEE Trans. Nucl. Sci.* **NS-19**(1), 270.
Teal, G. K., and Little, J. B. (1950). *Phys. Rev.* **78,** 647.
Thomas, H. G., Eberth, J., Decker, F., Burkhardt, T., Freund, S., Hermkens, U., Mylacus, T., Skoda, S., Teichert, W., Werth, A. V. D., Breniano, P. von, and Beral, M. (1993). *Nucl. Instr. Meth. Phys. Res.* **B74,** 523.
Trammell, R., and Walter. F. J. (1969) *Nucl. Instr. Meth.* **76,** 317.
Van der Pauw, L. J. (1958). *Philip. Res. Rep.* **13,** 1.
Van Goethem, L., Van Sande, M., De Laet, L., and Guislain, H. (1985). *Nucl. Instr. Meth. Phys. Res.* **A240,** 365.
Van Sande, M., Van Goethem, L., Delaet, L., and Guislain, H. (1986). *Appl. Phys.* **A40,** 257.
Varnell, L. S., Ling, J. C., Mahoney, W. A., Jacobson, H. S., Pehl, R. H., Goulding, F. S., Landis, D. A., Luke, P. N., and Madden, N. W. (1984). *IEEE Trans. Nucl. Sci.* **NS-31**(1), 300.
Zullinger, H. R., and Aitken, D. W. (1967). *IEEE Trans. Nucl. Sci.* **NS-14,** 563.
Zullinger, H. R., and Aitken, D. W. (1968). *IEEE Trans. Nucl. Sci.* **NS-15,** 466.
Zullinger, H. R., and Aitken, D. W. (1969). *IEEE Trans. Nucl. Sci.* **NS-16,** 47.

CHAPTER 3

Growth of Mercuric Iodide

A. Burger
NASA CENTER FOR PHOTONIC MATERIALS AND DEVICES
DEPARTMENT OF PHYSICS
FISK UNIVERSITY
NASHVILLE, TENNESSEE

D. Nason
EG&G ENERGY MEASUREMENTS, INC.
SANTA BARBARA OPERATIONS
SANTA BARBARA, CALIFORNIA

L. van den Berg
EG&G INSTRUMENTS, INC.
OAK RIDGE, TENNESSEE

M. Schieber
SCHOOL OF APPLIED SCIENCE AND TECHNOLOGY
HEBREW UNIVERSITY
JERUSALEM, ISRAEL

I. THE CRYSTAL STRUCTURE AND PHASES OF MERCURIC IODIDE	86
1. Crystal Structure	86
2. Phase Diagram	87
II. PHYSICAL PROPERTIES RELEVANT TO CRYSTAL GROWTH	87
1. Vapor Pressure	88
2. Stoichiometry	88
3. Thermal Properties	89
III. GROWTH OF HIGH PURITY MERCURIC IODIDE CRYSTALS	91
1. Purification of Starting Materials	91
2. Physical Vapor Transport Growth	93
3. Solution Growth	98
IV. CRYSTAL PERFECTION	98
1. Microscopy	98
2. Gamma Ray and X-Ray Diffraction Studies	99
V. RECENT DEVELOPMENTS	102
1. Seeded Growth	102
2. In-Situ Process Monitoring	103
3. Fundamental Studies	104
VI. CHALLENGES IN CRYSTAL GROWTH	106
1. Impurity Analysis	106

2. Stoichiometry 107
3. Bulk Growth 107
4. Film Growth 107
References 108

Single crystals of mercuric iodide have important applications as room temperature x-ray and gamma ray detectors and photocells with a wide variety of uses, as reviewed in Chapter 1. Although the crystal growth of mercuric iodide was reported as early as the beginning of the century (Bodroux, 1900) and drew some interest over the years in basic research, due mainly to its solid state phase transition, most of the progress in producing usable forms of the material has occurred in the past two decades. This chapter will review the progress on the material processing (synthesis, purification, and growth) of mercuric iodide and will describe some current developments. Unfortunately, not all the worthy contributions can be cited or discussed. It is intended that the references that are cited be consulted for further information.

I. The Crystal Structure and Phases of Mercuric Iodide

As with many other electronic materials, a knowledge of the crystal structure and phase diagram is necessary for controlling the properties of the material as it is processed through the various steps of purification and single crystal growth. This is even more important for HgI_2, due to a destructive phase transformation.

1. CRYSTAL STRUCTURE

Mercuric iodide belongs to the family of layer structured, heavy metal iodides. The α phase, in single crystal form, is of interest for electronic devices. Alpha-mercuric iodide is stable at room temperature and up to about 130°C. It has a tetragonal structure (P42 space symmetry) and is red colored. Two molecules are accommodated in the unit cell, which has parameters $a_0 = 4.372 \pm 0.002$ Å and $c_0 = 12.4399 \pm 0.0004$ Å at 25°C (NBS Monograph 1969). This results in a structural anisotropy of $c_0/a_0 = 2.85$ and leads to a theoretical density of 6.36 g/cm³. The β form is yellow and orthorhombic and stable from about 130°C to the melting point, 260°C. The α phase consists of layers of Hg and I atoms stacked in the sequence, I Hg I I Hg I I Hg I . . . along the [001] axis, with the iodine atoms placed in the corners of linked, distorted (HgI_4) tetrahedra. A covalent bonding between Hg and I atoms is consistent with the low melting point and high solubility in some organic solvents, including acetone, methanol, and DMSO, the latter being used for its crystal growth from solution (Nicolau, 1980). Van der Waals bonds occur between the planes of I atoms normal to [001] and account for the easy cleavage along these planes. As expected, the layered structure leads to anisotropy, which is manifested in many of the physical properties, including the thermal expansion and thermal conductivity, as will be discussed in detail in subsection 5.

2. PHASE DIAGRAM

High pressure ($p > 13$ kbar) and high temperature ($T > 130°C$) phases of mercuric iodide were studied (Bridgman, 1951; Newkirk, 1956). On cooling to 130°C, the high temperature orthorhombic phase, β-HgI_2, transforms (Newkirk, 1956; Rolsten, 1961) destructively to α-HgI_2, which limits the growth of useful crystals to temperatures below 130°C. The I–Hg phase diagram was established (Dworsky and Komarek, 1970) and shows eutectic transformations at 102°C on the iodine-rich side of HgI_2 and at 231°C on the mercury-rich side. There have been studies to determine the composition limits over which single phase tetragonal mercuric iodide can exist and whether mercuric iodide melts congruently. Some of these efforts will be discussed in Subsection 4. The Hg-I phase diagram also shows a mercurous iodide compound, Hg_2I_2, in addition to the mercuric iodide, HgI_2. Experiments have shown that this phase decomposes readily into mercury and mercuric iodide below its peritectic melting point at 235°C. This decomposition reaction has been used in the preparation of detector grade mercuric iodide crystals (Gospodinov, 1980). Mercurous iodide has a higher vapor pressure than mercuric iodide and is frequently found in commercially supplied HgI_2 reagents as a second phase (at a level of 50 ppm and higher). It can be removed as a yellow-greenish phase by vacuum sublimation (Burger et al., 1985).

The vapor phase HgI_2 molecule is linear, with a 2.59 Å distance between Hg and I atoms (Rolsten, 1961). In a vaporization study it was concluded (Piechotka, 1986) that mercuric iodide sublimes congruently and without appreciable dissociation according to the molecular reaction: $HgI_2(s) \rightarrow HgI_2(g)$. Although occasionally a pink color, characteristic of gaseous iodine, is observed during crystal growth from the vapor, this occurrence is usually attributed to excess iodine in the starting material rather than thermal decomposition.

II. Physical Properties Relevant to Crystal Growth

An ideal material for crystal growth in general should have, among other requirements, a high thermal conductivity and no solid phase transitions (Rosenberger, 1979). Neither of these requirements is fulfilled in mercuric iodide. The destructive solid state phase transformation precludes melt growth and limits the growth temperatures, T_g, to less than 125°C. During crystal growth by physical vapor transport some additional important requirements are a congruent sublimation and a sufficiently high vapor pressure occurring at a relatively low growth temperature. A low growth temperature has the advantages of minimizing container contamination and allowing the possibility of a transparent growth furnace. These latter advantages do exist for mercuric iodide.

1. Vapor Pressure

The equilibrium vapor pressure of mercuric iodide was calculated from thermodynamic data and the following relation was reported (Omaly, 1983):

$$\ln p_{HgI_2} = -11170/T + 25.586 - 1.63 \times 10^{-3} T \\ - 2.727 \times 10^{-7} T^2 - 0.883 \ln T$$

where the pressure is measured in Torr for temperatures, $T < 400$ K. The relatively high vapor pressure of 12.5×10^{-2} T around 120°C (Gmelins Handbuch, 1933) provides a satisfactory sublimation rate, which has facilitated the use of vacuum sublimation for both purification and growth processes. The vapor pressure is sufficient to purify, by evaporation at 150°C and recrystallization, approximately 500 g of material per week during current operations at EG&G/Energy Measurements, Santa Barbara Operations (EG&G/EM SBO). The rate is kept much slower (Lamonds, 1983) during single crystal growth to avoid parasitic nucleation.

2. Stoichiometry

A challenging and sometimes controversial issue in mercuric iodide growth is the subject of the stoichiometry. The range of composition over which mercuric iodide can exist without a change of phase is denoted by HgI_{2-x}, where $0 \leq |x| << 1$, $x > 0$ for mercury rich compositions and $x < 0$ for iodide rich compositions. Stoichiometric mercuric iodide is the term referring to the case of $x = 0$. The term "near stoichiometric" refers to the range of x over which the single phase is stable. This range is the "width" of the HgI_2 phase in the Hg–I diagram and may depend on temperature and other chemical components. The accurate measurement of composition, its control and relationship to electronic properties, has claimed the attention of several investigators. It was found (Beinglass et al., 1977) that iodine was being depleted during the prolonged purification process of repeated sublimations in a dynamic vacuum. They claimed that an improved device performance was achieved after reintroducing iodine during a subsequent closed sublimation in an iodine rich atmosphere, in an attempt to correct back toward the stoichiometric ratio. Such results were not confirmed by later studies. Numerous other experimental studies (Scharager et al., 1980; I. F. Nicolau and Rolland, 1981; Dishon et al., 1981; DeLong and Rosenberger, 1981; Burger,

Roth, and Schieber, 1982; Tadjine, 1983; Piechotka and Kaldis, 1986a; Burger *et al.*, 1990; Hermon *et al.*, 1993) particularly address the issue of the near stoichiometry range of the phase nominally denoted as HgI_2. The importance of this issue is in the possible formation, during mercuric iodide growth, of second phases such as iodine, mercurous iodide or even free mercury, yielding overall I/Hg atomic ratios sufficiently different from 2.00 (Burger *et al.*, 1990). Some of the difficulties in resolving this issue are the apparent narrowness of the single phase range at room temperature and the high vapor pressures of HgI_2 and I_2 which makes vacuum surface analysis difficult. The reactivity of HgI_2 with metals and certain polymers may also interfere with stoichiometry measurements. Other impurities may also interfere with some of the analytical determinations of the Hg/I atomic ratio. A recent study (Hermon *et al.*, 1993) has attempted to determine the single phase range by detecting the second phases that appear when mercuric iodide is progressively doped with various amounts of Hg_2I_2 and I_2, heated at various temperatures, and then quenched to room temperature. Solubility limits in the range of 100–200 ppm for Hg and I in mercuric iodide, at the temperature of approximately 110°C, have been proposed. The difficulty in such studies is the assumption that the original mercuric iodide prior to doping is strictly stoichiometric.

Evidence that deviations on both the iodine and mercury rich sides of the stoichiometric compositions were detrimental to nuclear radiation detector performance has been reported (Tadjine *et al.*, 1983), and it was concluded that the 2.00 iodine:mercury ratio was best. A recent analysis (Gerrish, 1993) based on many detectors fabricated at EG&G/EM SBO confirmed that intentional doping with iodine, apparently causing iodine rich compositions, had no significant effect on electrical transport properties or gamma ray detector performance. However, such iodine excess may introduce difficulties in the crystal growth process. Indeed, a recent review (Schieber *et al.*, 1994) concludes that any HgI_2 grown from the vapor phase is close to stoichiometry, even though free iodine may be detected during growth.

3. THERMAL PROPERTIES

Accurate measurements of the thermal diffusivity (Burger *et al.*, 1991a) have shown extremely low values of the thermal conductivity (among the lowest room temperature values found in a crystalline material) and demonstrated the high thermal anisotropy of mercuric iodide. The value of the thermal conductivities along the [100] and [001] are $k_{[100]} = 4.08 \times 10^{-3}$ J/(cm s K) and $k_{[001]} = 1.13 \times 10^{-3}$ J/(cm s K), respectively.

Thermal properties are important in controlling the morphology and quality of single crystals. A qualitative explanation of the crystal shape that develops during physical vapor transport (PVT) growth and an evaluation of the thermal response of the crystal during the growth process or cooldown were discussed in recent

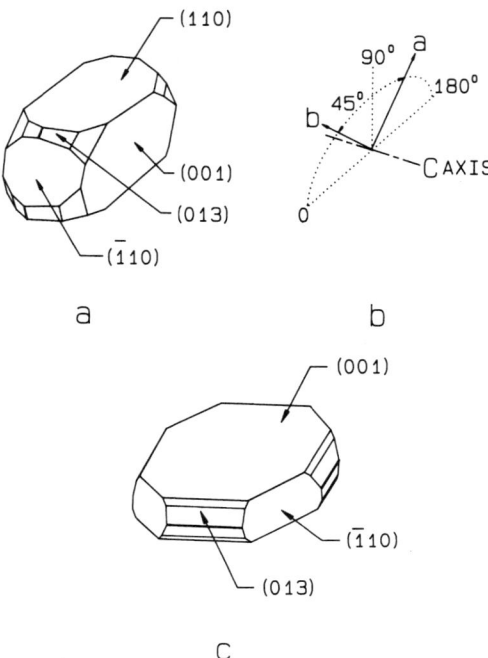

FIG. 1. Schematic drawing of the typical morphology of vapor grown, bulk mercuric iodide crystals grown with (a) [001] horizontal or (c) [001] vertical. The axes in (b) refer to orientation (a).

papers (Burger et al., 1992; Piechotka and Kaldis, 1992). The major faces of vapor grown, bulk mercuric iodide crystals are mainly the {001} and {110} faces, as illustrated in Fig. 1. Theoretical studies (Chernov et al., 1992; Roux et al., 1992) have concluded that a significant radial flow of heat from the furnace into the crystal and then out the bottom center establishes a significant temperature gradient in the crystal, a result also inferred from ampule perimeter temperature measurements (Burger et al., 1992). With the simplification that the impinging heat fluxes are the same over the various faces, the crystal should grow with dimensions inversely proportional to the axial thermal conductivities, and a length/width ratio of around 3.6 would be predicted for an orientation as in Fig. 1a. In fact, the ratio found is typically around 2, but this degree of agreement suggests that thermal conductivity anisotropy plays a major part in controlling growth morphology.

The thermal expansion coefficients along the principal crystallographic axes have been measured, and values of $\alpha_{11} = 11 \times 10^{-6}/°C$ and $\alpha_{33} = 54 \times 10^{-6}/°C$ were reported (Keller, Wang, and Cheng, 1993). Differences in thermal expansion coefficients between a growing mercuric iodide crystal and its glass growth substrate can in principle create significant thermal stresses at cooldown. An estimate (Burger et al., 1992) has concluded that the stresses developed could exceed by two orders of magnitude the value of the average yield stress of approximately 0.1 MPa necessary for shearing between (001) planes. Such stresses would result (Milstein, 1989a) in plastic flow and other extended defects. Extended defects produced during growth were investigated by x-ray topography (Gits, 1983), and their association with poor detector performance has been suggested (Georgeson and Milstein, 1989a).

III. Growth of High Purity Mercuric Iodide Crystals

Material of the highest purity and crystalline perfection is usually the goal in research and development of devices based upon high resistivity and stoichiometric crystals. Analytical methods for quantitative determination of impurities in mercuric iodide and their influence on detector performance are under development. It should be noted that, throughout the extended research and development period of this material since 1973, the yield of high quality detectors has constantly improved. The methods of purification and growth of HgI_2 will be described next. The crystal growth method most widely used is vapor phase growth, the method used to produce crystals for detectors. However, solution growth methods were also developed and will be briefly reviewed.

1. Purification of Starting Materials

The need for material with high purity and stoichiometric or near-stoichiometric composition became evident soon after the first report of the potential use of mercuric iodide single crystals as room temperature semiconductor nuclear radiation detectors (Willig, 1971). A comparative study (Schieber, 1976) of three methods of purification—zone refining, solvent extraction, and repeated sublimation—was undertaken. The conclusions of that study indicated that the latter was the most efficient technique for the purification of mercuric iodide. Since then, repeated sublimations constitute one of the main steps of the purification process. Improvements have been demonstrated (Gerrish, 1993) by variations in the synthesis (Skinner et al., 1988) and in the sequence of the purification steps. The commonly used preparation scheme developed at EG&G/EM SBO consists of (1) synthesis by precipitation from aqueous solutions of potassium iodide and mercuric chloride, (2) four cycles of vacuum sublimations, (3) melting and resolidification, (4) an additional vacuum sublimation, and (5) a closed system sublimation. This process has been assigned the 4XMXS notation, where "X" indicates the open tube, "S" is the closed tube sublimation, and "M" stands for melting.

The impurities most frequently reported (Muheim, Kobayashi, and Kaldis, 1983; R. B. James et al., 1989; Piechotka and Kaldis, 1992; Steinberg et al., 1989) in mercuric iodide are the hydrocarbons. Synthesis from the elements has been studied (Hermon, Roth, and Schieber, 1990) as a method of producing mercuric iodide of improved purity. However, where this method has been tried, there have often been difficulties in controlling the stoichiometry. Synthesis from the elements is also hazardous due to the high exothermicity and toxicity. A process (Hermon *et al.*, 1990) for the reduction of hydrocarbon impurities in the elements used in such a synthesis was studied, based on high temperature oxidation and cracking of elemental iodide.

The purification of mercuric iodide has been hampered by the lack of reliable analytical methods for measuring impurities in the matrix of this material. A current study uses a method based on persulfate oxidation and infrared detection of CO_2 to measure the carbon content in the ppm range (Cross, private communication, 1994). Cation impurities in mercuric iodide have been measured, and a procedure (Warburton 1985) to remove them by vapor phase oxidation has been investigated. Another study (Cross, Olivares, and Mroz, 1993) is underway to measure cation impurities using inductively coupled plasma mass spectroscopy.

Since its discovery in 1952 (Pfann, 1952), zone melting has become one of the most important methods of processing electronic materials. In particular, when used in the zone refining mode, it is a very efficient way of purifying some materials, especially elemental semiconductors. The method requires that the distribution coefficient of a specific impurity, k, be different than 1. When $k < 1$, as with most impurities, the result of the refining process is an accumulation of the impurities at the end where the molten zone travel terminates. The zone refining results from an early mercuric iodide study (Schieber, 1974) indicated that C and possibly Cu and Fe impurities were removable. However, this method has seldom been incorporated in the technology of mercuric iodide preparation (Pompon *et al.*, 1975; Stowell and Hutchinson, 1971) and particularly not in the preparation of significant amounts of material in the range of hundreds of grams.

A renewed effort in zone refining mercuric iodide has been initiated (Burger *et al.*, 1990). Using synthesized material that had already undergone 4*XMXS* purification processing, horizontal zone refining was done under conditions of 64 zone passes at a zone travel rate of 8.5 cm/hr. The molten zone length was typically 2.5 cm, which is about twice the quartz ampule diameter and one fifteenth the ingot length. The quartz ampules were fractured to recover the refined mercuric iodide, and there was more adherence to the quartz at the ends than at the center. A visual inspection was done along the refined ingot, and the front end was measurably darker than the central section. The rear end was slightly darker, though the color was not radially uniform. Samples taken along the ingot for differential scanning calorimetry measurements showed that the melting point was typically lowered 4°C at the front (last to freeze) end. These observations are consistent with the hypothesis that significant refining is occurring and that most but not all of the impurities have $k < 1$ and accumulate at the front end. For comparison, zone refining of the same material, but without the 4*XMXS* purification, produced much stronger darkening at the front end. Anion chromatography measurements showed an accumulation of 20–40 ppm of chloride ion at the front end, which can only partially explain a 4°C melting point depression there. It was concluded that the central region was the purest, in agreement with zone refining theory (Pfann, 1952).

Several single crystals of mercuric iodide have been grown from the large center sections of ingots of zone refined 4$XMXS$ material. Generally, the nucleation and growth were comparable to non-zone refined material. Performance gains in x-ray and γ-ray detectors fabricated from zone refined material have not yet been realized. There is a possibility that the local stoichiometry (Bohac and Kaufmann, 1975) is altered when zone refining is applied to a compound material. The large depression in the melting point and color variability also suggest this result. Current work is underway to evaluate zone refining as an intermediate, instead of terminal, stage in the purification scheme. This variation would replace the melting stage, M, in the 4$XMXS$ sequence; the overall scheme would still finish with the sublimation stage.

2. Physical Vapor Transport Growth

a. Large Crystals

The production of large, detector grade crystals of mercuric iodide in masses to 1 kg (van den Berg, 1992) follows a method that has evolved from the temperature oscillation method (TOM) developed originally by Scholz for Fe_2O_3 (Scholz and Kluckow, 1967; Scholz, 1974; Schieber, Schnepple, and van den Berg, 1976). TOM is a variation of the unseeded PVT method, in which the magnitude (and even the sign) of the external temperature gradient along the source crystal direction oscillates periodically to decrease the chances for additional (parasitic) nucleation. The use of the temperature oscillations was gradually abandoned (Lamonds, 1983) after the NASA-supported development of a static system having an improved crystal heating–cooling system produced similar or better results (van den Berg and Schnepple, 1982).

The vertical set up, the ampule geometry, and the seed selection process are some of the original features developed by Scholz still employed in the growth of large, prismatic mercuric iodide crystals at EG&G/EM SBO. A schematic drawing of the current system and typical temperature profiles are shown in Fig. 2. Single crystals of 250 g are typically grown in 8–12 weeks, and the typical growth orientation and morphology are as shown in Fig. 1a and Fig. 3. Variations include the use of an airstream cold finger, growing in an inverted vertical gradient from source material residing at the ampule bottom, and using implanted seeds (Subsection V.1). The growth of large crystals requires sawing, polishing, and etching to fabricate radiation detectors, whereas the smaller crystals such as platelets may be used as-grown, as will be discussed in Subsection III.2.c.

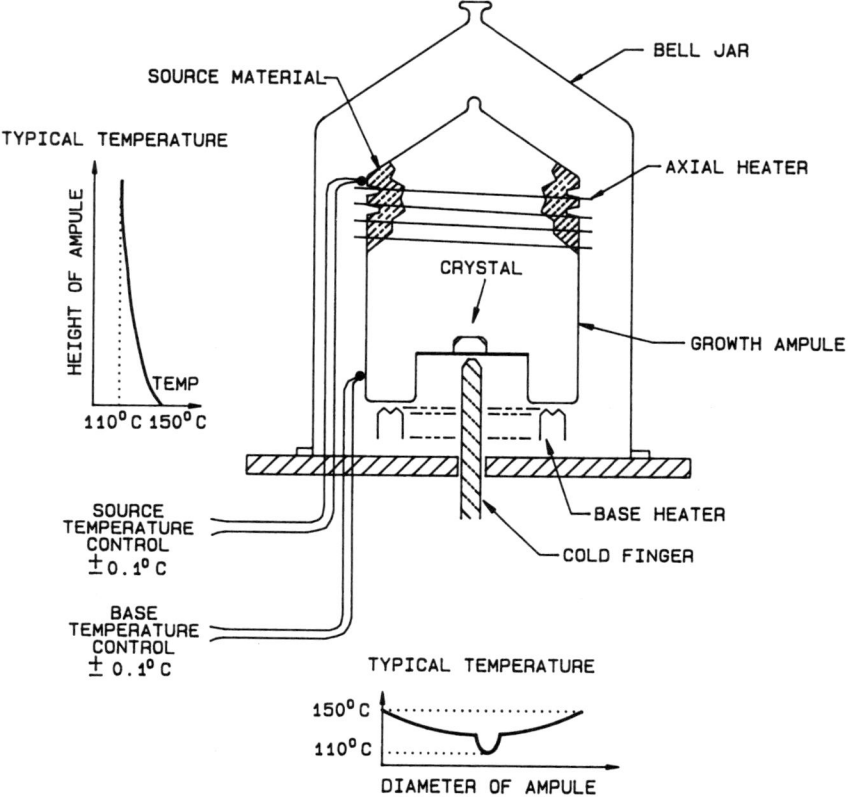

FIG. 2. Schematic drawing of the crystal growth furnace and temperature profiles employed in the growth of bulk mercuric iodide by physical vapor transport.

b. Space Flight Experiments

Since the early 1980s NASA has included crystal growth as an area of emphasis in the space program. In the selection of the materials to be grown, the scientific and technological importance of the crystal, together with a strong justification why microgravity conditions would be beneficial to the growth process, were taken into account. It has been generally assumed that vapor convective flow instabilities are linked to morphological irregularities and the incorporation of crystalline defects and even spurious nucleation. Convection flows have indeed been observed in the mercuric iodide growth system derived from the original Scholz design. Fine impurity particles have occasionally been seen rising from the bottom edge of the ampule toward the source location, which may suggest that the vapor flow passes over the source, becomes saturated, and returns downward through the center of the ampule toward the growing crystal. This type of flow would

FIG. 3. Photograph of a typical mercuric iodide crystal grown at EG&G/EM SBO.

greatly enhance the vapor transport as long as it remains regular and does not disturb the concentration field around the growing crystal. Irregularities in the convection flow are to be expected, causing changes of the concentration of mercuric iodide close to the surface of the growing crystal. Reducing these flows by minimizing gravity would be beneficial to the crystal quality.

An additional effect of gravity is that the crystal is soft enough at growth temperatures that internal slip can be caused by the weight of the crystal itself. This slip will cause an increased line defect density and ultimately may result in low angle grain boundaries, especially in the lower part of the crystal. These reasons and the technological interest in the material as a radiation detector has made mercuric iodide an attractive material for microgravity vapor growth experiments.

To investigate the effects of gravity on the crystal growth process, a vapor crystal growth system (VCGS) was designed. Two space experiments sponsored by NASA have been performed by van den Berg and coworkers in Spacelab 3 (van den Berg and Schnepple, 1989) and in the first International Microgravity Laboratory (IML-1) (van den Berg, 1993). The experiments were performed in a scaled down version of the ground based furnaces (Fig. 2) so the whole system could be accommodated in half a standard Spacelab rack. Several prototype furnaces were built to evaluate the thermal and mechanical performance of the system. For example, operating the system with the space between the ampule and the bell jar evacuated to simulate the no-convection conditions of space showed that an additional heater was required in the upper part of the furnace to maintain the temperature profile required for the crystal growth. These prototype furnaces were also used to prepare seeded ampules for the space experiment and grow crystals on the ground for comparison with the space crystal.

The ampules used in the space experiments had a specially designed pedestal with a cavity in the shape of a truncated cone. The base of the seeds grown into these ampules filled the cavity so the crystal was anchored to the pedestal and would not start to float around during the experiment in space. Approximately 20 g of purified mercuric iodide was loaded into each of six ampules. The ampules were evacuated and sealed off. The loaded ampules were then tested for pressure, heat, and vibration stability. Subsequently, seeds of approximately $5 \times 5 \times 3$ mm^3 were grown on the ground. The structural quality of the seeds was evaluated by means of gamma ray rocking curves (Yelon et al., 1981). Only seeds with a diffraction peak width equal to or less than 0.04° were accepted. Two of these ampules were installed in space rated furnaces.

The experiment timeline consisted of three parts: the heat up phase, the growth phase, and the cool down phase. During the heat up phase (4 hrs) the furnace with the ampule was gradually heated to the point where the temperature control points (source, pedestal, cooling sting) reached the preset values suitable for single crystal growth. At this point, the system transitioned automatically into the growth phase. During the growth phase the temperature of the bottom of the ampule (pedestal) was gradually decreased to promote continuing growth. If desired, it was also possible to increase the cooling of the crystal by reducing the temperature of the tip of the cooling sting. Changes in the preset growth program were made by a trained flight crew member in close coordination with the ground crew. The growth phase could be extended in real time to accommodate variations in the mission timeline. The cool down phase started automatically when the time for the growth phase had elapsed, nominally 4 hrs long. During this phase, the finished crystal was cooled down at a controlled, preset rate. The actual growth time for each experiment was approximately 100 hrs. During the growth, the crystals developed well-defined crystallographic planes intersecting at sharp edges. The temperature profile in the furnace could be increased without causing additional nucleation or breakdown of the surfaces. As a result, the growth rate increased to an average of 3 mm/day compared with 1 mm/day for ground based growth. This was a strong indication that diffusion based transport could be quite fast, as long as the system was not disturbed by convective currents.

The results of the Spacelab 3 and IML-1 experiments confirmed that the VCGS was able to grow crystals of unusually high quality (van den Berg and Schnepple, 1989; van den Berg, 1993). The results of the structural and electronic analysis of the crystal grown on the IML-1 flight, for example, showed that the space crystal in its entirety ($14 \times 12 \times 10$ mm^3) was several times more regular and homogeneous than the best earth grown crystal. Furthermore, the critical electronic parameter for detector purposes (mobility–lifetime product of the holes) was approximately ten times larger than the ground based average.

The next step in the space growth of HgI_2 crystals will be to instrument the growth ampule extensively so a detailed temperature profile of the ampule, and possibly the temperature of the surface of the growing crystal (Nason and Burger, 1991; Burger et al., 1992), can be obtained as well. This would provide data for detailed fluid dynamics calculations.

The forced flux method was originally developed for mercuric iodide as a growth system that could be easier to model theoretically. Independent control of the transport and kinetic parameters (Omaly et al., 1983; Burton, Cabrera, and Frank, 1951) was achieved in a two zone horizontal oven with a cylindrical tube (15 cm long, 1.5 cm diameter). The investigators applied the Burton, Cabrera, and Frank (BCF) (Burton et al., 1951) or Chernov (1961) analyses to explain the measured (001) parabolic and linear growth rates as a function of supersaturation in the temperature range of 100–120°C.

Growth experiments in space using the forced flux method (Cadoret, Brisson, and Magnan, 1989) to study the nucleation, vapor transport, and crystal growth rates (Coupat et al., 1994) of mercuric iodide have been performed during the flights of Spacelab 1, Spacelab 3, and IML-1. The more precisely known geometry of the growth system and the assumption that equilibrium conditions exist at the source and sink locations made it possible to calculate the supersaturation in the vapor phase surrounding the growing crystals. Experiments performed with pure mercuric iodide and additional inert gas in the growth tube indicated that solutal convection is predominant, although at a very low rate. The higher quality of the crystals grown in space is explained by the lack of disturbance, caused by gravity, of the concentration gradient layer around crystals growing at the low pressures used in the growth tubes. Perhaps an alternate transport mechanism needs to be formulated (Cadoret, private communication, 1994). Further experimental and theoretical studies may be able to clarify the transport issue. This issue is of great interest since crystal growth from the vapor is usually done more efficiently at low pressures. It is also believed that longer duration experiments on a platform like a space station will result in larger PVT grown crystals having higher quality than possible to achieve on the ground.

c. Small Platelet Growth

A simple yet efficient way of growing high quality thin (001) platelets of mercuric iodide was described (Faile et al., 1980). The platelets required no cleaving or polishing to be prepared as x-ray detectors. These crystals were grown by PVT from unpurified, commercial reagent grade starting material, in the presence of polyethylene. It was speculated that polyethylene influences the vapor transport and the mechanism of platelet formation by reacting with impurities or mercuric iodide at the crystal growth surface. One advantage of the platelet technique is that the crystals grow at an intermediate position along the temperature profile (120–80°C) in the growth furnace, allowing the more volatile species (Burger et al.,

1982, 1985) to separate and condense at the cold end of the ampule or tube. The platelet growth technique produces crystals in a short period of time that have a small face area (1–10 mm^2), which is sufficient for x-ray detectors. Small areas on the platelets were found (Burger *et al.*, 1982) to be dislocation free when etched in an ethanol–trichloroethylene solution (T. W. James and Milstein, 1981), with most of the dislocations occurring on the sides in contact with the ampule or tube wall. However, another study reports a dislocation density of 10^5 cm^{-2} was reported in another study (Przyluski and Laskowski, 1989). Variations of the platelet growth technique have been developed in which somewhat larger, more prismatic crystals are grown (still with addition of polyethylene). The crystals are then sectioned and fabricated into somewhat larger area wafers having a regular shape (Natarajan, private communication, 1994). Recent efforts have been made to grow thick plates of mercuric iodide without polymer additives, using various other transport gases (Lund, private communication, 1994).

3. Solution Growth

Research on the growth of mercuric iodide from liquid solutions was carried out (I. F. Nicolau, 1980). Optically clear, single crystals were grown using HgI$_2$ · DMSO molecular complex solvents. The crystals were grown at relatively low temperatures (30–50°C), and they exhibited low concentration of etch pits as revealed by methanol etching ($<10^2$/cm^2). The method allowed seeded growth and growth from iodine rich solutions. Although spectrometer grade material was obtained using solution growth, the method did not become widespread mainly because much higher quality detectors with much lower leakage currents were achieved with vapor grown material. The major problem seemed to be inclusion of solvents in the crystal.

IV. Crystal Perfection

Several methods have been employed to characterize bulk crystals or sections of crystals. The general purpose of these characterization studies has been to elucidate the bulk or subsurface defect structures to provide a better understanding of the crystal growth process and the electronic properties. In this review the evaluation of the crystalline perfection of the HgI$_2$ as determined by optical microscopy and by gamma and x-ray diffractions will be emphasized.

1. Microscopy

Optical microscopy studies have been performed (Milstein *et al.*, 1983; Milstein *et al.*, 1989b) on sections of mercuric iodide crystals, identifying dislocations and low angle boundaries that originated from microhardness measurements and seeking correlation with plasticity models (T. W. James and Milstein, 1981; Georgeson and Milstein, 1989b). Electron microscopy studies imaged (Y. F. Nicolau and Dupuy, 1989a) the surface morphology of mercuric iodide crystals that were grown on earth or in space and studied the effects of various dopants on the crystal microstructure (Y. F. Nicolau, Dupuy, and Kabsch, 1989b). A transmission optical microscopy study (K. James et al., 1992) was done to investigate sections of grown crystals of varying optical opacity, and evidence was found that the commonly seen growth bands are strings of discrete voids or semitransparent phases. An important impediment to studying mercuric iodide surfaces with higher resolution techniques (which require vacuum conditions), such as transmission electron microscopy (TEM) and scanning electron microscopy (SEM), has been the significant partial pressure at ambient temperature. The crystal tends to evaporate under vacuum conditions, leading to possible changes of the surface stoichiometry (Y. F. Nicolau, Moser, and Cobel, 1989c) and definite changes of the surface topography. The development of atomic force microscopy (Sarid, 1991; *Ultramicroscopy* 1992) has produced a technique with nanometer resolution that can be used in an ambient environment. Its first application for various surfaces of mercuric iodide, as-grown or fabricated, revealed cleavage steps, incomplete growth layer networks, and the nonuniform topography produced by aging and by vacuum, chemical or thermal etching (Azoulay et al., 1993).

Figure 4 shows atomic force microscopy images of the principal vapor growth faces of mercuric iodide: (a) the (001) face, (b) the (110) face, and (c) the (013) face. The as-grown (001) faces show a terraced structure and a network of growth ledges surrounding regular depressions, which are several molecular layers deep. In Fig. 4b the pits are regular, and the ledges follow strict crystallographic directions and are many molecular layers deep, with terracing which is visible even to the unaided eye. The (013) face is characterized by growth ledges and depressions along [100]. The features on as-grown natural (001) faces of mercuric iodide have been discussed in terms of the well-known phenomenon of segregation of impurities in front of advancing growth steps (Y. F. Nicolau, Dupuy, and Rolland, 1994). A correlation of the micro- and nanoscale structure of the surfaces of crystals grown under different conditions, including microgravity, should help elucidate mechanisms of incorporation of defects and their distribution along the various crystallographic directions.

2. Gamma Ray and X-Ray Diffraction Studies

Double crystal, gamma ray diffraction rocking curves are useful for studying the perfection of the crystal lattice in materials having a high atomic number. Gamma radiation having a wavelength 0.03 Å as obtained from ^{198}Au source is available and has an absorption coefficient of about 1 cm^{-1} in mercuric iodide, indicating that beam penetration can occur and information about the interior of a

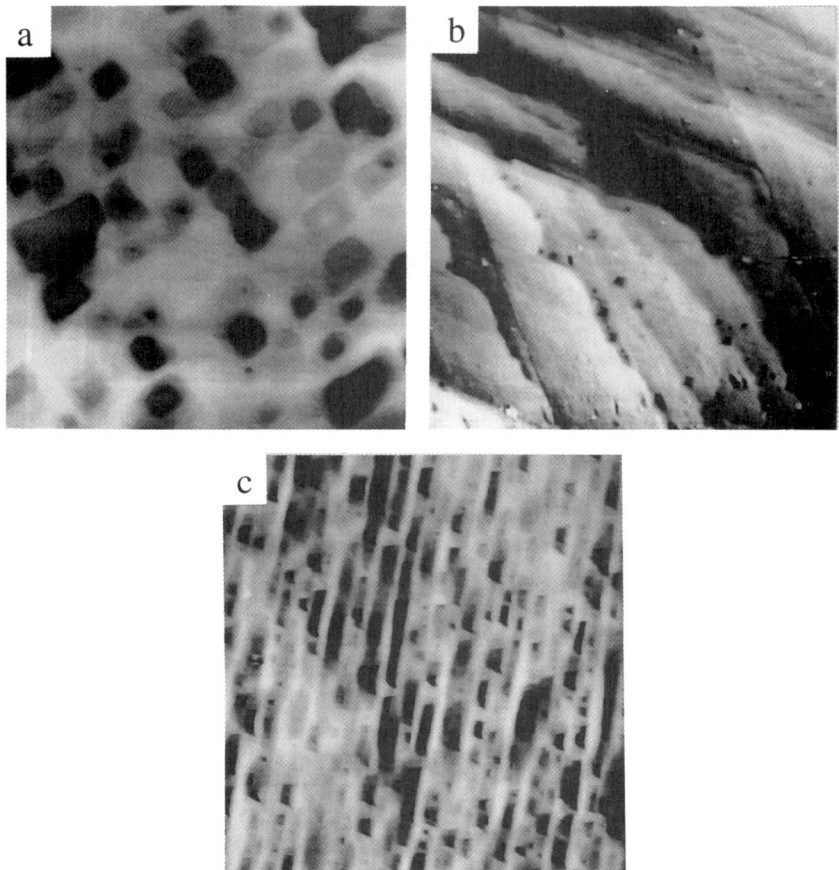

FIG. 4. Atomic force microscopy images of as-grown mercuric iodide crystal: (a) (001) face, scan size 14 μm × 14 μm, (b) (110) face, scan size 14 μm × 14 μm, (c) (013) face, scan size 5 μm × 5 μm, Z-scale is 250 nm (courtesy of Professor Michael A. George, Fisk University, Nashville).

crystal can be obtained. Using the rocking curve technique, lattice distortions such as tilts and twists of various sets of crystallographic planes can be measured by their broadening of the diffraction peaks. The extent to which a mosaic structure of subgrains, separated by low angle boundaries, constitutes the nominally "single" crystal is revealed by the peak splitting.

Gamma ray rocking measurements at the Missouri University Research Reactor (MURR) were used to determine the effects of various fabrication methods on the crystalline quality of slices of material fabricated from mercuric iodide crystals. The results (Yelon et al., 1981) showed that the mechanical cleaving is structurally damaging, and this has led to changes in fabrication methods. It was also concluded (Schieber et al., 1989) that detector performance correlates with crystal perfection. The gamma ray rocking method was used to preevaluate small seed crystals in ampules selected for mercuric iodide microgravity vapor growth experiments on Spacelab 3 (SL-3) and the International Microgravity Laboratory 1 (IML-1). This technique was subsequently used to evaluate the resultant space grown crystals. Currently, gamma rocking analysis is being investigated as a method to evaluate small single crystals that are intended to be used as implanted seeds in seeded mercuric iodide growth experiments (see the next subsection).

The spatial variability of the crystalline lattice perfection has been studied by scanned beam gamma rocking analysis on both ground and space grown crystals. In recent work at MURR, vertical scanning measurements were made on a grown crystal that had been cooled to ambient temperatures but otherwise undisturbed in the growth ampule. Some results are shown in Fig. 5 (Ross and Mann, private communication, 1994). The rocking curves were measured with 0.03 Å radiation diffracted from the (220) planes, at locations that range from near the growth base (2 mm) to near the upper grown surface (12 mm) along an approximate [110] lattice direction. The splitting and broadening of the peak is considerably larger near the growth base, showing decreased crystal perfection there. This result suggests either an effect caused by the mismatch of the thermal expansion between the HgI_2 crystal and the substrate, perhaps through thermal strain (Burger *et al.*, 1992), or a gravity effect caused by the crystal itself. The peak broadness at half-height (FWHM) of around 0.1° as seen in Fig. 5 is not unusual in this material, but FWHM values around 0.03° are not uncommon (Ross and Mann, private com-

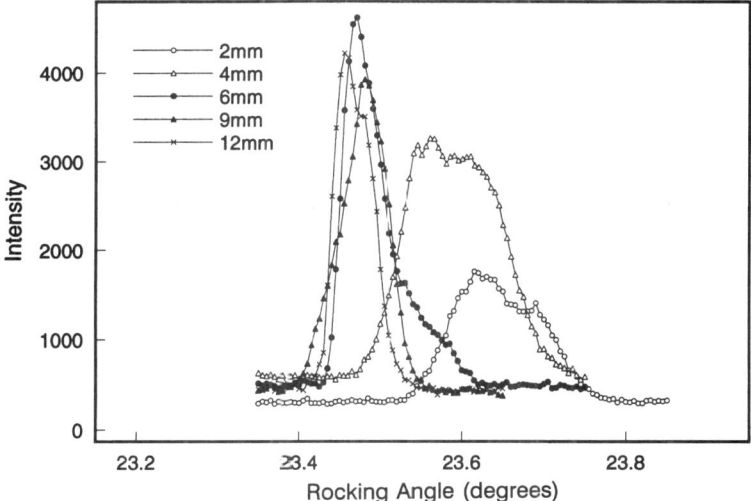

FIG. 5. Rocking curves using 0.03 Å gamma radiation for selected elevation from near the base (2 mm) to near the top surface (12 mm) in a mercuric iodide crystal (courtesy of J. Mann and Dr. F. Ross, Missouri University Research Reactor).

munication, 1994; Rossberg *et al.*, 1994) in local regions of single crystals of mercuric iodide.

Gamma rocking studies of mercuric iodide have also been done at Institute Max von Laue–Paul Langevin, Grenoble, France. Mapping measurements on crystals with large single grain segments revealed (Rossberg *et al.*, 1994) a considerable variation in crystalline perfection throughout the volume. The crystalline perfection was also reduced close to the crystal bottom, in agreement with the results shown in Fig. 5. The correlation of lattice structure and the nature of the distortions was discussed. Monochromatic syncrotron x-radiation (Steiner *et al.*, 1991) measurements at the U.S. National Institute of Standards and Technology (NIST) found evidence for micro inclusions in earth grown mercuric iodide crystals. X-ray reflection topography was used (Keller *et al.*, 1993) to map surface imperfections, to determine their correlation with fabrication methods and to characterize (Gits, 1983) the dislocation substructure.

The internal regularity of mercuric iodide has also been studied using synchrotron white beam x-radiation transmission. At CRMC2-CNRS, Marseille, France, volume topographies were obtained (Remy *et al.*, 1992; Remy, Gastaldi, and Le Lay, 1993) using thinned samples ($\sim 10^{-2}$ cm). These showed an extensive subgrain structure and plasticity related defects such as dislocations and kink bands.

More recent triaxial double crystal x-ray diffraction rocking curves showed that HgI_2 crystals that produced low grade detectors have both a large mosaicity and strain due to a spread in the unit cell dimensions (Schieber *et al.*, 1994). Such a strain contribution to the width of the rocking curves is absent in other bulk crystals such as GaAs.

V. Recent Developments

1. SEEDED GROWTH

Seeding is of interest in any crystal growth method and is widely used when appropriate techniques are available. One advantage is that the often slow and tedious process of initiating satisfactory spontaneous nucleation (self-seeding) can be bypassed.

A technique for implant seeding developed at EG&G/EM SBO (Nason, 1994) can be explained by reference to Fig. 2. The technique consists of (1) loading and sealing the growth ampule, (2) identifying the temperature settings appropriate for self-nucleation, (3) cooling the ampule, (4) opening the ampule neck, (5) implanting a seed at the selected orientation in the growth location, (6) resealing and reheating the ampule, and (7) finely adjusting the thermal conditions first for a small amount of evaporation and then (8) for the smooth growth of the seed. The seeds used to date have been pieces fabricated from previously grown, single crystals. They have been prepared by KI solution sawing, etched in KI solution and then rinsed in water before implantation. The seed orientation is known from the recognizable morphology of the parent crystal from which it was obtained. Several seeded crystals have been grown to date. By using the seeded growth there is a time savings of typically 2 weeks in an average growth time of 2–3 months for 250 g crystals over conventional, unseeded methods. However, careful monitoring is required at the time the seed is being stabilized. Sometimes a visible band occurs in the grown crystals, marking the original boundary of the seed, much as when any crystal is evaporated back and then regrown.

A particularly attractive advantage of seeded growth is the potential for superior crystals by using superior seeds. Initially, seeds have been cut from parent crystals that are simply visibly free of cracks, voids, and polycrystallinity. Even with this minimal selection standard, results from many detectors fabricated from these seeded crystals show performance that is comparable to self-nucleated crystals. Apparently, the seed fabrication, handling, and the growth process interruption do not degrade the resultant crystals. These results suggest there is potential for performance gains once a truly discriminating seed selection method is implemented. A study is underway in conjunction with MURR to grow crystals from seeds screened for their high crystalline perfection by prior gamma rocking analysis. Better optical methods for seed selection are also being investigated. More experimental data on comparison between seeded and self-nucleated crystals are needed to determine which method produces higher grade detectors.

2. In-Situ Process Monitoring

a. Iodine Concentration

The pink color of iodine vapor has been noticed occasionally in mercuric iodide growth experiments performed at EG&G/EM SBO that were not intentionally doped. When the color was observed during growth, its intensity was visually estimated with reference to saturated iodine at room temperature, which has a vapor concentration of 16 μmol/l. Even a trace of the pink color at growth temperatures of $\approx$100–120°C indicates an iodine partial pressure on the same order as the HgI_2 (0.1 Torr) vapor. This result has important consequences in modeling the process of crystal growth by vapor transport (Omaly et al., 1983). A method of measuring the significant concentrations of iodine, sometimes noticed during mercuric iodide growth, has been needed.

An optical technique based on absorption was established recently (Nason, Biao, and Burger, 1994) to determine iodine vapor concentrations in growth ampules. It was established first that iodine vapor at low concentrations obeys Beer's law of absorption and has an absorptivity of $7.7 \cdot 10^5$ cm^2 mol^{-1} at 514 nm. Iodine vapor concentrations can vary during growth and sometimes depend on the material purification processing. Differential scanning calorimetry (DSC) measurements of condensed iodine (Burger et al., 1990) have been correlated with these vapor absorption results and indicate (Nason et al., 1994) the presence of free crystalline iodine in vapor grown HgI_2. The variation in iodine vapor concentration monitored during growth indicates that iodine can incorporate unevenly into a crystal, creating sections of inhomogeneity with differing stoichiometry.

b. Surface Temperature

The surface temperature of mercuric iodide crystals as they grow is of fundamental importance because it affects the supersaturation, which controls the growth rate and the resultant crystal quality. Free standing crystals growing by PVT have significant temperature gradients (Subsection II.2), so methods that measure the overall temperature of the crystal bulk or nearby temperatures do not provide accurate surface temperatures. A novel, noncontact method has been developed for measuring the surface temperature of a growing mercuric iodide crystal. The method is particularly intended for situations where pyrometry is unsatisfactory, due to either low emittance or the combination of a temperature gradient and partial transparency in the solid. The method, called "reflectance spectroscopy thermometry" (RST) (Nason and Burger, 1991), is based on measuring the temperature dependence of the energy of a broad peak (the free exciton peak) in the reflection spectra, as detected with a low energy optical beam. The surface temperature, T, was found to be linearly dependent on the energy of the exciton peak, E_r:

$$T = \frac{2.236 - E_r}{4.3 \times 10^{-4}}$$

Where T is in °C and E_r in eV. An accuracy of ± 1.5°C for a slowly varying surface temperature has been demonstrated by this method.

c. Optical Growth Pattern Recognition

Progress has been made in developing an in situ optical monitoring method for the purpose of early detection of microscopic growth irregularities. The method allows growth parameters to be adjusted before macroscopic surface defects (voids, parasitic nuclei, etc.) develop. The method is based on real time images collected by a high resolution camera and analyzed by neural network processing with reference to images taken during satisfactory growth (Sawyer et al., 1991). Further development is needed before vapor growth applications are worthwhile.

3. Fundamental Studies

The commonly used method for growing mercuric iodide crystals of useful size by PVT in vertical furnaces is well established (Lamonds, 1983; Burger et al., 1991b; Zha, Piechotka, and Kaldis, 1991). Control of the growth process has been achieved largely by empirical control of the boundary heat flow parameters and the material purity. Crystals of satisfactory size and quality are being produced. However, it is generally acknowledged that further important gains in crystal quality may be possible if a more fundamental understanding of the mass and heat transport could be achieved and utilized.

The difficulties of a complete and rigorous analysis of PVT are numerous. However, significant progress has been made in identifying the key factors. A major motivation has been the space experiments with mercuric iodide, which made it possible to control gravity as an experimental variation (Subsection III.2). A study (Nadarajah, Rosenberger, and Alexander, 1992) has reviewed PVT theoretical work and investigated the coupling of mass transport and gravity in a two dimensional numerical analysis. Another study (Roux, Fedoseyev, and Roux, 1993) has used numerical methods to investigate the relationship of crystal morphology to the thermal radiation flux in the ampule. Analytical studies (Chernov et al., 1992; Nelson, 1993) of the morphological development of crystals under idealized geometrical conditions have also been done. A laser doppler system for measuring vapor convection in PVT systems, at much higher pressures than used for HgI_2 has been described (Jones et al., 1991). Such studies may help identify the critical ranges and necessary levels of control of the parameters (vapor phase composition and density, boundary temperatures, etc.) that most influence the crystals being grown.

On the experimental side, growth experiments at EG&G/EM SBO (Nason and Berry, 1994) have yielded useful information for optimizing the crystal growth geometry. Traditionally, crystals of a size useful for radiation detectors, 0.25–1.0 kg, have been intentionally grown with their [001] axis within 30° of horizontal, as illustrated in Fig. 1a (axes as in Fig. 1b), and shown in Fig. 3. The choice of orientation is made visually in the transparent ampule during self-seeding, where only a crystal with a desirable orientation is preserved for growth. The typical growth morphology in this orientation has a width/length ratio, relating the dimensions along [001] and [110], respectively, of 0.55–0.65.

Experiments were performed at EG&G/EM SBO in which crystals have been grown with their [001] axis vertical, as illustrated in Fig. 1b. These crystals grow flatter, with a [001]/[110] dimensional ratio of 0.3–0.4. These flatter crystals allow detectors of larger (001) face area to be fabricated from the same crystal mass. This difference in shape illustrates the sensitivity of the morphology to the thermal environment in the ampule. In experiments with the ampule modified from the cylindrical configuration of Fig. 2 toward a more thermally spherical configuration, the crystals with vertical [001] axes grew with a size ratio reduced to 0.2–0.25. A large crystal of 1.1 kg, and approximate dimensions, 100 × 100 × (12–20) mm^3, was grown in this orientation. These results illustrate how the morphology is far from inherent and is influenced strongly by controllable heat fluxes in the ampule.

The morphology is also influenced significantly by the growth rate. If the conventional rate of 1–2 mm/day is doubled or tripled, the minor faces such as (013) in Fig. 1 are greatly diminished, while the (110) area increases relative to (001). The crystal becomes generally bounded by fewer, larger faces. Rounded shapes result from a comparatively slow growth rate, with an elliptical profile as viewed along (110) and (001) faces that are small and circular and surrounded by a convex shoulder region. Quantitative morphological measurements may become crucial in validating models of mercuric iodide growth by PVT.

The accurate measurement of growth rates of crystals is important in understanding mechanisms of molecular attachment on the various crystal faces and their interplay with the mass transport and the thermal boundary conditions. Mercuric iodide growth rate measurements have been reported (Isshiki, Piechotka, and Kaldis, 1990) from in-situ optical measurements of crystal size changes as the ampule perimeter temperatures were changed incrementally. From the initial growth rates for different faces, these data have yielded useful kinetic values of $6-8 \times 10^{-3}$ cms^{-1}, where the kinetic coefficient is defined (Chernov et al., 1992) as the growth rate divided by the square of the supersaturation. In that study (Isshiki et al., 1990) the crystal dimensional changes were measured to ± 100 μm, and the actual face orientations were unknown. There may also be some uncertainty in the actual supersaturation values produced by the incremental changes in the boundary temperature, since it is difficult to thermally isolate the crystal surface from the heater that warms the source material to produce the supersaturation. Kinetic studies that have higher resolution of crystal diamensional changes and closer control of the supersaturation may be very important in developing an understanding of the growth process and controlling crystal development in these systems.

VI. Challenges in Crystal Growth

The goal of mercuric iodide crystal growth is to produce crystals that are easily grown, have a high crystalline perfection, and have a composition that is optimum for the fabrication of nuclear detectors. Four areas where advances would be especially beneficial are discussed in the following subsections. An earlier review (Piechotka and Kaldis, 1989) may be consulted for an additional perspective.

1. IMPURITY ANALYSIS

Mercuric iodide of purity adequate for successful crystal growth and useful devices can be produced by a certain sequence of processing steps (Subsection III.1). However, the optimization of that process is hampered by the lack of adequate information about the impurity level at the various stages of processing. An adequate and sensitive chemical analysis method, particularly for cations, is necessary for correlating the composition with device performance and identifying the levels of trace impurities that are either harmful or beneficial. An ultrapure crystal may be mechanically very soft and may lose its crystalline perfection during cutting, polishing, and other fabrication stages. Therefore, the knowledge of beneficial impurities to the electrical charge transport may lead to controlled doping or intensive purification steps to eliminate a specific type of undesirable component. It might also then be possible to distinguish between original impurities and those introduced during fabrication.

2. Stoichiometry

More accurate methods are needed to determine the near stoichiometric range of HgI_{2-x}. The intrinsic defect content, which is crucial to the electronic properties of compound semiconductors, depends critically on the stoichiometry parameter, x. Precise knowledge of the near stoichiometry range could allow the optimum compositions for device performance to be identified. The stoichiometry could eventually be controlled by appropriate selection of the growth temperature, control of the partial pressures of the vapor components, and perhaps by heat treatments.

3. Bulk Growth

A better fundamental understanding of the empirically successful process for growing free standing, bulk crystals would identify critical parameters and lead to crystals of superior crystallinity and optimized growth morphology. In particular, the effects of convective vapor transport and thermal and weight induced stresses are of interest. Additional microgravity experiments would be very helpful. Improvements in process control, such as direct monitoring and control of the growth rate or the crystal surface temperature, would be likely to increase crystal homogeneity.

4. Film Growth

There may be interest in thin layer mercuric iodide devices for photodetector applications. It would be worthwhile to develop such epitaxial film growth methods deposited on a suitable substrate.

Acknowledgments

The authors thank Dr. R. Monchamp, Dr. L. Franks, and Carol Ortale Baccash for their comments on the manuscript.

This work has been authored by a contractor of the U.S. government under Contract No. DE-AC08-93NV11265. Accordingly, the U.S. government retains a nonexclusive, royalty-free license to publish or reproduce the published form of this contribution, or allow others to do so, for U.S. government purposes.

REFERENCES

Azoulay, M., George, M. A., Biao, Y., Burger, A., Silberman, E., and Nason, D. (1993). *J. Vac. Sci. and Technol. B* **11**(5), 1782.
Beinglass, I., Dishon, G., Holzer, A., and Schieber, M. (1977). *J. Crystal Growth* **42**, 166.
Bodroux, M. F. (1900). *C. R. Acad. Sci.* **130**, 1622.
Bohac, P., and Kaufmann, P. (1975). *Mat. Res. Bull.* **10**, 613.
Bridgman, P. S. (1951). *Proc. Am. Acad. Arts. Sci.* **51**, 55.
Burger, A., Roth, M., and Schieber, M. (1982). *J. Cryst. Growth* **56**, 526.
Burger, A., Levi, A., Nissenbaum, J., Roth, M., and Schieber, M. (1985). *J. Cryst. Growth* **72**, 643.
Burger, A., Morgan, S. H., He, C., Silberman, E., van den Berg, L., Ortale, C., Franks, L., and Schieber, M. (1990). *J. Crystal Growth* **99**, 988.
Burger, A., Morgan, S. H., Henderson, D. O., Silberman, E., and Nason, D. (1991a). *J. Appl. Phys.* **69**, 722.
Burger, A., Morgan, S. H., Silberman, E., and Nason, D. (1991b). *Proceedings of SPIE Conference* **1557**, 245.
Burger, A., Morgan, S. H., Henderson, D. O., Silberman, E., and Nason, D., and Cheng, A. Y. (1992). *Nucl. Instr. and Meth. in Phys. Res.* **A322**, 427.
Burton, W. K., Cabrera, N., and Frank, F. C. (1951). *Phil. Trans. Royal Soc. A* **51**, 243.
Cadoret, R., Brisson, P., and Magnan, A. (1989). *Nucl. Instr. and Meth. in Phys. Res.* **A283**, 339.
Chernov, A. A. (1961). *Sov. Phys. Usp.* **4**, 116.
Chernov, A. A., Kaldis, E., Piechotka, M., and Zha, M. (1992). *J. Crystal. Growth* **125**, 627.
Cross, E., Olivares, J., and Mroz, G. (1993). *Semiconductors for Room Temperature Radiation Detector Applications.* MRS Symposium Proceedings **302**, 61.
Coupat, B., et al. (1994). *J. Crystal Growth* (Submitted).
DeLong, M. C., and Rosenberger, F. (1981). *Mater. Res. Bull.* **16**, 445.
Dishon, G., Schieber, M., Ben-Dor, L., and Halitz, L. (1981). *Mater. Res. Bull.* **16**, 565.
Dworsky, R., and Komarek, K. L. (1970). *Monatshefte Chem.* **101**, 966 and 985.
Faile, P., Dabrowski, A. J., Huth, G. C., and Iwanczyk, J. S. (1980). *J. Crystal Growth* **50**, 752.
Georgeson, G., and Milstein, F. (1989a). *Nucl. Instr. and Meth.* **A283**, 507.
Georgeson, G., and Milstein, F. (1989b). *Nucl. Instr. Meth. in Phys. Res.* **A285**, 488.
Gerrish, V. (1993). Materials Research Society Symposium Proceedings *Semiconductors for Room-Temperature Radiation Detector Applications,* **302**, 129.
Gits, S. (1983). *Nucl. Instr. Meth.* **213**, 43.
Gospodinov, M. (1980). *Kristal Technik* **15**, 263.
Gmelins Handbuch der Anorgan. Chemie. (1933). **8**(1), 78.
Hermon, H., Roth, M., and Schieber, M. (1990). *J. Crystal Growth* **106**, 68.
Hermon, H., Roth, M., Schieber, M., and Shamir, J. (1993). *Mat. Res. Bull.* **28**, 229.
Isshiki, M., Piechotka, M., Kaldis, E. (1990). *J. Crystal Growth* **102**, 344.
James, K., Gerrish, V., Cross, E., Markakis, J., Marschall, J., and Milstein, F. (1992). *Nucl. Instr. and Meth. in Phys. Res.* **A322**, 390.
James, R. B., Ottesen, D. K., Wong, D., Schlesinger, T. E., Schnepple, W. F., Ortale, C., and van den Berg, L. (1989). *Nucl. Instr. Meth. in Phys. Res.* **283**, 188.
James, T. W., and Milstein, F. (1981). *J. Mat. Sci.* **16**, 1167.

Jones, O. C., Glicksman, M. E., Lin, J. T., Kim, G. T., and Singh, N. B. (1991). *Proceedings of SPIE Conference* **1557**, 202.
Keller, L., Wang, E. X., and Cheng, A. Y. (1993). *Mat. Res. Symp. Proc.* **302**, 153.
Lamonds, H. A. (1983). *Nucl. Instr. and Meth. in Phys. Res.* **213**, 5.
Llacer, J. (1974). *J. Crystal Growth* **24-25**, 205.
Milstein, F., Farber, B., Kim, K., van den Berg, L., and Schnepple, W. F. (1983). *Nucl. Inst. and Meth.* **213**, 65.
Milstein, F., and Georgeson, G. (1989a). *J. Mater. Sci.* **24**, 328.
Milstein, F., James, T. W., and Georgeson, G. (1989b). *Nucl. Instr. Meth. in Phys. Res.* **A285**, 500.
Muheim, J. T., Kobayashi, T., and Kaldis, E. (1983). *Nucl. Instr. Meth.* **213**, 39.
Nadarajah, A., Rosenberger F., and Alexander, J. I. D. (1992). *J. Crystal Growth* **118**, 49.
Nason, D. (1994). Patent applied for.
Nason, D., and Burger, A. (1991). *Appl. Phys. Lett.* **59**(27), 3550.
Nason, D., and Berry, K. (1994). Unpublished results.
Nason, D., Biao, Y., and Burger, A. (1994). *J. Crystal Growth*.
NBS Monograph. (1969). *Standard X-ray Diffraction Patterns* **25**, Sect. 7, p. 32.
Nelson, J. (1993). *J. Crystal Growth* **132**, 538.
Newkirk, J. B. (1956). *Acta Metallurgica* **4**, 316.
Nicolau, I. F. (1980). *J. Cryst. Growth* **48**, 61.
Nicolau, I. F., and Rolland, G. (1981). *Mater. Res. Bull.* **16**, 759.
Nicolau, Y. F., and Dupuy, M. (1989a). *Nucl. Inst. and Meth. in Phys. Res.* **A283**, 355.
Nicolau, Y. F., Dupuy, M., and Kabsch, Z. (1989b). *Nucl. Inst. and Meth. in Phys. Res.* **A283**, 149.
Nicolau, Y. F., Moser, P., and Cobel, C. (1989c). *Nucl. Inst. and Meth. in Phys. Res.* **A283**, 167.
Nicolau, Y. F., and Dupuy, M., and Rolland, G. (1994). *J. Mater. Sci.: Mater. in Elect.* **4**.
Omaly, J., Robert, M., and Cadoret, R. (1981). *Mat. Res. Bull.* **16**, 785 and 1261.
Omaly, J., Robert, M., Brisson, P., and Cadoret, R. (1983). *Nucl. Instr. and Meth.* **213**, 19.
Pfann, W. G. (1952). *Trans AIME* **194**, 747.
Piechotka, M., and Kaldis, E. (1986a). *J. Less-Common Metals* **115**, 315.
Piechotka, M., and Kaldis, E. (1986b). *J. Crystal. Growth* **79**, 469.
Piechotka, M., and Kaldis, E. (1989). *Nucl. Instr. and Meth. in Phys. Res.* **A283**, 111.
Piechotka, M., and Kaldis, E. (1992). *Nucl. Instr. and Meth. in Phys. Res.* **A322**, 387.
Pompon, J. P., Stuck, R., Siffert, P., Meyer, B., Schwab, C. (1975). *IEEE Trans. Nucl. Sci.* **NS-22**, 182.
Przyluski, J., and Laskowski, J. (1989). *Nucl. Instr. Meth. in Phys. Res.* **A283**, 144.
Remy, F., Gastaldi, J., Grange, G., Jourdan, C., and Le Lay, G. (1992). *J. Crystal Growth* **121**, 243.
Remy, F., Gastaldi, J., and Le Lay, G. (1993). *Nucl. Instr. and Meth. in Phys. Res.* **B83**, 229.
Rolsten, R. F. (1961). *Iodide Metals and Metal Iodides*. John Wiley & Sons, New York, p. 153.
Rosenberger, F. (1979). *Fundamentals of Crystal Growth I*. Springer Series in Solid State Sciences **5**, Springer-Verlag, Berlin, p. 147.
Rossberg, A., Piechotka, M., Magerl, A., and Kaldis, E. (1994). *J. Appl. Phys.* **75**(7), 3371.
Roux, A., Fedoseyev, A., and Roux, B. (1993). *J. Crystal Growth* **130**, 523.
Sarid, D. (1991). *Scanning Force Microscopy*. Oxford Series in Optical and Imaging Sciences, Oxford University Press, New York.
Sawyer, C. R., Quach, V. T., Nason, D., and van den Berg, L. (1991). *Proceedings of SPIE Conference* **1567**, 1.
Scharager, C., Tadjine, A., Toulemonde, M., Grob, J. J., and Siffert, P. (1980). *Proc of Nucl. Physics, 7th Divisional Conference Nuclear Physics Methods in Material Research, Darmstadt, Germany*, 126.
Schieber, M., Schnepple, W. F., and van den Berg, L. (1976). *J. Crystal Growth* **33**, 125.
Schieber, M., Ortale, C., van den Berg, L., Schnepple, W. F., Keller, L., Wagner, C. N. J., Yelon, W., Ross, F., Georgeson, G. and Milstein, F. (1989). *Nucl. Instr. and Meth. in Phys. Res.* **A283**, 172.
Schieber, M., Carlston, R. C., Lamonds, H. A., Randtke, R. T., and Schnepple, F. W., and Gerrish, V., and van den Berg, L. (1990). *Nucl. Instr. Meth. in Phys. Res.* **A299**, 41.

Schieber, M., Roth, M., Yao, H. W., Devries, M., James, R. B., and Goorsky, M. J. *J. Crystal Growth*, in press.
Scholz, H. (1974). *Acta Electronica* **17,** 69.
Scholz, H., and Kluckow, R. (1967). *Crystal Growth,* ed. H. S. Peiser. Pergamon Press, Oxford, p. 117.
Skinner, N. L., Ortale, C., Schieber, M. M., and van den Berg, L. (1988). *J. Crystal Growth* **89,** 86.
Steinberg, S., Kaplan, I., Schieber, M., Ortale, C., Skinner, N., and van den Berg, L. (1989). *Nucl. Instr. Meth. in Phys. Res.* **A283,** 123.
Steiner, B., Dobbyn, R. C., Black, D., Burdette, H., Kuriyama, M., Spal, R., van den Berg, L., Fripp, A., Simchick, R., Lal, R. B., Batra, A., Matthiessen, D., and Dietcheck, B. (1991). *Proceedings of SPIE Conference* **1557,** 156.
Stowell, M. J., and Hutchinson, T. E. (1971). *Thin Solid Films* **835** and **411.**
Tadjine, A., Gosselin, D., Koebl, J. M., and Siffert, P. (1983). *Nucl. Instr. Meth.* **213,** 77.
Ultramicroscopy. (1992). **42–44,** articles and their references.
van den Berg, L. (1992). *Nucl. Instr. Meth. in Phys. Res.* **A322,** 453.
van den Berg, L. (1993). *Mat. Res. Soc. Symp. Proc.* **302,** 73.
van den Berg, L., and Schnepple, W. F. (1982). *Materials Processing in the Reduced Gravity Environment in Space,* p. 439, G. E. Rindone, Elsevier–North Holland, Amsterdam.
van den Berg, L., and Schnepple, W. F. (1989). *Nucl. Instr. and Meth. in Phys. Res.* **A283,** 335.
Warburton, W. K. (1985). Report IP-85-3, Institute of Physics, School of Medicine, University of Southern California.
Willig, W. R. (1971). *Nucl. Instr. and Meth.* **96,** 615.
Yelon, W. B., Alkire, R. W., Schieber, M. M., Rasmussen, S. E., Christensen, H., and Schneider, J. R. (1981). *J. Appl. Phys.* **52**(7), 4604.
Zha, M., Piechotka, M., and Kaldis, E. (1991). *J. Crystal Growth* **115,** 43.

CHAPTER 4

Electrical Properties of Mercuric Iodide

X. J. Bao

DEPARTMENT OF ANALYTICAL INSTRUMENTS
TN TECHNOLOGIES INC.
ROUND ROCK, TEXAS

T. E. Schlesinger

DEPARTMENT OF ELECTRICAL AND COMPUTER ENGINEERING
CARNEGIE MELLON UNIVERSITY
PITTSBURGH, PENNSYLVANIA

R. B. James

ADVANCED MATERIALS RESEARCH DEPARTMENT
SANDIA NATIONAL LABORATORIES
LIVERMORE, CALIFORNIA

I.	INTRODUCTION	112
II.	CARRIER TRANSPORT	113
	1. *Dark Resistivity and I–V Characteristics*	113
	2. *Mobility and Lifetime*	119
	3. *Effective Mass*	131
III.	DEEP LEVELS	132
	1. *Thermally Stimulated Current Technique*	133
	2. *Other Techniques*	144
IV.	PHOTOCONDUCTIVITY	148
	1. *Behavior of Photocurrent*	149
	2. *Photocurrent Quenching*	153
V.	SURFACE EFFECTS	154
	1. *Electrical Contacts*	154
	2. *Surface Recombination*	156
VI.	DETECTOR PERFORMANCE	158
	1. *Peak-to-Background Ratio*	158
	2. *Polarization Effect*	160
VII.	CONCLUSIONS	165
	References	165

List of Symbols

A	area of contact	$c_{e,h}$	capture probability of electron and hole
C	capacitance		

E	electric field	t_r	carrier transit time
E_{ph}	x-ray photon energy	V	applied voltage
e	2.718...	v_d	drift velocity
$e_{e,h}$	thermal emission probability of electron and hole	v_{th}	thermal velocity of free carrier
h	Planck's constant	β	heating rate
I_d	detector leakage current	ΔE_t	activation energy of carrier trap
i	transient current density		
J_{PF}	Poole–Frenkel current	ΔE	FWHM of x-ray peak
J_S	Schottky current	ϵ	average energy to generate one electron-hole pair
k	Boltzman's constant		
L	detector thickness	ϵ_0	permittivity in vacuum
ln	natural logarithm	ϵ_r	dielectric constant
log	logarithm with a base of 10	η	charge collection efficiency
m_0	electron rest mass	$\mu_{e,h}$	mobility of electrons and holes
$m^*_{e,h}$	effective mass of electron and hole		
		μ_{ph}	photoelectric absorption coefficient
$N_{c,v}$	effective density of states in the conduction and valence bands		
		ν	photon frequency
		ρ	resistivity
N_t	density of carrier trap	$\sigma_{e,h}$	carrier capture cross section of electron and hole trap
n_t	density of carrier trap occupied		
		τ_d	detrapping time
n	density of free electrons	$\tau_{\text{dielectric}}$	dielectric relaxation time
Q	collected charge	$\tau_{e,h}$	lifetime of electron and hole
Q_0	total charge generated	τ_t	trapping time
p	density of free holes	ϕ_{PF}	Poole–Frenkel barrier height
s	surface recombination velocity		
		ϕ_S	Schottky barrier height
T	absolute temperature		

I. Introduction

Even with the progress made since its first use as a nuclear spectrometer in 1971 (Willig, 1971), HgI_2 technologies are still far from being fully developed. Compared with other semiconductors commonly employed in various applications, the relationships between most defects and the electronic properties are still either poorly understood or unknown. The potential of HgI_2 nuclear detector technologies will certainly be more fully realized as more progress is made toward the understanding of the material properties responsible for the wide variability in detector performances. The electrical properties of HgI_2 in many instances determine the performance of the x-ray and γ-ray spectrometers and photodetectors fabricated from this material. Broadly speaking, the performance of a nuclear spectrometer is characterized by its energy resolution, peak to background ratio, efficiency, reliability, response speed, active volume, and the ease with which it can be handled. The electrical properties of importance include the dark resistivity of the material, mobility and lifetime of free carriers, effective masses of free carriers, behavior of traps, and surface effects. This chapter will be devoted to the description of the electrical properties of red HgI_2, particularly those important in using this material for x-ray and γ-ray detector applications. Other topics discussed in this chapter include the relationship of carrier traps to fabrication processes and device performance, measurement techniques used to characterize the electrical properties of HgI_2, and the theoretical background required to evaluate the data.

II. Carrier Transport

1. Dark Resistivity and I–V Characteristics

HgI_2 detectors are exclusively biased along the crystallographic c-axis for carrier collection, and transport properties do depend on the crystallographic direction. Therefore, the discussion of transport properties for crystals and detectors will refer to values along the c-axis, unless otherwise noted. Dark resistivity determines the leakage current in HgI_2 x-ray and γ-ray spectrometers. A large leakage current is undesirable mainly because it reduces the energy resolution of the detector. Leakage current contributes to the FWHM (full width at half maximum) of the detector system according to the relation (Dabrowski and Huth, 1978)

$$\Delta E_a = (2.355 \epsilon e) \sqrt{\frac{I_d \tau}{4q}} \tag{1}$$

where ΔE_d is the FWHM in eV due to leakage current, ϵ is the mean energy in eV required to create one electron–hole pair, e is 2.718, q is the electron charge in coulombs, I_d is the leakage current in amperes, and τ is the time constant of the shaping network in seconds.

One advantage of HgI_2 detectors is that their dark resistivity is many orders of magnitude greater than that of most other semiconductors, such as Si, Ge, III–V, and II–VI materials. While Si and Ge nuclear detectors are usually operated at cryogenic temperatures to reduce the leakage current, HgI_2 detectors have a high enough resistivity ($\sim 10^{13}$ Ω-cm) to be operated at room temperature. This advantage is not solely a matter of convenience, since it results in many new applications where cryogenic cooling is difficult or impossible and a compact portable system is essential (see the discussion in Chapter 1).

Dark resistivity in HgI_2 can be estimated by measuring the dark current and noting the sample geometry. Contacts are typically deposited on opposite faces of

HgI$_2$ slices to form a sandwich structure. The resistivity is then calculated according to the equation: $\rho = VA/I_d L$, where ρ is the resistivity in Ω-cm, V is the applied voltage in volt, I_d is the dark current in amperes, A is the area of the contact in cm^2, and L is the thickness in cm. Other commonly used techniques, such as the four point probe method or Hall effect measurement, are difficult to use on HgI$_2$ because of its very high resistivity.

When a high voltage is applied, the dark current in HgI$_2$ usually decreases slowly with time (Braatz and Zappe, 1984; Levi, Schieber, and Burshtein, 1985). The estimated dark resistivity at 295 K is about 10^{11} Ω-cm initially and reaches a saturation value of (0.6 to 20) $\times$ 10^{13} Ω-cm within 200 min. of applied bias (Braatz and Zappe, 1984). This decrease of dark current with time is believed to be due to the release of carriers from traps and is therefore described by a sum of exponentially decaying currents, each with a time constant corresponding to the detrapping time of a particular trap level. Investigations of this behavior have been used to understand carrier traps in HgI$_2$, and they will be discussed in Subsection III.2. The dependence of dark current on applied field is not linear over the entire range of voltages employed in detector operation (Braatz and Zappe, 1984; Mellet and Friant, 1989). The dark resistivity obtained, therefore, is only a rough estimate. Near room temperature, the dark current is very sensitive to temperature and increases rapidly with increasing temperature (Braatz and Zappe, 1984). For a HgI$_2$ detector of 1 cm^2 area, lowering the operating temperature to 0°C will reduce the dark current to below 1 pA compared to a dark current of about 10 pA at 300 K. The noise contribution from this source is then negligible when compared to the contribution from currently available preamplifiers (Levi et al., 1985).

The mechanism of dark conductivity in HgI$_2$ has not been thoroughly studied. Ionic conductivity due to the movement of iodine vacancies was observed in HgI$_2$ single crystals and polycrystals at a temperature of 120°C (Kitajima and Wagner, 1988). Since HgI$_2$ devices are usually operated near or below room temperature, ionic conductivity probably is not dominant under these conditions. The dark electronic conductivity in a dielectric material is generally due to thermal emission or tunneling, both of which may be further divided into either bulk controlled or electrode controlled (Yeargan and Taylor, 1968). Tunneling in HgI$_2$ is unlikely given the typical device configurations employed. For the case of thermal emission, the bulk controlled current is determined by the Poole–Frenkel effect, which is associated with barriers in the bulk of the material. The electrode controlled current is determined by the Schottky effect, which is associated with the barrier at the metal–semiconductor interface. Both of these effects are controlled by the lowering of the barrier by an applied electric field (Yeargan and Taylor, 1968). The barrier lowerings are given by

$$\Delta\phi_{PF} = \sqrt{\frac{q^3 E}{\pi \epsilon_e \epsilon_0}} = \beta_{PF} \sqrt{E} \qquad (2)$$

and

$$\Delta\phi_S = \sqrt{\frac{q^3 E}{4\pi\epsilon_r\epsilon_0}} = \beta_S\sqrt{E} \qquad (3)$$

and $\Delta\phi_{PF}$ and $\Delta\phi_S$ are the barrier lowerings due to the Poole–Frenkel and Schottky effects, respectively, E is the electric field at the barrier (which may be different from the applied field), ϵ_0 is the vacuum permittivity, and ϵ_r is the high frequency relative dielectric constant. The Poole–Frenkel and Schottky currents (for electrons) are given by (Yeargan and Taylor, 1968)

$$J_{PF} = q\mu_n n_o E \exp\left(-\frac{\phi_{PF} - \Delta\phi_{PF}}{rkT}\right) \qquad (4)$$

and

$$J_S = A_S T^2 \exp\left(-\frac{\phi_S - \Delta\phi_S}{kT}\right) \qquad (5)$$

where r is a parameter between 1 and 2 depending on the position of the Fermi level, and ϕ_{PF} and ϕ_S are the Poole–Frenkel and the Schottky barriers, respectively. In the Poole–Frenkel model, the concentration of free carriers generally takes the form

$$n = n_o \exp\left(-\frac{\phi_{PF}}{rkT}\right) \qquad (6)$$

where both n_o and ϕ_{FF} are functions of donor, acceptor and trap concentrations and their energy positions in the bandgap. The constant A_S is given by

$$A_S = \frac{2qk^3 m_e^*}{(2\pi)^2 h^3} \qquad (7)$$

where m_e^* is the electron effective mass, k is the Boltzman constant, and h is the Planck constant. Given the similarities in the functional forms of Eqs. (4) and (5), it is difficult to identify whether the dark current in HgI_2 is due to the Poole–Frenkel or the Schottky effects unless systematic studies are performed to measure the dependence of the dark current on the applied electric field and the effect of polarity, temperature, and contacts with different work functions (Yeargan and Taylor, 1968).

Braatz and Zappe (1984) have interpreted their measured dark currents in HgI_2 as due to the Poole–Frenkel mechanism, and a linear region in the plot of $\ln(J_{PF})$

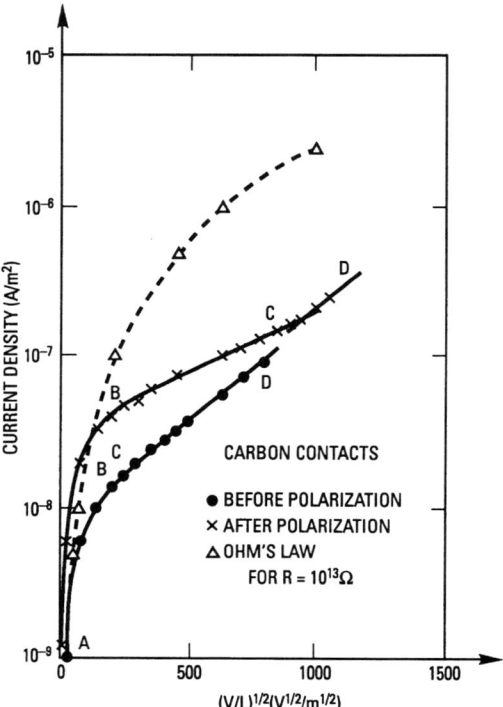

FIG. 1. Log(J) vs. $\sqrt{V/L}$ for a sample showing polarization after 24 hr under bias. (Reprinted from Mellet and Friant, 1989. ©1989 IEEE.)

vs. $\sqrt{E}$ was identified. From the slope of the linear region and the assumed Poole–Frenkel model, several parameters can be inferred such as trap concentration and trap energy levels. Figure 1 shows the plots of J_S vs. $\sqrt{V/L}$ obtained by Mellet and Friant (1989) for a sample before and after polarization. Ohm's law for 10^{13} Ω is also shown for comparison. The data can be divided into three regions. In the AB region, the curve follows Ohm's law. The BC and CD regions are linear and were interpreted as Schottky current. The polarization effect will reduce the slope in the BC region. The electric field in the expression for the Schottky current (Eq. (5)) was taken to be proportional to the applied bias

$$E = \gamma \frac{V}{L} \tag{8}$$

where V is the applied field, L is the sample thickness, and γ is a number that characterizes the electrode. An injecting contact has $\gamma < 1$, and a blocking contact has $\gamma > 1$. From the CD region in the plot of log(J_S) vs. $\sqrt{V/L}$, the parameter γ may be determined from the slope. The barrier height ϕ_S without bias can be calculated from the intercept with the y-axis. Further discussion of the electrical contacts can be found in Subsection V.1.

Another observation related to the I–V characteristics of HgI_2 was made by Levi *et al.* (1985). After a sample of HgI_2 has reached equilibrium, either by resting the sample in the dark under short circuit conditions for several hours or by illumination with an incandescent lamp for several minutes followed by a resting period in the dark (this latter procedure shortens the resting time to about half an hour), the dark current decays with time when a voltage is applied to this sample. This has also been observed by other workers (see Fig. 2a). If the applied voltage is then reversed, the dark current shows a current peak superimposed on a decaying current transient as shown in Fig. 2b. The position and magnitude of this current peak is a function of the preparing voltage V_P (the applied voltage before the reversal of bias), preparation time (the duration of the applied voltage before the reversal of the bias), and applied voltage V_R (the magnitude of the reversed bias). The explanation provided by Levi *et al.* (1985) is that the current peak is due to the trap controlled motion of holes. If the position of the peak t_m is assumed to be the transit time of the trap controlled charges, it will be related to the applied voltage by

$$t_m = \frac{L^2}{\mu_{\text{eff}} V_R} \qquad (9)$$

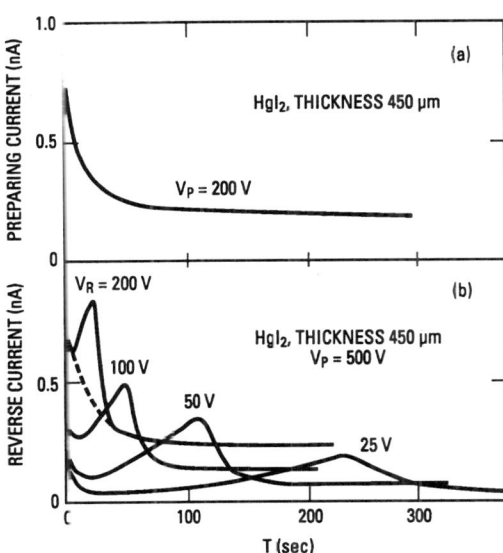

FIG. 2. Room temperature current vs. time transients under various conditions: (a) when the sample has reached thermal equilibrium before voltage application, (b) when the sample was subject to a voltage of 500 V for 10 min. before reversing the voltage. (Reprinted from Levi *et al.*, 1985, with permission.)

where μ_{eff} is the effective mobility of the trapped charges. The effective mobility is related to the trap parameters by (Nissenbaum, Schieber, and Burshtein, 1987)

$$\mu_{\text{eff}} = \mu_h \left(\frac{N_v}{N_t}\right) \exp\left(\frac{\Delta E_t}{kT}\right). \tag{10}$$

Indeed a plot of t_m vs. $1/V_R$ yields a straight line. The effective mobility determined from the slope is about 3.8×10^{-7} cm²/V-sec. This value is many orders of magnitude lower than the electron or hole mobilities and can be accounted for by trap controlled motion where the charged carrier undergoes multiple trapping and detrapping processes (Rose, 1951). The integrated amount of charge under the current peak was found to be much greater than the charge on the detector electrodes (capacitance times applied voltage). It was argued that this excess current superimposed on the decaying component of current is provided by free electrons that screen the space charges produced by the trap controlled motion of holes, instead of a space charge limited current of holes. Therefore, the average electron concentration is related to the total collected charge under the current peak by (Levi et al., 1985)

$$n_{\text{av}} = \frac{Q}{t_m A} \frac{L}{q\mu_n V_R} = \frac{Q}{qLA} \frac{\mu_{\text{eff}}}{\mu_e} \tag{11}$$

where Q is the total collected charge under the current peak obtained by integrating the current with time and subtracting the contribution due to the decay component and μ_e is the electron mobility. A plot of n_{av} vs. t_m is shown in Fig. 3. The time constant obtained from this plot is then the dielectric relaxation time, which is given by

$$\tau_{\text{dielectric}} = \frac{\epsilon_r \epsilon_0}{q\mu_e n} \tag{12}$$

where n is the dark electron density in HgI_2. The contribution of free holes to the conductivity is ignored here. The relaxation time thus obtained by Levi et al. (1985) at room temperature is about 25 sec. From Eq. (12), using $\epsilon_r = 9.75$ and $\mu_e = 100$ cm²/V-sec, an electron concentration of $n = 2 \times 10^3$ cm^{-3} was obtained, which corresponds to a resistivity of about 3×10^{13} Ω-cm. Similar measurements of dark current in SbI_3-doped HgI_2 crystals (Nissenbaum et al., 1987) resulted in a much smaller effective mobility of holes of 10^{-8} cm²/V-sec, which was attributed to a shallower trap with a higher concentration. The trap activation energy obtained for SbI_3-doped crystals was 0.5 eV compared with 0.7 eV for undoped HgI_2. These activation energies were obtained with Eq. (10) by using measured values for the effective mobility as a function of temperature. The trap density in the SbI_3-doped HgI_2 was related to the doping level and found to be

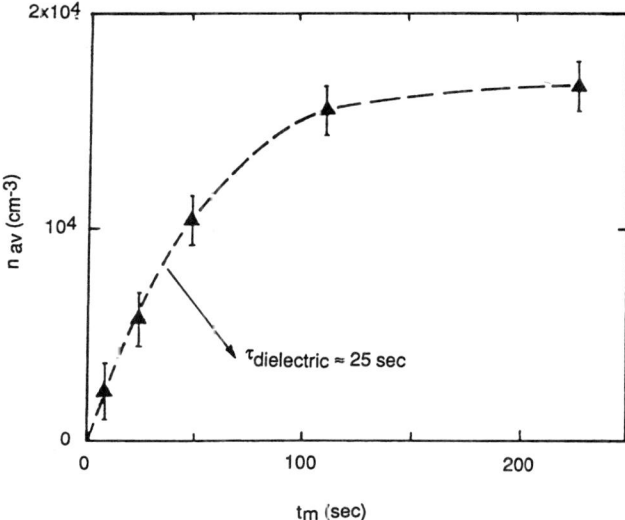

FIG. 3. Average electron density during a reversed current transient vs. reversed current peak time. (Reprinted from Levi et al., 1985, with permission.)

2×10^{19} cm^{-3} for the 1.5 weight% SbI$_3$-doped sample and 2×10^{18} cm^{-3} for the 0.1 weight% SbI$_3$-doped sample.

Marschall and Milstein (1991) have studied the behavior of dark conductivity during plastic deformation. An increase in dark current of 15 to 140% was observed during plastic deformation, and it was attributed to the freeing of trapped charges with the motion of glide dislocations.

2. MOBILITY AND LIFETIME

a. Definition of Lifetime

Several characteristic times will be defined here for clarity. Transit time is the time it takes for a carrier to traverse the entire thickness of the detector. Trapping time is the average time it takes for a free carrier to be captured by a trap. After a carrier is trapped, it may contribute to conduction when it is detrapped. Detrapping time is the average time it takes for a carrier to escape from the trap. Lifetime is the average time it takes for a free carrier to recombine with an electron–hole either directly or through recombination centers. In semiinsulating materials, trapping and detrapping very often dominate the carrier transport, and it is difficult to distinguish between trapping time and lifetime. For instance, if an electron is trapped and remains trapped longer than the time of interest, it will have the same effect as if it has recombined with a hole (except for some long term effects such as polarization and nonuniform electric fields). On the other hand, if the time of interest is longer than the detrapping time, then this electron is likely to be released into the conduction band and contribute to the conductivity. In the latter case, the effect of trapping is to produce an effective mobility of the carriers that is lower than the intrinsic mobility, as was discussed earlier in Section I. In this chapter, lifetime is loosely used as the average time it takes for the carriers to recombine or be trapped and remained trapped during the time of interest. In x-ray or gamma ray detection, the time of interest is usually the time constant of the shaping network, which defines the time the electronic system will spend collecting any charges induced by an incident x-ray or gamma ray photon.

b. Relationship with Detector Performance

The charge collection efficiency, η, is one of the most important factors to consider when designing x-ray or gamma ray spectrometers. It is related to peak broadening in x-ray spectra by the equation (Slapa et al., 1976)

$$\Delta E_{col} = \alpha(1 - \eta)(E_{ph})^{1/2} \tag{13}$$

where ΔE_{col} is the FWHM of the x-ray peak in eV contributed by incomplete charge collection, α is a constant equal to 4.7×10^5 $(eV)^{1/2}$, and E_{ph} is the energy of the incident x-ray photon in eV.

When nuclear radiations are detected by semiconductor detectors, carrier mobility and lifetime of the material determine the charge collection efficiency and carrier mobility and electric field determine the transit time. The charge collection efficiency is given by (Schieber et al., 1978)

$$\eta = \frac{\lambda_e}{L}\left[1 - \exp\left(-\frac{L-x}{\lambda_e}\right)\right] + \frac{\lambda_h}{L}\left[1 - \exp\left(-\frac{x}{\lambda_h}\right)\right] \tag{14}$$

where x is the distance from the negative electrode of the charge sheet generated by the incident energetic photon and $\lambda_{e,h}$ are the mean drift lengths of electrons and holes, respectively. For electrons, λ_e is defined as

$$\lambda_e = \frac{\mu_e \tau_e V}{L}. \tag{15}$$

For gamma rays that penetrate deeply in HgI_2 detectors (x may be anywhere between 0 and L), the charge collection efficiency is close to unity only when the mean drift lengths of both types of carriers are much longer than the detector thickness. For shallow penetrating x-rays, most of the carriers are generated close to one electrode and only one type of carrier must traverse the detector. In this case, the thickness of the detector has to be smaller than the mean drift length of the relevant carrier to achieve charge collection efficiency close to 1. In HgI_2 crystals, electrons have much higher mobilities than holes, and although electron lifetime is usually not very different from that of holes, the entrance electrodes of HgI_2 x-ray detectors are always biased negatively so that the electrons traverse the detector. The transit time (for electrons) is approximately given by

$$t_r = \frac{L^2}{\mu_e V}. \qquad (16)$$

To reduce peak broadening and shifting due to incomplete charge collection, either the applied voltage is increased or the detector thickness is reduced. Thus there is a tradeoff between detector active volume (or efficiency) and the energy resolution of the detector. Present day HgI_2 detectors have been limited to a thickness of about 0.4–5 mm primarily because of the short mean drift length of holes. Finally, the charge collection efficiency given in Eq. (14) assumes that the shaping network has a time constant longer than the transit time of electrons and holes.

c. Summary of Experimental Techniques

Several techniques have been employed to measure electron and hole mobilities, lifetime, and mobility–lifetime products; and some of the results are summarized here. The techniques can be divided into transient and nontransient methods. The transient methods are basically time of flight measurements, and they can be further categorized into current and charge transient techniques. Several sources of excitation have been used, such as α particles, x-ray photons, electron bursts, or photoexcitation by lasers. In the current transient method, high level excitation sources such as electron bursts or laser pulses are used, so it is possible to have space charge limited (SCL) current as well as space charge free (SCF) current. All the transient techniques can potentially determine the electron and hole mobilities and lifetimes.

The nontransient methods include x-ray or α particle spectra methods, photocurrent, photo-Hall effect measurements, and the photomagnetoelectric (PME) effect. The energy spectrum and photocurrent methods can only determine the mobility-lifetime product. Photo-Hall effect measurements can determine the Hall mobility of carriers if the conduction is primarily by one type of carrier. HgI_2 is highly resistive, and although Ag and Cu dopants can lower the resistivity by a few orders of magnitude, the resistivity is still too large for conventional Hall effect measurements. As a result, most studies of Hall mobility employ carrier generation by photons. This technique is thus limited by the availability of single carrier conduction, which can be achieved by suitable excitation wavelength and intensity. The photomagnetoelectric measurement can determine the dependence of carrier lifetime on the illumination intensity and anisotropy of carrier transport.

In the current transient method, an above bandgap excitation source is used to generate a certain amount of charge near one electrode. A current transient can then be measured as the charges traverse the thickness of the device under the influence of the applied voltage. If the generated charge is near the positively biased electrode, only holes will flow under the influence of the electric field and contribute to the current transient. In this case, the carrier transport properties of holes can be studied. Similarly, if the charge is generated near the negatively biased electrode, the transport properties of electrons can be studied. From an analysis of the waveform of the current transient, the carrier mobility and lifetime are determined. Currents in a semiinsulating material can be divided into space charge free and space charge limited currents. The criteria to determine whether a current is SCF or SCL is the total number of charges in the material. If the total charge in the material is much smaller than the charge on the electrode of the device (capacitance times applied voltage), it will not affect the electric field significantly and the current is considered space charge free. On the other hand, if the total charge in the material is much greater than the capacitor charge, the electric field in the device is determined mainly by the distribution of these charges and the current is space charge limited. Cho et al. (1975) have summarized the theoretical background which describes a current transient in a semiconductor under the assumption of uniform applied electric field.

  i. *Space Charge Free.* Without trapping,

$$i(t) = \begin{cases} \dfrac{Q_0}{t_r} & 0 < t < t_r \\ 0 & t > t_r \end{cases}. \tag{17}$$

With trapping but negligible detrapping,

$$i(t) = \begin{cases} \dfrac{Q_0}{t_r} \exp\left(-\dfrac{t}{\tau}\right) & 0 < t < t_r \\ 0 & t > t_r \end{cases}. \tag{18}$$

With both trapping and detrapping,

$$i(t) = \begin{cases} \dfrac{Q_0}{t_r} \left[\tau + \tau_d \exp\left(-\dfrac{\tau_d + \tau}{\tau_d \tau} t\right)\right] & 0 < t < t_r \\ \dfrac{Q_0 t_r}{2\tau_d \tau} \exp\left(-\dfrac{t}{\tau_d}\right) & t > t_r \end{cases} \tag{19}$$

where $i(t)$ is the current as a function of time, Q_0 is the total charge generated, τ_d is the detrapping time, and τ is the lifetime of the carriers.

Current pulse shapes for the three SCF cases are shown in Fig. 4. Since SCF current is easy to establish and interpret, most of the measurements of current transient belong to this regime. As can be seen from Fig. 4, it is possible to determine the carrier transit time if the lifetime is not very short compared with the transit time so that the leading edge of the pulse is well defined. From the transit

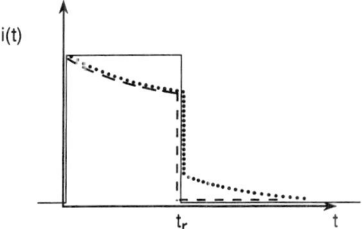

FIG. 4. Space charge free current pulse shapes: trap free case (solid line); trapping without detrapping (dashed line); trapping with detrapping (dotted line). (Reprinted with permission from Cho et al. 1975. ©1975 IEEE.)

time, the carrier drift velocity can be calculated by $v_d = L/t_r$. When the applied bias is varied, the drift velocity can be measured as function of the applied electric field. The slope of a plot of v_d vs. E, in the linear region, gives the carrier mobility. It is also possible to determine the lifetime from the current pulses if there is considerable decay of the current magnitude within the transit time. From Eq. (18), the current decays exponentially if detrapping is negligible. If $\ln[i(t)]$ vs. Δt is plotted, the lifetime can be determined from the slope.

ii. Space Charge Limited. The SCL currents can be divided into two cases, those with infinite reservoirs and those with finite reservoirs of carriers. Examples of waveforms are drawn in Fig. 5 for the cases of infinite and finite reservoirs. It

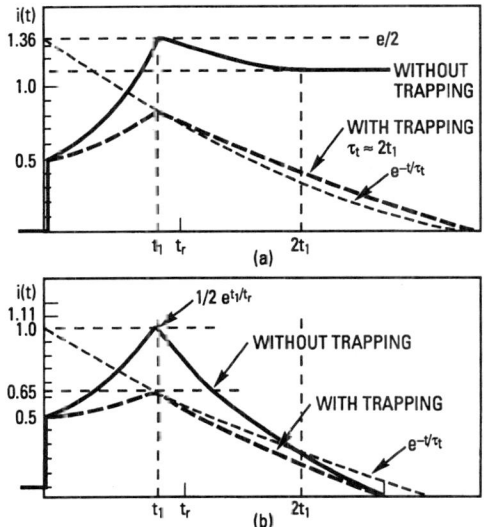

FIG. 5. Space charge limited current pulse shapes: trap free case (solid lines); trapping case (dashed lines); (a) infinite reservoir, (b) finite reservoir just sufficient to make the electric field zero at the electrode where the injection takes place at $t = 0$. (Reprinted with permission from Cho et al. 1975. ©1975 IEEE.)

should be noted that the current reaches maximum at a time of $t_1 \cong 0.787 t_r$. Under typical bias conditions, it is possible to satisfy the SCL condition only with intense excitation such as photoinjection with laser pulses or electron beam bursts. For example, for a detector with a capacitance of 10 pF and bias of 100 V, the number of charges on each electrode is about 10^{10}. If it is assumed that an energy of 5 eV is necessary to generate an electron–hole pair by an incident α particle, a 5.8 MeV (^{244}Cm) α particle can only generate about 1×10^6 electron–hole pairs. Thus typical x-ray and gamma ray photons to be detected by HgI_2 spectrometers can generate only SCF current pulses.

From the pulse shape of the SCL current, it is also possible to measure the transit time, since the current peaks at $t_1 \cong 0.787 t_r$. Determination of lifetime from SCL current is more difficult. It should be noted, however, that identifying whether a current is SCF or SCL is important, because if t_1 in the SCL current is mistaken as the transit time, an error of more than 20% may be introduced.

In the charge transient method, the charge induced by the transient current is measured. The properties of electrons and holes can be studied separately depending on the polarity of the bias in a manner similar to the current transient method. The advantage of this method is that it is closely related to x-ray detection. The technique simply uses a digital storage oscilloscope to monitor the output transient of a charge sensitive preamplifier. The excitation is usually either α particle or x-ray, and the current is SCF under typical experimental conditions. The charge transient is given by the integration of the current transient with respect to time. Using the current of the SCF case (without detrapping), the collected charge as a function of time is given by

$$Q(t) = \begin{cases} Q_0 \dfrac{\tau}{t_r}\left[1 - \exp\left(-\dfrac{t}{\tau}\right)\right] & t < t_r \\ Q_0 \dfrac{\tau}{t_r}\left[1 - \exp\left(-\dfrac{t_r}{\tau}\right)\right] & t > t_r \end{cases}. \qquad (20)$$

If the transit time is shorter than the lifetime, there is a sharp turn in the charge transient when the charges reach the electrode. Thus, the carrier drift velocity may be obtained by applying a voltage high enough so that transit time is short compared with lifetime and can be measured. On the other hand, if the applied field is small enough so that the lifetime is much shorter than the transit time, the lifetime may be determined by plotting $\ln[1 - Q(t)/Q(\infty)]$ vs t.

The x-ray spectra and photocurrent methods are able to give a rough estimate of mobility–lifetime product, either by analyzing the charge collection efficiency as a function of the applied electric field or the shape of x-ray photopeaks as discussed later. If the excitation is near one electrode, electrons or holes can be studied separately by varying the polarity of the applied bias. In an energy dispersive spectrum of x-rays, the position of the peak will initially increase with an increasing applied field because of increased charge collection efficiency and then saturate to a constant value as the charge collection efficiency approaches unity. The charge collection efficiency at a given applied field is calculated by taking the ratio of the peak position to the saturated peak position in the x-ray spectrum. In photocurrent measurements, the current is measured as a function of the applied voltage. The current is assumed to be directly proportional to the charge collection efficiency and will thus saturate as the applied field is increased, in a similar manner as the measured position of the x-ray photopeak increases with the applied field. If the charges are generated close to one electrode and surface recombination is negligible, the charge collection efficiency is given by

$$\eta = \frac{\mu\tau E}{L}\left[1 - \exp\left(-\frac{L}{\mu\tau E}\right)\right]. \tag{21}$$

Since L is known, the measured η as a function of applied bias can be fit with Eq. (21) using $\mu\tau$ as a parameter. The best fit will thus give an estimate of mobility–lifetime product. A simpler method can also be used to roughly estimate the mobility–lifetime product. If the applied electric field is $E_{\eta=0.5}$ when the charge collection efficiency is 0.5, the mobility–lifetime product is then

$$\mu\tau = 0.63\frac{L}{E_{\eta=0.5}}. \tag{22}$$

This method has been widely used to estimate mobility–lifetime product in HgI_2 detectors. However, if surface recombination is present and significant, the mobility–lifetime product obtained with this technique will be underestimated (Levi, Schieber, and Burshtein, 1983a). When surface recombination is present, the charge collection efficiency is given by

$$\eta' = \frac{\mu E}{\mu E + s}\frac{\mu\tau E}{L}\left[1 - \exp\left(-\frac{L}{\mu\tau E}\right)\right] \tag{23}$$

where s is the surface recombination velocity. In the general case, $E_{\eta=0.5}$ is related to the surface recombination velocity and mobility–lifetime by

$$E_{\eta=0.5} \approx \frac{s}{\mu} + 0.63\frac{L}{\mu\tau}. \tag{24}$$

If the surface recombination is negligible $L/\tau \gg s$, Eq. (24) reduces to Eq. (22). If the surface recombination dominates $L/\tau \ll s$, then

$$E_{\eta=0.5} = \frac{s}{\mu}. \tag{25}$$

It can be seen that, if surface recombination is significant, it is not possible to determine the mobility–lifetime product from the charge collection efficiency measured as a function of the applied field alone.

Another method to estimate the mobility–lifetime product from x-ray spectra is by examining the shape of an x-ray peak (Llacer et al., 1974). An x-ray peak in the energy dispersive spectrum can be calculated as follows. The charge collection coefficient is a function of the position of interaction x as shown in Eq. (14). The probability of an x-ray photon interacting with the detector at x is given by (assuming the x-ray is incident from the negative electrode)

$$P(x) = \mu_{ph} \exp(-\mu_{ph} x) \qquad (26)$$

where μ_{ph} is the photoelectric absorption coefficient. If $P(x)$ is plotted as function of η, one obtains the calculated x-ray spectrum (without pulse shaping). Note that in general η is a function of x (position) in contrast to the special case discussed in Eqs. (21) and (23). For HgI_2 under typical bias (Llacer et al., 1974), the electron is completely collected and the only parameter in the plot is the mobility–lifetime product of holes (assuming μ_{ph} is known). By fitting the calculated spectrum to the measured one, the $\mu\tau$ product for holes can be obtained. Since the calculation actually fits the energy peak broadening due to incomplete collection of holes, the x-ray used should have a high enough energy that it penetrates into the bulk so that surface recombination is small.

The conventional measurement of Hall mobility in materials of high resistivity is difficult in both measurement and data interpretation (Look, 1983). Modified techniques (photo-Hall effect) have been employed to study carrier mobility in photoconducting insulators such as diamond and alkali halides (Redfield, 1954; Klick and Maurer, 1951). In these techniques, illumination is used to generate charge carriers. However, measurements of mobilities are limited to those conditions where single carrier conduction can be achieved by the illumination. Photomagnetoelectric experiments involve the measurement of photocurrents with and without the presence of a magnetic field as a function of excitation wavelength (Roosbroeck, 1956; Gartner, 1957; Zitter, 1958).

d. Room Temperature Measurements and Effects of Excitation Intensity

Measurements of mobility, lifetime and their product for electrons and holes at room temperature are summarized in Table I. The mobility generally agrees among different workers and techniques. The lifetime measured by different techniques varies considerably, sometimes by orders of magnitude. The measured values of lifetime obtained from the current transient method are usually smaller than those measured by the charge transient method. One likely reason for the large variation in values for lifetime is the varying quality of the crystals used. Another more subtle explanation for this variation is that lifetime is a function of carrier

TABLE I

SUMMARY OF MEASUREMENTS OF MOBILITY AND LIFETIME IN HgI$_2$ AT ROOM TEMPERATURE

Reference	μ_e (cm^2/V-sec)	τ_e (μsec)	$\mu_e\tau_e$ (cm^2/V)	μ_h (cm^2/V-sec)	τ_h (μsec)	$\mu_h\tau_h$ (cm^2/V)	Method
Martin, Bach, and Guetin, 1974	75–90	>0.1	>6 × 10^{-6}	15	0.03	5 × 10^{-8}	a
Minder et al., 1974	100	0.1	1 × 10^{-5}	4	0.01	4 × 10^{-6}	a
Ottaviani, Canali, and Quaranta, 1975	100						a
Watt and Cho, 1976	100			3.5	0.15	5.3 × 10^{-7}	b
Malm, 1972	70		1–10 × 10^{-5}	4		1–10 × 10^{-6}	b
Ponpon et al., 1974	100			3			b
Swierkowski, Armantrout, and Wichner, 1974	94		1 × 10^{-4}	1		1 × 10^{-5}	b
Llacer et al., 1974	120				3	1.1 × 10^{-6}	b, c
Beyerle et al., 1983	100	>200	>2 × 10^{-2}	6	15	9.10^{-5}	b
Roth et al., 1987	97	3	2 × 10^{-4}	4	4.3	1.7 × 10^{-5}	b
Schieber et al., 1978			4 × 10^{-5}			4 × 10^{-6}	c
Burshtein, Akujieze, and Silberman, 1986			3 × 10^{-4}			2.8 × 10^{-4}	d

Methods: a. current transient method; b. charge transient method; c. energy spectra method; d. photocurrent method.

concentration, which varies significantly in semiinsulating materials where the densities of both types of carriers generated by extrinsic excitation are much higher than those at equilibrium. A strong dependence of lifetime on injection level has actually been confirmed by photomagnetoelectric measurements. The lifetime of electrons was found to depend on the excess electron density with an exponent varying between -0.33 and -0.63 (Manfredotti *et al.*, 1977a; Adduci *et al.*, 1977). The lifetime of holes was found to depend on the excess hole density with an exponent of -1 (Adduci *et al.*, 1977). In a current transient measurement, the number of charges generated is most likely larger than that generated by a single α particle or x-ray photo used in the charge transient method so that the preceding explanation seems reasonable. Since the measurements in which α particle or x-ray excitation is used are more closely related to the operation of HgI_2 detectors, the values obtained from these measurements are probably more meaningful.

e. *Effects of Temperature, Crystallographic Orientation, and Stoichiometry*

Minder *et al.* (1974) measured electron and hole mobilities as a function of temperature using an electron excited current transient method. Both electron and hole mobilities were found to decrease with increasing temperature, and the experimental results are shown in Fig. 6. For electrons, the temperature dependence

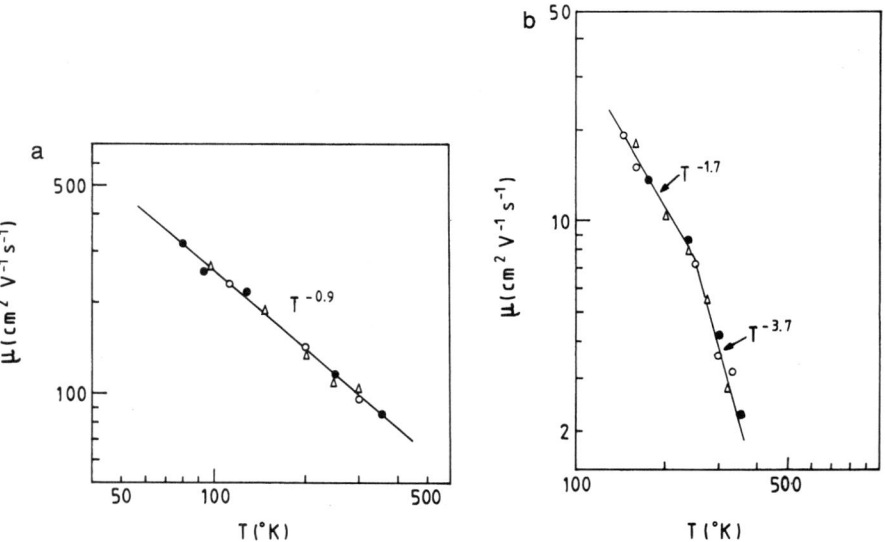

FIG. 6. Temperature dependence of the (a) electron [○, 243μ; △, 55μ; ●, 260μ] and (b) hole [○, 30μ; ●, 40μ; △, 55μ] mobility parallel to the c-axis. (Reprinted with permission from Minder *et al.*, 1974 and the American Institute of Physics.)

is given by

$$\mu_{e\|} \propto T^{-0.9} \qquad 114 < T < 300 \text{ K} \qquad (27)$$

where $\mu_{e\|}$ is the electron mobility in the direction parallel to the c-axis. For holes, the temperature dependence is given by

$$\mu_{h\|} \propto \begin{cases} T^{-1.7} & 140 < T < 240 \text{ K} \\ T^{-3.7} & 240 < T < 350 \text{ K} \end{cases} \qquad (28)$$

Using photo-Hall measurement, Bloch et al. (1977) found that above a temperature of 40 K, their samples always showed electron conduction irrespective of the wavelength of illumination. The Hall mobility of electrons measured as a function of temperature agrees well with the drift mobility of electrons measured by Minder et al. (1974). This indicates that the drift mobilities measured by the transient techniques probably reflect the intrinsic properties of the crystal with no trap controlled motion. The intrinsic mobilities of free carriers in a compound semiconductor like HgI_2 are typically controlled by scattering of carriers by non-polar optical phonons, polar optical phonons, and acoustic phonons. From the temperature dependence of the electron and hole mobilities, Minder, Ottaviani, and Canali, (1976) proposed that electrons are scattered mainly by polar optical phonons, while holes are mainly scattered by non-polar optical phonons. Below 15 K, single carrier conduction by holes was identified if the illumination wavelength was shorter than 5300 Å. Hole mobility decreases with temperature from 3 to 10 K by about one order of magnitude. Because of the low density of phonons at low temperatures, the scattering of carriers is either by neutral or ionized impurities. Hole mobility as high as 60,000 cm^2/V-sec was measured at low temperatures (Bloch et al., 1977).

HgI_2 crystals have a layered structure (perpendicular to c-axis) with very weak bonding between layers (T. W. James and Milstein, 1983). It is therefore expected that the carrier mobilities will also depend on the crystallographic orientation. Ottaviani et al. (1975) and Minder et al. (1976) measured electron mobility using electron excited current transients in the direction parallel and perpendicular to the crystallographic c-axis at various temperatures. The electron mobility parallel to the c-axis was found to be always greater than that perpendicular to the c-axis. The mobility of electrons perpendicular to the c-axis has a slightly different temperature dependence with a value of 65 cm^2/V-sec at room temperature (Minder et al., 1976). The data was well described by the expression

$$\mu_{e\perp} \propto T^{-0.8} \qquad 80 < T < 300 \text{ K} \qquad (29)$$

where $\mu_{e\perp}$ is the electron mobility in the direction perpendicular to the c-axis. From photomagnetoelectric measurements at room temperature, Manfredotti et al. (1977a,b) calculated the hole mobility perpendicular to the c-axis to be about

23 cm^2/V-sec, which is higher than the value of 4 cm^2/V-sec typically observed parallel to the c-axis. From Hall measurements at low temperature (below 15 K), Bloch et al. (1977) also found that, for specimens of high purity where the scattering is due to neutral impurities, the Hall mobility of holes in HgI$_2$ is higher in the direction perpendicular to the c-axis. Even though the absolute value varies considerably, the mobility of holes along the c-axis was always measured to be about a factor of 2 smaller than that perpendicular to the c-axis. The higher mobilities perpendicular to the c-axis suggests fabricating detectors with charge transport in this direction. However, we have found that this leads to prohibitively high leakage currents possibly due to a high level of injection from the contacts.

Using current transient methods and taking precautions to ensure a high field existing in the sample, Martin et al. (1974) observed a maximum electron drift velocity at an applied field of 80 kV/cm and a negative differential electron mobility of about -5 cm^2/V-sec after 80 kV/cm. The maximum differential electron mobility is at 50 kV/cm. However, Ponpon et al. (1974) observed no negative differential electron mobility up to 100 kV/cm in their measurements using an α particle charge transient method.

Electron and hole transport properties in HgI$_2$ with deviations from stoichiometry were studied in an attempt to establish correlations between detector quality and stoichiometry. Whited and van den Berg (1977) measured the mobility–lifetime product of electrons and holes in undoped HgI$_2$ and crystals doped with iodine or mercury. It was found that iodine doping deteriorates the mobility–lifetime product of electrons and mercury doping decreases the mobility–lifetime product of holes. The results are shown in Table II. It should be noted, however, that these composition variations are only relative. The exact stoichiometric point is not known. Hermon et al. (1991) have also attempted to correlate the lifetime

TABLE II

MOBILITY–LIFETIME PRODUCT OF ELECTRONS AND HOLES IN HgI^2 WITH VARIOUS DEVIATIONS FROM STOICHIOMETRY[a]

Stoichiometry	$\mu_n \tau_n$ (cm^2/V)	$\mu_p \tau_p$ (cm^2/V)
Heavily I$_2$	3×10^{-6}	3×10^{-7}
Lightly I$_2$	4×10^{-6}	8×10^{-7}
Typical	7×10^{-5}	4×10^{-7}
Lightly Hg	3×10^{-4}	$<10^{-7}$
Lightly Hg	5×10^{-4}	$<10^{-8}$
Heavily Hg	4×10^{-4}	$<10^{-7}$
Best	2×10^{-4}	1.5×10^{-5}
High Purity	1×10^{-4}	2.6×10^{-6}

[a]Reprinted with permission from Whited and van den Berg, 1977. ©1977 IEEE.

TABLE III
DEVIATIONS FROM STOICHIOMETRY AND CHARGE TRANSPORT PROPERTIES OF HgI_{2+x} CRYSTALS[a]

Hg Excess ± 7 (mole ppm)	I_2 Excess ± 5 (mole ppm)	τ_n (μsec)	τ_p (μsec)	FWHM at 6 keV (eV)	FWHM at 660 keV (keV)
0	143	3.0	2.4	660	>30
0	134	2.8	2.0	572	>30
0	55	6.6	2.6	425	>30
0	13	7.0	3.8	444	<30
0	0	5.4	1.7	393	>30
0	0	2.0	4.5	358	<30
0	0	5.0	3.5	385	<30
0	0	3.6	2.5	460	>30
14	0	6.3	6.1	402	<30
32	0	5.3	1.4	362	>30
619	0	1.2	0.65	876	>30
652	0	1.2	0.40	984	>30

[a] Reprinted with permission from Hermon et al., 1991 and Elsevier Science Publishers.

of electrons and holes with stoichiometry and energy resolution of detectors. The stoichiometry was measured by Raman spectroscopy for excess mercury and by spectrophotometric methods for excess iodine. Their findings are summarized in Table III. Excess mercury was found to reduce both electron and hole lifetimes. Excess iodine decreases the electron lifetime but does not have a significant effect on holes.

3. EFFECTIVE MASS

Effective mass is another important parameter pertaining to carrier transport. The calculation of effective conduction and valence band densities of states and carrier thermal velocities requires the knowledge of the effective masses of electrons and holes. Effective masses are also expected to depend on crystallographic orientation because of the anisotropy of HgI_2 crystals. Also related to the effective masses of electrons and holes is the reduced effective mass. Values of the effective masses of electrons, holes, and reduced mass vary considerably in the literature, depending on the methods used to obtain these values. Cyclotron resonance measurements, optical spectra of excitons, and band structure calculations have all been employed.

Using cyclotron resonance measurements at 1.2 K, Bloch et al. (1977, 1978) have obtained values of effective masses for electrons and holes in the direction perpendicular to the c-axis. The reported $m_{n\perp} = (0.37 \pm 0.02)m_0$, and $m_{p\perp} = (1.03 \pm 0.10)m_0$. From the anisotropy of the mobilities, the effective masses of electrons and holes parallel to the c-axis may be deduced, since the mobility and effective mass are related by

$$\frac{\mu_{\parallel}}{\mu_{\perp}} = \left(\frac{m_{\parallel}}{m_{\perp}}\right)^n \tag{30}$$

where n depends on the nature of scattering. Bloch et al. (1978) used $\mu_{h\parallel}/\mu_{h\perp} = 0.5$ (below 15 K). They also used a value of $n = -1$ for holes since the scattering by neutral impurities is dominant at low temperature and the matrix elements for neutral impurity scattering are isotropic. For electrons, $\mu_{e\parallel}/\mu_{e\perp} = 1.4$ (80 to 300 K) (Minder et al., 1976) was used. Since the temperature dependence of electron mobility suggests the presence of both polar optical phonon scattering and acoustic deformation potential scattering, n may range between -1.5 for the former and -2.5 for the latter. The inferred effective masses of electrons and holes in the direction parallel to the c-axis are $m_{e\parallel} = (0.31 \pm 0.03)m_0$ and $m_{h\parallel} = (2.06 \pm 0.50)m_0$.

Exciton masses may be deduced from optical spectra; however, these values vary among authors. This may be due to the different dielectric constant used in obtaining the exciton masses from the Rydberg constant or due to the polaron nature of the excitons in HgI_2, which makes the interpretation of the optical spectra more difficult. Anedda et al. (1977) performed wavelength modulated reflectivity spectra at 2 K to study the exciton properties of HgI_2. By using low frequency indices of refraction of $n_{\perp} = 2.39$ and $n_{\parallel} = 2.23$, exciton masses obtained from the optical spectra are $m_{r\perp} = 0.3m_0$, and $m_{r\parallel} = 0.51m_0$. Goto and Nishima (1979) obtained the following exciton masses: $m_{r\perp} = (0.68 \pm 0.1)m_0$, and $m_{r\parallel} = (1.2 \pm 0.1)m_0$ from resonant Brillouin scattering measurements. Novikov and Pimonenko (1971) obtained an exciton mass of $0.13m_0$ from ordinary $(E\|c)$ absorption spectra.

Turner and Harmon (1989) computed the effective masses to be $m_{e\perp} = 0.11m_0$, $m_{e\parallel} = 0.15m_0$, $m_{p\perp} = 0.17m_0$, and $m_{p\parallel} = 0.78m_0$ based on band structure calculations. Yee, Sherohman, and Armantrout (1976) obtained $m_{e\perp} = 0.4m_0$, $m_{e\parallel} = 0.4m_0$, $m_{p\perp} = 3.5m_0$, and $m_{p\parallel} = 1.2m_0$ from pseudopotential calculations of band structure, although an incorrect unit cell was used in their calculation. Bloch et al. (1978) pointed out that the measured effective mass may be different from those calculated because of the strong interaction between carriers and phonons to form polarons. Additional comments regarding the effective masses of HgI_2 as obtained from optical techniques are contained in Chapter 5 on optical properties of HgI_2.

III. Deep Levels

Since no dopants have been found that can lower the resistivity of HgI_2 considerably, shallow levels in HgI_2 are either highly compensated or they do not exist. Cu and Ag can reduce the resistivity of HgI_2 by a few orders of magnitude from 10^{13} Ωcm but this is still extremely high compared with resistivities on the order of 10^4 Ωcm or lower in typical doped semiconductors. Deep levels are therefore more important in affecting the performance of HgI_2 nuclear spectrometers. Recombination centers will reduce the lifetime of charged carriers while deep donor or acceptor levels may cause large variations in the electrical properties of HgI_2 from sample to sample. Defects in general also scatter carriers and reduce their mobility, especially at the lower temperatures, where carrier phonon scattering is smaller. Trap levels affect device performance in several important ways. First, if the trapped carrier remains trapped for longer than the charge collection time of the electronic system of the detector, then incomplete charge collection results. Second, even if the trapped carriers are released within the charge collection time and therefore collected, the effective mobility is lowered due to the trap controlled motion. This leads to an increase in the transit time, which in turn means that longer charge collection times are required to obtain complete charge collection. Third, trapped carriers modify the electric field in the device, which can cause polarization effects, giving rise to reliability problems in detectors. Polarization effects are typically observed as a change of efficiency, energy resolution, or charge collection efficiency with time. Finally, trapping near the metal contact–HgI_2 interface may alter the characteristics of the electrode and the charge transport across the electrode–HgI_2 interface.

A number of experimental techniques have been employed to study deep levels and their effects on device performance. A majority of these studies utilize thermally stimulated current (TSC) methods. Other techniques used include dark current transient methods, I–V measurements, photocapacitance measurements, photoresponse, and photoinduced transient spectroscopy (PITS).

1. Thermally Stimulated Current Technique

The TSC technique has been widely used to characterize carrier traps in semi-insulating or dielectric materials (Bube, 1960; Milnes, 1973; Look, 1983). The measurements are performed as follows. The sample is biased and cooled to a low temperature (usually to liquid nitrogen temperature of 77 K) while in the dark. The sample is then illuminated by an optical excitation source such as a laser or an incandescent lamp to generate electron–hole pairs and fill the carrier traps to be studied. After illumination, the temperature of the sample is raised gradually while the dark current is measured. As the temperature is increased, the trapped carriers are more likely to be released into their respective bands and contribute to the dark current, while at the same time, the number of trapped carriers is decreased. The competition of these two processes results in peaks in the dark current when measured as a function of temperature. The TSC processes can be described by carrier rate equations. If there is no direct transfer of carriers between traps, the equations are (for electrons) (Look, 1983)

$$\frac{dn_{ti}}{dt} = -n_{ti}\left(\frac{g_{i0}}{g_{i1}}\right)N_c \sigma_{ei} v_{th} \exp\left(-\frac{\Delta E_{ti}}{kT}\right) + n(N_{ti} - n_{ti})\sigma_{ei} v_{th} \quad (31)$$

and

$$\frac{dn}{dt} = -\frac{n}{\tau_e} - \sum_i \frac{dn_{ti}}{dt} \qquad (32)$$

where n_{ti} is the density of electrons trapped in the ith trap, N_{ti} is the density of the ith trap, σ_{ei} is the capture cross section of the ith trap, and g_{i0} and g_{i1} are the degeneracy factors of the ith traps when they are empty and occupied, respectively. These last parameters are generally assumed to be unity and thus ignored.

If the TSC peaks are well resolved so that each peak can be analyzed separately, the summation can be ignored (the index i can also be dropped), and simplified equations can be applied to each trap. Assuming the lifetime of the carrier is much smaller than the time scale over which the carrier density actually changes, i.e., $|dn/dt| \ll |n/\tau_n|$, the two simultaneous differential equations can be simplified to

$$n = \frac{\tau_e e_e n_t}{1 + \frac{\tau_e}{\tau_t}\left(1 - \frac{n_t}{N_t}\right)} \qquad (33)$$

$$\frac{dn_t}{dT} = -\frac{1}{\beta} \frac{e_e n_t}{1 + \frac{\tau_e}{\tau_t}\left(1 - \frac{n_t}{N_t}\right)} \qquad (34)$$

where β is the heating rate. The trapping and detrapping times are defined as the reciprocal of capture and emission probabilities (c_e and e_e), and they are given by

$$\tau_t = \frac{1}{c_e} = \frac{1}{N_t \sigma_e v_{\text{th}}} \qquad (35)$$

$$\tau_d = \frac{1}{e_e} = \frac{1}{N_c \sigma_e v_{\text{th}} \exp\left(-\frac{\Delta E_t}{kT}\right)}. \qquad (36)$$

a. Methods to Analyze TSC Spectra

Many methods have been developed over the years to analyze TSC data with the goal of obtaining relevant information such as activation energy, carrier capture rate, and trap concentration. A summary of these methods follows.

i. Heating Rate Method. Equations (33)–(34) may be solved for two limiting cases (Bao *et al.,* 1991a). Since the temperature dependence of τ_e, σ_e, and μ_e are generally not very strong, it can be assumed that their temperature dependences follow a power law; i.e., $\tau_e = \tau_{e0}T^\lambda$, $\sigma_e = \sigma_{e0}T^\alpha$, and $\mu_e = \mu_{e0}T^\gamma$. The temperature dependences of the effective density of states of the conduction band and the thermal velocity are $N_c = N_{c0}T^{1.5}$ and $v_{th} = v_{th0}T^{0.5}$, where N_{c0} and v_{th0} are functions of electron effective mass.

In the first case, if the trapping time is much longer than lifetime (slow trap), i.e., $\tau_t \gg \tau_e$, it can be shown that the temperature T_m at which the current peak occurs is related to the trap parameters and heating rate β by

$$\ln\left[\frac{T_m^{4+\alpha}}{\beta(T_m)}\right] = \frac{\Delta E_t}{kT_m} - \ln\left[\frac{kN_{c0}\sigma_{e0}v_{th0}}{\Delta E_e - kT_m(2 + \alpha + \gamma + \lambda)}\right]. \quad (37)$$

In the second case, if the trapping time is much shorter than the lifetime (fast trap), i.e., $\tau_t \ll \tau_e$, the solution for T_m can be shown to satisfy the following equation

$$\ln\left[\frac{T_m^{3.5-\lambda}}{\beta(T_m)}\right] = \frac{\Delta E_t}{kT_m} - \ln\left[\frac{\frac{kN_{c0}}{N_t\tau_{e0}}}{\Delta E_t + kT_m(2 + \alpha + \gamma + \lambda)}\right]. \quad (38)$$

To obtain Eqs. (37)–(38), the derivation does not require the heating rate to be constant. It suffices to know the heating rate where each current peak occurs, as long as the heating rate is not too irregular a function of temperature so as to give multiple solutions for T_m. To utilize Eqs. (37)–(38), several TSC curves have to be obtained with different heating rates. Since a constant heating rate for the entire TSC spectrum is not required, it is more practical to vary the heating power and record the temperature where the current peak occurs and the heating rate at this temperature. By plotting the left-hand side of Eqs. (37)–(38) vs $1/kT_m$, the activation energy of the trap may be obtained from the slope. From the intercept of the plot with the x-axis, σ_e and $N_t\tau_e$ can be obtained for the fast and slow traps, respectively. The temperature dependences of lifetime and capture cross section and whether a trap is fast or slow, however, are usually not known. If the temperature dependences of lifetime and capture cross section are not very strong, it is generally acceptable to ignore them altogether. It can be shown that if $\Delta E_t > 20\,kT_m$ (usually true), a variation of α or λ by 2 results in about a 10% uncertainty in ΔE_t. As for the determination of $N_t\tau_e$ and σ_e, they appear in Eqs. (37)–(38) in a logarithm term. The values obtained for $N_t\tau_e$ and σ_e, usually contain a very large uncertainty due to small errors in the temperature measurements, and the effects of temperature dependences are small and can be ignored. In general it is not known whether a trap is fast or slow. The best that can be done is to obtain both $N_t\tau_e$ and σ_e by assuming the trap is either slow or fast, respectively, and then check the values obtained. If one of the parameters obtained this way is unreasonable, then one trapping speed of the center can be ruled out and the other parameter can typically be accepted. This illustrates the difficulty of obtaining parameters such as $N_t\tau_e$ and σ_e from TSC data. In many cases, they are obtained under certain assumptions that may not be justified, and even if the assumptions are justified, the values contain large uncertainties simply due to the nature of the mathematical equations governing the TSC process.

ii. Delayed Heating Method. In this method, the traps are initially filled and the sample is held at some temperature for varied lengths of time so that some number of the traps are emptied. The TSC measurements are then performed (Stuck *et al.*, 1976) as a function of time after excitation. The total charge under the TSC peak is then a function of the waiting time and related to the detrapping time by

$$Q_{\text{TSC}} \propto \exp\left(-\frac{t}{\tau_d}\right). \tag{39}$$

Since the detrapping time is related to the activation energy by Eq. (36), the activation energy and carrier capture cross section can be obtained by measuring τ_d at different temperatures and plotting $\ln(\tau_d)$ vs. $1/kT$.

iii. Initial Rise Method. From Eq. (33), it can be seen that if the trapping time and lifetime are not strong functions of temperature, then the carrier density is a strong function of temperature through the emission rate. Therefore, at the onset of a TSC peak, when n_t has not changed very much, the activation can be obtained by plotting $\ln(n)$ vs $1/kT$ regardless of whether the trap is fast or slow (Gelbart *et al.*, 1977). When using this method, it is sometimes necessary to employ a "cleaning" technique. This technique is useful when two TSC curves are close together in temperature so that the initial rise of the high temperature peak is superimposed with the falling edge of the lower temperature peak. It is possible to significantly reduce the interference of the lower temperature peak by holding the sample at a temperature just below the high temperature peak for a period of time, so that the carriers trapped in the centers related to the low temperature peak are released. The TSC spectrum taken after this cleaning procedure will include less of the low temperature peak and the initial rise of the high temperature peak can be measured with greater accuracy.

iv. Fermi Level Method. In this method, the trap is assumed to be in equilibrium with its corresponding band (fast trap). The trap energy level is assumed to be at the Fermi level and the activation energy can then be obtained by (Bube, 1957)

$$\Delta E_t = kT_m \ln\left(\frac{N_c}{n_{\max}}\right) \tag{40}$$

where n_{max} is the carrier density at the current peak. After the activation energy is obtained this way, the capture cross section can then be calculated using Eq. (37).

v. Grossweiner's Method. If retrapping is ignored (slow trap) and $\Delta E_t/kT_m > 20$, $N_c \sigma_n v_{th} > 10^7/\text{sec}$, the trap parameters can be estimated by (Grossweiner, 1953)

$$\Delta E_T = \frac{1.51 kT_m}{T_m - T_-} \tag{41}$$

where T_- is the temperature where the TSC is half of its maximum on the low temperature side. In this method, β should be constant over the TSC current peak. The capture cross section can then be calculated by

$$\sigma_e = \frac{3T_- \beta \exp\left(\frac{\Delta E_t}{kT_m}\right)}{2N_c T_m (T_m - T_-)}. \tag{42}$$

vi. Luschick's Method. If retrapping is assumed dominant, the trap activation energy may be estimated by (Stuck et al., 1976)

$$\Delta E_t = \frac{kT_m^2}{T_+ - T_m} \tag{43}$$

where T_+ is the temperature where TSC is half of its maximum on the high temperature side.

vii. Numerical Method. From Eqs. (33)–(34), it can be seen that the shape of TSC curve for n/N_t is a function of only the heating rate β, activation energy ΔE_t, capture cross section σ_e, and trap density and lifetime product $N_t \tau_e$. Since a TSC curve is not symmetric about its maximum, T_-, T_m, and T_+ can be measured and used to obtain ΔE_t, σ_e and $N_t \tau_e$ numerically. In this method, the heating rate is measured as a function of temperature and used in Eqs. (33)–(34) to solve the TSC curve numerically with three parameters: ΔE_t, σ_e, and $N_t \tau_e$. The temperatures at which the TSC current reaches maximum (T_m^c), and half-maximum on the low (T_-^c) and high (T_+^c) temperature side of the peak can be obtained from this calculation as a function of the trap parameters (Bao et al., 1992):

$$\begin{aligned} T_-^c &= T_-^c (\sigma_e, \Delta E_t, N_t \tau_e) \\ T_m^c &= T_m^c (\sigma_e, \Delta E_t, N_t \tau_e) \\ T_+^c &= T_+^c (\sigma_e, \Delta E_t, N_t \tau_e). \end{aligned} \tag{44}$$

From the TSC spectrum, temperatures at which the measured TSC current is maximum and half-maximum can also be obtained and are labeled as T^m_-, T^m_m, and T^m_+. With these values, a three dimensional root finding problem is set up using the Newton–Raphson method (Press et al., 1988) with three unknowns: σ_e, ΔE_t, and $N_t \tau_e$. The three equations are

$$T^c_- (\sigma_e, \Delta E_t, N_t \tau_e) - T^m_- = 0$$
$$T^c_m (\sigma_e, \Delta E_t, N_t \tau_e) - T^m_m = 0 \qquad (45)$$
$$T^c_+ (\sigma_e, \Delta E_t, N_t \tau_e) - T^m_+ = 0.$$

It should be noted that if a trap is either slow or fast, the TSC curve is dependent on only two trap parameters, ΔE_t and σ_e for the former and ΔE_t and $N_t \tau_e$ for the latter. In these situations, Eq. (45) will result in overdetermination, and care should be taken to identify which of the three trap parameters cannot be determined. A limit can then be established for this parameter, and the nature of the trap (i.e., fast or slow) can also be identified.

b. TSC Spectra of Mercuric Iodide

Stuck et al. (1976) made TSC measurements on samples prepared by chemical etching in an aqueous KI solution and cleaving. In the samples prepared by etching, one peak at 217 K was found and attributed to a hole trap since it was present only when the bias polarity of the illuminated contact was positive (see Fig. 7a). After comparing several techniques, a mean value of 0.45 eV for the activation energy and 4×10^{-15} cm^2 for the capture cross section were obtained. In the samples prepared by cleaving, two more peaks at 114 and 145 K were observed under both polarities of bias as shown in Fig. 7b. The activation energies for these two traps were estimated to be about 0.19–0.22, and 0.22–0.24 eV, respectively.

The effects of iodine and mercury doping on the TSC spectra of HgI$_2$ were investigated by Whited and van den Berg (1977). Figure 8 shows TSC spectra from undoped (a), iodine doped (b), and mercury doped (c) samples. Several traps were observed at 81 K, 98 K (electron), 102 K (hole), 138 K (hole), 162 K (hole), and 180 K (hole). TSC peaks that appear in only one kind of sample were then attributed to native defects due to deviation from stoichiometry. A peak at 81 was observed in undoped samples, and it was attributed to iodine vacancy V_I. Two peaks at 89 and 220 K observed in heavily Hg doped samples were attributed to iodine vacancy V_{I2} and mercury interstitial I_{Hg}. A peak at 214 K in iodine doped samples was assigned to mercury vacancy V_{Hg}.

Tadjine et al. (1983) found seven peaks in their TSC spectra at temperatures of 100, 110, 125, 155, 173, 205, and 225 K. By correlating with the detector performance, it was found that the samples with an intense peak at 173 K exhibit fast polarization and poor resolution. A TSC spectrum is shown in Fig. 9a for such a

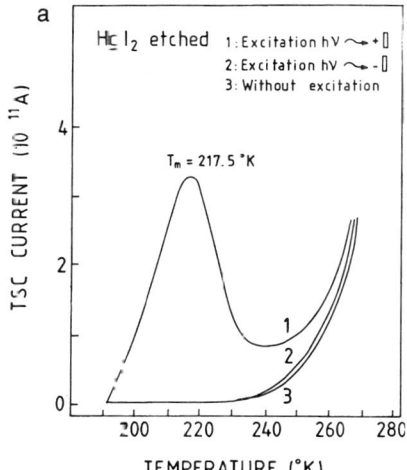
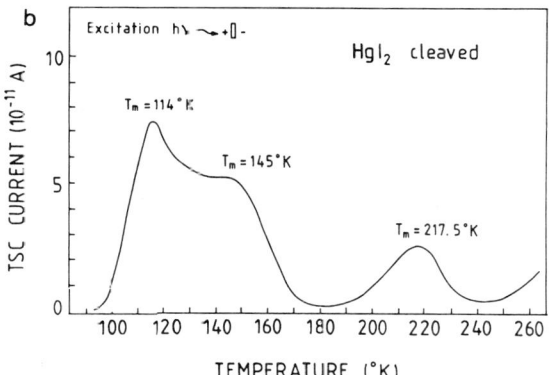

FIG. 7. TSC current in (a) etched and (b) cleaved HgI$_2$ samples. (Reprinted with permission from Stuck et al., 1976 and the American Institute of Physics.)

sample. Samples with a low intensity peak at 173 K were found to produce good detectors (see the TSC spectrum shown in Fig. 9b).

Suryanarayana and Acharya (1989) have studied the effects of aging, baking, storage in I$_2$ vapor, and application of bias on TSC spectra of HgI$_2$. From Fig. 10a, as-grown crystals show two TSC peaks at 170 and 230 K. The crystals show two peaks at 170 and 252 K for aged samples and two peaks at 170 and 280 K for the baked samples. As can be seen in Fig. 10b, if a bias of 500 V is applied to the aged sample for 12 hr, the peak at high temperature disappears and the intensity of the peak at low temperature increases. For samples stored in iodine vapor, there is only one peak at 180 K in the TSC spectrum. The activation energies and capture cross section were obtained using the Fermi level method.

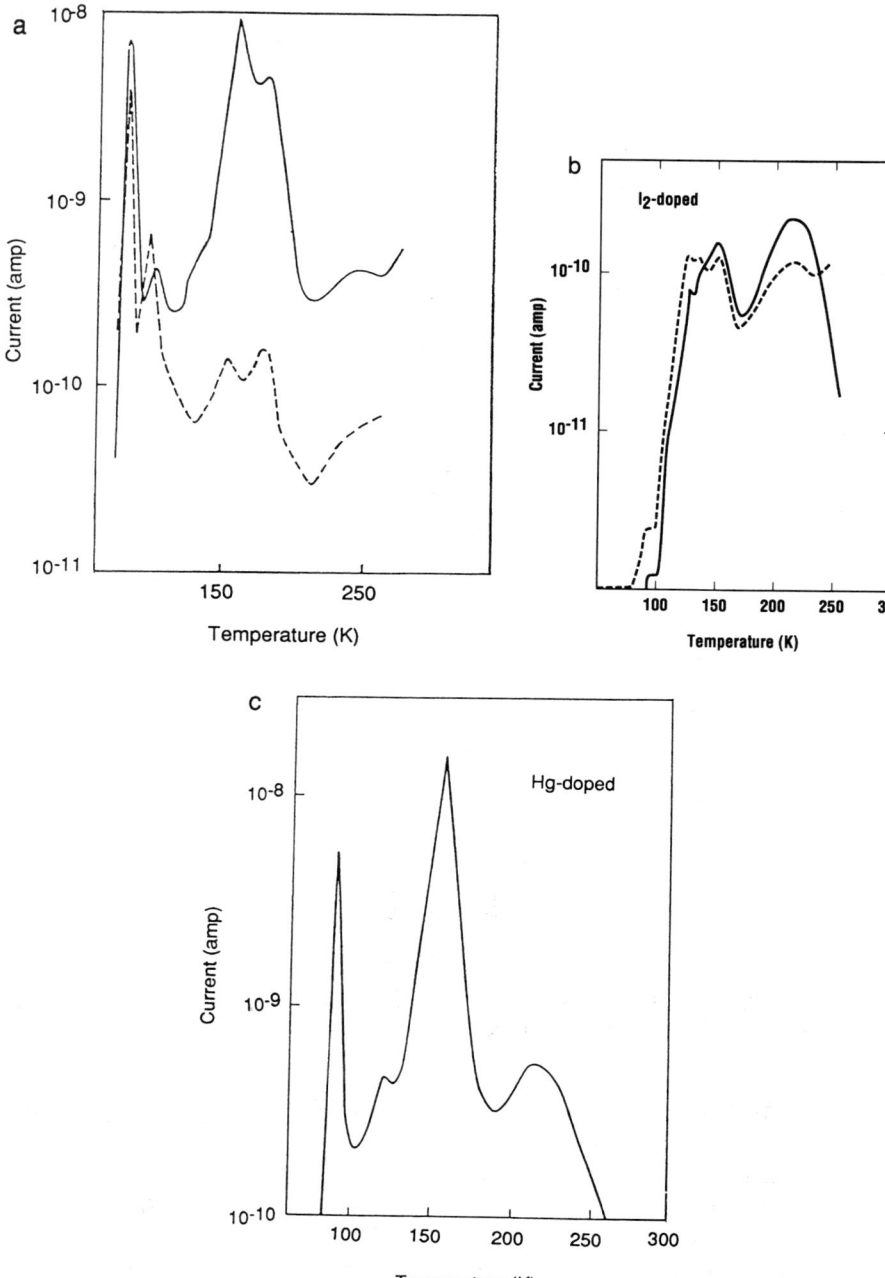

FIG. 8. (a) TSC of undoped crystal for both hole and electron traps (—, hole traps; ---, electron traps); (b) TSC of heavily iodine doped crystals for both hole and electron traps (---, hole traps; —, electron traps); (c) TSC of heavily mercury doped crystal. (Reprinted with permission from Whited and van den Berg, 1977. ©1977 IEEE).

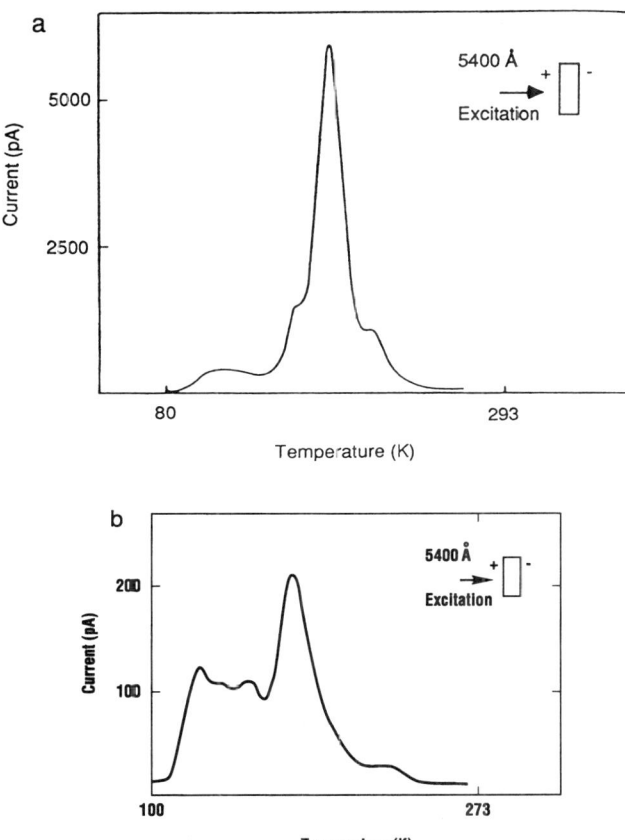

FIG. 9. (a) TSC spectrum of a crystal that makes detectors with fast polarization and very poor energy resolution; (b) TSC spectrum of a crystal that makes detectors with no polarization effect and very good energy resolution at values below 60 keV. (Reprinted with permission from Tadjine et al., 1983 and Elsevier Science Publishers.)

Prolonged (~24 hr) illumination by monochromatic light (~5900 Å) at room temperature reduced the 203 K peak and increased the 170 K peak dramatically in the TSC spectra measured by Mohammed-Brahim (1981) (see Fig. 11). Accompanied by the emergence of the peak at 170 K, the x-ray detector will resolve the 59.6 keV photon from an Am-241 source. It was concluded that a direct correlation between the 170 K peak and improvement of nuclear detection exists.

Gelbart et al. (1977) have tried to identify whether a trap is fast or slow. The activation energy was obtained by the initial rise method and then compared with those obtained assuming retrapping is fast (Fermi level method) or slow (Grossweiner's method). Among the traps observed, those at 95, 132, and 178 K were

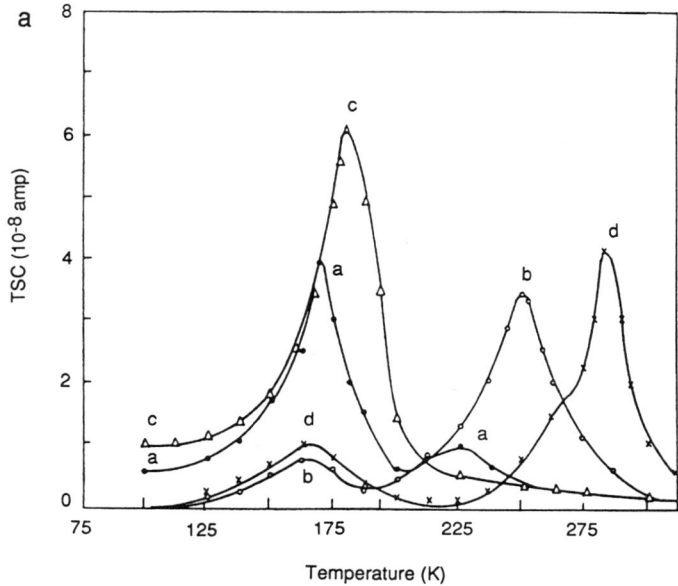
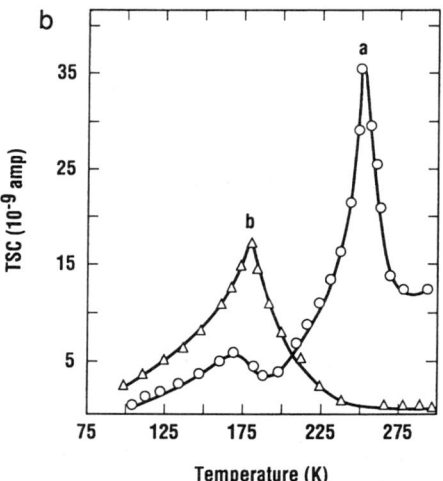

FIG. 10. (a) TSC spectra of HgI_2 crystals: *a* as-grown crystals, *b* aged samples, *c* as-grown crystal stored in iodine chamber, *d* as-grown crystal after baking at 350 K for 30 minutes; (b) TSC spectra of HgI_2 crystals: *a* aged samples, *b* same as *a* after application of dc bias 500 V for 12 hr. (Reprinted with permission from Suryanarayana and Acharya, 1989 and JOURNAL OF ELECTRONIC MATERIALS, a publication of the Minerals, Metals & Materials Society, Warrensdale, PA 15086.)

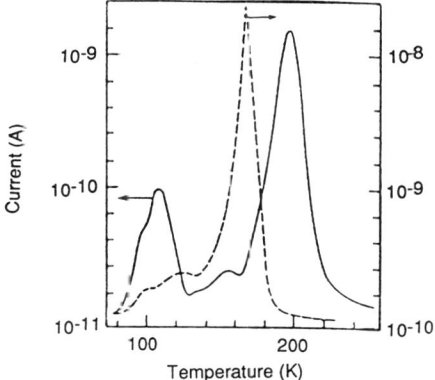

FIG. 11. Illumination effect on TSC spectra of a solution grown sample (from Mohammed-Brahim, 1981).

identified as slow traps, and those at 116, 144, and 200 K were identified as fast traps.

Blasi *et al.* (1978) have used the integrated area under each TSC peak to estimate the concentration of traps. If it is assumed that the traps are completely filled at low temperature, then the trap density may be estimated by

$$N_t = \frac{A_c}{qVG} \quad (46)$$

where A_c is the total integrated charge under a TSC peak in coulomb, V is the volume of the sample, and G is the photoconductive gain calculated for the steady state excitation of a photocurrent with magnitude equal to the TSC average current. The trap densities estimated for the two traps observed by the authors at 112 and 212 K are 3.5×10^{15} and 8.3×10^{15} cm^3. Hyder (1977) observed two TSC peaks at about 167 and 233 K in samples doped with Ag (about 3 ppm).

The effects of different contact materials on TSC were studied by Bao (1991). As can be seen in Table IV, the contacts have a dramatic effect on the TSC spectra. With the exception of samples contacted with Cu and Ag, where fast diffusion into HgI_2 bulk has been observed even at room temperature and therefore the effect may be due to the bulk, the variation of the TSC spectra with different contact materials is most likely due to the difference in the defects introduced by the interaction of the contact material with HgI_2 in the near-surface region. In fact, low temperature photoluminescence spectroscopy studies (Bao *et al.*, 1990, 1991a; R. B. James *et al.*, 1989, 1990) have shown that most of these contact layers modify the defect structure in the near surface region.

TABLE IV

EFFECT OF CONTACTS ON TSC SPECTRA OF HgI_2 (from Bao, 1991)

Contact		T_1	T_2	T_3	T_5	T_6	T_8	T_{10}
Cu	$T_m(K)$			116		161		
Ag	$T_m(K)$	90	96				212	
Al	$T_m(K)$	83		119	146	170		
Mg	$T_m(K)$				147	162	206	273
Au	$T_m(K)$						193	
Ni	$T_m(K)$	89			147	166	202	
In	$T_m(K)$	87		119	150	168		
Sn	$T_m(K)$				149	163		
Pd	$T_m(K)$				149	163		
Cr	$T_m(K)$				144	157	211	
ITO	$T_m(K)$		95	133	144	165	213	

Because the TSC data can be analyzed so many ways and TSC measurements themselves are very sensitive to experimental conditions such as excitation power, wavelength, sample geometry, illumination time, waiting time, heating rate, sample preparation, or contacts, it is difficult to compare results obtained from different workers. In addition, among the four parameters of activation energy, capture cross section, trap density, and lifetime product, it is possible to determine activation energy with only reasonable accuracy. The other parameters depend exponentially on activation energy and thus small uncertainties in the value of this parameter lead to large uncertainties in the other parameters. The reported values should be viewed as rough estimates in terms of order of magnitude and hence be used with caution. Peaks in TSC spectra obtained from several workers are summarized in Table V.

2. Other Techniques

Several other techniques have been employed to study deep levels in HgI_2. These techniques have not been correlated as well as TSC measurements to detector fabrication and performance. Brief reviews of isothermal current, photo-induced current transient spectroscopy, and photocapacitance spectroscopy follow.

a. Isothermal Current

When a voltage is applied to a HgI_2 sample, the dark current will decrease with time (Braatz and Zappe, 1984). This decay of current has been explained as due to the release of trapped carriers, which then drift under the applied electric field and are collected by the electrodes (Mellet and Friant, 1989). If retrapping is ignored, the current is then given by (Micocci et al., 1983)

$$I_d = \sum_i A_i \exp\left(-\frac{t}{\tau_{di}}\right). \quad (47)$$

By best fitting the decay of the dark current with time using Eq. (47), several detrapping times may be obtained. From the detrapping time, and assuming a typical capture cross section, the trap activation energy (assuming electron traps) can be estimated by

$$\Delta E_t = kT \ln(N_c \sigma_e v_{th} \tau_d). \quad (48)$$

Using this technique, Mellet and Friant (1989) studied the detrapping time in HgI_2 crystals grown either by solution or vapor phase and also crystals prepared with various surface treatments and deposited with a variety of contacts. Detrapping times from 0.68 to 40,000 seconds have been observed in these crystals, which correspond to trap activation energies from 0.67 to 0.96 eV.

b. Photoinduced Current Transient Spectroscopy

Alvarez and Saura (1990) have employed PITS to study traps in HgI_2. The technique and the interpretation of the data are very similar to deep level transient spectroscopy (DLTS), used widely in characterizing carrier traps and recombination centers in semiconductors. Instead of the typical boxcar integrators employed in DLTS, a two phase lock-in amplifier was used (Pons, Mooney, and Bourgoin, 1980). The basic idea, however, is still the same, in that a rate window is created and the transients are measured as a function of temperature. A peak will occur in the spectrum when the emission probability corresponding to the rate window was reached at a certain temperature. This corresponds to a measurement of the emission probability (related directly to the rate window) at the peak temperature. The position of the peak in terms of temperature will vary as the rate window is changed. A measurement of the shift in peak temperature as a function of rate window gives the emission probability as a function of temperature. By proper data analysis using Eq. (36), the trap activation energy and the capture cross section can be obtained. Using a modulated light source to create the necessary trap filling, Alvarez and Saura (1990) have observed five traps with activation energies 0.19, 0.22, 0.32, 0.44, and 0.49 eV. The capture cross sections range from 3.4×10^{-16} to 3.3×10^{-13} cm^2. The densities of the traps estimated from the amplitude of the current decay are from 4×10^{13} to 8×10^{14} cm^{-3}.

c. Photocapacitance Spectroscopy

Photocapacitance spectroscopy involves the measurement of ac capacitance under illumination as a function of the illumination wavelength. When the photon

TABLE V
TSC RESULTS FROM VARIOUS WORKERS

Reference		T_1	T_2	T_3	T_4	T_5	T_6	T_7	T_8	T_9	T_{10}	T_{11}
Bube, 1957	$T_m(K)$			129	142		172	196		222	267	293
	ΔE_t(eV)			0.34	0.38		0.45	0.51		0.59	0.77	
	Comments											
Whited and van den Berg, 1977	$T_m(K)$	81, 89	98	102	138		162	180	214	220		
	ΔE_t(eV)	V_1 & V_{22}	e	h	h		0.32	h	V_{Hg}, e	I_{Hg}, h		
	Comments						h					
Blasi et al., 1978	$T_m(K)$		112						212			
	ΔE_t(eV)		0.17						0.45			
	Comments		h						h			
Gelbart et al., 1977	$T_m(K)$	95	116		132	144	168	178	200			
	ΔE_t(eV)	.15–.23	.19–.28		.12–.29	.27–.35	.36–.49	.37–.51	.43–.54			
	Comments	Slow	Fast		Slow	Fast	Slow	Slow	Fast			
Suryanarayana and Acharya, 1989	$T_m(K)$						170	180		230	252	280
	ΔE_t(eV)						0.46	0.46		0.63	0.63	0.70
	Comments							I-strg			Aged	Baked

Reference													
Hyder, 1977	T_m(K)						167				233		
	ΔE_t(eV)						0.26				0.73		
	Comments												
Bornstein and Bube, 1987	T_m(K)		102		143		159	170	196				
	ΔE_t(eV)		0.28		0.36		0.40	0.43	0.51				
	Comments												
Muller, Friant, and Siffert, 1978	T_m(K)	107	112	125	145		170	190	200		217	245	
	ΔE_t(eV)												
	Comments		e		h						h		
Tadjine et al., 1983	T_m(K)	100	110	125	155		173		205		225		
	ΔE_t(eV)	0.16	0.28	0.32	0.40		0.45		0.50		0.59		
	Comments						Rad						
Stuck et al., 1976	T_m(K)		114		145						217		
	ΔE_t(eV)		0.19		0.22						0.45		
	Comments		clv		clv						h		
Mohammed-Brahim, Friant, and Mellet 1983	T_m(K)	96(8)	112(5)	125	140	155(60)	170	185	204				252	275
	ΔE_t(eV)	.10–.13	.16–.19	.20–.23	.21–.26	.25–.30	.50–.59	.40–.44	0.43					
	Comments						Good						

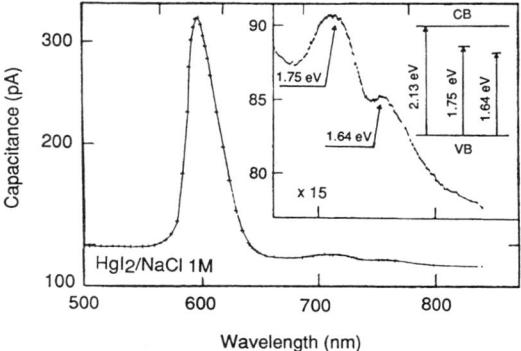

FIG. 12. Electrochemical photocapacitance spectroscopy spectrum and associated energy transitions for single crystal HgI_2 in 1 M NaCl. The back electrode is biased at +1 V. (Reprinted with permission from Burger et al., 1989 and Elsevier Science Publishers.)

energy is equal to the activation energy of a deep center, the carriers are liberated from this deep center into the corresponding band edge. These carriers will redistribute under the bias either as free carriers or as trapped carriers in a different center or location. This changes the polarization of the sample and a capacitance peak is observed. Burger et al. (1989) observed three peaks using electrochemical photocapacitance spectroscopy (EPS). A NaCl electrolyte cell was used and the back carbon contact was biased +1 V. The measurement frequency was 1 kHz. The spectrum is shown in Fig. 12 along with an interpretation of the levels and transitions involved. The three peaks appear at photon energies of 2.13, 1.75, and 1.64 eV. Photocapacitance in a longer wavelength region was measured by Zolotarev et al. (1970). Three peaks at 0.54, 0.69, and 0.89 eV were observed. Since photo quenching also occurs at 0.69 and 0.89 eV (see the next section), these peaks are explained as due to the liberation of holes from traps.

IV. Photoconductivity

HgI_2 was one of the first materials used for studies of photoconductivity study (Nix, 1935). Bube (1957) reported a systematic investigation of photoconductivity of HgI_2. The high resistivity of HgI_2 makes many typical semiconductor characterization techniques difficult to perform and many measurements of the electrical properties of HgI_2 have relied on the application of light. These experimental techniques include current transient measurements, photocurrent, internal photoemission, photoresponse, TSC, PITS, photomagnetoelectric effect, photo-Hall effect, cyclotron resonance measurements, photoelectrochemical measurements, and photocapacitance measurements, many of which have been discussed above. In addition, illumination by light affects the performance of detectors, the length of time for HgI_2 to reach equilibrium, and the photocurrent generated by a primary

1. BEHAVIOR OF PHOTOCURRENT

Photocurrent measurements of HgI_2 show a maximum at wavelengths near the bandgap (see Fig. 13). Several explanations for this peak have been reported. The peak may be a result of significant surface recombination (Adduci et al., 1977; Burshtein et al., 1986; Bornstein and Bube, 1987) that reduces the photocurrent for the highly absorbing photon energies. As the wavelength of the incident light decreases near the bandgap, the photocurrent will first increase due to an increase in the absorption, and as the wavelength continues to decrease, more and more carriers will be generated near the surface region. Here they recombine quickly through surface states due to the high carrier density and do not contribute to the photocurrent. A second explanation is that the peak in photocurrent is due to exciton formation, which causes a resonant absorption at a wavelength slightly below the bandgap (Manfredotti et al., 1977a,b). The exciton energy measured by

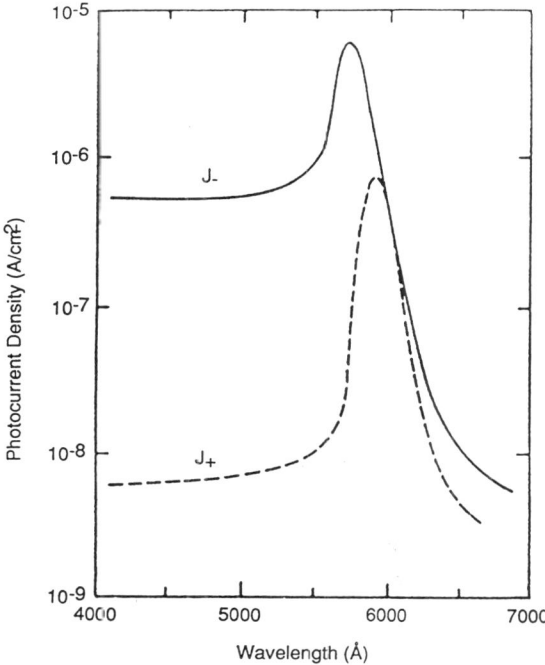

FIG. 13. Spectral response of photocurrent density for the photodetector cell with transverse geometry in which light and electric field are parallel; J^- is the photocurrent density when the illuminated surface is negative, and J^+ is the photocurrent density when the illuminated surface is positive. (Reprinted with permission from Borstein and Bube, 1987 and American Institute of Physics.)

Novikov and Pimonenko (1971) is about 29 meV, which is comparable to thermal energy at room temperature. However, for the excitons to contribute to the photocurrent, they have to dissociate and produce free carriers. The most likely process is to give their energy to trapped carriers that are then liberated to the corresponding band edge to contribute to the photocurrent.

Light spot scanning measurements conducted by Bube (1957) found that the photocurrent is limited mainly to a small region at or near the cathode; i.e., as the light spot is moved between the cathode and anode, the current is maximum when the spot is near the cathode. The voltage drop also occurs mostly near the cathode. This probably indicates that the photocurrent is due mainly to an electron contribution. When the light spot is near the cathode, more electrons will move toward the anode, resulting in more electrons in the active region and a higher photocurrent.

When photovoltaic short circuit current is measured as a function of wavelength, another interesting phenomenon has been observed (Bao, 1991). The setup for this measurement is shown in Fig. 14, where the direction of the current is also displayed. This configuration is used to measure the barrier height of the back contact and will be discussed in Subsection V.1. The current spectral response is shown in Fig. 15. It can be noted that the short circuit current reverses its direction twice as the wavelength of the excitation light increases. At short wavelengths, the current is negative (region I); as the wavelength increases, the current reverses its direction to positive (region II); and then changes back to a negative direction for all long wavelength responses (regions III and IV). These observations can be explained in terms of the dependence of carrier lifetime on the excess carrier density (or excitation intensity) (also refer to Subsection II.2.d). If the lifetimes of electrons and holes depend on the excess carrier with the form $\tau_e \propto \Delta n^\alpha$ and $\tau_h \propto \Delta p^\lambda$, then the excess carrier density will be related to the carrier generation rate per unit volume G by

$$\Delta n \propto G^{1/(1-\alpha)} \quad (49)$$
$$\Delta p \propto G^{1/(1-\lambda)}.$$

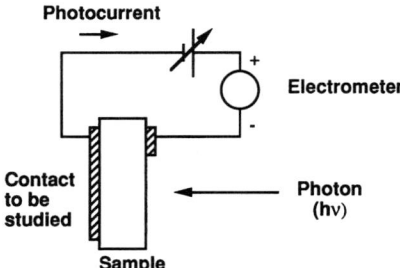

FIG. 14. Experimental setup for the photoresponse measurements. The configuration is designed to measure the barrier height of the back contact.

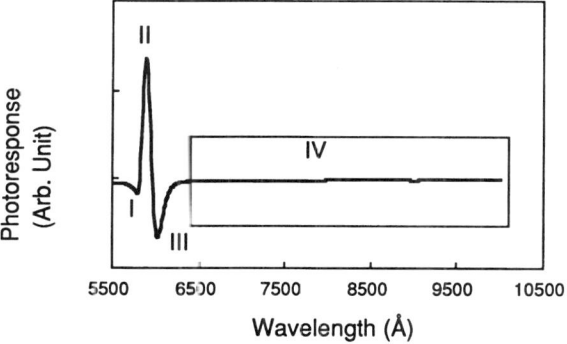

FIG. 15. Photoresponse from a palladium contacted sample measured at room temperature. The four regions labeled I, II, III, and IV are believed to arise from different processes in the sample.

If $\alpha > \lambda$, then $\Delta n < \Delta p$ at low G, and $\Delta n > \Delta p$ at high G. If $\alpha < \lambda$, the opposite is true. From photomagnetoelectric measurements by Adduci et al. (1977) and Manfredotti et al. (1977a), $\alpha = -0.33$ to -0.63, and $\lambda = -1$, thus, the contribution to the photocurrent is expected to be by holes in the low intensity excitation region and by electrons in the high intensity excitation region. Figure 15 can be explained as follows, using the band diagram shown in Fig. 16. In region I, absorption is primarily near the surface because of the above bandgap excitation; thus the generation rate is relatively high and the conduction is primarily by electrons. The sample then behaves like an n-type semiconductor and a positive space charge will be formed near the front surface. This results in a band bending as shown in Fig. 16 and causes the current to be negative. As the wavelength increases to region II, the absorption is still mainly near the front surface, but the generation rate per volume decreases as the absorption coefficient is reduced. The conduction becomes p-type and the band bending and current reverses its direction. As the wavelength is further increased to region III and IV, the excitation is more uniform across the sample, and the conduction is p-type. Because of the configuration of the measurements setup shown in Fig. 14, the back contact is more efficient than the front contact, which is away from the illumination spot. The short circuit current is then a result of the band bending near the back contact as shown in Fig. 16, and the current will reverse its direction again. At even longer wavelengths in region IV, the carriers are no longer generated from band to band transition but from transition between the band edge and the electrical contact as shown in Fig. 16. A similar observation was made when the intensity of the illumination is changed (Blasi et al., 1978). The current reverses its sign as the intensity of the illumination is increased. The conduction was found to be p-type at low intensity and n-type at high intensity.

When photocurrent is measured (with above bandgap light) as a function of temperature, several things were observed. Below 15 K, the photocurrent de-

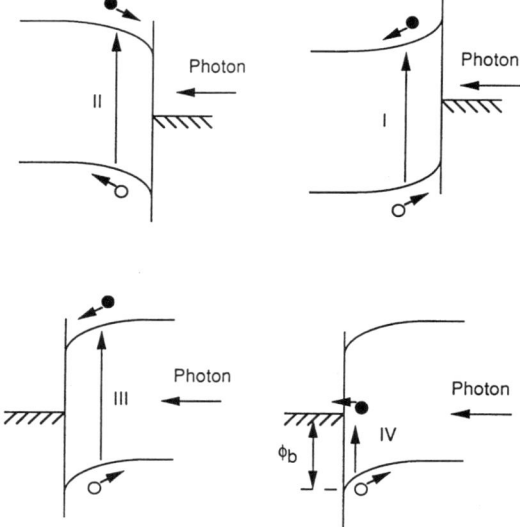

FIG. 16. Band diagrams illustrating the four processes that result in the short circuit current photocurrent.

creased rapidly with increasing temperature. At this temperature, the photocurrent was due mainly to holes, whose mobility also decreases rapidly with increasing temperature. Thus, the decrease in photocurrent is controlled by mobility, not carrier lifetime (Bloch *et al.*, 1977). At higher temperatures (>15 K), the photocurrent usually shows current peaks at several temperatures. These changes can not be accounted for by the temperature dependence of the mobility only, rather, the lifetime of carriers also changes. Bube (1957) has attributed the current peaks as due to the changeover of certain centers from the role of recombination centers to trapping centers as the Fermi level moves toward the center of the bandgap and crosses these centers. This changeover may decrease the hole occupancy on this level and the holes are redistributed to other more efficient recombination centers, thereby reducing the lifetime of the conducting carriers (electrons in this case). The decrease in photocurrent with increasing temperature is referred to as a "thermal quenching" effect. The activation energy of the center and the temperature of the peaks is related to the carrier density by

$$\ln(n_{\max}) = \ln\left(\frac{N_v \sigma_n}{\sigma_e}\right) - \frac{\Delta E_t}{kT_m}. \tag{50}$$

It can be seen that, if the carrier density is varied, the current maximum will appear at a different temperature. Since the carrier density can be varied by changing the excitation power, it is possible to obtain information of the trap by measuring the current peak position as a function of illumination power. The photocurrent peaks are shifted to higher temperatures as the excitation power is increased. Bube (1957) obtained an excitation energy of 0.51 eV and a σ_h/σ_e ratio of 10^3 for the photocurrent peak at about 193 K. Mohammed-Brahim et al. (1983) observed several photocurrent peaks at 80, 120, and 200 K with corresponding excitation energies of 0.21, 0.36, and 0.5 eV.

2. PHOTOCURRENT QUENCHING

Photoquenching was observed in HgI_2 by Zolotarev et al. (1970), Mohammed-Brahim et al. (1983) and Saura and Tognetti (1975). In these studies, a photocurrent was first measured using a primary excitation near the bandgap. A secondary illumination with below bandgap photon energy was then applied, and the photocurrent was measured and compared with the one obtained without secondary illumination. For some photon energies of the secondary illumination, the photocurrent will be quenched. If the photon energy of the secondary illumination is varied, the spectral response of photo quenching (percentage of photocurrent loss due to the secondary illumination) can be measured.

As shown in Fig. 17, Saura and Tognetti (1975) observed a sharp decrease as the photon energy of the secondary light source increases to 0.56 eV. A second quenching was observed as the photon energy of the secondary light source approaches 0.75 eV. The quenching disappeared when the photon energy of the secondary source was increased beyond 1.75 eV. The authors explained the spectra

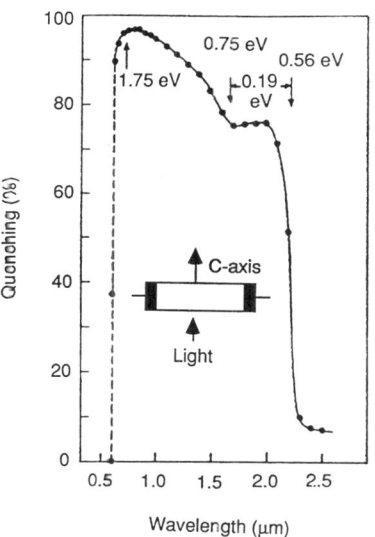

FIG. 17. Infrared quenching spectrum of HgI_2 at 77 K. The number of quenching photons was the same at all wavelength: $V = 70$ V, $\lambda_{prim} = 5400$ Å (from Saura and Tognetti, 1975).

as due to a center 0.56 eV above the valence band edge. The first rise in quenching is due to the transfer of a hole from this trap to the valence band. The second rise is attributed to the transfer of holes from the same center to the second valence band located 0.19 eV below the valence band maximum. The onset of the disappearance of photo quenching at 1.75 eV corresponds to the beginning of transfer of electrons from this center to the conduction band edge. A similar quenching effect was observed by Mohammed-Brahim *et al.* (1983), except for the location of the onset of quenching. Mohammed-Brahim *et al.* (1983) observed the first rise in quenching at 0.4 eV and the second one at 0.48 eV. The second quenching was present up to photon energies of about 1.77 eV. The quenching peaks observed by Zolotarev *et al.* (1970) are located at 0.60 and 0.89 eV. These two photon energies correspond to two of the three photocapacitance peaks observed by the same authors.

The photo quenching effect may be explained in a similar way as the thermal quenching effect. It is due to the transfer of trapped carriers into the corresponding band edges and the capture by a more efficient recombination center, which leads to a reduction in the lifetime of the free majority carriers of the other type. Since the photoconductivity is most likely to be due to electron conduction (especially when the excitation power is not too low), the energies levels obtained by these quenching measurements are most likely relative to the valence bands.

V. Surface Effects

Surface effects play an important role in the applications of HgI_2 nuclear detectors. Injecting contacts and surface conduction increase the leakage current and are undesirable for HgI_2 detectors. The recombination rates near the surface are generally very different from the bulk of a crystal because of the presence of surface states. This effect influences performance considerably, since high recombination rates result in incomplete charge collection. Therefore, proper choice of the electrical contact material, method of contact deposition, and surface treatment are extremely important in the fabrication of HgI_2 detectors.

1. Electrical Contacts

The nature of the charge transport across the contact–semiconductor interface depends on the work functions of the metal and semiconductor. It is also very sensitive to the presence of surface states, which are often created by chemical interactions between the metal and the semiconductor and by various processing procedures used to attach the electrical contacts. Contacts made from a variety of materials have been tried on HgI_2 detectors. Several of them have been successfully used for fabricating HgI_2 nuclear detectors or photodetectors. These include palladium, colloidal carbon, gold, germanium, transparent indium–tin oxide (ITO), conducting hydrogel, and saltwater. Palladium and carbon are usually the contact materials of choice for most x-ray and gamma ray detector spectrometers. With the exception of carbon, which is often applied to HgI_2 in the form of colloidal carbon, hydrogel, or saltwater, these contact materials are deposited either by thermal evaporation or sputtering. During thermal evaporation or sputtering a HgI_2 crystal is exposed to vacuum and heat treatment, and it often interacts with the contact material of choice during the deposition procedures. Because HgI_2 is a soft, reactive material with a high vapor pressure, all of the processing conditions are potential sources of defects in the near-surface region, leading to the incorporation of undesirable surface states.

Metals that form barriers to carrier transport on HgI_2 have been measured by two methods, internal photoemission and a dark current method. Other techniques such as the capacitance method are difficult to apply to HgI_2 due to its high resistivity.

The experimental setup for the internal photoemission measurement has already been shown in Fig. 14. To avoid absorption in the contact, the illumination is shone through the opposite site of the contact to be studied. The photocurrent under a bias or in a short circuit condition is measured as function of incident photon energy. In the long wavelength region (labeled as IV in Fig. 15), the photocurrent is due to the photoelectric effect where carriers are excited from the electrical contact into the semiconductors (see Fig. 16). The photoresponse per photon is related to the incident photon energy and the barrier height by (Sze, 1981)

$$\sqrt{R} \sim h\nu - \phi_b \quad \text{if} \quad h\nu - \phi_b > 3kT \tag{51}$$

where R is the photocurrent per incident photon, $h\nu$ is the energy of the photon, ϕ_b is the barrier height.

Figure 18 shows the photoresponse as a function of photon energy for a palladium contacted HgI_2 sample for photon energies in the range of 1.1 to 2.0 eV (i.e., using the long wavelength photoresponse from about 0.5 to 1 μm, region IV of Fig. 15). The current was measured in a short circuit configuration to avoid any influence of the dark current. The measured value of barrier height of the palladium contact was found to be 1.1 eV. From the direction of the current flow in the short circuit configuration, the barrier was found to be a hole barrier (p-type conduction). A current peak due to photoionization of a defect level occurs at 1.38 eV (see Fig. 18).

Similar measurements have also been performed on contacts made from a variety of other metals. Almost all of these showed some degree of polarization effect, that is, the results of measurements depended on time and sample history, thereby making the measured value unreliable. This is not surprising since palladium contacts are now widely used as the contact material of choice, due to its relatively good performance and the high quality of detectors that can be fabricated with Pd contacts.

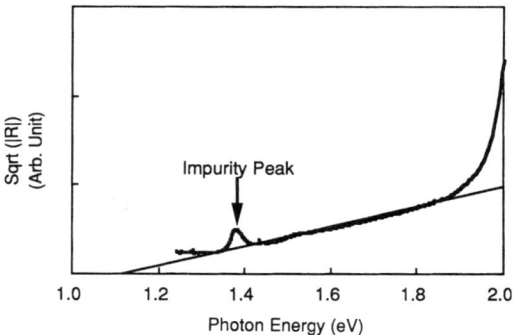

FIG. 18. Plot of the square root of the photoresponse as a function of photon energy to obtain the barrier height. The barrier height measured for the palladium contacted is 1.05 eV from the intercept of the straight line with the x-axis. A defect peak at 1.38 eV can also be seen.

The dark current technique has also been used to measure the barrier height, by assuming that the dark current in HgI_2 arises from thermal emission from the contact (Mellet and Friant 1989). By measuring the dark current as a function of applied electric field and using Eqs. (3) and (5), the barrier height was determined to be about 1.2 eV, for both palladium and carbon contacts. Mellet and Friant (1989) also studied the correlations between the contact parameter γ (defined in Eq. (8)) and the detector performance. It was found that γ (from Eq. (8)) ranges from 26 to 70 and higher values are usually related to higher energy resolution in the detectors. Crystals grown from the vapor phase tend to have higher γ values than those grown from solution.

2. SURFACE RECOMBINATION

Surface recombination also influences the performance of the HgI_2 devices. It directly affects the amount of charges that are separated and eventually collected by the electrical contacts. The surface recombination is characterized by a surface recombination velocity and is defined as $s = N_{st}v_{th}\sigma$, where N_{st} is the area density of recombination centers near the surface. For the purposes of this discussion, recombination also includes any trapping processes that hold charge for a time longer than the charge collection time of the detector electronics.

Using Eqs. (23)–(24), Levi et al. (1983a) have studied surface recombination in HgI_2 detectors. As can be seen from Eq. (24), if the electric field $E_{\eta=0.5}$ (when the charge collection efficiency is half) is plotted vs. the detector thickness, it is possible to determine whether the surface recombination is significant. Figure 19 shows such plots for electrons and holes (Levi et al., 1983a). From the plot for electrons, there is hardly any dependence on the detector thickness. Therefore, Eq. (25) holds and the surface recombination for electrons is significant and

4. ELECTRICAL PROPERTIES OF MERCURIC IODIDE

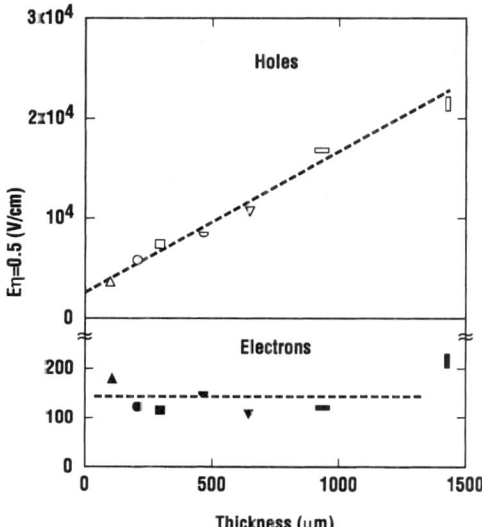

FIG. 19. Half-maximum collection field vs. detector thickness for holes (top) and electrons (bottom). (Reprinted with permission from Levi *et al.*, 1983a and the American Institute of Physics.)

greater than bulk trapping (i.e., $s_n \gg L/\tau_n$). The value of s_n was calculated to be about 1.5×10^4 cm/sec using an electron mobility of 100 cm^2/V-sec and lifetime of electrons $\tau_n \gg 10^{-5}$ sec. The plot for holes is quite different from that for electrons. From the intercept with y-axis and using a hole mobility of 4 cm^2/V-sec, the surface recombination for holes (s_p) is about 10^4 cm/sec. From the slope a lifetime τ_p for holes of about 10^{-6} sec is obtained. In this case, the hole surface recombination is negligible.

This method used to determine the surface recombination velocity requires the fabrication of several detectors with different thicknesses. It is possible, however, to measure surface recombination velocity using only one device. In their measurements, Roth *et al.* (1987) first used the charge transient method to determine the mobilities and lifetimes of electrons and holes, and the charge collection efficiency η (Eq. (21)) was calculated assuming no surface effect. The actual charge collection efficiency η' is then related to the surface recombination velocity by

$$\frac{\eta}{\eta'} - 1 = \frac{s}{\mu E}. \qquad (52)$$

The surface recombination velocity may be obtained from the slope of the plot ($\eta/\eta' - 1$) vs. $1/\mu E$. For electrons, the surface effect was also found to be important and $s_n = 2 \times 10^4$. The hole surface recombination was found to be negligible, in agreement with Levi *et al.* (1983a).

The effect of surface treatment on surface recombination was studied by Levi *et al.* (1983b). The surface recombination velocity of electrons was measured with the sample treated with potassium iodide (KI) aqueous solutions at different temperatures, concentrations, and waiting times before contact deposition. It was found that the lowest surface recombination was achieved by etching with 20% KI solution at temperatures between 0 to 10°C, and the contact deposition made about 24 hr after etching. Mellet and Friant (1989) have found that nitric acid treatment of solution grown HgI_2 improved the low-energy response of the detectors considerably.

VI. Detector Performance

The qualities of radiation detectors are characterized by their energy resolution, peak to background (peak to valley) ratio, efficiency, active volume, response time, radiation damage resistance, and reliability. Most of these qualities are closely related to the electrical properties of the detector material, as discussed in various sections of this chapter and related chapters in this book. This section will be devoted to the discussion of background counts in HgI_2 x-ray and gamma ray detectors, and polarization effects.

1. Peak-to-Background Ratio

When an x-ray spectrum is taken from a HgI_2 detector, a certain amount of background counts occurring in channels below the photo peak can be observed even though the incident x-rays are monoenergetic. In spectroscopy applications, these background counts may impose a detection limit when x-rays of different energies are present. These background counts cover all energies below the photo peak and are not due to x-ray escapes, which usually produce sharp peaks in the low energy region. To quantify this continuous background, the peak to background ratio can be measured and used as a figure of merit to describe this aspect of the detector performance. Continuous backgrounds have been observed in Si and Ge detectors and have traditionally been related to a dead layer beneath the entrance contact.

The origins of the dead layers for Si and Ge detector with x-ray energies below a few keV have been associated with photogenerated carriers moving to the entrance surface against the applied field and not contributing to the charge collection. The carriers may move to the entrance either through diffusion, which is a multiple scattering process, or as hot carriers that escape to the surface without scattering (Caywood, Mead, and Mayer, 1970; Llacer, Haller, and Cordi, 1977). Estimated dead layer thicknesses for Si and Ge x-ray detectors are not very different from measured values at x-ray energies below a few keV (Goulding, 1977; Musket and Bauer, 1973; Llacer *et al.,* 1977). In the case of HgI_2 detectors, the estimated dead layer thickness due to hot carriers is much less than the measured value and can be ignored. The dead layer thickness due to diffusion is a function of the applied field and is approximately given by (Dabrowski et al., 1981)

$$d = \frac{kT\mu}{4v_d} \qquad (53)$$

where μ is the carrier mobility and v_d is the carrier drift velocity. Since the electron drift velocity will not saturate under bias typically applied to HgI_2 detectors (also see Subsection II.2), the dead layer due to the carrier diffusion is a function of the electric field through the drift velocity. It should be pointed out that the electric field used to calculate the drift velocity should be that near the entrance contact, which may not be equal to the applied electric field.

Dabrowski et al. (1981) measured the dead layer thicknesses for HgI_2 detectors using Al and Mg x-rays. The dead layer thicknesses were found to depend both on the x-ray energy and on the contact materials. For a detector with palladium contacts, the dead layer thicknesses are 0.24 μm for the 1.49 keV Al x-rays and 0.15 μm for the 1.25 keV Mg x-rays. For a detector with carbon contacts, the dead layer thicknesses are 0.08 μm for the Al x-rays and 0.05 μm for the Mg x-rays. The large difference between the detectors with different contacts was attributed to the difference in band bending near the entrance electrodes, resulting in different total electric field strength even under the same applied field.

Dead layer thicknesses measured from a large number of HgI_2 detectors are about 0.091 μm at 5.89 keV (Mn x-rays from a Fe-55 radioactive source) and 1.6 μm at 22.16 keV (Ag x-rays from a Cd-109 radioactive source). These values do not vary very much among good detectors (good energy resolution, minimum tailing, etc.) and are not very sensitive to the applied electric field above a certain bias. Figure 20 shows the peak to background ratio (inversely proportional to the dead layer thickness) at 5.89 keV measured as a function of the applied bias. As can be seen, the peak to background ratio changes very little with the applied field above about 300 V. The rapid decrease of peak to background ratio below 300 V is due to the incomplete depletion of the detector at low bias. If Eq. (53) is used to estimate the electric field near the entrance electrode, the obtained value will be much smaller than a typical applied bias, which is difficult to account for even if there is certain band bending near the surface. It appears then that the origins of the dead layer may be more than just the diffusion mechanism in HgI_2. One likely source of the background counts is the energetic electrons generated by the incident x-rays through the photoelectric effect. These electrons will gradually lose their energy by ionization of more electrons along the path. If the energetic photoelectrons are generated near the entrance surface, they may deposit only part of their energy before reaching the surface and thus product counts in lower energy channels. The ranges of these electrons can be estimated from the Katz–Penfold formula (Katz and Penfold, 1952; Manfredotti and Nastasi, 1984). They are about 0.082 μm and 1.3 μm at 5.89 keV and 22.16 keV, respectively, for HgI_2. These

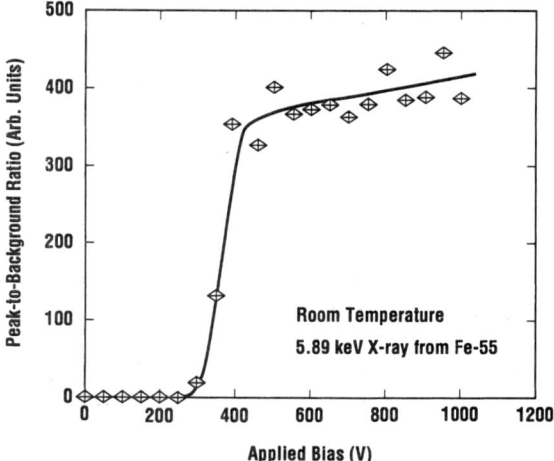

FIG. 20. The peak to background ratio measured at 5.89 keV as a function of the applied bias for a HgI$_2$ detector. The thickness of the detector is about 400 μm.

values are reasonably close to the measured values of dead layer thicknesses given the simplicity of the calculation. The dead layer due to such an effect will not depend on the electric field and will provide the lower limit for the thickness (more accurate values may be obtained with more elaborate calculations taking into account the details of interaction and the directional dependence of photoelectrons). Therefore dead layer in HgI$_2$ x-ray detectors may come from both carrier diffusion and energetic photoelectrons. At low electric fields, the dead layer due to diffusion may dominate and is a function of the electric field. At higher electric fields, the dead layer comes mainly from the energetic photoelectron escape and is an intrinsic property of the material and independent of the electric field.

2. POLARIZATION EFFECT

In the application of semiconductors in x-ray and gamma ray detections, the polarization effect is defined as the time dependence of the detector performance. Possible causes of polarization include trapping, detrapping, and change of defect structure in the detector. The trapping and detrapping of carriers will change the space–charge distribution in the detector and thus modify the electric field profile. The charge collection efficiency will be altered correspondingly through the changes in the average drift length of carriers (see Eq. (14)) or the effectiveness of moving carriers away from surface recombination centers (see Eq. (23)). Since the time constants such as trapping and detrapping times may vary considerably among different traps, time dependent detector performances can be observed with quite different time scales. The polarization may have drastically different effects on the x-ray spectra. The x-ray spectra may either improve or degrade, and the time constant may vary from seconds to several years. These differences come from the different defect structures in the detectors, which may be related to different crystal growth techniques, different impurities and native defect in the detectors, different contact materials and deposition techniques, different detector fabrication procedures, etc. The detector behaviors can be further complicated by the simultaneous presence of both trapping and detrapping, detrapping through other mechanisms such as Auger recombination (Gerrish 1992), and other factors such as radiation energy, flux, penetration depth, and detector temperature. Therefore, the observed behaviors of polarization effects in HgI_2 can be quite different among researchers (Whited, Schieber, and Randtke, 1976; Holzer and Schieber, 1980; Mohammed-Brahim, Friant, and Mellet, 1985; Squillante, Shah, and Moy, 1990; Gerrish, 1992; Natarajan, Scoopo, and Henderson, private communication, 1992).

a. Trapping

The effect of carrier trapping on the x-ray spectra is shown in Fig. 21 (Natarajan, Scoopo, and Henderson, private communication, 1992). The gradual degra-

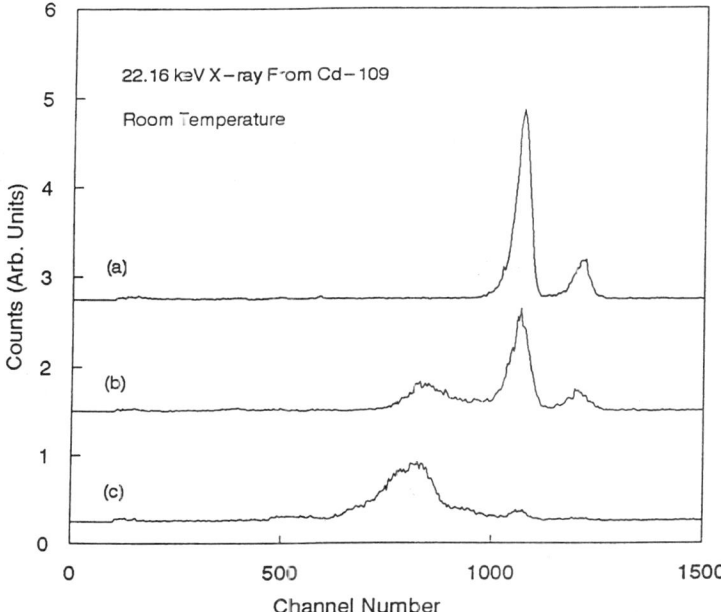

FIG. 21. X-ray spectra of 22.16 keV showing the polarization effect due to electron trapping in the HgI_2 detectors; spectrum (a) was taken before polarization and spectra (b) and (c) were taken at increased levels of polarization (trapping) (from Natarajan, Scoopo, and Henderson, private communication, 1992).

dation of the x-ray spectra form (a) to (c) is caused by increased concentration of trapped electron carriers. This polarization effect is attributed to trapping for several reasons. First, the changes in the x-ray spectra can be induced only in the presence of excitation. Second, the changes are completely reversed if the excitation source is removed. Third, the reversal in the absence of excitation is slowed down considerably if the detector is cooled to about $-5°C$, indicating that the reversal process is thermally activated (release of the trapped carriers from the defect states responsible for the polarization effect). In Fig. 21, the trapped carriers are believed to be electrons because the polarization effect degrades the x-ray spectra. This is due to reduced electric field near the entrance electrode (negatively biased), which can be produced only by the accumulation of more negative space charge in the detector. It is entirely possible, however, for a trapping process to improve the x-ray spectra, if the trapped carriers are holes instead of electrons. Such a polarization effect will then have all the characteristics behaviors as described for Fig. 21, except that the x-ray spectra will improve as the trapping (degree of polarization) increases (Mohammed-Brahim et al., 1985). Holzer and Schieber (1980) studied the effects of radiation on the count rate, energy resolution, and the position of the photo peak. As the radiation flux or energy increases, detectors that polarize will deteriorate faster indicating a trapping process is responsible for the observed degradation.

b. Detrapping

Since the electric field in the detector can also be modified by a change in space charge due to detrapping of carriers, detrapping may also lead to a polarization effect. The reversal process described in the previous paragraph is one form of such a polarization effect. If polarization is due to detrapping, it will not need excitation and the time constant related to the polarization will become very temperature sensitive (for thermally activated detrapping) as observed for the reversal process just described. Another polarization effect in HgI_2 detectors due to detrapping is observed after a bias is first applied to the detector. Figure 22 shows a group of 5.89 keV x-ray spectra taken at different time intervals after the bias is applied. Even after 8 hr, the spectrum is still improving. Spectrum (e) has a peak-to-background ratio about twice as large as spectrum (d). The changes in the x-ray spectra shown in Fig. 22 were attributed to the detrapping process after the bias was applied and the depletion region was being established. This detrapping process is also responsible for the dark current transient as described by Eq. (47). In reality, it is more likely to have several different traps detrapping at the same time after the application of bias. Some of these traps may be electron traps and some may be hole traps. Thus, even though Fig. 22 shows a case in which the electron detrapping dominates over the hole detrapping because the x-ray spectrum improves, it is also possible to have a dominant hole detrapping process; in which case, the x-ray spectrum will degrade. The detector may reach an optimal

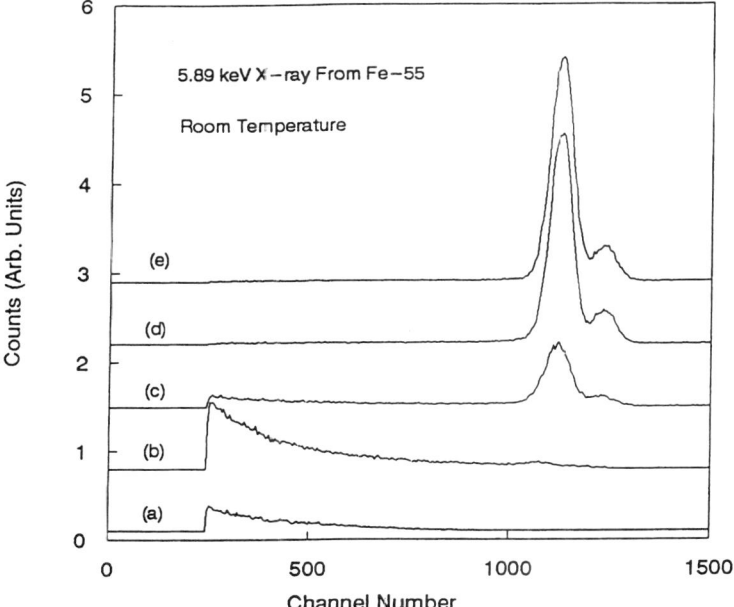

FIG. 22. X-ray spectra at 5.89 keV showing polarization effect due to electron detrapping in the HgI$_2$ detectors after the application of the bias. (Spectra (a) to (e) were taken 0, 1, 2, 8, and 24 hr after the bias was applied. The detector was not irradiated with the x-rays between these measurements.

(or worst) performance at a certain time if both electron and hole detrapping is present and these can dominate at different times because of differing detrapping times.

c. *Others*

Auger recombination of holes in HgI$_2$ gamma ray detectors has been observed by Gerrish (1992). In this process, trapping of the holes results in the ejection (detrapping) of electrons. The ejection of electrons is not thermally activated and may show different behaviors as described in Subsection VI.1.b. In this process, the trapping affects the detection of the energy of the incident x-rays because of the gain created by the ejected electrons (Gerrish 1992). On the other hand, as the electron trap empties, the space charge will be more positive and the rate of the Auger recombination will be less. Hence, a time dependent detector performance (improvement of energy resolution for gamma rays in this case) was observed that may take up to several weeks to stabilize after application of bias.

Polarization effects due to trapping and detrapping are reversible processes. Permanent changes in defect structure may also result in time-dependent changes

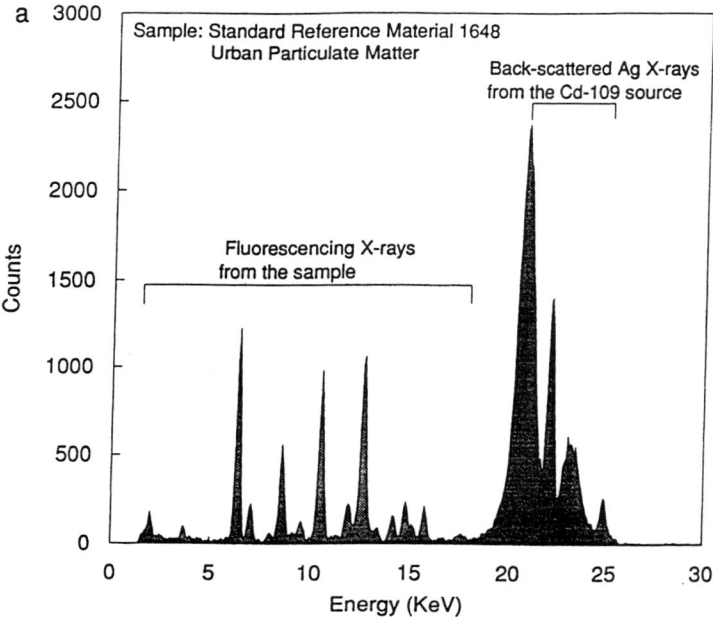
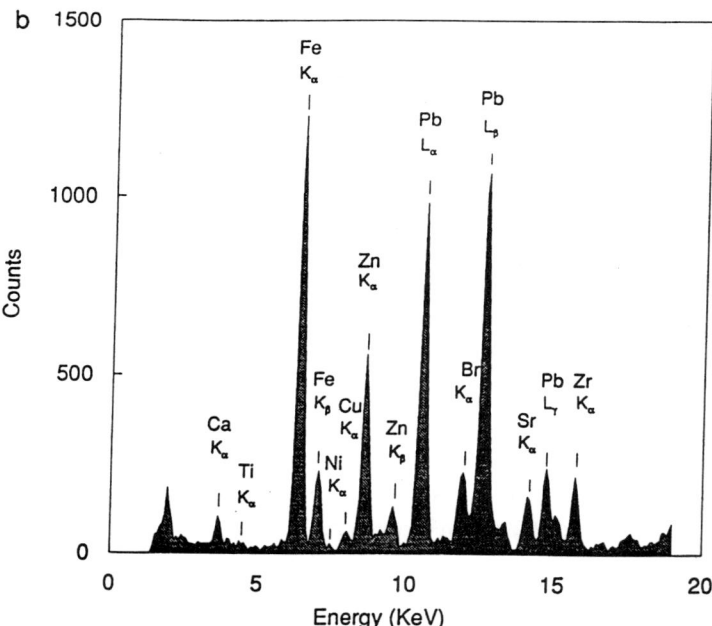

FIG. 23. (a) An x-ray spectrum taken with the battery-powered field portable XRF system: Spectrace-9000, manufactured by TN Technologies, Inc.; (b) the fluorescence portion of the spectrum with identifications of the observed elements by their characteristics x-ray emissions (from Berry and Voots, private communication, 1992).

in the detector performance. Most of the permanent degradations observed in HgI_2 detectors are related to its relatively high vapor pressure (Swierkowski et al., 1974; Squillante et al., 1990). Advances in the encapsulation techniques for HgI_2 detectors, especially with the use of parylene coatings, have greatly alleviated these problems (Iwanczyk et al., 1990).

VII. Conclusions

The electrical properties of mercuric iodide most relevant to x-ray and gamma ray detection were discussed in this chapter. Several techniques that have been extensively used to characterize HgI_2 crystals and detectors were discussed in detail. These techniques include current and charge transient measurements, TSC, and photoconductivity. Two of the present challenges in HgI_2 sensor technology are to increase control of the detector fabrication processes and to modify the detector processing strategy in such a way so as to increase the manufacturing yield of high quality detectors. HgI_2 detectors have been successfully utilized in commercial field portable x-ray fluorescence (XRF) systems used for alloy analysis and other general-purpose applications (TN Technologies 1992). These XRF systems are the first field portable XRF instruments with energy resolution approaching that of a laboratory (Si) system and far exceeding that of a gas-filled proportional analyzer. The full-width at half-maximum of these systems at 5.89 keV employing HgI_2 detectors cooled to $-5°C$ by Peltier coolers is about 270 eV as compared with 150 eV for a Si detector (cooled by liquid nitrogen) and 800 eV for a gas-filled proportional counter. Figure 23a shows an x-ray spectrum taken with such a portable system, Spectrace-9000 (TN Technologies 1992) from a standard reference material and excited by one of the three built-in excitation sources (Cd-109 radioactive source emitting 22.16 keV Ag x-rays) (Berry and Voots, private communication, 1992). Figure 23b shows the fluorescence portion of the spectrum in Figure 23a with the identification of the observed elements (Berry and Voots, private communication, 1992). The superb spectroscopic capabilities are demonstrated in Fig. 23b with the well-resolved characteristic x-rays very close in energies. With the three excitation sources employed by the Spectrace-9000, the 19 lb. battery powered field portable instrument is capable of performing quantitative analysis of elements from S to U. Progress in optimizing the various material parameters and processing steps, such as crystal growth, stoichiometry, impurity control, surface treatment, choice of contact, deposition method, and encapsulation, will determine the future of this technology.

REFERENCES

Adduci, F., Cingolani, A., Ferrara, M., Lugara, M., and Minafra, A. (1977). *J. Appl. Phys.* **48,** 342.
Alvarez, F., and Saura, J. (1990). *J. Mat. Sci. Lett.* **9,** 569.

Anedda, A., Raga, F., Grill, E., and Guzzi, M. (1977). *Il Nuovo Cimento* **38B**, 439.
Bao, X. J. (1991). Ph.D. thesis, Carnegie Mellon University.
Bao, X. J., Schlesinger, T. E., James, R. B., Stulen, R. H., Ortale, C., and van den Berg, L. (1990). *J. Appl. Phys.* **67**, 7265.
Bao, X. J., Schlesinger, T. E., James, R. B., Gentry, G. L., Cheng, A. Y., and Ortale, C. (1991a). *J. Appl. Phys.* **69**, 4247.
Bao, X. J., Schlesinger, T. E., James, R. B., Cheng, A. Y., Ortale, C., and van den Berg, L. (1991b). *Mat. Res. Soc. Symp. Proc.* **170**, 541.
Bao, X. J., Schlesinger, T. E., James, R. B., Cheng, A. Y., Ortale, C., and van den Berg, L. (1992). *Mat. Res. Soc. Symp. Proc.* **242**, 767.
Beyerle, A. G., Hull, K. L., Markakis, J., Schnepple, W., and van den Berg, L. (1983). *Mat. Res. Soc. Symp. Proc.* **16**, 191.
Blasi, C. D., Galassini, S., Manfredotti, C., Micocci, G., Ruggiero, L., and Tepore, A. (1978). *Nucl. Instr. Methods* **150**, 103.
Bloch, P. D., Hodby, J. W., Jenkins, T. E., Stacey, D. W., and Schwab, C. (1977). *Il Nuovo Cimento* **38B**, 337.
Bloch, P. D., Hodby, J. W., Schwab, C., and Stacey, D. W. (1978). *J. Phys. C* **11**, 2579.
Bornstein, J., and Bube, R. H. (1987). *J. Appl. Phys.* **61**, 2676.
Braatz, U., and Zappe, D. (1984). *Phys. Stat. Sol. (a)* **86**, 407.
Bube, R. (1957). *Phys. Rev.* **106**, 703.
Bube, R. (1960). *Photoconductivity of Solids*. John Wiley & Sons, New York.
Burger, A., Shi, W., Silberman, E., Franks, L., and Schnepple, W. F. (1989). *Nucl. Instr. Methods Phys. Res. A* **283**, 232.
Burshtein, Z., Akujieze, J. K., Silberman, E. (1986). *J. Appl. Phys.* **60**, 3182.
Caywood, J. M., Mead, C. A., and Mayer, J. W. (1970). *Nucl. Instr. Meth.* **79**, 329.
Cho, Z. H., Watt, M. K., Slapa, M., Tove, P. A., Schieber, M., Davies, T., Schnepple, W., Randtke, P., Carlston, R., and Sarid, D. (1975). *IEEE Trans. Nucl. Sci.* **NS-22**, 229.
Dabrowski, A. J., and Huth, G. C. (1978). *IEEE Trans. Nucl. Sci.* **NS-25**, 205.
Dabrowski, A. J., Iwanczyk, J. S., Barton, J. B., Huth, G. C., Whited, R., Ortale, C., Economou, T. E., and Turkevich, A. L. (1981). *IEEE Trans. Nucl. Sci.* **NS-28**, 536.
Gartner, W. (1957). *Phys. Rev.* **105**, 823.
Gelbart, U., Yacoby, Y., Beinglass, I., and Holzer, A. (1977). *IEEE Trans. Nucl. Sci.* **NS-24**, 135.
Gerrish, V. (1992). *Nucl. Instr. Methods Phys. Res.* **A322**, 402.
Goto, T., and Nishima, Y. (1979). *Solid State Commun.* **31**, 751.
Goulding, F. S. (1977). *Nucl. Instr. Meth.* **142**, 213.
Grossweiner, L. I. (1953). *J. Appl. Phys.* **24**, 1306.
Hermon, H., Roth, M., Nissenbaum, J., and Schieber, M. (1991). *J. Cryst. Growth* **109**, 376.
Holzer, A., and Schieber, M. (1980). *IEEE Trans. Nucl. Sci.* **NS-27**, 266.
Hyder, S. B. (1977). *J. Appl. Phys.* **48**, 313.
Iwanczyk, J., Barton, J. B., Dabrowski, A. J., Kusmiss, J. H., and Szymczyk, W. M. (1983). *IEEE Trans. Nucl. Sci.* **NS-30**, 363.
Iwanczyk, J., Dabrowski, A. J., Markakis, J. M., Ortale, C., and Schnepple, W. F. (1984). *IEEE Trans. Nucl. Sci.* **NS-31**, 336.
Iwanczyk, J. S., Wang, Y. J., Bradley, J. G., Albee, A. L., and Schnepple, W. F. (1990). *IEEE Trans. Nucl. Sci.* **NS-37**, 2214.
James, R. B., Bao, X. J., Schlesinger, T. E., Markakis, J. M., Cheng, A. Y., and Ortale, C. (1989). *J. Appl. Phys.* **66**, 2578.
James, R. B., Bao, X. J., Schlesinger, T. E., Ortale, C., and Cheng, A. Y. (1990). *J. Appl. Phys.* **67**, 2571.
James, T. W., and Milstein, F. (1983). *J. Mat. Sci.* **18**, 3249.
Kanel, H. V., Wachter, P., and Gerischer, H. (1984). *J. Electrochem. Soc.* **131**, 77.
Katz, L., and Penfold, A. S. (1952). *Rev. Mod. Phys.* **24**, 28.

Kitajima, K., and Wagner, J. B. Jr. (1988). *Solid State Ionics* **28–30,** 1146.
Klick, C. C., and Maurer, R. J. (1951). *Phys. Rev.* **81,** 124.
Levi, A., Schieber, M. M., and Burshtein, Z. (1983a). *J. Appl. Phys.* **54,** 2472.
Levi, A., Burger, A., Nissenbaum, J., Schieber, M., and Burshtein, Z. (1983b). *Nucl. Instr. Methods* **213,** 35.
Levi, A. Schieber, M. M., and Burshtein, Z. (1985). *J. Appl. Phys.* **57,** 1944.
Llacer, J., Watt, M. M. K., Schieber, M., Carlston, R., and Schnepple, W. (1974). *IEEE Trans. Nucl. Sci.* **NS-21,** 305.
Llacer, J., Haller, E. E., and Cordi, R. C. (1977). *IEEE Trans. Nucl. Sci.* **NS-24,** 53.
Look, D. C. (1983). *Semiconductors and Semimetals* **19,** 75.
Malm, H. L. (1972). *IEEE Trans. Nucl. Sci.* **NS-19,** 263.
Manfredotti, C., and Nastasi, U. (1984). *Nucl. Instr. Meth. Phys. Res.* **A225,** 138.
Manfredotti, C., Murri, R., Quirini, A., and Vasanelli, L. (1977a). *IEEE Trans. Nucl. Sci.* **NS-24,** 158.
Manfredotti, C., Murri, R., and Vasanelli, L. (1977b). *Solid State Commun.* **21,** 53.
Markakis, J. M. (1988a). *IEEE Trans. Nucl. Sci.* **35,** 356.
Markakis, J. M. (1988b). *Nucl. Instr. Methods Phys. Res.* **A263,** 499.
Marschall, J., and Milstein, F. (1991). *Appl. Phys. Lett.* **58,** 1422.
Martin, G. M., Bach, P., and Guetin, P. (1974). *Appl. Phys. Lett.* **25,** 286.
Mellet, J., and Friant, A. (1989). *IEEE Trans. Nucl. Instr. Methods Phys. Res.* **A283,** 199.
Micocci, G., Rizzo, A., Tepore, A., and Zuanni, F. (1983). *Phys. Stat. Sol. (a)* **80,** 263.
Milnes, A. G. (1973). *Deep Impurities in Semiconductors.* John Wiley & Sons, New York.
Minder, R., Majni, G., Canali, C., Ottaviani, G., Stuck, R., Ponpon, J. P., Schwab, C., and Siffert, P. (1974). *J. Appl. Phys.* **45,** 5074.
Minder, R., Ottaviani, G., and Canali, C. (1976). *J. Phys. Chem. Solids* **37,** 417.
Mohammed-Brahim, T. (1981). *Phys. Stat. Sol. (a)* **65,** K1.
Mohammed-Brahim, T., Friant, A., and Mellet, J. (1983). *Phys. Stat. Sol. (a)* **79,** 71.
Mohammed-Brahim, T., Friant, A., and Mellet, J. (1985). *IEEE Trans. Nucl. Sci.* **NS-32,** 581.
Muller, J. C., Friant, A., and Siffert, P. (1978). *Nucl. Instr. Methods* **150,** 97.
Musket, R. G., and Bauer, W. (1973). *Nucl. Instr. Meth.* **109,** 593.
Nissenbaum, J., Schieber, M., and Burshtein, Z. (1987). *J. Appl. Phys.* **61,** 2921.
Nix, C. (1935). *Phys. Rev.* **47,** 72.
Novikov, B. V., and Pimonenko, M. M. (1971). *Sov. Phys. Semicond.* **4,** 1785.
Ottaviani, G., Canali, C., and Quaranta, A. A. (1975). *IEEE Trans. Nucl. Sci.* **NS-22,** 192.
Ponpon, J. P., Stuck, R., Siffert, P., and Schwab, C. (1974). *Nucl. Instr. Methods* **119,** 197.
Pons, D., Mooney, P. M., and Bourgoir, J. C. (1980). *J. Appl. Phys.* **51,** 2038.
Press, W. H., Flannery, B. F., Teukoslky, S. A., and Vetterling, W. T. (1988). *Numerical Recipes in C.* Cambridge University Press, New York.
Redfield, A. (1954). *Phys. Rev.* **94,** 526.
Rose, A. (1951). *RCA Rev.* **12,** 362.
Roosbroeck, W. V. (1956). *Phys. Rev.* **101,** 1713.
Roth, M., Burger, A., Nissenbaum, J., and Schieber, M. (1987). *IEEE Trans. Nucl. Sci.* **NS-34,** 465.
Saura, J., and Tognetti, N. P. (1975). *Phys. Stat. Sol. (a)* **31,** K125.
Schieber, M., Beinglass, I., Dishon, G., Holzer, A., and Yaron, G. (1978). *IEEE Trans. Nucl. Sci.* **NS-25,** 644.
Slapa, M., Huth, G. C., Seibt, W., Schieber, M. M., and Randtke, P. T. (1976). *IEEE Trans. Nucl. Sci.* **NS-23,** 102.
Squillante, M. R., Shah, K. S., and Moy, L. (1990). *Nucl. Instr. Meth. Phys. Rev.* **A288,** 79.
Stuck, R., Muller, J. C., Ponpon, J. P., Scharager, C., Schwab, C., and Siffert, P. (1976). *J. Appl. Phys.* **47,** 1545.
Suryanarayana, P., and Acharya, H. N. (1989). *J. Electr. Mat.* **18,** 481.
Swierkowski, S. P., Armantrout, G. A., and Wichner, R. (1974). *IEEE Trans. Nucl. Sci.* **NS-21,** 302.
Sze, S. M. (1981). *Physics of Semiconductor Devices,* 2nd ed. Wiley-Interscience, New York.

Tadjine, A., Gosselin, D., Koebel, J. M., and Siffert, P. (1983). *Nucl. Instr. Methods* **213,** 77.
TN Technologies. (1992). Product brochures for Metallurgist-XF and Spectrace-9000.
Turner, D. E., and Harmon, B. N. (1989). *Phys. Rev. B* **40,** 10516.
Van den Berg, L. (1992). *Nucl. Instr. Methods Phys. Res.* **A322,** 453.
Watt, M. K., and Cho, Z. H. (1976). *IEEE Trans. Nucl. Sci.* **NS-23,** 124.
Whited, R. C., and van den Berg, L. (1977). *IEEE Trans. Nucl. Sci.* **NS-24,** 165.
Whited, R. C., Schieber, M. M., and Randtke, P. T. (1976). *J. Appl. Phys.* **47,** 2230.
Willig, W. R. (1971). *Nucl. Instr. Methods* **96,** 615.
Yeargan, J. R., and Taylor, H. L. (1968). *J. Appl. Phys.* **39,** 5600.
Yee, J. H., Sherohman, J. W., and Armantrout, G. A. (1976). *IEEE Trans. Nucl. Sci.* **NS-23,** 117.
Zitter, R. N. (1958). *Phys. Rev.* **112,** 852.
Zolotarev, V. F., Kikineshi, A. A., Semak, D. G., Fedak, V. V., and Chepur, D. V. (1970). *Sov. Phys. Semicond.* **4,** 802.

CHAPTER 5

Optical Properties of Red Mercuric Iodide

X. J. Bao

DEPARTMENT OF ANALYTICAL INSTRUMENTS
TN TECHNOLOGIES, INC.
ROUND ROCK, TEXAS

R. B. James

ADVANCED MATERIALS RESEARCH DEPARTMENT
SANDIA NATIONAL LABORATORIES
LIVERMORE, CALIFORNIA

T. E. Schlesinger

DEPARTMENT OF ELECTRICAL AND COMPUTER ENGINEERING
CARNEGIE MELLON UNIVERSITY
PITTSBURGH, PENNSYLVANIA

I. INTRODUCTION	169
II. BAND STRUCTURE	170
III. EXPERIMENTAL TECHNIQUES AND MEASURED VALUES FOR OPTICAL CONSTANTS	177
1. *Measurement of Bandgap and Shift with Temperature*	177
2. *Absorption*	183
3. *Ellipsometry Measurements of Optical Constants*	185
4. *Optical Properties near and below the Bandgap*	187
5. *Phonon Structure of Red Mercuric Iodide*	192
6. *Radiative Recombination of Nonequilibrium Electron–Hole Pairs*	195
7. *Phonon-Assisted Electron–Hole Pair Photoluminescence*	202
8. *Effects of Geometrical Configuration on Photoluminescence Spectra*	202
9. *Temperature Dependence of the Photoluminescence Spectrum*	204
IV. STUDY OF PROCESSING BY PHOTOLUMINESCENCE SPECTROSCOPY	205
1. *Purification and Stoichiometry*	205
2. *Etching and Vacuum Exposure*	207
3. *Contacts*	210
4. *Detector Performance*	214
V. CONCLUSIONS	216
References	216

I. Introduction

Mercuric iodide (HgI_2) in its red tetragonal form has many properties that make it well suited for fabricating x-ray and gamma ray detectors that can be operated without cryogenic cooling. These properties include a high bulk resistivity ($\sim 10^{13}$ Ω-cm), which ensures a low dark current during detector operation; large atomic masses of the constituent components of HgI_2 (i.e., $Z = 80$ and 53 for Hg and I, respectively), which allows for a high stopping power for x-rays and gamma rays; and high photosensitivity so that the number of electron–hole pairs generated in the crystal is proportional to the energy of the incident photon. Although the potential for manufacturing high resolution spectrometers has been clearly demonstrated, problems continue to be associated with incomplete charge collection, in which case the amount of charge collected for each photon is no longer uniquely proportional to its energy, and with device polarization, in which case the channel numbers corresponding to particular photopeaks or count rate are found to change with time. These problems with charge transport have motivated considerable research in the intrinsic and extrinsic optical properties of HgI_2 crystals, particularly those properties that appear to be related to detector quality.

Optical spectroscopy is a powerful technique to characterize the intrinsic and extrinsic energy levels in many semiconducting crystals. It has been shown that many of the detrimental carrier traps in mercuric iodide crystals and detectors can be probed using optical techniques. However, one difficulty frequently encountered in interpreting the optical measurements is the lack of knowledge of the electronic band structure of the undoped material and the energy levels corresponding to particular contaminants. The next part of this chapter focuses on calculations of the energy bands of mercuric iodide, density of states, momentum matrix elements, and anisotropic dielectric functions. Comparisons with measured values of the bandgap, effective masses, and absorption are made. The remaining portion of the chapter emphasizes numerous experimental reports on the optical properties, such as absorption, reflection, phonon spectrum, and defect states in as-grown and doped mercuric iodide crystals that can be studied by optical spectroscopy.

II. Band Structure

Angular resolved photoemission can be used to probe the electronic levels of HgI_2. By studying the energy of the photoelectrons, it is possible to obtain the energy of the initial (occupied) and final states, the wave vector **k**, and the spin. However, the relatively high vapor pressures of Hg and I complicate use of a photoemission technique and require that the material be cooled to low temperatures to eliminate loss of material from the surface via sublimation. Since photoemission measurements have not yet been used to map the dispersion of the energy bands of HgI_2, the current state of knowledge of the band structure relies primarily on theoretical work, and these investigations are reviewed in this section.

The space group for mercuric iodide is D_{4h}^{15}. The primitive tetragonal unit cell consists of two Hg atoms and four I atoms, as shown in Fig. 1. The nearest neighbor Hg-I spacing is about 2.78 Å. The crystal has inversion symmetry about the

5. OPTICAL PROPERTIES OF RED MERCURIC IODIDE 171

FIG. 1. The primitive tetragonal unit cell for HgI_2. (Reprinted with permission from D. E. Turner and B. N. Harmon, *Phys. Rev. B* **40**, 10516. ©1989 The American Physical Society.)

midpoint between the two Hg atoms. If one chooses the midpoint between the two Hg atoms as the origin of the coordinate system, the atomic positions of the two Hg atoms are $(-a/4, -a/4, -c/4)$ and $(a/4, a/4, c/4)$ and those of the four I atoms are $(-a/4, a/4, -0.111c)$, $(a/4, -a/4, 0.111c)$, $(-a/4, a/4, 0.389c)$, and $(a/4, -a/4, -0.389c)$. The values of a and c are 4.37 Å and 12.44 Å, respectively. The tetrahedral coordination of the mercury atoms in the unit cell is more clearly shown in Fig. 2.

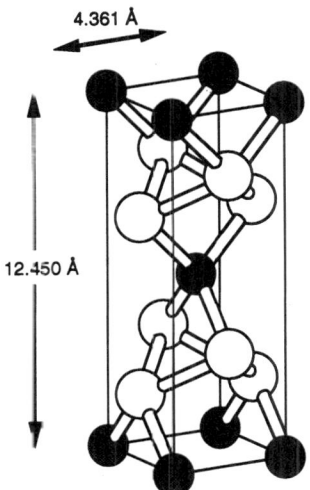

FIG. 2. The primitive HgI_2 unit cell drawn to emphasize the tetrahedral coordination of the mercury atoms. (Reprinted with permission from D. E. Turner and B. N. Harmon, *Phys. Rev. B* **40**, 10516. ©1989 The American Physical Society.)

Yee, Sherohman, and Armantrout, (1976) reported the first empirical pseudopotential calculation of mercuric iodide. In their calculation there was no self-consistency to allow for charge transfer, and no relativistic effects were included. Yee *et al.* (1976) predicted that the bottom of the conduction band did not occur at the same point in k-space as the top of the valence band. The fundamental bandgap was calculated to be indirect with the smallest energy gap occurring in the Γ–M direction of [110]. The prediction of an indirect fundamental bandgap was noted in several experimental reports, and it was often used to interpret optical measurements. Several years later Turner and Harmon (1989) pointed out that the crystal structure used in the work of Yee *et al.* (1976) was incorrect and the calculated results bear little relation with reality.

The report by Turner and Harmon (1989) was the first self-consistent calculation of the electronic band structure of HgI_2. A local density approximation was used, and both spin–orbit interaction and relativistic effects were included. Significant differences in the electronic bands were found between Yee *et al.* (1976) and Turner and Harmon (1989); for example, a direct rather than an indirect bandgap was found. Turner and Harmon also predicted that covalent bonding was dominant in HgI_2 crystals, and an ionicity of less than $Hg^{+0.1}$ was assigned. Figure 3 shows the calculated bands near the energy gap along high symmetry lines. The direct bandgap was calculated to have a value of 0.52 eV and found to be located at the zone center. The calculated value of 0.52 eV is much less than the experimental value of 2.37 eV (Novikov and Pimonenko, 1971), and the predicted bare electron and hole effective masses were also too small. Most of the discrepancy in the bandgap and effective masses is due to self-energy corrections for the excited states of the material, and the overall band structure calculated in Turner and Harmon (1989) appears qualitatively correct.

A few years after the calculations of Turner and Harmon (1989), an empirical nonlocal pseudopotential calculation of mercuric iodide in its red tetragonal form was reported by Chang and James (1992). Spin–orbit interaction and relativistic effects were also included, and the correct unit cell was used. Values for the electron and hole effective masses, optical matrix elements, absorption spectra, and complex dielectric function were reported. The local and nonlocal pseudopotentials were adjusted to fit the experimental value of the bandgap and qualitatively agree with the overall band structure obtained by the first principles calculation of Turner and Harmon (1989). Approximately 620 planewaves with energies less than 8 Rydberg were included in the diagonalization procedure. Figure 4 shows the calculated band structure of HgI_2. The calculated fundamental bandgap is direct at the zone center with a value of 2.37 eV, and the splitting between the first two valence bands (i.e., the heavy and light hole bands) is about 0.2 eV, both of which agree with the measured values at 4.2 K (Novikov and Pimonenko, 1971; Kanzaki and Imai, 1972).

Figure 5 shows a magnified view of the band structure of Chang and James (1992) near the zone center. The left side shows the electronic structure with wave vector **k** along the [110] direction, and the right-hand side shows the bands along

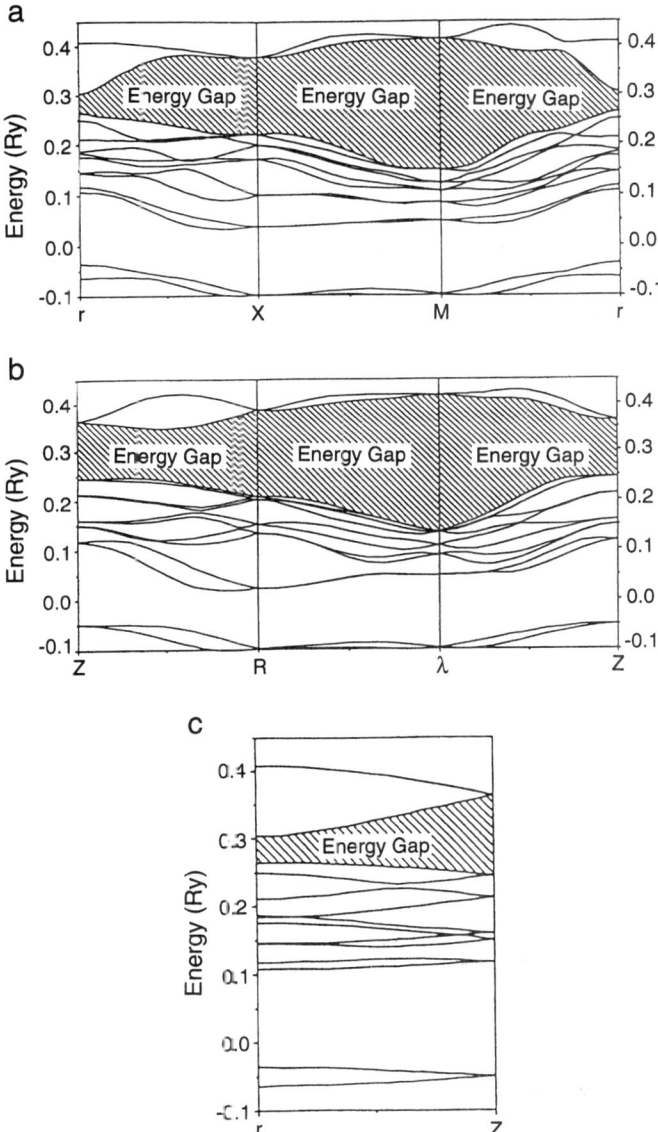

FIG. 3. Calculated bands of mercuric iodide near the energy gap along high symmetry lines. (Reprinted with permission from D. E. Turner and B. N. Harmon, *Phys. Rev. B* **40**, 10516. ©1989 The American Physical Society.)

the [001] direction. The unit of k is in $2\pi/a$. The conduction band is found to be nearly isotropic and the valence band to be much more anisotropic. The effective masses and reduced effective masses are shown in Table I, along with the mea-

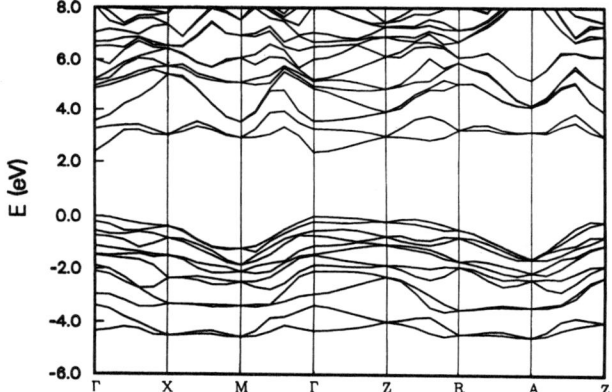

FIG. 4. Calculated band structure of mercuric iodide. (Reprinted with permission from Y. C. Chang and R. B. James, *Phys. Rev. B* **46**, 15040. ©1992 The American Physical Society.)

sured values. Table I also shows the polar masses (m^e and m^h), which were calculated by Chang and James using an intermediate coupling polaron theory.

Figure 6 shows the squared optical matrix elements P^2 for transitions between

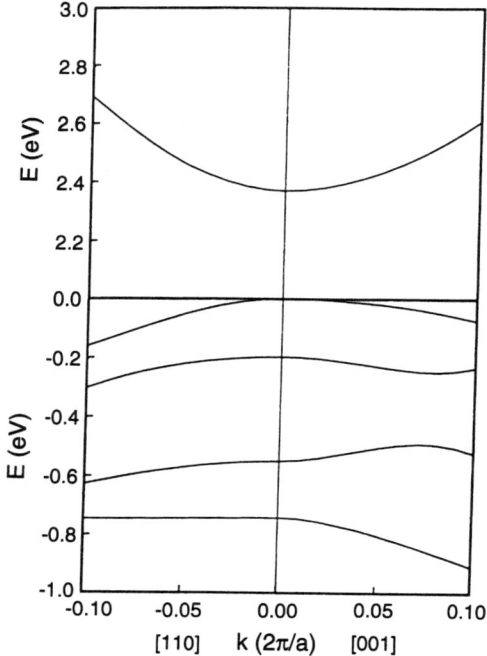

FIG. 5. Magnified view of the band structure of Fig. 4. (Reprinted with permission from Y. C. Chang and R. B. James, *Phys. Rev. B* **46**, 15040. ©1992 The American Physical Society.)

TABLE I

EFFECTIVE MASSES FOR MOTION PERPENDICULAR TO ($m_\perp$) AND PARALLEL TO THE C-AXIS ($m_\parallel$). THE SUBSCRIPTS AND SUPERSCRIPTS e AND h DENOTE ELECTRON AND HOLE, RESPECTIVELY. THE SUBSCRIPTS APPLY TO THE BARE EFFECTIVE MASSES AND THE SUPERSCRIPTS APPLY TO THE POLARON EFFECTIVE MASSES

	$m_{e,\perp}$	$m_{e,\parallel}$	$m^e_\perp$	$m^e_\parallel$	$m_{h,\perp}$	$m_{h,\parallel}$	$m^h_\perp$	$m^h_\parallel$	$\mu_\perp$	$\mu_\parallel$
Theory	0.22[a]	0.30[a]	0.29[a]	0.37[a]	0.59[a]	1.02[a]	0.89[a]	1.43[a]	0.22[a]	0.29[a]
Experiment	0.29[b]	0.25[b]	0.37[b]	0.31[b]	0.56[b]	1.72[b]	1.03[b]	2.06[b]	0.24[c]	0.31[c]

(Reprinted with permission from Y. C. Chang and R. B. James, *Phys. Rev. B* **46**, 15040. © 1992 The American Physical Society.)

[a] Chang and James (1992)
[b] Schluter and Schluter (1974)
[c] Goto and Kasuya (1981)

the valence and conduction bands near the zone center (Chang and James, 1992). The curves shown in the figure are for transitions from the topmost five valence bands to the lowest energy conduction band. Here, P^2 is defined as

$$P^2_{ij} = 2/m \, |\langle k, i | e \cdot p | k, j \rangle|^2$$

The labels $v1$, $v2$, $v3$, $v4$, and $v5$ denote transitions between the first to fifth valence bands, respectively, and the lowest conduction band. The solid curves are for the in-plane polarization (i.e., **e** perpendicular to the **z** direction) and the dashed curves are for polarization along the c-axis (i.e., **e** parallel to **z**). For optical

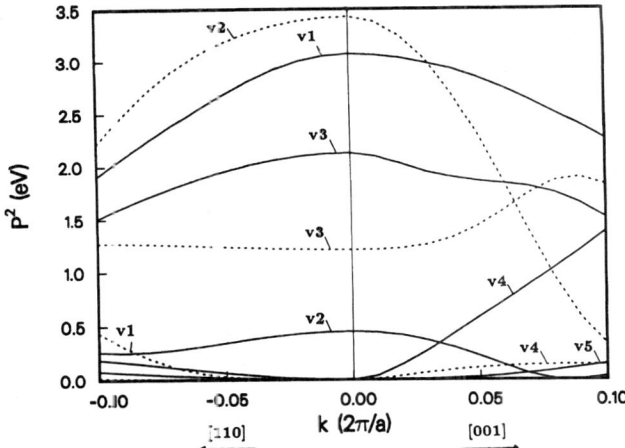

FIG. 6. Calculated squared optical matrix elements of P^2 for transitions between the valence and conduction bands near the zone center. (Reprinted with permission from Y. C. Chang and R. B. James, *Phys. Rev. B* **46**, 15040. © 1992 The American Physical Society.)

absorption at wavelengths at the fundamental band edge, only the $v1$ transitions are resonant. The matrix elements for linearly polarized light perpendicular to the z direction are large at the zone center, whereas for z polarized light the matrix elements vanish at the zone center and are very weak for all values of **k**. These results for P^2 are consistent with the character of the heavy hole band, since the heavy hole state transforms like $(x + iy)\uparrow$. Here, the symbol $\uparrow$ denotes a spin up. For the $v2$ matrix elements, light polarized along the z direction is strongly absorbed, and the matrix elements are about six times larger at the zone center than the matrix elements for polarization perpendicular to **z**. This is consistent with the character of the light hole band, since the light hole state is a linear combination of states that transform like $(x + iy)\downarrow$ and $z\uparrow$ with the majority component having $z\uparrow$. Given the dependence of P^2 on the direction of light polarization, it is clear that the depth of penetration of light for photon energies near the fundamental bandgap depends strongly on the orientation of the crystal with respect to the direction of propagation of the linearly polarized light.

Using calculated values for the electronic band structure and optical matrix elements, one can model the absorption coefficient as a function of the incident photon energy. Chang and James (1992) used an effective mass approximation to obtain values for the real and imaginary parts of the complex dielectric constant of HgI_2 for photon energies near the fundamental bandgap. The calculation also included effects of excitons on the absorption spectrum. The solid curve in Fig. 7 shows the values for the absorption coefficient for unpolarized light, and the dotted curve shows the absorption coefficient for incident light polarized parallel to the z direction. Fig. 7 also shows the experimental absorption coefficient for thin HgI_2 films at temperatures of 4.2 and 79 K (Kanzaki and Imai, 1972). Since the films were polycrystalline, comparisons should be made between the experimental data at 4.2 K and the calculated results for the case of unpolarized light. The calculations by Chang and James (1992) predict that the heavy hole exciton absorption is significant only for light polarized perpendicular to the c-axis, and the light hole exciton absorption is nonzero for both polarizations with the absorption for light polarized parallel to **z** much stronger than for light polarized perpendicular to **z**. These predictions are also in agreement with reflectivity measurements of Kanzaki and Imai (1972).

The calculations by Turner and Harmon (1989) and Chang and James (1992) have significantly advanced the level of knowledge of the electronic bands and absorption spectrum of HgI_2. Using measured values for the phonon dispersion curves, it is now possible to model the electron–phonon and hole–phonon matrix elements and determine the scattering rates for carriers as a function of their wave vector **k**. By integrating the carrier–phonon scattering rates over the distribution of free carriers, one can in principle determine the electron and hole mobilities and their dependences on the lattice temperature. Although these calculations have not been reported to date, the task is a tractable one with modern supercomputers. The values for mobilities due to intrinsic scattering mechanisms (i.e., carrier–phonon) can then be used to determine the fundamental limits for the carrier mo-

5. OPTICAL PROPERTIES OF RED MERCURIC IODIDE

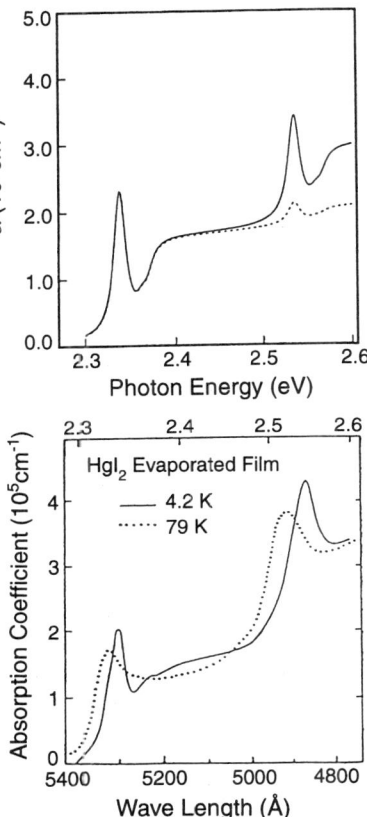

FIG. 7. Absorption spectra of HgI_2 near the fundamental absorption edge. Top, theoretical prediction; the solid curve is for unpolarized light and the dotted curve is for light polarized perpendicular to the c-axis. (Reprinted with permission from Y. C. Chang and R. B. James, *Phys. Rev. B* **46**, 15040. ©1992 The American Physical Society) Bottom, experimental absorption coefficient of thin mercuric iodide films at 4.2 and 79 K, both curves are for unpolarized light (Kanzaki and Imai, 1972).

bilities and ideally explain the wide variation in the reported values of the hole mobility of HgI_2 at 300 K.

III. Experimental Techniques and Measured Values for Optical Constants

1. MEASUREMENT OF BANDGAP AND SHIFT WITH TEMPERATURE

The energy of the bandgap and its temperature dependence is vital to understanding the performance of the radiation detectors fabricated from mercuric iodide. When HgI_2 photocells and x-ray and gamma ray spectrometers are used over a range of temperatures, the performance of the devices will shift slightly due to changes in the optical and electrical properties of HgI_2. In some cases this change in the amount of photogenerated charge makes it difficult to spectrally resolve elements that have x-rays of similar energies, and temperature control of the detectors is typically required for stable signals. The temperature controllers reduce slightly the portability and compactness of the instruments, which may limit their usefulness in certain applications. Independent of the demands on temperature control for each application, it is clear that the understanding of the temperature dependence of detectors requires knowledge of the change in the bandgap, E_G. Measurements of $E_G(T)$ have been reported by a number of investigators using a variety of techniques. In one of the earliest studies, Bube (1957) used reflection measurements on polycrystalline samples to obtain a value of 2.14 eV for E_G at room temperature. Burger and Nason (1992) used excitonic reflection to obtain a direct bandgap of 2.292 eV at 300 K, which is considerably larger than the bandgap reported by Bube (1957). Since the measurements of Burger and Nason do not rule out the existence of an indirect bandgap at lower energy, it is possible that the measurements of Bube (1957) and Burger and Nason (1992) are not in conflict. At 77 K Chester and Coleman (1971) measured a value for E_G of 2.331 eV, and Bube (1957) obtained a value of 2.334 eV. At 4.2 K Goto and Nishina (1978) obtained a value of 2.3707 eV by emission, and Novikov and Pimonenko (1971) measured 2.369 eV by absorption. Anedda *et al.* (1977) reported 2.397 eV using wavelength modulated reflection at 2 K. Lopez-Cruz (1989) performed an analysis based on photoconductivity measurements and concluded that at 0 K, HgI_2 has a direct bandgap at 2.27 eV and an indirect gap of 2.203 eV. The measurements by Lopez-Cruz (1989) and Burger and co-workers (1992) suggest that the indirect and direct bandgaps of mercuric iodide are close together. If this is indeed the case, it might account for the photoluminescence (Novikov and Pimonenko, 1971) data, absorption (Kanzaki and Imai, 1972) measurements, and band structure calculations (Turner and Harmon, 1989; Chang and James, 1992) that predict a direct gap material, and the photoconductivity (Lopez-Cruz, 1989) and electroabsorption measurements (Chester and Coleman, 1971) that suggest an indirect gap material.

The temperature coefficient of the bandgap has also been reported by several investigators. Bube (1957) reported a coefficient of -14×10^{-4} eV/K in the temperature range of 330 to 400 K, Burger and Nason (1992) observed a value of -3.9×10^{-4} eV/K from 285 to 400 K, and Harbeke and co-workers (1974) reported -6.5×10^{-4} eV/K in the same temperature range. At low temperatures, Novikov and Pimonenko (1972) quoted a value of -2.6×10^{-4} eV/K from 20 to 77 K, and Merz *et al.* (1983) measured -1.13×10^{-4} eV/K from 32 to 45 K. Nason and Burger (1991) used the location of peaks in the reflectance spectrum to infer a value of the surface temperature during the growth of HgI_2 crystals.

Significant differences in the values for E_G and its temperature dependences are apparent. Some of the discrepancy may be attributed to the different techniques used to measure E_G and differing role of excitonic effects in the experiments. Further work needs to be done to more fully understand the details of the band

TABLE II

REFRACTIVE INDICES OF HgI_2 FOR POLARIZATION PERPENDICULAR TO C

Wavelength	n	R
Na D-line 589 nm	2.71 ± 0.04	21.4%
He–Ne laser 632 nm	2.62 ± 0.01	20.0%

(Reprinted with permission from Kanzaki and Imai, 1972 and the Physical Society of Japan.)

structure of HgI_2 and role of crystalline perfection and purity on the measured results for $E_G(T)$.

a. Excitation near and above the Fundamental Band Edge

The reflectance spectrum of single crystal and polycrystalline mercuric iodide over the region of 2 to 6 eV was studied by Kanzaki and Imai (1972), and several peaks were detected. Ellipsometry measurements have also been performed by Yao, Johs, and James, (1993), and values for the optical properties of HgI_2 have been reported.

i. The Ordinary Spectrum. Refractive indices were measured at room temperature using incident photon energies of 2.10 and 1.959 eV from Na and He–Ne lasers, respectively. Table II shows the measured values for the refractive indices of single crystal HgI_2 for the case of polarization perpendicular to the c-axis.

Figure 8 shows the reflectivity spectrum of HgI_2 at 4.2 K for photon energies

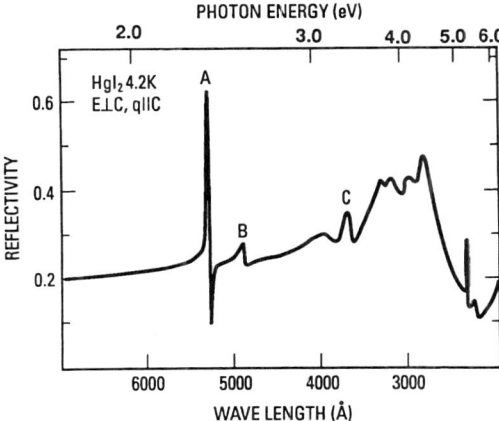

FIG. 8. Reflectance measurements at 4.2 K. (Reprinted with permission from Kanzaki and Imai, 1972 and the Physical Society of Japan.)

TABLE III

The Sharp Reflectivity Peaks near the Fundamental Edge of HgI_2.
A Dash Indicates Values Not Measured by Kanzaki and Imai (1972)

	$q \parallel c$	$E \perp c$	$q \perp c$	$E \perp c$	$q \perp c$	$E \parallel c$
	4.2 K	79 K	4.2 K	79 K	4.2 K	79 K
A	2.339	2.320	2.338	2.320	—	—
B	2.538	2.505	2.538	2.505	2.538	2.505
C	3.35	3.31	—	—	—	3.32

between 2 and 6 eV. Here the polarization of the light is perpendicular to the c-axis and the photon wave vector **q** parallel to c. The peaks at 2.339 (5301 Å) are sharp and large. For photon energies less than 2.339 eV, the material is transparent, indicating that the peak at 2.339 eV is due to formation of an exciton. The peaks of the reflectance spectrum are listed in Tables III and IV.

Figure 9 shows the reflectance spectrum at 4.2 K in the region near the bandgap. The circles show the measured reflectivities in a configuration with the light polarization perpendicular to the c-axis and the photon wave vector parallel to c, and the dotted line shows the measured results for light polarized perpendicular to the c-axis and the photon wave vector perpendicular to c. As shown in the figure, the observed values of reflectivities in both configurations agree with each

TABLE IV

The Reflectivity Peaks of HgI_2
for Polarization Perpendicular to the c-Axis
and Photon Wave Vector Parallel to c

$T = 4.2$ K			$T = 79$ K	
eV	Å	Remarks	eV	Å
2.339	5301	Very sharp A	2.320	5343
2.368	5236	Very weak		
2.538	4885	Sharp B	2.505	4955
3.10	4000	Broad	3.06	4050
3.351	3700	Sharp C	3.306	3750
3.73	3325		3.71	3340
3.84	3230		3.83	3240
4.13	3000			
4.38	2830		4.35	2850
5.29	2345	Very sharp	5.25	2360
5.38	2305			
5.46	2270	Broad	5.44	2280

(Reprinted with permission from Kanzaki and Imai, 1972 and the Physical Society of Japan.)

5. OPTICAL PROPERTIES OF RED MERCURIC IODIDE

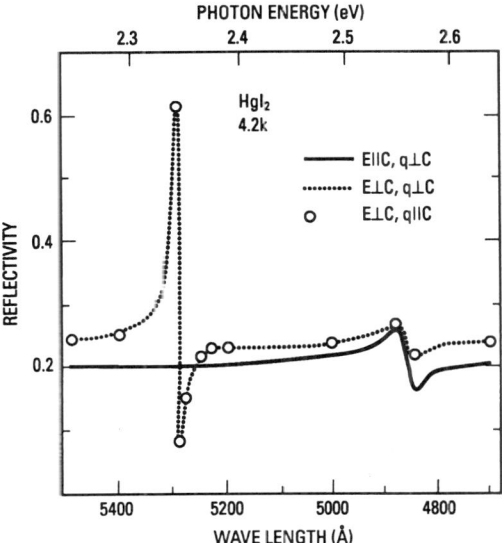

FIG. 9. Reflectivity spectra of HgI_2 at 4.2 K near the fundamental absorption edge: $E \perp c$, $q \perp c$ (dotted line); $E \perp c$, $q \| c$ (circles); $E \| c$, $q \perp c$ (solid line). (Reprinted with permission from Kanzaki and Imai, 1972 and the Physical Society of Japan.)

other. The configurations with $E \perp c$ correspond to the ordinary spectrum of HgI_2. Figure 10 shows the details of the reflectivity spectrum at 4.2 K for $E \perp c$ and $q \| c$ for photon energies in the 5 to 6 eV range.

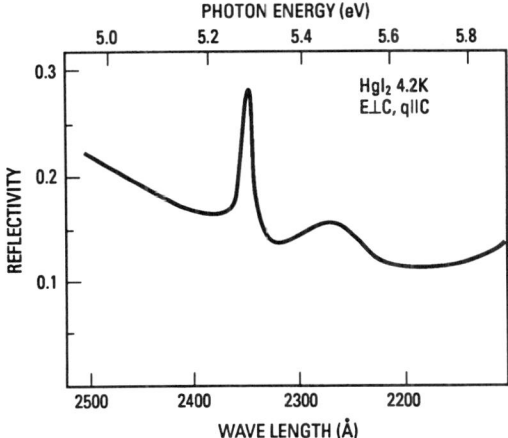

FIG. 10. Reflectivity spectrum of H_gI_2 at 4.2 K at the 5 ~ 6 eV region, $E \perp c$, $q \| c$. (Reprinted with permission from Kanzaki and Imai, 1972 and the Physical Society of Japan.)

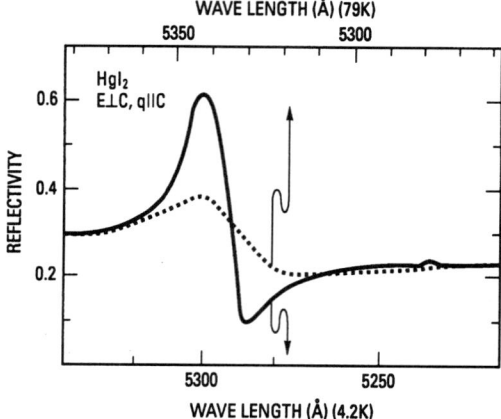

FIG. 11. Reflectivity spectra of HgI$_2$ at 4.2 K (solid line) and at 79 K (dotted line) near the fundamental edge in enlarged scale, $\mathbf{E}\perp\mathbf{c}$, $\mathbf{q}\|\mathbf{c}$. (Reprinted with permission from Kanzaki and Imai, 1972 and the Physical Society of Japan.)

Figure 11 shows reflectance spectra on an enlarged scale for wavelengths close to the band edge and temperatures of 4.2 and 79 K. The configuration used in this experiment is $\mathbf{E}\perp\mathbf{c}$ and $\mathbf{q}\|\mathbf{c}$. As the temperature is increased, the spectrum is shifted to longer wavelengths and becomes less sharp, which is primarily due to the reduction of the bandgap at higher temperatures. Figure 12 shows the reflectance of HgI$_2$ at 100 K for $\mathbf{E}\perp\mathbf{c}$ and for photon energies up to 10 eV, as measured by Anedda and coworkers (1977).

ii. The Extraordinary Spectrum. The extraordinary reflectance spectrum (i.e, $\mathbf{E}\|\mathbf{c}$ and $\mathbf{q}\perp\mathbf{c}$) was also measured by Kanzaki and Imai (1972). The results

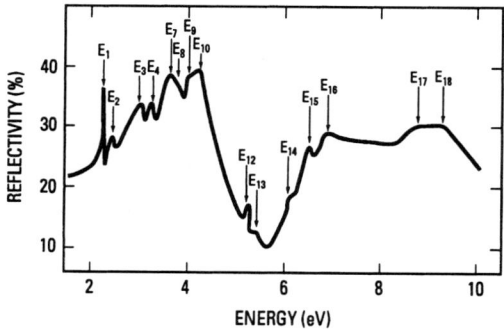

FIG. 12. Reflectivity spectrum of red HgI$_2$ at 100 K ($\mathbf{E}\perp\mathbf{c}$ and $\mathbf{k}\|\mathbf{c}$). (Reprinted from Anedda *et al.*, *Solid State. Commun.* **39**, 1121 ©1981 with kind permission from Elsevier Science Ltd., The Boulevard, Langford Lane, Kidlington, OX5 1GB.)

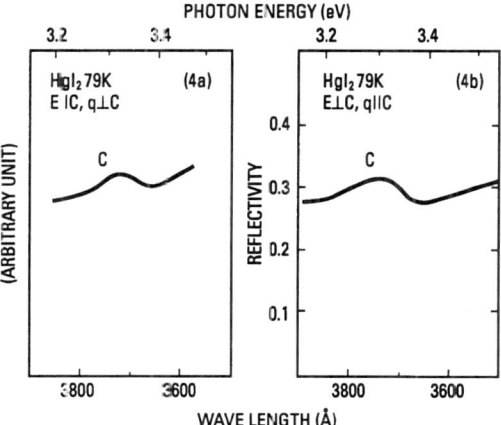

FIG. 13. Reflectivity spectra of HgI_2 at 79 K near 3.3 eV: $\mathbf{E}\perp\mathbf{c}$, $\mathbf{q}\perp\mathbf{c}$ (a); $\mathbf{E}\perp\mathbf{c}$, $\mathbf{q}\|\mathbf{c}$ (b). (Reprinted with permission from Kanzaki and Imai, 1972 and the Physical Society of Japan.)

are shown in Figure 9 by the solid curve. The extraordinary spectrum of HgI_2 has no peak near 5300 Å, and the first peak occurs at about 4885 Å (2.538 eV), which coincides with the second peak in the ordinary spectrum. The absence of the peak near 5300 Å in the extraordinary reflectivity spectrum is explained by the near zero optical matrix elements calculated by Chang and James (1992) (see Fig. 6).

Figure 13 shows the reflectance spectrum at 79 K for photon energies near 3.3 eV (Kanzaki and Imai 1972). A peak at about 3730 Å (3.32 eV) is observed in both the ordinary and extraordinary configurations. This peak corresponds to the peak at 3700 Å (3.351 eV) observed at 4.2 K (see Figure 12).

2. ABSORPTION

For photon energies exceeding the bandgap of HgI_2, the absorption coefficient is quite large ($>10^5$ cm^{-1}), and transmission measurements must be performed on thin films. It is difficult to polish and etch HgI_2 to submicron thicknesses, but thin films can be easily prepared by evaporating mercuric iodide in vacuum and depositing the evaporated material onto a cooled substrate. Kanzaki and Imai (1972) prepared thin polycrystalline films by evaporation in vacuum onto glass plates. Optical microscopy revealed that polycrystals of about 10 μm size were formed, and the c-axes of the domains were typically random in direction. The polycrystalline nature of the films requires that both the ordinary and extraordinary absorption occur in the film. The results of Kanzaki and Imai (1972) are shown in Fig. 7, along with the theoretical predictions for the absorption coefficient. The features of the transmittance measurements are consistent with reflectivity measurements (Kanzaki and Imai, 1972) and theory (Chang and James, 1992; Chang, Sim, and James, 1993).

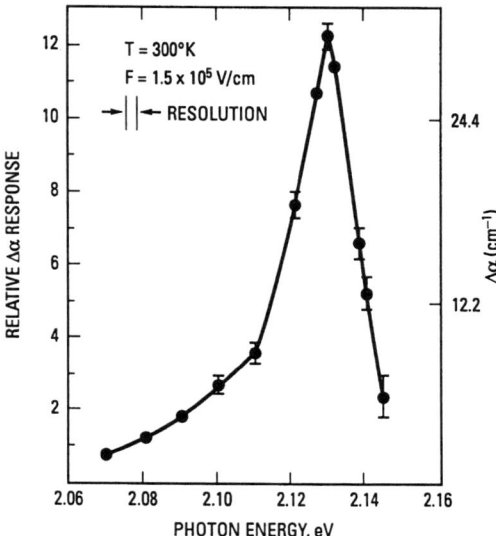

FIG. 14. Electroabsorption signal in HgI_2 at 300 K. (Reprinted from M. Chester and C. C. Coleman, *J. Phys. Chem. Solids* **32**, 223 ©1971 with kind permission from Elsevier Science Ltd., The Boulevard, Langford Lane, Kidlington, OX5 1GB.)

The electroabsorption spectrum of HgI_2 has been investigated by Chester and Coleman (1971) as a function of temperature and electric field. The physical quantity determined by the measurements of Chester and Coleman (1971) is the change in the optical absorption coefficient induced by an electric field. Figure 14 shows typical results at 300 K for a field of 1.5×10^5 V/cm. The electroabsorption peak occurs at a spectral energy of 2.127 eV, which is slightly less than the bandgap of 2.14 ± 0.005 eV at 300 K.

Electroabsorption (EA) data were taken by Chester and Coleman (1971) at several temperatures. Figure 15 shows the dependence of the electroabsorption peak on the applied field for photon energies of 2.127 eV (coinciding with peak at 300 K) and 2.313 eV (location of peak at 88 K). The form of the EA spectrum did not change with the applied field, and the magnitude of the peak varies approximately as the 4/3 power of the field. A 4/3 power of the EA signal with the electric field is expected for an indirect bandgap material (Chester and Coleman, 1971), whereas a 1/3 power is expected for an allowed direct bandgap excitation. However, the theory that predicts a 4/3 power dependence for indirect bandgap materials does not include the case when exciton absorption is involved. Moreover, the EA peak observed by Chester and Coleman (1971) always occurred approximately 0.010 eV below the bandgap. There exists an exciton at an energy of about 0.010 eV below the bandgap (Bao *et al.*, 1990b), which suggests strongly that the EA signal at each temperature is associated with exciton absorption.

5. OPTICAL PROPERTIES OF RED MERCURIC IODIDE

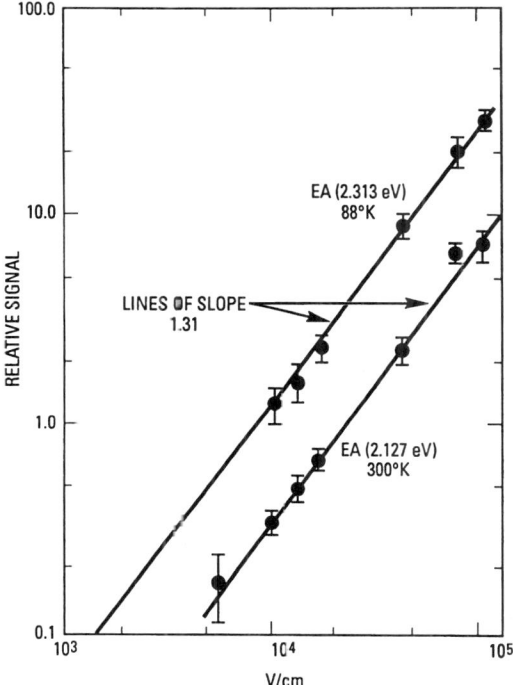

FIG. 15. Electroabsorption signal voltage dependence. (Reprinted from M. Chester and C. C. Coleman, *J. Phys. Chem. Solids* **32**, 223 ©1971 with kind permission from Elsevier Science Ltd., The Boulevard, Langford Lane, Kidlington, OX5 1GB).

3. ELLIPSOMETRY MEASUREMENTS OF OPTICAL CONSTANTS

The imaginary part of the dielectric constant was first measured by Anneda and co-workers (1977), and the results are shown in Fig. 16 for energies between 2 and 10 eV. More recent measurements of the optical properties of HgI_2 were measured by Yao and Johs (1993), Schieber et al. (1994), and Yao et al. (1994) using variable angle spectroscopic ellipsometry (VASE). Figure 17 shows values for the room temperature anisotropic dielectric functions, ϵ_1 and ϵ_2, as measured by VASE (Schieber et al., 1994; Yao et al., 1994) for electric fields perpendicular and parallel to the c direction. Figure 18 shows measured values for the absorption coefficient in the energy range of 2–5 eV at 293 K, and Fig. 19 shows measured values for the reflectance in the 1–5 eV range (Yao et al., 1994) at 293 K. Based on the VASE measurements, one sees that the depth of penetration of light and its reflectance are strongly dependent on the polarization of the radiation and the orientation of the crystal. Surface aging effects of HgI_2 crystals have also been studied by monitoring the changes in the dielectric constants after chemical etch-

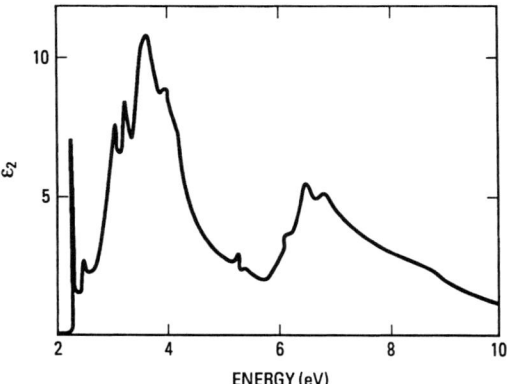

FIG. 16. Imaginary part of the dielectric constant versus energy. (Reprinted from Anedda *et al.*, *Solid State. Commun.* **39,** 1121 ©1981 with kind permission from Elsevier Science Ltd., The Boulevard, Langford Lane, Kidlington, OX5 1GB.)

ing and as a function of time (Schieber *et al.*, 1994; Yao and Johs, 1993). Some of the surface aging effects appear to be related to excess iodine contained in the crystals.

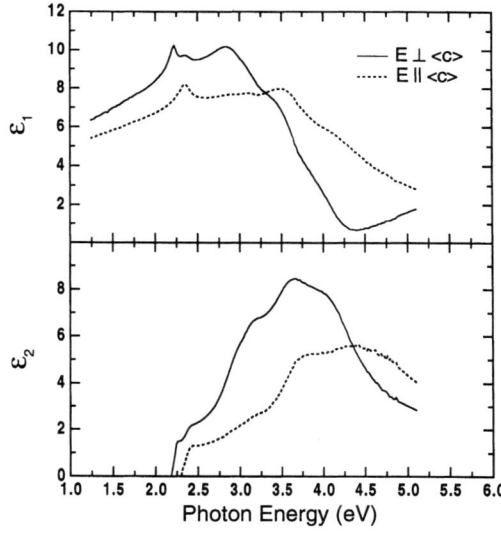

FIG. 17. Anisotropic optical constants of HgI_2 as measured by VASE (from Yao *et al.*, 1994).

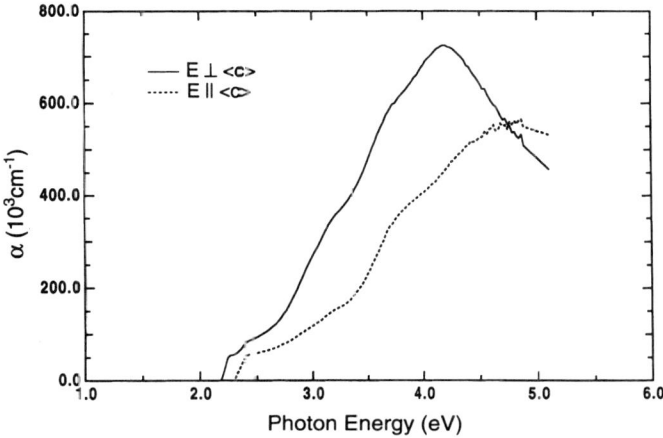

FIG. 18. Absorption of HgI$_2$ as measured by VASE (from Yao *et al.*, 1994).

4. Optical Properties near and below the Bandgap

Since HgI$_2$ is more optically transparent at photon energies near and below the bandgap, studies of the absorption coefficient at these wavelengths can be performed on thicker single crystals. The nature of the absorption still depends on the orientation of the light polarization with respect to the c axis of HgI$_2$. Figure 20

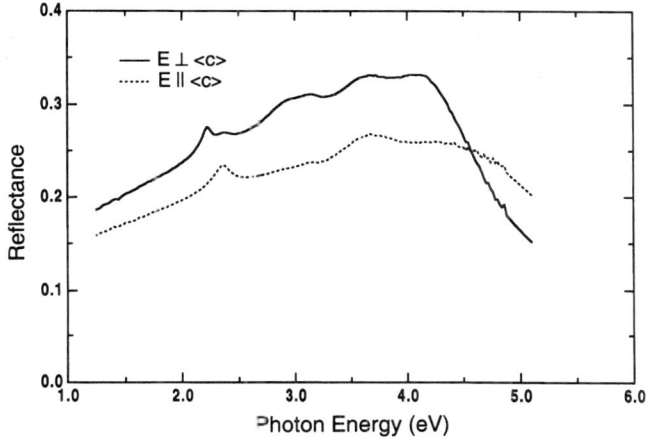

FIG. 19. Polarized reflectance of HgI$_2$ (from Yao *et al.*, 1994).

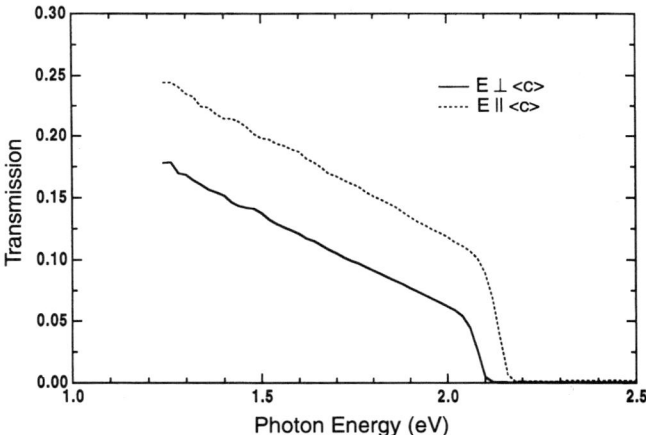

FIG. 20. Polarized transmission of HgI_2 (from Yao et al., 1994).

shows room temperature values for the polarized transmission of HgI_2 for photon energies in the 1.2 to 2.2 eV range (Yao et al., 1994).

Reflectivity in the ordinary configuration (i.e., $\mathbf{E}\perp\mathbf{c}$) at 4.2 K has been reported by Akopyan et al. (1975). At this temperature, the ordinary absorption ($\mathbf{E}\perp\mathbf{c}$) shows continuous absorption (Novikov and Pimonenko, 1971). However, in the extraordinary absorption spectrum (i.e., $\mathbf{E}\|\mathbf{c}$) at 4.2 K, the absorption consists of discrete resonant peaks (Novikov and Pimonenko, 1971) (see Fig. 21). The strongest peak (labeled as λ_0) occurred at 5297 Å and two weaker lines were measured at 5246 Å and 5239 Å. The continuous absorption at shorter wavelengths was attributed to interband transitions, and the bandgap was inferred to be 5230 Å (= 2.369 eV). On the long wavelength side of the large peak at 5297 Å, two narrow lines were observed at 5310 and 5321 Å, whose intensities varied from sample to sample. These same lines are also observed with varying intensities in the photoluminescence spectrum of HgI_2 (Novikov and Pimonenko, 1971) and have been described as excitonic in nature.

Akopyan et al. (1975) also observed these absorption peaks and assigned slightly different values for the peak positions and linewidths. They measured an absorption peak at 5309.1 Å and found the FWHM of the peak to be much less than that of the emission line at 5308 Å ($\mathbf{E}\perp\mathbf{c}$). To further investigate this line, the authors (Akopyan et al., 1975) measured its behavior under a uniaxial stress with $\mathbf{P}\perp\mathbf{c}$. The shift of the bandgap with pressure was reported earlier by Zahner and Drickamer (1959). As stress was applied, the absorption line at 5309 Å became more intense than the line at 5297 Å. Such behavior is characteristic of a forbidden exciton transition, with the increase in the absorption at 5309 Å being due to the mixing of forbidden and allowed transitions by the uniaxial stress. A similar stress dependence was observed for the A_F line in CdS crystals (Hopfield and Thomas,

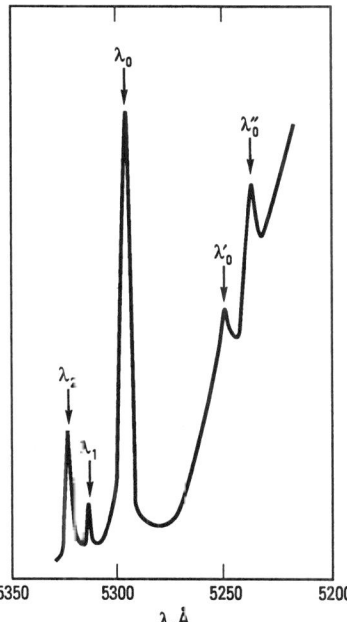

FIG. 21. An absorption spectrum of HgI$_2$ single crystal with its optic axis in the plane of the platelet. The spectrum was recorded at 4.2 K. The scale of the long wavelength part of the spectrum is magnified along the ordinance to show the structure. (Reprinted with permission from Novikov and Pimonenko, 1971 and the American Institute of Physics.)

1960, 1961; Thomas and Hopfield, 1959). The line in CdS was interpreted to result from transitions into an optically forbidden component of the ground state exciton level, which is split from the main allowed component by exchange interaction.

In a magnetic field ($H = 30$ kOe) in the orientation **H**∥**c**, the 5309 Å absorption line splits into a doublet, but it is not affected for **H**⊥**c**. The A_F line in CdS shows the same magnetic field dependence (Hopfield and Thomas, 1960), which gives further confidence in the assignment of the 5309 Å of HgI$_2$ to transitions into an optically forbidden component of the ground exciton state. The amount of exchange splitting depends on the ratio of the exciton Bohr radius to the lattice constant; and for exciton states of large radius (relative to the lattice constant), the exchange splitting is typically only a few tenths of a meV (Akopyan et al., 1975).

At higher temperatures, the large absorption peak at 5297 Å broadens, and it gradually shifts to lower energies because of bandgap narrowing. At 77 K this line is located near 5328 Å and its full-width at half-maximum has increased by greater than 4. At this temperature the continuous absorption attributed to band to band transitions began at 5260 Å (Novikov and Pimonenko, 1971). The two narrow absorption lines (located at 5310 and 5321 Å at 4.2 K) disappear at about 30 K.

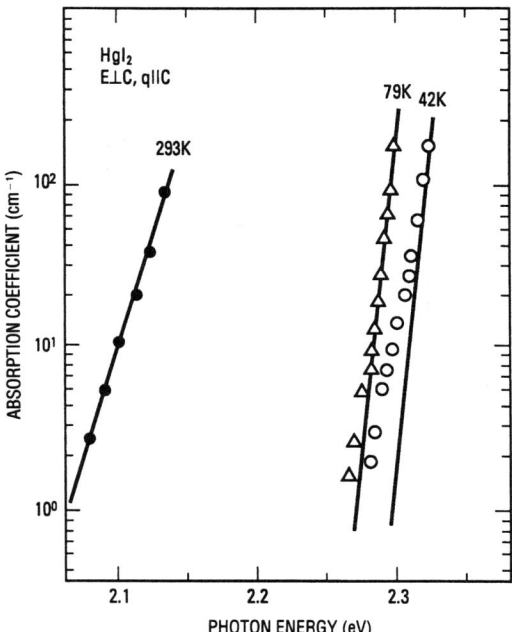

FIG. 22. Absorption spectra of HgI_2 single crystals in the tail part of the fundamental absorption edge. (Reprinted with permission from Kanzaki and Imai, 1972 and the Physical Society of Japan.)

Figure 22 shows data by Kanzaki and Imai (1972) on polycrystalline samples for the absorption coefficient on the low energy side of the fundamental bandgap at temperatures of 4.2, 79, and 293 K. Since the absorption coefficient has approximately an exponential dependence on the incident photon energy, the spectrum of HgI_2 is well described by Urbach rule (Urbach, 1953). At lower temperatures the deviations from Urbach rule are larger, which may be due to the excitonic effects (Novikov and Pimonenko, 1971). In addition, there may be increased contributions at lower temperatures from photoionization of carriers bound to impurities. As pointed out by Chester and Coleman, internal electric fields may be present in HgI_2, which can produce an internal Franz–Keldysh shift and lead to a larger tail in the absorption coefficient (Redfield, 1965).

The refractive index for ordinary absorption (i.e., $E \perp c$) for photon energies less than the bandgap is shown in Fig. 23. For even longer wavelengths, the material becomes much more transparent and thicker crystals are preferred for transmission measurements. Photoconductivity measurements have been performed on palladium contacted HgI_2 detectors, and an absorption peak was observed at about 9373 Å (Bao, 1991). This peak was attributed to photoionization of an electron or hole carrier bound to an impurity. Since mercuric iodide crystals do not naturally contain palladium (Soria *et al.*, 1994) and this peak was not seen on detectors with

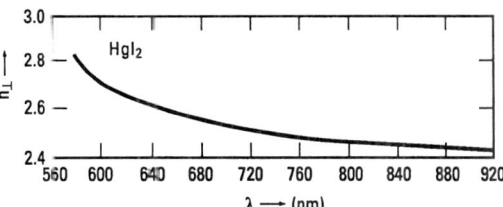

FIG. 23. The refractive index for ordinary absorption (i.e., $\mathbf{E} \perp \mathbf{c}$) for photon energies less than the bandgap. (Reproduced with permission from Seiskind, 1960.)

contacts other than Pd, it was believed due to Pd impurities. If this belief is indeed correct, it indicates that Pd either diffuses into HgI_2 during detector processing or aging or that Pd can migrate into the bulk due to the influence of an applied bias. Since Pd is the contact material of choice for most HgI_2 detector applications, the presence of diffusion or electrodrift phenomenon of Pd into HgI_2 is the subject of further investigation. Moreover, the movement of carrier traps not limited to Pd related defects in the material can lead to polarization effects, which are detrimental to device performance. Recent optical and electrical measurements have revealed that some elements (e.g., Cu and Ag) do indeed move under the application of an electric field (James et al., 1993; Van Scyoc et al., 1993).

For infrared wavelengths, most of the absorption is associated with impurities, although lattice absorption is significant for photon energies in the restralen region. For photon energies greater than the restralen region, two possible mechanisms for infrared absorption are the excitation of vibrational modes associated with contaminants and excitation of free carriers in the material. Several absorption bands associated with the presence of impurities have been observed in transmission infrared spectra experiments (James et al., 1989a) (see Fig. 24). The strength of the absorption in these bands varies between samples due to differences in the amounts of impurities contained in the crystals. Attempts by James et al. (1989a) to identify the impurities associated with the absorption have been partly successful. For example, some specimens displayed a narrow absorption peak at about 1600 cm^{-1}, which corresponds to the measured energy for absorption by the H-O-H bending mode of water (Nyquist and Kagel, 1971). The absorption measured at about 2850, 2915, and 2970 cm^{-1} is due to the presence of aliphatic hydrocarbons in the bulk material. A broad absorption band peaked at about 3240 cm^{-1} was also detected by James et al. (1989a) and attributed to presence of free water or a water of hydration. Since HgI_2 has not been shown previously to form a hydrate, the water is probably bonded to a contaminant. Both calcium and sodium iodate are possible choices because they have infrared absorption spectra that resemble the measured results of James et al. (1989a) and Ca and Na are present in the HgI_2 crystals in the 1–10 ppm range (Schieber, Roth, and Schnepple, 1983; Soria, Natarajan, and James, 1994). Another absorption band observed by James et al. (1989a) was peaked at about 3500 cm^-. Since

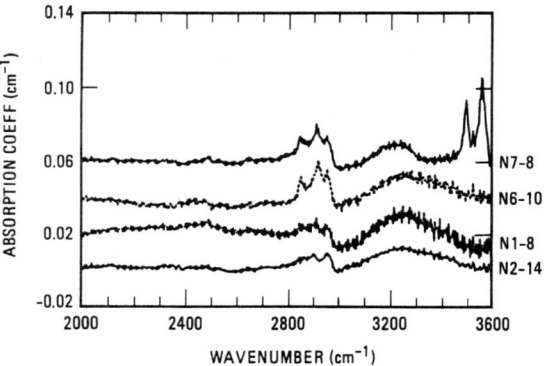

FIG. 24. FTIR spectra of four HgI_2 samples in the 2000–3600 cm^{-1} range. (Reprinted with permission from James et al., 1989a and Elsevier Science Publishers.)

strong absorption associated with the OH stretching vibrations of water and the hydroxyl group occur between 3200 and 3700 cm^{-1} (Nyquist and Kagel, 1971), it was concluded by James et al. that the mercuric iodide samples studied in their work contained a significant amount of both oxygen and hydrogen and that much of the oxygen and hydrogen was chemically bonded as H_2O and possibly OH. For excitation in the 600–900 cm^{-1} range, several weak narrow absorption bands were detected by James et al. (1989). The mostly likely candidates for these weak absorption bands included mercurous iodide, calcium iodate, sodium iodate, and compounds containing a carbonate anion (Schieber et al., 1983; Nyquist and Kagel, 1971).

The contribution from free carriers depends on the carrier density, and since undoped HgI_2 is an insulating material, negligible free-carrier absorption is expected. Furthermore, few impurities have been demonstrated to be electrically active in the material. However, it was recently shown that Cu and Ag dopants are electrically active in HgI_2 (Bao et al., 1990a; James et al., 1993). At present, no infrared transmission measurements have been reported on Cu- or Ag doped HgI_2, and the role of free-carrier absorption has not been determined. It is also possible that the large increase in the conductivity of Cu and Ag doped HgI_2 is due to ionic conduction instead of Cu or Ag acted as an electrically active substitutional impurity. Since many electron and ion probe diagnostic techniques demonstrate charging problems on insulating materials like HgI_2, further investigations on HgI_2 crystals that have been intentionally doped with Cu or Ag (to greatly lower the sample resistivity) are expected in the near future.

5. Phonon Structure of Red Mercuric Iodide

The vibrational frequencies of the transverse and longitudinal modes of HgI_2 have been obtained from the polarized far-infrared reflection spectrum (Prevot,

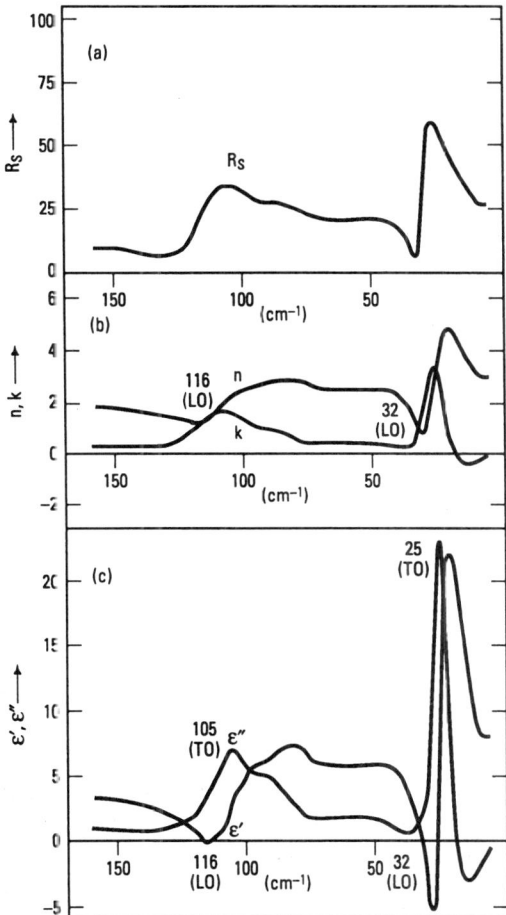

FIG. 25. Reflectivity and the optical and dielectric constants. (Reprinted from Y. Ogawa et al., Spectrochimica Acta **32A**, 49 ©1976 with kind permission from Elsevier Science Ltd., The Boulevard, Langford Lane, Kidlington, OX5 1GB.)

Schwab, and Dorner, 1978). Figure 25 shows the reflectivity and the optical and dielectric constants as measured by Mikawa, Jacobsen, and Brasch, (1966). The reflectivity R_s was measured for light polarized perpendicular to the c-axis. The incidence angle was fixed at 20° in the experiment. The refractive index n, the absorption coefficient κ and the real and imaginary parts of the complex dielectric constant, ϵ' and ϵ'', were obtained according to the model given by Roessler (1965, 1966) and Nakagawa (1971).

Raman measurements have also been conducted to measure the energies of the LO_1, LO_2, and LO_3 longitudinal optical phonon energies near the zone center (Nakashima, Mishima, and Mitsuishi, 1973; Adams and Hooper, 1970; Haisler

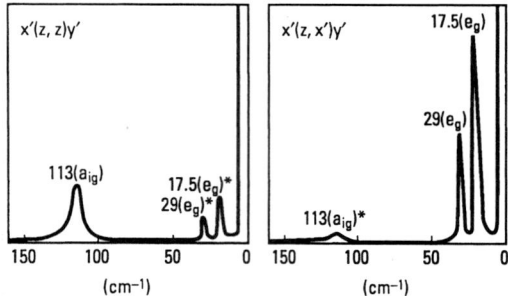

FIG. 26. The polarized Raman spectra. The experimental axes x' and y' are related to the crystal x- and y-axes by a 45° rotation about z(c). The appearance of the bands with asterisks is ascribed to imperfections in the crystal. (Reprinted from Y. Ogawa et al., Spectrochimica Acta **32A,** 49 ©1976 with kind permission from Elsevier Science Ltd., The Boulevard, Langford Lane, Kidlington, OX5 1GB.)

et al., 1984; Prevot and Biellmann, 1979). Strong Raman bands are observed by several investigators at 17.5, 29, and 113 cm^{-1} (see Fig. 26, Ogawa et al., 1976). Weaker bands were also reported by Melveger et al. (1968) at 46, 142 and 246 cm^{-1}. Further discussions on the assignments for the Raman bands are contained in Biellmann and Prevot (1980) and the references noted therein.

Phonons with a nonnegligible wave vector **q** can also participate in radiative transitions, such as excitation and photoluminescence. For indirect bandgap materials, phonon participation is typically required to conserve crystal momentum for photon energies near the bandgap. Radiative recombination can result in either the creation or destruction of a phonon, although phonon absorption can be usually ignored for measurements conducted at low temperatures. The energy of the photon emitted by a radiative transition that involves one or more phonons will be altered from the "no phonon" energy by the amount of energy corresponding to the phonon(s).

The lower section of Fig. 27 shows phonon dispersion curves of HgI_2 at room temperature (Prevot et al., 1978). The full (open) symbols show acoustic (optic) vibrations either in transverse (squares) or longitudinal (circles) geometries for polarizations within the layer plane. The triangles in the figure refer to modes with polarization parallel to the principal axis. The upper section of Fig. 27 shows calculations of the phonon modes using a rigid ion model (Sim, Chang, and James, 1994). Additional phonon dispersion curves are predicted by Sim et al. (1994) that were not measured by Prevot et al. (1978). The absence of these curves in the experimental measurements has been explained based on the crystal symmetry of HgI_2. Higher energy phonon dispersion curves, along with values for the sound velocities and frequencies of infrared active modes have been predicted, and they are in accord with available experimental data (Sim et al., 1994). Further discussions of the angular dispersion of the phonon frequencies and the infrared active modes can be found in the paper by Sim et al. (1994).

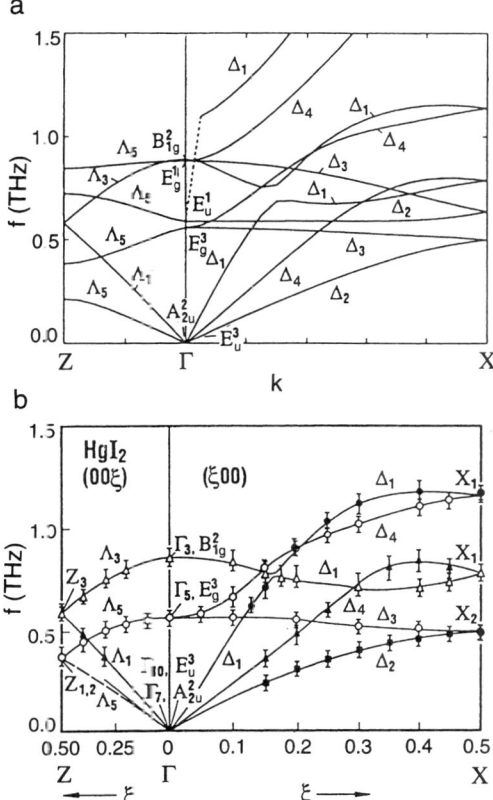

FIG. 27. Low energy phonon dispersion curves measured in red HgI$_2$ at room temperature. The full (open) symbols show acoustic (optic) vibrations either in transverse (squares) or longitudinal (circles) geometries for polarizations within the layer plane. The triangles in the figure refer to modes with polarization parallel to the principal axis. (From Prevot et al., reprinted with permission by VCH Publishers ©1978.)

6. Radiative Recombination of Nonequilibrium Electron–Hole Pairs

Excitation of mercuric iodide detectors by above band gap excitation leads to the creation of nonequilibrium electron–hole pairs. These photogenerated carriers move under the influence of the applied bias and are scattered by phonons, impurities, native defects, and dislocations. At zero bias and open circuit conditions, the electrons and holes eventually recombine to establish equilibrium. The recombination can be divided into three primary categories: (1) direct recombination in the bulk material, (2) recombination by way of traps, and (3) surface recombination. For direct recombination a free electron in the conduction band recombines with a hole state in the valence band in a single transition. Energy conservation considerations requires that an amount of energy be given up that corresponds to the energy difference between the conduction and valence band states. One mechanism for disposing of this excess energy is by emission of radiation. In some cases the emitted radiation will involve the creation or annihilation of one or more phonons.

Prior to recombination, a free electron and hole can first be bound together by their electrostatic attraction to form an exciton. All excitons are unstable against a recombination process in which the electron drops into the hole. This process often involves the emission of radiation at an energy slightly less than the bandgap of the material, because the formation of excitons leads to a lowering of energy by an amount corresponding to the binding energy of the exciton. Free excitons are electrically neutral and move through the crystal transporting their excitation energy. At low temperatures free excitons in HgI_2 are trapped at a rate that exceeds the free exciton lifetime. This trapping further lowers the excitation energy of the excitons by an amount equal to the binding energy of the free exciton to its trapping sites. These trapping sites can be impurities, stoichiometric deviations, and extended defects. By monitoring the energy of the emitted recombination luminescence, one can determine information on the energy levels of the electrons, holes, free excitons, and bound excitons involved in the recombination process. Free and bound exciton luminescence are often in the form of sharp lines that are clearly separated in energy from the continuous band to band radiation and from other excitonic lines. Since the information gained on bound–exciton recombination assists in constructing an energy level diagram of the dominant traps, photoluminescence is a particular useful technique in characterizing defects in HgI_2 crystals. The exciton lifetimes in HgI_2 can also be studied by measuring the rate of decay of the emitted photoluminescence with fast exciting pulses (Williams, Anderson, and Banet, 1991).

Another recombination mechanism involves an electron that combines with a hole in the valence band in a two step process, which may be separated by a relatively long amount of time. Here, an electron can become captured by a deep trap; and some time later, a hole is influenced by the potential of this occupied electron trap. The trap can also capture the hole leading to the eventual recombination of the two carriers. This type of process can be more probable than a single step one, because the probability of a radiative recombination process increases greatly as the radiated energy is decreased.

The third area for recombination mentioned previously also involves traps, but in this case, the traps are located in the near-surface region of the crystal or detector. In this case, electrons and holes generated in the bulk of the material diffuse to the surface. The recombination rate depends on the quality of the surface, and it is often much larger than the recombination rate in the bulk. In some cases surface recombination may dominate and thereby reduce the advantages associated with long carrier lifetimes in bulk material.

Recombination at surfaces often involves only phonons to conserve energy, in which case no radiation is emitted. This process has been shown to be important in mercuric iodide crystals and devices (Levi et al., 1983, 1985), although the precise mechanism for recombination (multiphonon or Auger) has not been conclusively determined. It has been shown that the techniques employed for treating HgI_2 surfaces have an important influence on defects at the surface (David et al., 1993), on leakage currents (Jayatirtha et al., 1993), and on the rate of surface recombination (Levi et al., 1983).

Photoluminescence and cathodoluminescence spectroscopy has been extensively used to study the radiative recombination of HgI_2 crystals and detectors (Novikov and Pimonenko, 1971, 1972; Merz et al., 1983; Akopyan et al., 1987; Bao et al., 1990a, b, c; James et al., 1989b; Petroff, Hu, and Milstein, 1989). Although the origin and nature of some of the emission lines have been identified, a complete description of all of the photoluminescence features is yet to emerge. Several reasons explain the difficulties in identifying each defect manifested in the photoluminescence spectrum. First, the absorption of electromagnetic waves in HgI_2 depends on the geometric relationship between the direction of the light polarization, the wave vector of the incident wave, and the c-axis of the crystal. Thus, the light penetrates to different depths in the material, depending on the geometrical configuration of the experiment. The reabsorption of the emitted photons depends on their wavelengths, so the relative intensities of the photoluminescence peaks will change as the depth of penetration of the exciting beam changes. Second, the crystals used in various studies were grown using different purification procedures and growth techniques, and the crystals contained different levels of impurities and deviations from stoichiometry. Third, handling practices have a significant effect on the dominant recombination centers, because HgI_2 is soft, highly reactive, sensitive to temperature changes, and has a high vapor pressure. It also develops a degraded surface layer for samples stored in air over a period of a few days, and this aged layer is different from a freshly etched HgI_2 surface. Fourth, HgI_2 is difficult to systematically dope with impurities, and efforts to intentionally dope a crystal during growth often lead to other changes in the sample, such as deviations from stoichiometry. Finally, the photoluminescence spectrum of HgI_2 at energies near the fundamental band edge consists of many sharp spectral lines that are spaced close together and often difficult to resolve. These lines vary in intensity due to differences in crystalline perfection and surface conditions. Confusion arises when comparing reported values for peak positions and relative magnitudes by different groups of investigators, especially when these groups also have different spectral resolutions for their experimental setups.

The remainder of this section focuses on measurements of the radiative recombination in HgI_2 crystals and the effects of the electrical contacts on the incorporation of new carrier traps. Bube (1957) reported the first low temperature photoluminescence spectrum of HgI_2, and three emission bands with peaks located at 5360, 5675, and 6200 Å were reported at 77 K. These bands were labeled 1, 2, and 3, respectively. By cooling the crystal to 4.2 K, Novikov and Pimonenko (1971) found that the band at 5360 Å observed by Bube actually consisted of many narrow lines. All of the lines reported in by Novikov and Pimonenko (1971) were assigned as free or bound exciton emissions and their phonon replicas. Merz et al. (1983) reported similar results for the photoluminescence spectrum, but with different assignment to a few lines. As more data became available and the quality of the crystals continued to improve, Wong et al. (1988a) pointed out that some of the emission lines assigned earlier as phonon replicas did not have the correct peak positions. Goto and Nishina (1978) was the first to observe several weak steplike emissions between 5230 and 5290 Å, which they designated phonon assisted indirect recombination of free electron–hole pairs. Akopyan et al. (1975) carefully studied many properties of the first three exciton lines, including their polarization and changes under the influence of magnetic field and uniaxial stress.

The highest resolution photoluminescence spectrum existing today on undoped HgI_2 was obtained by Bao et al. (1990b). Although their experiments were conducted with relatively high instrument resolution, the primary reason that more emission peaks were resolved by Bao et al. (1990b) is probably due to additional improvements in the crystalline purity and quality, thus allowing the dominant peaks to be much narrower in energy and the relatively weaker peaks nearby in energy to be more clearly resolved from the background. Figure 28 shows the 4.2 K photoluminescence for photon energies near the fundamental bandgap. Bao et al. (1990b) found that many lines previously reported as singlets were actually composed of more than one line, which explained some of the earlier inconsistencies in peak positions. A total of 26 photoluminescence lines were reported between 5290 and 5400 Å, and several lines were observed for the first time. Some of the peaks are not well resolved in Fig. 28, but were much more clearly resolved

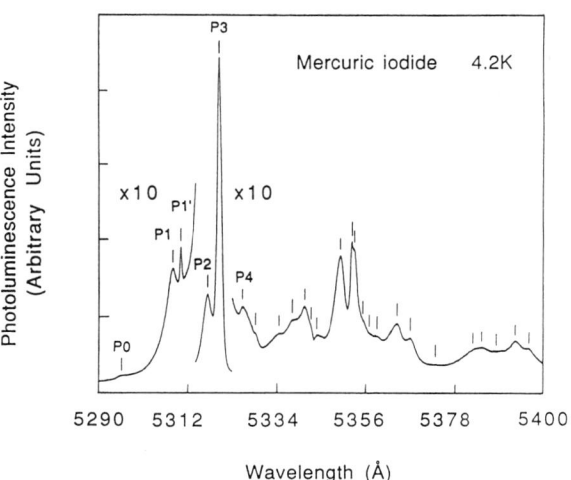

FIG. 28. A 4.2 K near-band edge photoluminescence spectrum of HgI_2 between 5290 and 5400 Å. Vertical bars indicate the positions of the emission lines. The spectral resolution of the spectrometer was 0.35 Å. (Reproduced with permission from Bao et al., 1990b and the American Institute of Physics.)

TABLE V

WAVELENGTHS AND PHOTON ENERGIES OF BAND EDGE PHOTOLUMINESCENCE PEAKS OF HgI_2 BETWEEN 5290 AND 5400 Å

	4.2 K[a]		4.2 K[b]		1.6 K[c]	
Wavelength (Å)	Energy (eV)	Notation	Energy (eV)	Notation	Energy (eV)	Notation
5296.9	2.3400	P0	2.340	λ_0	2.3402	1A*
5309.1	2.3347	P1	2.335	λ_1	2.3349	1A
5311.2	2.3337	P1'			2.3332	
5317.9	2.3308	P2	2.331	λ_x	2.3300	1B
5320.8	2.3296	P3	2.329	λ_2	2.3286	1C
5326.4	2.3271	P4	2.326	λ_0-LO	2.3262	1A*-LO
5329.6	2.3257					
5335.3	2.3232		2.323	λ_0-LO-TO		
5338.8	2.3217				2.3214	1A-LO
5341.6	2.3205					
5343.3	2.3197				2.3198	
5344.7	2.3191		2.319	λ_1-LO		
5350.8	2.3165		2.317	λ_x-LO	2.3162	1B-LO
5353.5	2.3153					
5354.1	2.3150		2.315	λ_2-LO		
5356.3	2.3141				2.3144	1C-LO
5358.4	2.3132					
5359.8	2.3126		2.313	λ_0-2LO	2.3120	1A*-2LO
5364.5	2.3106		2.310	λ_0-2LO-TO	2.3105	1A-2LO
5367.8	2.3092		2.308	λ_1-2LO	2.3083	
5374.2	2.3064					
5383.5	2.3024		2.303	λ_2-2LO	2.3022	1B-2LO
5383.3	2.3016				2.3012	1C-2LO
5390.1	2.2997					
5394.1	2.2979		2.298	λ_0-3LO	2.3977	1A*-3LO
5397.3	2.2966		2.296	λ_0-3LO-TO	2.2959	1A*-3LO-TO

Reproduced with permission from Bao et al., 1990 and the American Institute of Physics.
[a] Bao et al., 1990b
[b] Novikov and Pimenko, 1971
[c] J. L. Merz et al., 1983

in other spectra. Table V lists these 26 peaks along with their wavelengths and photon energies. Peaks and their assignments reported by Novikov and Pimonenko (1971) and Merz et al. (1983) are shown for comparison.

The first five lines shown in Fig. 28 with wavelengths of 5296.9, 5309.1, 5311.2, 5317.9, and 5320.8 Å (labeled P0, P1, P1', P2, and P3) are generally accepted to be free or bound exciton emission lines, although there is some uncertainty on the exact assignments (Bao et al., 1990b). Table VI shows the FWHM of these five lines (Bao et al., 1990b), along with their observations in the ordinary and extraordinary absorption and emission spectra (Novikov and Pimonenko, 1971; Akopyan et al., 1975; Tubbs 1972). From the table one finds that the lines

TABLE VI

Observations of the First Five Lines between 5290 and 5400 Å in the Ordinary ($E \perp c$) and Extraordinary ($E \parallel c$) Absorption and Emission Spectra with Their FWHM as Measured in the 4.2 K Emission Spectra from This Work

Lines[a] (4.2 K)	5296.9 Å P0	5309.1 Å P1	5311.2 Å P1'	5317.9 Å P2	5320.8 P3
Absorption (4.2 K) $E \perp c$		Continuous, strong absorption[c,d]			
$E \parallel c$	5297 Å[b,d]		5310 Å[c,d]		5321 Å
Emission (4.2 K) $E \perp c$		5308 Å[b,c]	5318 Å[b]		
$E \parallel c$	5297 Å[c]			5320 Å[b]	
FWHM (4.2 K)	0.88 meV	1.6 meV	0.19 meV	1.3 meV	0.61 me

Reproduced with permission from Bao et al., 1990b and the American Institute of Physics.
[a] Bao et al., 1990b
[b] Tubbs, 1972
[c] Akopyan et al., 1975
[d] Novikov and Pimenko, 1971

at 5296.1 ($P0$), 5311.2 ($P1'$), and 5320.8 Å ($P3$) can be grouped together because each has relatively narrow line widths and they appear as well resolved peaks in both the emission and absorption spectra ($E \parallel c$). In addition, the relative intensity of the line at 5320.8 Å ($P3$) follows closely the line at 5311.2 Å ($P1'$), suggesting that the origin of the two lines is the same or strongly correlated. On the other hand, the lines at 5309.1 Å ($P1$) and 5317.9 Å ($P2$) fall into a separate group, in which they both appear in the ordinary emission ($E \perp c$) spectra while the ordinary absorption spectra show only a continuum. Furthermore, the two lines have distinctly larger FWHMs than lines $P0$, $P1'$, and $P3$. It has also been clearly demonstrated that the intensity of the line at 5317.9 Å ($P2$) follows the intensity of the line at 5309.1 Å ($P1$), which further supports the idea of a relationship between the two lines. The energy difference between $P2$ and $P1$ is approximately equal to the LO_2 phonon at 3.9 meV, although $P2$ has not been previously reported as a phonon replica of $P1$.

Akopyan and co-workers (1975) concluded that $P0$ is due a longitudinal free exciton transition, $P1$ is associated with a transverse free exciton transition, and $P1'$ is due to a spin forbidden free exciton transition (i.e., electron and hole spin in the same direction). The longitudinal–transverse exciton splitting would then be given by the difference in energy between the peaks of $P0$ and $P1$, which is approximately equal to 5.3 meV. Akoypan and co-workers (1975) conclude that continuum absorption is observed in the $E \perp c$ configuration because the absorption is too strong to observe the weaker excitonic effects. Since the absorption is much less in the $E \parallel c$ configuration, Akoypan et al. (1975) argue that the $P0$ resonant excitation can be observed. A pure longitudinal exciton cannot be excited by an electromagnetic wave, so it is speculated that $P0$ is due to the interaction of the

the transverse exciton with the incident electromagnetic wave to form a polariton in the crystal. If these assignments of $P0$, $P1$, and $P1'$ are correct, it is likely that $P2$ is related to a transverse exciton and $P3$ to a longitudinal exciton. The temperature dependence of the $P2$ and $P3$ emissions suggests that they are bound exciton emissions, in which case it can be postulated that $P2$ is a transverse bound exciton transition and $P3$ is a longitudinal bound exciton. Further discussions of the bound excitons $P2$ and $P3$ will be presented later in this section.

The peaks on the long wavelength side (i.e., between 5321 and 5400 Å) will now be discussed. The three dominant optical phonons in the Raman spectra have zone center phonon energies of 2.3, 3.9, and 14.3 meV. Novikov and Pimonenko (1971) and Merz et al. (1983) have attributed the rest of the lines on the long wavelength side of $P3$ as phonon replicas of 2.3 and 14.3 meV phonons. No phonon replicas have been assigned to the 3.9 meV. Wong et al. (1988a) argued that some of the lines identified as phonon replicas cannot be so, because the energy separation from their no phonon parent line was not quite correct. Bao et al. (1990b) later explained much of the discrepancy between different authors regarding the peak positions of several photoluminescence emissions. Bao et al. (1990b) showed that many of emission lines that were previously assumed to be singlets were actually composite lines. Moreover, many of the other photoluminescence peaks reported by Bao et al. (1990b) had asymmetric shapes suggesting that they too had multiline compositions.

Several difficulties are encountered when one tries to assign numerous lines to exciton emissions and their phonon replicas. First, if only the 2.3 and 14.3 meV phonons are used in assigning phonon replicas, many lines simply cannot be explained because their peak positions do not correspond to a combination of zone center phonons. Second, if the phonon at 3.9 meV is also included, the assignments become somewhat arbitrary because many of the emission lines are closely spaced, and they could be assigned to either different parent lines or different combinations of emitted phonons. In theory, the ratio between the different bound exciton lines and their phonon replicas should be constant, so that measurements of the relative intensities should be helpful in organizing all of the lines into groups and thus lead to their identification as phonon replicas. However, in practice, one observes that the bound exciton emissions vary in intensity from sample to sample due to different concentrations of impurities or defects to which the excitons are bound. Moreover, the potential list of phonon replicas are so closely spaced that there is significant overlap between adjacent emissions, which makes it impossible to precisely determine the relative intensities of most of the photoluminescence peaks. Given these difficulties, assignment of most of the weaker emissions as phonon replicas is highly speculative. Nonetheless, there appears to be general agreement in the literature that the intensities of the lines at 5350.8 and 5353.5 Å (see Table V) follow closely in intensity the lines at 5317.9 ($P2$) and 5320.8 Å ($P3$), respectively. In addition, the two sets of peaks differ from each other by 14.3 meV, which corresponds to the LO_3 optical phonon energy. The evidence seems convincing that the lines at 5350.8 and 5353.5 Å are indeed the first LO_3 phonon replica of the lines at 5317.9 ($P2$) and 5320.8 Å ($P3$).

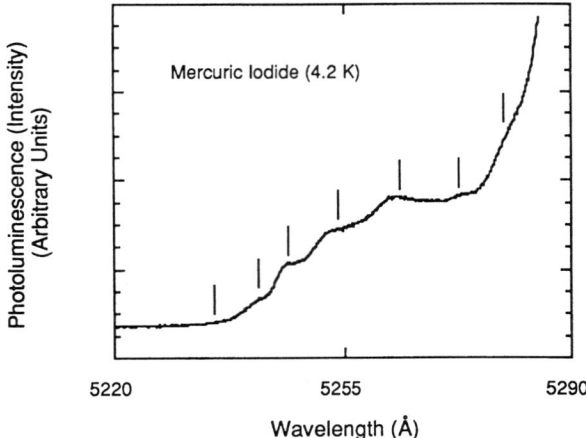

FIG. 29. A 4.2 K near-band edge photoluminescence spectrum of HgI_2 between 5220 and 5290 Å. Vertical bars indicate the positions of the emission lines. The spectral resolution of the spectrometer was 2.0 Å. (Reproduced with permission from Bao *et al.*, 1990b and the American Institute of Physics.)

7. Phonon–Assisted Electron–Hole Pair Photoluminescence

Figure 29 shows a typical 4.2 K spectrum at wavelengths between 5220 and 5290 Å. These lines are usually about three orders of magnitude weaker in intensity than the emission lines between 5290 and 5400 Å discussed earlier. They have been attributed to phonon assisted indirect radiative recombination of free electron–hole pairs. The peak positions were first measured by Goto and Nishina (1978) and later by Bao and co-workers (1990b). Table VII shows the measured peak position by the two groups of investigators. It is obvious that there is considerable difference between the peak positions reported. Bao and co-workers (1990b) observe two more peaks on the longer wavelength side, and Goto and Nishina (1978) report two more peaks on the shorter wavelength side. Part of the difficulty in assigning peak positions results from the weak intensities and stepwise nature of these lines, which makes it very difficult to determine the peak positions to better than about 1 Å.

8. Effects of Geometrical Configuration on Photoluminescence Spectra

Although most photoluminescence measurements on HgI_2 have been conducted in a reflection mode, one measurement has been made in which the laser beam excited the back side of the slab and the luminescence was collected from the front surface in a direction parallel to the normal (Bao *et al.*, 1990c). Figure 30

5. OPTICAL PROPERTIES OF RED MERCURIC IODIDE

TABLE VII

WAVELENGTHS AND PHOTON ENERGIES OF BAND EDGE PHOTOLUMINESCENCE PEAKS OF HgI_2 BETWEEN 5220 AND 5290 Å

Wavelength (Å)[a] 4.2 K	Energy (eV)	Energy (eV)[b] 4.2 K	Notation
		2.3707	E_g
5235.0	2.3677	2.3684	$-LO_1$
		2.3662	$-2LO_1$
5241.5	2.3648	2.3639	$-3LO_1$
5246.4	2.3626	2.3614	$-4LO_1$
5253.0	2.3596	2.3589	$-5LO_1$
5263.1	2.3551	2.3562	$-LO_3$
5273.3	2.3505		
5280.7	2.3472		

Reproduced with permission from Bao et al., 1990b and the American Institute of Physics.
[a] Bao et al., 1990b
[b] Goto and Nishina, 1978

shows 4.2 K PL spectra taken in the reflection and transmission modes for a slice with the c-axis perpendicular to the normal of the sample. Comparisons between the spectra reveal that $P0$, $P1$, $P1'$ and $P3$ are almost completely reabsorbed for spectra taken in the transmission mode. This observation is consistent with absorption measurements.

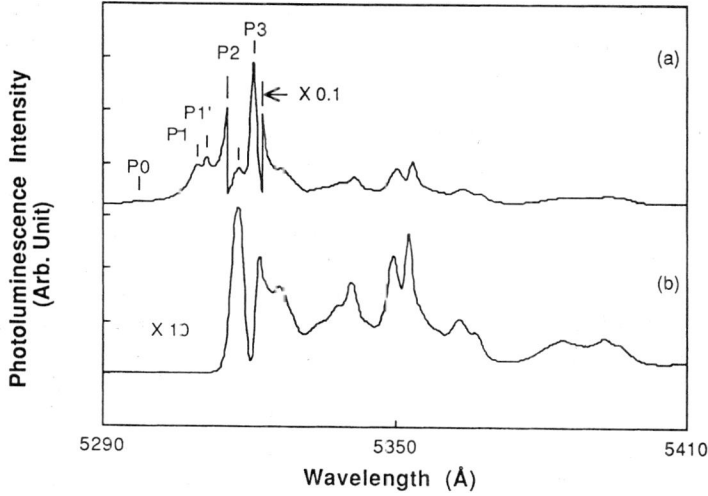

FIG. 30. 4.2 K photoluminescence spectra obtained in (a) reflection and (b) transmission modes. (Reproduced with permission from Bao et al., 1990c and the American Institute of Physics.)

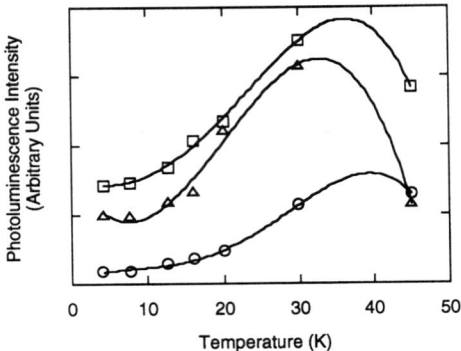

FIG. 31. Photoluminescence intensities of $P0$, $P1$, and $P5338.8$ as a function of temperature (from Bao, 1991). ⊙, $P0$; ⊟, $P1$; △, $P5338.8$.

9. Temperature Dependence of the Photoluminescence Spectrum

The temperature dependence of photoluminescence spectrum may provide clues for the identification of spectral lines in addition to bandgap shift with temperature (see the previous subsections). The temperature behaviors of the first several photoluminescence lines were studied by several investigators (Novikov and Pimonenko, 1972; Merz et al., 1983; Bao, 1991). It was found that the intensities of $P0$ and $P1$ increase with temperature below about 30 K (Bao, 1991; Novikov and Pimonenko, 1972; Merz et al., 1983). Above 30 K, the intensities of $P0$ and $P1$ start to decrease with temperature. The intensities of $P0$ and $P1$ as a function of temperature are shown in Figure 31. This temperature behavior of $P0$ and $P1$ leads to their assignments as free exciton lines, which is consistent with discussions in the previous subsections on the origin of these emission lines. The intensities of $P2$ and $P3$, on the other hand, decrease with temperature, and their thermal activation energies can be obtained from the temperature dependence. Since these lines were believed due to bound excitons, the thermal activation energies were then the binding energies of these bound excitons. The measured thermal activation energies for $P2$ and $P3$ are tabulated in Table VIII. The peak positions of $P2$ and $P3$ relative to $P1$ and $P0$ are also listed. These measurements seem to agree with the assignment of these lines as discussed in the previous sections.

Figure 32 shows the temperature dependence of the three commonly observed broad photoluminescence bands label $B2$ (5595 Å), $B3$ (6200 Å), and $B4$ (7550 Å) (Merz et al., 1983; Bao et al., 1990a; Bao, 1991). The intensity of $B2$ decreases with temperature. The intensities of $B3$ and $B4$ increase with temperature below about 50 K and decrease above 50 K. The behavior of $B3$ and $B4$ has yet to be explained. More discussion on these bands will be provided in the next section on processing of HgI_2 detectors.

TABLE VIII

Summary of Thermal Activation Energies for $P2$ and $P3$ by Different Workers
(Novikov and Pimonenko, 1972; Merz et al., 1983; Akopyan et al., 1987; Bao, 1991)

Peak	E_T (meV)[a]	E_T (meV)[b]	E_T (meV)[c]	E_T (meV)[d]	Peak Position
$P2$	4.9	4		2.4	$P1$, 3.9 meV
$P3$	9.3	6	12	11.4	$P0$, 10.4 meV

[a] Bao, 1991
[b] Akopyan et al., 1987
[c] Novikov and Pimenko, 1972
[d] Merz et al., 1983

IV. Study of Processing by Photoluminescence Spectroscopy

1. Purification and Stoichiometry

Using low temperature photoluminescence spectroscopy, Merz et al. (1983) first studied the processes commonly employed to purify raw materials for vapor phase crystal growth. These processes include open tube sublimation, closed tube sublimation, and melting (Merz et al., 1983). Figure 33 shows the effect of sublimation on the 77 K photoluminescence spectra of HgI_2 powders (Merz et al., 1983). It can be seen that the intensity of band 2 relative to band 1 increases with each sublimation run. Band 3 practically disappeared after the first sublimation run. Chemical analyses show that the sublimation process significantly reduces the concentration of impurities in HgI_2, indicating that Band 3 is probably associated with impurities (Soria et al., 1994). It is also suspected that sublimation reduces the iodine concentration by preferential removal of the more volatile species, which suggests that the band 2 to band 1 ratio is an indicator of the relative

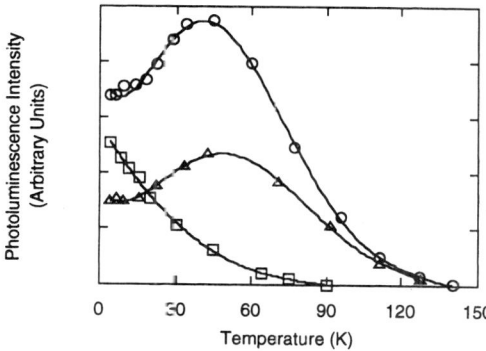

FIG. 32. Photoluminescence intensities of bands 2 (⊟), 3 (⊖), and 4 (△) as a function of temperature (from Bao, 1991).

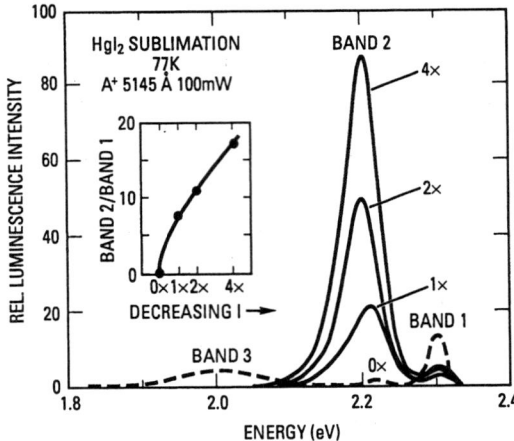

FIG. 33. Photoluminescence spectra of HgI$_2$ powder at 77 K, after various sublimation steps. The designations $N\times$ ($N = 0, \ldots, 4$) refer to the number of times the powder is passed along the sublimation tube. The inset shows the intensity ratio of band 2/band 1 for successive sublimation runs, using an arbitrary scale along the horizontal axis. (Reprinted with permission from Merz et al., 1983 and Elsevier Science Publishers.)

concentration of iodine in the material. Subsequent doping experiments with iodine or mercury appear to agree with the hypothesis (Bao et al., 1992a). Figure 34 shows 77 K photoluminescence spectra from six single crystals grown from materials doped with either iodine or mercury. Crystal A is heavily doped with iodine, B is lightly doped with iodine, C is undoped, D is lightly doped with mer-

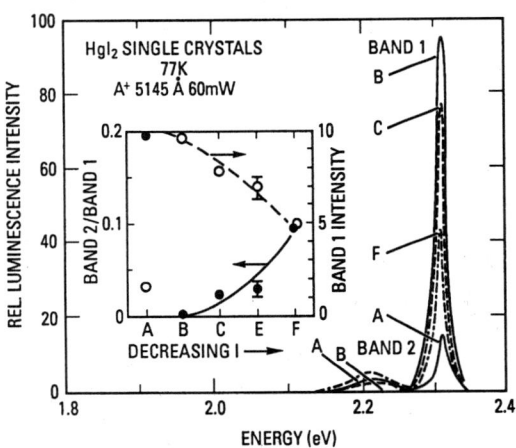

FIG. 34. Photoluminescence of doped HgI$_2$ single crystals at 77 K. Inset shows band 2/band 1 ratios, and intensity of band 1 for samples doped with Hg and I$_2$. (Reprinted with permission from Merz et al., 1983 and Elsevier Science Publishers.)

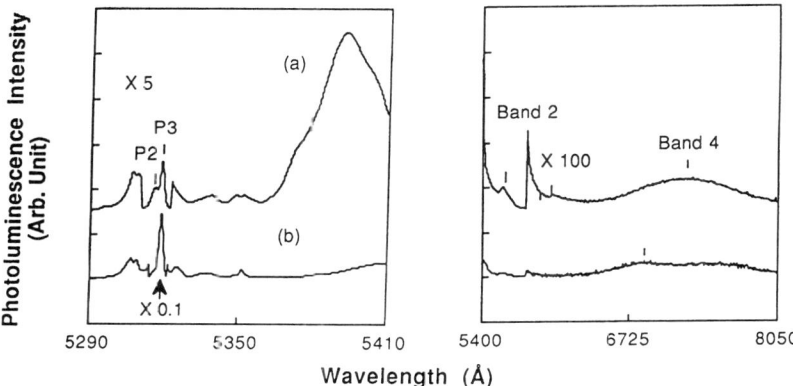

FIG. 35. 4.2 K photoluminescence spectra taken from a HgI_2 crystal (a) before and (b) after a six day storage with vapor of iodine. The sample was etched with 10% KI before storage. (Reprinted with permission from Bao et al., 1992a and the American Institute of Physics.)

cury, E is moderately doped with mercury, and F is heavily doped with mercury. As can be seen from the figure, with the exception of sample A, which shows abnormally weak luminescence intensity, the band 2 to band 1 ratio increases with decreasing iodine content, and the absolute intensity of band 2 also increases with decreasing iodine content.

The effect of stoichiometry was further studied by Bao et al. (1992a). HgI_2 samples were stored in the presence of either iodine or mercury vapor and photoluminescence spectra were taken before and after the storage. Figure 35 shows the effect of storage in iodine vapor for six days. It is observed that the intensity of $P3$ will increase in the presence of iodine vapor. The effect of storage with mercury is shown in Fig. 36. The sample was sitting on a glass slide with the front surface exposed. After six days, the front surface looked metallic. And as can be seen from Fig. 36, spectrum (c) taken from the front surface has degraded dramatically with little intensity in the band 1 region. The back surface, which was not directly exposed to the vapor, however, changed less (spectrum (b) of Fig. 36). Nevertheless, there is a drastic decrease in the intensity of $P3$ after storage (compare spectra (a) and (b)). Further storage of the sample with iodine vapor recovers some of the luminescence intensity, but the spectrum (d) is still of low quality. The results of these experiments seem to be consistent with results of Merz et al. (1983) in that a low band 1 intensity relative to band 2 is an indication of iodine deficiency, since $P3$ usually dominates the band 1 region.

2. ETCHING AND VACUUM EXPOSURE

Etching of HgI_2 by aqueous solution of KI has been used extensively in the fabrication of HgI_2 detectors for sawing, polishing, and reducing crystal size.

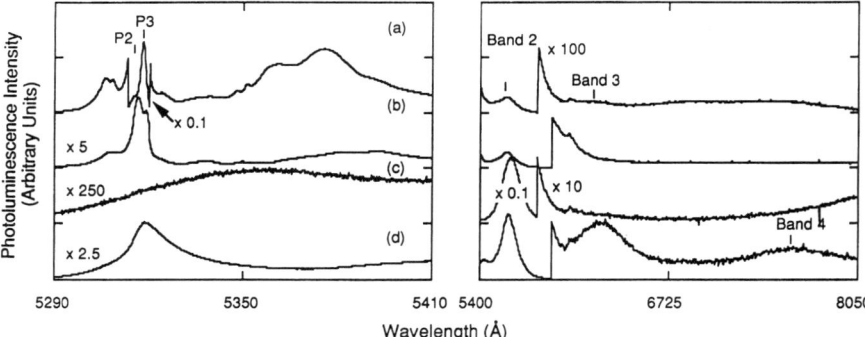

FIG. 36. 4.2 K photoluminescence spectra taken from a HgI_2 sample. They were obtained (a) before storage (b) from back of the sample after storage (c) from the front surface after storage with mercury vapor for six days, and (d) from front surface after storage with iodine vapor for two days following the storage with mercury. (Reprinted with permission from Bao et al., 1992a and the American Institute of Physics.)

Figure 37 shows the effect of chemical etching by 10% KI aqueous solution (Bao et al., 1990d). The intensity of $P3$ is enhanced after the etching. As a matter of fact, $P2$ was the dominating line before the etch, but $P3$ became dominant after the etch. The explanation is that the etching removes a surface layer that has degraded during storage. This surface layer will form in about 24 hr under typical storage conditions, such as in a desiccator (more discussion later with aging). It is further speculated that the degraded surface is due primarily to preferential removal of iodine species because of its relatively high vapor pressure. Therefore, a spectrum with relatively low intensity of $P3$ is probably related to iodine deficiency, which is consistent with stoichiometry study in the previous subsection. Several other etchants have also been studied such as nitric acid and methanol (Bao et al., 1991; David et al., 1993). It was found that in general all etchants have similar effect on the near-bandgap photoluminescence. However, there may

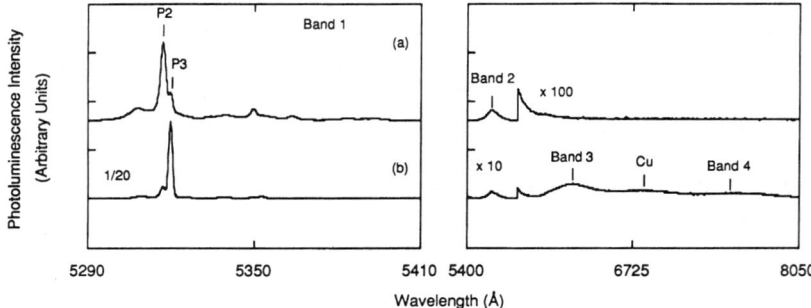

FIG. 37. A 4.2 K photoluminescence of a HgI_2 single crystal before and after a 10% KI etch for about 3 min. (Reprinted with permission from Bao et al., 1990b and the American Institute of Physics.)

be variations in the emission lines in the long wavelength region, which are associated with the purity of the etchant and contaminations of the crystal during etching. The possible incorporation of impurity defects during etching has been verified for Ag and Cu by doping experiments, where a known amount of the impurities was intentionally added into the etching solution (Van Scyoc et al., 1993).

Mercuric iodide surfaces change upon exposure to air (Schieber et al., 1994; Yao et al., 1994). The top photo in Fig. 38 shows the surface of a HgI_2 sample

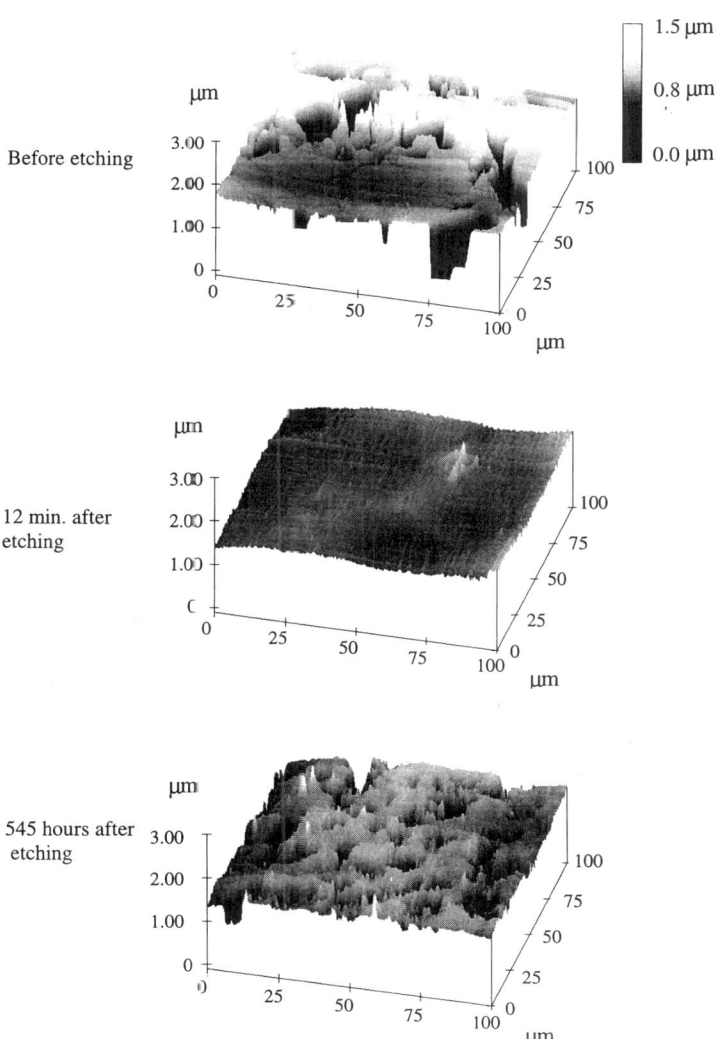

FIG. 38. Atomic force microscopy images showing effects of chemical etching and aging on the roughness of HgI_2 surfaces (from Yao et al., 1994).

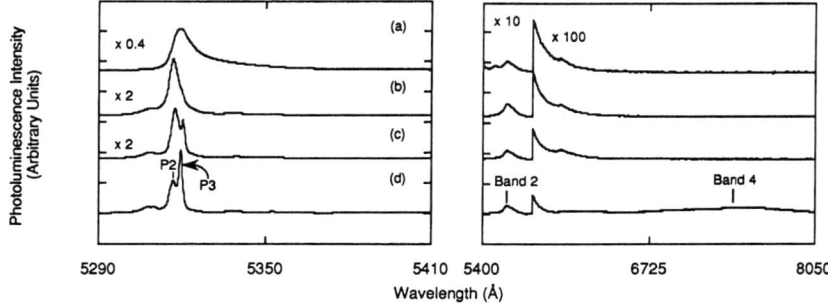

FIG. 39. The effect of aging when the crystal is exposed to air. The spectra were taken (a) 20 days, (b) 7 days, (c) 20 hr, and (d) minutes after a 10% KI etch. (Reprinted with permission from Bao *et al.*, 1992a and SPIE.)

that has been stored in a dessicator for a few months. The surface is quite rough with ridges that are several microns higher than the valleys. The middle photo shows the same surface after it has been chemically etched for 2 min in a 10% KI solution. The chemical etching removes most of the surface roughness, but this roughness returns with continued exposure of the sample to air. The bottom photo of Fig. 38 shows the surface at 545 hr after etching (Yao *et al.*, 1994).

The effect of aging on the low temperature photoluminescence of HgI_2 is shown in Fig. 39 (Bao *et al.*, 1992a). As aging progresses, $P3$ decreases and gradually disappears while $P2$ becomes dominant. This is exactly the reverse process as seen in the etching study. After about 20 days, the spectrum is predominantly a featureless broad strong peak slightly shifted away from $P2$ toward $P3$. From the gradual change of the spectra, it appears that this broad peak derives from $P2$. From a large number of measurements, it was found that typically the effect of aging can be observed within about 24 hr for a freshly etched sample stored in air. Changes in the stoichiometry of the near-surface region and roughening upon aging are expected to contribute to the photoluminescence spectra.

The effect of vacuum exposure on the near-bandgap photoluminescence spectra of HgI_2 is very similar to that of chemical etching (Bao *et al.*, 1990d). Unlike aging during storage, where there seem to be a preferential loss of iodine, the vacuum exposure removes the surface layer congruently. This is probably due to the high rate of removal of the HgI_2 molecules under vacuum. The effect of vacuum exposure is thus more like an etching than an aging effect.

3. Contacts

Currently the most widely used contact materials are Pd and colloidal carbon. Transparent contacts such as indium–tin oxide and hydrogel contacts have also been used for fabrication of HgI_2 photodetectors. This is primarily because most metals react with HgI_2 (Cheng, 1993). Since the ideal contact material for each

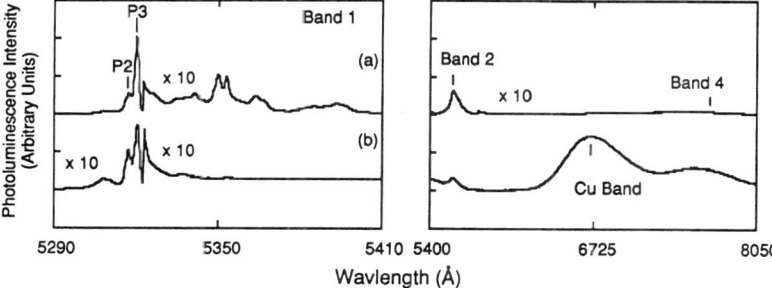

FIG. 40. A 4.2 K photoluminescence spectra of HgI_2 taken from the same spot (a) before and (b) a semitransparent Cu layer was deposited. (Reprinted with permission from Bao et al., 1990a and the American Institute of Physics.)

detector application has yet to be identified, measurements of contact chemistry and device performance for detectors employing different electrode materials are currently of interest. In some cases, investigations of HgI_2–contact interfacial region also increase our understanding of the effect of impurities on the charge transport in devices, particularly for contact materials that diffuse into HgI_2 and dope the materials.

Several contact materials were studied carefully with optical and electrical techniques. They include materials that are currently in use such as Pd, C, In, and Sn and materials of potential use such as Cu, Ag, W, Au, Pt. Results for Cu and Ag will be discussed here. Interested readers can refer to the references for more details and for discussions on other contact materials (James et al., 1989b, 1990, 1992; Schlesinger et al., 1992; Wong et al., 1988b; Bao et al., 1990a, c, d, 1991; Bao, 1991; George et al., 1993). Figure 40 shows two photoluminescence spectra taken from the same spot on a HgI_2 crystal obtained before and after a semitransparent Cu layer is deposited (Bao et al., 1990a). Several changes can be observed. The most dramatic difference is the introduction of a new broad band centered at about 6720 Å between band 3 and band 4. This band has never been observed before and is apparently due to the incorporation of Cu into the HgI_2 crystal and is thus labeled the "Cu" band. Photoluminescence spectra were also taken from spots that were masked off during the Cu deposition. The spots were 2 mm away from the Cu layer. All of these spectra showed emission associated with the Cu band, which indicates that Cu diffuses easily along the crystal surface. A back-doping experiment was performed to study the bulk diffusion of Cu in HgI_2. Figure 41 shows the absolute intensities of the Cu band as a function of position. Four spectra were taken along each of the three directions A, B, and C. The excitation spots were first moved toward the Cu dot from position 1 to 2 and then away from the Cu dot from position 3 to 4. From the plot, it can be seen that the spots close to the Cu dot have stronger Cu band intensity than those farther away. Because the Cu dot was deposited on the other side of the sample (the sample is about 1 mm thick), it is concluded that Cu also diffuses readily through the

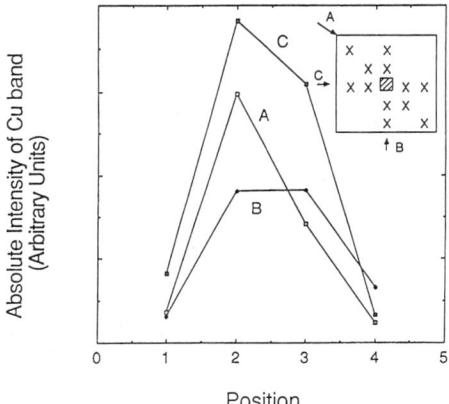

FIG. 41. Absolute intensities of the Cu band (at about 6720 Å) in the 4.2 K photoluminescence spectra as a function of the approximate position from the back-doping experiment. The insert shows the approximate location of points on a HgI_2 substrate (about 1 cm × 1 cm in size) from which the spectra were taken. The small square in the middle indicates the Cu film deposited on the back side. (Reprinted with permission from Bao et al., 1990a and the American Institute of Physics.)

bulk of the material. It should be noted that the HgI_2 samples were never intentionally annealed to enhance the diffusion. The only time the sample may have been expose to momentary heating is during the contact deposition by thermal evaporation.

Nuclear detectors were also made with Cu as electrodes (Bao et al., 1990a). These detectors had very poor performance and high leakage current. Further electrical measurements of Cu contacts also showed fast diffusion of Cu in HgI_2. These investigations revealed and that Cu atoms in HgI_2 are electrically active since the resistivity of HgI_2 was lowered by several orders of magnitude by Cu doping (Van Scyoc et al., 1993). Copper is also believed to significantly degrade the hole transport in devices.

Similar experiments were performed for Ag (James et al., 1993). The top curve of Figure 42 displays a typical photoluminescence spectrum from a point 2 mm away from the silver film. This spectrum was taken 3 days after deposition of the Ag film, and it closely resembles spectra taken from other slabs of the same crystal that were not deposited with Ag. For points beneath the silver film, the spectra were significantly different. Figure 42b shows the spectrum from a typical point beneath the Ag film. The emission in the band 1 spectral region is dominated by a single feature peaked at about 5321.2 ± 1.5 Å. The intensity of band 3 was greatly reduced. Another important difference between the photoluminescence spectra from points beneath the Ag film and away from it is the presence of a new broad feature located at about 5490 Å. This emission peak typically appears as a distinct shoulder on the short wavelength side of band 2 (see Fig. 42b). Since this

5. OPTICAL PROPERTIES OF RED MERCURIC IODIDE 213

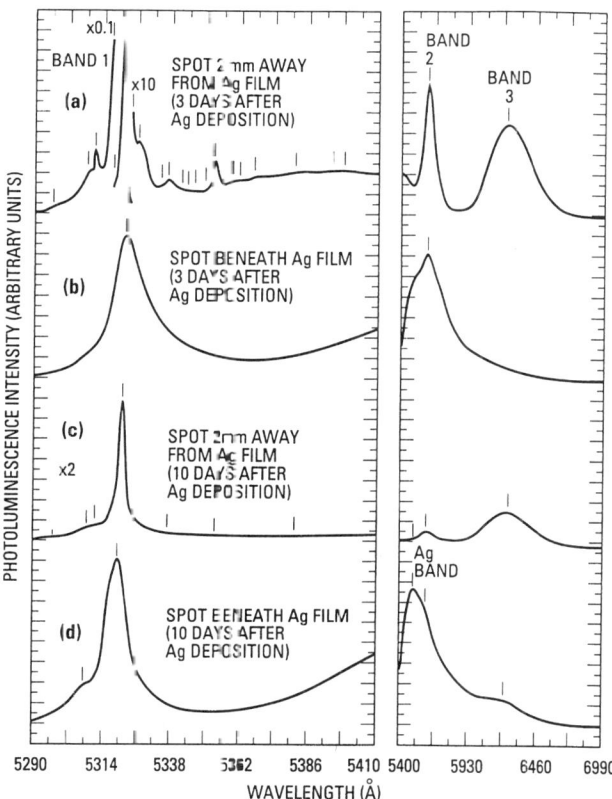

FIG. 42. A 4.2 K photoluminescence spectra from spots that are either beneath or 2 mm away from the Ag film. Spectra in 1a and 1b were taken 3 days after the Ag film was deposited, and the spectra in 1c and 1d were taken 10 days after the film was deposited. (Reprinted with permission from James et al., 1993 and the Materials Research Society.)

peak is only observed in crystals with Ag impurities and it is the dominant photoluminescence feature in regions with high silver concentrations, it is labelled "Ag band." A separate experiment was performed on a 1 cm × 1 cm × 0.06 cm HgI_2 slab with a semitransparent 0.4 cm diameter Ag film deposited on one side only. Photoluminescence spectra were obtained from several spots on the back side that was separated from the circular Ag film by 0.06 cm of bulk HgI_2. Spectra were taken 3 days after the Ag deposition, and it looks indistinguishable from undoped HgI_2. However, after a period of 10 days, the spectrum from the same region exhibited the presence of Ag dopants. In particular, the emission observed at 5490 Å, which is an indicator of Ag impurities, is clearly resolved in the spectrum (see Fig. 43a and b). Figure 43c displays a typical spectrum from a point on the

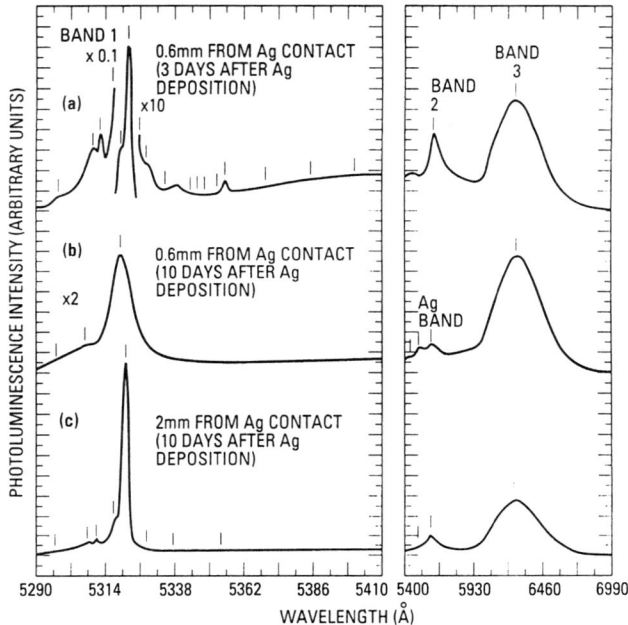

FIG. 43. A 4.2 K photoluminescence spectra taken from different points on the backside of a Ag contacted HgI$_2$ sample. The points were separated from the Ag film by either 0.6 or 2 mm of bulk HgI$_2$. (Reprinted with permission from James *et al.*, 1993 and the Materials Research Society.)

back side of the sample that was separated from the Ag film by 0.2 cm of the bulk HgI$_2$. The spectrum was also taken 10 days following the Ag deposition. This spectrum showed a smaller intensity of the "Ag band" for regions that were separated farther from the Ag film by bulk HgI$_2$. These measurements reveal that silver diffuses through the bulk of the material instead of surface diffusion only. The rate of diffusion at room temperature is estimated to be on the order of 10 μm/hr.

4. Detector Performance

For gamma ray applications, the detector performance is usually quantified by the photopeak resolution and peak to valley ratio (Gerrish, 1993). Figure 44 shows a pulse height spectra of the 662 keV gamma ray from a Cs-137 source. The resolution is 1.7% and the peak to valley ratio is 15:1. Merz *et al.* (1983) first observed that crystals that produce better detectors tend to have higher intensity of luminescence in the band 1 region. Since the band 1 region corresponds to the near-band edge wavelength, it is reasonable to believe that higher intensities from this region indicate higher quality crystals. Photoluminescence spectra taken from

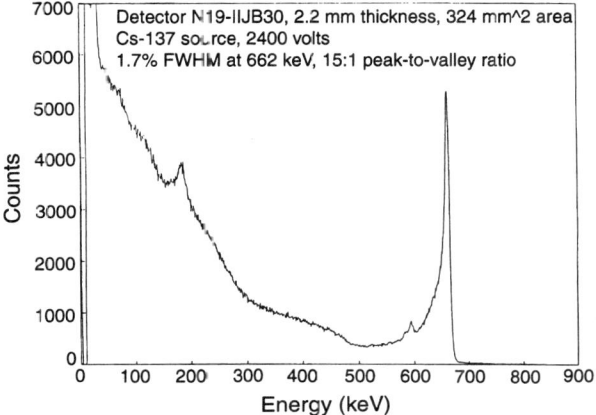

FIG. 44. A spectrum of the 662 keV gamma ray from a Cs-137 source taken with a HgI$_2$ detector at room temperature.

working detectors also show that better detectors have higher band 1 to band 2 ratio, as shown in Fig. 45. In addition, it is found from a large number of measurements that band 3 is also detrimental for detector performance (Bao et al., 1992; James et al., 1992). Based on these measurements it appears likely that larger emission from the deeply bound states (i.e., at wavelengths greater than ~5400 Å) are indicators of poor detector grade crystals.

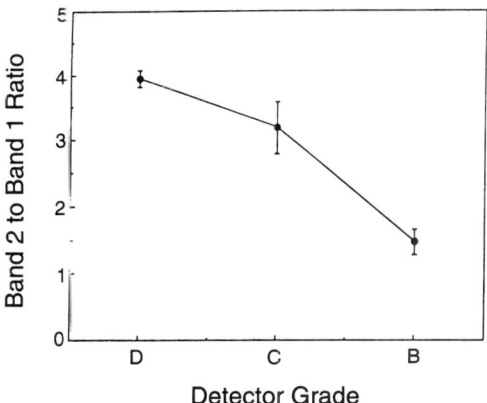

FIG. 45. Band 2 to band 1 ratios of three working detectors graded B, C, and D. The intensities of band 1 and band 2 were obtained by integrating from 5290 to 5410 Å and from 5410 to 5690 Å, respectively using 4.2 K photoluminescence spectra. The error bars are the standard deviation of the means from several measurements on each sample. (Reproduced with permission from Bao et al., 1992b and Elsevier Science Publishers.)

V. Conclusions

Since the fabrication of the first HgI_2 nuclear detectors more than two decades ago, there have been much progress in several major aspects of HgI_2 sensor technologies. With advances in device processing (including purification, growth, fabrication, contacts, passivation, and testing) of HgI_2 and associated pulse processing electronics, the first commercial products were realized in 1988 in the form of field portable x-ray fluorescence systems designed and manufactured by TN Technologies, Inc. At the same time, fundamental research in the material properties of HgI_2 has increased the knowledge base of this material tremendously, which in turn helps the development of improved crystal growth and device processing procedures. In addition, developments of applications for HgI_2 in many other areas are keenly pursued by many researchers in the world. Better understanding of the optical properties of HgI_2 and development of new techniques will certainly become more important in the future. First, the application of HgI_2 in nuclear detection is fundamentally related to its photosensitivity. Therefore HgI_2 device behavior and performance will be better understood with greater knowledge of its fundamental electrical and optical properties. Second, given the high resistivity of HgI_2, the majority of characterization techniques used to study the electrical properties of HgI_2 utilize light sources for excitation. Third, as the technology is becoming more commercialized, rapid and cost effective testing in the manufacturing environment will be in demand for nondestructive characterization of materials, crystals, and detectors for sorting, grading, and process control. These analytical techniques will be derived from the research efforts that emphasize the relationships of crystalline purity, structural perfection, contact quality, with measurements of detector performance and manufacturing yield.

References

Adams, D. M., and Hooper, M. A. (1970). *Astr. J. Chem.* **24**, 885.
Anedda, A., Raga, F., Grill, E., and Guzzi, M. (1977). *Il Nuovo Cimento* **38B**, 439.
Anedda, A., Grilli, E., Guzzi, M., Raga, F., Serpi, A. (1981). *Solid State. Commun.* **39**, 1121.
Akopyan, I., Novikov, B., Permogorov, S., Selkin, A., and Travnikov, V. (1975). *Phys. Stat. Sol. (b)* **70**, 353.
Akopyan, I. K., Bondarenko, B. V., Kazennov, B. A., and Novikov, B. V. (1987). *Sov. Phys. Solid State* **29**, 238.
Bao, X. J. (1991). Ph.D. thesis, Carnegie Mellon University.
Bao, X. J., Schlesinger, T. E., James, R. B., Stulen, R. H., Ortale, C., and van den Berg, L. (1990a). *J. Appl. Phys.* **67**, 7265.
Bao, X. J., Schlesinger, T. E., James, R. B., Ortale, C., and ven den Berg, L. (1990b). *J. Appl. Phys.* **68**, 2951.
Bao, X. J., Schlesinger, T. E., James, R. B., Cheng, A. Y., and Ortale, C. (1990c). *Mat. Res. Soc. Symp. Proc.* **163**, 1027.
Bao, X. J., Schlesinger, T. E., James, R. B., Stulen, R. H., Ortale, C., and Cheng, A. Y. (1990d). *J. Appl. Phys.* **68**, 86.

Bao, X. J., Schlesinger, T. E., James, R. B., Gentry, G. L., Cheng, A. Y., and Ortale, C. (1991). *J. Appl. Phys.* **69,** 4247.
Bao, X. J., James, R. B., Hung, C. Y., Schlesinger, T. E., Cheng, A. Y., Ortale, C., and van den Berg, L. (1992a). *SPIE Proc.* **1736,** 60.
Bao, X. J., Schlesinger, T. E., James, R. B., Harvery, S. J., Cheng, A. Y., Gerrish, V., and Ortale, C. (1992b). *Nucl. Instr. Meth. Phys. Res.* **A317,** 194.
Biellman, J., and Prevot, B. (1980). *Infrared Phys.* **20,** 99.
Bube, R. (1957). *Phys. Rev.* **106,** 703.
Burger, A., and Nason, D. (1992). *J. Appl. Phys.* **71,** 2717.
Burger, A., Morgan, S. H., Silberman, E., Nason, D., and Cheng, A. Y. (1992). *Nucl. Instr. Methods. Phys. Res.* **A322,** 427.
Chang, Y. C., and James, R. B. (1992). *Phys. Rev. B* **46,** 150404.
Chang, Y. C., Sim, H. K., and James, R. B. (1993). In *Semiconductors for Room-Temperature Radiation Detector Applications,* ed. R. B. James et al. Mat. Res. Soc. Symp. Proc. **302,** 121.
Chelikowsky, R. J., and Cohen, M. L. (1976). *Phys. Rev. B* **14,** 556.
Cheng, A. Y. (1993). In *Semiconductors for Room-Temperature Radiation Detector Applications,* ed. R. B. James et al. Mat. Res. Soc. Symp. Proc. **302,** 141.
Chester, M., and Coleman, C. C. (1971). *J. Phys. Chem. Solids* **32,** 223.
David, D. C., Van Scyoc, J., Khudatyan, M., James, R. B., Anderson, R. J., and Schlesinger, T. E. (1993). In *Semiconductors for Room-Temperature Radiation Detector Applications,* ed. R. B. James et al. Mat. Res. Soc. Symp. Proc. **302,** 147.
Gerrish, V. M. (1993). In *Semiconductors for Room-Temperature Radiation Detector Applications,* ed. R. B. James et al. Mat. Res. Soc. Symp. Proc. **302,** 129.
George, M. A., Azoulay, M., Burger, A, Biao, Y., Silberman, E., and Nason, D. (1993). *Thin Solid Films* **236,** 180.
Goto, T., and Kasuya, A. (1931). *J. Soc. Jpn.* **50,** 520.
Goto, T., and Nishina, Y. (1978). *Sol. Stat. Commun.* **25,** 123.
Haisler, V. A., Zaletin, V. M., Kravchenko, A. F., and Yashin, G. Y. (1984). *Phys. Stat. Sol. (b)* **121,** K13.
Harbeke, G., and Tosatti, E. (1974). *Proceedings of the 12th International Conference on the Physics of Semiconductors,* Stuttgart, 626.
Hopfield, J. J., and Thomas, D. G. (1960). *J. Phys. Chem. Solids* **12,** 276.
Hopfield, J. J., and Thomas, D. G. (1961). *Phys. Rev.* **122,** 35.
James, R. B., Ottesen, D. K., Wong, D., Schlesinger, T. E., Schnepple, W. F., Ortale, C., and van den Berg, L. (1989a). *Nucl. Instr. Meth. Phys. Res.* **A283,** 188.
James, R. B., Bao, X. J., Schlesinger, T. E., Markakis, J. M., Cheng, A. Y., and Ortale, C. (1989b). *J. Appl. Phys.* **66,** 2578.
James, R. B., Bao, X. J., Schlesinger, T. E., Ortale, C., and Cheng, A. Y. (1990). *J. Appl. Phys.* **67,** 2571.
James, R. B., Bao, X. J., Schlesinger, T. E., Chang, A. Y., Ortale, C., and van den Berg, L. (1992). *Nucl. Instr. Meth. Phys. Res.* **A322,** 435.
James, R. B., Bao, X. J., Schlesinger, T. E., Cheng, A. Y., and Gerrish, V. M. (1993). In *Semiconductors for Room-Temperature Radiation Detector Applications,* ed. R. B. James et al. Mat. Res. Soc. Symp. Proc. **302,** 103.
Jayatirtha, H. N., Azoulay, M, George, M. A., and Burger, A. (1993). In *Semiconductors for Room-Temperature Radiation Detector Applications,* ed. R. B. James et al. Mat. Res. Soc. Symp. Proc. **302,** 161.
Kanzaki, K., and Imai, I. (1972). *J. Phys. Soc. Jnp.* **32,** 1003.
Levi, A., Burger, A., Nissenbaum, J., and Schieber, A. (1983). *Nucl. Instr. Meth.* **213,** 35.
Levi, A., Schieber, M., and Burshtein, Z. (1985). *J. Appl. Phys.* **57,** 1944.
Lopez-Cruz, E. (1989). *J. Appl. Phys.* **65,** 874.
Melveger, A. J., Khanna, R. K, Guscott, B. R., and Lippincott, E. R. (1968). *Inorganic Chem.* **7,** 1630.

Merz, J. L., Wu, Z. L., van den Berg, L., and Schnepple, W. F. (1983). *Nucl. Instr. Meth.* **213,** 51.
Mikawa, Y., Jakobsen, R. J., and Brasch, J. W. (1966). *J. Chem. Phys.* **45,** 4528.
Nakagawa, I. (1971). *Bull. Chem. Soc. Japan* **44,** 3014.
Nakashima, S., Mishima, H., and Mitsuishi, A. (1973). *J. Raman Spectr.* **1,** 325.
Nason, D., and Burger, A. (1991). *Appl. Phys. Lett.* **59,** 3550.
Novikov, B. V., and Pimonenko, M. M. (1971). *Sov. Phys. Semicond.* **4,** 1785.
Novikov, B. V., and Pimonenko, M. M. (1972). *Sov. Phys. Semicond.* **6,** 671.
Nyquist, R. A., and Kagel, R. O. (1971). *Infrared Spectra of Inorganic Compounds.* Academic Press, New York.
Ogawa, Y., Harada, I., Hiroatsu, M., Takehiko, S., and Hiraishi, J. (1976). *Spectrochimica Acta* **32A,** 49.
Petroff, P. M., Hu, Y. P., and Milstein, F. (1989). *J. Appl. Phys.* **66,** 2525.
Prevot, B., and Biellmann, J. (1979). *Phys. Stat. Sol. (b)* **95,** 601.
Prevot, B., Schwab, C., and Dorner, B. (1978). *Phys. Stat. Sol. (b)* **88,** 327.
Redfield, D. (1965). *Phys. Rev.* **140A,** 2056.
Roessler, D. M. (1965). *Brit. J. Appl. Phys.* **16,** 1359.
Roessler, D. M. (1966). *Brit. J. Appl. Phys.* **17,** 1313.
Schieber, M., Roth, M., and Schnepple, W. F. (1983). *J. Cryst. Growth* **65,** 353.
Schieber, M., Roth, M., Yao, H., DeVries, M., James, R. B., Goorsky, M. (1994). *J. Cryst. Growth,* in press.
Schlesinger, T. E., Bao, X. J., James, R. B., Cheng, A. Y., Ortale, C., and van den Berg, L. (1992). *Nucl. Instr. Meth. Phys. Res.* **A322,** 414.
Schluter, I. C., and Schluter, M. (1974). *Phys. Rev. B* **9,** 1652.
Sieskind, M. (1960). *Rev. Opt. Theor. Intrum.* **39,** 239.
Sim, H. K., Chang, Y. C., and James, R. B. (1994). *Phys. Rev. B* **49,** 4559.
Soria, E., Natarajan, M., and James, R. B. (1994). Unpublished.
Thomas, D. G., and Hopfield, J. J. (1959). *Phys. Rev.* **116,** 573.
Tubbs, M. R. (1972). *Phys. Stat. Sol. (b)* **49,** 11.
Turner, D. E., and Harmon, B. N. (1989). *Phys. Rev. B* **40,** 10516.
Urbach, F. (1953). *Phys. Rev.* **92,** 1324.
Van Scyoc, J. M., Schlesinger, T. E., James, R. B., Cheng, A. Y., Ortale, C., and van den Berg, L. (1993). In *Semiconductors for Room-Temperature Radiation Detector Applications,* ed. R. B. James *et al.* Mat. Res. Soc. Symp. Proc. **302,** 115.
Williams, L. R., Anderson, R. J. M., and Banet, M. J. (1991). *Chem. Phys. Lett.* **182,** 422.
Wong, D., Schlesinger, T. E., James, R. B., Ortale, C., van den Berg, L., and Schnepple, W. F. (1988a). *J. Appl. Phys.* **64,** 2049.
Wong, D., Bao, X. J., Schlesinger, T. E., James, R. B., Cheng, A., Ortale, C., and van den Berg, L. (1988b). *Appl. Phys. Lett.* 53, 1536.
Yao, H., and Johs, B. (1993). In *Semiconductors for Room-Temperature Radiation Detector Applications,* ed. R. B. James *et al.* Mat. Res. Soc. Symp. Proc. **302,** 341.
Yao, H., Johs, B., and James, R. B. (1994). Unpublished.
Yee, J. H., Sherohman, J. W., and Armantrout, G. A. (1976). *IEEE Trans. Nucl. Sci.* **NS-23,** 117.
Zahner, J. C., and Drickamer, H. G. (1959). *J. Phys. Chem. Solids* **11,** 92.

CHAPTER 6

Growth Methods of CdTe Nuclear Detector Materials

Makram Hage-Ali and Paul Siffert

CENTRE DE RECHERCHES NUCLÉAIRES
LABORATOIRE PHASE
STRASBOURG, FRANCE

I. INTRODUCTION	219
II. PHASE DIAGRAM	220
1. *T(x) Projection*	220
2. *P(T) Projection*	221
3. *Field of Existence*	225
III. SYNTHESIS AND PURIFICATION	229
1. *Purification of Elements*	229
2. *Tube Graphitization*	231
3. *Synthesis*	231
IV. GROWTH OF BULK CDTE	232
A. STOICHIOMETRIC GROWTH METHODS: CONGRUENT OR NEAR-CONGRUENT MELTS	233
1. *Zone Melting*	233
2. *Bridgman Methods*	234
B. SOLVENT GROWTH METHODS	239
1. *Te-Rich Solution Bridgman*	239
2. *Traveling Heater Method*	240
3. *Direct THM Synthesis and Growth*	244
V. HIGH RESISTIVITY MATERIALS	245
1. *High Purity Materials*	245
2. *Compensation*	246
VI. EXPERIMENTAL RESULTS AND CONCLUSION	254
References	255

I. Introduction

The various material preparation techniques for applications of cadmium telluride (CdTe) can be found in many books, treatises, proceedings and articles, depending on the specific application, including (Zanio, 1978; Siffert and Cornet, 1971; *Rev. Phys. Appl.*, 1977) infrared windows, electro-optical modulators, luminescent diodes, photorefractive solar cells or nuclear detectors. Each application requires a specific preparation method. Cadmium telluride can be produced in low or high electrical resistivity, n- or p-type, bulk or thin layer material, in polycrystalline or monocrystalline ingots. We restrict ourselves here to those methods suitable for nuclear detection because this constitutes the most promising application to date.

For this purpose, one needs a reasonable mobility lifetime product for both electrons and holes, high resistivity, a low number of grain boundaries and a low concentration of trapping centers. At the same time, production rate and reproducibility are of interest.

The quality of the material and its electrical characteristics are dependent on structural defects and chemical dopants: vacancies, interstitials, loops, and associations. The behavior of these defects is determined by composition, temperature and pressure, which are the main phase diagram parameters.

II. Phase Diagram

To justify the choice of crystallization method, one must first describe the thermodynamic properties of the cadmium–tellurium system (phase diagram and equilibrium relationship between components), which summarizes the various combinations of composition, temperature and partial pressure under which the three phases—solid, liquid, and gas—coexist in equilibrium. One must remember that, in the growth of CdTe crystals, the solid CdTe is formed under quasi-equilibrium conditions.

The phase diagrams are a function of three variables: temperature, T; pressure, P; and molar fraction, x. In practice, the $T(x)$ and $P(T)$ projections are of main interest.

1. $T(x)$ PROJECTION

Figure 1 gives the projection of temperature–composition. We see that the systems have only one compound, with equiatomic proportion CdTe that melts congruently at 1092°C. The two liquidus lines indicate the temperature at which a liquid with a given composition will crystallize in the absence of supercooling. At the same time, the figures shows the composition of the liquid phase in equilibrium with solid CdTe. This $T(x)$ projection was obtained from the fit of many results and data from Kobayashi (1911), De Nobel (1959), Lorenz (1962a), Steininger, Strauss, and Brebrick (1970). The shape of the liquidus curve, studied by Jordan (1970), is quite asymmetric: the temperature decreases more rapidly on the nonmetallic Te rich side. One can see an inflection point on each side of the diagram and a rise to a peak at the maximum melting point with an acute angle. This is a general trend for other II–VI compounds as well, and it is due to the strong

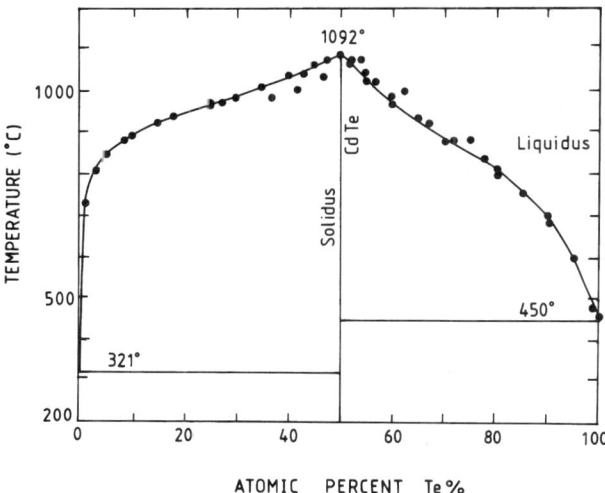

FIG. 1. CdTe phase diagram $T(x)$ temperature-composition projection.

deviation from ideality in the liquid phase, associated with the ionic bonding contribution for the Cd–Te system, even if CdTe is predominantly covalent. The ionic portion is enough to produce a maximum at the melting point. This behavior is not present in III–V compounds where the bonding is still more covalent: for InSb, which is in the same row as CdTe in the periodic table, the melting point is 525°. Due to the large ionic contribution, an increase of the free energy of the Cd–Te system is obtained by ordering and solidification near the stoichiometric composition, rather than far from stoichiometry. One can introduce the interchange energy in calculations by using random pairing or the quasi-chemical approach (Steininger et al., 1970). However, non-random pairing seems to be present and a thermodynamic model for the liquidus of binary system A–B in equilibrium with AB compounds and congruent melting point was developed by Jordan (1970). It is well described by Zanio (1978), but one should note that all models fail to differentiate the metal rich from chalcogen rich side. They predict a symmetrical shape of the liquidus while all the II–VI phase diagrams are unsymmetrical and require different treatments for every part of the diagram, metal or chalcogen.

2. $P(T)$ PROJECTION

To determine the best conditions of CdTe crystal growth, one has to take into account the properties of the vapor phase in equilibrium with liquid and solid phases. It will be seen later that vapor pressure can be a determinant parameter during growth or annealing treatments.

With regards to the composition of the vapor phase in equilibrium with the condensed phase for CdTe, mass spectrographic measurements (Drowart and Goldfinger, 1958; Goldfinger and Jeunehomme, 1963) show that CdTe molecules do not exist in vapor form, only $Cd(g)$ atoms and $Te_2(g)$ diatoms exist as gaseous species in the vapor phase, and this seems to be the general rule for all II–VI systems. However, the partial pressure of Cd and Te are not independent and are related by an equilibrium constant. This means that only one pressure of Cd or Te defines the system from which the second partial pressure can be directly deduced.

To begin, consider the qualitative behavior of the Cd partial pressure (P_{Cd}) and of the Te partial pressure (P_{Te2}) along an isotherm during the variation of composition of Figure 2a (note that this figure is a schematic diagram, only for explanation).

The chemical potential of one constituent rises uniformly with the atomic fraction of this constituent. As a result, the partial pressures of the constituents (P_{Cd} and P_{Te2}) change with the average composition (x) when only one condensed phase is present, but remain constant when two phases are present. This can be seen in Figure 2b, where it is seen that when P_{Cd} increases, P_{Te2} decreases.

One point at the solid line Cd rich side is in equilibrium with Cd vapor at a pressure $P(A)$ and Te vapor at a pressure $P(A')$; the same behavior is seen for the Te rich side at a point with corresponding equilibrium values $P(B)$ and $P(B')$.

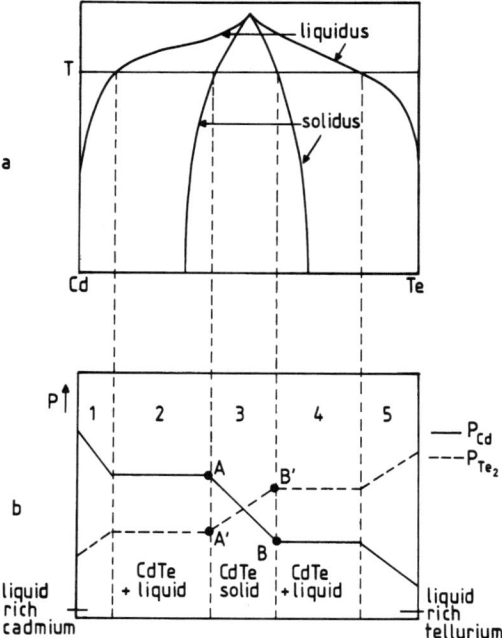

FIG. 2. Schematic behavior of (Cd) and (Te) partial pressure with composition x.

6. GROWTH METHODS OF CdTe NUCLEAR DETECTOR MATERIALS

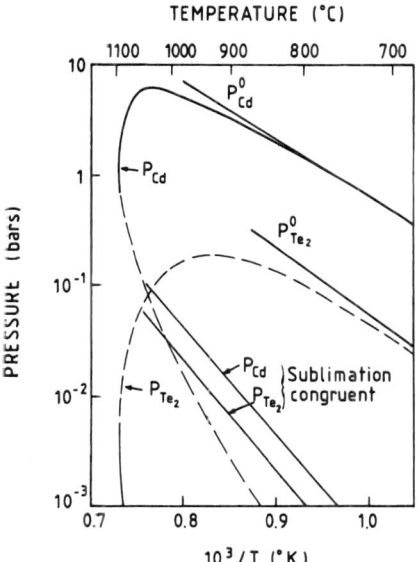

FIG. 3. CdTe phase diagram pressure–temperature, $P(T)$, projection.

The $P(T)$ diagram locus of points A, A' and B, B' when the temperature changes was determined by De Nobel (1959), Lorenz (1962a), Brebrick and Strauss (1964), and theoretically by Jordan and Zupp (1969) (Figs. 3 and 4). Strauss (1971) shows a synthesis of these results. First, consider the solid line of the curve $P_{\text{Cd}A}$. It represents the Cd vapor pressure at equilibrium with the solid Cd rich side (s_A), while the dashed line of the same curve $(P_{\text{Cd}B})$ gives the Cd vapor pressure at equilibrium with the solid Te rich side (s_B) (Fig. 4). The same connection exists between the curve representing the Te pressure $(P_{\text{Te}2})$. One can

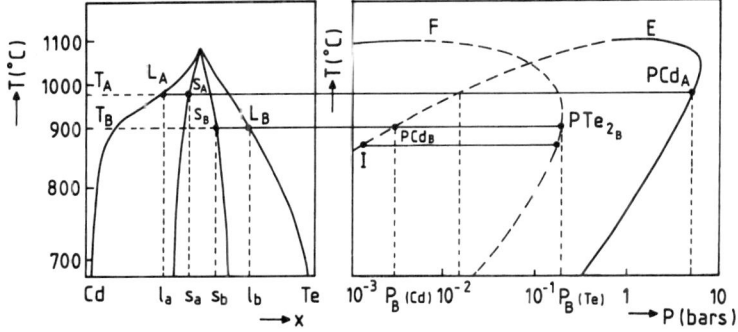

FIG. 4. Correspondence of phase diagram $T(x)$ and $P(T)$ projections.

see the correspondence between the $P(T)$ projection and $T(x)$ projection with the solid Cd and Te rich sides (Fig. 4).

As can be seen in Figure 3, at low temperatures, P_{Cd} and P_{Te2} tend to the lines of P^0_{Cd} and P^0_{Te2}, which are respectively pure Cd and Te_2 saturated vapor pressure. Points E and F represent the pressure of these elements at the maximum fusion temperature $1092°C$. One can see that, when all three phases coexist at equilibrium, the maximal pressure is $P_{Cd} = 6.6$ bars.

Quantitatively speaking, when a Cd–Te system is annealed under vacuum, partial pressures are related by the reaction equation and the corresponding mass action law:

$$CdTe(s) \leftrightarrow Cd(g) + \frac{1}{2}Te_2(g) \tag{1}$$

$$K_{CdTe}(T) = P_{Cd} P_{Te2}^{1/2} = \exp\frac{\Delta G}{RT} \tag{2}$$

where s = solid, g = gas, K = equilibrium constant of the reaction, ΔG = free energy formation, R = perfect gas constant. Brebrick and Strauss (1964) have determined these expressions:

$$\log K = -15 \cdot \frac{10^3}{T} + 9.82 \tag{3}$$

$$\Delta G(Kcal/mole) = -68.64 + 44.94 \, 10^{-3} \, T \tag{4}$$

Sublimation is a special case of equilibrium between the solid and vapor phases. When the composition x is the same in both cases, the locus of these points is represented by the two "lines of congruent sublimation," which must satisfy the stoichiometric condition:

$$P_{Cd} = 2 \, P_{Te2} \tag{5}$$

In this case, the total pressure P_T is minimum and congruent sublimation lines between 780 and 940° are given by Brebrick and Strauss (1964):

$$\log P_{Te2}(sub) = -10 \cdot \frac{10^3}{T} + 6.346 \tag{6}$$

$$P_{Cd}(sub) = 2P_{Te2}(sub) \tag{7}$$

If CdTe is annealed under high temperature near these last conditions without

6. Growth Methods of CdTe Nuclear Detector Materials   225

imposed P_{Cd} or P_{Te2}, intense sublimation occurs following the relation (1) toward the right side. In this case, one must note that, theoretically, an applied high pressure of inert gas will not change the situation, knowing that the diffusion rate of a chemical species in the vapor phase is directly proportional to its partial pressure. In this case, P_{Cd} and P_{Te2} are still equal, both Cd and Te diffuse normally, and sublimation is still present. In contradiction, Vandekerkof (1971) reports the necessity of a high inert gas pressure, higher than the dissociation pressure P_d, to avoid sublimation. This is the case in the Czochralski method, for example.

Annealing of CdTe under a high pressure of Cd or Te avoids sublimation because, following Eq. (2), an increasing pressure of one constituent results in a decrease in the pressure of the others. This is the reason for the necessity of imposing a pressure of Cd or Te during annealing, to avoid sublimation and the resulting defects.

3. Field of Existence

To this point, a CdTe crystal was considered as an equiatomic compound: 50% Cd, 50% Te (stoichiometric). However, at a given temperature, one component is still stable in a composition range, the range of homogeneity. The boundaries of this field on the Te and Cd rich sides are defined as the solidus lines. However, this range is far too narrow (central line in Fig. 1) to be measured by classical chemical techniques. The most useful information is derived from electrical measurements on crystals treated to obtain deviations from the stoichiometry. If the defects resulting from these deviations act like a donor or acceptor, electrical measurements are then possible and allow the detection of minute deviations of the original constituent if there is no other external influences such as impurities or complexes.

This means that, to establish solidus lines from electrical measurements, one must first define a *correct model of defects* present in the materials used but keep in mind that such a determination can be easily and always *contested* due to the large number of parameters and the low concentration of these defects. Some restrictive hypothesis are needed: (1) every stoichiometric deviation leads to an active electrical defect; (2) only structural defects are present, impurities are neglected; (3) crystal cooling after annealing can modify the results by associations, the only reliable results are those made directly at high temperature; and (4) at lower temperatures, the study of native defects and their associations is needed.

  i. Native Defects. In a MD (M = metal, D = nonmetal) compound, the simplest defects are vacancies (V_M, V_D) and interstitials (M_i, D_i). Others are possible, too. Here one must note that different types of defects are not independent; reactions occur, leading to an equilibrium where different associations are possible.

ii. Frenkel Defects. M leaves its site to an interstitial site, creating V_M and M_i. The related equilibrium equation and the applied mass action law are

$$M_M \leftrightarrow M_i + V_M \tag{8}$$

$$[V_M][M_i] = K_F = C_F \exp(-W_F/KT) \tag{9}$$

where K_F, W_F = Frenkel constant and energy, K = Boltzmann constant, C_F = variable, depending on the number of possible interstitial sites and their density (the same things can be applied to D).

iii. Schottky Defects. When displaced atoms move to available vacancies at the surface (S), leaving only vacancies in the bulk, again

$$(M_M + D_D) + (2V_S) \leftrightarrow (M_S + D_S) + (V_M + V_D) \tag{10}$$

$$[V_M][V_D] = K_S = C_S \exp(-W_S/KT) \tag{11}$$

where K_S = Schottky constant and C_S = variable, depending on the site concentration.

a. CdTe Case

On the Cd side, when annealing under increasing Cd pressure (decreasing P_{Te}), CdTe material begins as p-type and changes to n-type (De Nobel, 1959). The conclusion was that the p-type is associated with a lack of Cd (V_{Cd} or Te_i) while the n-type is associated with an Cd excess (Cd_i or V_{Te}). Cd radiotracer diffusion experiments show very high diffusion, interpretable only as interstitial bonds, which means that one is observing Frenkel type defects ($V_{Cd} + Cd_i$).

Consider now the case of donors. Their number increases with P_{Cd}. Two cases are possible: single ionization and double ionization.

With single ionization, the equilibrium equation and mass action law are

$$Cd(g) \leftrightarrow D^+ + e^- \tag{12}$$

$$\frac{[N_D^+][n]}{P_{Cd}} = K_1 \tag{13}$$

where D^+ = singly ionized donor, N_D^+ = D^+ concentration, n = electrons concentration.

If there is no other donor, the neutrality equation becomes

$$N_D^+ = n \tag{14}$$

6. GROWTH METHODS OF CdTe NUCLEAR DETECTOR MATERIALS

solving Eqs. (13) and (14):

$$n = K'_1 \cdot (P_{Cd})^{1/2} \tag{15}$$

With double ionization, if the donors are doubly ionized, in the same manner,

$$Cd(g) \leftrightarrow D^{++} + 2e^- \tag{16}$$

$$\frac{[N_D^{++}][n^2]}{P_{Cd}} = K_2 \tag{17}$$

$$N_D^{++} = \frac{n}{2} \tag{18}$$

$$K_2 = \frac{n^3}{2P_{Cd}} \tag{19}$$

$$n = K'_2 \cdot (P_{Cd})^{1/3} \tag{20}$$

Consequently, in principle, it is easy to see the charge state of the structural ionized donors by looking at the log $n = f(\log P_{Cd})$ function. If the gradient is 1/2, the donor is singly ionized. If it is 1/3, it is doubly ionized, provided that no other donor (impurities) are present.

Many authors (Whelan and Shaw, 1968; Zanio, 1970; Smith, 1970) found 1/3 but Matveev et al. (1969) and De Nobel (1959) found 1/2. After quenching to RT, Smith (1970) has established the carrier concentration on the n-type side versus P_{Cd} by Hall measurements at high temperature:

$$n = 6.9 \ \partial \ 10^{18} \exp\frac{W_{inc}}{3KT} \cdot (P_{Cd})^{1/3} \tag{21}$$

where n is in cm^{-3}, $W_{inc} = 1.7$ eV, and P_{Cd} is in Torr.

The intersection of the isothermal lines of Eq. (21) with the Cd saturated leg of the $P(T)$ projection in Figure 3 gives the concentration of $[n]$ along the Cd rich side of the solidus and under the cited restrictions. Smith (1970) found

$$n_S = 3 \cdot 10^{21} \exp\left(-\frac{0.91 \text{ eV}}{KT}\right) \tag{22}$$

This shows that the solubility of the Cd excess in CdTe is retrograde, and Smith (1970) supposes that the melting point is on the Cd rich side, although other authors put it in the other side. Theoretically, the situation is not clear.

By temperature dependent Hall effect measurements in undoped crystals annealed at low Cd pressure (or high Te_2 pressure), De Noble (1959) found p-type conductivity on the Te rich side. He assumed that a native acceptor is responsible for that situation and he measured an energy activation for this defect of 0.15 eV, supposed to be the first ionization level. Strauss (1971) however contested this conclusion, due to the limited hole concentration, and suggested the influence of impurities. Lorenz and Segall (1963b) observed an acceptor level at 0.05 eV in n-type crystals converted to p-type by annealing under Te pressure and quenching. They proposed this level as the first ionization level, but without the study of the hole concentration as a function of pressure. Smith (1970) performed this study by measurement of the Hall coefficient as a function of P_{Te_2} at high temperatures, up to 1000°C. He observed that the hole concentration was constant over one magnitude of P_{Te_2} variation, which excluded as a first approximation the native defect acceptor to be responsible for the p-type conductivity. He concluded that impurities were responsible. It will be seen later that both of these associations account for this hole concentration and p-type conductivity.

As a conclusion of this section concerning the determination of solidus lines (the field of existence or the homogeneity region), electrical measurements on Cd rich side are well enough accepted to be representative of a solidus line, assuming a doubly ionized defect for the deviation from the stoichiometry. On the Te rich side, the situation is more complicated. The electrical activity of the impurities do not allow us to substitute the carrier concentrations and deviation from the stoichiometry. If one assumes that the measured carrier concentrations are those of acceptor native defects, these carrier concentrations represent only the upper limit to the deviation from the stoichiometry on the Te rich side of the solidus line. The real line is certainly located at a lower concentration if we assume again no compensation by impurities. However, this is a starting point and the final diagram of the solidus lines can not be established without knowledge of the correct model of the native defects, which will be discussed later. However, we show in Figure 5 the solidus lines calculated by De Nobel (1959) and Smith (1970).

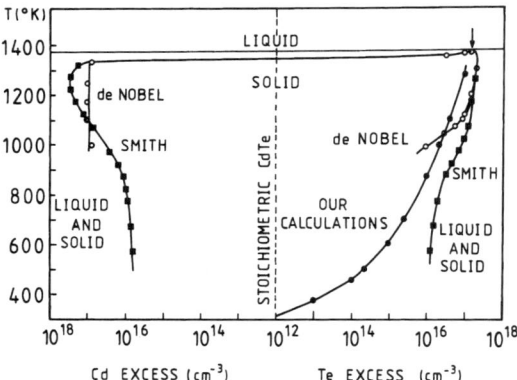

FIG. 5. Detailed "existence region" of solid CdTe.

In the following sections, the methods of purification, synthesis, and crystal growing, devoted especially to the elaboration of high resistivity materials of good crystallinity characteristics, are presented, prior to discussing special cases of these materials and the compensation models.

III. Synthesis and Purification

Purification and crystal growing generally run parallel. The quest for electronic grade purity materials demands the strictest precautions against impurities, and if one can start with 99.9999% ($6N$) elements, one can avoid many sources of subsequent problems.

1. PURIFICATION OF ELEMENTS

Cd and Te elements with 99.999% ($5N$) and $6N$ purity are commercially available, but even though this normally can be considered as rather pure from a chemical point of view, it is still of poor quality for semiconductors, where one or to decades more are needed. This can be reached actually, but the prices and handling problems limit the acquisition of such elements. The use and handling of $7N$ elements or more, without contamination, is extremely difficult. For this reason, many crystals growers prefer to do purification in situ to avoid this contamination problem. Some of these purification processes, generally used for high resistivity materials growth are discussed in the following.

a. Cadmium

The segregation coefficient for the majority of impurities in Cd is around 1, making the purification of Cd by zone melting not very useful. While it is possible to reduce the concentration of Cu, Pb, Sn, and Na, the Au, Ag, and Li concentration remains constant. Many authors (Aleksandrov, 1960; Wernick, 1960; Mochalov, Urubkova, and Lepis, 1966; Schaub and Potard, 1971) have studied the classical purification by zone melting. Under Ar, He, H_2, N_2, or mixed gas, the concentration of P, S, As, and O or other volatile impurities are reduced, but metallic ones, as seen before, remain constant. Resistivity measurements show purities of $6.5N$.

Good results have been obtained by distillation also (Silvey, Lyons, and Silvestri, 1961; Aleksandrov and Udovikov, 1973), generally under vacuum, but H_2 can be used to reduce O. Cu and Pb are difficult to remove.

However, processes combining both methods are generally used, and purities of $7N$ have been achieved (Schaub and Potard, 1971). Purities are measured by resistivity, spark mass spectrometry, and atomic absorption. Many results are reported in the literature (Zanio, 1978).

b. Tellurium

Segregation coefficients of more than 32 impurities in Te (Zanio, 1978; Zaiour, 1988) are well known to be favorable for zone melting purification, with the exception of Cd with $K_0 > 1$ and for Se and Na with $K_0 \approx 0.2-0.5$. The presence of H_2 during the refining seems to be effective in removing impurities, due to the evaporation coefficient of the volatile elements. The general form of the distribution impurities (C_{ng}) along the ingot after n passes through the oven and taking into account this evaporation (Zaiour, 1988) is given by the following formula:

$$\frac{C_{ng}}{C_0} = \left[\frac{K}{K+g}\right]^n \left\{1 - (1 - K - g)\exp\left[-K + g\frac{x}{b}\right]Z_n\right\} \quad (23)$$

with b = zone length, C_0 = original impurities concentration, n = crossing number, K = effective liquid segregation coefficient, g = effective evaporation coefficient, x = length position of C_{ng}, $Z_n = f(n, K, g) \to 1$ when $x \to \infty$.

In this latter case, the stationary state occurs at saturation with

$$C_{ng} = \left[\frac{K}{K+g}\right]^n \quad (24)$$

when $g = 0$, Eq. (23) gives the classic Pfann's formula for solid–liquid segregation. One can determine the number of passes for good purification of C_{ng}/C_0 as

$$n = \theta \cdot V \cdot \log\left(\frac{C_0}{C_{ng}}\right) \quad (25)$$

$$\theta = \frac{nL}{V\left[L \cdot \log\left(\frac{C_0}{C_{ng}}\right)\right]} \quad (26)$$

with V = zone melting speed and L = ingot length.

However, after a number of passes, one has a certain competition between purification and contamination, which leads to an equilibrium state. With additional passes, which do not help in additional purification, it seems that for three zones–oven systems, 10 passes is the optimal value (Zaiour, 1988).

Purification of Te by zone melting was the subject of many studies. Baïmakov and Petrova (1960), Schaub and Potard (1971), and others reached a purity of $6N$.

Another method of purification was used also, namely the distillation of Te under vacuum or Ar–H_2 atmosphere. This is also effective for purification, where purities of $6N$ were reached (Vanyukov, Koshitov, and Bulanov, 1969; Abdullaev et al., 1965).

However, the best results are obtained by combining both methods. From Kujawa (1963) to Zaiour (1988), carrier concentration of $1-4 \times 10^{14}$ cm^{-3}, equivalent to 7N, have been achieved.

A nonclassical method of purification consists of the addition of a few 100 ppm of CdTe, prior to the refining by zone melting (Redden, Bult, and Bullong, 1986). Such an approach, in contradiction with all other theories, seems to be effective. This may be due to the formation of complexes with better K_{eff} or by gettering.

2. Tube Graphitization

The melting temperature of CdTe is 1092° C, but Cd sublimes at above 320° C. That means that CdTe cannot be grown without liquid or solid encapsulation to prevent the loss of volatile components. Vitreous carbon, boron nitride, and silica tubes can be used. The last one is the most useful for complex facilities.

The following are a few remarks regarding silica tubes: the quality of the silica tubes depends on their origins. They can contain many impurities, especially

Al	up to 100	ppm.W	Li	up to 2	ppm.W
Fe, Ti	up to	0.8 ppm.W	Na	up to 1	ppm.W
K, Ca	up to	3 ppm.W	Mn, Mg	up to 0.2	ppm.W

At high temperatures, these impurities can diffuse into CdTe and severely contaminate it. Water, in the form of OH ions, at high concentrations, can decrease the temperature of softening of silica up to $\approx 1300°$ C and less, leading to the deformation and the explosion of the tube. The OH ion concentration can easily be measured by IR absorption at 2.7 μ in a classical spectrometer at room temperature.

Another problem arises from the fact that oxygen is always present in elements in oxide form. These oxides, especially CdO, react with silica to form Cd silicate (SiO_3Cd) compounds, and lead to the breakage of the tube. Coating the inner surface of the tube with a thin layer of graphite is needed to avoid contact between metals and the tube. This layer must be formed with great care. The tube must be successively cleaned by trichlorethylene, acetone, methylic alcohol, and a 10% HF solution, rinsed in pure water, evacuated to 10^{-6} Torr, and heated at 1000° C for 1 day. The tube is then slowly filled to 150 Torr of pure methane, which is cracked at $\approx 1000°$ C for 1 hr on the inner surface. This layer of carbon formed a few microns thick is then baked at 1000° C for a few hours to increase the adherence to the tube. The rupture of the vacuum is done under H_2 or Ar pressure (methane cracking can be replaced by benzene or alcohol).

3. Synthesis

To avoid contamination, all preparations must be performed under a laminar flow hood with high efficiency filtration. Cd and Te elements (5N–6N) as purchased or formerly purified and are used in bars to limit the surfaces available for oxidation. Cd is etched in a "nitrol" solution (4% NO_3H in methylic alcohol), because this has been considered to be the most important oxide source. But, as will be seen later, tellurium may be the major source of oxide. Tellurium can be etched in a Br (2–10%)–methanol solution. However, Br can be found later as a trace impurity in the compound with important electrical consequences (see later). In some cases, other solutions such as dilute HCl are preferable. The elements are quickly introduced into the prepared tube in equiatomic proportion within the weighing precision.

Depending on the required vapor pressure, several options are possible:

1. A synthesis can be performed in a two zone Bridgman type oven, where the pressure is determined by the first zone temperature (Cd or Te container), while in the second zone, the synthesis takes place.

2. A Te excess of 1–5% can be added to follow the Te equilibrium loop in Figure 3 and the maximum pressure of Cd that can be reached in the tube is around 0.8 bar.

3. A Cd excess, even at trace levels allows the pressure to reach values as high as 6–8 bars, and higher if one is out of equilibrium by rapid heating.

The choice is determined by the subsequent application and the required characteristics.

The tube is then exhausted to 10^{-6} Torr, with liquid nitrogen traps used to trap contamination from oil pumps and sealed. Strictly speaking, it is preferable to perform synthesis in a horizontal position at the beginning and to manage the free surface over the liquid to avoid, for instance, local overpressure. In this kind of synthesis, the tubes have a diameter ranging from 15 to 25 cm, and a length up to 50 cm, allowing charges up to 500 g of CdTe. In a well-regulated furnace, the temperature rises very slowly (thanks to automatic control from room temperature (RT), 1100°C in 24 hr), to about 800°C where the exothermic reaction of the formation of CdTe takes place. The temperature then rises rapidly up to the melting point of CdTe owing to the heat of reaction. The major opportunities for explosion occur at this moment if the pressure equilibrium conditions are not respected, due to the rapid temperature increase of the mixture. When the synthesis is achieved, the furnace is tilted to the vertical position to obtain cylindrical ingots and the temperature is slowly lowered to RT.

Growth of CdTe crystals and purification of the material go together. Purification will be treated simultaneously during the discussion of various growth methods.

IV. Growth of Bulk CdTe

Monocrystals of CdTe have been grown since 1958 by De Nobel, Kyle, Triboulet, Bell, Maximowski, and others. A complete bibliography up to 1978 is given by Zanio (1978).

The metallurgy of this material uses classical methods of growth: vapor phase transport, Bridgman and related methods, zone melting, growing from solvents, Chzochralski, and many other methods. However, these methods have required adaptations particular to CdTe, which is a very volatile compound and consequently unstable at high temperatures.

The analysis of the published experimental results allows one to see that generally every method leads to a special kind of material, with specific, well-determined characteristics. Zone melting induces n or p low resistivity, THM leads to a p-type material with high resistivities, and so on.

In this chapter, the discussion is restricted to the methods allowing for the fabrication of detectors working at RT. However, simplified methods or the combination of a few methods can help one to reach the required characteristics.

A. Stoichiometric Growth Methods: Congruent or Near-Congruent Melts

1. ZONE MELTING

The zone melting method is well known for purification as seen before. The principles were well established by Pfann (1958). This method is based on the fact that the solubility of a given impurity is generally higher in the liquid C_L than in the solid C_S. The more $K = (C_S/C_L) < 1$, the more efficient is the purification. The coefficient K is given by Pfann's formula:

$$C_S = C_0 \left[1 + (K - L) \exp\left(-K\frac{x}{L}\right) \right] \tag{27}$$

where C_S is the concentration of the impurities after one pass, x is the position where C_S is measured, and b is the length of the melted zone.

Pfann established this theory for a horizontal tube setup. De Nobel (1959) has used this configuration as well. However, the vertical arrangement is more generally employed since it avoids contamination by vapor phase impurities through the free surface at the top of the horizontal melt. The vertical tube use was first suggested by Heumann (1962) and used on CdTe by Lorenz and Halsted (1963a).

Zone melting can be used as a congruent or near congruent method, either for synthesis by introducing Cd and Te elements, for purification as indicated before, or for crystal growth. In the two last methods, one must avoid a free volume on the top of the ingot for CdTe vapor decomposition. This is done by sealing with double plugs as shown in Figure 6. All three purposes require temperatures around 1150°C in a very small space. This is reached by an RF (1 MHz) induction coil and a susceptor (ring of graphite) that have two configurations:

1. For purification, the ring of graphite is split on one side. In this case, the external surface of the ring is the primary of a transformer and coupled to the

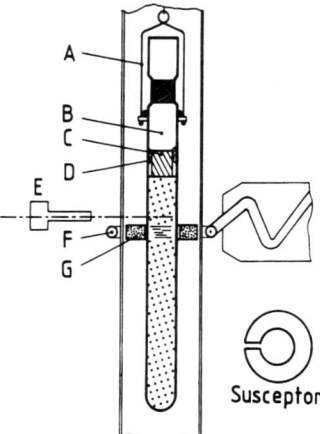

FIG. 6. Schematic of melting zone technique. A, Holder; B, silica tube; C, silica stopper; D, condensed CdTe; E, pyrometer; F, inductor; and G, susceptor.

inductor coil. The internal surface is the secondary of the same transformer and coupled to the CdTe in the tube. It provides, first, the needed preheating of CdTe to assist the coupling and, second, some mixing favorable to the purification in the melted zone by means of the high frequency currents in the melt. The thermal regulation is accomplished by an optical pyrometer focused on the melted zone. The complete cycle of purification includes 20 passes at a speed of 2–3 cm/hour.

2. For crystal growth, one must avoid the induced HF currents, to ensure better stability in the liquid and at the interface. In this case, the split graphite ring is replaced by an unsplit ring, which acts as a normal furnace. The optical pyrometer or thermocouple measures the temperature of the material within the ring. The crystal growth is then done by the last pass at a lower speed of 5 mm/hour.

Zone melting allows growth of single crystals up to 10 cm^3 in volume. The doping type remains the same as in the initial type after synthesis and imposed pressure. Generally n-type material is found with the resistivity around 10–500 $\Omega \cdot$ cm and the best measured values of electron mobility are $\mu_e \approx 1.5 \cdot 10^5$ cm^2/volt $\times$ second at 35 K and 10^3 cm^2/volt $\times$ second at RT (Triboulet, 1971; Woodbury, 1974).

Overpressure of Cd or Te or doping can be employed to change the conduction type or the resistivity within certain limits. This will be discussed later.

2. BRIDGMAN METHODS

This method is the most often employed technique in crystal growth: liquid material crosses over a negative temperature gradient where solidification occurs.

When the container of liquid is moving it is called the "Stokberger method," while when the temperature gradient is moving this is the Bridgman method. Presently, "Bridgman" is the name generally used for both methods.

This method can be used for synthesis by the introduction of elemental Cd and Te prior to heating or for crystal growth, but less so for purification even if a certain degree of purification occurs during the growth. From the technical and physical points of view, two kinds of methods must be considered: vertical and horizontal Bridgman

a. Vertical Bridgman

This technique is most extensively used, because it produces large crystals of uniform size and characteristic ingots, while it also allows some freedom in controlling the stoichiometry, the type, the conductivity, and the imposed overpressure. It can be seeded to favor monocrystals. Appearing in Figure 7, the furnace includes three zones of temperature: $T1$ to impose Cd or Te overpressure following $P(T)$ projection (Fig. 3); $T2 > 1092°C$ for the melt; $T3 < 1092°C$ with the gradient between $T2$ and $T3$ allowing for the solidification.

Many authors attach great importance to the value of this gradient while Kyle (1971) states the need for a sharp gradient $> 20°C/cm$. Tanaka *et al.* (1987) and

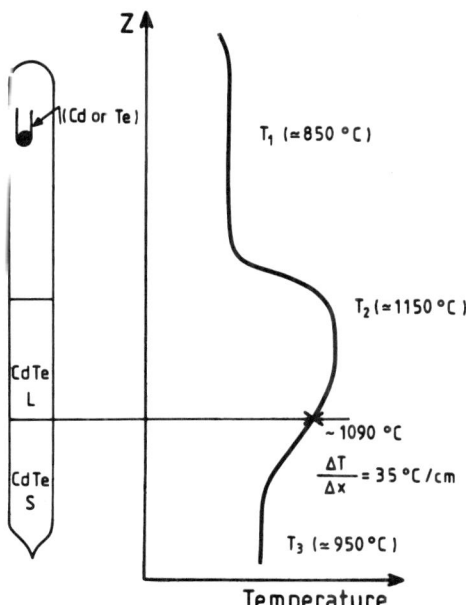

FIG. 7. Schematic of Bridgman crystal growth technique.

Pfeiffer and Mühlberg (1992) state the opposite opinion with a gradient as soft as possible, 4° C/cm or less. In any case, the most important parameter is the liquid–solid interface shape and not the gradient. Both depend on the furnace design and the heat exchange. The convex interface pointed in the direction of growth leads to monocrystals.

Effectively, thermal conductivity is a determining factor for growth rate and interface shape. A speed much more than 0.5 cm/hr leads to a concave shape for a tube of 2.5 cm diameter because it assists transverse heat flow and decreases axial heat flow. The reverse leads to a convex shape. To be exact the interface shape is a function of k_s/k_l: solid conductivity over the liquid zone at the interface melting temperature (Pfeiffer and Mühlberg, 1992) with the ratio ≥ 1 leading to a convex shape and the ratio <1 leading to a concave one that produces polycrystals.

Forced cooling of the ingot nose can accelerate the axial flow of heat with subsequent improvement in crystal size. This forced cooling can be simulated by increasing the length of the ingot (or decreasing the aspect ratio $a = d/l$). The length of the ingot acts as a heat sink cooler.

The pressure above the melt greatly affects the crystallinity with Cd pressure around 1 atm appearing to be the best value.

Rotation should, in principle, smooth out asymmetrical furnaces and minimize radial fluctuation, but no evidence of this has been observed, both rotating or not rotating variants appear to be effective.

i. Temperature of the Melt. This temperature is of great importance but it is a subject of controversy and debate. Crystal growers such as Kyle (1971) found experimentally that melt temperatures well above the melting point of CdTe promote a concave interface and polycrystals. These results were confirmed later by Steer *et al.* (1992) with finite element simulation of the temperature gradient and with experimental results.

On the other hand, finite element simulations show the influence on the ratio k_s/k_l of the geometric aspect ratio and of the relative position of the furnace tube (Pfeiffer and Mühlberg, 1992) in accordance with experimental results. At the same time, these experimental results show the importance of superheating with $\Delta T > 10°C$ (Rudolph and Mühlberg, 1992). According to these authors, due to the covalent character of CdTe in *low heated* ($<$ superheating temperature) melt, an arrangement of atoms in "tetrahedral" clusters is predicted; while in *superheated* melt, these clusters are dissociated into molecules or at least in to rings, chains, or atoms. Tetrahedra or highly organized clusters are assumed to reduce the nucleation energy of polycrystals; this means that superheating avoids multiple nucleation. This can be demonstrated experimentally by showing the supercooling–superheating relation. In Figure 8, it can be seen that low heating $\Delta T^+ < 10°C$ for the melt results in a total absence of supercooling, while superheating of $\Delta T^+ > 10°C$ shows an abrupt supercooling up to $\Delta T^- = 25°C$, which demonstrates the absence of undesirable nucleation. Rudolph and Mühlberg (1992) also

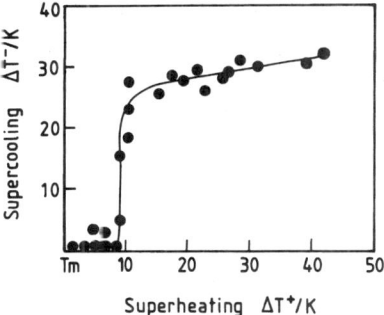

FIG. 8. Supercooling as a function of superheating. (Reprinted with permission from Rudolph and Mühlberg, 1993 and Elsevier Science S. A.)

have shown experimentally that highly superheated (>10°C) melt leads to monocrystals.

ii. Cd Overpressure. It was seen before that it is possible to establish a Cd overpressure in the first furnace ($T1$). Here one has two kinds of equilibria: between vapor and liquid and between liquid and solid. The liquid phase can be considered as a communication medium between vapor and solid. Many authors have shown that overpressures around 1 atm (or 800–850°C temperature of the Cd source) corresponding to the Cd pressure of conversion from n- to p-type (Kyle, 1971; Rudolph and Mühlberg, 1992) leads to better crystallinity, lower carrier concentrations, higher resistivity, homogeneous characteristics, high IR transmission, and some purification.

iii. In Doping. The Bridgman method is well known in terms of supplying good nuclear detector material when doped with In. If the solidification follows the cooling slowly, the electrical properties of the material are determined by the In concentration and vapor overpressure, while if the ingot is rapidly cooled (quenched), the electrical properties are dominated by the overpressure or quench temperature. As examples (Kyle, 1971): for $2.5 \cdot 10^{17}$ In, 1 atm Cd, quench $T > 1000°C$, $\rho = 10^6 - 10^7 \, \Omega \cdot cm$; for $2.5 \cdot 10^{17} - 2 \cdot 10^{18}$ In, 1–2 atm Cd, quench $T < 1000°C$ or slow cooling, $\rho < 1 \, \Omega \cdot cm$; for $2.5 \cdot 10^{17}$ In, 0.84 atm Te, slow cooled, $10^8 - 10^9 \, \Omega \cdot cm$.

b. *Horizontal Bridgman*

As in the vertical Bridgman method, a sealed tube with three horizontal temperature profiles is used. The main difference is that the vapor phase pressure is in equilibrium with both the liquid and solid phases at the same time. However, this can be an inconvenience for contamination from the vapor phase. This method was used by Lorenz (1962b) and Medvedev *et al.* (1968) and crystals doped with Cl for high resistivity for nuclear detectors were obtained (Arkadyeva and Matveev, 1977) and Matioukhin (Riga). Generally, however, it seems that using the horizontal Bridgman method with an excess of one of the components, to prevent the decomposition of the molten zone, is well adapted to nuclear detection material.

c. High Pressure Bridgman

The growth of CdTe crystals from a stoichiometric melt is the ideal solution to avoid inclusions on the one hand, or vacancies on the other hand, and is a classical situation during crystal growth. At 1100°C or more, it is difficult to prevent contamination of the melt by O, Si, Na, and many other impurities generally contained in the silica. One can achieve growth at lower temperatures by using a Te rich melt that gives the best reported detectors, as will be seen later. But the difficulty, by these methods, is in achieving large homogeneous crystals with inclusion free volumes. To overcome this difficulty, Raiskin (1970) and Raiskin and Butler (1988) developed a high pressure vertical Bridgman method, with pressures up to several hundred atmospheres at temperatures up to 1600°C. This method eliminates the need for high strength sealed crucibles like silica and allows the use of high purity containers such as boron nitride or graphite, which is an excellent oxygen reducer. In addition, high pressure allows for the imposition of temperature profiles that promote the growth of monocrystals. The elements are introduced, alloyed, lowered out of the heater, and program cooled to RT. Raiskin and Butler (1988) have grown 2 kg ingots, 7.5 cm in diameter. The electrical characteristics were poor, but uniform. These results led the authors to another approach, to improve lattice perfection by alloying ZnTe with CdTe to form the ternary crystal $Cd_{1-x}Zn_xTe$. Butler, Lingren, and Doty (1991) thus combined the structural perfection, achieved by Zn atoms on Cd site, and the advantage of the higher bandgap of CdZnTe, leading to lower leakage currents and higher operating temperature and resulting in a high degree of uniformity and in principle a better $\mu\tau$ product. Experimentally announced results were satisfactory: very high resistivity $5 \times 10^{10} - 10^{11} \, \Omega \cdot cm$, without intentional compensation, uniformity, and monocrystals. The stated absence of polarization is typical with the type of contacts used. The advantages of Zn introduction were explained by an enhancement of the covalent component of the interatomic bonding in the crystal (Sher *et al.*, 1985). But the role of the high pressure is a little blurred, with no indication of whether it is Cd, Te, or inert gas overpressure given! If it is Cd or Te pressure, high resistivities cannot be expected, due to the excess of one of the species. If it is an inert gas pressure, one cannot avoid evaporation of Cd or Te but may avoid the rapid evaporation of volatile impurities of column II or VI in the periodic table, leading to a compensation by species more difficult to hold back. This problem is still under study.

6. GROWTH METHODS OF CdTe NUCLEAR DETECTOR MATERIALS

B. Solvent Growth Methods

Growing CdTe from solutions presents many advantages, in particular, lowering the growth temperature as can be seen in the $T(x)$ projection phase diagram, of Cd or Te solution. Consequently, many parameters are more favorable:

- Reduction of the impurity acquired from the crucible,
- Lower solubility of the impurities in the solid phase,
- Better segregation coefficients,
- Reduction of the native defects in the crystal.

One immediately thinks of Te or Cd as solvents but other elements are possible also, such as Sn or Si (Rubenstein, 1968) or from $CdCl_2$ by Taguchi. However, for nuclear detector material, solvents other than Te have not been reported or successfully and repeatly used, so this discussion is restricted to Te rich solution methods.

1. TE-RICH SOLUTION BRIDGMAN

Called the "solution depletion" or "solution impoverishment" method, these are well represented by two approaches.

The first one, from Zanio (1974), consists of a flat bottom quartz ampule cooled by liquid circulation cold fingers (two concentric tubs). A typical starting charge contains 27 Wt % Cd, with an initial interface temperature of 900° C. Rather than lowering the ampule through the Bridgman-type furnace, its temperature is reduced by 4° C/hr. Nucleation occurs at the tip of the cold finger. The solid–liquid interface is convex, due to the sharp temperature gradient imposed by the finger. Crystals up to 20 cm^3, precipitate free, have been grown by this method. However, the method suffers from strict volume limitation because a 100 g ingot corresponds to a final solid–liquid interface temperature of 650° C, which is a limit for growth.

The second approach is that of Schaub et al. (1977), which can be summarized as follows: a sealed tube, ≈5 cm in diameter, filled by Cd and Te, with $x_{Te}^\circ = 0.6$, corresponding to 960° C is lowered in a vertical Bridgman furnace. Both synthesis and dissolution are simultaneous during the heating of the furnace. During the growth, the ingot decreases in length. The concentration of the CdTe in the Te solvent decreases continuously, leading to an increase x_{Te} and the corresponding interface temperature decreases along the liquidus line. For example, when starting with $y = 0$ mm, $x_{Te}^\circ = 0.5$ and $T = 960°$ C; and when $y = 138$ mm, $x_{Te} = 0.9$ and $T = 700°$ C. This results in two observations:

- The interface moves along the furnace out of the *suitable* axial and transverse gradient points.
- The crystallization speed is then different from the container speed.

These two problems need to be, and are, solved as follows. During the growth, the temperature T of the furnace is lowered as a function of the grown length (y), according to the liquidus phase diagram. This function, $T = T(y)$ can be calculated and imposed by a temperature regulator. Also, there exists a critical maximum speed of growth rate, under which constitutional supercooling and consequently concave interface are avoided. On the other hand, this growth rate decreases continuously with the interface temperature, which itself is decreasing with the length y, and normally one must fix this speed at the required minimum at the end of the growth. The maximum speed can be calculated with the gradient as a parameter that is a logarithmic function of y. One can select a linear function for the speed kept lower than this maximum speed; however, if (v) is linear in (y), it is exponential in t (time of growth). One can select parametric functions like

$$v = a + by = ae^{bt} \qquad (27)$$

with $a = 0.92$, $b = -5.8 \cdot 10^{-3}$ (experimental choice). Then one can have different curves $v(y)$, $v(t)$, $T(t)$, and $y(t)$. In this latter case, one can see that the imposed variable speed (0.92 to 0.22 mm/hr) leads to an appreciable gain in time of 2 compared to a fixed one at 0.22 mm/hr.

This method allows for the growth of large monocrystals with high purity and ingots up to a few kg, but like all Te rich methods, it presents precipitation and inclusion problems. With convenient doping, good nuclear detector materials are obtained.

2. Traveling Heater Method

The traveling solvent method (TSM) was first used by Weinstein and Mlavsky (1964) for the growth of GaAs. It consists of a temperature gradient across the solvent that dissolves the solute at the hot end and deposits it in the cold end. The traveling heater method (THM) is based on the same principle but the gradient is established by the relative motion of rod and furnace. An extended THM bibliography can be found in the work of Wolff and Mlavsky (1974). This method was widely developed for CdTe preparation by the Tyco group (Bell, Hemmat, and Wald, 1970; Wald, Bell, and Menna, 1973). To date, the best materials for spectrometric nuclear detectors have been produced by this method, and it deserves some attention.

If one considers the phase diagram (Fig. 1), it can be seen that the melting temperature can be decreased drastically by increasing the Te fraction (off stoichiometry) and with 80% Te melting occurs at 800°C (as opposed to 1092°C). The same thing can be done for Cd but with more technical difficulties (pressure, solubility). Te as a solvent is a better choice; at the same time, it is a "homosolvent" that avoids contamination.

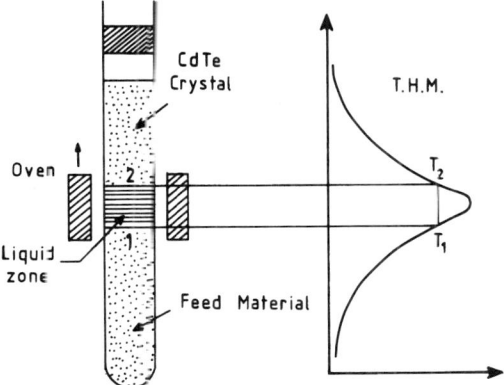

FIG. 9. Schematic of THM crystal growth technique.

Figure 9 shows the THM principle. The Te melted zone crosses through a CdTe ingot. The melt is generated by a small annular furnace. The imposed temperature profile is schematized on the right.

If the heating coil is arranged such that $T_2 > T_1$ in Figure 9, then the solubility $S_2(T_2)$ is higher than $S_1(T_1)$. With $S_2 > S_1$, this leads to an excess of CdTe at the hotter interface T_2, corresponding to a gradient of CdTe concentration from T_2 toward T_1 and a consequent diffusion of Te from the hotter interface (T_2) to the cooler one (T_1). CdTe is thus deposited on this side as epitaxial growth. But, as will be seen later, T_2 can be made $>T_1$ by continuous motion of the ampoule in the furnace. Thus, two parameters are important:

- The solvent optimum temperature is determined from $T(x)$ phase diagram projection (Fig. 1) to the first order.
- The growth rate is determined mainly by the liquid diffusion rate of CdTe in melted Te at the former temperature.

However, problems arise from solutions that are *constitutionally supercooled*, which must be eliminated. This can be done by taking into account the following considerations (Wald and Bell, 1972, 1975):

1. When the liquid–solid interface is convex, convention current depletes the periphery of Cd and enriches it with Te, which is more desirable than the reverse in a concave situation.
2. The growth takes place by material transport in the zone. The analysis of the transport process is complicated and two limiting cases can be considered to simplify it (Fig. 10): *diffusion only*, if the diffusion of Cd atoms through the Te zone

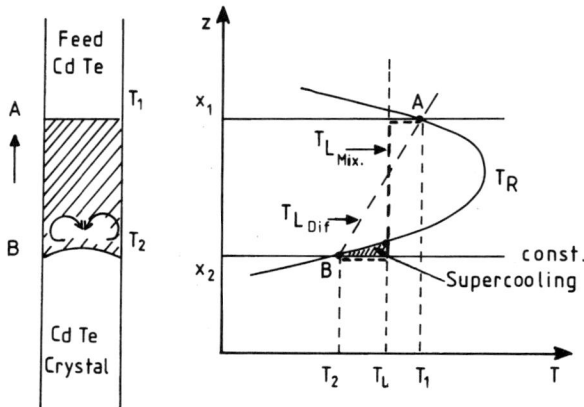

FIG. 10. Material transport in the THM method following complete mixing or only diffusion model.

is quasi-constant and quasi-independent of the temperature, the concentration gradient is constant from A to B in Figure 10, and then the actual temperature T_R within the whole melted zone is higher than the liquidus line temperature T_L and no supercooling occurs ($T_R > T_{LD}$); and *complete mixing*, natural convection and mixing can be complete, and then the temperature and concentration in the melted zone are constant. The liquidus line T_L crosses over T_R at some point (Fig. 10), leading to supercooling and unstable conditions with multiple nucleation taking place. During actual crystal growth, the liquidus line is intermediate between both and supercooling can be avoided, especially by adjusting the temperature of the melted zone and its temperature profile so that the CdTe solubility (or the saturation concentration) gradient at the middle of the zone is higher than the average value. This means one needs a continuous temperature "bump" in the middle of the melted zone. It is quite difficult experimentally to determine the temperature distribution in both longitudinal and radial directions, and a theoretical simulation is easier.

3. Longitudinal temperature distribution: Figure 11 shows the geometry employed. A rod of CdTe with a Te zone in the middle is assumed to be in thermal equilibrium with its surrounding by conduction, convection, and radiation. Heating is supplied by a hot Te zone at T_1 and the loss of heat to the outside at T_2. By balancing the heat supplied and lost, the equation governing the temperature as a function of the position x in the rod can be written as a differential equation. Using known constants and boundary conditions, solutions of this system can be found (Bell and Wald, 1971; Bell, 1974a). In Figure 11, we show the case of a symmetrical geometry, and in the displacement case, an increase of 14°C is observed in the middle of the zone. This was basically a one dimensional approach. The temperature distribution in the other direction will now be considered.

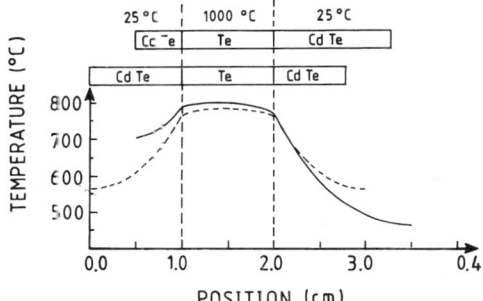

FIG. 11. Temperature profile in the THM Te zone in the symmetrical (– – –) and 0.5 cm displacement (———) cases (according to Wald et al. (1973)).

4. The Temperature: For crystal growth radial as well as longitudinal temperature profiles must be considered. Thus, assuming homogeneous and isotropic thermal properties at quasi-equilibrium, the equation of the heat distribution is Laplace's equation: $\nabla^2 T = 0$. The numerical and analytical solution has been done by Bell and Wald (1971) and Bell (1974). In a symetrical rod, Figure 12 shows the plots of isotherms for two temperature values of the solid–liquid interface. At 750°C, the interface is convex and within the hot zone, which is desirable for monocrystal growth.

The longitudinal temperature distribution profile on the axis and at the periphery of the ingot can be calculated, too, but it is preferable to measure these important features experimentally. This can be done by measuring the temperature pro-

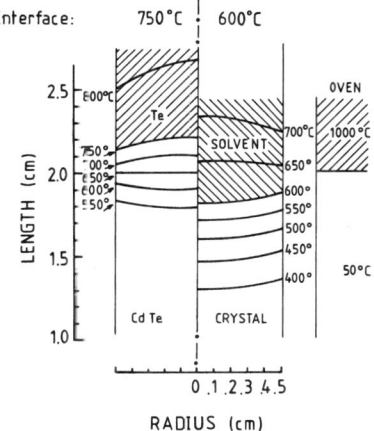

FIG. 12. Isotherms plot in two THM cases 750°C and 600°C interface temperatures following calculation of Bell et al. (1971).

files at the two cited points in CdTe rod or, better, in an ingot of volcanic stone (stumatite) that has approximately the same thermal conductivity. Such experimental axial and lateral profiles can be compared to the theoretical points in Figure 12. One can note that outside of the oven the axial temperature T_A is higher than the lateral one T_L, due to the conductivity of the rod and cooling through the surface. The interfaces are *flat* when $T_A = T_L$, *convex* for $T_A < T_L$, and *concave* for $T_A > T_L$. This means that the best situation is when the Te melted zone length is equal to or smaller than the intersection line of the two profiles; however, this is not so easy to realize.

5. Growth speed can be limited. It was mentioned before that growth is limited by the diffusion rate of CdTe across the Te solvent. This coefficient is given by Brice (1973) as

$$\eta D = 8.2 \times 10^8 \cdot T(1 + 3\ V_2/V_1)^{2/3} \times V_1^{-1/3} \tag{28}$$

where η is the viscosity of the solvent and V_1 and V_2 are the molar volume of CdTe and Te. The speed v_m then follows:

$$v_m \leq \frac{-\dfrac{dx}{dT} \times DG}{x(2x - 1)} \tag{29}$$

where G is the temperature gradient, x is the molar part of Te. The maximal speed between 800–1000°C is around 0.2–1 mm/hr for $G = 50°C/cm$.

As a consequence of various calculations and measurements, one can see that the choice of temperature, geometry (mainly the ingot length), diameter, and heater dimension are of importance and every small change must be taken into account.

The optimum configuration must be determined for every case. For example, for crystallinity and a convex interface, it is better to have shorter solvent zones and a longer heater; but to obtain fast growth and avoid constitutional supercooling, a large gradient and short heater is better and a compromise must be found.

It is the same for precipitation. Low temperature, in principle, decreases the number of constitutional Te precipitates due to the retrograde solidus, but low temperature also decreases the Te diffusion and increases precipitation. THM is still the best method to produce high material quality for spectrometric nuclear detectors, even if problems of homogeneity and precipitates are still present.

3. Direct THM Synthesis and Growth

Because of purity considerations, to date materials for nuclear detectors were purified prior to and after synthesis, but before crystal growth. However, purifi-

6. GROWTH METHODS OF CdTe NUCLEAR DETECTOR MATERIALS   245

cation and especially different handling is a source of new contamination, and experience shows that impurities such as Cu, Ca, K, Na, Si, C, O, Fe, or Cr can easily be introduced, up to $10^{14}-10^{15}$ atoms/cm^3 level during the various growth steps. Recently it appears that all these steps can be easily shortened by the direct synthesis and growth of CdTe and many other II–VI materials by THM, even $5N$ elements. The method consists of the following steps: A rod of Cd is set in the center of a normally prepared tube with the selected Te zone with equiatomic quantities of Te pieces, squeezed all around. The stoichiometry is not so critical and within the weighing precision, growth can take place. In the usual THM manner, the synthesis, purification, and growth are done simultaneously. Performed equiatomic cylindrical sections of Cd and Te can be used, too, but are more difficult to realize.

Good results have been obtained (Wald et al., 1973) by this method, owing to the low temperature of the whole process; the limited number of steps, only one; and the self-purification of THM. These materials can be used also as feed for other growth methods

V. High Resistivity Materials

Before talking about high resistivity semiinsulating material and compensation of cadmium telluride (CdTe) a few facts must be noted. The major applications of cadmium telluride such as nuclear detectors, epitaxy substrates, and electrooptic devices cannot allow more than 10^{-7} Amp. current leakage under an electric field of 1000–3000 V/cm and require an active volume, generally a depleted zone free of charge carrier, as large as possible in the mm–cm range. These requirements explain the necessity of high resistivity semiinsulating material in the range of $\rho = 10^8-10^9$ Ωcm.

1. HIGH PURITY MATERIALS

As will be seen later in this subsection, theoretical calculation shows that the limit of resistivity obtainable by chemical and physical purification of CdTe materials lies around 10^8 Ωcm but experimental measurement shows a maximum resistivity of 10^6 Ωcm and commonly around 10^4 Ωcm. For the best purified CdTe grown by THM improved methods and with very good nuclear detection properties (Cornet, 1976; Wald et al., 1973), it was shown that resistivity increases with decreasing temperature of the zone. The best results are obtained at 700° C with a suitable gradient. However, this is the minimum below which the density of precipitates become prohibitive at the speed considered. This resistivity is high enough to be considered by some authors as a consequence of compensation by residual impurities in Cd, Te, or quartz. On the other hand, it is few orders of magnitude less than theoretical expectations and values resulting in thin sensitive regions. This is probably due to structural defects and residual impurities; however, if extended purification can remove a large fraction of the chemical impurities, structural defects are more difficult to eliminate, and the ultimate values that can be reached lie around 10^6 Ωcm for ρ and 10^{-3} for $\mu\tau$. Therefore, further improvement of semiinsulating CdTe cannot be done directly without chemical compensation by an appropriate element.

2. Compensation

a. Compensation Elements

Common donors have been considered for compensation: group III elements like In substitutional on Cd sites have been used with some success by the Hughes group (Kyle, 1971) and Al was suggested but without evidence; on the other side, group VII (halogen) substitutionals on Te sites were successfully used by many authors (Cornet, 1976; Wald et al., 1973; Siffert et al., 1975).

Compensation by elements such as Ge (Scharager et al., 1975), Mg, Se, or Zn have been reported, but without extensive studies. Recently, the possibility of complementary compensation by unexpected elements like Cu (Biglari et al., 1988; Biglari et al., 1989), Fe and V (Moravec et al., 1992) appears to be possible and the extension of work done for Si and GaAs shows a possibility of compensation by hydrogen (Mergui, 1991).

Au seems to have some possibility as a compensation element like Cu, while the role of O is more contradictory. Generally it appears as an acceptor in CdTe but some results suggest the possibility of amphoteric behavior (Mergui, 1991.

b. Compensation Models

Spectroscopic methods like thermostimulated current (TSC), photoinduced current transient spectroscopy (PICTS), photoluminescence (PL), and others show more than 20–30 different defect levels in the forbidden gap of CdTe with different ionization states and signs. This leads to a general decrease of the electrical characteristics: first, a decrease of resistivity by increasing the density of carriers; second, a decrease of the $\mu\tau$ product owing to the time constant of trapping and detrapping of carriers, which decreases the mobility and the lifetime of these carriers.

Thus by increasing resistivity ρ and the $\mu\tau$ product the device quality material can be enhanced less by purification and more by compensation as mentioned before. However, experimental studies of compensation are particularly difficult due to the large number of parameters. In this case, the modeling of compensation allows for a better understanding of the material behavior and processing.

6. GROWTH METHODS OF CdTe NUCLEAR DETECTOR MATERIALS

The first theoretical study was performed by De Nobel (1959). He considered the presence of Frenkel structural defects on Cd sites as a cadmium vacancy V_{Cd} and interstitial Cd_i, which can be ionized and which introduce a localized level in the bandgap. De Nobel established the concentration of these defects at 700° C as a function of Cd pressure, for "undoped" material. However, he used a quite different energy level diagram than is accepted now; Kröger (1965) generalized this model by considering the compensation by group III donors with association between V^{--} and D^+, without taking into account these associations in his model. Both found nonlinear equilibrium equations. Using mass action and solving only for some particular cases, Bell et al. (1974) introduced the previously cited chemical associations in the model as neutral complexes despite the evidence for ionized complexes. Agrinskaya, Arkadeva, and Matveev (1970), Agrinskaya et al. (1971), and Höschl et al. (1975) studied the same problems in the case of THM, while Selim, Swaminathan, and Kröger (1975) applied Kröger's calculation to the In compensation case. But with De Nobel's level diagram, Marfaing (1977) assumed the existence of neutral V_{Cd} in large concentration and described the compensation process by a donor–neutral V_{Cd} interaction with stabilisation by electron trapping.

These models all have a partial view of the problem, and for this reason, a rather complete, self-consistent model was developed that synthesizes the former models but takes into account the noticed deficiencies (Stuck, Cornet, and Siffert, 1977), the majority of known levels in the band gap and the different ionization states of defects and complexes. Moreover, for calculation complexity reasons, this model was directed only at halogen compensation, but in view of recent work and results, a new reading of this model should be of great help, first to understand some unexpected results and general compensation mechanisms, second to improve this model and modeling in general.

However, in the model simulation, one must introduce supposed defect parameters like the type and the energy level with respect to bands: precise knowledge of these parameters allows for better agreement between model and experiment. Thus after reviewing and talking about defects and their corresponding level diagrams, the case of the pure crystal model is discussed and then the compensated model introduced.

i. Defects. Defects in CdTe are mainly structural defects, impurities and complexes of the two; the excess, the lack or equilibrium between these different levels determines the absolute electrical behavior of the material.

Such defects can be classified in few major families or bands (TSC, PICTS, PL) <0.14 eV, 0.14–0.20 eV, 0.20–0.40 eV, 0.40–0.70 eV. Every level or band has its own specific action on the electrical characteristics of the material.

Detailed discussion of the defect levels is well reviewed (Kröger, 1977; Cornet, 1976; Bell et al., 1970), but prior to the study of their correlation, we give a brief overview of these defects and their simplified level diagrams as shown in Figure 13. It should be noticed that assignment of levels is rather an assumption than an absolute proof, thus the authors reveal here their general belief.

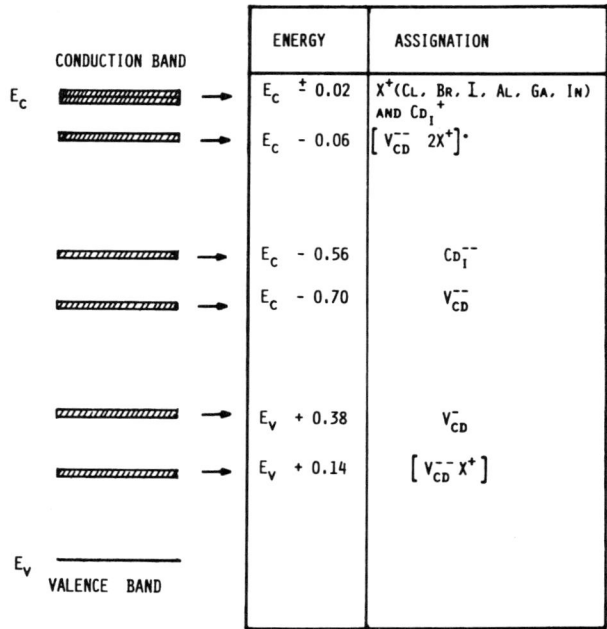

FIG. 13. Simplified model of the more common defect levels in a high ρ CdTe bandgap.

- The level at $E_v + 0.14$ eV or band $0.14-0.20$ eV is an important level, extensively studied and attributed to many defects, but it is now generally assumed that this is a cadmium vacancy–donor ionized complex $(V_{Cd^{--}}, X^+)^-$.
- $E_v + 0.38$ eV is attributed to a simply charged cadmium vacancy (V_{Cd^-}). The authors believe that 0.43 eV is more appropriate but, for reasons of convenience, 0.38 eV is still used,
- $E_c - 0.56$ eV is generally attributed to interstitial cadmium (Cd^{++}).
- $E_c - 0.6$ to 0.7 eV is due to a well-established doubly charged cadmium vacancy.
- $E_c - 0.06$ is due to a neutral cadmium vacancy–donors complex $(V_{Cd^{--}}, 2x^+)^0$.
- $E_c - 0.02 \pm$ is due to many donors doping such as Cl, In, Br, or Al.
- Other levels have been found in some special cases and are not introduced in the model, as will be discussed later.

ii. Model. Pure crystal The crystal is assumed to be free of dopants, present are only structural defects such as Frenkel disorders:

$$V_{Cd}^0 + Cd_i^0 = Cd_{Cd}$$
$$KF = [V_{Cd}^0][Cd_i^0]$$

6. GROWTH METHODS OF CDTE NUCLEAR DETECTOR MATERIALS

which can be neutral, singly, or doubly ionized. Equilibrium between those states can be written

$$Cd_i^0 = Cd_{i+} + e^- \tag{30}$$

$$Cd_{i+} = Cd_{i++} + e^- \tag{31}$$

and the same equilibrium for the V_{Cd}^0. (32)(33)

The law of mass action can be applied to this equilibrium if $[n]$ and $[p]$ are electron and hole concentration:

$$K_1 = [Cd_i^+][n]/[Cd_i^0] \tag{34}$$

$$K_2 = [Cd_i^{++}][n]/[Cd_i^+] \tag{35}$$

$$K_3 = [V_{Cd^-}][p]/[V_{Cd}^0] \tag{36}$$

$$K_4 = [V_{Cd^{--}}][p]/[V_{Cd^-}]. \tag{37}$$

The same treatment must be performed for the intrinsic thermal native electronic disorder:

$$e_i^- + h_i^+ = 0$$

$$K_i = [n][p] = n_i^2 \tag{38}$$

where $(n_i) = K^{1/2}$ is the intrinsic carrier concentration. Finally, the required electrical neutrality is expressed by

$$[n] + [V_{Cd^-}] + 2[V_{Cd^{--}}] = [p] + [Cd_{i+}] + 2[Cd_{i++}] \tag{39}$$

One has a nonlinear system of six Eqs. (34) to (39) that have been solved by De Nobel for particular cases. For a more general approach, some additional developments are required, concerning the cadmium pressure–temperature relation in the THM case, Frenkel constant KF, intrinsic constant K_i, and the distribution of defects that are supposed to follow the Boltzmann statistics (employed for simplicity). In this case the equilibrium constants are given by

$$K_x = \alpha Nc \, \exp(-Ex/KT),$$

where α is the number of possible configurations of one electron in the defect

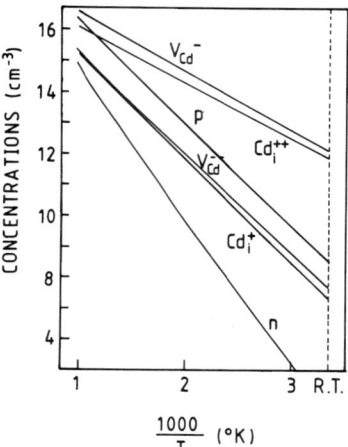

FIG. 14. Temperature dependence of defect concentrations in pure crystal.

(1/2 or 2), Nc is the density of state in the conduction band (partition function), and Ex is the energy level.

Iterative solutions can then be done with the temperature as a parameter (see Fig. 14). One can see that all defects decrease drastically and uniformly from 1000°C to RT, that $[V_{Cd^-}] - [Cd^{++}] - [p]$ are four orders of magnitude greater than $[V_{Cd^{--}}] - [Cd^+] - [n]$, that (V_{Cd^-}) and (Cd^{++}) are the major carriers with $10^{13}/cm^3$, this THM material is of the (p)-type at RT, and that its ultimate resistivity at RT under the assumptions made is given by ($[n]$ and $[p]$ from Fig. 14) are $\rho = [e(n\mu_n + p\mu_p)] - 1 \approx 4.10^8 \, \Omega \cdot cm$. This means that theoretically it is possible to have semiinsulating material by purification. However, experimental measurements show the ultimate resistivity of highly purified crystals around $10^6 \, \Omega \cdot cm$, indicating the necessity of compensation.

Furthermore, starting from the knowledge of V_{Cd} and Cd_i concentrations, one can calculate the Te excess concentration present at every temperature, and this allows one to construct, point by point, the solidus line of the $T(x)$ projection in the extended phase diagram. This calculated line can be seen in Figure 5 in comparison with Smith's (1970) and De Nobel's (1959) data. It seems that the values obtained are a little lower than the points of Smith, which must be considered as an upper limit as has been mentioned before.

Compensated crystals Bell et al. (1970) introduced the V_{Cd} and halogen associations first, but as neutral complexes without self-compensation. Here, the former model is extended to the case of the presence of donor impurities (denoted X), mainly the more common ones, halogen atoms, because they are supposed to be shallow enough to be fully ionized. In addition to the previous equations, equi-

6. GROWTH METHODS OF CDTE NUCLEAR DETECTOR MATERIALS

libria due to associations between V_{Cd} and X are introduced but here their ionization state is taken into account:

$$X \Rightarrow X^+ + e^-$$

$$V_{Cd^-} + X^+ \Rightarrow (V_{Cd}X)^0 \tag{40}$$

$$(V_{Cd^-} - X)^0 + e^- \Rightarrow (V_{Cd}X)^- \tag{41}$$

$$(V_{Cc}X)^- + X^+ \Rightarrow (V_{Cd}X)^0 \tag{42}$$

Applying again the law of mass action on these equilibria

$$K_5 = [(V_{Cd}X)^0]/[V_{Cd^-}][X^+] \tag{43}$$

$$K_6 = [(V_{Cd}X)^-]/[(V_{Cd}X)^0][n] \tag{44}$$

$$K_7 = [(V_{Cd}2X)^0]/[(V_{Cd}X)^-][X^+] \tag{45}$$

the electrical neutrality is now modified:

$$[n] + [V_{Cd^-}] + [(V_{Cd}X)^-] - 2[V_{Cd^{--}}]$$
$$= [p] + [Cd^+] + 2[Cd^{++}] + [X^+] \tag{46}$$

and the conservation of total impurity number $[X]_T$ requires

$$[X]_T = [X^+] + [(V_{Cd}X)^0] + [(V_{Cd}X)^-] + 2[(V_{Cd}2X)^0] \tag{47}$$

The new nonlinear system of 10 equations to be solved is (34) to (38) and (43) to (47) where $[X]_T$ is the free parameter. Here again, one needs additional development for K. This can be written as follows:

$$K = C \cdot \exp\frac{\Delta H}{KT}$$

where C and ΔH are the entropy and enthalpy variations of the reaction; C corresponds to the normalized number of possible association sites (four in a cubic crystal), and ΔH can be found by comparing the heat of formation of CdX_2 and CdTe (Bell et al., 1970). One can also suppose that the ΔH needed to form a simple or double association is the same, which means that $K_5 = K_7$.

Starting from these considerations, this system has been solved by an iterative process again. The first set of results is given in Figures 15 and 16.

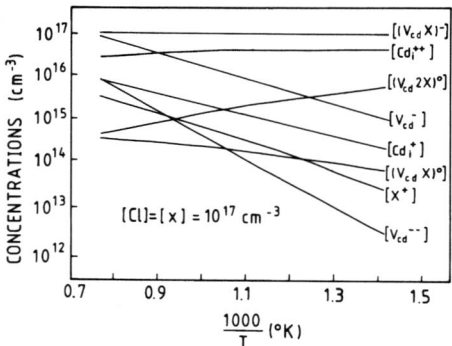

FIG. 15. Temperature dependence of defects concentration in (Cl) compensated CdTe.

With the temperature as a parameter, with $[X]T = 10^{17}$ at/cm^3 in the case of chlorine and bromine donors impurities, Figure 17 gives the concentration of the defects as a function of chlorine concentration at 400° C (below this, the system is "frozen"; Agrinskaya *et al.* (1971).

A few remarks can be deduced from these figures:

• Cadmium vacancies and interstitials are independent of the nature of the halogen impurity,
• Concentration of V_{Cd} in a compensated material are 10 times lower than in the pure one whereas Cd$_i$ are 10 times higher!
• $[X^+]/[X]_T = 10^{-3}$ for chlorine but 10^{-1} for bromine,
• $(V_{Cd}X)^-$, (Cd_i^{++}) and $(V_{Cd}2X)^0$ are predominant at RT and $[X]_T = 10^{17}$ cm^{-3},

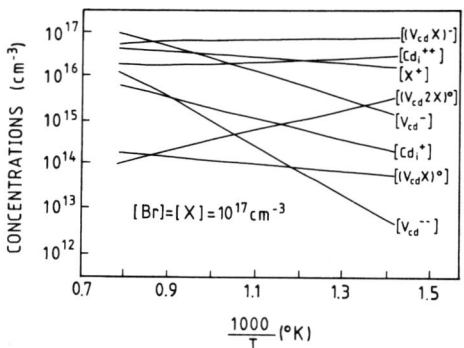

FIG. 16. Temperature dependence of defects concentration in (Br) compensated CdTe.

6. GROWTH METHODS OF CdTe NUCLEAR DETECTOR MATERIALS

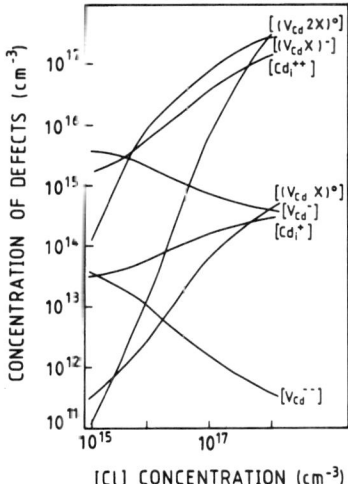

FIG. 17. Defects concentration as a function of (Cl) concentration.

- Compensation is effective (Fig. 17), both $(V_{Cd}X)^0$ and $(V_{Cd}2X)^0$ increase while $(V_{Cd}-)$ and $(V_{Cd}--)$ decrease, with the increasing $[X]_T$, especially both Cd_i increase as well, but we must notice that the solubility limit of $[X]$ is not introduced in the model.

On the other hand, one can see that for a very active $[X]_T$ of 10^{16} cm^{-3}

- The acceptor $(V_{Cd}-) \approx (E_v +0.4\text{ eV})$ is still there with 10^{15} cm^{-3},
- And the acceptor $(V_{Cd}--) \approx (E_c -0.66\text{ eV})$ with $\approx 10^{13}$ cm^{-3},
- The donor $(Cd^{++}) \approx (E_c -0.56\text{ eV})$ with $\approx 10^{16}$ cm^{-3},
- And the donor $(Cd^+) \approx (E_c -0.02\text{ eV})$ with $\approx 10^{14}$ cm^{-3},
- The acceptor $(V_{Cd}X)- \approx (E_v +0.14\text{ eV})$ with $\approx 10^{16}$ cm^{-3}.

This means that, even in the halogen compensated material with $[X]_T = 10^{16}$ cm^{-3}, one still has up to 10^{16} cm^{-3} ionized centers of both signs that can act as trapping and recombination centers with severe electrical effects like tailing, polarization, and large time constant decay.

Höschl et al. (1992) present a new model based on Berding et al.'s (1990) calculations and using Fermi–Dirac statistics (rather than Boltzman's in the former model). This model shows practically the same results for $V_{Cd}--$, $(V_{Cd}, Cl)^-$ and $(V_{Cd}, 2Cl)^0$ probability, and it seems that there is now some agreement about the general behavior of complexes.

VI. Experimental Results and Conclusion (Hage-Ali and Siffert, 1991)

In highly purified CdTe, a wide range of added 10^{16}–$10^{18}/\text{cm}^3$ Cl leads to high resistivity materials, with only 2–10^{16} of levels related to Cl. This was confusing at first, but finally Cl excess was found in the Te zone and the increase of $\mu\tau$ product with Cl can then be explained as impurity purification by Cl.

In Figure 18, the resistivity behavior (Biglari *et al.*, 1988, 1989) of a CdTe Cl doped material is seen in terms of Cu concentration, and it shows a sharp maximum at 100–250 ppb (Cu) with $\rho = 5 \cdot 10^9 \, \Omega \cdot \text{cm}$, which is much higher than CdTe compensated by Cl only. This can be explained as complementary compensation of Cd^+ if Cu is an acceptor or $(V_{Cd}, Cl)^-$ if Cu is a donor. The same behavior for V and Fe with resistivity reaching a few $10^9 \, \Omega \cdot \text{cm}$ has also been observed (Moravec *et al.*, 1992).

H introduced by diffusion or ion implantation in atomic or molecular form shows a drastic increase of resistivity in low resistive material. Again the compensation effect is clear. Au, Mg, Zn, Se show the same effect as well.

Looking at Figure 17 and considering these experimental results, one can conclude that compensation is a more general process than believed before and can and must be understood in terms of general equilibria between structural defects, donor and acceptor impurities and their different neutral and ionized complexes.

The compensation is limited in the case of any particular element only by the solubility limit, the structural defects created for each specific system, and the creation energy of each complex.

In pure CdTe at RT, the dominant defect is (V_{Cd^-}) and $(Cd_{i^{++}})$. Adding Cl, leads to dominant defects $(V_{Cd^{--}}, Cl)^-$ and $(Cd_{i^{++}})$ that need Cu, V, or Fe

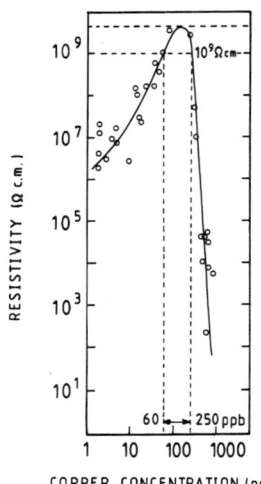

FIG. 18. Resistivity of CdTe material as a function of copper concentration.

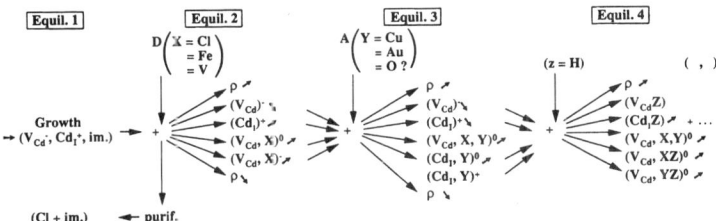

FIG. 19. Schematic succession of equilibria following the addition of few elements.

for $(Cd_{i^{++}})$ and H for $(V_{Cd}, Cl)^-$. Adding these elements probably changes the equilibrium again and leads to other defects such as $(V_{Cd^{--}}, Cu)^-$ and so on (Fig. 19).

Thus it becomes clear that self-compensation is an equilibrium problem. CdTe is more self-compensated than other semiconductors, owing to the high mobility of defects in this material. During ion implantation experiments in CdTe (Cornet *et al.*, 1972), it was impossible to make CdTe *amorphous* even with 70 keV 10^{17} Bi implantation. This means that defects are relatively free and can be partially annealed at room temperature. The same can be extended to dopant defects. If one introduces an impurity, one is then out of equilibrium. CdTe adjusts to the situation by creating the opposite sign defect (thanks to the high mobility), and generally speaking, doping is an unstable situation (out of equilibrium). Foreign atoms need to be frozen for doping, and this is generally done by quenching. For CdTe the "freeze" temperature seems to be at very low very temperature, resulting in *high self-compensation*. Self-compensation viewed in this way is easy to understand. However, one still needs to understand why there is this high defect mobility.

REFERENCES

Abdullaev, G. B., Shilkin, A. I., Shakhalakhinskii, M. G., and Kuliev, A. A. (1965). *Selen. Tellur. Ihk. Premen* **42**.

Agrinskaya, N. V., Arkadeva, E. V., and Matveev, O. A. (1970). *Sov. Phys. Semicond.* **4**, 347.

Agrinskaya, N. V., Arkadeva, E. N., Matveev, O. A., and Sladkova, V. A. (1971). In *Proceedings of the First International Symposium on CdTe: A Material for Gamma-Ray Detectors*, ed. P. Siffert and A. Cornet. Centre de Recherches Nucléaires, Strasbourg.

Aleksandrov, B. N. (1960). *Fiz. Metal i Metalloved* **9**, 362.

Aleksandrov, B. V., and Udovikov, V. I. (1973). *Isv. Akad. Nauk. SSSR Met.* **2**, 17.

Arkadyeva, E. N., and Matveev, O. A. (1977). *Rev. de Phys. Appl.* **12**(2), 239.

Baïmakov, A., and Petrova, Z. N. (1960). *Tsvetn. Met.* **33**, 43.

Bell, R. O. (1974). *J. Electrochem. Soc.* **121**, 1366.

Bell, R. O., and Wald, R. V. (1971). In *Proceedings of the First International Symposium on CdTe: A Material for Gamma-Ray Detectors*, ed. P. Siffert and A. Cornet. Centre de Recherches Nucléaires, Strasbourg.

Bell, R. O., Hemmat, N., and Wald, F. (1970). *Phys. Stat. Sol. (a)* **1**, 375.

Bell, R. O., Wald, F. V., Canali, C., Nava, F., and Ottaviani, G. (1974). *IEEE Trans. Nucl. Sc.* **NS-21,** 331.
Berding, M. A., Van Schilfgaarde, M., Paxton, A. T., and Sher, A. (1990). *J. Vac. Sc. Techn. A* **8,** 1103.
Biglari, B., Samimi, M., Hage-Ali, M., Koebel, J. M., and Siffert, P. (1988). *J. Cryst. Growth* **89,** 428.
Biglari, B., Samimi, M., Hage-Ali, M., Koebel, J. M., and Siffert, P. (1989). *Nucl. Instr. and Meth. in Phys. Res. A* **283,** 249.
Brebrick, R. F., and Strauss, A. J. (1964). *J. Phys. Chem. Solids* **25,** 1441.
Brice, J. C. (1973). *Growth of Crystal from Liquid.* North-Holland Press, Amsterdam.
Butler, J. F., Lingren, C. L., and Doty, F. P. (1991). IEEE Symposium on Nuclear Science, Nucl. Sc. Symp. Proc., Santa Fe.
Cornet, A. (1976). Thesis, Université Louis Pasteur, Strasbourg.
Cornet, A., Hage-Ali, M., Grob, J. J., Stuck, R., and Siffert, P. (1972). *IEEE Trans. Nuc. Sc.* **NS-19,** 358.
De Nobel, D. (1959). *Philips Res. Rep.* **14,** 361.
Drowart, J., and Goldfinger, P. J. (1958). *J. Chim. Phys.* **55,** 721.
Goldfinger, P. J., and Jeunehomme, M. (1963). *Trans. Faraday Soc.* **59,** 2851.
Hage-Ali, M., and Siffert, P. (1992). *Nucl. Instr. Meth. Phys. Res.* **A322,** 313.
Heumann, K. F. (1962). *J. Electrochem. Soc.* **109,** 345.
Höschl, P., Polivka, P., Piosser, V., and Sakalas, A. (1975). *Czech. J. Phys. B* **25,** 585.
Höschl, P., Grill, R., Franc, J., Moravec, P., and Belas, E. (1993). *Mat. Sc. and Eng. B.* **16,** 215.
Jordan, A. S. (1970). *Met. Trans.* **1,** 239.
Jordan, A. S., and Zupp, R. R. (1969). *J. Electrochem. Soc.* **116**(9), 1285.
Kobayashi, M. (1911). *Z. Anorg. Chem.* **69,** 1.
Kröger, F. A. (1965). *J. Phys. Chem. Sol.* **26,** 1717.
Kröger, F. A. (1977). *Rev. de Phys. Appl.* **12**(2), 205.
Kujawa, R. V. (1963). *Phys. Stat. Sol.* **3,** 1089.
Kyle, N. R. (1971). In *Proceedings of the First International Symposium on CdTe: A Material for Gamma-Ray Detectors,* ed. P. Siffert and A. Cornet. Centre de Recherches Nucléaires, Strasbourg.
Lorenz, M. R. (1962a). *J. Phys. Chem. Solids* **23,** 939.
Lorenz, M. R. (1962b). *J. Appl. Phys.* **33,** 3304.
Lorenz, M. R., and Halsted, R. E. (1963a). *J. Electrochem. Soc.* **110,** 343.
Lorenz, M. R., and Segall, B. (1963b). *Phys. Lett.* **7,** 18.
Marfaing, Y. (1977). *Rev. de Phys. Appl.* **12**(2), 211.
Matveev, O. A., Rud', Y. V., and Sanin, K. V. (1969). *Sov. Phys. Semicond.* **3,** 779.
Medvedev, S. A., Maksimovskii, S. N., Koevkov, Y. V., and Shapkin, P. V. (1968). *Inorg. Mat.* **4,** 1759 (English translation).
Mergui, S. (1991). Thesis, Université Louis Pasteur, Strasbourg.
Mochalov, A., Urubkova, E., and Lepis, V. (1966). *Cadmium,* ed. D. M. Chizhikov. Pergamon Press, Oxford.
Moravec, P., Hage-Ali, M., Chibani, L., and Siffert, P. (1993). *Mat. Sc. and Eng. B.* **16,** 223.
Pfann, W. G. (1958). *Zone Melting.* John Wiley & Sons, New York.
Pfeiffer, M., and Mühlberg, M. (1992). *J. of Crystal. Growth* **118,** 269.
Raiskin, E. R. (1970). *Crystal Growing by High Pressure and Temperature.* Iformazia SEV O Nauchno-techiche-skom sotrudnichestve, Ministry of Chemical Industry, Moscow.
Raiskin, E. R., and Butler, J. F. (1988). *IEEE Trans. Nuc. Sc.* **NS-35,** 81.
Redden, R. F., Bult, R. P., and Bollong, A. B. (1986). *TMS* (chemical paper of Cominco) **A86–60.**
Rev. de Phys. Appl. (1977). Vol. 12(2). Proceedings of the Second International Symposium on CdTe: Physical Properties and Applications.
Rubenstein, M. (1968). *J. Cryst. Growth* **3,** 309.
Rudolph, P., and Mühlberg, M. (1993). *Mat. Sc. and Eng. B.* **16,** 8.
Scharager, C., Muller, J. C., Stuck, R., and Siffert, P. (1975). *Phys. Stat. Sol. (a)* **31,** 247.

6. GROWTH METHODS OF CDTE NUCLEAR DETECTOR MATERIALS

Schaub, B., and Potard, C. (1971). In *Proceedings of the First International Symposium on CdTe: A Material for Gamma-Ray Detectors* ed. P. Siffert and A. Cornet. Centre de Recherches Nucléaires, Strasbourg.

Schaub, B., Gallet, J., Brunet-Jailly, A., and Pelliciari, B. (1977). *Rev. de Phys. Appl.* **12**(2), 147.

Selim, F. A., Swaminathan, V., and Kröger, F. A. (1975). *Phys. Stat. Sol. (a)* **29**, 465.

Sher, A., Chen, A., Spicer, W. E., and Shil, C. (1985). *J. Vac. Sc. Technol. A* **3**, 105.

Siffert, P., and Cornet, A. (eds.). (1971). *Proceedings of the First International Symposium on CdTe: A Material for Gamma-Ray Detectors*. Centre de Recherches Nucléaires, Strasbourg.

Siffert, P., Cornet, A., Stuck, R., Triboulet, R., and Marfaing, Y. (1975). *IEEE Trans. Nucl. Sc.* **NS-22**(1), 221.

Silvey, G. A., Lyons, V. I., and Silvestri, U. J. (1961). *J. Electrochem. Soc.* **108**, 653.

Smith, F. T. J. (1970). *Met. Trans.* **1**, 617.

Steer, C., Hage-Ali, M., Koebel, J. M., and Siffert, P. (1993). *Mat. Sc. and Eng. B.* **16**, 48.

Steininger, J., Strauss, A. J., and Brebrick, R. F. (1970). *J. Electrochem. Soc.* **117**, 1305.

Strauss, A. J. (1971). In *Proceedings of the First International Symposium on CdTe: A Material for Gamma-Ray Detectors*, ed. P. Siffert and A. Cornet. Centre de Recherches Nucléaires, Strasbourg.

Stuck, R., Cornet, A., and Siffert, P. (1977). *Rev. de Phys. Appl.* **12**(2), 218.

Tanaka, A., Musa, Y., Seto, S., and Kawasaki, T. (1987). *Mat. Res. Soc. Symp. Proc.* **90**, 111.

Triboulet, R. (1971). In *Proceedings of the First International Symposium on CdTe: A Material for Gamma-Ray Detectors*, ed. P. Siffert and A. Cornet. Centre de Recherches Nucléaires, Strasbourg.

Vanyukov, A. V., Koshitov, O. A., and Bulanov, A. I. (1969). *J. Appl. Chem.* **42**, 1920 (English translation).

Venderkerkov, J. (1971). In *Proceedings of the First International Symposium on CdTe: A Material for Gamma-Ray Detectors*, ed. P. Siffert and A. Cornet. Centre de Recherches Nucléaires, Strasbourg.

Wald, F. V., and Bell, R. O. (1972). *Nat. Phys. Sc.* **273**, 13.

Wald, F. V., and Bell, R. O. (1975). *J. Cryst. Growth* **30**, 29.

Wald, F. V., Bell, R. O., and Menna, A. A. (1973). *Int. Tyco Technical Report*.

Weinstein, M., and Mlavsky, A. I. (1964). *J. Appl. Phys.* **35**, 1892.

Wernick, J. P. (1960). *Trans. AIME* **218**, 763.

Whelan, R. C., and Shaw, D. (1968). *Phys. Stat. Sol.* **29**, 145.

Wolff, G. A., and Mlavsky, A. I. (1974). *Crystal growth, theories and techniques*, **1**, ed. C. H. L. Goodman. Plenum Press, New York.

Woodbury, H. H. (1974). *Phys. Rev. B* **9**, 5188.

Zaiour, A. (1988). Thesis, Université Louis Pasteur, Strasbourg.

Zanio, K. (1970). *J. Appl. Phys.* **41**, 1935.

Zanio, K. (1974). *J. Electron. Mat.* **3**, 327.

Zanio, K. (1978). In *Cadmium Telluride*, ed. Willarson-Bear. Semiconductors and Semimetals **13**, Academic Press, New York.

CHAPTER 7

Characterization of CdTe Nuclear Detector Materials

Makram Hage-Ali and Paul Siffert

CENTRE DE RECHERCHES NUCLÉAIRES
LABORATOIRE PHASE
STRASBOURG, FRANCE

I. INTRODUCTION	259
II. IMPURITIES ANALYSIS	260
1. *Spark Mass Spectrography Analysis*	260
2. *Atomic Absorption*	261
3. *Nuclear Activation*	261
4. *X-Ray Fluorescence*	263
5. *Secondary Ion Mass Spectrometry*	265
6. *Ion Chromatography—IR Absorption*	265
III. SURFACE ANALYSIS	266
1. *Lapped and Polished Surfaces*	267
2. *Chemically Etched Surfaces*	268
3. *Oxidized Surfaces*	270
4. *Interfaces and Contact Analysis*	270
IV. ELECTRICAL AND OPTICAL CHARACTERIZATION	277
1. *Resistivity*	277
2. *Mobility, Lifetime, and $\mu\tau$ Product*	279
3. *Photoluminescence Analysis*	281
4. *ODMR and Related Methods*	282
5. *TSC and PICTS Measurements*	283
V. DISCUSSION AND CONCLUSIONS	286
References	287

I. Introduction

For conventional semiconductor materials, it is relatively easy or at least possible to prepare crystals with predetermined properties chosen in advance. Up to now, unfortunately, this has not been the case for II–VI materials and especially for CdTe. General rules are sometimes established, but with great variability.

Inspection and investigation, wafer by wafer, is still, generally, the only way to select material with desired properties. This demonstrates the importance of characterization of CdTe material. The choice of characterization methods and their combination determines the success or failure of device applications.

In our case, the devices being considered are nuclear detectors, which are a sensitive application. Uses of these detectors include nuclear plants, safety, spectrometry, and nuclear medicine. There is no room for error; thus characterization and knowledge of precise properties becomes imperative: properties of the bulk, surface, contacts and interfaces, devices and their processes. Each of these studies is a research field in itself. Here we present a summary of the more important measurement methods and properties for nuclear detector applications.

It must be noted that conventional analysis methods for semiconductors generally fail in the case of high resistivity CdTe, due to the difficulty in obtaining ohmic contacts, due to the thermal instability even at rather low temperature, and the list can be longer. The greatest part of the analysis can and must be done through detector operation itself. We will see this later, in the chapter on detectors.

The literature contains a considerable amount of characterization data, results and experiments concerning II–VI compounds crystals and especially CdTe material. However, in the following sections, the review of methods and characteristics is restricted to subjects of interest to people concerned with nuclear detectors.

II. Impurities Analysis

Generally speaking, impurity analysis is a very wide field of investigation, searching for most of the elements of the periodic table. Nearly every element in every matrix is the subject of extensive studies, and researchers are always looking for higher and higher sensitivity and accuracy. In the case of CdTe, this problem is particularly acute owing to difficulty of purification and the great action of some trace impurities in doping or compensation.

A number of methods capable of detecting impurity traces have been employed. It is generally preferable to use methods covering wide range of masses but always, gaps and interferences lead to the use of complementary methods.

1. SPARK MASS SPECTROGRAPHY ANALYSIS

For a very long time, spark mass spectrography analysis (SMSA) was one of the more sensitive methods covering a wide range of elements. It is now outshined only by inductively coupled plasma mass spectrometry (ICPMS) or substituted by atomic absorption (AA) for cost and volume consideration.

Spark mass spectrography analysis is a mass spectrometry method with electric and magnetic analyzers. The specificity comes from the ion source principle. The material to be analyzed is used as electrodes in a spark chamber with various kinds of sparks: high voltage discharge, dc low voltage arc, and the more popular high frequency (20–30 MHz) and voltage spark. The spark is used for evaporation and ionizing the elements of the sample. This is then followed by acceleration and analysis.

The detection limit in this method is theoretically rather low: the average is about 0.1 ppm atoms but can reach 0.01 ppm for Ag or Sn and 0.002 for Li in Te and is limited to 1 ppm for Fe or Ni in Si. Schaub and Potard (1971) have used this method to measure impurity concentrations in Cd and Te prior to the synthesis. It is a standard method for Cd and Te suppliers; but if this method is highly recommended for elemental materials, in compound materials interferences and double ionization are sources of errors and moreover this method is a rather destructive one.

2. ATOMIC ABSORPTION

SMSA is a spectroscopic method, where all masses are scanned at the same time, but atomic absorption is a more targeted analysis. Basically, every element can absorb the light spectra it is able to emit owing to electronic transitions between its levels. The element to be measured is not excited or ionized but rather driven to its fundamental atomic state, free of chemical bonds, by two methods after dissolution in a convenient solvent, by acetylene flame torch or by graphite oven heating. Light, characteristic of the element being searched for, is emitted by a hollow cathode lamp directed onto the atomized material, resulting in absorption if the element is present. The absorption is proportional to the concentration of the element, and the measurement is done by comparison between the initial light intensity and the transmitted one, with the pure solvent absorption and with absorption of standard samples with known concentrations. This method is very popular, owing to its low cost and precision. Although, volatile materials and light elements are not measurable by this method, the sensitivity can reach few ppb for many elements. Figure 1 shows an example of an analysis by this method of different THM CdTe ingots.

This method has allowed the measurement of the segregation coefficient K_{eff} and K_0 of more than 20 elements in Te. It has also allowed the study of different purification processes on nuclear detector characteristics through measurement of more than 24 elements in the ppb range (Zaiour, 1988) and has permitted the first demonstration of CdTe(Cl) compensation by copper at a 200 pbb level (Biglari et al. 1988, 1989a).

AA is a destructive method, which needs minimum samples of 1 g.

3. NUCLEAR ACTIVATION

Nuclear activation can be performed in two ways: by charged particles, such as deuterium or proton charged particle activation in an accelerator or by neutron reaction in an atomic reactor plant. In each case, the elements to be detected and

Ingots	Elements	Fe	Al	Cu	Ni	Cr	Mn	Ag	Au	Mg	Ca	K
	Te pure	1.5	2.88	0.2	≤0.02	0.06	0.08	0.035	≤0.02	0.3	5	1.5
	Cd pure	0.2	0.100	0.1	0.03	≤0.01	0.02	≤0.01	≤0.02	≤0.1	≤1	≤0.1
1370	N°2	0.622	0.7	≤0.03	0.05	≤0.01	≤0.02	≤0.01	≤0.02	≤0.1	1.1	≤0.1
	N°24	≤0.05	≤0.15	≤0.02	≤0.02	≤0.01	≤0.02	≤0.01	≤0.02	≤0.1	0.9	≤0.1
	N°39	≤0.05	0.9	≤0.03	≤0.02	≤0.01	≤0.02	≤0.01	≤0.02	≤0.1	2.14	≤0.1
	N°43	≤0.05	0.98	0.04	≤0.02	≤0.01	≤0.02	≤0.01	≤0.02	≤0.1	0.85	≤0.1
1336	N°3	0.28	≤0.1	0.06	≤0.005	≤0.01	≤0.005	0.1	≤0.02	≤0.05	≤1	≤0.1
	N°5	0.09	≤0.1	0.06	≤0.005	≤0.01	≤0.005	69	≤0.02	≤0.05	≤1	≤0.1
	N°20	0.03	≤0.1	0.09	≤0.005	≤0.01	≤0.005	≤0.02	≤0.02	≤0.05	≤1	≤0.1
	N°22	0.1	≤0.1	0.095	≤0.01	≤0.01	≤0.005	≤0.02	≤0.02	≤0.05	≤1	≤0.1
	N°35	0.3	≤0.1	0.06	≤0.01	≤0.01	≤0.005	26	≤0.02	≤0.05	≤1	≤0.1
	N°39	0.845	≤0.1	0.017	≤0.005	≤0.01	≤0.005	≤0.05	≤0.02	≤0.05	≤1	≤0.1
	N°42	0.17	≤0.1	0.005	≤0.02	≤0.01	≤0.005	≤0.02	≤0.02	≤0.05	≤1	≤0.1
	N°47	0.1	≤0.1	0.009	≤0.02	≤0.01	≤0.005	≤0.02	≤0.02	≤0.05	≤1	≤0.1
1350	N°4	2.14	0.2	0.7	≤0.1	≤0.01	≤0.02	≤0.01	≤0.02	≤0.1	5.7	4.59
38	N°45	0.06	0.15	0.045	≤0.02	≤0.01	≤0.02	≤0.01	≤0.02	≤0.1	0.8	≤0.1

FIG. 1. Atomic absorption of trace elements in different CdTe ingots and wafers.

interferences are different, but the data analysis is the same. Both are nondestructive methods, in principle.

a. Neutron Activation

Samples are introduced in special locations of an atomic reactor where they are irradiated with $10^{11}-10^{15}$ neutrons. *Slow neutrons* include (n, γ) reactions, where impurity elements are activated, generally in the same chemical species, but with specific γ emission. For example, ^{59}Co (n, γ) ^{60}Co can be detected by classical spectrometric chain detection. If the activated nucleus is a β emitter, this results in new chemical emitter nucleus. However, the analysis process is still the same. *Fast neutrons* induce (n, p) and (n, α) reactions and are useful for light elements such as Li, N, K, Ca, and S, for example.

Neutron activation was used to determine the concentration of S and Br in CdTe, in the ppm range or less. One must note that a neutron generator (14 MeV) can be used, too, with less sensitivity.

b. Charged Particle Activation

This method uses charged particle beam delivered by accelerators at a sufficient energy to overcome the coulomb barrier of the nucleus and make possible the nuclear activation reaction. This methods has been used in CdTe to measure the absolute concentration of C (Chibani *et al.*, 1991) and to measure the segregation coefficient of C in THM CdTe, using the (d, n) reaction):

$$^{12}C\ (d,\ n)\ ^{13}N \xrightarrow{\beta^+} {}^{13}C$$

^{13}N has a half life of 10.05^m. During activation, a layer of deposited carbon, from the cracking of vacuum system oil, must be removed by a highly controlled chemical etch, in order to know the thickness removed. The activity is then measured by a β^+ spectrometer, such as two NaI gamma ray detectors or Ge-well detectors. The radiative decay is compared to a standard to give the absolute concentration of carbon. The sensitivity can reach a few ppb, but etching uncertainty and inteference of other impurities limit the sensitivity to 0.1–1 ppm. Elements such as Si, B, or Al can also be detected. The advantage of these methods is the high sensitivity and the absolute measurement but it is limited by the possible reactions.

4. X-RAY FLUORESCENCE

Particle induced x-rays have been used for various kinds of analytical problems with MeV accelerators. The major advantage of particles, compared to electrons,

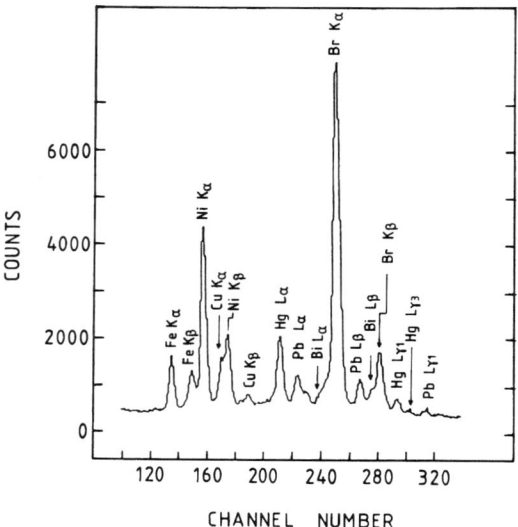

FIG. 2. Proton induced x-ray (PIXE) spectra of some impurities in CdTe.

is the reduced background, thus yielding higher sensitivity in trace elements analysis. The samples to be analyzed are irradiated by the particle beams of an MeV accelerator in a vacuum setup. X-rays are generally detected by Si(Li) solid state detectors. However, a distinction must be made between two different approaches: particle induced x-ray emission and heavy ion induced x-ray emission. PIXE (particle induced x-ray emission) use light elements such as protons and α particles as a projectile. This technique is widely used to determine the concentrations and presence of heavy element traces in CdTe (Al Neami, 1988; Al Neami et al., 1992). Fe, Ni, Ca, Hg, Pb, Bi, and Br can be measured in CdTe ingots (Fig. 2), especially the accumulation of these elements toward the end of the ingot and in the Te zone. HIXE (heavy ion induced x-ray emission) uses heavy elements as the projectile (Heitz, 1984). This method allows measurements of light elements in a heavy matrix, even if interferences exist. Following the Barat–Lichten rule (Barat and Lichten, 1972), an adequate choice of the ion–matrix–element system allows the concentration measurement of the last (element) one, in the best conditions. Iturbe-Garcia (1981), Al Neami (1988), and Al Neami et al. (1992) have measured the concentration profiles of Cl, S, and Si in Te and CdTe ingots as well as the presence of elements due to solvents after cleaning (Fig. 3), like Cl from trichlorethylene. Chibani et al. (1992) have used this method to establish correlations between Si concentration in CdTe with electrical measurements. C and Si are critical elements in CdTe, owing to the growth method in graphite covered quartz tube leading to possible contamination by these two elements. Both approaches are non destructive analysis methods.

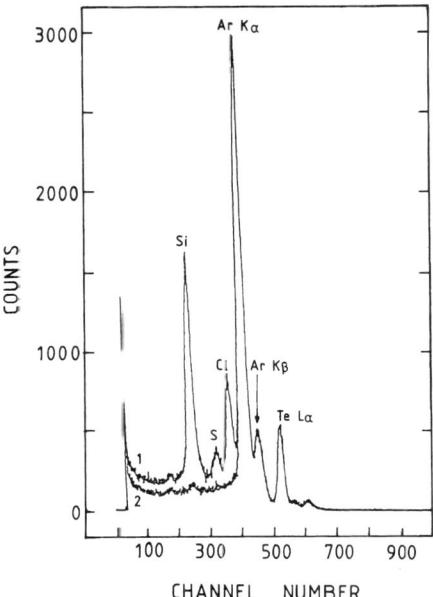

FIG. 3. Heavy ion induced x-ray (HIXE) spectrum of (Si) and (Cl) in CdTe.

5. Secondary Ion Mass Spectrometry

The very popular secondary ion mass spectrometry (SIMS) is more a surface analysis method. However, by adequate etching and surface milling it can serve as a concentration profiler and bulk analysis method. The principle is well known. Sputtering is by a low energy Ar, O, or Cs ion gun, then naturally or laser ionized species are analyzed by a quadrupole spectrometer and detected. SIMS has been widely used to observe contamination of CdTe by impurities and follow their profiles from the surface into the bulk. Figure 4 shows bulk spectra of THM CdTe. However, even if the sensitivity reached is many times less than 1 ppm, direct measurement of impurity concentration is still difficult. This is due to the large difference of the ionization of certain elements in the presence of other elements; for example, Cd ionization is enhanced by a factor of 10^3 in presence of Br. SIMS is more often used for profiling, in the near surface or for the detection of impurities to direct atomic absorption spectrometry (AAS) or nuclear methods toward specific impurity concentrations.

6. Ion Chromatography—IR Absorption

The recent introduction in the market of high sensitivity ion chromatography apparatus, with a wide range of mass detection, permits the fast analysis of mono-

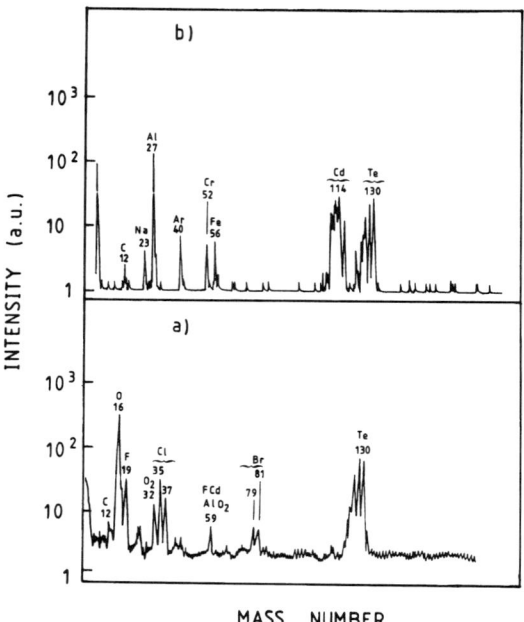

FIG. 4. SIMS spectrum of bulk CdTe that shows common impurities. (a) CdTe etched Br_2-methanol bulk, negative ions; $j = 0.75$ µA/cm². (b) CdTe etched Br_2-methanol bulk, positive ions; $j = 0.75$ µA/cm².

and divalent anions and cations in UV, conductivity, or IR detection, especially for low mass elements such as Cl. This of great interest in doping and impurity measurements. Recently, it was successfully used by Ohmori, Iwase, and Ohno (1992) in detecting Cl in compensated CdTe up to 0.5 ppm and less. However, ion chromatography is still a destructive method while IR absorption is not. This last method can be useful to measure light bonded elements such Li, Na, P, Ag, or Cu (Finkielsztejn-Milchberg, 1983) but with the advances in FTIR spectroscopy, the far- and near-IR levels are now available and levels related to Fe diffused in CdTe have been reported by Carmo and Soares (1992).

Another useful method is IR microscopy ($\lambda \approx 1.3$ µm). It permits the localization of metallic inclusions and precipitates up to 1 µm and less, especially those aligned in chains along grain boundaries and twins and can, therefore, constitute a spark path between the two contacts (Fig. 5).

III. Surface Analysis

It is well known that surface properties determine semiconductor device properties to a great extent. CdTe is even more sensitive to the surface than other

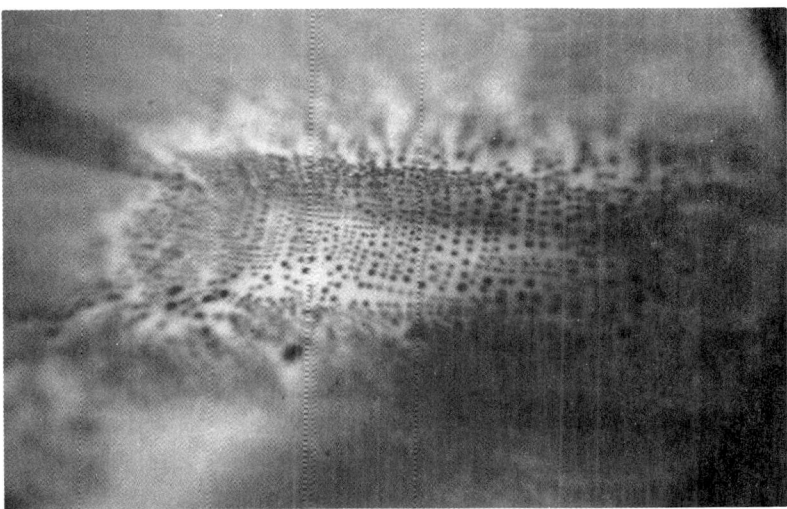

FIG. 5. IR microscopy of chain aligned (Te) inclusion around a small twin.

semiconductors, and small changes, of a few surface layers, is enough to make devices such as detectors vary in quality from the best to the worst. At the same time the surface is the external part of the material: It is subject to external attack and contamination and is the interface between the bulk and the external circuitry. All these reasons have led to extensive and thorough investigation of all surface behavior and characteristics.

A large variety of surface treatments is possible. However, just a few of them have been used for nuclear detector applications. Thus, only these surface treatments are considered in order of priority.

1. LAPPED AND POLISHED SURFACES

Lapped and polished surfaces are similar in most aspects, only the polishing grain size makes a difference, as well as the thickness of the damaged layer. This thickness is generally 2 to 20 times the diameter of the polishing grain, and variation is due mainly to the polishing pressure differences; the average lays in the 5 to 20 μm range.

Measurements by grazing incidence electron diffraction performed by Marple (1966), showed that this layer appears almost amorphous, but it is more a system of disoriented blocks than a glass, even for an optically polished surface with 0.03 μm grains. Ellipsometry measurements do not show extended oxidation (<20 Å). Even after 3 mon of air exposure, the Δ ellipsometry parameter is still rather constant over time. The sheet resistance was measured to be $3 \cdot 10^{11}$ Ω/sq (Ponpon, 1982) but we believe that is is an exaggerated value, due mainly to the difficulty in obtaining ohmic contacts. In fact, Kuzel and Lucas (1966) deduced, by field effect measurements, that the polished surfaces show an inversion layer of p-type material and only a small band bending in n-type. This fact was explained as Cd vacancies and CdO formation. However, CdO has not been detected and TeO_x is more probable. SIMS measurements on these surfaces show generally highly levels of contamination by many impurities present in the abrasive powders (Hage-Ali et al., 1979), such as $C.H_n$, Na, Al, SiC, Cr, and Fe and also Ti and Ba. These facts are of importance for future devices.

Part of the applied voltage on these contacted surfaces can be lost depending on the model adopted:

- The disturbed surface is considered a high resistivity layer and the voltage loss is ohmic.
- The disturbed surface is an inverted layer leading to a kind of heterojunction barrier with large dead zones.

But, in conclusion, even though this surface is stable in time, it is still far from a clean or perfect surface.

2. CHEMICALLY ETCHED SURFACES

There exist large numbers of CdTe etching solutions, producing different kinds of surface stoichiometries, but again only a few are adapted to nuclear detector surface treatments. The discussion is restricted to these treatments.

In fact, etching is used not only for cleaning, but also for bright surface polishing, for etch pits to reveal Cd or Te termination, and for surface enrichment by Cd, Te, or oxides—all depending on the expected application. An extended review of etching solutions can be found in Mergui (1991).

A Te rich layer can be obtained by an oxidizing solution like $7(K_2Cr_2O_7)$–$3(H_3SO_4)$, $3(HF)$–$2(H_2O_2)$–H_2O, $2(HCl)$–$2(HNO_3)$–H_2O, $4(KrCr_2O_7)$–$10\ HNO_3$–H_2O, and such as EDTA. For a few solutions, etch pits appear at the same time.

More solutions devoted to etch pit formation have been studied in nuclear detector materials. Such studies give an idea of the concentration of precipitates, inclusions, and defects. The most often used in $0.5\ Br_2 + 10\ AgNO_3 + CH_3OH$, but the standard one is 3 (HF) (conc.)–$2(H_2O_2)$ (30%) $2(H_2O)$ (Nakagawa, Maeda, and Takenchi, 1980), which reveals clear etch pits on the Cd ("A") face.

A face staining etch is used to identify the faces on [111] oriented wafers. The most useful is a mixture of 50 vol (lactic acid 85%)–8 vol (HNO3) (70%)–2 vol (HF) (conc.). After a 10 sec immersion, rinsing, and drying, the Cd side of the [111] crystal (face A) shows a *black color,* while the Te side (face B) shows a *gray color.* As mentioned before, the Cd (A) face can reveal etch pits; and both sides, immediately, reveal any extra grains or twins in the crystal, as the opposite color.

Many other etching solutions for CdTe surfaces may be employed to create a bright surface for polished and mirror finishes (Gaugash and Milnes, 1981; Ponpon, 1985; Ricco, White, and Wrighton, 1984; Feldman, Opila, and Bridenbough, 1985). However, the one most often used as a standard treatment for nuclear detector surface treatment is Br-methanol.

Br-methanol (BM) etching was first reported by Fuller and Allison (1962) for III–V materials and was widely used after that for II–VI materials. The concentrations used vary from 0.5 to 20% Br in methanol, generally around 2–5%. It was reported to create pits or orange peel surfaces. It leaves a nonstoichiometric surface composition (atomic ratio Cd/Te ≠ 1). Here the situation is more complicated. While several authors (Amirtharaj and Pollak, 1984, Danaher et al., 1986, Feldman et al. 1985, Haring et al. 1983, Solzbach and Richter, 1980, Werthen et al. 1983) reported Te rich surfaces, another group (Hage-Ali et al. 1979, Patterson and Williams, 1978) reported Cd rich surfaces. A third group of authors (Dharmadasa, Roberts, and Petty 1982, Gaugash and Milnes, 1981, Zitter, 1971) obtained stoichiometric surfaces; and finally, Konova, Shepov, and Nedev (1986) obtained Cd rich surfaces for n-type and Te rich for p-type materials. Bowman (1988) reports a drastic decrease in the Cd/Te ratio with increasing water proportion and slight crystallographic dependence.

Few conclusions can be extracted from these results: the stoichiometry of Br-methanol etched surfaces is type (n or p) and hygrometry dependent. The great distribution of the results depends on the groups, material origin, and growth and analysis methods. Many other parameters, such as temperature, impurity concentration at the surface, and resistivity, for example, are not well studied. However, a consensus exists that aging this surface increases its Te proportion, probably in the form of TeO_x. SIMS measurements show clearly a peak of TeO, and ellipsometry measurements show after etching logarithmic time dependence of growth in air of a layer of Te oxide (50 Å in 14 days) (Hage-Ali et al. 1979). It must be noticed that the methods employed for determination of stoichiometry are mainly XPS and sometimes Auger, one time Raman, and one time high resolution RBS. Notice that these two last ones are nondestructive methods and are an absolute and simultaneous mass and concentration versus depth profiling methods. It uses the fact that Cd or Te excess layers are not crystallized to analyze CdTe wafers in channeling along a crystalline direction, in comparison with cleaved surfaces (as a reference, see Figure 6).

SIMS is useful and sensitive as a technique to measure surface stoichiometry. However, ionization enhancements between species like Cd by Br and Cl or Si by O lead to profiles that are somewhat difficult to interpret (Hage-Ali et al., 1979). With some experience, one can extract useful information from the profiles to construct the real surface component structure.

Knowledge of the surface stoichiometry is of great importance, because it determines the property of contacts in both cases: of deposited metal (evaporated or sputtered), where the species represent the real contact, Cd, Te or their oxides; and of chemical deposited contacts, where the presence of one or other species change completely the chemical reaction behavior.

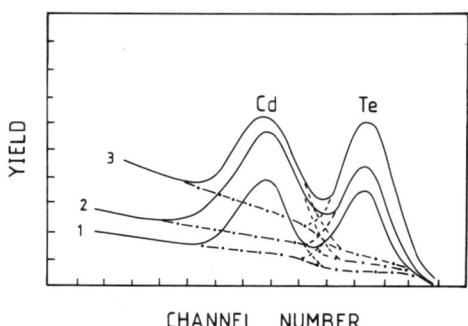

FIG. 6. Channeled RBS spectrum of a (1) cleaved, (2) freshly etched, and (3) aged CdTe surface.

3. Oxidized Surfaces

Native oxidized layers always exist on CdTe surfaces exposed to air, especially etched surfaces. This effect is enhanced by metal deposition via an electrically assisted diffusion process. This oxide layer determines widely the value and quality of the electrical surface barrier. As was shown before, this oxide is TeO_x–TeO_2, rather than CdO. Generally, these oxide layers show p-type conduction, probably due to a Cd deficiency (Dubbrovskii 1961). However, this layer is around 50 Å thick after few weeks. To increase this thickness, intentional oxidation has been made, mainly by dipping in H_2O_2 + NH_4(OH) sol. at RT. Depending on the reaction time, a thickness of up to 500 Å has been reached, and by boiling up to 800 Å, and the oxide is more stable. Figure 7 shows a SIMS analysis of these layers. However, ellipsometry studies show a long term evolution of these surfaces (Hage-Ali 1980).

4. Interfaces and Contact Analysis

Contacts and interfaces determine to a large degree the electrical properties of devices. Even if the material is of the highest quality, bad contacts or interfaces result in bad devices. A wide variety of contacts on CdTe are possible. However, only a few give useful results on high resistivity nuclear detector materials.

a. Evaporated Contacts

Al evaporated contacts have been widely used on lapped or polished surfaces. This results in a bias at higher voltage with lower dark current, which sometimes gives rise to a polarization effect. The effect appears in the form of a count rate decrease with time after the bias is applied. An extensive interface study of this contact has not been pursued, owing to the nonuniformity of the lapped coated

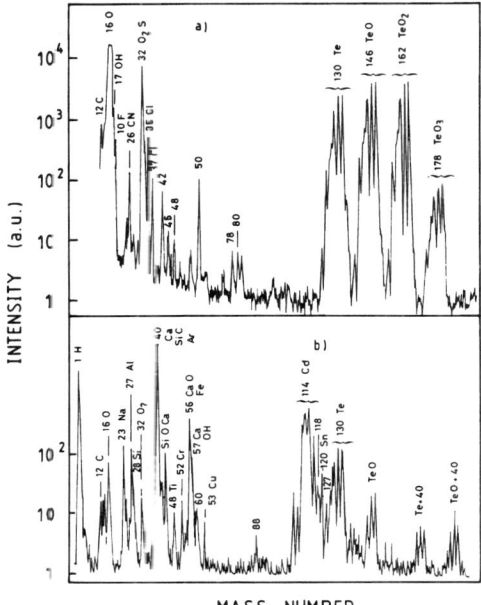

FIG. 7. SIMS spectrum of an H_2O_2 oxidized CdTe surface. Note the exclusive presence of TeO_x. (a) Negative ions; (b) positive ions.

surface, even though such a study can explain many of the problems connected to this kind of contact. However, we can notice that aluminium is known to be easily and quickly oxidized in air, and the presence of Cd, Te, TeO_2, and H_2O on the CdTe surface can lead to a variety of electrically active compounds. Despite the difficulties, the profiles and characteristics of these thin layers need to be further studied.

b. *Electroless Contact Au, Pt*

Electroless deposition of Ag, Ir, Rh, Au, Pt has been used by Wald and Bell (1974). The most useful is certainly Au and Pt electroless contacts, employing gold or platinum chloride, widely used now for CdTe nuclear detectors. According to De Nobel (1959), during reaction with chloride, Cd leaves the material to the solution as Cd^+ and Te precipitates on the CdTe surface as a p-type film, mainly at the interface. Part of the Te is still in the Au or Pt deposited layer, allowing the continuation of the reaction that occurs with Te rather than the Cd. Figure 8 shows SIMS results for Pt, Cl, Cd, and Te profiles of platinum chloride dipped samples. Enhancement of Cd by Cl is clearly seen. Auger profiles for platinum chloride (Fig. 9) are more demonstrative. Figure 10 shows RBS profiles of electroless deposited gold with an evaporated contact as reference.

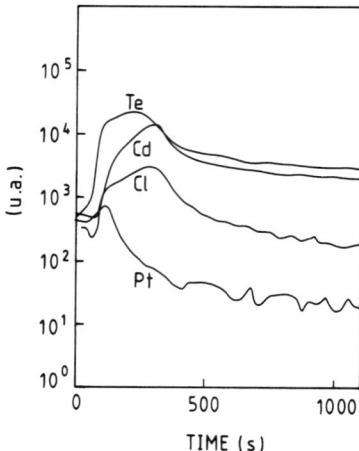

FIG. 8. SIMS profile of electroless deposited platinum contact and interface.

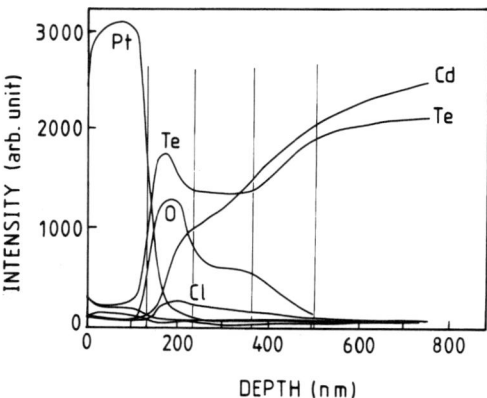

FIG. 9. Auger profile of electroless deposited platinum contact and interface. (Reprinted with permission from Elsevier Science.)

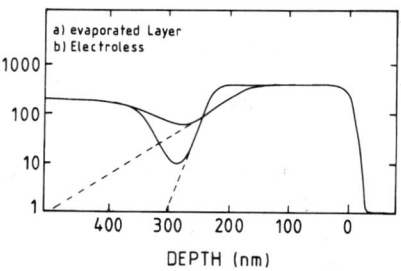

FIG. 10. RBS spectra of electroless and evaporated Au contact on CdTe.

Electroless profiles show a diffusionlike shape. However, diffusion is unlikely at this temperature and over this depth range, and the most likely behavior is that of immediate CdTe dissolution up to 500 nm by aurochloric acid (one should realize that, within a second in this etching solution, microns are dissolved). This dissolution is followed by redeposition of a succession of different layers from the bulk to the surface region:

- A graded layer with decreased Cd and Te and increased O and Cl,
- A layer with a quasi-constant amount of Te, O, Cl, and decreased Cd,
- A layer of TeO_x (TeO_2?) decreasing Cd with small increase of Cl at the interface with the preceding, and containing small Pt concentrations,
- A thin layer of decreasing Te, increasing Pt, depleted in Cd,
- The last layer is a quasi single Au or Pt mixture, with 10% of Te up to the surface, which ensure the continuation of the reaction as noted before.

A few words have to be said about these dissolution and redeposition processes and their conditions. Despite the commonly used "electroless" description of this process, it is a pure "electrolysis" process in fact. Two electrodes of Cd and Te or CdTe dipped in Au or Pt chloride provide excellent battery behavior at ≈ 1.4 V and ≈ 1 A/cm^2. Unfortunately it polarizes with time. If the two electrodes are adjoining, a violent reaction results due to the short circuit of the battery. Cd and CdTe electrodes show only 0.1 V and a very weak reaction. In fact, the dissolution and redeposition here are pure electrolysis reactions in a closed circuit, leading to a permanent change of the solution composition that results in different composition equilibria and a consequent deposition in every layer.

Like all equilibria problems, the mechanism of this reaction in solution is dependent on many parameters: temperature, concentration, pH, time, carrier concentration, and defects and impurities at the CdTe–solution interface (n or p type) (Musa et al. 1983a, Musa, Ponpon, and Hage-Ali, 1983b). Figure 11 gives the variation of the Pt thickness as a function of the bath solution temperature, and

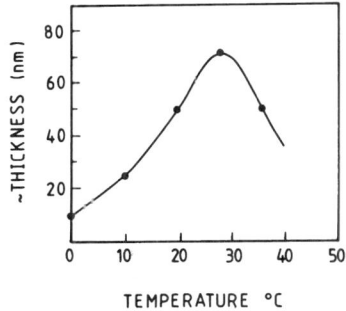

FIG. 11. Deposited (Pt) layers thickness evolution with bath temperature.

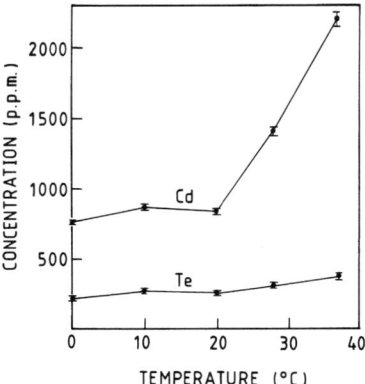

FIG. 12. (Cd) and (Te) AA concentration evolution with temperature in the bath after CdTe deposition.

the study shows a maximum around 28°C. Figure 12 gives the Cd and Te concentrations by AAS in the bath solution after deposition, as a function of the temperature. Simon (1991) shows that free Cd is present in the solution and increases in concentration with temperature.

In fact, the reaction (Au, Pt thickness and related deposition) is limited only by the amount of the CdTe initially dissolved in the solution in the few first seconds. The reaction kinetic is slightly different for Au and Pt, but the general behavior is certainly the same. The TeO_x layer and the first layer, presumably p-type due to the lack of Cd, determine to a large part the electrical behavior of the contact. TeO_2 is a semiconductor, and, in fact, this contact is a heterojunction between TeO_2 and CdTe with highly disturbed interfacial regions. If there are some tunneling effects, they can be attributed to Au or Pt, O, or Cl doping in the layers, also to V_{Cd}, V_{Te} and their complexes and carrier density. One must note, too, that the formerly mentioned graded profiles can be attributed also to nonuniformity in the thickness of layers as reported by Fowell *et al.* (1990) and Williams (1991) using STM and BEEM measurements.

In conclusion, despite the good results obtained with this contact, it is now clear that it is a more complex process than believed before. A complete and comprehensive study of this contact is still needed.

c. *Diffused Contacts*

For the elements classically used in semiconductor diffusion, results in CdTe were generally disappointing. The formation of layers with electrically active impurities is made difficult for two reasons: first, during the doping process, lattices can be compensated by induced defects as mentioned in the previous chapter;

second, heating CdTe at a temperature higher than 160°C without some precautions leads to drastic changes in the electrical properties of the bulk material, due to the formation of V_{Cd} and the breaking of some complex bonds, resulting in resistivity decrease and enhanced trapping. Li was studied by Vul and Chapnin (1966) and by Arkadeva, Matveev, and Sladkova (1969), and P by Hall and Woodbury (1968). Au has been studied by many authors and a diffusion coefficient was extracted (Hage-Ali et al. 1973, Akutagawa et al. 1975) by RBS and mass spectrometry. Bi was studied, too, owing to its easy evaporation and analysis in CdTe and to its appearance in the volume V, like P, As, and Sb (Hage-Ali et al. 1973). However, for the two latter elements, Au and Bi, fast diffusion saturation was found, attributed to compounds or complex (Cd, Te–Au, Bi) formation starting at 390°C for Au and less than 350°C for Bi. Cu diffusion was studied by Mann, Linker, and Meyer (1972) and by Woodbury and Aven (1968). For all these elements, the extracted diffusion coefficients present a wide distribution among different authors, since surface state preparation, and crystal quality seem to be dominant.

Although these last diffused elements provided inferior results, another element, In, seemed to be more promising. It has been widely used as an ohmic contact on n-type low resistivity material, by alloying In in a neutral Ar atmosphere at 300°C for a few minutes (Segall, Lorenz, and Halstead 1963), leading to good x-ray and α particle detectors (Cornet, Hage-Ali, and Siffert 1971). Typical CdTe gamma detectors use M-S-M structures with Au or Pt as contacts on p-type high resistivity material. With a medium barrier, this results, generally, in low applied voltages and relatively high leakage currents that limit the useful electric fields in the device to 400–600 V/cm. Consequently, sensitive volume charge collection and speed are very limited. The solution is well known for Ge and Si. The p-i-n structure approximated by M-π-n (π high, ρ p-type) structure for CdTe if M is easy to do. Many studies have been performed to realize an n region in p-type high resistivity CdTe, with different degrees of success (Squillante et al. 1989, Khusainov, 1991, Mergui 1991). The process is in principle simple: The surface is etched or polished and a layer of In or In_2O_3 source is deposited on the surface. In is then thermally diffused within the temperature limits indicated before for CdTe under low Cd pressure. For measurements of the thickness effects on the properties of the detector, a thick detector is made, at the beginning, 1.5–2 mm, and the p contact is an electroless Au contact. After data collection, this contact is carefully removed from the detector to make it thinner and thinner. The best thickness depends on the $\mu\tau$ product of the material, which is ≈ 300 μ for $\mu\tau = 10^{-4}$ material. The diffused n region formation requires knowledge of the In diffusion coefficient. As for other metals, this coefficient is widely varying, depending on the crystal surfaces and external pressure. A few significant papers describe the situation well (Mergui 1991, Watson and Shaw, 1983, Brown and Willoughby, 1982, Yokota et al. 1991) and the results can be summarized as follows: In is very soluble in CdTe, even at 200°C, the solubility increases from a Cd rich to Te overpressure, below 400°C in the interesting region of detector

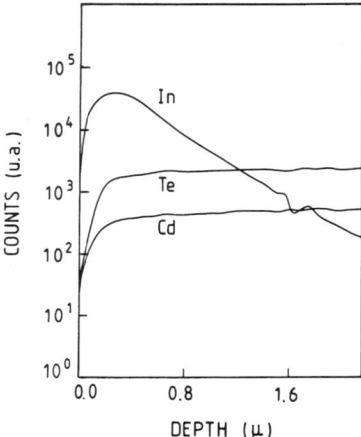

FIG. 13. SIMS profile of diffused In in CdTe (160°C-30'-Ar-H$_2$ flux).

processing. D seems to be contant at $D = 2.10^{-13}$ cm^{2}s^{-1}, with In + Te source but this is obtained for only one condition. Figure 13 shows SIMS profiles of diffused In (500 Å, $T = 160°$C, $t = 30'$ ArH$_2$).

We must notice that after evaporation, and even before annealing, diffusion up to 3000 Å can be observed. However, despite these difficulties, diffused In is still the more promising candidate for p-i-n or M-π-n structures, as it will be seen in the next chapter.

d. Implanted Contacts

High resistivity CdTe material is a quenched material where associations are frozen at temperatures ranging from 100 to 400°C. Ion implantation and subsequent annealing can disturb these complexes and break existing needed bonds. This explains the little work done in this field. Starting with p-type CdTe, n conversions were obtained by 2–40 keV In ion implantation (Kachurin et al. 1967a). The doping efficiency was found to increase with energy up to 40 keV, but decrease with increasing temperature and become ineffective after 150°C. This suggests a doping by complex centers, as noted in the previous chapter. The same group (Kachurin et al., 1967b) introduced 5–40 keV Ag and Ga with the same behavior, P also, but Al and Sb have been found ineffective. Some p-n junctions by $3 \cdot 10^{15}$ As/cm^2 (500°C) implantation at 400 keV in p-type CdTe have been reported by Donnelly et al. (1968). CdTe was preannealed in Cd pressure (650°C) and postannealed after SiO$_2$ encapsulation at 650°C and Cd pressure again. The electrical activity was estimated to be 0.1–1% of As. Many authors have studied implantation of P, Bi, Tl, Xe, Ar, and In (Meyer and Lang 1971, Gettings and Sephens, 1974, Norris et al. 1981) with different degrees of sophistication and

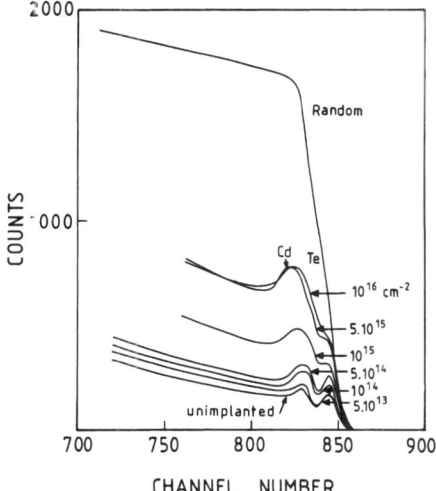

FIG. 14. Random and aligned along ⟨110⟩ direction RBS spectra of CdTe 40 Kev $5 \cdot 10^{13} - 10^{16}$ Bi/cm² ion implantation.

success. However, nuclear radiation detectors have been prepared (Cornet *et al.* 1972) using 30 keV, $5 \cdot 10^{14}$ Bi/cm² implantation 200°C–30′ and vacuum annealing. Rutherford Backscattering (RBS) x-ray fluoresence shows saturation of the lattice defect at 47% of amorphization for $5 \cdot 10^{15}$ Bi/cm², and even $3 \cdot 10^{17}$ Bi/cm² was not successful to overrun this limit up to amorphization (Fig. 14). This detectors shows a resolution of 23 keV for ^{241}Am α particles at 5.484 MeV. Finally, one must note that defect annealing can be accomplished by laser pulses (Norris *et al.* 1981) or preferably by 0.4 J/cm² of Nd:YAG laser enhanced diffusion of adequate layer thickness (200 Å of In) (Musa 1983).

IV. Electrical and Optical Characterization

Up to now chemical impurity analysis and surface-interface characteristics have been discussed. However, the properties of devices are strongly affected, too, in a large extent by bulk characteristics: defects, impurities and their associations, which can be measured or estimated by different complementary electrical and optical methods.

1. Resistivity

Resistivity is one of the most important characterization measurements in nuclear detector materials for various reasons. First, it is a global indication of purity and defect concentration in the material. Second, the sensitive effective thickness of x-ray and gamma ray detectors is a direct function of the resistivity and, as will be seen later in the chapter on detectors, affects directly the detector radiation efficiency.

Directly after crystal growth, the resistivity is measured in different sections of the ingot by the well-known four point probe (Van der Pauw 1958) method. The difficulty here is that materials that are semiinsulating, in the 10^8-10^{10} Ωcm range require high impedance ($10^{13}-10^{14}$ Ω) electrometers, ammeters, or synchronous detection instruments to evaluate very low currents in resistivity or Hall mobility measurements. These are in principle the more precise methods. The problem is that contacts are supposed to be ohmic, and as is known, it is not possible to have such a contact on high resistivity p-type material without damaging treatment. A possible contact is one that would have a low barrier of about 0.2 eV (Musa 1983). Many authors reported ohmic contact on low resistivity n-type materials by In alloying (Segall et al. 1963). For p-type material, contacts are made by electroless deposition (De Nobel 1959). For medium resistivity, material (ρ = 100–500 Ωcm), Musa et al. (1983a) found that in the electroless contact case the specific contact resistance (ρ_c) change, like $\rho^{1.13}$ ($\rho = \alpha\rho_c^{1.13}$) and it is seen that low ρ_c cannot be achieved by this way. Other metals like In alloys, Cu or Ag were used with some success. Electroless Au on LiNO$_3$ treated surfaces (Triboulet and Rodot, 1968) provide good results, but at the price of thermal annealing in excess of 200–400°C with the predicted consequences in material quality. This problem is still open, and for actual nuclear detector materials, electroless Au is used most. Figure 15 shows the resistivity behavior of different kinds of THM CdTe ingots

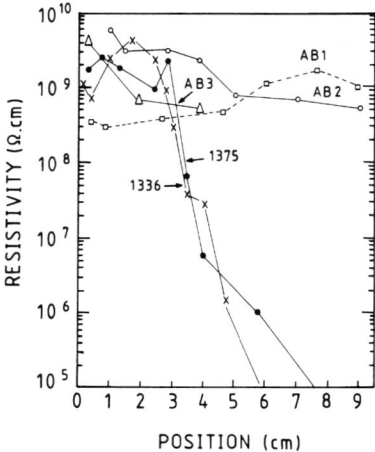

FIG. 15. Resistivity (ρ) evolution along different THM CdTe ingots.

with typical materials and highly purified materials. As can be seen, resistivity generally decreases from the beginning of the ingot toward the end, owing to the saturation of the Te zone by impurities along the ingot. For highly purified elements, this saturation arose farther along the ingot. As a final remark, during measurement, samples must be stored in the dark for the same time to obtain reproducible results.

The simplest way to measure resistivity is to use the $I(V)$ characteristic $R = V/I$ and $R = \rho\, 1/s$. That requires ideal ohmic contacts, which means that this method can give only qualitative and indicative results.

There exist more realistic resistivity measurement methods based on the depleted sensitive zone calculations in nuclear detector devices, and we will see it later in the chapter on detectors. As a final remark, high resistivity is an essential prerequisite condition but not a sufficient one. Materials can reach 10^7–$10^8\ \Omega \cdot$ cm (Schaub et al. 1977) in Cl compensated Te rich solutions, or up to $10^{10}\ \Omega \cdot$ cm in Bridgman group V doped ingots, without good nuclear detection properties, probably owing to a low $\mu\tau$ product caused by a high density of deep levels and recombination centers.

As was mentioned in the previous chapter, resistivity can be increased by compensation, halogen, and complementary Cu (Fig. 18 in Chapter 6) or by hydrogen treatment with annealing or implantation of H. Figure 16 shows the variation of the relative resistivity of materials after implantation with 2 MeV H^+ versus the initial resistivity of the material. It has been seen that H^+ is effective to obtain low ρ and ineffective for high $\rho\ 10^9\ \Omega \cdot$ cm.

2. MOBILITY, LIFETIME, AND $\mu\tau$ PRODUCT

The $\mu\tau$ product is a global measure of the material and electric charge carrier (electrons or holes) transport properties, such as defects and impurities concentra-

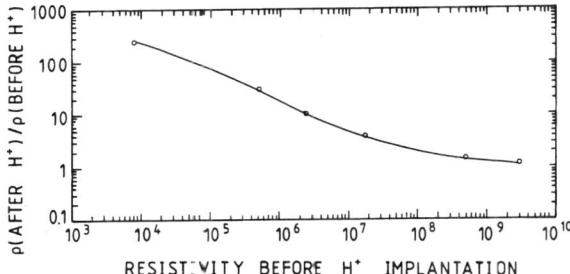

FIG. 16. The ratio of the resistivity (ρ') after (H^+) implantation (2 MeV) on (ρ) before hydrogen implantation, evolution with (ρ) previously.

tion, trapping and detrapping centers, their behavior and cross sections, and consequently of detection quality as well as collection and detection efficiency and resolution. The mobility was well studied experimentally and theoretically by a number of authors (Segall *et al.* 1963, Triboulet and Marfaing 1973). However, as mentioned previously for the resistivity measurements, mobility cannot be easily measured by the classic Hall method under electric and magnetic fields owing to the high resistivity material needed for nuclear detectors. A dynamic method such as "time of flight" is indicated. This consists in deducing the drift velocity from the collection time of carriers generated at one side of a planar detector. Carriers can be generated by different kinds of radiation, α particles of ^{241}Am at 5.5 MeV, pulsed electron beam ($\approx$40 keV) (Alberigi Quaranta, Canali, and Ottaviani 1970, Stuck 1975), or low energy gamma rays from natural sources, with the more suitable and flexible source being the pulsed electron beam. Depending on the side irradiated, the parameters of electrons or holes can be studied with electrons for the negative side and holes for the positive side. With a fast oscilloscope, one can measure the charge collection time in current mode or the potential risetime T_R, which is related to carrier drift velocity v by $v = d/T_R$, where d is the detector thickness. The mobility is then $\mu = v/\mathscr{E}$, where $\mathscr{E}$ is the electric field $\approx V/d$, and where V is the applied voltage. Figure 17 shows the mobility of some high resistivity In and Cl doped material.

The carrier lifetime τ_c before trapping can be deduced from the collection time and efficiency and the Hecht relation (Hecht 1932). The product of τ_c and $\mu\tau_c$ will be seen in detail in the next chapter.

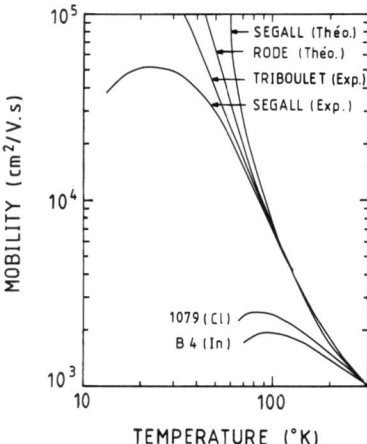

FIG. 17. Mobility of carrier evolution with temperature for low resistivity (experimental and theoretical), (Cl) and (In) doped (exp.) CdTe.

3. PHOTOLUMINESCENCE ANALYSIS

Photoluminescence (PL) and cathodoluminescence (CL) are the most widely used methods for CdTe materials characterization (Norris and Barnes 1977). Despite expensive experimental setup and the need for liquid He at 1.8–4.2 K, these methods need no electrical contact or particular electrical behavior, thus leading to very sensitive and repetitive measurements independent of resistivity, contact quality, or leakage current. While this is an advantage of these techniques, the drawback of these methods is that they do not measure the irreversibility of the electrical behavior of materials. In these experiments, samples are excited by a laser (typically an Ar ion) for PL and by an electron gun for CL. Luminescence is analyzed by a spectrometer in the 0.8–1.6 eV range. Many authors have studied CL and PL spectra in CdTe (Norris and Barnes 1977, Halsted and Segall 1963, Pautrat *et al.* 1985, Lischka *et al.* 1985) and generally found levels lying from 0.05 up to 0.15 eV with this last one dominating. Longer wavelengths lead to the IR domain, which is more difficult for analysis. PL is widely used to characterize CdTe samples qualitatively. In the case of nuclear detector materials, comparative results have been obtained for THM (CdTe : Cl) ingots grown at different temperatures (Cornet 1976). The PL band at 1.42 eV increases sharply from 800 to 950° C (Fig. 18) and is interpreted as higher dissolution of Cl in CdTe with temperature. The same result was found for increasingly In doped material (Barnes and Zanio 1975) and the PL intensity increases as compared to undoped THM material as one moves from Al, In up to Cl doped material (Furgolle *et al.* 1974). Many authors believe they have identified Li, N, Cu, Ag, Au, N, P, and As acceptor sites

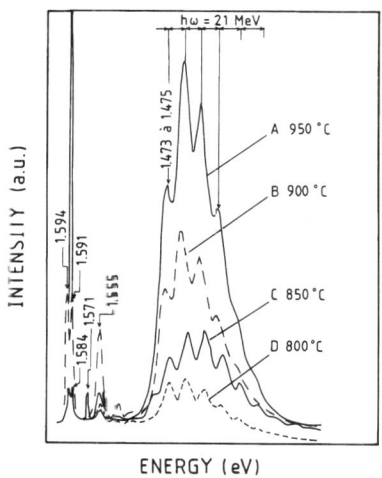

FIG. 18. Evolution of the 1.4 eV photoluminescence band for different crystal growth temperature.

and up to six donor lines attributed to Ga, In, and Cl. However, it is not always easy to identify unambiguously a doping site from complex defect sites. If some correlation is found with impurities, the only possible conclusion is that these impurities are involved in the centers but are not necessary "the doping" center. The same impurity can be related to two or more different levels: one doping and another complex related to a center. Such is the case in the PL work of Lischka et al. (1985) on Fe in CdTe, where two levels at ≈ 0.14 eV and ≈ 0.47 eV are clearly apparent. The same behavior is apparent for Cu, as will be seen later. This shows the need for complementary measurements for precise assignment and concentration, such as those of Zimmermann et al. (1992) by RT diffusion of Ag combined with electrical resistivity and various doping measurement (Seto et al., 1992) or combined with ODMR, for example, as can be seen in the next subsection.

4. ODMR AND RELATED METHODS

Optically detected magnetic resonnance (ODMR) allows for the combination of the potential of electron spin resonance (ESR) for structural identification with the high sensitivity of PL when combined with magnetic circularly polarized emission (MCPE), which permits better resolution. The higher resolution is achieved by frequency modulation leading to selection by lifetime (Meyer et al. 1992). A number of analyses using the combination of PL and ODMR (Hofmann et al. 1990, Benz et al. 1991, Stadler et al. 1991) have led to very precise results concerning many PL lines or levels that can be seen in CdTe:Cl compensated high resistivity material. They are summarized as follows. The A center acceptor with binding energy of $E_a = 0.12 \pm 0.003$ eV has been identified using PL, ODMR as a complex $(V_{Cd'} Cl_{Te})^-$, and interactions with three neighbors T_e are resolved. In ternary systems such as CdZnTe, the A center cannot be the major defect responsible for self-compensation in these materials since it is only a small fraction of the defects observed, and it cannot be responsible for the p-type conversion in ZnTe. Extrinsic impurities, not detectable by PL, are the 3d transition metal ions such Fe, Ni, Co, and Mn. ESR identified these defects even in undoped CdTe materials, and a trap at $\approx E_v + 0.2 \pm 0.005$ eV was found by constant photocurrent method (CPM) and attributed to a level where $Fe^{3+/2+}$ and $Ni^{2+/+}$ are involved. Ni seems to be involved with another level at $E_C - 0.64$ eV. ODMR studies on recombination luminescence involving a no phonon line at 1.473 eV show that both band–acceptor and donor–acceptor transitions contribute to this luminescence, and this explains the contradiction between emission decay time and thermal activation energy of the luminescence quenching. Each one is related to two different transitions, with an acceptor taking part in the recombination. One can see by the observation of three resonnance signals that the symmetry is lower than tetrahedral. This excludes the common extrinsic impurities like Ag or Cu acting as a direct doping acceptor for this level at $\approx E_V + 0.14$ eV. This settles the question of the role of extrinsic impurities in shallow levels at ≈0.14 eV (Molva and Le Si Dang 1983), and this is in accordance with results found by TSC and PICTS discussed in the next subsection.

5. TSC AND PICTS MEASUREMENTS

As has been seen before, purely optical methods present many advantages in precision, sensitivity, and repeatability but do not give a realistic image of electrical behavior or good estimates of concentration or cross section. Despite all the contact and current leakage problems, thermostimulated current (TSC) and photoinduced current transient spectroscopy (PICTS) are of great help to study and evaluate parameters of these levels.

a. TSC

TSC is well described in the literature (Grossweiner 1953, Braunlich 1979). In principle, the material to be studied is cut into wafers on which detectors are prepared by classical methods. The device is then cooled to $L.N_2$ or L.He temperature, resulting in lowering the Fermi level position toward the valence band edge. Then carriers of both signs are generated by laser, lamps, or radiation. Under a bias electric field, carriers cross through the material and fill up trap centers, where they are frozen in these deep levels. Excess energy is needed for emission, which is provided by progressive temperature increase at a constant rate and leads to the emission of trapped carriers when the Fermi level crosses a certain defect level. Thus carrier emission is recorded versus temperature, and adequate models allow the calculation of the energy and cross section as well as concentration of levels present. The TSC emission process involves two differential rate equations that cannot be solved directly without approximation. This has been done simply by Grossweiner (1953). Exact mathematical and numerical resolutions of TSC equations for slow and fast retrapping was done by Elkomoss et al. (1985, 1987). These allow more precise calculation of the parameters if added to the "thermal cleaning" procedure (Samimi, 1992) for filtering of peaks in the spectra.

b. PICTS

PICTS has evolved from the well-known deep level transient spectroscopy (DLTS) (Lang 1974) where levels are electrically excited by a pulse in previously depleted detectors, resulting in trapping at the beginning and emission at the end of the pulse. In low resistivity material, the pulse is a voltage variation and emission is detected by capacitance measurements. In high resistivity CdTe for nuclear detector, this is done by light pulses and transient current or charge measurements are performed (Biglari et al. 1989).

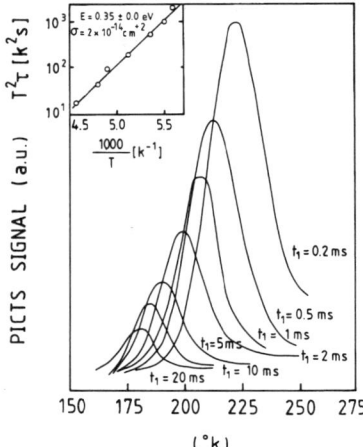

FIG. 19. PICTS spectra of (Cu) doped CdTe with corresponding Arrhenius plot for level energy determination (0.35 eV).

c. Results

In light of these two methods, a number of conclusions have been reached:

• Cu doping of CdTe results in a level at $E_V + 0.35 \pm 0.03$ eV (Fig. 19) shown by PICTS, but this is not the only level related to Cu in CdTe. Figure 20 shows

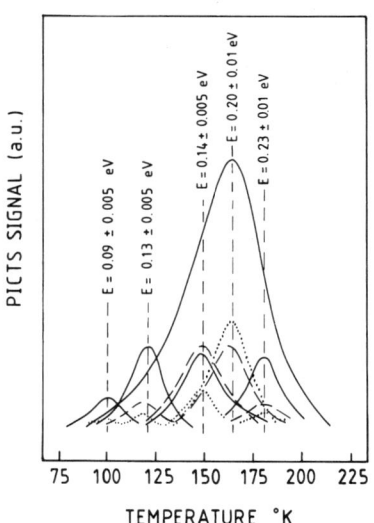

FIG. 20. Fine structure of the 0.16 eV (0.10–0.23) energy band (PICTS). (—), $p = 10^8$ Ω cm; (···), $p = 5 \cdot 10^6$ Ω cm; (--), $p = 1.2 \cdot 10^5$ Ω cm.

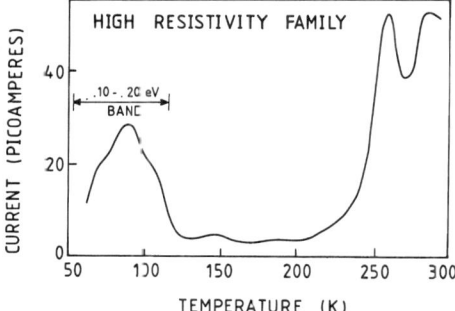

FIG. 21. Characteristic thermostimulated (TSC) spectrum of high resistivity THM CdTe material.

the behavior of two levels at 0.35 ± 0.03 eV and 0.16 ± 0.02 eV (supposed to be above E_V) as a function of copper concentration of the material. This behavior is quite complex and depends on whether the material is Cl compensated or not. In the uncompensated CdTe the 0.16 eV level increases up to 200 ppb corresponding to the curve/maximum of Figure 18, in Chapter 6. It then decreases in favor of Cu doping level at 0.35 eV, while in the compensated CdTe, both increase in the same fashion but with ratios of two orders of magnitude greater of 0.16 eV which is in fact a band of a few levels corresponding to different complementary compensation (Fig. 20).

• High resistivity material is characterized by an intense band at 0.10–0.20 eV, which corresponds to the different compensation such as (Fig. 20) $(V_{Cd}^{--}, Cl)^-$ or (Cu_{Cd}^-, Cd_i^+) or $(V_{Cd}, Cl, Cu^=)^0$ and similar associations with Fe, V, Ni, and others. Figure 21 shows high resistivity families characterized by an intense 0.10–0.20 eV band, around 80 K and a depleted one between 120 K up to 140 K. Figure 22 shows a low resistivity family, with an intense peak in the 0.27–0.35 eV region. One can see that *high resistivity* material is characterized by a *high ratio between the 0.10–0.20 eV and the 0.3 eV band, and a low ratio results in low resistivity material.*

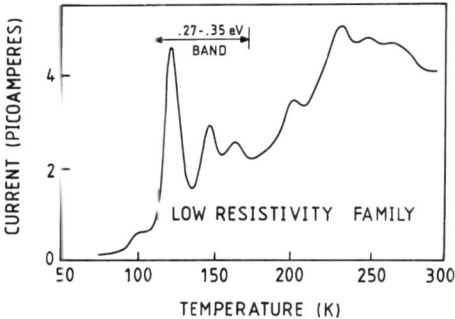

FIG. 22. Characteristic (TSC) spectrum of low resistivity CdTe.

- Irradiation by gamma rays of intense ^{60}Co radiosource or hydrogen treatment by annealing in H_2 or by low energy implantation at a few keV or 2 MeV implantation of H^+ result in increasing the resistivity (under same conditions). Following the ratio of the two bands rule as indicated before, extended numbers of results can be seen in a number of articles, which cover defects, irradiation, H, Cu, and resistivity behavior in CdTe (Samimi et al. 1985, 1987, 1989, Biglari et al. 1987, 1989a, Biglari 1989, Hage-Ali et al. 1991).

V. Discussion and Conclusions

In view of the different optical and electrical CdTe characterizations, some remarks can be made in summary:

- Contrary to earlier PL results, residual impurities cannot alone determine in major part the conductivity of high resistivity materials, as suggested by some authors. That can be true for concentration in excess of $10^{17}-10^{18}$ cm^{-3} but for impurity concentrations lower than 10^{16} cm^{-3} (high ρ materials), the association and complexes with at least two shallow acceptor and donor levels and one deep donor level, involving maybe some residual impurities, seem to determine the conductivity (Meyer et al. 1993) and this is generally the case for nuclear detector materials.
- Very pure materials made from $7N$ elements result in low resistivity materials as determined for pure materials in the last chapter, in the range from 10^6 to 10^8 $\Omega \cdot$ cm.
- Cu, Fe, or V added in excess of 0.3–2 ppm to pure elements increase appreciably the resistivity up to the higher values attained if precipitation is not present, and this results in good detectors.
- At equilibrium, self-compensation still determines the behavior of materials. The possibility of doping n- or p-type depends on the mobility of the defect at the temperature employed. Doping is possible outside of equilibrium as in MBE growth or quenching grown layers. In general, it depends on the material one considers: every material, every treatment, every temperature, and every condition is a special case that must be treated knowing all the parameters involved including, temperature, pressure, diffusion, contaminants, and kinetics.
- Silicon and carbon, which were used to explain the role of extrinsic impurities in PL experiments, do not seem to play a great electrical role in CdTe (Chibani et al. 1992). No levels in the 0.14–0.7 eV band related to Si and C are observed and they have no action in the resistivity of the material, even at a few tens ppm level.
- The final remark concerns the agreement between optical and thermal activation energies, especially in controversial material like CdTe and discussed in an excellent series of articles (see, for example, Grimmeiss and Klevermann, 1991).

They show that optical results if not taken at very low temperatures must be compared with the Gibbs free energy ΔG_n values and not with enthalpies. Note that generally one has

$$\Delta G_n = \Delta H_n - T \cdot \Delta S_n$$

where ΔH_n is the enthalpy variation and ΔS_n the total entropy variation. However, energies deduced from thermal activation still provide a relatively good indication when comparing results from the same method. For comparison with other methods, a precise approach is required and calculations of Gibbs free energy is indicated by Grimmeiss and Klevermann (1991).

REFERENCES

Akutagawa, W., Turnbull, D., Chu, W. K., and Mayer, J. W. (1975). *J. Phys. Chem. Solids* **36**, 521.
Alberigi Quaranta, A., Canali, C., and Ottaviani, G. (1970). *Rev. Sc. Instr.* **41**, 1205.
Al Neami, A. (1988). Thesis, Université Louis Pasteur, Strasbourg.
Al Neami, A., Bordas, A. K., Hage-Ali, M., Larcher, J., Siffert, P., and Heitz, C. (1992). *Nucl. Instr. and Meth. in Phys. Res* B **63**, 71.
Amirtharaj, P. M., and Pollak, F. H. (1984). *Appl. Phys. Lett.* **45**, 789.
Arkadeva, E. N., Matveev, O. A., and Sladkova, V. A. (1969). *Sov. Phys. Semicond.* **2**, 1264.
Barat, M., and Lichten, W. (1972). *Phys. Rev. A* **6**, 211.
Barnes, C. E., and Zanio, K. (1975). *J. Appl. Phys.* **46**, 3959.
Benz, K. W., Sinerius, D., Stadler, W., Meyer, B. K., Hofmann, D. M., and Omling, P. (1991). Proceedings of the European Conference on Crystal Growth, Budapest.
Biglari, B. (1989). Thesis, Université Louis Pasteur, Strasbourg.
Biglari, B., Samimi, M., Koebel, J. M., Hage-Ali, M., and Siffert, P. (1987). *Phys. Stat. Sol. (a)* **100**, 589.
Biglari, B., Samimi, M., Hage- Ali, M., Koebel, J. M., and Siffert, P. (1988). *J. Cryst. Growth* **89**, 428.
Biglari, B., Samimi, M., Hage-Ali, M., Koebel, J. M., and Siffert, P. (1989a). *Nucl. Instr. and Meth. in Phys. Res. A* **283**, 249.
Biglari, B., Samimi, M., Hage-Ali, M., Koebel, J. M., Siffert, P. (1989b). *J. Appl. Phys.* **65**, 1112.
Bowmann, P. T. (1988). Electrochem. Soc. Meeting, Wawaï, USA.
Braunlich, P. (1979). *Thermally Stimulated Relaxation in Solids*. Springer-Verlag, Heidelberg.
Brown, M., and Willoughby, A. F. W. (1982). *J. Cryst. Growth* **59**, 27.
Carmo, M. C., and Soares, M. J. (1993). *Mater. Sci and Eng.* **B16**, 246.
Chibani, L., Stoquert, J. P., Hage-Ali, M., Koebel, J. M., Abdesselam, M., and Siffert, P. (1991). *Appl. Surf. Sc.* **50**, 177.
Chibani, L., Hage-Ali, M., Stoquert, J. P., Koebel, J. M., and Siffert, P. (1993). *Mater. Sci and Eng.* **B16**, 202.
Cornet, A. (1976). Thesis, Université Louis Pasteur, Strasbourg.
Cornet, A., Hage-Ali, M., and Siffert, P. (1971). In *Proceedings of the First International Symposium on CdTe: A material for gamma-ray detectors*, ed. P. Siffert and A. Cornet. Centre de Recherches Nucléaires, Strasbourg
Cornet, A., Hage-Ali, M., Grob, J. J. Stuck, R., and Siffert, P. (1972). *IEEE Trans. Nuc. Sc.* **NS-19**, 358.
Danaher, W. J., Lyons, L. E., Marychurch, M., and Morris, G. C. (1986). *Appl. Surf. Sc.* **27**, 338.
De Nobel, D. (1959). *Philips Res. Rep.* **14**, 361.
Dharmadasa, I. M., Roberts G. G., and Petty, M. C. (1982). *J. Phys. D: Appl. Phys.* **15**, 901.

Donnelly, J. P., Foyt, A. G., Hinkley, E. D., Lindley, W. T., and Dunmock, J. O. (1968). *Appl. Phys. Lett.* **12**, 303.
Dubrovskii, G. B. (1961). *Sov. Phys. Sol. State* **3**, 431.
Elkomoss, S. G., Samimi, M., Hage-Ali, M., and Siffert, P. (1985). *J. Appl. Phys.* **57**, 5313.
Elkomoss, S. G., Samimi, M., Unamuno, S., Hage-Ali, M., and Siffert, P. (1987). *J. Appl. Phys.* **61**, 2230.
Feldman, R. D., Opila, R. L., and Bridenbough, P. M. (1985). *J. Vac. Sc. Technol. A* **3**, 1988.
Finkielsztejn-Milchberg, G. (1983). Thesis, Université Grenoble.
Fowell, A. E., Williams, R. H., Richardson, B. E., and Shen, T. H. (1990). *Semic. Sc. Technol.* **5**, 348.
Fuller, C. S., and Allison, H. W. (1962). *J. Electrochem. Soc.* **109**, 880.
Furgolle, B., Hoclet, M., Vandevyver, M., Marfaing, Y., and Triboulet, R. (1974). *Sol. Stat. Comm.* **14**, 1237.
Gaugash, P., and Milnes, A. G. (1981). *J. Electrochem. Soc.* **128**, 924.
Gettings, M., and Sephens, K. G. (1974). *J. Cryst. Growth* **22**, 50.
Grimmeiss, H. G., and Klevermann, M. (1991). *Appl. Surf. Sc.* **50**.
Grossweiner, L. I. (1953). *J. Appl. Phys.* **24**, 1306.
Hage-Ali, M. (1980). Thesis, Université Louis Pasteur, Strasbourg.
Hage-Ali, M., Mitchell, I. V., Grob, J. J., and Siffert, P. (1973). *Thin Sol. Films* **19**, 409.
Hage-Ali, M., Stuck, R., Saxena, A. N., and Siffert, P. (1979). *Appl. Phys.* **19**, 25.
Hage-Ali, M., Yaacoub, B., Mergui, S., Samimi, M., Biglari, B., and Siffert, P. (1991). *Appl. Surf. Sc.* **50**, 377.
Hall, R. B., and Woodburry, H. H. (1968). *J. Appl. Phys.* **39**, 5361.
Halsted, R. E., and Segall, B. (1963). *Phys. Rev. Lett.* **10**, 392.
Haring, J. P., Werthen, J. G., Bube, R. H., Gulbrandsen, L., Jansen, W., and Lusher, P. (1983). *J. Vac. Sc. Technol. A* **1**, 1469.
Hecht, K. (1932). *Zeitschr. Phys.* **77**, 235.
Heitz, C. (1984). *Progress in Cryst. Growth and Charact.* **8**, 131.
Hofmann, D. M., Meyer, B. K., Probst, V., and Benz, K. W. (1990). *J. of Cryst. Growth* **101**, 536.
Iturbe-Garcia, J. L. (1981). Thesis, Université Louis Pasteur, Strasbourg.
Kachurin, G. A., Gorodetskii, A. E., Lohurets, Y. V., and Smirnov, L. S. (1967a). *Sov. Phys. Sol. State* **9**, 375.
Kachurin, G. A., Gorodetskii, A. E., Zelevinskaya, V. M., and Smirnov, L. S. (1967b). *Sov. Phys. Semic.* **1**, 1187.
Khusainov, A. R. (1992). *Nucl. Instr. and Meth. in Phys. Res* **A322**, 335.
Konova, A. A., Shepov, A., and Nedev, I. (1986). *Thin Sol. Films* **140**, 189.
Kuzel, R., and Lucas, V. (1966). *Phys. Stat. Sol.* **14**, 169.
Lang, D. V. (1974). *J. Appl. Phys.* **45**, 3023.
Lischka, K., Brunthaler, G., and Jantsch, W. (1985). *J. Cryst. Growth* **72**, 355.
Mann, H., Linker, G., and Meyer, O. (1972). *Sol. State Comm.* **11**, 475.
Marple, D. T. F. (1966). *Phys. Rev.* **150**, 728.
Mergui, S. (1991). Thesis, Université Louis Pasteur, Strasbourg.
Meyer, B. K., Stadler, W., Hoffmann, D. M., Omling, P., Sinerius, D., and Benz, K. W. (1992). *J. of Cryst. Growth* **117**, 656.
Meyer, B. K., Hoffmann, D. M., Stadler, W., Salk, M., Eiche, C., and Benz, K. W. (1993). *Mat. Res. Soc. Symp. Proc.* **302**, 189.
Meyer, O., and Lang, E. (1971). In *Proceedings of the First International Symposium on CdTe: A Material for Gamma-Ray Detectors*, ed. P. Siffert and A. Cornet. Centre de Recherches Nucléaires, Strasbourg.
Molva, E., and Le Si Dang (1983). *Phys. Rev. B* **27**, 6222.
Musa, A. (1983). Thesis, Université Louis Pasteur, Strasbourg.
Musa, A., Ponpon, J. P., Grob, J. J., Hage-Ali, M., Stuck, R., and Siffert, P. (1983a). *J. Appl. Phys.* **54**, 3260.
Musa, A., Ponpon, J. P., and Hage-Ali, M. (1983b). *Mat. Res. Soc. Symp. Proc.* **16**, 225.

Nakagawa, K., Maeda, K., and Takenchi, S. (1980). *J. Phys. Soc. Jpn* **49**, 1909.
Norris, C. B., and Barnes, C. E. (1977). *Rev. Phys. Appl.* **12**, 219.
Norris, C. B., Westmark, C. I., Entine, G., Lis, S. A., and Serreze, H. B. (1981). *Rad. Effect. Lett.* **58**, 11.
Ohmori, M., Iwase, Y., and Ohno, R. (1993). *Mater. Sci and Eng.* **B16**, 283.
Patterson, M. H., and Williams, R. H. (1978). *J. Phys. D: Appl. Phys.* **11**, L83.
Pautrat, J. L., Francon, J. M., Magnea, N., Molva, E., and Saminadayar, K. (1985). *J. Cryst. Growth* **72**, 355.
Ponpon, J. P. (1982). *Appl. Phys. A* **27**, 11.
Ponpon, J. P. (1985). *Sol. Stat. Electron.* **28**, 689.
Ricco, A. J., White, H. S., and Wrighton, M. S. (1984). *J. Vac. Sc. Technol. A* **2**, 910.
Samimi, M. (1992). Thesis, Université Louis Pasteur, Strasbourg.
Samimi, M., Biglari, B., Hage-Ali, M., and Siffert, P. (1985). *J. of Cryst. Growth* **72**, 213.
Samimi, M., Biglari, B., Hage-Ali, M., and Siffert, P. (1987). *Phys. Stat. Sol. (a)* **100**, 251.
Samimi, M., Biglari, B., Hage-Ali, M., Koebel, J. M., and Siffert, P. (1989). *Nucl. Instr. and Meth. in Phys. Res. A* **283**, 243.
Schaub, B., and Potard, C. (1971). In *Proceedings of the First International Symposium on CdTe: A Material for Gamma-Ray Detectors*, ed. P. Siffert and A. Cornet. Centre de Recherches Nucléaires, Strasbourg.
Schaub, B., Gallet, J., Brunet-Jailly, A., and Pelliciari, B. (1977). *Rev. de Phys. Appl.* **12**(2), 147 (Proceedings of the Second International Symposium on CdTe: Physical Properties and Applications).
Segall, B., Lorenz, M. R., and Halstead, R. E. (1963). *Phys. Rev.* **129**, 2471.
Seto, S., Tanaka, A., Masa, Y., and Kawashima, M. (1992). *J. Cryst. Growth* **117**, 271.
Simon, L. (1991). Internal report, CRN-PHASE, Strasbourg.
Solzbach, U., and Richter, H. J. (1980). *Surf. Sc.* **97**, 191.
Squillante, M. R., Entine, G., Frederick, E., Cirignano, L., and Hazlett, T. (1989). *Nucl. Instr. and Meth. in Phys. Res. A* **283**, 323.
Stadler, W., Wang, F., Schwarz, R., Oettinger, K., Meyer, B. K., Hofmann, D. M., Sinerius, D., and Benz, K. W. (1991). Presented at ICAM-91, Strasbourg, France.
Stuck, R. (1975). Thesis, Université Louis Pasteur, Strasbourg.
Triboulet, R., and Marfaing, Y. (1973). *J. of Electrochem. Soc.* **120**, 1260.
Triboulet, R., and Rodot, M. (1968). *CR Acad. Sc. Paris B* **266**, 498.
Van der Pauw, L. G. (1953). *Philips Res. Rept.* **1**, 13.
Vul, B. M., and Chapnin, V. A. (1966). *Sov. Phys. Sol. State* **8**, 206.
Wald, F. W., and Bell, R. O. (1974). *Tyco Report US AEC AT (11-1)* **3**, 545.
Watson, E., and Shaw, D. (1983). *J. Phys. C: Solid State Phys.* **16**, 515.
Werthen, J. G., Haring, J. G., Fharenbruch, A. L., and Bube, R. H. (1983). *J. Phys. D: Appl. Phys.* **16**, 2391.
Williams, R. H. (1991). *Surf. Sc.* **251–252**, 12.
Woodbury, H. H., and Aven, M. (1968). *J. Appl. Phys.* **39**, 5485.
Yokota, K., Nakai, H., Satoh, K., and Katayama, S. (1991). *J. Cryst. Growth* **112**, 723.
Zaiour, A. (1988). Thesis, Université Louis Pasteur, Strasbourg.
Zimmermann, H., Boyn, R., Rudolph, P., Bollmann, J., and Klimakow, A. (1993). *Mater. Sci and Eng* **B16**, 139.
Zitter, R. N. (1971). *Surf. Sc.* **28**, 335.

CHAPTER 8

CdTe Nuclear Detectors and Applications

Makram Hage-Ali and Paul Siffert

CENTRE DE RECHERCHES NUCLÉAIRES
LABORATOIRE PHASE
STRASBOURG, FRANCE

I. INTRODUCTION	291
II. DETECTION PARAMETERS	292
III. CdTe DETECTORS	293
1. *Starting Materials*	293
2. *Detector Devices*	294
3. *Electrical Characteristics*	296
4. *Main Detector Properties*	305
IV. IMPROVEMENT OF DETECTOR QUALITY	316
1. *New Growth Process Materials*	316
2. *Single Type Carrier Collection*	317
3. *(n-i-p) (M-π-n) Structures and Cooling*	320
4. *Electronic Treatment*	322
V. APPLICATIONS OF CdTe DETECTORS	326
1. *Spectrometers*	327
2. *Safeguard Systems*	327
3. *Dosimetry*	328
4. *Space and Astrophysics*	328
5. *Medical Applications*	328
6. *Industrial Applications*	330
References	331

I. Introduction

Since 1879 many papers have reported numerous properties of CdTe. In the 1950s detailed investigations of CdTe were made, in particular by Kroeger and De Nobel (1955) and by Boltaks, Konorov, and Matveev (1955) and again by De Nobel (1959). De Nobel's thesis describes the behavior of this compound and especially the behavior of dopants in terms of a coherent model that established the basis for various uses of the compound. One cannot forget the following sentence from a patent filed by De Nobel and Kroeger (1962):

    As is well known, CdTe is a semiconductor which has very advantageous properties as compared to the other semi conductor chalcogenides of Cd, such as comparatively great mobility, a simple controllability of the conductivity from (*n*) to (*p*) type and conversely, so that CdTe may be used in semi conductor devices such as crystal diodes or transistors. It is also known that CdTe is photosensitive to many kind of radiation, for example, to infrared and visible radiation and X-radiation, so that it may be used in photosensitive devices such, for example, as photodiodes or as photoconductive bodies or infrared telescopes, image intensifiers, camera tubes and photoelectric cells, X ray dosimeters and the like.

It was a very promising list to which more possible applications have already been or will be added, especially in the x-ray and gamma ray detector field and their new extended applications.

II. Detection Parameters

Among the large family of binary compounds, only a few can potentially be used as nuclear radiation detector material. Table I shows a few selected semiconductors with their selection criteria. These criteria include the following.

- Atomic number (z): Only photoelectric interactions are of interest, and these increase with the atomic number z (power 4 to 5) of the absorbant material, leading to the preferential use of those with a higher z. CdTe with an average z of 50 (Cd 50–Te 52) is well situated.
- Bandgap E_g: For room temperature operation with low noise level due to thermal generation, high bandgaps of more than 1.3 eV are needed. However,

TABLE I

OTHER PROMISING BINARY COMPOUNDS AND THEIR PROPERTIES

S.C.	E_g(300 K) eV	μ_e(300 K) cm²·v⁻¹·s⁻¹	μ_h(300 K) cm²·v⁻¹·s⁻¹	τ_e (s)	τ_h (s)	Z	creat. pa eV
Si	1.12	1500	600	3×10^{-3}	3×10^{-3}	14	3.61(300
Ge	0.67	3900	1800	10^{-3}	10^{-3}	32	2.96 (90
C(diam.)	5.47	2000	1550	10^{-8}	10^{-8}	6	13.2(300
GaAs	1.43	8500	420	10^{-7}	10^{-7}	31–33	4.27
GaP	2.25	300	100	10^{-8}	10^{-8}	31–15	7.8
CdS	2.42	300	50	10^{-8}	10^{-8}	48–16	6.3
CdTe	1.5	~1000	~80	10^{-6}	10^{-6}	48–52	4.43(300
InSb	0.17	78000	750	10^{-7}	10^{-7}	49–51	1.2
GaSb	0.67	4000	1400	10^{-8}	10^{-8}	31–51	
InAs	0.36	~33000	460			49–33	
InP	1.27	4600	150			49–15	
AlSb	1.52	200	550			13–51	
HgI₂	2.1	100	4	10^{-7}	10^{-8}	80–53	4.15(30

when this energy is high enough, the probability of trapping and the electron–hole pair creation energy becomes prohibitive. Optimal values are situated between 1.5 and 2.2 eV CdTe with 1.5 eV at room temperature is acceptable.

- Resistivity: For materials with $Z > 40$, a thickness of about 2 mm can absorb a appreciable fraction of gamma rays in the 100 KeV range. Such a space charge must be obtained with a reasonable applied voltage (100–500 V). One can perform calculations to demonstrate that the semiconductor must have $N_A - N_D < 10^{11}$ free carriers/cm^3 at room temperature, equivalent to a resistivity of 10^8–10^9 $\Omega \cdot$ cm.

- Transit time: The transit time of carriers across the sensitive zone of the detector is inversely proportional to their mobility. With $\mu_h = 100$ cm^2/V $\cdot$ sec, this leads to a transit time on the order of a microsecond. Greater times lead to higher trapping probability.

- Trapping center density: Semiconductor materials, and particularly binary compounds, are affected by the trapping centers associated with the presence of imputities, stoichiometric defects, and complexes. Defects that introduce discrete levels in the bandgap lead to the trapping of carriers for variable times, resulting in drastic effects on various detection parameters, including average carrier lifetime, collection efficiency, mobility, and electron–hole pair creation energy. Two parameters are used to characterize these phenomena—trapping center density, N_T, and carrier lifetime before trapping, τ. It is easy to prove that trapping is negligible when $N_T < 10^{10}$ cm^{-3} and $\tau < 10^{-6}$ sec.

- The last, and fundamental, detector material selection criterium is the ability to prepare the material with a sufficient degree of purity and crystalline perfection.

Up to now, except for the classical materials Si and Ge, only two materials satisfy the majority of these criteria: CdTe mainly for gamma ray detection and HgI$_2$ mainly for x-ray detection. In this chapter special attention is devoted to CdTe detectors.

III. CdTe Detectors

1. STARTING MATERIALS

One must keep in mind that the description of high quality crystals of compound semiconductors is generally more subtle than that of Si or Ge. The reasons for this include, as mentioned, the binary nature of the material, in which any deviation from perfect stoichiometry affects strongly the electrical as well as the transport properties, and the easy decomposition of the compound when heated (for example, during crystal growth). This high sensitivity to growth conditions leads to crystals having properties that can be quite different. These experimental difficulties explain, partly, the apparently slow progress noted for these counters. Furthermore, this material does not represent the same economical importance as Si or Ge.

Since photon counters need rather thick sensitive zones, only growth methods that lead to high resistivity crystals is of real interest. Although many techniques have been considered in the past, essentially three groups of methods are employed today: zone melting, the traveling heater method (THM), and the Bridgman method of normal freezing of a tellurium rich solution. With the exception of the zone melting technique, which gives crystals of high quality but low resistivity, just sufficient for x-ray spectroscopy, the other methods operate in a tellurium solvent. This is to reduce the growth temperature to about 700 to 900°C rather than to go through CdTe melting at about 1050°C. The temperature reduction has a decisive influence on the material contamination by impurities from quartz mixtures and the furnace elements. The existence domain of the compound (solidus line) on the tellurium rich side of the temperature–composition projection of the phase diagram is strongly retrograde. When going from the growth temperature down to the ambience, the concentration of Te has to be reduced by several orders of magnitude. This can produce tellurium precipitates if the conditions are not strictly controlled.

It has been shown that it is, theoretically, possible to reach very high resistivity (10^8 Ω · cm) without any external compensation. In fact, singly charged cadmium vacancies V_{Cd}^-, doubly charged cadmium vacancies V_{Cd}^{--}, singly charged cadmium interstitials Cd^+, and doubly charged cadmium interstitials are always present. These give rise to several levels in the forbidden gap and thus have to be compensated for either by indium or mainly by halogens, primarily chlorine. The generally accepted model gives to chlorine a similar role as lithium has in Si or Ge compensation in the presence of residual acceptors, where the compensation of residual zinc by Li follows:

$$Zn^{--} + n\,Li^+ \rightarrow (Li^+, Zn^-)^- \text{ and } (2Li^+, Zn^{--})^0 \text{ neutral triplet}$$

In a similar manner, the doubly charged vacancy, V_{Cd}^{--} that acts as an acceptor in CdTe can be compensated for by a donor X from group III (Al) or group VII (Cl, Br, I) as follows:

$$V_{Cd}^{--} + n\,Cl^+ \rightarrow (Cl^+, V_{Cd}^{--})^- \text{ and } (2Cl^+, V_{Cd}^{--})^0 \text{ neutral triplet}$$

In the same manner residual chemical impurities like, Cu, V, Fe, seem to perform the same compensation of Cd_i^+, Cd_i^{++} and other ionized complexes.

In conclusion, it should be noted that the fundamental understanding of CdTe is still under discussion. The crystallographic quality has greatly improved in recent years, and large crystals are now obtained by zone solvent growth and Bridgman methods. Furthermore, the small angle deviations of various domains within a single crystal (mosaic effect) have been greatly reduced.

2. Detector Devices

Crystals are sliced in 1–4 mm thick wafers and cleaned by conventional semiconductor procedures. Then the handling depends upon the nature of the material and the kind of detector wanted. Essentially, three categories of devices can be prepared:

1. For low energy, high resolution x-ray spectrometry, low resistivity n-type samples, from zone melting grown crystals, are used. A gold Schottky barrier is realized on a bromine–methanol etched surface, having further received some "kitchen"-type surface treatment. The active dots are less than 6 mm in diameter to keep the diode's capacitance small, as discussed by Arkadeva, Matveev, and Melnikova (1980), Dabrowski, Iwanczik, and Triboulet (1975), and Akobirova et al. (1975).

2. For gamma ray spectrometry, both THM and solvent grown crystals are used, with or without Cl or In compensation. Due to the rather long time needed for crystal growth, these methods offer the advantage of lower temperature, leading to much less contamination by the quartz ampoule.

Essentially three kinds of structures have been investigated: Schottky diodes, quasi-ohmic contacts, and n-i-p structures. Schottky diodes are mainly on lapped surfaces, since the barrier height on etched wafers is generally below 1 eV, giving rise to excess current and noise. On these lapped surfaces, the active area diameter is limited to about 4–5 mm. This kind of contact is reported by Hage-Ali et al. (1979a), Patterson and Williams (1978), Mead and Spitzer (1964), Akobirova et al. (1975), and Seraphy (1980). Quasi-ohmic contacts are employed on semiinsulating materials, produced by electroless gold (platinum) chloride deposition on chemically etched surface. The electroless contact deposition techniques are of interest, since the reaction proceeds significantly below the surface, thus avoiding many of the surface complications. However, in addition to the redeposition of Au(Pt) tellurium, and the possible formation of tellurides, a migration of Au (or Pt) occurs. When increasing the bias voltage, the noise quickly increases, limiting the electric field and, consequently, the charge collection efficiency. Large current reductions are possible however by cooling. The n-i-p structures would be best and have also been considered but problems arise during the heat treatment for diffusion or annealing of damage after ion implantation (Chu et al., 1978; Bean, 1976). Pulsed annealing (Norris et al., 1981) offers certain advantages even though the first results are not fully satisfactory, due to defects resulting from the high thermal gradient; much better results have been observed on ohmic contacts (An, Tews, and Cohen, 1982). However the problem of n-i-p structures was elegantly solved in great part by the Riga group (Khusainov, 1991). One must note that spectrometers prepared on crystals having a resistivity in excess of $10^7 \, \Omega \cdot$ cm (Van Der Pauw) exhibited strong polarization when rectifying contacts were used (Malm and Martini, 1974; Siffert et al., 1976). In the case of n-i-p this problem is partially solved by cooling to low temperature.

3. For x-ray or gamma ray counting, without energy resolution, CdTe can be used as a solid state ionization chamber. Generally, electroless contacts on etched surfaces are used, as already discussed. For high flux measurements these counters have been employed in the photovoltaic mode, without any external bias, in a

manner similar to solar cells (Fox and Agouridis, 1978; Entine, Squillante, and Serreze, 1981).

3. ELECTRICAL CHARACTERISTICS

a. *I–V Characteristic*

Both leakage current and breakdown voltage are strongly dependent on the nature of the material and the diode manufacturing process. Typical results are as follows.

On low resistivity n-type material with a gold Schottky barrier, currents as low as 0.2 nA at 30 V and 3 nA at 100 V have been reached at room temperature. The breakdown voltage is in the range to 100–125 V.

On high resistivity p-type crystals with lapped surfaces, currents of 10^{-8}–10^{-9} A have been obtained at 500–1000 V (3–4 mm diameter). The breakdown voltage exceeds 1000 V.

On the same material but with electroless gold or platinum contacts, the currents are higher by a factor 10E2–10E3 and the breakdown is often around 100 V. This explains why these devices cannot be employed as spectrometers. In their paper, Agrinskaya and Matveev (1980) found a correlation between the deviation from Ohm's law and resolution of detectors that they attributed to homogeneity, however currents decrease drastically with cooling (Fig. 1).

b. *Capacitance*

Many authors (Kacherininov, Matveev, and Maslova, 1969; Cornet *et al.,* 1970; Rabin, Tabatabai, and Siffert, 1978) have found that, for diodes prepared on low

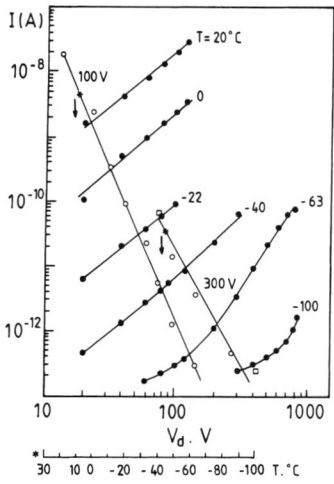

FIG. 1. $I(v)$ characteristic of Au–CdTe–Au detectors, and $I(T)$ at 100 V and 300 V.

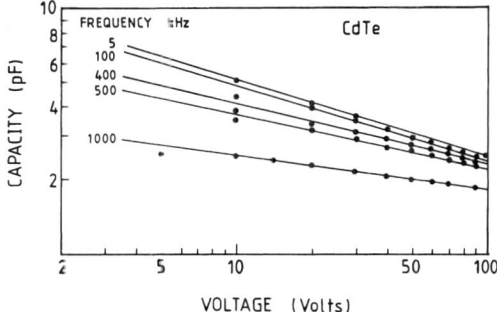

FIG. 2. Capacitance evolution of pure Al–CdTe–Al with voltage and frequency.

resistivity n-type material, the capacitance C changes with bias voltage V as $C \propto V^{-n}$ with $n \simeq 0.2-0.3$. This deviation from the classical Schottky diode is interpreted as a consequence of the presence of deep levels around $E_C - 0.55$ eV (Siffert et al., 1978).

For high resistivity crystals, the capacitance, for high frequencies, is independent of voltage. For the undoped crystals, variations of C are observed with both temperature and frequency, due to the presence of a deep level at $E_V + 0.15$ eV. In chlorine compensated samples, a time dependent variation of C can be found under some particular conditions, due to polarization by a very deep level at about $E_V + 0.7$ eV. Figure 2 shows the evolution of the capacitance versus voltage, with the frequency as a variable parameter, for high resistivity uncompensated p-type material, where $C \propto V^{-2.7}$. Figure 3 shows the capacitance evolution for Cl com-

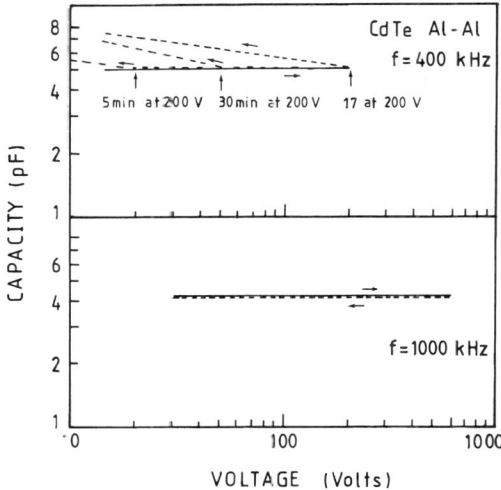

FIG. 3. Capacitance evolution of Cl compensated Al–CdTe–Al (no variation at 1000 kHz, but voltage and time dependent variation at 400 kHz).

pensated high resistivity p-type material with Al contacted lapped surface, which is supposed to be a surface barrier contact, at 1000 kHz. The capacitance is constant, but at 400 Hz, when the detector is maintained for a long time under bias, the capacitance increases for a decreasing voltage, which results from long term polarization effects. As a consequence, capacitance measurements cannot be employed for resistivity evaluation, as is usual for Si diodes, since the expected change with voltage is not (in general) obeyed. However, one must note that many discrepancies observed in capacitance measurement are attributed to the presence of levels in the bandgap, and several authors have considered theoretically the influence of levels (Schibli and Milnes, 1968; Rabin et al., 1978).

c. Depletion Layer Depth—Sensitive Volume

Owing to the extreme importance of this part, we treat it in some detail. In an abrupt junction, the thickness of the depletion layer under reverse bias V is expressed by

$$X \approx (2\epsilon\mu\rho V)^{0.5} \tag{1}$$

where X = depletion depth, ϵ, μ is the dielectric constant and the mobility of the material, $\rho = \dfrac{\mu}{e|N_A - N_D|}$, the resistivity, and V is the applied bias:

$$\text{for } p\text{-type CdTe,} \quad X(\mu) = 0.124\sqrt{\rho V} \ (\Omega\text{cm} \cdot V) \tag{2}$$

$$n\text{-type,} \quad X(\mu) = 0.4\sqrt{\rho V} \ (\Omega\text{cm} \cdot V) \tag{3}$$

Figure 4 allows the calculation of the depletion depth from the resistivity and the applied bias, and the evaluation of efficiency for different gamma energies.

However, the determination of ρ in high resistivity p-type material is certainly one of the more difficult tasks in CdTe. One can take advantage, however, of the property of nuclear radiation detection to measure this depletion layer independently. In fact, detectors are rarely completely depleted, the leakage current limiting the applied electric field results in a limited depleted sensitive zone less than the real physical detector thickness.

Figure 5 shows the attenuation coefficient of x-ray and gamma rays in CdTe, Si, Ge and HgI$_2$, from which one can extract the number of absorbed photons in a given thickness x of the material used. Conversely, if one knows the number of absorbed photons one can extract the sensitive depleted thickness of the detector. The absorbed number of photon I_m in this layer is

$$W = \frac{I_m}{I_0} = 1 - \exp(-\alpha x) \tag{4}$$

where I_0 is the initial number of photon, α is the attenuation coefficient extracted

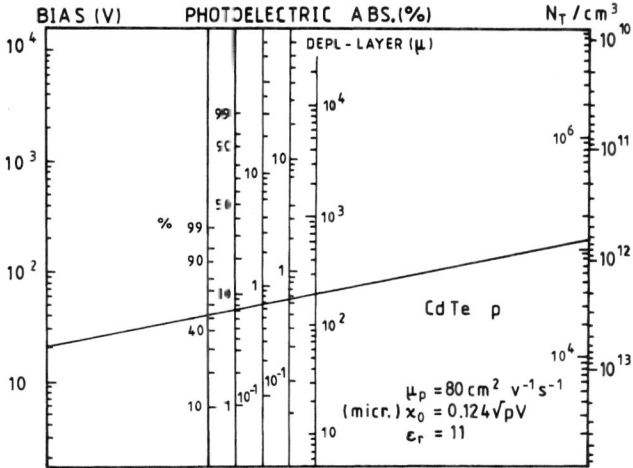

FIG. 4. Nomograph for depleted zone and photoelectron absorption as a function of voltage and resistivity for *p*-type CdTe.

from Figure 5, and x is the thickness of the absorption layer. Here the sensitive depleted layer is

$$x = -\frac{\ln(1 - W)}{\alpha} \qquad (5)$$

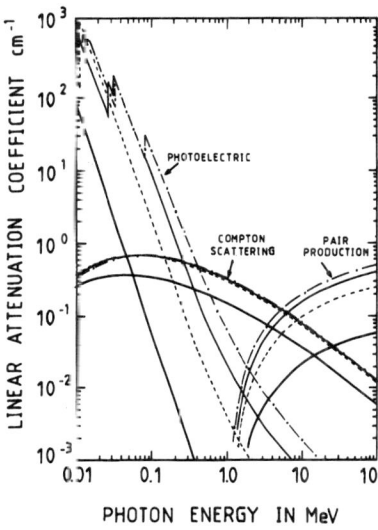

FIG. 5. Attenuation coefficient for different semiconductors as a function of x-ray and gamma ray energy. (—), Si; (--), Ge; (—), CdTe; (-·-), HgI$_2$.

This can be obtained by using a calibrated gamma ray source or any radioactive source (generally ^{57}Co or ^{241}Am) that can be calibrated by a known fully depleted detector controlled by classic methods. I_0 is then the theoretical emitted number of photons within the solid angle of the detector in relation to the radioactive source. I_m is the total integrated number of counts in the detected spectrum for a particular photon. Knowing then α and W, x is easily calculated by formula (5). This is the more precise method, when one has detectors with the same resolution.

In practice, the situation today, is as follows. For diodes prepared on low resistivity (100–500 Ω · cm) n-type materials, the depletion width increases as expected up to a maximum value of about 100 μ, which gives an active volume of 10^{-4} cm^3. This value constitutes a serious handicap for spectroscopy applications.

For spectrometers fabricated on lapped high resistivity p-type materials, the depletion depth is limited to a few mm. The apparent resistivity ρ' of a biased diode has been found to be 500–1000 times lower than the value measured by the Van der Pauw method. This limitation of the extension of the depletion depth has (probably) several possible origins, but two are dominant: fast polarization appearing as soon as the bias voltage is applied increasing the ionized $[N_A - N_D + N_T]$ concentration; and a high series resistance due to the lapped film (about 10 μm in thickness) reduces the effective applied voltage.

Therefore, the maximum effective depletion depth is limited to 1–2 mm. This limit seems sufficient for most applications, as it is equivalent to more than 1 cm of germanium (photoelectric). Since the active area is limited to about 5 mm in diameter, as already indicated, the volume of these spectrometers is also limited. Diode structures on etched materials are necessary for any further progress.

For counters prepared on etched high resistivity p-type crystals on which the contacts are applied by electroless deposition of gold or platinum, the full space between electrodes is sensitive. By electrooptical measurements depletion zones up to 6–7 mm have been observed. However, since the breakdown voltage is rather low, the possible applied electric field within the material is weak, thus efficient charge collection can be achieved only near the negative biased electrode (Fig. 6) due to the difficulties holes have in reaching the collecting electrode before trapping. As a result, strong degradation of the spectrometric properties appears when the thickness increases, and in practice, the spectroscopic properties are poorer than on lapped structures, where the effective sensitive zone is small.

d. Resistivity

This is one of the most important parameters to be determined. Several techniques may be used to measure the resistivity.

i. Van der Pauw (dc) Technique (V.D.P.). The classical procedure uses four point contacts made by the electroless deposition of gold chloride. One can observe 10^7 Ω · cm on highly purified uncompensated material, and up to

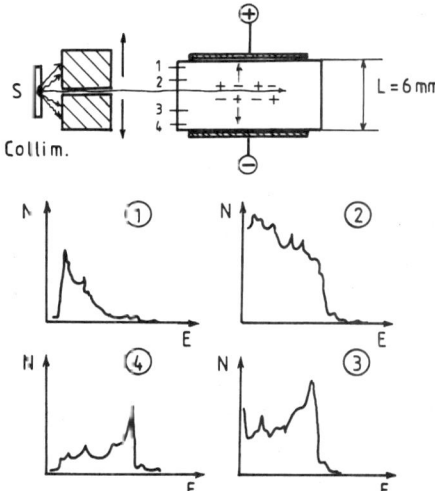

FIG. 6. ^{57}Co gamma ray spectra at different irradiation depth showing the electric field and electron–hole mobility effect.

10^9–10^{10} $\Omega \cdot$ cm for Cl compensated samples. However, there are doubts whether these values really correspond to bulk resistivities or are related to contact issues. To clarify this point complementary methods are needed.

ii. Hall (ac) Effect. On high resistivity samples, the Hall signal is of small amplitude close to the noise level. This necessitates the use of synchronous detection of alternating current flow of frequency f_1 and ac magnetic field of frequency f_2. When the same samples as before are used, comparison between the two dc and ac results is then possible. Both values of resistivity are generally on the same order of magnitude (Siffert *et al.*, 1978), but doubts still exist about contact problems.

iii. Use of Detector Properties. It was seen in Subsection III.3.c that the sensitive depleted layer x is expressed by the formula (2) for p-type CdTe as

$$X(\mu) = 0.124\sqrt{\rho \cdot V} \; \rho(\Omega \cdot \text{cm}), \text{V}$$

Knowing X from formula (5) leads to

$$\rho = \frac{X^2}{(0.124)^2 \cdot V} = \frac{[\ln(1 - W)]^2}{(0.124 \cdot \alpha)^2 \, V} \quad \text{for } p\text{-type} \tag{6}$$

The same section also showed how to measure X independently, by using only the attenuation coefficient and the detection spectra of the detector. This allows for the measurement of the effective resistivity of the detector by direct measurement and a simple model. This effective resistivity is called the "nuclear apparent resistivity ρ'." Siffert (1978) shows a comparison between the classic V.D.P. resistivity ρ and this nuclear resistivity ρ'. The problem is that the latter is 10^2–10^3 lower than the former, leading to a complete review of results and calculations concerning CdTe nuclear detectors with ρ as a parameter. The issue is to understand the reason for this discrepancy; and again fast polarization defect levels, high series resistance, and contact barriers are put forward, probably all of them are involved. At this time, with no simple explanation, this discrepancy must be accepted as fact, and in the future, for any work making use of the value of resistivity, only ρ' should be considered.

iv. Photopeak Escape. Several authors (Fioratti and Piermatty, 1971; Hansen, Feund, and Finck, 1970) have calculated the probability for a given fluorescence X ray to escape from the material during the absorption of a photon in a solid medium. In particular (Jäger and Thiel, 1977) using calculations of (Sherman, 1955) has extended this theory to the case of CdTe counters when the sensitive depleted layer is smaller than the absorption length of the primary photon and the contact size. In this case, the escape of x-rays occurs essentially throughout the two contact areas in an amount proportional to the supposed depletion thickness X (itself directly dependent on the resistivity ρ' as has been seen before), and it then becomes possible to evaluate the ratio photopeak to escape peak for any x-ray. This calculation is reported as a function of (ρ') with some measured experimental points (Fig. 7) as deduced from ^{57}Co gamma ray spectra, and reasonable agreement is obtained. However, if this approach seems to be sufficiently general to be used for the evaluation of the nuclear resistivity (ρ'), it should be remembered that the detector must have sufficiently good energy resolution to permit precise evaluation of both peaks (as shown in the figure).

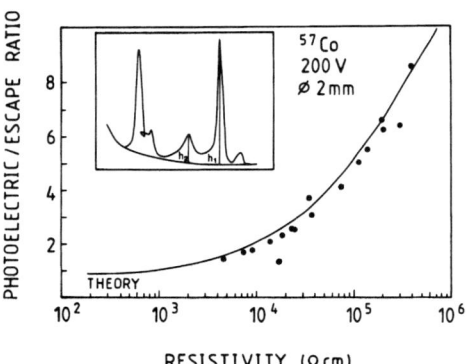

FIG. 7. Photopeak to escape ratio h_1/h_2 as a function of resistivity.

e. $\mu\tau$ Product

This parameter can be classified along with the electrical parameters even though it is a detector and material property. It is a quality index of the material and the detector, and it can be determined by several methods.

Separate measurement of μ by the Hall effect is generally possible. However, materials for nuclear detectors have very high resistivity, leading to difficulty in measurements using this method.

Easier evaluation can be done by making use of detection characteristics, using low energy gamma ray or α particles at one side ($+$ or $-$) of the detector, and allowing one carrier species to drift along the detector thickness (d) to the other contact under the electric field $\epsilon = V/d$. The transit time t_r, can be measured by the width of the current pulse $i(t)$ or from the rise time of the charge waveforms $Q(t)$. The carrier speed is then

$$v = \frac{d}{t_r} \quad (7)$$

The mobility μ is

$$\mu = \frac{v}{\epsilon} = \frac{d^2}{V_t} \quad (8)$$

where V is the bias voltage.

The lifetime of the carrier can be deduced from the shape of the current pulse, which is supposed to be exponential and of the form shown in Figure 8:

$$i = i_0 \, e^{-t/\tau} \quad (9)$$

$$\text{with} \quad i_0 = \frac{N_0 \, ev}{d} \quad (10)$$

The $\mu\tau$ product can be evaluated directly using detection properties and the Hecht relation as follows. The collection efficiency η is defined as:

$$\eta = \frac{Qm}{Qo} = \frac{hm}{ho} \quad (11)$$

where Qm, hm are the measured charge collection and pulse height, and Qo, ho are the theoretical values. The parameter hm is deduced directly from the spectrum by the channel number of the detected gamma ray peak (generally [241]Am (60 Kev) or [57]Co (122 Kev)). The parameter ho is more difficult to derive: it is obtained by comparison with a good Si surface barrier detector. If Si and CdTe

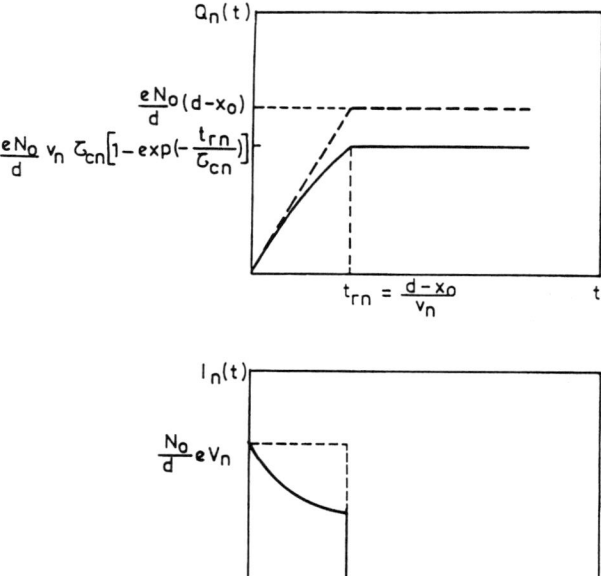

FIG. 8. Collected charge and current electron pulses, with (solid line) and without (dashed line) trapping.

detectors are connected in parallel under the same bias, with the same radioactive source, one can compare both pulse heights:

$$\frac{ho(\text{CdTe})}{h_{\text{Si}}} = \frac{E(\text{Si})}{E(\text{CdTe})} = \frac{3.61(\text{eV})}{4.43(\text{eV})} \quad (12)$$

where $E(\text{Si})$ and $E(\text{CdTe})$ are, respectively, the pair creation energy of Si and CdTe. The efficiency is then

$$\eta = \frac{hm}{ho} = \frac{hm}{h_{\text{Si}}} \times \frac{E_{\text{CdTe}}}{E_{\text{Si}}} \quad (13)$$

The Hecht relation (Mergui, 1991) is

$$\eta = \frac{\tau}{t}\left(1 - \exp\frac{-t}{\tau}\right) \simeq 1 - \frac{X^2}{2V\mu\tau} \quad (14)$$

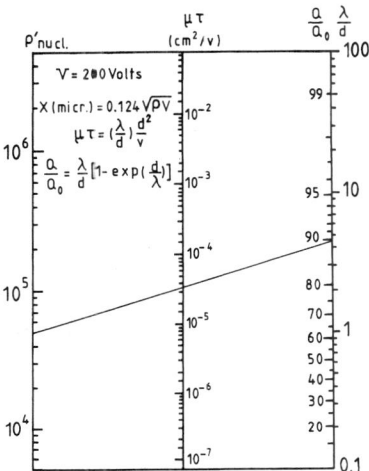

FIG. 9. Nomograph for ($\mu\tau$) product determination as a function of a charge collection efficiency and nuclear resistivity ρ'.

then

$$\mu\tau = \frac{X^2}{2V(1-\eta)} = \frac{X^2}{2V} \times \frac{1}{1 - \frac{h_{CdTe}}{h_{Si}} \times \frac{E_{CdTe}}{E_{Si}}} \quad (\text{cm}^2 \cdot \text{V}^{-1}) \quad (15)$$

where X is the sensitive depleted zone as measured before under V bias voltage. This calculation can be done using the nomograph in Figure 9, using known values of ρ' (nucl.) and η.

The $\mu\tau$ product for electrons and holes can be easily estimated by this method. For a THM (traveling heater method) grown, p-type high resistivity detector material, the $\mu\tau$ product is between $(0.8-3)\ 10^{-3}\ \text{cm}^2 \cdot \text{V}^{-1}$ for electrons and less than 10^{-4} for holes. By using another approach based on the dependence of gamma ray counting rate versus applied voltage (Kacherininov, Matveev, and Matyukhin, 1979), $\mu\tau$ values as large as $5 \cdot 10^{-2}$ for electrons have been reported.

4. MAIN DETECTOR PROPERTIES

a. *Pulse Amplitude—Charge Collection Efficiency*

The energy required per electron–hole pair generation is 3.61 eV (300 K) in Si, 2.98 eV (77 K) in Ge and 4.46 eV (300 K) in CdTe. Therefore, in an ideal situation, the pulse amplitude of a CdTe diode corresponds to 81% of that exhibited by a Si counter. However, in practice, these amplitudes are observed only for the purest low resistivity materials, whereas for the THM or solvent grown crystals, the pulse amplitude reaches 80–90% of the ideal height. This imperfect collection efficiency may have two origins, which in fact have some correlation:

1. Either too small a carrier mobility–lifetime product, as discussed before (starting from the Hecht relation, it is possible to determine this quantity for both electron and hole collection, by analyzing the pulse height of a counter when bombarded with shallow penetrating particles or gamma rays) or pulses with a long rise time, leading to incomplete charge collection during the clipping time of the amplifier.

2. Surface preparation and contacts. Table II gives the influence of surface treatment conditions on pulse amplitude delivered for gamma rays and α-particles, (Hage-Ali et al., 1979b). These conditions are of importance for the barrier height, the dead zone, and the consequent electric field extension.

The total charge collection efficiency η depends on the point x_0 (measured from negative entrance electrode of the radiation) at which the electron–hole pair has been generated. For a uniform field, the value is given by

$$\eta(x_0) = \frac{\lambda n}{d}\left[1 - \exp\left(-\frac{d - x_0}{\lambda n}\right)\right] + \frac{\lambda p}{d}\left[1 - \exp\left(-\frac{x_0}{\lambda p}\right)\right] \quad (16)$$

where d is the detector thickness, λ is the $\mu\tau\, V/d$ for electrons and holes (Day, Dearnaley, and Palms, 1967; Siffert et al., 1974). Similar calculations have been performed for a nonuniform field distribution.

b. Detection Efficiency

i. Sensitive Volume. Efficiency is directly related to the depletion layer, and as has been seen before, in an abrupt junction, the thickness of this sensitive de-

TABLE II

RELATIVE PULSE AMPLITUDE FOR DIFFERENT SURFACE TREATMENTS AND IRRADIATION DIRECTIONS

Source	Irradiation Side	Lapped Al Evap.	Etched Electroless Pt, Au	Etched Oxide + Al
α^{241}Am	+	0.35	0.20	0.85
	−	0	0.96	1
γ^{57}Co	+	0.90	0.97	0.95
	−	0.90	0.99	0.97

pletion layer is expressed by

$$X(\mu) = a(\rho \cdot V)^{0.5} \qquad a = 0.124 \text{ for } p\text{-type and } 0.4 \text{ for } n\text{-type}$$

Figure 4 allows an easy determination of X as a function of the applied volatge and the base material resistivity, for the most common case of p-type substrates. On the basis of the resistivity values generally published in the literature for compensated crystals, a few volts would be sufficient to achieve rather thick sensitive zones. In practice, however, this is not true for most materials, and the apparent resistivity is 500 to 1000 times lower than expected. Therefore, in these ingots, the maximum depletion layer thickness is limited to about 2 mm in the p-type samples and a little more in the n-type materials. The precise knowledge of X is therefore fundamental for efficiency calculation, and in Subsection III.3.c practical measurement methods for X were discussed.

ii. Theoretical Efficiency. The nuclear spectroscopist is interested mainly in the intrinsic peak efficiency, which is the probability that the photon interacts in the medium by producing counts in the full energy peak. Photoelectric absorption along with multiple Compton scattering contribute to the full energy peak formation. Using the attenuation coefficient (Fig. 5), the calculated pure photoelectric efficiency is tabulated in Table III as a function of photon energy for various sensitive thickness. It should be mentioned that for medium ($\sim$500 keV) energy, the gain in efficiency, when compared to a Ge (Li) diode, is 5–6, as a result of the higher atomic numbers in the compound.

Multiple scattering of the photons resulting in the total energy transfer to the sensitive medium can further increase this efficiency. For thin counters ($X = $ 1 mm), about half the events in the peak can be due to the Compton scattering contribution. However, the intrinsic peak efficiency is not sufficient to give a real idea of the CdTe detector performance, the solid angle of the detector not being considered. The evolution of the calculated relative efficiency of CdTe planar structures as a function of sensitive thickness and active area show that a gain in efficiency could be achieved much faster by an increase of the active area rather than by a thicker stopping medium, making necessary the growth of larger size uniform crystals.

iii. Real Efficiency. Evaluation of the real efficiency of spectrometers can be done by various measurements. One can determine the full energy peak efficiency relative to a 1.5" $\times$ 1.5" NaI (Tl) scintillator as for Ge detectors. The absolute full energy peak efficiency can be measured by using calibrated sources, which is the more precise approach. One can determine the real efficiency through the resistivity measurement, using all methods mentioned earlier for ρ measurement, mainly the real resistivity (ρ') (nuclear sensitivity). One can evaluate the depletion layer X as shown before and then estimating the absorbed number of photon by calculation or by nomograph (Fig. 4).

TABLE III

ABSORBED FRACTION OF X-RAYS AND GAMMA RAYS IN CdTe AS A FUNCTION OF THICKNESS AND PHOTON ENERGY

Sensitive Thickness (mm)	Fraction Absorbed (%) of Photon Energy (keV)												
	20	50	100	200	300	400	500	600	700	800	900	1000	1500
0.5	99.9	95	39.3	9.1	4.9	3.2	2.8	2.2	2	1.9	1.7	1.6	1.2
0.75		98.9	52.8	13.3	7.2	4.8	4.1	3.3	3	2.9	2.5	2.4	1.9
1		99.8	63.2	17.3	9.5	6.3	5.4	4.4	3.9	3.8	3.3	3.3	2.5
2		99.9	77.7	24.8	13.6	9.3	8	6.5	5.8	5.7	5	4.7	3.7
3			86.5	31.6	18.1	12.2	10.6	8.6	7.7	7.5	6.6	6.2	4.9
4			95	43.4	25.9	17.7	15.4	12.6	11.3	11	9.7	9.2	7.2
5			98.2	53.2	33	22.9	20	16.5	14.8	14.4	12.7	12	9.5
6			99.3	61.3	39.3	27.7	24.3	20.1	18.1	17.7	15.6	14.8	11.8
7			99.8	68	45.1	32.3	28.5	23.7	21.3	20.9	18.5	17.5	13.9
8			99.9	73.6	50.3	36.6	32.3	27	24.4	23.9	21.2	20.1	16.1
9				78.1	55.1	40.6	36	30.2	27.4	26.8	23.8	22.6	18.1
10				81.9	59.3	44.3	39.5	33.3	30.2	29.6	26.4	25.1	20.1
				85	63.2	48	42.8	36.2	33	32.3	28.8	27.5	22.1

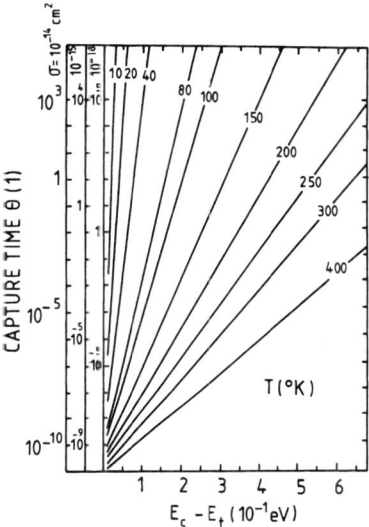

FIG. 10. Capture time as a function of level position for different temperature and capture cross sections.

If detection methods are used for depletion thickness, and ρ' measurement, in high quality detectors the real efficiencies can reach theoretical ones (Table III). However, if V.D.P. resistivity is taken into account, it is not surprising to find very low efficiencies (up to 0.001%). It should be mentioned that usually several effects can lead to a loss of efficiency. The escape of x-rays gives rise to a photopeak 23–27 keV lower than the expected photopeak of the line. This affects the efficiency as a function of the resistivity. However, this escape decreases with increasing sensitive volume and increasing gamma energy. Owing to the trapping phenomena, poor charge collection can occur. In fact, if the responsible defect traps the carriers for a time θ expressed by (Fig. 10)

$$\theta^{-1} = \sigma v_{\text{th}} + N_t \qquad (17)$$

where v_{th} is the thermal velocity of carriers, σ is the capture cross section, N_t is the trap or defect concentration. If the time θ is larger than the pulse shaping time constants of the amplifier (clipping time), the generated pulse is lost leading to one count less and a decrease in efficiency. The presence of deep traps can lead to decreasing efficiency with time: the polarization effect that will be treated later.

c. *Energy Resolution*

Spectroscopists working with Si or Ge detectors are used to energy resolutions determined essentially by statistical fluctuations in the electron–hole pair generation. Exceptions occur only for very low energy photons, where noise plays a role, and for charged particles, where nuclear energy loss fluctuations contribute to peak broadening. However, in Si and Ge counters, both used at 77 K in photon spectroscopy, the thermal noise becomes small. With CdTe, the goal being to operate at room temperature, this contribution is no longer negligible. The various contributions to peak broadening are these:

- Amplifier and detector noise: Diode leakage current and the noise of the amplifier produce line broadening, which can be calculated. When aluminium is used on small area surfaces, values around 1 keV are observed, becoming much larger for electroless gold contact devices. As long as the current is kept in the 10^{-8} to 10^{-9} A range, the contribution of these effect is not predominant.
- Ballistic defect: Due to the rather low drift velocity of the carriers, ballistic defects occur for thick detectors. For the thickness considered (2 mm), this effect is no longer dominant.
- Trapping–detrapping: This constitutes the most important contribution to peak broadening. Several authors have considered this effect on peak shape; their models predict either Gaussian broadening or a tailing on the low energy side of the peak. For a single level, uniformly distributed through the detector, when one type of carrier has a diffusion length much greater than that of the other and also in excess of X, the FWHM L_t is given by

$$L_t = K(1 - \eta)E_0 \qquad (18)$$

where K is a constant, η the pulse height defect of a photon of energy E_0. In the case of Ge spectrometers, the constant K was found to be approximatively the same for all detectors, indicating that the trapping distribution is uniform through the sensitive region. However, it was observed experimentally that L_t increases with $E_0^{1/2}$ rather than with E. In CdTe the situation is as follows for high resistivity material with lapped contacts. For a given detector at a certain energy E_0, the preceding relation is followed—the peak broadening is proportional to pulse height defect η. However, for various counters, the value of K changes from detector to detector, which is an indication of nonuniformity and nonconstant trap distribution. When the photon energy is changed, the FWHM is degraded more for small than large detectors. The following conclusions can be drawn from these results: at very low energy the resolution is limited by detector noise level; and when energy increases, incomplete charge collection is the most pronounced contribution, but the evolution vs. energy varies from detector to detector. However, the degradation in resolution vs. energy is much faster for the smaller detectors due to escape of photoelectrons out of the sensitive volume.

For electroless gold or platinum contacts, due to the small electric field, $1 - \eta$ becomes very important and the resolution can be degraded; multiple peaking appears and very short amplifier time constants must be used due to the high

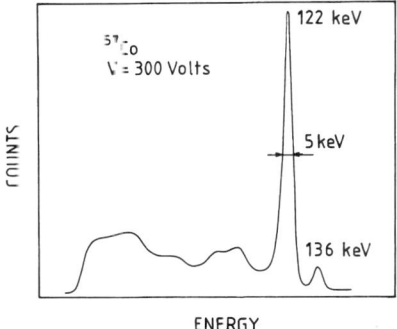

FIG. 11. ^{57}Co spectrum of Al contact on lapped surfaces for nonpolarizing (Cl) compensated CdTe material.

leakage current. However, improvement of surface treatments and the quality of these contacts have now led to detectors better in both resolution and efficiency.

Figures 11 and 12 show some results of detectors with Al lapped surface fabricated from CdTe (Cl) THM material. Figures 13 and 14 show results for electroless contacts on etched surface with the same material, where one sees that the price of energy resolution improvement for thin detectors is an important loss of efficiency due to the sensitive volume.

d. Stability

Polarization effects are considered short term changes, even if this problem is solved in many cases, but at one time it was the most severe problem encountered in CdTe detectors. Until recently, all crystals exceeding some critical resistivity

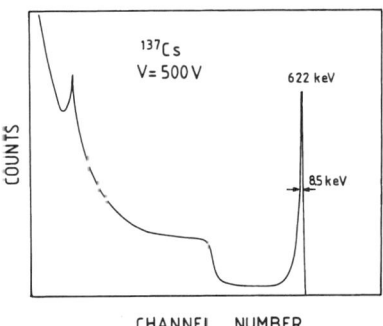

FIG. 12. ^{137}Cs spectrum of Al contact on lapped surfaces for nonpolarizing (Cl) compensated CdTe material.

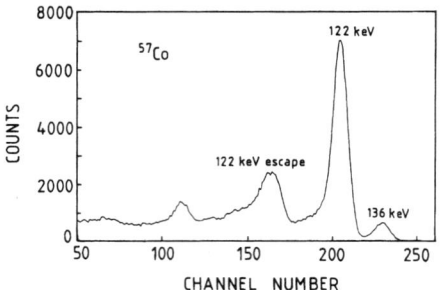

FIG. 13. ^{137}Cs spectrum of electroless Au contacts for thin (130 μ) CdTe detector.

value ($\rho \simeq 7 \cdot 10^7 \, \Omega \cdot$ cm or $\rho' \simeq 7 \cdot 10^4 \, \Omega \cdot$ cm), exhibited a progressive reduction of both signal amplitude and couting rate as a function of time after the detector was biased. Several models have been proposed (Malm and Martini, 1973; Bell, Entine, and Serreze, 1974; Siffert et al., 1976, 1977) to explain this effect. They consider either the trapping or detrapping of carriers on a deep level as a function of time. This level is located at about midgap at $E_v + 0.7$ eV (± 0.1 eV) and has a capture cross section of $\sigma = 10^{-16}$–10^{-17} cm^2 and a concentration N_T varying from 10E10 to 10E14 cm^{-3}. Its origin is probably related to the presence of doubly charged cadmium vacancies $[V_{Cd}]^{--}$ that release weakly bonded holes when the bias voltage V is applied on the detector. The original depletion layer width is expressed by

$$X^2 = \frac{2\epsilon_V}{q(N_A - N_D)} \tag{19}$$

where all symbols have their usual meaning.

Due to the emission of holes, a progressive variation of ionized centers occurs

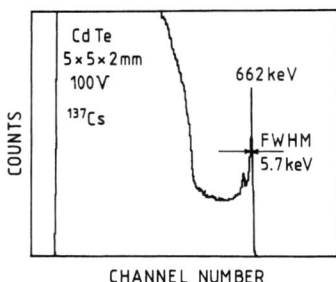

FIG. 14. ^{137}Cs spectrum of electroless Au contact for 5 × 5 × 2 mm CdTe detector, 5.7 Kev resolution is measured.

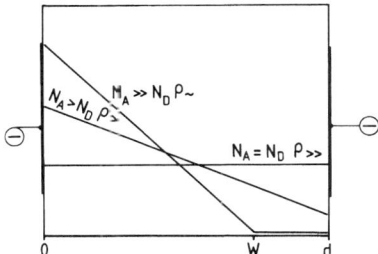

FIG. 15. Electric field profiles in CdTe junction for different resistivities or carrier concentration.

with time: $N_A - N_D + N_T(t)$. Therefore, the depletion width decreases with time as indicated in Figure 15. If all the centers are uniformly distributed, it becomes

$$\frac{X^2(0) - X^2(t)}{[X(t) - \alpha X(0)]^2} = \frac{N_T}{N_A - N_D}[1 - \exp(-e_p t)] \tag{20}$$

where $\alpha = (E_F - E_T/qV)^{1/2}$.

Here $\alpha \ll 1$, therefore,

$$\frac{X^2(0)}{X^2(t)} - 1 \simeq \frac{N_T}{N_A - N_D}\{1 - \exp(-e_p t)\} \tag{21}$$

where e_p is the deep level emission coefficient given by

$$e_p = \sigma v_{\text{th}} N_V \exp\left(-\frac{E_T - E_V}{kT}\right) \tag{22}$$

in which v_{th} is the carrier thermal velocity and N_V the number of states in the valence band. This model is in good agreement with experimental results, even in the case of very severe polarization (Fig. 16).

To overcome this problem, four possibilities exist:

1. Use lower resistivity material to keep the Fermi level away from the deep level E_T so that, when the band is bent, E_T does not cross E_F. This approximation has been used in the past for THM "pure" crystal growth.
2. A similar result is achieved by shining light on the contact electrode, increasing the free carrier concentration.
3. Reduce the band bending in such a way that E_T cannot cross E_F; in other words, apply quasi-ohmic contacts and use low electric fields. Such a method has been developed by Wald and Bell (1975), with electroless gold or platinum contacts and is now largely employed.

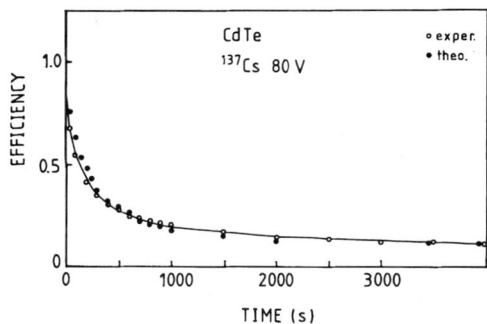

FIG. 16. Polarization effect, which is the evolution of efficiency with the time.

4. Suppress the polarization level E_T by a correct choice of the crystal growth conditions. This can be achieved sometimes even on lapped surfaces with aluminium electrodes. This opens new possibilities to the high resistivity CdTe crystals.

However, if the hole deep level is well accepted as an origin of the polarization, one is still faced with many questions. If the polarization is a bulk problem, why should surface treatments or contact changes, such as electroless or oxides on etched surfaces versus aluminum contacts on lapped surface, overcome this problem? If polarization is a problem of mechanical stress only on a lapped surface why does this problem not exist in the n-type (In) doped material? One first answer can be given by an equilibrium reached between the two parts of the responsible level bent above and below the Fermi level, at a given band bending, owing to the interface contact such as electroless (Au) or (Pt) (Fig. 17a). But this must

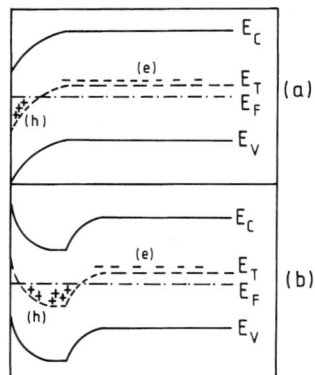

FIG. 17. Band diagram of (a) contact on an etched surface (b) contact on a lapped surface or adequate interfacial layer.

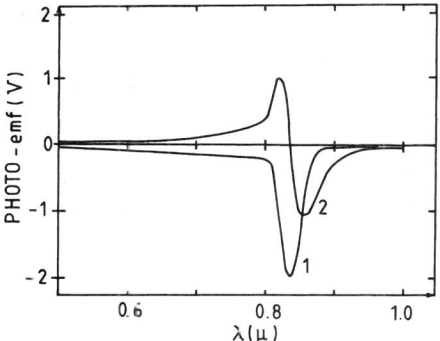

FIG. 18. Photo-emf spectra as a function of wavelength for Au/CdTe on (1) etched surface and (2) lapped surface.

be correct only for one given bias and such is not the case: polarization vanishes at every bias. It seems that another model is more realistic: the p-type high resistivity (CdTe:Cl) material becomes n^+-type in the lapped surface layer due to mechanical defects, dislocations, and other stresses related to the process, scratching, and the low mechanical resistance of CdTe. This leads to a deep (n^+, p) junction interpreted sometimes as a high resistivity series layer, (Al) being the ohmic contact on this n^+-layer. Detrapped holes are then gradually trapped in this "well" under the surface, leading to the polarization effect. Electroless Au contacts can reverse the situation (Fig. 17b). This unusual model is well supported by many results and facts such as the photo-emf spectral response of Akobirova et al. (1977) (Fig. 18) and by the occasional growth of very high resistivity, polarization free, materials (Hage-Ali et al., 1980), probably n-type doped, or the lapped surface layer still p-type for some chemical or thermic reason. In any event, further investigation are still needed in this field. Subjects of such investigation could include, for example,

- Long term evolution. Counters prepared by lapping and polishing the surface, which have received aluminium contacts, show no evolution over several years.
- Detectors realized with etched surfaces and electroless gold contacts. These can, sometimes, give rise to problems, depending on the preparation procedure, since the etched surface can exhibit evolution with time.
- Polarization. This can be produced when very high electron–hole pair generation is used, as has been shown.
- Irradiation effects. Irradiation of both chlorine and indium doped materials by 33 MeV protons has been considered, along with strong gamma ray, electron, or neutron bombardment (Norris, Barnes, and Zanio, 1977). For example with ^{60}Co gamma rays, an improvement is first observed followed by degradation for fluxes higher than 10^5 rad./hr. Annealing at 200°C restored the original performance.

IV. Improvement of Detector Quality

When the CdTe material is high resistivity, CdTe nuclear detectors can be of the homojunction type with two similar contacts, ohmic or not. A better detector can be realized with a n-i-p structure. In the lower resistivity case, n-p junctions or barriers are necessary. In all cases one needs low leakage current for low noise, a high electric field for better collection efficiency and a wider depleted zone, a high electron and hole $\mu\tau$ product for better charge collection, low trapping, and good resolution. As has been seen before, when photons interact with the detector material at a certain depth x from the (+) contact, a number of pairs of electron–hole are generated. Under the influence of the electric field, electrons are directed to the (+) contact with mobility around $\mu_e = 10^3$ cm^2/volt $\times$ second, low probability of trapping, and time $t = 60$ ns, even with 2 mm thickness, while for holes, $\mu_h = 80$, and a high probability of trapping leading to a large distributed time constant that can reach 2 μsec or more under the same conditions. The subsequent current pulse, which is the sum of both parts, is strongly dependent on the behavior of the holes. For electrons the $\mu\tau$ product can easily reach $5 \cdot 10^{-3}$ cm^2 V^{-1} easily, while for holes this product is rarely better than $5 \cdot 10^{-4}$, with a large variation.

Amelioration of this situation can certainly be realized by the perfection of the material itself. However, we are in the presence of a few contradictory parameters, in which improvement of one is made at the expense of the others, and some compromise must be generally found for the optimal performance. This means that one is now faced with some limitations in a classic material point of view, and improvement must be found presently in other fields.

1. New Growth Process Materials

The chapter on crystal growth mentioned the emergence of high pressure Bridgman (HPB) CdTe material followed quickly by CdZnTe (HPB) material.

For CdTe materials the reported announced performance (Raiskin and Butler, 1988) is large volume, uniform crystals up to 2 kg. If the spectrometric properties are poor, the uniform characteristic and low price of the crystal still make possible the fabrication of large numbers of counting detectors with similar properties for many applications. Dose rate response appears to be linear over five decades, but no indication about the resistivity or the conductivity type of the material is reported; however, more recent results (Johnson et al., 1993) seem to sometimes find better detection qualities.

Improvement of this latter material was done by alloying CdTe and ZnTe to obtain $Cd_{0.8} Zn_{0.2}$ Te, again by HPB (Butler et al., 1991, 1992) (Johnson et al., 1993), which results in a quartz free, high purity growth process, up to 5 kg of homogeneous crystal, reported lower lattice defect densities with consequent improvement of the $\mu\tau$ product, and higher resistivity up to 10^{11} $\Omega \cdot$ cm. Without compensation or doping, lower leakage current, and higher temperature operation (up to 100°C) owing to the larger bandgap, the crystal homogeneity allows the realization of arrays up to 3.7 cm on a side. However, the best announced resolution is about 7% for 122 keV of ^{57}Co, at 100°C and so this detector is a counter more than a spectrometer.

Even if the performance of such a material is moderate, it is still a promising detector material, and in the light of the last publication, better results are expected.

2. SINGLE TYPE CARRIER COLLECTION

It has been discussed previously that decreasing detection qualities are essentially attributed to the variation in hole properties, while electron properties seem to be more uniform thanks to their higher mobility. The evident solution is to detect only electrons if that is possible. Planar, thick, detectors can do that (Zanio, 1977) for low penetrating x-rays exciting the negative electrode. In fact, the advantage of cylindrical and spherical configurations have long been realized in ion chambers and other gas tubes, where the electron mobility greatly exceeds the mobility of positive ions, which are atoms or nucleii. One can establish a parallel between these ions and holes from the mobility point of view and use this configuration in semiconductor detectors, which present differences between the two carriers' mobility.

The electric field in planar detector is that of planar imperfect capacitor expressed as

$$\mathcal{E}_1 = \frac{V}{d} + \frac{Ne}{\epsilon}\left(\frac{d}{2} - x\right) \simeq \frac{d-x}{\mu\rho\epsilon} \tag{23}$$

where $\mathcal{E}_1$ is constant or linear function of the thickness. For a cylindrical capacitor, it is

$$\mathcal{E}_2 = \frac{V}{r \ln\left(\frac{r_2}{r_1}\right)} \tag{24}$$

and for spherical one it is

$$\mathcal{E}_3 = \frac{V}{r^2\left(\frac{1}{r_1} - \frac{1}{r_2}\right)} \tag{25}$$

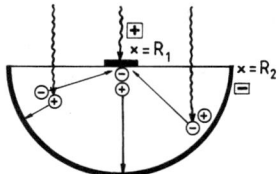

FIG. 19. Behavior of charge carriers in a spherical or cylindrical configuration.

d is the detector thickness or depleted zone, r_1 is the inner radius r, x is the current point, r_2 is the outer radius or radius of depletion.

It is clear that the electric field $\mathscr{E}_{2,3}$ is nonuniform in both the cylindrical and spherical cases and presents a much higher field region near the inner contact at (r_1). Several authors have probed these structures (Fig. 19), first Zanio (1972, 1977) with a more theoretical and experimental study done by Malm et al. (1975, 1977) and later by Alekseeva et al. (1985). The earlier work of Zanio since 1972 has shown the practical interest of these configurations with good resolution in n-type (In) doped 10^7 $\Omega \cdot$ cm CdTe. This work was extended experimentally and theoretically by Malm et al. (1975, 1977) especially by using the general form of the Ramo theorem as

$$dq = -q \frac{\partial \mathscr{E}}{\partial V} \cdot dx \qquad (26)$$

Starting from this form the charge and current of electrons or holes, in planar and spherical detectors is derived, allowing for the theorical plotting of the spectra, as $dN/dQ = f(Q/Q_0)$, as a function of the $(\mu\tau)$ product and the geometry. Results can be summarized as follows:

- The best results (resolution) are obtained when the less trapped carriers (electrons) are collected toward the positive (+) biased center contact—the condition is that $(\mu\tau)_e \gg (\mu\tau)_h$.
- Best resolution occurs at less than full collection in contrast to two carrier collection in planar detectors.
- Good resolution is obtained even if the electron drift length is much less than the electrode spacing.
- Charge collection efficiency was found to cross by a maximum versus $\mu\tau V$ and change with r position.

Experimentally resolution of 5.5% for ^{137}Cs was found, and spatial sensitivity was found to be more pronounced around the negative contact for this n-type semi conductor.

Hemispherical detectors can be simulated by a cubic five face contact, and the central contact in the sixth face (called "U" shape) with quasi similar results.

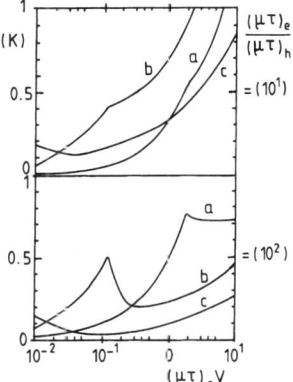

FIG. 20. Behavior of the detector quality factor (K_i) as a function of $(\mu\tau)_e V$ product, $(\mu\tau)_e/(\mu\tau)_h$ ratio, and the geometry. (a), Spherical; (b), cylindrical; (c), planar.

These results and studies were repeated by (Alekseeva et al., 1985) in more detail, including comparative analysis of detector quality in the planar, cylindrical, and spherical cases and more convenient choice of the parameters studied. The detector quality parameter chosen was the factor (K) equal to the ratio of the number of pulse detected in a well-defined peak divided by the total number of detected pulses in the sensitive volume (spectra area). This factor (K) was studied as a function of the product $(\mu\tau V)_e$ and change with r position. For each set of the $\mu\tau$ product and bias V, K is maximal for the planar, cylindrical, or spherical configuration. In the case of spherical detectors there exists an optimum of r_2/r_1 ratio for every $(\mu\tau V)$ product. Knowing these relations it is possible to make the best choice (Figs. 20 and 21) for an experimental resolution of 6 keV for ^{57}Co, which was found with $(\mu\tau)_e > 10^2 (\mu\tau)_h$ at ($-30°$C).

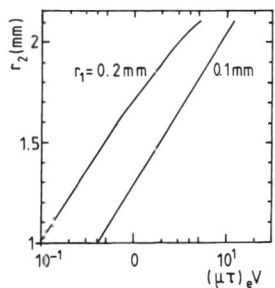

FIG. 21. Evolution of the optimal sphere radius (r_2) with internal contact radius (r_1) and $(\mu\tau)_e V$ product in a spherical detector.

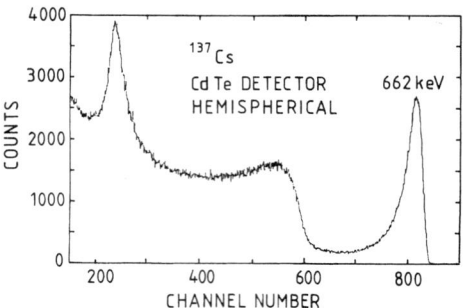

FIG. 22. ^{137}Cs spectrum for a hemispherical CdTe detector.

Relatively good detectors can be made in this way on p-type high ρ material even with medium quality materials. A semicylindrical detector of more than 600 mm^3 was fabricated with 26 keV resolution for ^{137}Cs (Fig. 22), and in hemispherical configuration, a resolution of 6 keV for ^{57}Co at $r_2 = 2-3$ mm was found with positive (+) central contact. However, mechanical shaping and contact quality are the real limits of these configurations, even for the "U" shape configuration.

3. (n-i-p), (M-π-n) STRUCTURES AND COOLING

Typical CdTe detectors at the present time generally use a metal–semiconductor–metal (M-S-M) structure where the metal is a high work function material, such as gold or platinum, which forms a low surface barrier on p-type CdTe. This results in relatively high leakage currents under moderate bias, and although structures present relatively good properties, they exhibit some limitations in that applied bias voltage is limited to 600–1500 V/cm, for thicknesses up to 2 mm. Other limitations in speed and number of counts also exist, and there are a few applications where these limitations are unsuitable. In spectrometry, for example, where better resolution is needed, the first idea that occurs is certainly the (n-i-p) structure as suggested by, among others, Lis et al. (1979), and Squillante et al. (1989).

The (n-i-p) structure uses low work function metal contact like (In) to provide a high barrier height and form the (n) side of the (n-i-p). The other side must be the reverse situation. In principle this structure offers many advantages: high bias voltage operation leading to very large active volumes and uniform electric field (Fig. 23) over the extended depletion volume.

However, at present In compensated high resistivity CdTe material is n-type, and halogen compensated high ρ CdTe is p-type, and neither is exactly an intrinsic material. The p-type contact becomes a tricky problem. Fortunately, many advantages of this structure can be achieved with the so-called M-π-n structure (Fig. 23) where, in the p-type CdTe, the (p) contact is replaced by a low barrier metallic contact like Au or Pt, which forms essentially a quasi-ohmic contact.

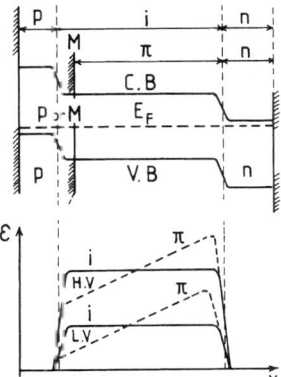

FIG. 23. Band diagrams and corresponding electric field profiles for both $(n\text{-}i\text{-}p)$ and $(M\text{-}\pi\text{-}n)$ structures.

The reported device fabrication steps (Squillante et al., 1989) are as follows: surfaces of (THM) grown CdTe (Cl) are polished and etched with Br–methanol; electroless Au was chosen for the (M) region; and the (n) region was formed by heavy thermal (In) diffusion in the opposite surface. A $\phi 14$ mm active surface on a $\phi 16$ mm wafer has been achieved, and the thickness is chosen by repeated lapping and redeposition of the gold contact.

Bias voltages up to 2000 V for 2 mm thickness are possible, allowing timing resolution better than 1 nsec with a FWHM distribution of 14.5 nsec, and a moderate reported energy resolution (good enough for a specific application).

On the other hand, the presence of a good $(n\text{-}p)$ barrier in such a structure, but in a Pt–CdTe–In (diff.) configuration, allows the use of these detectors in the photovoltaic mode. Internal field is provided by the junction barrier to collect the charge over six decades of flux starting at 10^6 phot./cm$^2 \cdot$ s.

However, the major problem still is the degree of success in the $n\text{-}i\text{-}p$ or $M\text{-}\pi\text{-}n$ processing procedure:

- This concept was reviewed successfully by the Riga group (Khusainov, 1991) in improving the choice of the material and the diffusion process of In. To increase the degree of success, both horizontal Bridgman and (THM) CdTe (Cl) methods were used, with $(\mu\tau)_e = (0.5-1)\, 10^{-3}$ cm^2/V for the former and 10 times less for holes. After diffusion of In, the thickness is adjusted again by reprocessing, with the metal contact being gold. Dark leakage current (10^{-9} A/cm^2) is two decades less than in the $M\text{-}S\text{-}M$ structure at room temperature.
- A new approach was the introduction of small *thermoelectric cooling* Peltier elements to $-40°$ C with a number of improvements. Even for $M\text{-}S\text{-}M$ structure detector of 30 mm^2, leakage currents reduce from 10^{-7} at $20°$ C to 10^{-11} at $-40°$ C. For $M\text{-}\pi\text{-}n$, the ratio is the same down to 10^{-13} A/cm^2.

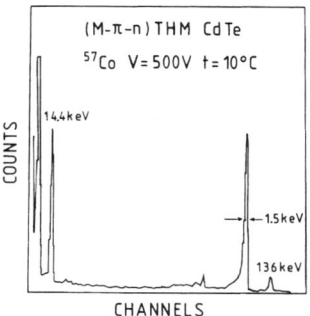

FIG. 24. ^{57}Co spectrum with Peltier cooler element and $(M\text{-}\pi\text{-}n)$ CdTe detector.

- The polarization effect can be decreased. In fact, at $-35°$ C trapped carriers are frozen. At higher temperatures, even at 20°C, detrapping or trapping occur again, leading to polarization. Earlier studies with cooling have shown confusing results, but the new results showed that temperature and level configuration are critical thus optimal conditions must be found for every set of detectors to avoid polarization at low temperatures.

The study of the field distribution and depleted region using IR investigation has shown that the depleted depth is limited by the concentration of deep centers. Strong correlation was found between the depleted zone thickness and the lifetime τ_h of holes at room temperature; in other words, with the $\mu\tau$ products of both carriers. The main limit for large active volume it still the presence of deep acceptor centers and the possibility of achieving a large $\mu\tau$ product ($\mu\tau = (1-5) \cdot 10^{-3}$ cm^{2}V) allows sensitive thickness of 0.3–1.5 mm and high energy resolution (Fig. 24) even with small cooling.

It appears that an $(n\text{-}i\text{-}p)$ or $(M\text{-}\pi\text{-}n)$ structure combined with small thermoelectric cooling allows for the construction of highly improved CdTe detectors. These detectors are able to compete with Ge spectrometric possibilities with a resolution of 1.5 keV for ^{57}Co at 122 keV. This is especially true when enough good material with $\mu\tau = 5 \cdot 10^{-3}$ cm^2/V or more is available.

4. Electronic Treatment

a. Rise Time Selection

Common techniques to enhance resolution of planar CdTe detectors include the pulse shape discrimination (PSD) first described by (Jones (1975, 1977)). Various kinds of PSD techniques are reported, characterized by two parallel paths: the

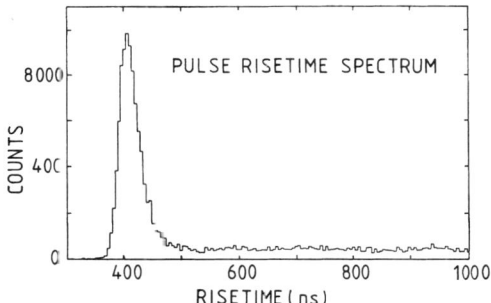

FIG. 25. Pulse rise time distribution of 2 mm thick detector at 90 V.

energy or pulse height path occurs mostly in a common spectroscopic amplifier, while the second path is a complicated system that measures and selects the rise time of the charge collection as described by a few authors (Jones and Woolam, 1975; Hagemann *et al.*, 1988, 1990; Squillante *et al.*, 1989). The signal of this second path is used to gate the pulse height path to select pulses with the right rise time as defined by suitable criteria.

Two methods have been used first the zero crossing method used by Jones and Squillante, which is well suited to usual comparators with delay times around 20 nsec. The second method is a modified constant fraction discriminator with two trigger points. This method can be used in the case of CdTe detectors thanks to its relatively long (>100 nsec) charge collection times. Figure 25 shows this rise time distribution of a good detector. This principle was successfully used by Richter and Siffert (1991), see Figure 26. Using a very good CdTe detector and

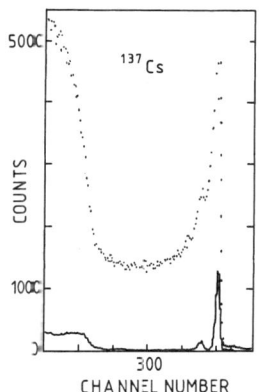

FIG. 26. ^{137}Cs spectra with and without rise time selection around the distribution peak of Fig. 25.

the same method, Hagemann, Berndt, and Arlt (1988) have reported 4.3 keV resolution with the 662 keV line of ^{137}Cs.

b. Compensation of Trapping Losses (correction)

More than a signal treatment method, PSD is a powerful "analysis method" that allows for the understanding of the detector physics especially in the case of CdTe detectors. Figure 25 shows the distribution of rise times of charge collection. If we select pulse heights with the shortest rise time, as seen in Figure 26, more than 80% of all detected pulses are lost if good resolution is required. It is well known that the low energy peak tail in the energy spectrum of a line such as the 662 keV of ^{137}Cs is attributed to trapping of carriers in deep centers. A distribution of detrapping time leads to a wide distribution of rise times and consequent incomplete charge collection during the clipping time of the amplifier (as shown by the ^{137}Cs spectrum in Fig. 26). If one investigates the variation of the spectrum shape as a function of the charge collection by taking different spectra, each in coincidence with one window of the rise time distribution of Figure 25, the results can be seen in Figure 27. One sees that for longer collection times the photopeak moves to lower energies as expected. Figure 28 shows the peak shift versus the rise time from 400 nsec up to 2 μsec. The initial spectrum is the summation of all subspectra, and the peak shift function seem to be rather linear. This opens up the possibility of easier charge loss correction. Rather than excluding 80% of counts one can correct every pulse height following a linear fractional loss function. However, spectra measured with longer rise times have the worst resolutions, and those limit the possible corrections. Also, they demand a pulse preselection to suppress pulses with overly long collection times, and this limits the method. A description of the correction system can be found in Richter and Siffert (1991), Richter, Siffert, and Hage-Ali (1992), and Eurorad (1992). The principle can be summarized as follow. The charge generated in the detector is preamplified, converted, and amplified in the normal fashion. The signal is then dispatched in two paths: first to a gated integrator to store information; second, a constant fraction discriminator measures the rise time of the pulse, which is converted to voltage V by a time amplified converter (TAC) (a window can be determined there to command the gated output). An adjustable level corresponding to the minimum rise time is subtracted from V, and the result is multiplied by the integrated signal. This represents the correction to be added to the signal, and the final corrected signal is gated for possible rise time selection.

This system works well enough for detectors of reasonable quality. Where the holes still play some role in making selection possible and generate correction diagnostics, results are impressive. Figures 29 and 30 show ^{137}Cs and ^{57}Co spectra before and after correction. In Figure 29 we can compare selection and correction to estimate the gain in efficiency along with improved resolution.

8. CdTe Nuclear Detectors and Applications   325

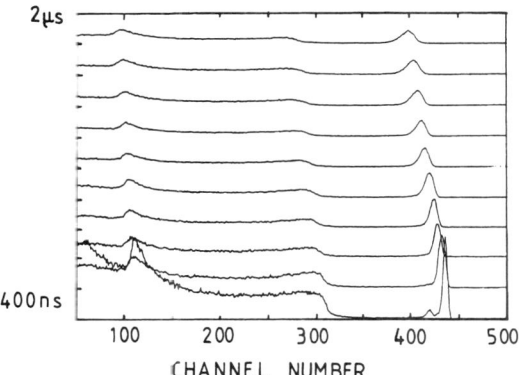

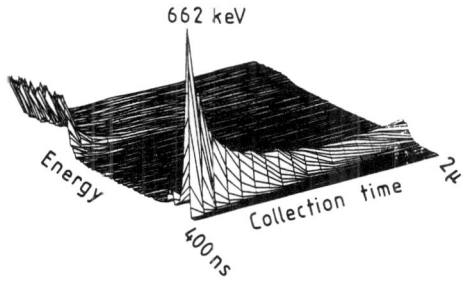

FIG. 27. ^{137}Cs spectra measured for different rise time windows along Fig. 25 distribution.

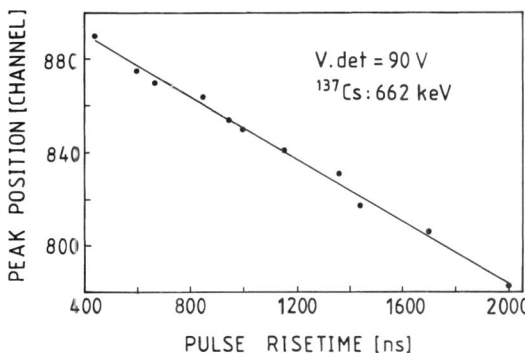

FIG. 28. ^{137}Cs 622 keV peak position as a function of the pulse rise time window (from Fig. 27) (note the linear shift!).

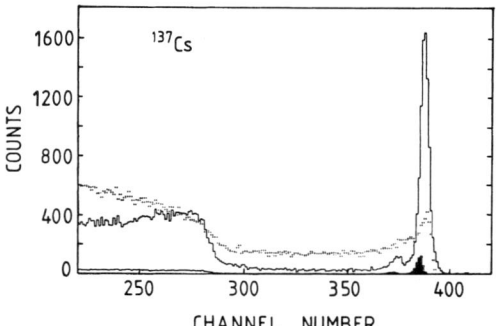

FIG. 29. ^{137}Cs spectra for CdTe detector: without electronic treatment, with rise time selection, and with pulse height compensation.

V. Applications of CdTe Detectors

CdTe detectors have not yet been employed in the large number of applications that they can, in principle, address. Nobody doubts their potential, but up to the 1980s, they have still enjoyed much less success than detectors such as Geiger–Muller tubes, scintillator–photomultiplier systems, and Si–Ge semiconductor detectors. However, step by step, they have begun to find their own field of application especially in advanced technologies: the nuclear domain, space, and medicine. The situation is now quite different, with a rich bibliography and many industrial detector and systems producers all around the world.

Several categories of the various applications can be considered: chronological, detector type, or subject of interest. The last classification, mixing with some of the others, seems to be the more appropriate.

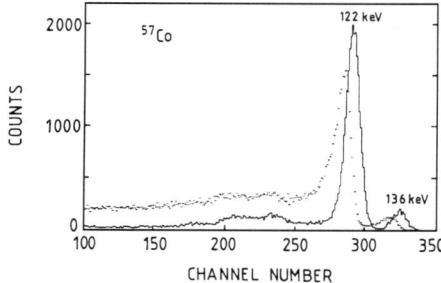

FIG. 30. ^{57}Co spectra with and without pulse height compensation.

1. SPECTROMETERS

Both scientific and industrial fields offer applications to CdTe spectrometers when room temperature operation, small size, low voltage requirements, and reasonable resolution are needed. To date, CdTe has not found many applications in the laboratory spectrometric fields, where detector volume and liquid Nitrogen cooling are not so problematic for Si and Ge detectors. However the recent development in hemispheric, n-i-p, thermoelectric Peltier cooling, and electronic processing, with resolutions of 1.5 keV for ^{57}Co and 5–8 keV for ^{137}Cs, allow the use of these detectors for measurement "in the field" of different materials concerning critical atomic activity (Arlt, 1991).

a. Uranium and Spent Fuel

Unirradiated power reactor fuel is checked by the uranium x-ray or gamma rays at 186 keV using instrumentation holes to introduce the small probe. Enrichments of 1.6–3.6% in the WWER 440 fuel is resolved easily in the case of irradiated fuel, which is stored in bundles with difficult access. Small collimated CdTe detectors allow the evaluation of ^{137}Cs activity for every bundle and to check its number.

b. Classification of Plutonium as Weapons Grade (low burnup) or Not (high burnup)

The first, with 93% enrichment, shows a prominent ^{239}Pu gamma peak at 129 keV, which is easy to see, while this peak vanishes at 63% of ^{239}Pu. It seems that it is possible to see mixed oxide material of U and P in the same manner.

c. Chemical Analysis

Another exciting spectrometric application is chemical analysis directly in bore holes. Minerals like uranium, as well as the presence of oil, can be detected directly on the wall of a bore hole either by fluorescence x-ray excitation using a gamma ray emitter, or by (n, γ) reaction using a neutron generator. A CdTe based system was built and has shown real capabilities (Siffert, 1982). Mention should also be made of results of Dabrowski et al. (1977) with low resistivity CdTe surface barrier in x-ray detection with 1.1 keV resolution of 6 keV.

2. SAFEGUARD SYSTEMS

Individual safeguard systems using CdTe detectors have appeared since 1975 (Cornet et al., 1975). They start at the size of a large watch (Siffert, 1982), with a buzzer like the Radiabip of Eurorad (1992) up to a more sophisticated personal dosimeter with display, integrated doses, and computer processing possible (RSK 30 from Eurorad) from 50 keV to 3 MeV.

3. Dosimetry

Dosimetry is a more sophisticated extension of the preceding application, which is generally a safety directed system. For many years Si detectors were felt to be the best choice owing to the low atomic number (Z) of Si, which is the closest to the low Z of human tissue in comparison to the high Z of CdTe. On the other hand, this high Z allows high x-ray and gamma ray absorption and thus higher sensitivity to lower doses. The absolute dose measurement required correction of the absorption in relation to the energy and then conversion to actual tissue dose (Kronenberg, 1985). This can be done easily now using microprocessors, analog circuits, or more simply by filter correction (Fassassi, 1987). Such dosimeters are now commonly present in the market. A complete dosimeter was constructed with particle–photon differentiation capability (Nagarkar et al., 1991) to monitor the irradiation of space flight crew members, who are exposed to natural radiation x-rays, gamma rays, and protons.

4. Space and Astrophysics

Dyes' (1971) paper describes a CdTe based system for rocket propellant gauging at zero gravity using gamma ray absorption of a set of gamma ray sources on one side and a set of CdTe detectors strategically arranged on the other side. This system showed the feasability of such a gauging principle over a wide range of working temperatures of $\pm 125°C$. Many other space applications have been found, especially compact CdTe detector hybrid electronics packages for use in the severe environments associated with rocket reentry into the atmosphere (Lyons, 1977). Special attention must be paid to the elimination of microphonics and shock. CdTe is highly piezoelectric. A special mounting package was tested from 300–2000 Hz without noise, and it allowed operation at temperatures up to 70°C. Such a design opens the way for more complicated space applications, such as the use of CdTe arrays for astrophysical gamma ray imaging systems. Evaluation of (2 × 2 × 2 mm) to (2.5 × 2.5 × 20 mm) detectors show resolution of 4.5 keV for ^{57}Co and 10 keV for ^{22}Na, with timing possibilities as good as 45 nsec, allowing the mapping of the sky in the high gamma energy domain (0.5–10 MeV) with good angular resolution, of at least 0.5° (Baldazzi et al., 1991; Caroli et al., 1991). One must mention the work of Zanio (1977) and Droms et al. (1976) in the field of monitoring the thickness of shields and nosetips of rockets and space vehicles.

5. Medical Applications

Up to now, the medical field represents certainly the best opportunity for CdTe detectors thanks to their high absorption cross section, ability to work at rather high temperatures >70° C, low applied bias voltage, and small size (<1 mm possible), allowing precise localization. Since the earlier development by Meyer, Martini, and Sternberg (1972) of CdTe probes for ^{75}Se gamma rays in rat's eyes for diagnosis, a great many studies, instruments, and production of CdTe based probes have appeared on the market and in the literature for various medical applications. These cover diagnosis of dental infections and cardiovascular problems such as venous thrombosis, control of pre- or postsurgical diagnosis (Entine, Garcia, and Tow, 1977), bone mineral measurement by gamma ray transmission (Vogel, Ullman, and Entine, 1977) for bone mineral losses in osteoporosis, lung densitometry for diagnosis of pulmonary edema (Kaufman et al., 1977), and biotelemetry for physiological parameter study, such as the clearance of injected ^{133}Xe in the tibial muscle or bloodflow in the heart using 99 m − Tc as described in the excellent work of Bojser et al. (1977). More recently, a large number of publications have appeared concerning new medical uses of CdTe detectors and arrays (Scheiber and Chambron, 1991; Entine et al., 1989; Michael, Squillante, and Entine, 1991) for tumor localization, cerebral bloodflow, the classic bone density, gastric bile reflux, monitoring kidney function, and myocardial bloodflow. Radiotherapy has found CdTe arrays the best monitoring systems for precise beam positioning on tumors within 4 sec versus 40 sec for Si diodes. Instruments are now currently available in the market from handheld, simple, low cost probes (Fig. 31), to sophisticated computer aided diagnosis for preoperative tumor local-

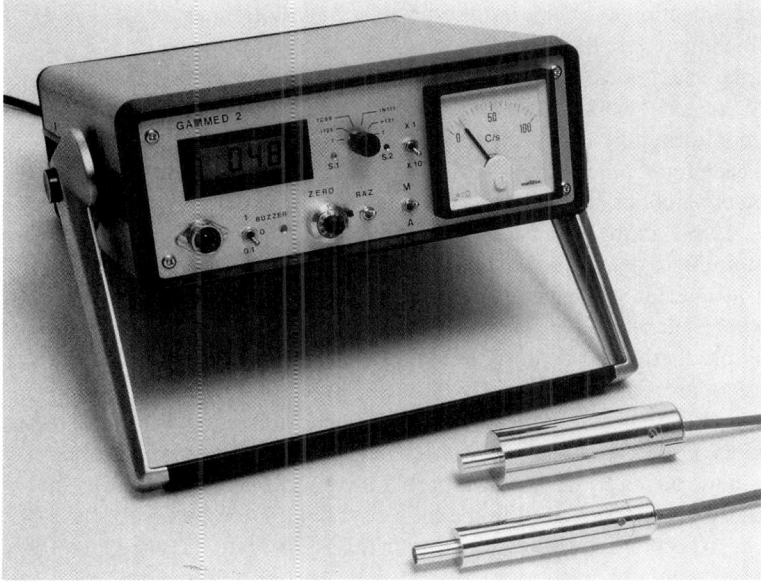

FIG. 31. Surgical CdTe probe based system, with a handheld probe and analogic–digital readout module. (Courtesy of Eurorad Co., with permission.)

ization (Hossein-Foucher et al., 1992). In conclusion medical applications are an extremely promising area for CdTe detectors.

6. INDUSTRIAL APPLICATIONS

To this point only one kind of CdTe detector has been considered: the ionization chamber. But there exists another possibility for CdTe; namely, as a photoconductive cell. Even if this detector is not useful in spectrometry and requires more powerful radiation, it can be interesting in industrial applications, especially in the high energy radiation field. Another, and more common, use for CdTe detectors checks the integrated pulse current when the number of counts exceeds $3-6 \cdot 10^5$ c/sec and pulse counting electronics cannot follow the rate; here the current mode is more suitable. This has the disadvantage of leakage current, which is integrated at the same time.

Studies and applications of photoconductive CdTe cells are well explained in Cuzin (1991) and Verger et al. (1991). Applications are in x-ray flash radiography, in laser fusion experiment control, in single shot synchrotron bunch x-ray monitoring, in power x-ray flash radiography, and in detonics experiments (Mathy, Cuzin, and Gagelin, 1991). On the other hand, there is a great interest in integrated current mode CdTe detectors owing to the simplicity of the electronics: in the pulse mode, problems arise from tailing in the pulse decay (photomemory), which reduces the pulse rate response. It appears that there is some possibility in the double injection contacts on n-type materials (Zelenina et al., 1989). Tomographic images are then possible.

In the usual pulse height counting mode, industrial application are numerous. Since the early work of Jones (1977) in the inspection of nuclear power stations and monitoring the deposited activity on parts, many other industrial applications have been found: the narcotics scanner is able, by the backscatter of ^{57}Co photons, to detect narcotics and explosives, even in hollow body panels or car tires (Entine et al., 1989); in the beta backscatter mode, it can determine atomic number Z in coal mining to determine the coal seam and the amount of gas in it; and it can be used in nuclear waste processing and management. CdTe detectors and arrays have been successfully used in an automatic system for quality control in ammunition production, to determine the level of gunpowder, voids in lead tips, and obstructions of the 2 mm fire hole (Eisen, 1991).

A more important application of CdTe detectors should also be mentioned: scanners. Even though they have not found great success in medical applications up to now, owing to well known problems (polarization, electronics, and uniformity), they have been used in the industrial field. Exciting results have been obtained. Arrays of a few hundred detectors, 1.5 mm wide, biased up to 300 V, with electronics able to stand more than $3 \cdot 10^5$ c/sec in pulse height mode, under x-rays or gamma rays of few hundred keV, can scan a whole car with its passengers, and a 1 cm metallic rod can be seen even behind a few mm of steel, thanks to the

8. CdTe Nuclear Detectors and Applications

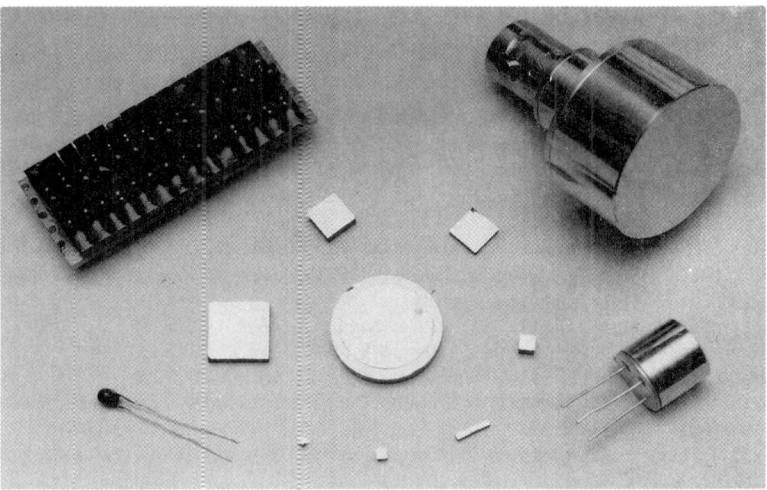

FIG. 32. Classic array of 16 detectors for scanner purpose with other geometry detectors. (Courtesy of Eurorad Co., with permission.)

one photon sensitivity in this mode. Such very promising results can be extrapolated to the medical field with some effort in analog electronics.

The same results are found in dc integrated current mode scanners with roughly the same number and dimension of array detectors (Fig. 32). Two layers, with detectors biased with a few volts used for energy separation to differentiate low Z from high Z materials. The main application is in luggage inspection in airports. The current generated is three times greater than the CsI–Si photodiode currently used and extension to medical applications is easier, due to the simplicity of the electronics (Hage-Ali et al., 1989).

This system can be used to scan an entire container but needs an 8–15 MeV gamma ray. That was achieved using 25(25 × 15 × 0.8 mm^3) edge detectors operating in the photoconduction mode, operating over five decades (Glasser et al., 1991) and was able to measure a 100 μm delamination in a solid rocket motor.

In conclusion, the great number of potential applications of CdTe detectors have been shown. The quality and number of exciting studies in this field is great. Despite a few limitations in the material growth, including uniformity and defects and in contact deposition, with some clever solutions, new systems can be found and achieved in spite of the wait for perfect material, which we hope is not so far off.

References

Agrinskaya, N. V., and Matveev, O. A. (1980). *Sov. Phys. Semicond.* **14**, 611.
Akobirova, A. T., Maslova, L. V., Matveev, O. A., and Khusainov, A. K. (1975). *Sov. Phys. Semicond.* **8**, 1103.

Akobirova, A. T., Maslova, L. V., Matveev, O. A., Ryvkin, S. M., and Khusainov, A. K. (1977). *Rev. Phys. Appl.* **12**, 331.
Alekseeva, L. A., Dogarov, P. G., Ivanov, V. I., and Khusainov, A. R. (1985). *Pribory I Technika Eksper.* **1**, 54.
An, C., Tews, H., and Cohen, G. (1982). *J. Cryst. Growth* **59**, 289.
Arkadeva, E. N., Matveev, O. A., and Melnikova, E. V. (1980). *Sov. Phys. Semicond.* **14**, 424.
Arlt, R., Gsock, K. H., and Rundquist, D. E. (1991). Proceedings of the Seventh Int. Workshop on R.T.S.C. X and γ Ray Detectors, Ravello, Italy, to be published in *NIM in Phys. Res. 1992*.
Baldazzi, G., Bollini, D., Caroli, E., Casali, F., Chirco, P., Di Cocco, G., Donati, A., Dusi, W., Landini, G., Malaguti, G., Rossi, M., and Stephan, J. B. (1992). *Nucl. Instr. Meth. Phys. Res* **A322**, 644.
Bean, J. C. (1976). Thesis, Stanford University.
Bell, R. O., Entine, G., and Serreze, H. B. (1974). *Nucl. Instr. Meth.* **117**, 267.
Bojsen, J., Rossing, N., Soeberg, O., and Vadstrup, S. (1977). *Rev. Phys. Appl.* **12**, 361.
Boltaks, B. I., Konorov, P. P., and Matveev, O. A. (1955). *Z. Tekh. Fiz.* **25**, 2329.
Butler, J. F., Lingren, C. L., and Doty, F. P. (1991). Conference Record Symp. Santa Fe, *IEEE Trans. Nucl. Sci.*
Butler, J. F., Doty, F. P., Apotovsky, B., Lajzerowicz, and Verger, L. (1993). *Mat. Sc. and Eng.* **16**, 291.
Caroli, E., Baldazzi, G., Di Cocco, G., Donali, A., Dusi, W., Malaguti, G., Rossi, M., and Stephan, J. B. (1992). *Nucl. Instr. Meth. Phys. Res.* **A322**, 639.
Chaubey, A. K., and Gupta, H. V. (1977). *Rev. Phys. Appl.* **12**, 313.
Chu, M., Fahrenbruck, A. L., Bube, R. H., and Gibbons, J. F. (1978). *J. Appl. Phys.* **49**, 322.
Cornet, A., Siffert, P., Coche, A., and Triboulet, R. (1970). *Appl. Phys. Lett.* **17**, 432.
Cornet, A., Regal, R., Ponpon, J. P., and Siffert, P. (1975). Second ISPRA Nucl. Elect. Symp. Stresa, Italy, p. 169.
Cuzin, M. (1987). *NIM in Phys. Res.* **A253**, 407.
Cuzin, M. (1992). *Nucl. Instr. Meth. Phys. Res.* **A322**, 347.
Dabrowski, A. J., Iwanczik, J., and Triboulet, R. (1975). *Nucl. Instr. Meth.* **126**, 417.
Dabrowski, A. J., Chwaszczewska, J., Iwanczyk, J., Triboulet, R., and Marfaing, Y. (1977). *Rev. Phys. Appl.* **12**, 297.
Day, R. B., Dearnaley, G., and Palms, J. M. (1967). *IEEE Trans. Nucl. Sci.* **NS-14**, 487.
De Nobel, D. (1959). *Philips Res. Rept.* **14**, 361 and 430.
De Nobel, D., and Kroeger, F. A. (1962). U.S. Patent No. 3,033,791 May.
Droms, C. R., Langdon, W. R., Robinson, A. G., and Entine, G. (1976). *IEEE Trans. Nucl. Sci.* **NS-23**, 498.
Dyes, O. (1971). In *Proc. of First Int. Symp. on CdTe γ Ray Detectors*, ed. P. Siffert and A. Cornet. Centre de Recherches Nucléaires, Strasbourg.
Eisen, Y. *Nucl. Instr. Meth. Phys. Res.* **A322**, 596.
Entine, G., Garcia, D. A., and Tow, D. E. (1977). *Rev. Phys. Appl.* **12**, 355.
Entine, G., Squillante, M. R., and Serreze, H. B. (1981). *IEEE Trans. Nucl. Sci.* **NS-28**, 558.
Entine, G., Waer, P., Tiernan, T., and Squillante, M. R. (1989). *NIM in Phys. Res.* **A283**, 282.
Eurorad. (1992). *Charge Loss Corrector (CLC) User Manual.* Eurorad Co (CTT), Strasbourg, France.
Fassassi, K. (1987). Thesis, Université Louis Pasteur, Strasbourg.
Fioratti, M. P., and Piermattey, S. R. (1971). *Nucl. Instr. Meth.* **96**, 605.
Fox, R. J., and Agouridis, D. C. (1978). *Nucl. Instr. Meth.* **157**, 65.
Glasser, F., Thomas, G., Cuzin, M., and Verger, L. (1992). *Nucl. Instr. Meth. Phys. Res.* **A322**, 619.
Hage-Ali, M., Stuck, R., Saxena, A. N., and Siffert, P. (1979a). *Appl. Phys.* **19**, 25.
Hage-Ali, M., Stuck, R., Scharager, C., and Siffert, P. (1979b). *IEEE Trans. Nucl. Sci.* **NS-26**, 281.
Hage-Ali, M., Scharager, C., Koebel, J. M., and Siffert, P. (1980). *Nucl. Instr. Meth.* **176**, 499.
Hage-Ali, M., Koebel, J. M., Ritt, C., and Siffert, P. (1989). Brite-Euram Contract No. PL-BE-3036-89.

Hagemann, U., Berndt, R., and Arlt, R. (1988). *Kern energie* **31**.
Hagemann, U., Richter, M., Berndt, R., and Stephan, M. (1990). *ZfK Report* **706**, ZfK., Rossendorf, Germany.
Hansen, J. S., Feund, H. U., and Finck, R. W. (1970). *Nucl. Phys.* **A142**, 104.
Hossein-Foucher, C., Rousseau, J., Bedoui, H., Regal, R., Carnaille, B., Venel, H., Proye, C., and Marchandise, X. (1992). *Innov. Tech. Biol. Med.* **13**, 57.
Jäger, H., and Thiel, R. (1977). *Rev. Phys. Appl.* **12**, 293.
Johnson, C. J., Eissler, E. E., Cameron, S. E., Kong, Y., Fan, S., Jovanoric, S., and Lynn, K. G. (1993). *Mat. Res. Soc. Symp. Proc.* **302**, 463.
Jones, L. T. (1977). *Rev. Phys. Appl.* **12**, 379.
Jones, L. T., and Woolam, P. B. (1975). *Nucl. Instr. Meth.* **124**, 591.
Kacherininov, P. G., Matveev, O. A., and Maslova, L. M. (1969). *Sov. Phys. Semicond.* **3**, 451.
Kacherininov, O. G., Matveev, O. A., and Matyukhin, D. G. (1979). *Sov. Phys. Semicond.* **13**, 756.
Kaufman, L., Gamsu, G., Savoca, C., and Swann, S. (1977). *Rev. Phys. Appl.* **12**, 369.
Khusainov, A. K. (1992). *Nucl. Instr. Meth. Phys. Res.* **A322**, 335.
Kroeger, F. A., and De Nobel, D. (1955) *J. electron* **1**, 190.
Kronenberg, S. (1985). *IEEE Trans. Nucl. Sci.* **NS-32**, 945.
Lis, S. A., Ellis, R., Shuman, R., Serreze H. B., and Entine, G. (1979). R.M.D., Inc. contract no. BNL 436743-S.
Lyons, R. B. (1977). *Rev. Phys. Appl.* **12**, 385.
Malm, H. L., and Martini, M. (1973). *Can. J. Phys.* **51**, 2336.
Malm, H. L., and Martini, M. (1974). *IEEE Trans. Nucl. Sci.* **NS-21**, 322.
Malm, H. L., Canali, C., Mayer, J. W., Nicolet, M. A., Zanio, K., and Akutagawa, W. (1975). *Appl. Phys. Lett.* **26**, 344.
Malm, H. L., Litchinsky, D., and Canali, C. (1977). *Rev. Phys. Appl.* **12**, 303.
Mathy, F., Cuzin, M., and Gagelin, J. J. (1992). *Nucl. Instr. Meth. Phys. Res.* **A322**, 615.
Mead, C. A., and Spitzer, W. G. (1964). *Phys. Rev.* **134**, 935.
Mergui, S. (1991). Thesis, Université Louis Pasteur, Strasbourg.
Meyer, E., Martini, M., and Sternberg, J. (1972). *IEEE Trans. Nucl. Sci.* **19**, 237.
Nagarkar, V., Squillante, M. R., Entine, G., Stern, I., and Sharif, D. (1992). *Nucl. Instr. Meth. Phys. Res.* **A322**, 623.
Norris, C. B., Barnes, C. E., and Zanio, K. (1977). *J. Appl. Phys.* **48**, 1659.
Norris, C. B., Westmark, C. I., Entine, G., Lis, S. A., and Serreze, H. B. (1981). *Rad. Effects Lett.* **53**, 115.
Patterson, M. H., and Williams, R. H. (1978). *J. Phys. D. Appl. Phys.* **11**, L83.
Rabin, B., Tabatabai, H., and Siffert, P. (1978). *Phys. Stat. Sol.* **a49**, 577.
Raiskin, E., and Butler, J. F. (1988). *IEEE Trans. Nucl. Sci.* **NS-35**, 81.
Richter, M., and Siffert, P. (1991). *Nucl. Instr. Meth. Phys. Res.* **A322**, 529.
Richter, M., Siffert, P., and Hage-Ali, M (1993). *Mat. Sc. and Eng.* **B16**, 296.
Scheiber, C., and Chambron, J. *Nucl. Instr. Meth. Phys. Res.* **A322**, 604.
Schibli, E., and Milnes, A. G. (1968). *Solid State Elects.* **11**, 323.
Seraphy, M. (1980). Thesis, Université Louis Pasteur, Strasbourg.
Sherman, J. (1955). *Spectrochim. Acta* **7**, 283.
Siffert, P., Gonidec, J. A., Cornet, A., Bell, R. O., and Wald, F. (1974). *Nucl. Instr. Meth.* **115**, 13.
Siffert, P., Berger, R., Scharager, C., Cornet, A., Stuck, R., and Bell, R. O. (1976). *IEEE Trans. Nucl. Sci.* **NS-23**, 159.
Siffert, P., Hage-Ali, M., Stuck, R., and Cornet, A. (1977). *Revue de Phys. Appl.* **12**, 335.
Siffert, P. (1978). *Journal de Phys.* **39**, C3.
Siffert, P., Rabin, B., Tabatabai, H., and Stuck, R. (1978). *Nucl. Instr. Meth.* **150**, 31.
Siffert, P. (1982). MRS Meeting, Boston.
Squillante, M. R., Entine, G., Frederick, E., Cirignano, L., and Hazelett, T. (1989). *NIM in Phys. Res.* **A283**, 323.

Squillante, M. R., and Entine, G. (1992). *Nucl. Instr. Meth. Phys. Res.* **A322,** 569.
Verger, L., Cuzin, M., Gande, G., Glasser, F., Mathy, F., Rustique, J., and Schaub, B. (1992). *Nucl. Instr. Meth. Phys. Res.* **A322,** 357.
Vogel, J., Ullman, J., and Entine, G. (1977). *Rev. Phys. Appl.* **12,** 375.
Wald, F. V., and Bell, R. O. (1975). Contract Report AT (11-1) 3545 Tyco-Waltham.
Zanio, K. (1972). *Isot. Radiat. Tech.* **9,** 456.
Zanio, K. (1977). *Rev. Phys. Appl.* **12,** 343.
Zelenina, N. K., Ignatov, S. M., Karpenko, V. P., Maslova, L. V., and Matveev, O. A. (1989). *NIM Phys. Res.* **A283,** 274.

CHAPTER 9

$Cd_{1-x}Zn_xTe$ Spectrometers for Gamma and X-Ray Applications

R. B. James
ADVANCED MATERIALS RESEARCH DEPARTMENT
SANDIA NATIONAL LABORATORIES
LIVERMORE, CALIFORNIA

T. E. Schlesinger
DEPARTMENT OF ELECTRICAL AND COMPUTER ENGINEERING
CARNEGIE MELLON UNIVERSITY
PITTSBURGH, PENNSYLVANIA

Jim Lund and Michael Schieber
ADVANCED MATERIALS RESEARCH DEPARTMENT
SANDIA NATIONAL LABORATORIES
LIVERMORE, CALIFORNIA

I. INTRODUCTION	336
II. GROWTH OF $CD_{1-x}ZN_xTE$ CRYSTALS	337
III. MATERIAL PROPERTIES OF $CD_{1-x}ZN_xTE$	339
1. *Resistivity*	340
2. *Alloy Composition*	341
3. *Photoluminescence Spectrum*	342
4. *Charge Transport*	342
5. *Absorption Coefficient for X-Rays and Gamma Rays*	345
IV. DEFECT CHARACTERIZATION AND EFFECTS ON DEVICE RESPONSE	346
1. *Etch Pit Densities*	346
2. *X-Ray Rocking Curves*	346
3. *Precipitates*	348
4. *Impurities*	349
V. DETECTOR CHARACTERIZATION AND EFFECTS ON DEVICE RESPONSE	350
1. *Detector Fabrication*	350
2. *Nuclear Spectroscopic Data at Room Temperature*	350
3. *Detector Current–Voltage Characteristics*	353
4. *Large-Volume Gamma Ray Spectrometers*	355
5. *p-i-n Gamma Ray Detectors*	362
6. *X-Ray Detector Response*	364
7. *Detector Polarization*	368
8. *Temperature Dependence*	369
9. *Pulse Risetime Discrimination and Compensation*	372

VI. Imaging Applications 375
VII. Future Work 378
   References 378

I. Introduction

Most of the efforts on $Cd_{1-x}Zn_xTe$ (CZT) gamma ray spectrometers have been motivated by the benefits and commercial success of CdTe devices and the possibility of manufacturing relatively large homogeneous CZT crystals at an acceptable cost. CdTe detectors have been shown to operate at room temperature, have good stopping power for energetic photons, produce reliable spectra, and be fairly rugged in field applications (see the chapters on CdTe in this text and the references contained there). CdTe devices and instruments are sold today for environmental monitoring, waste remediation, industrial gauging, monitoring nuclear materials, and medical instrumentation. Although CdTe has shown great promise as a gamma ray detector, it does have some drawbacks that limit the widespread use of these devices. In particular, CdTe spectrometers are limited to small sizes, which makes the instruments less sensitive to gamma emitting sources, and the leakage currents are too large for many x-ray applications. Depending on contaminant levels and the choice of dopant for compensation, CdTe detectors sometimes display polarization effects, in which case the counting rates or peak positions change with time (Siffert *et al.*, 1976; Johnson *et al.*, 1993; Sato *et al.*, 1995). CdTe detectors are also relatively expensive, which increases the difficulty in manufacturing and marketing large volume and large area CdTe detector arrays. Most of the problems with CdTe are related to the growth of detector crystals of suitable size and quality, and to the relatively high leakage currents in the devices. By increasing the homogeneity of CdTe ingots and yield of acceptable material, one can produce larger detectors and lower the unit price of commercial devices. Furthermore, by reducing the defect density in the crystals, the remaining concerns with polarization effects will diminish or may disappear altogether, in much the same way that polarization problems with CdTe:Cl gamma ray detectors have been resolved in the past. The growth of detector grade CZT crystals has been pursued because CZT offers a more immediate solution to the problem of noise due to leakage current, and the crystals can be grown to larger sizes at a reduced cost compared to CdTe. Although problems with purity and homogeneity of CZT continue to exist, significant progress is expected over the next few years.

Most spectrometer grade $Cd_{1-x}Zn_xTe$ crystals are grown by a high pressure Bridgman (HPB) technique (see, for example, Kimura and Komiya, 1973; Fitzpatrick, 1988; Raiskin and Butler, 1988; Kikuma *et al.*, 1979), and therefore the emphasis of the work reported in this chapter is on HPB-grown crystals. This approach has the potential of solving many of the technical problems associated with production of large volume $Cd_{1-x}Zn_xTe$ crystals. Application of the high pressure Bridgman technique for the growth of $Cd_{0.96}Zn_{0.04}Te$ and $Cd_{0.8}Zn_{0.2}Te$

9. $Cd_{1-x}Zn_xTe$ Spectrometers for Gamma and X-Ray Applications

crystals led to the first reports of CdZnTe gamma ray spectrometers (Doty et al., 1992; Butler, Lingren, and Doty, 1992), and some distinct advantages of alloying CdTe and ZnTe to produce $Cd_{1-x}Zn_xTe$ crystals were identified. For example, the bandgap of CZT crystals is increased compared to CdTe, thereby greatly reducing the leakage current (and noise) of the detectors and allowing for their use at either lower photon energies or higher temperatures.

The focus of this chapter is on $Cd_{1-x}Zn_xTe$ detector technology, but discussions of CdTe (i.e., $x = 0$) will be presented when appropriate. This chapter will discuss growth of CZT crystals, followed by a review of the properties of the material, particularly those properties that pertain to detector performance. A review of the performance of CZT gamma and x-ray spectrometers and imaging arrays will be presented, and the chapter will conclude with a brief discussion of future work needed to advance the technology.

II. Growth of $Cd_{1-x}Zn_xTe$ Crystals

Single crystals of CZT were originally grown to replace CdTe as better matched substrates for $Hg_{1-x}Cd_xTe$ epilayers used as infrared (IR) detectors. The first reports on CZT bulk crystals already indicated an apparent improvement in the crystallinity compared to the earlier substrate materials.

Although there are many reports of CZT crystal growth in the literature, bulk crystals are grown primarily by the same methods as CdTe; namely, the Bridgman method (see Zanio 1978). A crucible containing the melt is moved slowly through an axial temperature gradient, in a vertical or horizontal furnace. Alternatively, rather than moving the crucible or furnace, both can be kept stationary and the temperature profile adjusted. Since the vapor pressure of Cd is much larger than that of Te, the melt is usually enclosed in a crucible such as quartz, carbon-coated quartz, or a graphitic carbon crucible. Various techniques are used to prevent or at least reduce the evaporation of Cd. The addition of Zn appears to improve the properties of the material. Since Zn has a lower vapor pressure than Cd, it can compensate for losses of Cd during melt. For the preparation of CZT for detector applications, the use of the vertical Bridgman method under high pressure of ~100 atm seems to be very promising and will be discussed further in this review.

There is an extensive literature regarding crystal growth of CZT. We shall first review some of the work that discusses crystal growth of CZT as a substrate material, and afterwards we shall concentrate on high pressure Bridgman of CZT for radiation detectors. Sen et al. (1988) reported 5 cm diameter crystals of CZT in a computer controlled vertical Bridgman furnace. A similarly grown CZT crystal was later used as a gamma ray detector in a p-i-n configuration, where the positive and negative electrodes were p-doped and n-doped epilayers of mercury cadmium telluride deposited by liquid phase epitaxy on $Cd_{1-x}Zn_xTe$ bulk substrate materials (Hamilton et al., 1994). Other vertical Bridgman crystals were grown by

Bruder et al. (1990, 1993). A stationary horizontal Bridgman method was used by Trivedi and Rosemeier (1989). The cooling of the melt was controlled electronically. Higher purity CZT crystals were reported by Cheuvart et al. (1990). However, no radiation detectors were fabricated from these crystals. The liquid encapsulated Czochralski (LEC) method, using B_2O_3 as an encapsulant, was also reported to grow CZT crystals. An inert gas partial pressure of 1–10 atm and a partial pressure of Cd vapor of 1–5 atm were used by Kotani and Tatsumi (1987). No detectors were fabricated from the LEC crystals.

It is interesting to note that CdTe crystals grown as substrates never worked as high resolution radiation detectors. Since the substrate material must have a high degree of crystallinity for epitaxial matching purposes and high purity material is required, it may be speculated that defects arising from deviations from stoichiometry, which are less critical for substrates, deteriorate the CdTe charge transport properties to the point that the material is unacceptable as a radiation detector.

To achieve a stoichiometric CZT crystal, the composition of the melt must be controlled by partial pressure control of its constituent elements. Since Cd has the highest vapor pressure, it is expected to have the greatest influence on the stoichiometry of the growing boule. Here, stoichiometry refers to the balance between the metal and anion species in the ternary II–VI compound [i.e., the ratio of (Cd+Zn)/Te)]. Without control of the Cd partial pressure, the melt will lose Cd and therefore become Te rich. Furthermore, as the crystal growth progresses, the melt become continuously enriched with excess Te as the Cd preferentially evaporates. Deviations from stoichiometry in the growing crystal would be expected to give rise to native point defects such as vacancies, interstitials, and possibly antisite defects.

The high pressure (~ 100 atm) of inert gas above the melt in HPB grown material is expected to suppress some of the Cd loss, but it is probably not sufficient to maintain an equilibrium stoichiometric composition in the melt throughout the growth cycle. At this time there has not been a detailed comparative study of the stoichiometric deviations in CZT crystals grown by HPB and conventional Bridgman methods, and it is still uncertain to what extent the high pressures of argon reduce the Cd vaporization from the melt. There is, however, a significant body of evidence suggested that CZT crystals grown by a HPB method have much higher resistivities and produce better large-volume (> 0.1 cm^3) nuclear detectors. Thus, this section will primarily concentrate on the high pressure growth method and review some of the pertinent work in this field.

Significant strides in the commercialization of $Cd_{1-x}Zn_xTe$ room temperature radiation detectors have occurred since the first reports by Doty et al. (1992) and Butler et al. (1992). Most of the successes are linked to continued improvements in the purification of starting materials and the HPB growth of relatively large uniform crystals designed for fabrication into gamma ray detectors. Although the demands on the properties of $Cd_{1-x}Zn_xTe$ as a substrate material were much less stringent than those for a bulk CZT spectrometers, it is almost certain that the rapid improvements seen recently in the performance of CZT gamma ray spectrometers would not have been possible without the earlier work on growth and characterization of CZT substrates for IR detector applications.

In general, the growth of high-purity stoichiometric $Cd_{1-x}Zn_xTe$ alloys from the melt is complicated by the relatively high melting temperatures and vapor pressures at the melting point. The melting temperature of $Cd_{1-x}Zn_xTe$ is comparable to or exceeds the softening point of quartz, thus the growth of CZT crystals in sealed quartz ampoules typically leads to the incorporation of significant amounts of oxygen and other impurities, even when carbon coatings are applied to the crucible (Butler et al., 1993b). These impurities are often detrimental to the crystallinity of the material (Sen and Stannard, 1993) and to the carrier lifetimes in fabricated detectors (Bube, 1967). Thus, it is advantageous to avoid the use of quartz ampoules in the growth of $Cd_{1-x}Zn_xTe$. One promising technique involves growth in a high pressure chamber and uses an inert atmosphere to reduce the mean free path (and diffusion) of the vapor species thereby eliminating the need for sealed ampoules.

In practice, high purity graphite components are encased in a large steel shell designed to withstand internal pressures of over 100 atmospheres (Doty et al., 1992). A graphite heater provides a temperature profile that has been optimized for Bridgman growth. High purity graphite crucibles are fitted with caps to minimize evaporative loss. The growth furnaces are designed to accommodate crucibles with up to 10 cm diameters and growth charges of up to 10 kg. The starting elements for crystal growth are multiply purified and weighed to a stoichiometric composition. The entire growth cycle is completed in a time period of about 1 mon (Doty et al., 1992; C. J. Johnson et al., 1993).

This high pressure Bridgman approach allows for $Cd_{1-x}Zn_xTe$ crystals to be grown near stoichiometry with respect to the metal–chalcogenide ratio and to have a high resistivity without the introduction of compensating dopants, such as chlorine (Butler et al., 1992; C. J. Johnson et al., 1993; Meyer et al., 1993). The crystals have a high degree of uniformity as evidenced by the relative performance of detectors fabricated from samples taken along the total length and diameter of a boule (Butler et al., 1992) and by the performance of large volume and large area CZT gamma ray spectrometers (Glick et al., 1994; Olsen et al., 1994a).

III. Material Properties of $Cd_{1-x}Zn_xTe$

This section discusses some of the material properties of CZT crystals, especially those directly related to the response of gamma ray spectrometers. In selecting a suitable semiconductor for a room temperature gamma or x-ray detector, it is imperative that the material have high bulk resistivity, good photoconductive properties, acceptable electron and hole transport, and uniformity. High bulk resistivities are required to maintain low noise associated with leakage current in the detector and ensure the ability for the device to operate at room temperature. The material must be a photoconductor so that the number of electron–hole pairs generated in the detector is proportional to the energy of the incident gamma ray. Ideally, the number of electron–hole pairs would be high to allow for low spectral broadening due to the statistics in the photogenerated charge produced for each photon. High charge collection of the electrons and holes is also important, which means that carrier trapping should be understood and minimized. The need to maximize the transport of electrons and holes requires an understanding of the defect states limiting the carrier mobilities and lifetimes. The important defect states can be extrinsic point defects (e.g., impurities), native defects (e.g., vacancies), extended defects (dislocations and stacking faults), and stoichiometry variations. Thus, optical, electrical, and structural measurements of CZT alloys are important in understanding the defect states in detector grade crystals. In addition, the crystals must be uniform with respect to the charge generated by the absorption of an x-ray or gamma ray. Thus, one must be concerned about local variations in the bandgap of the detector, which would lead to a variation in the number of electron–hole pairs produced by absorption of an energetic photon. Since the Cd:Zn ratio has a strong effect on the bandgap of the alloy, it is imperative that this ratio be well controlled during growth and be uniform throughout the active area of the detector. Variations in the Cd:Zn composition within the growth ingot will be particularly troublesome in the production of large area and large volume CZT detectors. Brief discussions on the resistivity, Cd:Zn alloy composition, radiative recombination centers, carrier mobilities and lifetimes, absorption coefficient, etch pit densities, x-ray rocking curves, precipitates, and impurities follow.

1. Resistivity

The bandgap of ZnTe is substantially larger than the bandgap of CdTe; thus, the resistivity of intrinsic $Cd_{1-x}Zn_xTe$ crystals is expected to be much larger than that of CdTe. Intrinsic resistivities at room temperature are predicted to range from about 10^{10} Ω-cm for CdTe to over 10^{16} Ω-cm for ZnTe. The measured resistivities of $Cd_{1-x}Zn_xTe$ crystals for $x = 0$, 0.04, and 0.20 are shown in Table I (Butler et al., 1993a). By increasing the fraction of x from 0.0 to 0.20, there is an increase

TABLE I

RESISTIVITIES OF $Cd_{1-x}Zn_xTe$ CRYSTALS GROWN BY A HIGH PRESSURE BRIDGMAN METHOD (Butler et al. 1993a)

Composition	Resistivity (Ω-cm)
CdTe	3.0×10^9
$Cd_{.96}Zn_{.04}Te$	2.5×10^{10}
$Cd_{.8}Zn_{.2}Te$	2.5×10^{11}

9. $Cd_{1-x}Zn_xTe$ Spectrometers for Gamma and X-Ray Applications

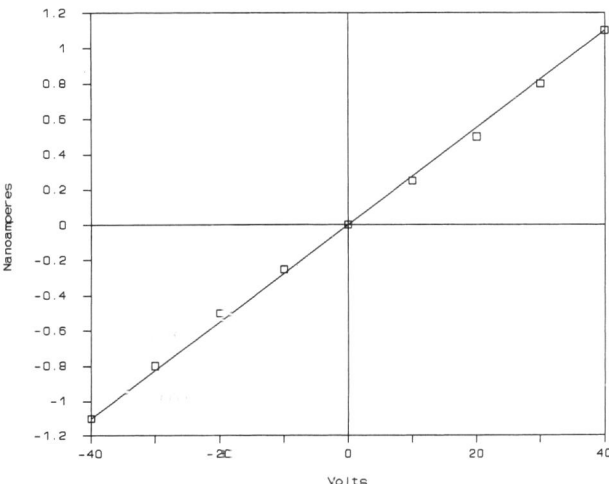

FIG. 1. Typical current–voltage curve at room temperature; detector dimensions: 10 × 10 × 1.5 mm. (Reprinted with permission from Butler et al., 1993c and Materials Research Society.)

of almost two orders of magnitude in the resistivity of the crystals, which is equivalent to a reduction of about two orders of magnitude in the leakage current.

Typical room temperature current–voltage curves for a $Cd_{0.8}Zn_{0.2}Te$ crystal having dimensions of 10 × 10 × 1.5 mm are shown in Fig. 1. The reduction in the leakage current by the addition of Zn to CdTe is important because two of the dominant factors limiting the performance of CZT radiation detectors are (1) incomplete charge collection of mobile carriers and (2) electronic noise associated with leakage current and the first amplification stage. If other factors are held constant, the spectral broadening due to incomplete charge collection decreases with increasing bias voltage, whereas the contribution due to leakage current increases. Because of these competing factors, there is an optimal bias voltage if one is most interested in minimizing the spectral linewidths of the detector. The higher resistivity of $Cd_{1-x}Zn_xTe$ compared to CdTe (for nonzero x) shifts the optimal bias voltage to higher values, thus allowing for narrower linewidths in the photopeaks.

2. Alloy Composition

For $Cd_{1-x}Zn_xTe$ crystals grown from the melt, the mole fraction of Zn has been shown to vary with respect to a sample's location within the boule (Sen and Stannard, 1993; Schlesinger et al., 1995). The Zn variation is due to the fact that the segregation coefficient of Zn has a value of 1.3; consequently, there is a gradual change in the value of x from the tip to the tail of the solidified boule. Typically, the Zn content for a nominal $Cd_{0.96}Zn_{0.04}Te$ boule can vary from about 0.03 to 0.05 from one end to the other (Sen and Stannard, 1993). In practice, about 1 in. of each end of the boule is typically discarded, and the remaining portion has a much more uniform alloy composition.

The gradual change in alloy composition along a boule has dire consequences for producing large volume $Cd_{1-x}Zn_xTe$ gamma ray spectrometers, because the variation in Zn with position causes a variation in the bandgap and in the number of photogenerated electron–hole pairs per incident gamma ray. In addition, the leakage current is much higher in regions having lower Zn content, which is particularly important in cases where shot noise is the dominant noise mechanism. The variation in bandgap within the active volume of the detector is manifested by a different nuclear spectroscopic response (i.e., different location of the photopeaks), depending on where in the device the electron–hole pairs were created. By integrating counts from all regions within the detector, one obtains photopeaks that are substantially broader than those observed for a detector that has a more uniform alloy composition, assuming all other factors have been held constant. Thus, it is imperative that $Cd_{1-x}Zn_xTe$ crystals selected for device fabrication have a uniform or nearly uniform Zn content.

3. Photoluminescence Spectrum

The value for the Zn composition can be determined to better than 0.01% by measuring the position of the free exciton peak at low temperatures (Schlesinger et al., 1995). Using photoluminescence spectroscopy the free exciton is well resolved at 4.2 K, and for small values of x, one can use the empirical expression

$$FE(x) = 1.5964 + 0.455x + 0.33x^2 \tag{1}$$

to determine approximately the mole fractions of Cd and Zn (Doty et al., 1992). Here, $FE(x)$ is the position of the free exciton in electron volts. In addition, one can use the measured energies of the photoluminescence peaks together with the sharpness, intensity, and linewidths of the near-band edge emission lines to determine the presence of shallow and deep level defects that act as radiative recombination centers and to obtain information on the crystallinity and alloy broadening of the $Cd_{1-x}Zn_xTe$ crystal (Doty et al., 1992; Butler et al., 1993c).

4. Charge Transport

To fabricate high resolution $Cd_{1-x}Zn_xTe$ gamma ray spectrometers, there are several constraints on the bulk material properties in addition to those placed on resistivity and uniform alloy composition. For example, detectors with the narrowest linewidths are obtained under conditions where the charge carriers' mean free paths are much longer than the detector thicknesses, preferably by more than an order of magnitude (Knoll and McGregor, 1993). Under these circumstances, practically all of the carriers generated by absorption of the incident x-rays or gamma rays are collected at the electrodes, and the degradation due to incomplete charge collection is small (see, for example, Dabrowski et al., 1983). Thus, the average carrier mean free paths for electrons and holes are important material properties in an assessment of the potential of $Cd_{1-x}Zn_xTe$ radiation detectors.

The mean free path of each carrier is given by the product of the lifetime, mobility, and electric field. C. J. Johnson et al. (1993) used a transient charge technique to measure the carrier lifetimes. Room temperature values for the electron and hole lifetimes were measured to be 0.57 and 0.33 μsec, respectively.

The charge carrier mobilities are also important parameters in determining the mean free paths of carriers, although mobilities of detector grade crystals may be limited by intrinsic carrier–phonon scattering. If this is the case, the carrier mobilities cannot be further increased by improving the material. Ideally, the room temperature electron and hole mobilities should be as large as possible, thereby maximizing the mean free paths for fixed carrier lifetimes and electric field.

Charge carrier mobilities of $Cd_{0.8}Zn_{0.2}Te$ have been reported using a time of flight measurement (Burshtein et al., 1993). The crystals used in this experiment were grown by a high pressure Bridgman method, and they were not intentionally doped with impurities. The electron mobility was found to have a $T^{-1.1}$ temperature dependence in the range 200–320 K, with a room temperature mobility value of 1350 cm²/V-sec. The hole mobility displayed a $T^{-2.0}$ dependence with lattice temperature over the range 200–320 K, with a room temperature mobility of 120 cm²/V-sec (Burshtein et al., 1993). Burshtein et al. (1993) argued that the hole mobility is a dispersive trap controlled one, so that the measured values for holes may not be indicative of intrinsic lattice scattering events. Readers may refer to discussions by Sher and Montroll (1975) and Tiedje and Rose (1980) for more information on dispersive trap controlled mobility.

Similar measurements of charge carrier mobilities of CZT grown by the high pressure Bridgman method have been reported by C. J. Johnson et al. (1993) for specimens with $x = 0.0$ and by Polichar, Schiarato, and Reed (1992) for $x = 0.2$. The values they obtained are slightly smaller than those reported by Burshtein et al. (1993).

The mobility values obtained by Burshtein et al. (1993) are higher than those reported for other $Cd_{0.8}Zn_{0.2}Te$ crystals grown by conventional Bridgman methods (Triboulet, Neu, and Fotouhi, 1983). The lower values of mobilities obtained by Triboulet et al. (1983) may be related to higher levels of impurities due to the use of quartz containers for their $Cd_{0.8}Zn_{0.2}Te$ crystal growth.

One of the fundamental figures of merit for a semiconductor x-ray or gamma ray spectrometer is the mobility–lifetime product. This product can often be obtained by measuring the photopeak shift as a function of the applied bias (Doty et al., 1994). Using excitation that is predominantly in the near-surface region (e.g., alpha particles), the induced pulses can be fit to the Hecht equation to determine the mobility–lifetime product of each charge carrier (Hecht, 1932). C. J. Johnson et al. (1993) studied the transport properties of high pressure Bridgman CdTe and obtained values of 6×10^{-4} cm^2/V for electrons, and 3×10^{-5} cm^2/V for holes. These mobility–lifetime products for electrons and holes generally agree with previous measurements (Siffert, 1978; Whited and Schieber, 1979; Sakai, 1982; Ohmori, Iwase, and Ohno, 1993). Similar investigations on $Cd_{0.8}Zn_{0.2}Te$ crystals were performed by Butler et al. (1993b), and mobility–lifetime products of 8×10^{-4} cm^2/V were reported for electrons and 3×10^{-6} cm^2/V for holes. The low mobility–lifetime product for holes was attributed to small hole lifetimes, approximated to be 30 nsec (Butler et al., 1993b). This hole lifetime is much smaller than the value obtained on chlorine doped CdTe (Ohmori et al., 1993). Measurements have also been performed on ZnTe crystals grown by high pressure Bridgman techniques (Doty, 1994). The highest values obtained were 2×10^{-4} cm^2/V for electrons and 2.5×10^{-5} cm^2/V for holes.

More recent measurements by Lund et al. (1994a) on $Cd_{0.9}Zn_{0.1}Te$ crystals have shown mobility–lifetime products of up to 6×10^{-3} cm^2/V for electrons and 9×10^{-5} cm^2/V for holes (see Fig. 2 for the mobility–lifetime product for electrons). In addition, Lund et al. (1994a) have performed fits of the charge collection for several CZT detectors. For these fits the mobility–lifetime product of holes is used as the only adjustable parameter. Hole mobility–lifetime products of a few times 10^{-5} cm^2/V are typically found to give the best fit to the data.

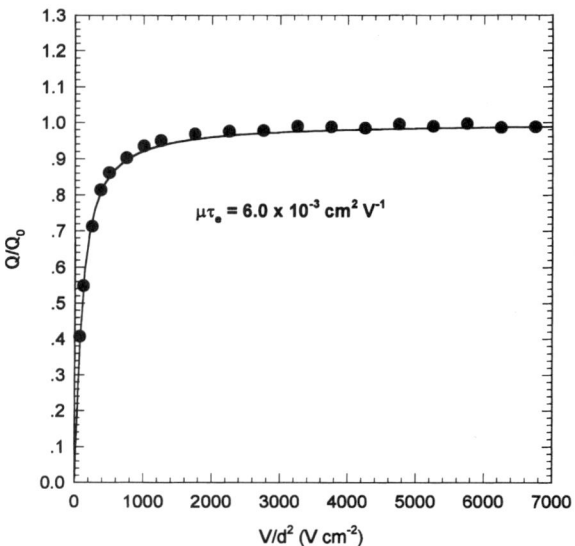

FIG. 2. Mobility–lifetime measurement of electrons on a $Cd_{0.9}Zn_{0.1}Te$ detector (from Lund et al., 1994a).

Most of the recent enhancements in carrier lifetimes have resulted from improvements in the purity and crystallinity of $Cd_{1-x}Zn_xTe$ crystals. The larger values for the mobility–lifetime products (especially for the electrons) indicate that continued increases in the size and detectability of CZT spectrometers are forthcoming, particularly if hole lifetimes can be further increased or there are developments in the electronics to compensate for the hole collection problem.

5. Absorption Coefficient for X-Rays and Gamma Rays

The values for the mean free paths determine the maximum thickness allowed for a high resolution $Cd_{1-x}Zn_xTe$ gamma ray spectrometer. Thus, another important material property is the efficiency of a $Cd_{1-x}Zn_xTe$ detector for stopping incident x-rays and gamma rays. For example, if the product of the linear absorption coefficient and thickness of a detector is much less than unity, then the sensitivity of the detector for measuring incident energetic photons is poor. The linear absorption coefficient of $Cd_{0.8}Zn_{0.2}Te$ has been calculated by Butler et al. (1993a), and the values are shown in Figure 3. Based on the lattice parameter values for CdTe and ZnTe (S. M. Johnson et al., 1990), the density of $Cd_{0.8}Zn_{0.2}Te$ was calculated to be 5.811 g/cm^3. Using these calculated values together with the thickness of a detector, one can determine the stopping power of

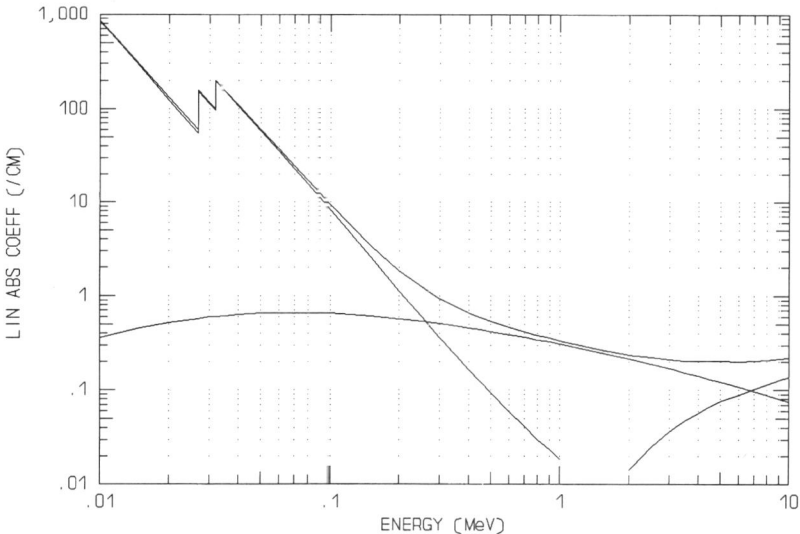

FIG. 3. Linear absorption coefficient curves for $Cd_{0.8}Zn_{0.2}Te$ showing photoelectric, Compton, and pair production values. (Reproduced with permission from Butler et al., 1993a and Materials Research Society.)

$Cd_{0.8}Zn_{0.2}Te$ for x-rays and gamma rays in the 0.01–10 MeV range. For example, for a gamma ray with an energy of 100 keV, the linear absorption coefficient is about 10 cm^{-1}, and one would prefer a detector having a thickness of greater than 0.1 cm.

IV. Defect Characterization and Effects on Device Response

Other techniques that have been used to characterize $Cd_{1-x}Zn_xTe$ crystals to determine if they are of spectrometer grade quality include etch pit densities, x-ray diffraction rocking curves, infrared microscopy, and impurity analyses. Brief comments will be made on each of these measurements, and interested readers can refer to the references for details.

1. Etch Pit Densities

A low value of etch pit density is typically measured for samples having low defect densities. In addition, some correlations have been established between specific dislocation features and epitaxial $Hg_{1-x}Cd_xTe$ IR detector performance (C. J. Johnson et al., 1993). Using a Nakagawa etch on (111) $Cd_{1-x}Zn_xTe$ surfaces, etch pit densities have been measured and reported by C. J. Johnson et al. (1993) and Butler et al. (1993c) for crystals grown by HPB methods. C. J. Johnson et al. (1993) reported a variety of features observed on CdTe crystals grown at eV Products. Some regions were found to have uniform etch pit densities of about $4 \times 10^4/cm^2$, others showed graded etch pit densities, and some areas near grain boundaries had etch pit densities of about $1 \times 10^6/cm^2$. Butler et al. (1993c) also reported values for $Cd_{1-x}Zn_xTe$ samples, which were grown at Aurora Technologies. Etch pit densities of $1.8 \times 10^5/cm^2$, $1.0 \times 10^4/cm^2$, and $0.5 \times 10^4/cm^2$ were reported for crystals having a Zn content of 0.0, 0.04, and 0.20, respectively. These values for etch pit densities are significantly lower than those found for $Cd_{1-x}Zn_xTe$ crystals grown in sealed quartz ampoules (Butler et al., 1993c), which is further evidence that crystals grown by high pressure Bridgman methods have less defect densities than those grown by conventional Bridgman methods.

2. X-Ray Rocking Curves

Butler et al. (1993c), C. J. Johnson et al. (1993), and Goorsky et al. (1995) reported x-ray diffraction measurements on $Cd_{1-x}Zn_xTe$ crystals. Butler et al. (1993c) measured (111) oriented samples obtained from nine $Cd_{0.96}Zn_{0.04}Te$ boules and from one $Cd_{0.8}Zn_{0.2}Te$ boule, all grown by high pressure Bridgman methods. The surfaces were chemically polished and subjected to an etch prior to exposure. Typical FWHMs of their rocking curves were measured to be about

14 arc-seconds for specimens taken from the $Cd_{0.96}Zn_{0.04}Te$ boules and 12 arc-seconds for the $Cd_{0.8}Zn_{0.2}Te$ boule. The narrowest linewidth observed by Butler et al. (1993c) was on a $Cd_{0.96}Zn_{0.04}Te$ sample (FWHM of 10 arc-seconds), which equaled the intrinsic resolution of their monochromator.

C. J. Johnson et al. (1993) reported x-ray diffraction measurements on one CdTe detector. The crystal used to fabricate the detector was grown by high pressure Bridgman methods. Although the detector displayed high resolution for 59.5-keV incident gamma rays, a great deal of spot to spot variation in the FWHMs of the double crystal rocking curves was observed. Using CuKα radiation and an x-ray spot size of 0.2 × 0.2 cm, the FWHMs obtained from four different regions on the 1 × 1 × 0.2 cm sample ranged from 44 to 120 arc-seconds. Each region showed some degree of lattice disruption as indicated by the presence of multiple peaks in the rocking curves.

Triaxial x-ray diffraction (TAXRD) rocking curves were also performed for $Cd_{1-x}Zn_xTe$ crystals and detectors by Goorsky et al. (1995). TAXRD is capable of distinguishing between the broadening of the x-ray diffraction peaks due to mosaicity and strain. Figure 4 shows a comparison between two $Cd_{0.9}Zn_{0.1}Te$ detectors, one having good overall performance and the other one having poor

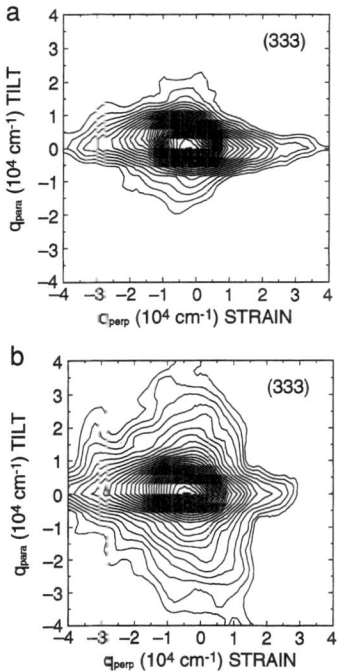

FIG. 4. Comparison of good (top) and bad (bottom) $Cd_{1-x}Zn_xTe$ detectors using (333) reflections x-ray reciprocal space maps.

performance. The bad detector exhibited a more extensive mosaic spread, indicated by the increased scattering in the q_{para} direction. These results are similar to those obtained by Schieber et al. (1994) for mercuric iodide gamma-ray detectors, in which case the TAXRD curves were much broader for bad detectors than for good ones.

Based on these x-ray diffraction measurements, it is important that similar measurements be conducted on samples that have a wide range of detector quality and the linewidths of the rocking curves correlated with the device performance. Given the spatial variation in the rocking curves observed for some detectors, one should also use a collimated gamma ray source and spatially map the device response over the entire active area of the detector, then correlate these measurements with maps of the rocking curve measurements performed on the same detector. Ideally, these measurements would be conducted using several different gamma ray energies to further determine the relationship between the detector characteristics and the bulk and near-surface structural properties of $Cd_{1-x}Zn_xTe$ crystal.

3. Precipitates

Infrared microscopy has been used to identify numerous inclusions and precipitates in $Cd_{1-x}Zn_xTe$ crystals, particularly in the study of substrates for $Hg_{1-x}Cd_xTe$ IR detectors (see, for example, Vydyanath et al., 1992; Johnson et al., 1992; Sen and Stannard, 1993). The presence of Te precipitates in $Cd_{1-x}Zn_xTe$ material is found to be variable, although there is a strong tendency for precipitates to occur on or near grain boundaries (C. J. Johnson et al., 1993). Generally, it is accepted that the occurrence of Te precipitates is unavoidable for $Cd_{1-x}Zn_xTe$ crystals grown from the melt and cooled to room temperature (Sen and Stannard, 1993). Although the Te precipitates appear to be present in all $Cd_{1-x}Zn_xTe$ boules grown from the melt, the shape of some of the precipitates appears to be unique to the high pressure Bridgman growth method (C. J. Johnson et al., 1993). At present full control of the Te precipitates in $Cd_{1-x}Zn_xTe$ is not possible; however, a technique for reducing the precipitates has been reported (Vydyanath et al., 1992). In practice, postgrowth thermal annealing of the crystals has proven to be the best way to affect the size, density, and distribution of Te precipitates (Sen and Stannard, 1993). However, if the precipitates act as gettering sites for HPB grown boules, then it is possible that the post-growth annealing can destroy the detector performance by creating new electrically-active states in the device.

Although the effect of Te precipitates on the performance of $Cd_{1-x}Zn_xTe$ gamma ray detectors is poorly understood, it has been shown that in some crystals the Te precipitates act as effective gettering sites for impurities and that they are associated with a high density of dislocation clusters (Sen and Stannard, 1993; C. J. Johnson et al., 1993). Consequently, $Cd_{1-x}Zn_xTe$ crystals with relatively high concentrations of precipitates are usually rejected for fabrication of x-ray and

9. $Cd_{1-x}Zn_xTe$ Spectrometers for Gamma and X-Ray Applications

gamma ray spectrometers. An improved control of the partial pressures of the constituents during the growth cycle and a better understanding of the role of precipitates on device performance should allow for larger homogeneous volumes and a higher manufacturing yield of detector grade material.

4. Impurities

Although impurities are not intentionally introduced into $Cd_{1-x}Zn_xTe$ crystals grown by high pressure Bridgman methods, impurities still are present in the crystals. These impurities can influence the electrical properties as trapping sites or compensating defects. Given that the current–voltage curves are well described by a Fermi level near the midpoint of the energy gap (Butler et al., 1993c; Egarievwe et al., 1994), one might assume that the material is intrinsic; however, impurities that are unintentionally introduced into the $Cd_{1-x}Zn_xTe$ boule are most likely acting to highly compensate the material. Much work is yet to be done on the role of impurities on the material properties and detector characteristics of $Cd_{1-x}Zn_xTe$ devices.

There are several potential sources of impurities in $Cd_{1-x}Zn_xTe$ boules, some of the most notable being components of the graphite furnace, quartz ampoules, carbon coatings, and especially from the Cd, Zn, and Te source materials. The purity of Te is especially important, and contaminants contained in the Te can lead to a higher density of twins and microtwins (Bell and Sen, 1985; Sen and Stannard, 1993). Contaminants in the Te include CO_2, H_2O, N_2 and O_2, and attempts are currently made to rid Te of its gaseous impurities. Some of the techniques employed to minimize the contaminants in Te include distillation, zone refining of Te, reaction and sublimation of CdTe and ZnTe, and the growth of crystals without exposure of the material to air after each step (Sen and Stannard, 1993).

A few cations are also detected in $Cd_{1-x}Zn_xTe$ crystals in relatively high concentrations ($\sim 10^{16}$ cm^{-3}). These include Ni, Pb, Al, Fe, Cu, and C (Sen and Stannard, 1993; Soria and James, 1995). These elements are also routinely found in the commercial starting materials. Unfortunately, the concentrations of contaminants in the starting materials vary between different suppliers and also tend to vary from batch to batch for the same supplier (Sen and Stannard, 1993; Soria and James, 1995).

Effective quantitative screening techniques of the source and purified materials must be established and implemented to reproducibly achieve impurity levels of less than 10 ppb in $Cd_{1-x}Zn_xTe$ crystals and detectors. Care must always be taken to avoid contamination of the crystals and detectors during handling. Based on studies of the performance of other room temperature gamma detector materials (e.g., CdTe and HgI_2), an order of magnitude reduction in the concentration of impurities in $Cd_{1-x}Zn_xTe$ will have a substantial effect on the mobility–lifetime products of electron and hole charge carriers, and it will increase the yield of detector grade material.

V. Detector Characterization and Effects on Device Response

1. DETECTOR FABRICATION

$Cd_{1-x}Zn_xTe$ detectors are usually cut in the form of square slabs with areas ranging from a few square millimeters up to more than 200 mm^2 and thicknesses ranging between 1 mm and 3 mm. Experimental devices have been produced with cross-sectional areas of over 600 mm^2 and thicknesses of over 25 mm (Glick et al., 1994).

In the detector fabrication process, $Cd_{1-x}Zn_xTe$ boules are grown and then cut into slices having the desired dimensions. The slicing and dicing are typically performed using a diamond, inside diameter saw. Br:MeOH etching in a 5% Br solution can be used to remove the surface damage from the sawing step. To minimize surface oxidation, one should attach the electrodes immediately after the final etch. Gold contacts are most often applied using the electroless AuCl method (see, for example, Musa and Ponpon, 1983); although, evaporated Au (Schirato, Polichar, and Reed, 1994), sputtered Pt, and electroless Pt contacts have also been used (Butler et al., 1993b). The detector surfaces are cleaned after the electroding process to remove residual salts and other byproducts of the electroless AuCl deposition process. After the contact deposition, colloidal graphite is frequently used to attach a thin (~1-mil) Pd, Pt, or Au wire to the electrodes for application of an electrical bias. In some cases the surfaces are next covered with a thin coating of Humiseal to help with surface passivation and mechanical stability (Burshtein et al., 1993).

In the electroding process caution must be exercised to use a fresh AuCl solution and allow an optimal time for the solution to be in contact with the $Cd_{1-x}Zn_xTe$ substrate. However, independent of efforts to standardize this AuCl contact deposition technique, the process is still somewhat variable, and it is dependent on the person performing the processing step (C. J. Johnson et al., 1993). Figure 5 shows an example of a proton-induced x-ray emission spectrum showing the presence of Au and Cl after the AuCl contact deposition step. The concentrations of Au and Cl were found to vary significantly from spot to spot.

2. NUCLEAR SPECTROSCOPIC DATA AT ROOM TEMPERATURE

The energy resolution of $Cd_{1-x}Zn_xTe$ detectors is limited primarily by electronic noise (e.g., detector leakage current, detector capacitance, and preamplifier input) and by incomplete charge collection, particularly for the hole carriers. For small detectors and relatively low quantum energies (less than about 30 keV), electronic noise due to the leakage current and from the preamplifier input tend to be the dominant noise factors. For detectors with cross-sectional areas exceeding about 1 cm^2, the detector capacitance is also a significant factor to the total elec-

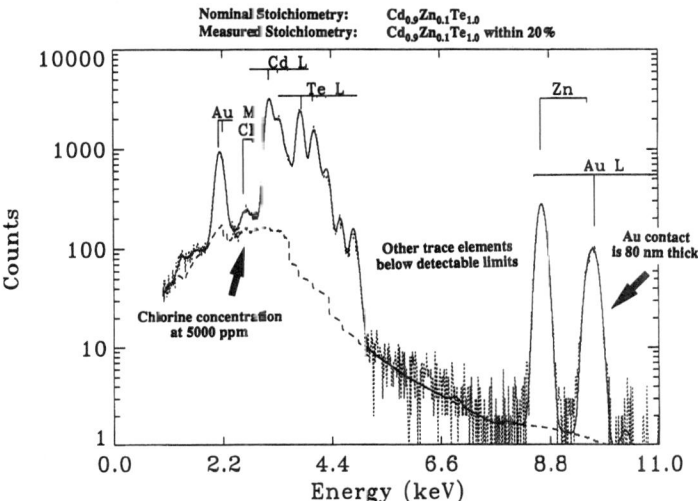

FIG. 5. Proton-induced x-ray emission spectrum for a $Cd_{0.9}Zn_{0.1}Te$ detector after the AuCl contact deposition process (from Antolak et al. 1995).

tronic noise. The influence of incomplete charge collection is more important at the higher quantum energies, in which case the absorption coefficient is smaller and electron–hole pairs are created deeper in the detector. At these higher energies (greater than about 60 keV), hole carriers can become easily trapped within the detector volume, in which case the pulse output for an incident gamma ray is dependent on the location where the gamma ray was absorbed.

The effect of these broadening mechanisms is seen in the spectra shown in Figures 6–8 taken with a $0.5 \times 0.5 \times 0.15$-$cm^3$ $Cd_{0.8}Zn_{0.2}Te$ detector biased at 230 V (Butler et al., 1993c). These data were taken at room temperature using a Cd-109, Am-241, and Co-57 source, respectively. High charge collection is seen at the lower energies, as evidenced by the 17:1 peak to valley ratio for the 22.1 keV photopeak of Cd-109. At higher energies the incomplete charge collection is more severe, as evidenced by the low energy tailing of the 60 keV and 122 keV photopeaks.

Similar pulse height spectra are observed for $Cd_{0.9}Zn_{0.1}Te$ detectors fabricated by eV Products. A spectrum taken at room temperature using Am-241 is shown in a paper by Hess et al. (1994). The FWHM of the 59.5 keV photopeak is about 4.2%, which is comparable to the spectra shown in Figure 7 for a $Cd_{0.8}Zn_{0.2}Te$ detector fabricated by Aurora Technologies. A spectrum taken using Cs-137 as a radioactive source shows a photopeak at 662 keV, but given that the $Cd_{0.9}Zn_{0.1}Te$ detector was only 0.2 cm thick, the efficiency of photopeak at 662 keV was quite small (Hess et al., 1994).

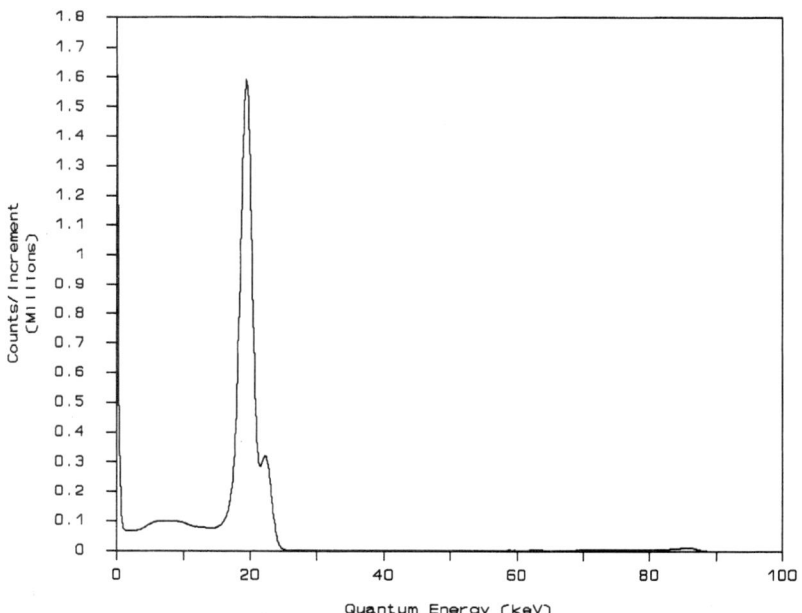

FIG. 6. Energy spectrum of Cd-109 measured with a $Cd_{0.8}Zn_{0.2}Te$ detector at room temperature. Resolution (FWHM) of the 22.1 keV photopeak is 2.24 keV. (Reproduced with permission from Butler et al., 1993c and Materials Research Society).

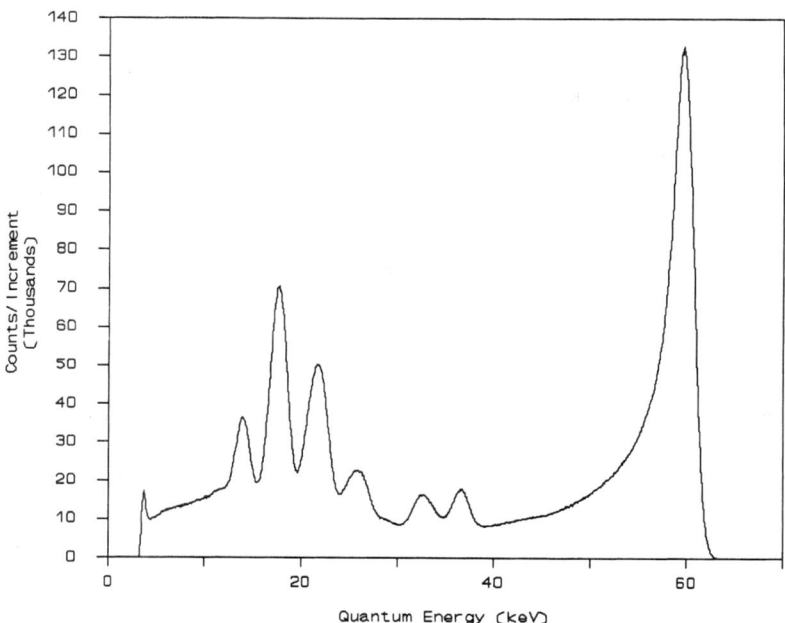

FIG. 7. Energy spectrum of Am-241 measured with a $Cd_{0.8}Zn_{0.2}Te$ detector at room temperature. Resolution (FWHM) of the 59.5 keV photopeak is 2.89 keV. (Reproduced with permission from Butler et al., 1993c and Materials Research Society).

9. $Cd_{1-x}Zn_xTe$ Spectrometers for Gamma and X-Ray Applications

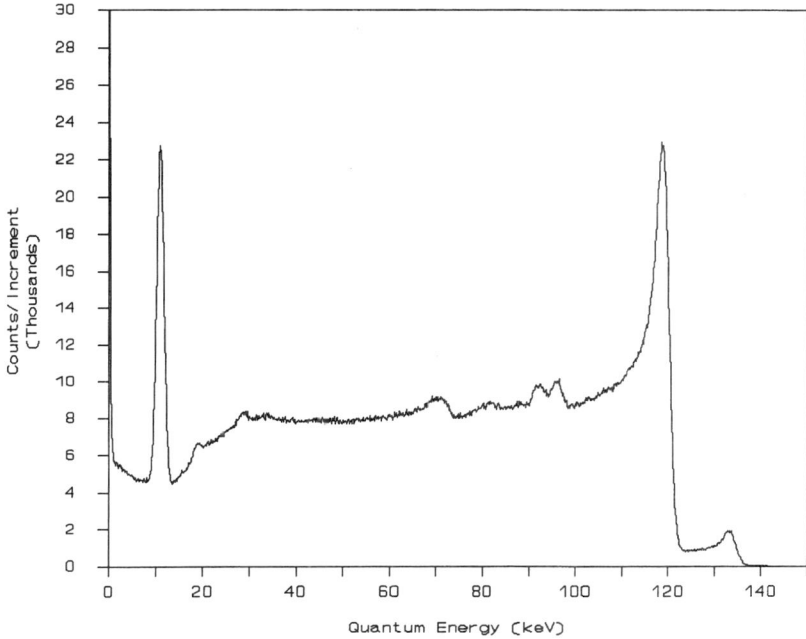

FIG. 8. Energy spectrum of Co-57 measured with a $Cd_{0.8}Zn_{0.2}Te$ detector at room temperature. Resolution (FWHM) of the 122 keV photopeak is 2.9 keV and 2.16 keV for the 14.5 keV photopeak. (Reproduced with permission from Butler et al., 1993c and Materials Research Society).

3. Detector Current–Voltage Characteristics

C. J. Johnson et al. (1993) reported three characteristic current–voltage curves that were found to correlate to the Am-241 gamma performance of CdTe detectors grown by high-pressure Bridgman methods (see Fig. 9). The characteristic I–V curves were labeled Types I, L, and D, respectively, for increasing, linear, and decreasing resistance shapes. Two trends were found in these I–V studies on graded CdTe detectors. First, the apparent resistivity was largest for the Type I detectors, with these devices having resistivities of about 10^{11} Ω-cm. The resistivities decreased to a low 10^{10} Ω-cm for the Type L detectors, and further decreased to mid 10^9 Ω-cm for the Type D detectors. Second, the Type I detectors tended to display good 59.5 keV photopeaks with low noise thresholds. The Type L detectors showed poorer photopeaks and higher noise levels than the Type I detectors, and the Type D detectors yielded no 59.5 keV photopeak and had significantly higher noise thresholds Among the 50 detectors tested in this study, 82% of the high resolution detectors were Type I, 55% of the low resolution detectors were Type L, and 80% of the detectors showing no photopeak were Type D. A strong correlation between detector response and the I–V curves was evident from this

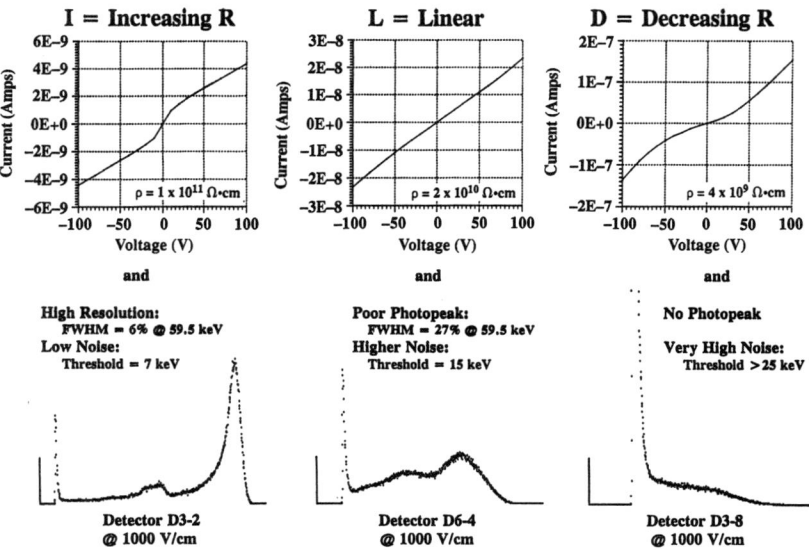

FIG. 9. Current–voltage characteristics of HPB-grown CdTe detectors and Am-241 spectra. (Reproduced with permission from C. J. Johnson et al., 1993 and Materials Research Society.)

work (see Fig. 10). Although C. J. Johnson et al. (1993) reported no model to attempt to explain this correlation, a couple of points are fairly obvious. First, the increase in the noise thresholds with decreasing resistivity is most likely explained by the larger detector leakage currents. Second, the increased trapping as one goes from the Type I to Type L to Type D detectors suggest that the defect contributing to the observed reduction in the resistivity of the crystal also plays a critical role as a performance controlling defect, and it is most likely an electrically active trapping site. This implies that the Type I detectors have a much lower concentration ($\sim 1-2$ orders of magnitude) of these electrically active defects than the Type D detectors. Further work should be completed to provide additional tests on other $Cd_{1-x}Zn_xTe$ detectors with $x > 0$.

C. J. Johnson et al. (1993) also reported a correlation between the type of measured I–V curve (i.e., Type I, L, or D) and the location in the ingot from which the crystals were taken. A strong correlation between the number of Type I (good) detectors and location above the ingot bottom was found. Here, the ingot bottom represents the part of the CdTe boule that is first to freeze. This study showed that 85% of the units located 2 cm from the bottom were Type I detectors, and 0% of the units located 6 cm above the bottom were Type I. Furthermore, the percentages of Type L and Type D units increased monotonically from 15% and 0%, respectively, at 2 cm from the bottom to 50% each at the 6 cm location. These data strongly suggest that some performance limiting defect is monotonically increasing from the bottom toward the top of the ingot axis. This defect may be associated with an impurity that either increases from the bottom to the top of the boule and

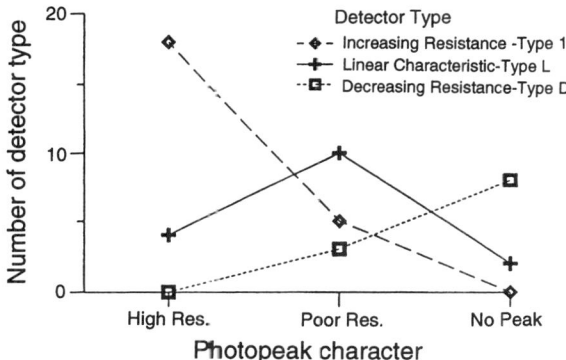

FIG. 10. Correlation of current–voltage characteristic types and ingot location for HPB-grown CdTe crystals. (Reproduced with permission from C. J. Johnson et al., 1993 and Materials Research Society.)

acts as an electrically active charge trap, or one that decreases from bottom to top of the boule and acts to compensate other defects present in the material. Additional testing programs to monitor the defect states, contaminants levels, and Zn content and their variations along the ingot axis are greatly needed.

Other current–voltage measurements have been reported by Butler et al. (1993a, 1993c; Parsons et al., 1994) on $Cd_{0.8}Zn_{0.2}Te$ detectors grown by HPB. The typical shape of the I–V curves for detectors studied by Butler et al. (1993a) was fairly linear (see Fig. 1). A value of 2×10^{11} Ω-cm was deduced for the bulk resistivity. If the measurements by C. J. Johnson et al. (1993) are accurate and values of 10^{11} Ω-cm are achievable on HPB CdTe, then a higher value of resistivity would be expected for intrinsic $Cd_{0.8}Zn_{0.2}Te$.

Other issues clearly remain regarding the current–voltage measurements on CZT detectors and their relationship to device performance. For example, further investigations should be conducted to understand the "intrinsic" nature of the undoped HPB $Cd_{1-x}Zn_xTe$ material, the use of electrode materials or contact deposition processes that can lead to blocking contacts, the possibility of p-i-n structures to lower leakage currents, and the relationship between the shape of the current–voltage curves with detector performance for gammas that excite primarily the near-surface region and for more energetic gammas that uniformly excite the bulk of the detector.

4. LARGE VOLUME GAMMA RAY SPECTROMETERS

There is a widespread need for lightweight, portable gamma ray spectrometers suitable for detection of nuclear proliferation, treaty verification, special nuclear materials control, and IAEA-like applications. In many cases the signals associated with the particular radioactive materials of interest are barely distinguishable from the noise or other naturally occurring radiation. Consequently, the detectors should have high sensitivity and high energy resolution (i.e., substantially better than scintillators). In addition, the constraints on portability and maintenance often require that the devices operate at room temperature.

Progress on increasing the detectability of CZT detectors to weak gamma emitting signals will be brought about by increasing the energy resolution or increasing the active volume of the detectors. Increases in the active volume of CZT spectrometers can be achieved with several approaches: by increasing the mobility–lifetime product of the charge carriers so that thicker detectors (>1 cm) can be produced, by increasing the homogeneity of crystalline slabs to allow for thin (<0.3 cm) detectors with much larger surface areas, by engineering of alternative detector designs to obviate the problem with poor hole collection, or by arraying several smaller detectors to produce a detector assembly with a much larger cross-sectional area and thickness.

This subsection reviews recent work on the fabrication and characterization of CZT detector arrays and large volume spectrometers. Discussions on CZT arrays will be presented first, followed by a review of recent work on relatively thick planar CZT detectors, closing with a brief presentation on new developments in the area of large volume coaxial detectors.

Figure 11 shows spectra for two $5 \times 5 \times 1.5$ mm $Cd_{0.9}Zn_{0.1}Te$ detectors using a Barium-133 source (from Olsen *et al.,* 1994a). The detectors were connected in a side by side lateral array geometry without an energy correction circuit. Each detector had a bias voltage of 250 V, and the entrance electrode was negatively biased. The data collection time for the experiment was selected to be 2000 sec. The four dominant Ba-133 gamma ray peaks at energies of 276–384 keV are easily observed in the figure. These peaks are observed as double peaks, because

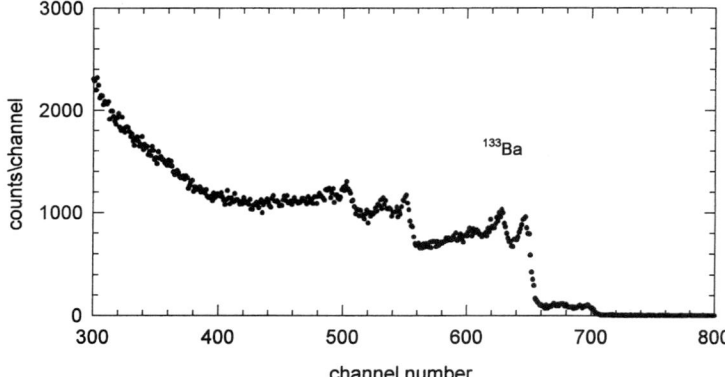

FIG. 11. Spectrum for two $5 \times 5 \times 1.5$ mm CZT detectors connected in parallel using a Ba-133 radioactive source and no energy correction circuit. Note the presence of the double peaks associated with two different detector responses to the incident radiation (from Olsen *et al.,* 1994a).

of the different response of the two detectors in the assembly. Although this two element detector array increases the sensitivity of the detector to radiation, it can actually lower the detectability of the instrument, because of the difficulty in uniquely determining the energy of the incident gamma ray based on the position of the photopeaks. In the event of other background radiation, significant problems with false alarms would be expected.

Data taken on other commercial CZT detectors show that, in general, the detectors have different carrier collection efficiencies, which cause the location of the photopeaks to vary between different detectors (Olsen et al., 1994a). This variation in peak positions complicates the use of vertical and horizontal stacking of detectors as a way to greatly increase the effective sensitivity of the detector assembly by arraying several individual detectors.

An electronic scheme was developed by Olsen et al. (1994a) to adjust for the different charge induced at the preamp input from the different detectors configured in an array, so that the response from each detector is identical. This allows a number of small semiconductor detectors to be stacked horizontally or vertically to greatly increase the overall system sensitivity (by a factor of 10 or more). Moreover, the output of each detector can be added prior to the preamp, and multiple preamps are not required. There is a small penalty in using the electronic scheme developed by Olsen et al. (1994a) to array several small detectors, in that it contributes to increases in the overall leakage current and capacitance of the system. This increased leakage current and capacitance leads to an increase in the system noise, which slightly degrades the energy resolution of the detector assembly. However, in most cases, the small decrease in energy resolution is more than offset by the simplification in the electronics and the increased overall sensitivity of the detector array. The increased sensitivity is manifested by a much lower counting time required to acquire an energy spectrum from a weak source and a reduced false alarm rate in picking out a weak signal from the background radiation.

Figure 12 shows a spectrum for the same CdZnTe detector assembly shown in Fig. 11, except that an RC corrective circuit has been connected to one of the detectors (from Olsen et al., 1994a). Here, the RC circuit on one of the detectors has been designed so that the positions of the photopeaks are aligned with the other detector. Using this approach, a number of smaller detectors can be connected together, and effective cross-sectional areas of 50 cm^2 are achievable. The limit on cross-sectional area will be determined by the capacitance and leakage current of the system. The performance of an array of small detectors would be approximately equal to that of a single detector with the same surface area. Moreover, if inhomogeneities in the single large area detector produce spot to spot variations in the spectra, then the performance of a number of small detectors would be expected to be significantly better than that of the single large detector. At present, high quality detectors having cross-sectional areas greater than about 5 cm^2 are not available, but in the event they are produced in the future, it is still possible to array these detectors to further enhance the sensitivity of a CZT gamma sensing instrument.

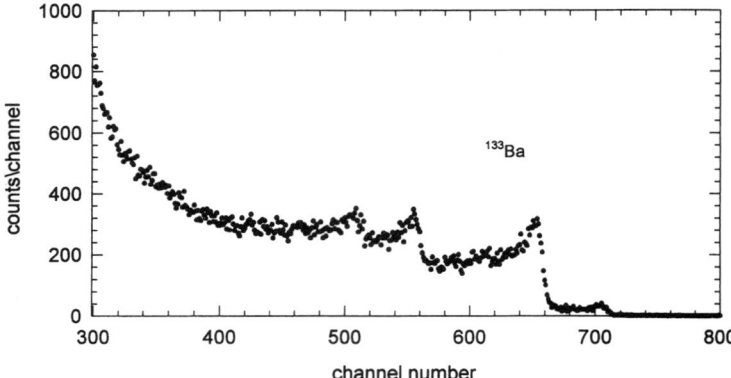

FIG. 12. Spectrum for two 5 × 5 × 1.5 mm CZT detectors connected in parallel using a Ba-133 radioactive source with an energy correction circuit. Note the double peaks as seen in Fig. 11 are now aligned to produce the same photopeaks (from Olsen *et al.*, 1994a).

Figure 13 shows a schematic of a vertical stacking of room temperature semiconductor detectors, which is useful for building an array for detecting high energy gammas. In this case, the entrance electrode is negatively biased for each detector in the array. This scheme causes the positive electrode of one detector to face the negative electrode of the adjacent detector. A thin insulator (or air gap) must be placed between the detectors to avoid a short circuit. Thin insulating layers are typically effective capacitors, and they can be the source of a large amount of unwanted noise. Olsen *et al.* (1994a) showed that much better spectra can be

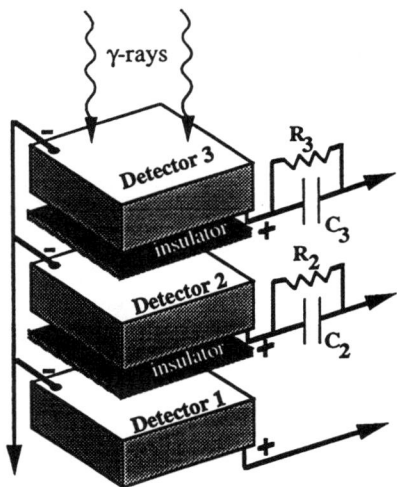

FIG. 13. Schematic drawing showing the detector assembly and energy shift circuit for multiple detectors arranged in a vertical stack (from Olsen *et al.*, 1994a).

9. $Cd_{1-x}Zn_xTe$ Spectrometers for Gamma and X-Ray Applications

obtained by wiring the detector assembly so that the positive face of one detector faces the positive face of the adjacent one. A thin low-Z insulator is still placed between detectors so that the energy corrective RC circuit will work. In this setup the voltage difference between the faces of adjacent electrodes is small, and it does not significantly contribute to the input capacitance as seen by the preamp.

For low energy gammas incident from the top of the vertical cascade, most of the excitation will take place in only the top detector, but for high energy gammas the sensitivity of the vertical array will be much greater than a single detector due to the increased absorption of the incident gammas. An alternative geometry is to illuminate the detector cascade from the side so that radiation impinges equally on each detector.

Using the approach by Olsen *et al.* (1994a), it is now possible to continue adding several detectors into a vertical stack to effectively absorb gamma rays with energies of 1000 keV and larger. For example, the top spectrum in Fig. 14 shows

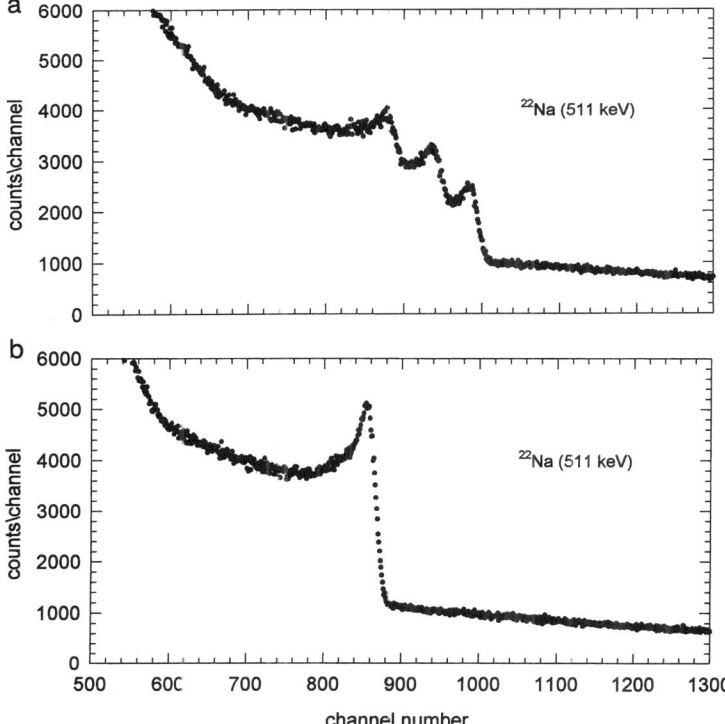

FIG. 14. The data at the top show a spectrum taken with Na-22 radiation incident on three 5 × 5 × 2 mm $Cd_{0.9}Zn_{0.1}Te$ detectors, which have been stacked vertically and connected in parallel. The spectrum at the bottom shows the results for the same vertical cascade, except in this case the energy correction circuit has been switched on. Here, the energy correction switch has been designed so that the photopeaks associated with each detector are aligned (from Olsen *et al.*, 1994a).

the results for three 5 × 5 × 2 mm CdZnTe detectors connected in a vertical cascade, but in this case no energy corrective circuit has been added. (Olsen *et al.*, 1994a). This particular spectrum was taken with Na-22 radiation, which has a characteristic gamma photopeak at 511 keV. Three distinct photopeaks are observed due to the differing response of each detector in the cascade. Without changing any other experimental conditions, the RC energy corrective circuit is switched on, and some of the charge normally induced at the preamp input is now dissipated into the RC networks. In this setup the detector having photopeaks at the lowest channel numbers is connected directly between the preamp and ground, and the other two detectors are connected to RC circuits so that all detectors have photopeaks in the same positions. The bottom curve in Fig. 14 shows a Na-22 spectrum for the same cascade, but in this measurement the energy corrective circuit has been switched on. Both the sensitivity and peak to valley ratio of the detector assembly is better than any one of the smaller detectors, and the rate of counts into the channel number associated with the photopeak is greatly increased.

Individual CZT detectors that are much thicker than those commercially available are still desirable for high gamma ray energies, particularly if the sources of radiation at the measurement location are weak. Attempts have been made to fabricate thick (>10 mm) experimental CdZnTe detectors (see Glick *et al.*, 1994; Olsen *et al.*, 1994a). Figure 15 shows a spectrum using Na-22 radiation for a 10 × 10 × 10 mm $Cd_{0.9}Zn_{0.1}Te$ detector, which was irradiated from the negatively biased entrance electrode (from Olsen *et al.*, 1994a). The FWHM of the 511-keV photopeak is less than 3%. Glick *et al.* (1994) reported measurements on a 10 × 10 × 25 mm $Cd_{0.9}Zn_{0.1}Te$ detector. Figure 16 shows spectra for the 59.5-keV gamma radiation emitted from Am-241 and the 662-keV gammas emitted from Cs-137 (from Glick *et al.*, 1994). The FWHM of the 59.5 keV photopeak is 8%, and the FWHM for the 662 keV photopeak is about 5%. These thick experimental

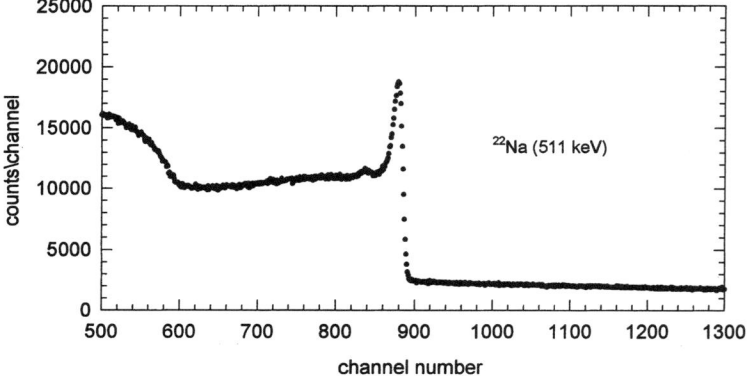

FIG. 15. Spectrum taken with Na-22 radiation incident on a 10 × 10 × 10 mm $Cd_{0.9}Zn_{0.1}Te$ detector (from Olsen *et al.*, 1994a).

9. $Cd_{1-x}Zn_xTe$ Spectrometers for Gamma and X-Ray Applications

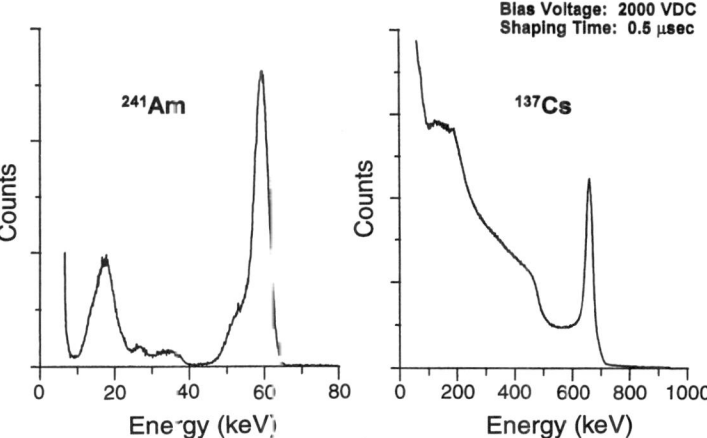

FIG. 16. Spectra taken with a 10 × 10 × 25 mm $Cd_{0.9}Zn_{0.1}Te$ detector showing the gamma radiation emitted from Am-241 and from Cs-137. Note the photopeaks at 59.5 and 662 keV (from Glick et al., 1994).

CZT detectors are not typical of those readily available today, but given the rapid improvements in the development of this technology, thick spectrometers should be available within two years. At that time, it should be straightforward to array these large volume detectors using the electronic scheme discussed earlier in this subsection.

Thick planar detectors require relatively high voltages. For example, a 1-cm thick CZT detector requires about 1 kV for good electron charge collection. Thus, a 4-cm thick CZT planar detector would require a bias voltage of over 10 kV. These power supply requirements can present difficulties for lightweight, portable instruments. For applications in which thick CZT spectrometers are needed, a new detector geometry is desired, preferably one that allows for a large active volume while maintaining modest demands on the power supply (i.e., batteries). One option is to fabricate a coaxial CZT detector, analogous to the cryogenically cooled coaxial germanium detectors available commercially. In consideration of the recent progress in the development of high resolution planar CZT detectors with detection volumes greater than 1 cm^3 (Glick et al., 1994; Olsen et al., 1994a), the production of a coaxial CZT spectrometer appears to be technologically feasible.

Work is now underway to develop a coaxial detector with a detector volume of 10 cm^3, which will further extend the sensitivity of the CZT detector to higher energy gammas. Moreover, the distance from the inner to outer walls of the coaxial detector will only be about 1 cm, so that the bias voltage required will be less than 1 kV. If efforts to develop a large volume coaxial spectrometer are successful, then a new instrument will be available that meets many of the desired attributes of a portable instrument—greatly increased sensitivity to high energy

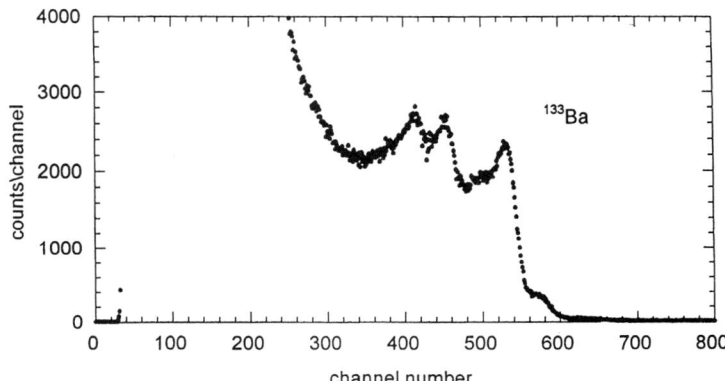

FIG. 17. Spectra with a 2.6 cm³ open-end coaxial CZT detector using a Ba-133 radioactive source (from Olsen *et al.,* 1993b).

gammas, high temperature operation, light weight, and energy resolutions much better than those available with scintillator systems.

A prototype 2.6 cm³ CZT coaxial detector has been recently fabricated by eV Products (Eissler *et al.,* 1994). The detector has been tested by Eissler *et al.* (1994) and Olsen *et al.* (1994b) for several radioactive sources (see Fig. 17 for results using Ba-133). This new work will advance the understanding of problems in manufacturing coaxial CZT detectors and most likely lead to higher quality large volume devices. Further effort is now underway to resolve problems with crystalline perfection, ingot cracking, and device processing steps. Characterization studies will also be completed within the coming months to measure the electric field inside the detector and the role of the open ends. These studies are motivated by the unusual dependence of the nuclear spectroscopic tests on the bias voltage. The commercial viability of coaxial CZT detectors is still uncertain, and barriers associated with homogeneity of large blanks, cutting of blanks from screened CZT ingots, drilling of open- and closed-end holes while maintaining crack free detector volumes, chemical etching to remove damage, mechanical polishing of sharp edges, attachment of gold or other acceptable electrodes, and passivation of surfaces must be resolved.

5. *P-I-N* GAMMA RAY DETECTORS

Reductions in the charge carrier injection from the metallic contacts of silicon and germanium detectors have been achieved by fabricating *p-i-n* diode structures (Woldseth, 1973). This earlier work on Si and Ge devices provided a model for alternative approaches to improve the quality of CdTe and CdZnTe detectors. Hazlett *et al.* (1986) used diffusion of Au and In into CdTe to provide *p*- and *n*-type regions. Although the charge injection by the *p* and *n* contacts was re-

9. $CD_{1-x}ZN_xTE$ SPECTROMETERS FOR GAMMA AND X-RAY APPLICATIONS

duced, doping with chlorine was still needed to reach resistivity values high enough for gamma ray detector applications. At about the same time, Shin (1985) and Ryan et al. (1985) used mercury cadmium telluride for construction of CdTe p-i-n detectors. The CdTe crystals used to produce the p-i-n detectors had a resistivity of less than 10^5 Ω-cm, and higher resistivity substrates would probably have yielded improved devices. Although the CdTe substrates were not ideal, reasonable energy resolutions were still measured (e.g., 8% for the Co-57 line at 122 keV).

Cadmium zinc telluride p-i-n gamma detectors were recently constructed by growing p- and n-type HgCdTe epitaxial layers on opposite faces of bulk $Cd_{0.96}Zn_{0.04}Te$ wafers (Hamilton et al., 1994). The p-i-n diode structure was found to reduce the difficulty in contacting CZT and provide for barriers against charge injection. In this study the substrates were grown by vertical Bridgman methods and the HgCdTe epitaxial layers were grown by liquid-phase epitaxy. The detectors had a typical thickness of 0.09 cm and cross-sectional area of 0.25 cm². Although the detectors tested in this study were fairly small in area, the bulk growth and LPE process is capable of producing crystals with areas up to 35 cm². The structure of the resulting device is shown schematically in Fig. 18. Figure 19 shows a typical energy spectrum for U-235 using a CZT p-i-n detector. Spectra were also measured for photon energies in the range of 20 to 400 keV, and high energy resolution was obtained. For example, a resolution of 1.82% FWHM was reported for the 356 keV line of Ba-133 with a peak to valley ratio of 3.4 (Hamilton et al., 1994).

The results of these measurements have demonstrated promise for CZT p-i-n diode structures, at least in the case of thin detectors. For thicker detectors, several questions need to be resolved regarding the depletion region in the devices and its dependence on applied bias. In addition, the spectroscopic behavior of CZT p-i-n structures, which have been fabricated using high resistivity substrates grown by high pressure Bridgman methods, should be determined. The widespread use of p-i-n CZT gamma detectors will most likely require that thicker detectors be fabricated in the future.

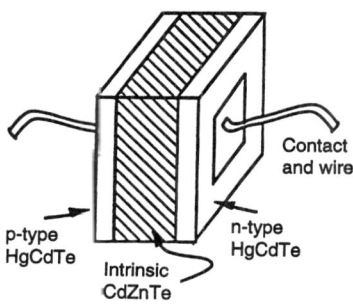

FIG. 18. Drawing of p-i-n CZT detector structure (from Hamilton et al., 1994).

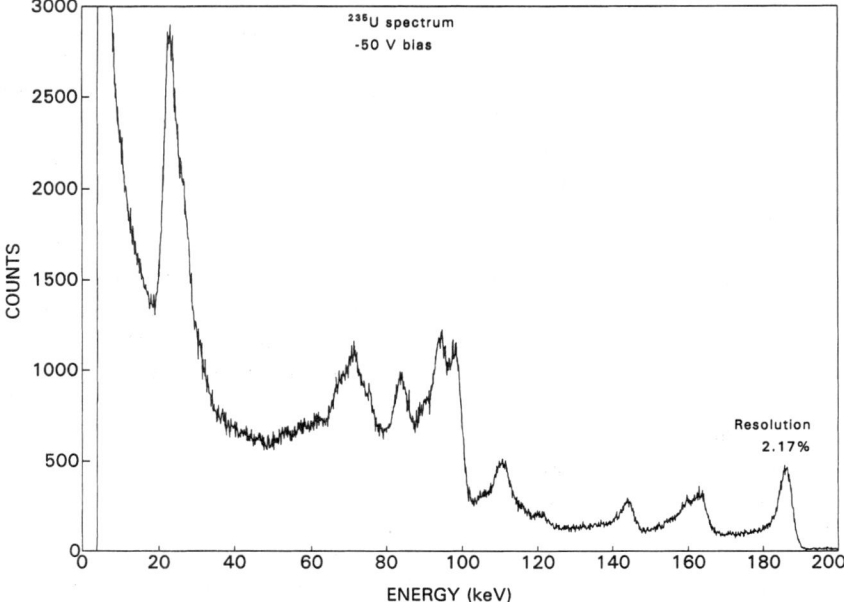

FIG. 19. Energy spectrum for U-235 using a *p-i-n* CZT detector (from Hamilton *et al.*, 1994). The width of the 186-keV photopeak is 4.03 keV, and the peak-to-valley ratio is 5.3.

6. X-Ray Detector Response

This subsection reviews the recent results on the use of $Cd_{1-x}Zn_xTe$ crystals to detect lower photon energies relevant to x-ray fluorescence measurements and the application of these detectors for measuring elements through their inner shell x-ray emissions. For soft x-ray applications, it is desirable for the leakage current to be in the tens of picoamp range or lower. This requirement on leakage current means that large bandgap $Cd_{1-x}Zn_xTe$ material must be used (i.e., material having as large a value for the Zn mole fraction as possible, without a significant loss in the carrier lifetimes).

An alternative scheme is to reduce the leakage current from the room temperature values by cooling the detectors. Since the Fermi level of CZT crystals grown by high-pressure Bridgman methods is pinned near the middle of the bandgap (Egarievwe *et al.*, 1994; Butler *et al.*, 1993b), the lattice temperature has a strong effect on the leakage current. Figure 20 shows the variation in the leakage current with temperature at a bias voltage of 50 V for a 10 × 10 × 1.5 mm $Cd_{0.8}Zn_{0.2}Te$ detector (from Butler *et al.*, 1993a). To reduce the leakage current to a few picoamps or less for measuring soft x-rays, one needs to cool the detector to about −30°C or below. This level of cooling can be easily achieved using low power Peltier refrigerators. These Peltier cooling stages are commercially available and

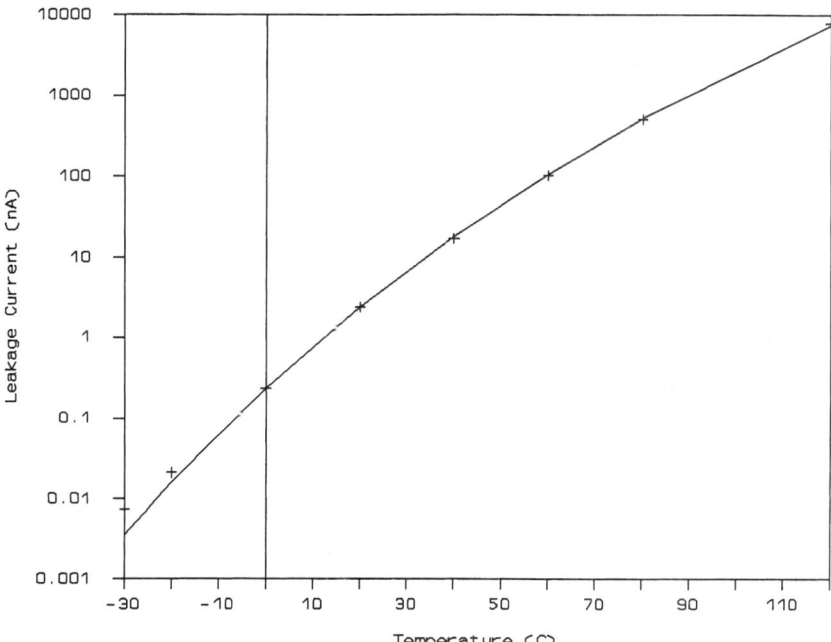

FIG. 20. Dependence of leakage current on temperature. Detector dimensions: 10 × 10 × 1.5 mm; bias voltage: 50 V. (Reproduced with permission from Butler et al., 1993c and Materials Research Society).

have been incorporated into HgI_2 based x-ray fluorescence instruments without difficulty (Natarajan, Bao, and Henderson, 1994, private communication). It is also straightforward to cool the preamplification stage with the same Peltier refrigerator, which further reduces the noise.

By reducing the leakage current to values of a few tens of picoamps or less, one can replace the resistive feedback preamplifiers with pulsed optical feedback. This retrofit further reduces the system noise, thus improving the energy resolution. In addition, the lowering of the leakage current permits the use of higher bias voltages. The higher voltages allow for more efficient charge collection since the transit time of the carriers is inversely proportional to the electric field inside the detector. The high electric field is particularly important for the hole carriers, which have a mobility–lifetime product that is typically about two orders of magnitude lower than that for the electron carriers (Lund et al., 1994a).

The x-ray entrance electrode for a CZT x-ray detector would be negatively biased, so that the slow and easily trapped holes have a shorter distance to traverse, thus improving the overall charge collection properties of the device. Figure 21 shows a schematic of the measurement setup for a cooled CZT x-ray detector with pulsed optical feedback (from Niemela and Sipila, 1994). With this type of configuration, x-ray spectra can be obtained and values have been reported for Fe-55

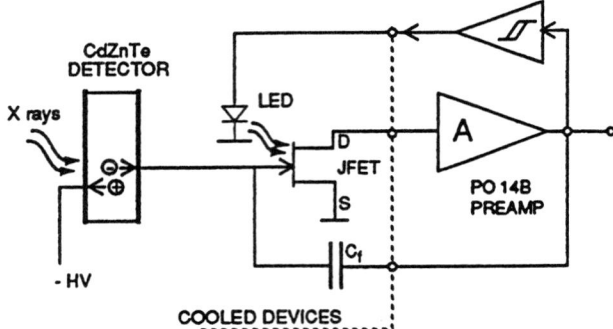

FIG. 21. Setup for CZT x-ray detector measurements (from Niemela and Sipila, 1994).

excitation at 5.9 keV (Butler et al., 1993a; Niemela and Sipila, 1994). Figure 22 shows a spectrum taken with a $2 \times 2 \times 2$ mm $Cd_{0.8}Zn_{0.2}Te$ detector produced by Aurora Technologies (Niemela and Sipila, 1994). The detector was cooled to $-40°C$, and a bias voltage and shaping time of -1000 V and 6 μsec were used. The Fe-55 FWHM was measured to be 240 eV, and the test pulser FWHM was 187 eV. An average peak to background ratio of 218:1 was obtained. The crystal contribution to the electronic noise was calculated to be 150 eV.

The energy resolution of the Fe-55 photopeak (at 5.9 keV) degrades with increasing temperature of the detector (Butler et al., 1993a). For example, at $-30°C$ the energy resolution increased to 282 eV at the optimal shaping time of 3 μsec (Niemela and Sipila, 1994). The pulser resolution was 231 eV, from which the crystal contribution was deduced to be 162 eV. There was also a reduction in the peak to background ratio to 170:1.

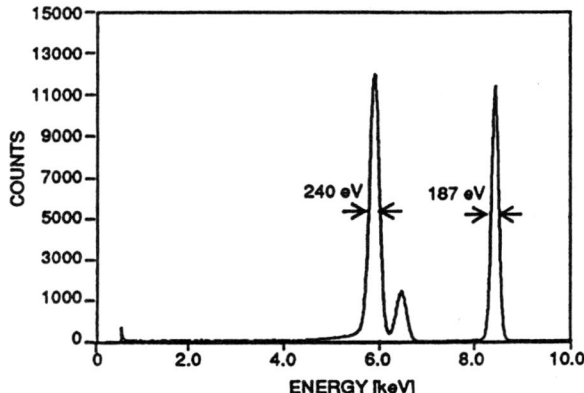

FIG. 22. A spectrum taken at $-40°C$, 6 μsec shaping time, and -1000 V bias. The Fe-55 FWHM is 240 eV, and the test pulser FWHM is 187 eV. The average peak to background ratio is 218:1 (from Niemela and Sipila, 1994).

9. CD$_{1-x}$ZN$_x$TE SPECTROMETERS FOR GAMMA AND X-RAY APPLICATIONS   367

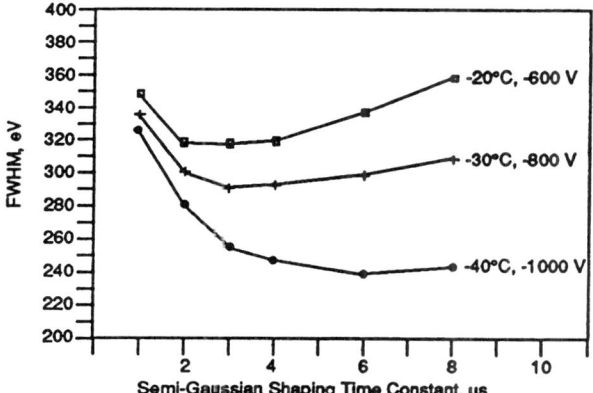

FIG. 23. Resolution of the Fe-55 line measured at several shaping time constants and temperatures (from Niemela and Sipila, 1994).

Figure 23 shows the energy resolution at temperatures of $-40°C$, $-30°C$, and $-20°C$ at several shaping time constants (Niemela and Sipila, 1994). Based on the measured results, it is clear that the optimal shaping time is longer at lower temperatures. Furthermore, by using even lower temperatures and corresponding reductions in the leakage current, still better values for the FWHMs are obtainable.

The soft x-ray detection capability of CZT was further demonstrated by placing samples of titanium and potassium in front of a $2 \times 2 \times 2$ mm^3 detector and using an Fe-55 source to excite inner shell transitions of the materials (Niemela and Sipila, 1994). The fluorescent spectrum of Ti and K are shown in Figs. 24 and 25, respectively, using a CZT detector cooled to $-30°C$. The emission lines at 3–5 keV are well resolved in the figures, and together with suitable software, quan-

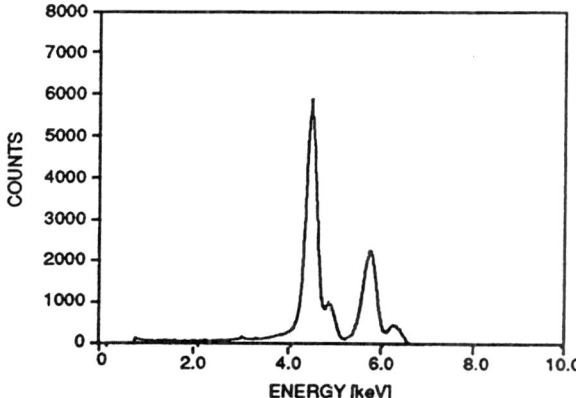

FIG. 24. Titanium K-lines (4.51 and 4.93 keV), obtained at $-30°C$ with Fe-55 excitation (from Niemela and Sipila, 1994).

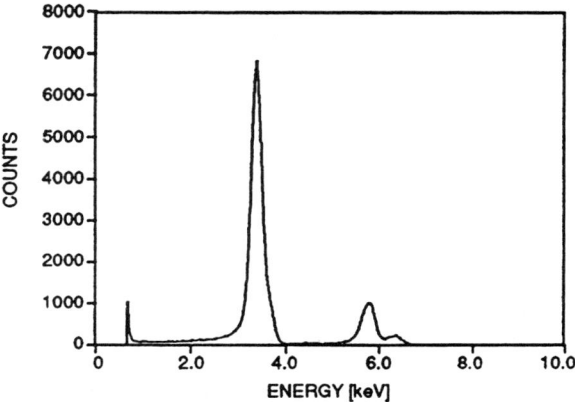

FIG. 25. Potassium K-lines (3.31 and 3.59 keV), obtained at $-30°C$ with Fe-55 excitation (from Niemela and Sipila, 1994).

titative analyses of the compositions of materials can now be achieved using cooled CZT detectors. The low energy x-ray cutoff depends on the system noise, which is below about 400 eV at operating temperatures of $-30°C$ or less.

These studies show that high resistivity CZT is a promising material for development of a portable, high resolution x-ray instrument that can operate without liquid nitrogen cooling. Some cooling of the detectors is needed to reduce the leakage current and enable the use of low noise pulsed optical feedback preamplifiers. These compact detectors have a disadvantage compared to mercuric iodide x-ray devices in that a greater degree of cooling is required; but CZT detectors also have some distinct advantages associated with their higher chemical stability, negligible room temperature vapor pressures, no low temperature phase transition, and potential for photolithographic device processing. Further work will determine the commercial viability of CZT detectors for measuring soft x-rays.

7. Detector Polarization

There exists a large number of reports concerning the polarization behavior of CdTe and HgI_2 detectors (see, for example, Malm and Martini, 1974; Siffert *et al.*, 1976; Zanio, Krajenbrink, and Montano, 1974; Hodgkinson, Howes, and Totterdell, 1979; Hage-Ali *et al.*, 1979; Sato *et al.*, 1995). Additional discussions on detector polarization can be found in this text within the chapters on characterization of CdTe and HgI_2.

During the 1970s and 1980s, considerable work was done on the polarization of CdTe:Cl gamma ray detectors. In most cases, the counting rates or the peak positions of the devices were found to change in time, and this aging was detrimental to the commercialization of CdTe:Cl detectors. Most of the problems with polarization were attributed to space charge effects associated with crystal growth and detector fabrication processes, and through improvements in CdTe:Cl technology, the polarization problems have for the most part been solved.

In view of the importance of polarization to commercialization, a thorough study of its possible occurrence in CZT detectors is needed. A report by Butler et al. (1992) showed that the counting rates for $Cd_{0.8}Zn_{0.2}Te$ detectors did not change with time over a period of 1 week. A similar measurement was performed by Lund et al. (1994a) on $Cd_{0.8}Zn_{0.2}Te$ and $Cd_{0.9}Zn_{0.1}Te$ detectors grown by high pressure Bridgman methods, and no polarization effects were exhibited. It appears that CZT detectors do not exhibit short term polarization problems; however, additional work must be conducted to investigate concerns with long term polarization behavior.

8. Temperature Dependence

Because of the wider bandgap of $Cd_{1-x}Zn_xTe$, one expects better high temperature performance of $Cd_{1-x}Zn_xTe$ gamma ray and x-ray spectrometers compared to CdTe. A comparative study of the temperature dependence of $Cd_{0.8}Zn_{0.2}Te$ and CdTe gamma ray detectors was completed by Egarievwe et al. (1994). Figure 26 shows the results of the logarithm of leakage current (I_F) divided by the square of the temperature (T) versus the reciprocal of the temperature for a $Cd_{0.8}Zn_{0.2}Te$ and a CdTe detector (Egarievwe et al., 1994). The resistivity of the $Cd_{0.8}Zn_{0.2}Te$ and CdTe detectors at 23°C is measured to be 8×10^{10} and

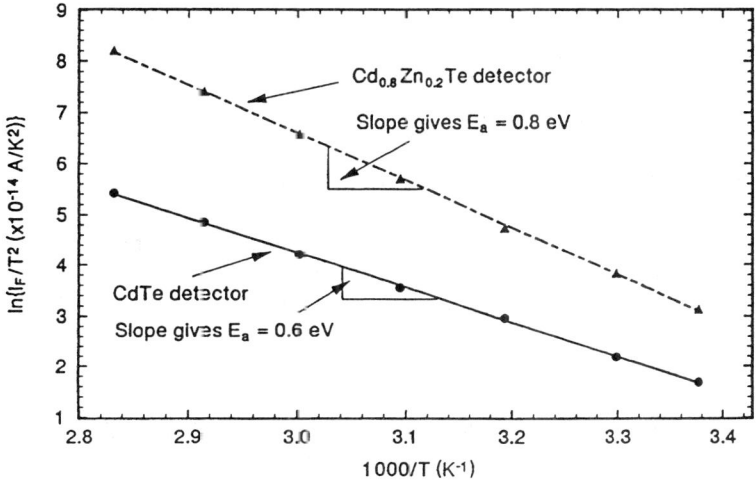

FIG. 26. Plot of $\ln(I_F/T^2)$ versus the reciprocal of the temperature for CdTe and $Cd_{0.8}Zn_{0.2}Te$ detectors from which the activation energies E_a can be obtained. The value of the external applied electric field was 250 V/cm for both detectors (from Egarievwe et al., 1994).

0.2 × 10^{10} Ω-cm, respectively. Activation energies of 0.8 and 0.6 eV are deduced for the $Cd_{0.8}Zn_{0.2}Te$ and CdTe detectors, respectively. An activation energy of 0.8 eV was also measured by Butler et al. (1993b) for $Cd_{0.8}Zn_{0.2}Te$ detectors with metallized Au and Pt contacts. Since the activation energy for an intrinsic semiconductor is one half of its bandgap, the $Cd_{0.8}Zn_{0.2}Te$ detector, which has a bandgap of about 1.6 eV, exhibits an intrinsic-like Fermi level, although it is most likely a highly compensated material. The bandgap of CdTe is about 1.45 eV, and since the activation energy is measured to be about 0.6 eV from the band edge, the measurement indicates that the Fermi level of the CdTe detector is governed by a deep level defect.

The performance of $Cd_{0.8}Zn_{0.2}Te$ and CdTe detectors was also reported by Egarievwe et al. (1994). For temperatures in the range of 25–70°C, the peak positions of $Cd_{0.8}Zn_{0.2}Te$ were found to shift much less than for CdTe. In addition, the $Cd_{0.8}Zn_{0.2}Te$ detectors exhibited little deterioration in energy resolution (for an 81 keV photopeak) for heating up to about 60°C. The current–voltage curves were nearly identical after cycling the detectors from room temperature up to the maximum temperature used in this study (about 85°C).

High temperature measurements of the response of $Cd_{0.9}Zn_{0.1}Te$ gamma ray detectors were conducted by Glick et al. (1993). In this study, a nuclear spectrum was taken on a 10 × 10 × 2 mm detector at a temperature of 27°C using an Am-241 gamma ray source. The detector was next heated to 85°C for a period of 15 hr. The detector was allowed to cool to room temperature, and a spectrum was again taken with an Am-241 source. The spectra were nearly identical before and after the heating at 85°C, indicating that the detectors can survive temperature cycling of this degree.

Similar experiments were performed by James et al. (1995) on CZT, CdTe, GaAs, PbI_2, and HgI_2 gamma ray detectors. For these candidate materials the results showed that CZT and PbI_2 detectors were the most stable at higher temperatures. Upon heating up to 80°C, only CZT was capable of good energy resolution at the higher energy photons (i.e., photopeaks above about 70 keV). For photon energies below about 70 keV, the noise associated with leakage current at 80°C significantly degraded the energy resolution in the CZT detectors, and poor or no energy resolution was typically observed at this elevated temperature (James et al., 1995). CdTe detectors were also found to show little degradation after temperature cycling, and fair energy resolution was obtainable at temperatures up to near 80°C, provided the shaping time was substantially reduced (Squillante, Olschner, and Lund, 1995).

The temperature response of a CdTe detector grown by HPB has been reported by C. J. Johnson et al. (1993). Nuclear spectra were obtained on a 10 × 7 × 2 mm^3 detector in the temperature range of −60 to 40°C. Figure 27 shows a three dimensional plot of pulse height analysis versus temperature using gamma rays from Am-241. A drastic reduction in the pulse height of the 59.5 keV photopeak is observed at temperatures less than about −25°C. The FWHM is found to be optimal at about −20°C, where the peak to valley ratio is also maximized. The unusual temperature dependence was associated with defect trapping properties

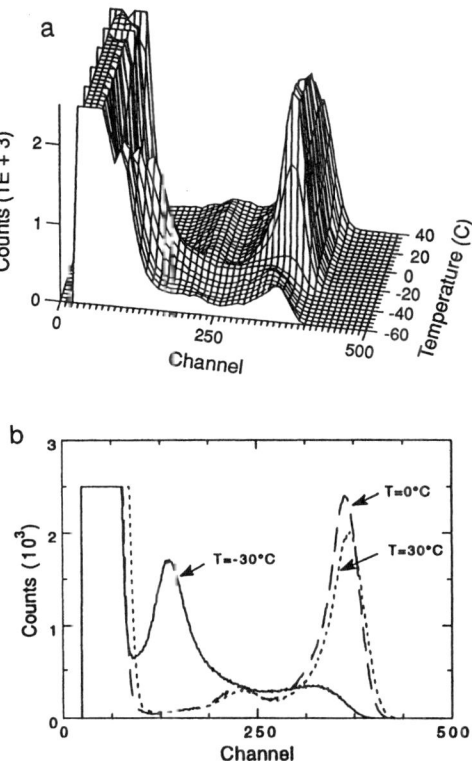

FIG. 27. ^{241}Am pulse height analysis from −60 to 40°C for a CdTe detector grown by high pressure Bridgman methods. The detector dimensions were 10 × 7 × 2 mm, and the bias voltage was 200 V. (Reprinted with permission from C. J. Johnson et al., 1993 and Materials Research Society.)

that were affecting the carrier lifetimes, mobilities, and other temperature dependent material properties. In that respect, other analytical techniques such as photoionization of trapped charge, photocapacitance, and thermally stimulated current may provide information on the binding energy, the capture cross section, and the concentration–lifetime product of the particular defect responsible for the reduced counts in the photopeak. Another possibility is that the internal field distribution in the CdTe detector is temperature dependent, and the voltage drop does not extend across the entire detector at temperatures below about −25°C. Figure 28 shows the temperature dependence of the electron and hole charge collection on voltage under LED illumination. Signal amplitudes are also plotted as a function of temperature for a 200 V bias. The optimal charge collection for electrons is found to be about −20°C, which is in agreement with the pulse height analysis investigations. At −60°C the maximum collected electron charge has decreased by about a factor of 2. The signal associated with the hole carriers has a maximum value between 0 and −20°C, and the amplitude of the hole signal is

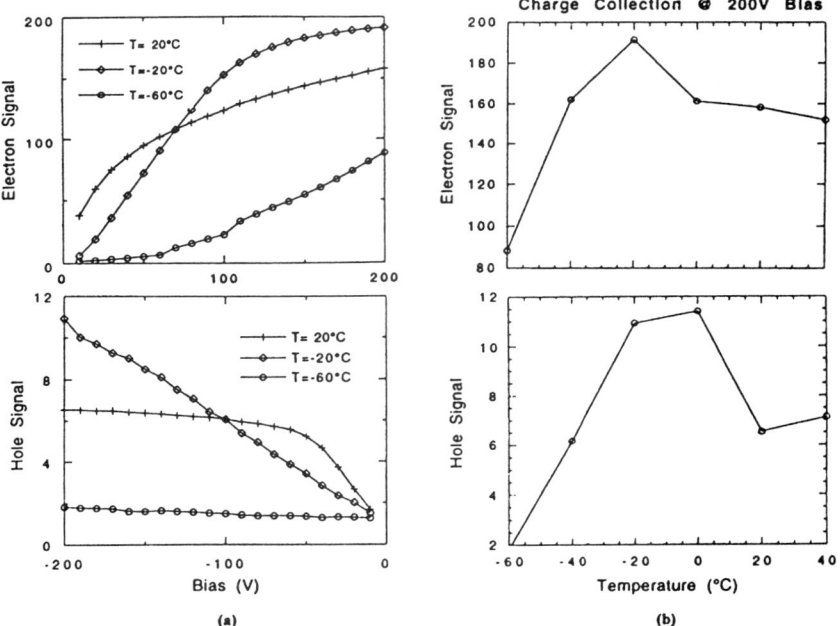

FIG. 28. Electron and hole signal versus bias from −60 to 40°C with LED excitation. Detector dimensions: 10 × 7 × 2 mm. (Reprinted with permission from C. J. Johnson *et al.*, 1993 and Materials Research Society.)

about an order of magnitude smaller at −60°C. The large reduction in the signal amplitude of holes in the temperature range of −40 to −60°C is most likely attributed to a deep hole trap in the bulk of the material. Further discussions are contained in the paper by C. J. Johnson *et al.* (1993).

In general, the noise threshold for CZT detectors, as well as the noise for other important semiconductor nuclear detectors, is a strong function of temperature. For most candidate materials, the high temperature performance is limited by the leakage current in the device, although other material properties can affect the stability and survivability after temperature cycling. For high temperature applications, it is clear that CZT detector technology has a strong potential, particularly for crystals with an even higher bulk resistivity. One way to achieve higher values for the resistivity is to increase the Zn content from 0.1–0.2 to even higher values. Work in this area is expected over the next few years.

9. Pulse Risetime Discrimination and Compensation

Gamma ray interactions in CZT detectors excite electron-hole pairs that are subsequently measured as they transit through an electric field applied externally across the device. Electrons have a higher drift mobility than holes, and as a result, electrons are swept from a detector much quicker than holes. Additionally, the holes have considerably shorter minority carrier lifetimes than electrons. Together, the lower velocities and short lifetimes create a situation in which electrons are "collected" much more efficiently than holes, resulting in position-dependent charge collection and deterioration in resolution (Knoll and McGregor, 1993). The energy resolution is degraded by the asymmetry of the photopeaks, the reduced peak-to-valley ratio, and the counts on the low-energy side of the photopeaks.

Since the mobility of electrons is much higher than holes, the risetime associated with motion of the electron carriers is much shorter than for hole carriers in CZT detectors. The difference in risetimes allows one to separate the components of each carrier type. Thus, corrective rise time discrimination and compensation by electronic means is a viable method to reduce the effects of hole trapping losses and thereby produce improved energy resolution.

A method using the shape of the detector output pulse has been employed to correct for "hole tailing" in the gamma ray signal of CdTe devices (Jones and Woollam, 1975; Richter and Siffert, 1992). Lund et al. (1994b) applied this technique to CZT gamma-ray detectors and was able to discriminate events associated with a small hole component. By discarding events associated with incomplete hole collection, exceptionally high energy resolutions were measured throughout the 60–1300 keV range. Figures 29, 30, and 31 shows spectra for Cs-137,

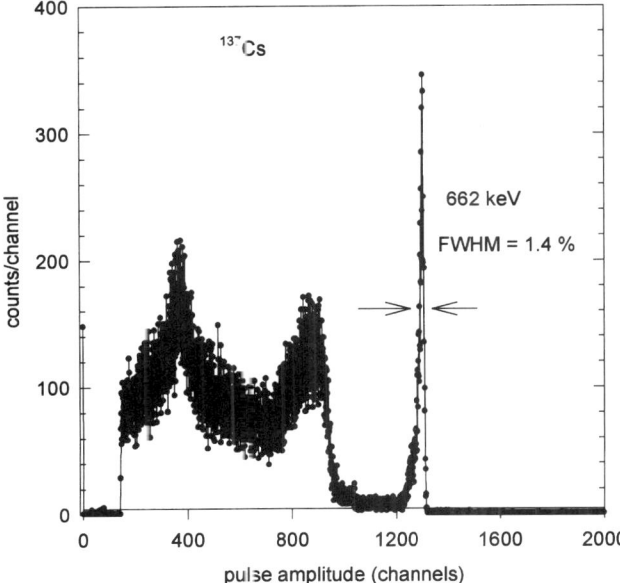

FIG. 29. Energy spectrum of Cs-137 with a $Cd_{0.9}Zn_{0.1}Te$ detector at room temperature using pulse risetime discrimination. Detector dimensions: $18 \times 18 \times 2$ mm; bias voltage: -1200 V (from Lund et al., 1994b).

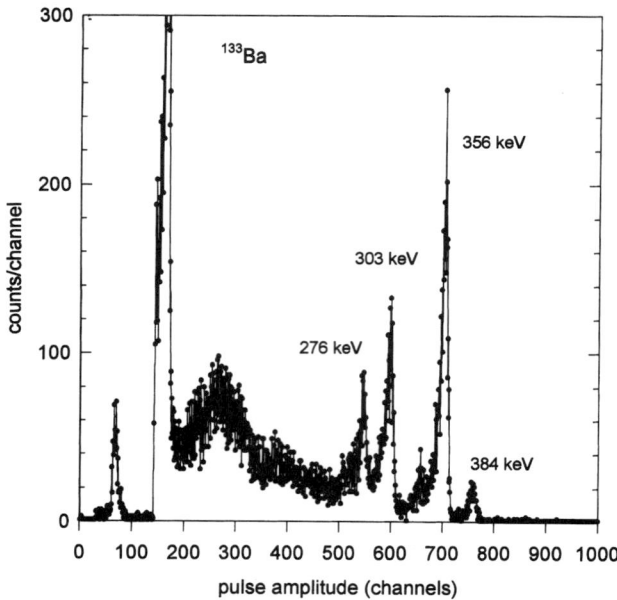

FIG. 30. Energy spectrum of Ba-133 with a $Cd_{0.9}Zn_{0.1}Te$ detector of room temperature using pulse risetime discrimination. Detector dimensions: $18 \times 18 \times 2$ mm; bias voltage: -1200 V (from Lund et al., 1994b).

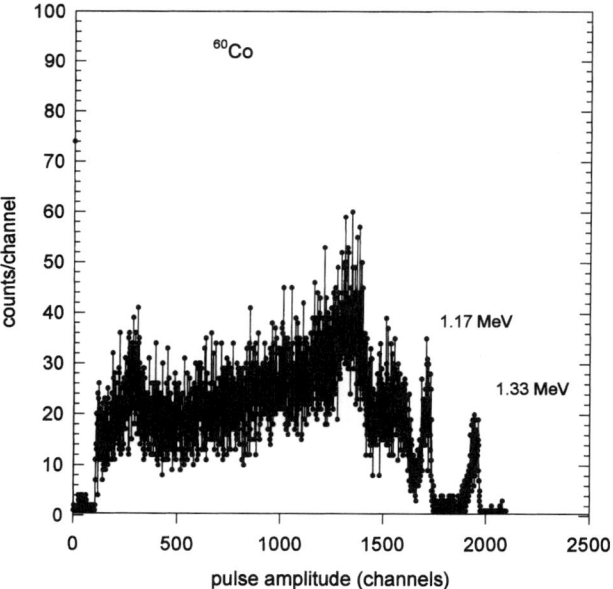

FIG. 31. Energy spectrum of Co-60 with a $Cd_{0.9}Zn_{0.1}Te$ detector at room temperature using pulse risetime discrimination. Detector dimensions: $18 \times 18 \times 2$ mm; bias voltage: -1200 V (from Lund et al., 1994b).

Ba-133, and Co-60, respectively (Lund et al., 1994b). The technique employed by Lund et al. (1994b) relies on the fact that holes travel at much lower velocities than the electrons, and thus the signals with a significant hole component can be rejected.

In situations where the counting rates are low, one would like to correct the pulses having incomplete hole collection, rather than reject them. The correction is achieved by measuring the ratio of the fast signal component (composed of both electron and hole motion) and the slow component (composed of only hole motion). The ratio reveals the interaction depth in the detector and the initial gamma ray energy, thus allowing for proper scaling of a correction factor that amplifies the reduced output pulse and compensates for the incomplete charge collection. In the case of poor electron-to-hole signal rise-time ratios, the entire gamma ray event will be rejected in order to preserve the energy resolution of the detector. Using a compensation algorithm for the pulse height which is dependent on location of the gamma event, one can correct the pulse height and achieve greatly improved gamma ray energy resolution without a significant loss in detector efficiency (Richter and Siffert, 1992; Eisen and Horovitz, 1994). There are numerous examples where an improvement in the energy resolution and a reduction in the low energy counts of the photopeaks allow for new radiation sensing applications (see, for example, Ruhter and Gumnick, 1994).

VI. Imaging Applications

Semiconductor arrays for direct detection and imaging of x-rays and gamma rays offer some distinct advantages for applications in nuclear medicine, simultaneous dual energy radiography, space sciences, national security, nonproliferation inspections, and other areas. For example, the use of CZT detector arrays offers the potential of improved spatial resolution, long term stability, spectrometer mode imaging, direct digitization, signal processing at the imaging plane, and other benefits. In addition, CZT has some major advantages over Si and Ge in that the devices operate at room temperature and the detectors and individual pixels can be made relatively small while maintaining interaction efficiency and energy resolution. Several monolithic x-ray and gamma ray imaging arrays have been fabricated and tested using CZT crystals (Schirato et al., 1993, 1994; Doty et al., 1993, 1994; Butler et al., 1992, 1993c, 1994; Lorenz et al., 1994; Polichar et al., 1992, 1994). These CZT arrays have been produced in linear and area geometries, and the results of each type of imager will be reviewed.

A linear x-ray scanner has been produced by Schirato et al. (1994) using a $Cd_{0.9}Zn_{0.1}Te$ crystal. Each array was 25.3 mm long with 32 independent detector elements, yielding a gap spacing of 0.8 mm from center to center. Gaps of less than 100 μm separated the individual pixels. Two of the 25.3 mm long linear arrays were mounted end to end in a printed circuit board header package, resulting in a 64-element detector module. Each pixel was directly coupled to the input

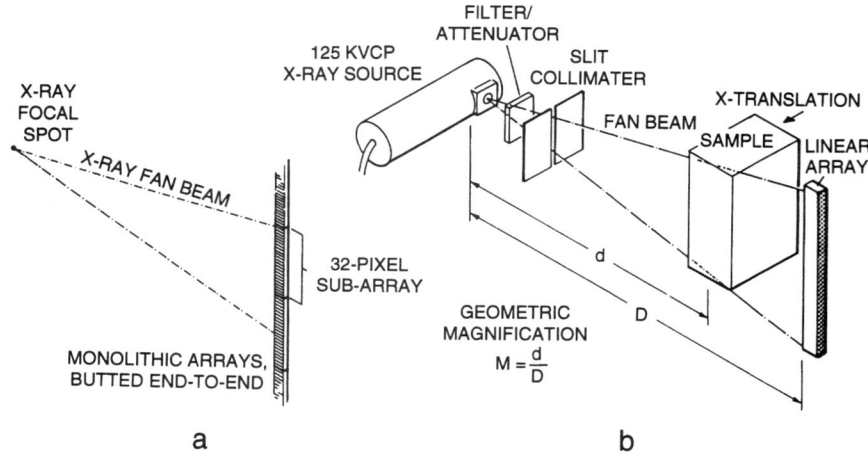

FIG. 32. Linear scan imaging: (a) monolithic arrays are butted end to end to produce a long linear imaging array, (b) imaging setup and relationship for geometric magnification for direction normal to sample translation (from Schirato et al., 1994).

FET of an independent charge sensitive preamplifier channel. Linear amplifiers shaped the pulses into bipolar pulses with a peaking time of 1 μsec. Pulses that exceeded a preset lower level discriminator were counted and used to produce transmission x-ray images. Figure 32a shows the linear scan x-ray imager, and Fig. 32b shows how transmission images were generated by translating a sample in front of the linear scanner at a controlled speed. The spatial resolution of the arrays and readout system for x-rays was measured by scanning a finely collimated fan beam along the array. The array exhibited low crosstalk or spatial spread and good uniformity of response (Schirato et al., 1994). Transmission images were taken using the x-ray linear scanner. The dose for the images ranged from 70 to 150 μR. Figure 33 shows an image of a mechanical timer with various materials having different thicknesses adjacent to the timer, and Fig. 34 shows a view of a medical foot phantom. These photographs show a considerable amount of detailed structure using a relatively low dose rate, demonstrating the photon counting capability of a CZT linear detector array system. Schirato et al. (1994) also used the energy discrimination ability of the system to produce dual energy images of samples without changing the source potential. This dual energy capability can be used to determine the material composition or to eliminate particular materials from the image. Readers interested in obtaining additional information on linear CZT detector arrays should refer to Polichar et al., 1992, 1994; Schirato et al., 1993; Butler et al., 1992, 1993c; Doty et al., 1993.

For many imaging applications one would prefer a two dimensional monolithic CZT detector array. Butler et al. (1993c) developed a 32 × 32 monolithic array of $Cd_{0.8}Zn_{0.2}Te$ detectors, each having an area of 1 mm^2. All amplifying and data processing electronics were installed into a small and portable box with dimen-

FIG. 33. X-ray transmission image of a mechanical timer made with 0.8 mm pitch CZT linear arrays (from Schirato et al., 1994).

sions of 25.4 × 25.4 × 10.2 cm. A laptop computer provided data control, image processing, visual display, and storage. Images were obtained using a phantom of the human thyroid gland (Butler et al., 1994). The study incorporated simulations of four abnormalities that could represent cancer within the gland. The smallest

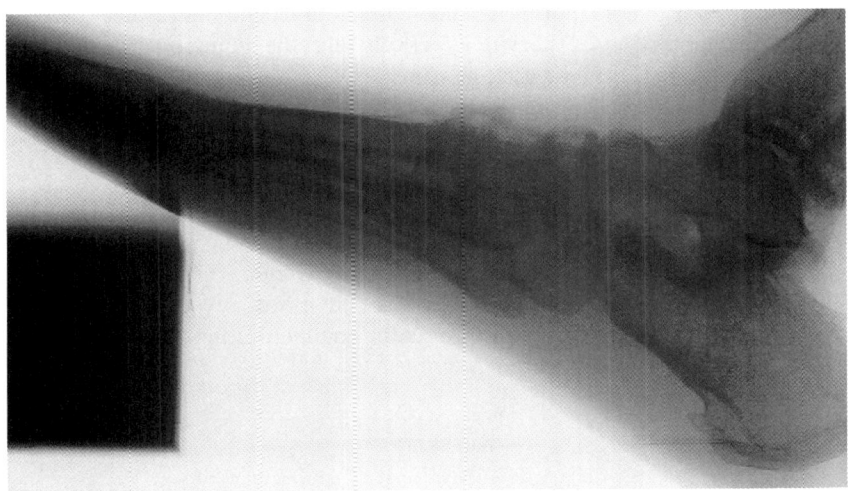

FIG. 34. X-ray transmission image showing side view of foot phantom with penetration of ankle (from Schirato et al., 1994).

abnormality was 6 mm in diameter. The imaging experiment was performed using the CZT detector array and then repeated with an Anger camera in place of the array. Butler *et al.* (1994) concluded that the CZT detector array was at least equal in performance to that of the Anger camera.

More recently, CZT detector arrays based on 48×48 monolithic elements have been fabricated (Doty *et al.*, 1994). The array was designed for use with silicon multiplexer IC readouts, which was first suggested by Gaalema (1985) and recently demonstrated by Barber *et al.* (1994) with cooled Ge arrays and single element CdTe and CZT detectors. The 2304 element array fabricated by Doty *et al.* (1994) represents a significant reduction in pixel dimensions over previous CZT arrays, and it is compatible with mature silicon readout technology. X-ray and gamma images acquired with the 48×48 array are anticipated in the near future.

VII. Future Work

CZT detector technology has been rapidly evolving since the first spectra was reported by Doty *et al.* (1992) and Butler *et al.* (1992). Although CZT crystals are clearly promising for many gamma ray detector and imaging applications, the energy resolution of devices is still limited by the transport of holes. Materials work should focus on the defects responsible for the relatively low hole–lifetime product, compared to that observed for the electron carriers. Once the particular defects limiting the hole transport have been identified, the crystal growth or detector processing procedures should be modified to reduce or eliminate the detrimental defects. At the same time, researchers should also be examining ways to obviate the problems with poor hole transport by modifying the pulse processing electronics. For example, pulse rise time correction techniques or other detector designs should be explored that will effectively relax the requirements on the hole charge collection. The fabrication of detectors and imaging arrays from future crystals with improved hole transport properties will greatly expand the applications of CZT sensor technology.

Acknowledgments

We would like to thank Richard Olsen, D. S. McGregor, Pat Doty, Jack Butler, Carl Johnson, Elgin Eissler, Richard Schirato, John Van Scyoc, Arnold Burger, Bill Hamilton, Ching Wang, and Ed Soria for communicating their data with us prior to the publication of their manuscripts.

References

Antolak, A. J., Bench, G. S., Morse, D., Pontau, A. E., Lund, J., and James R. B. (1995). Unpublished data.
Barber, H. B., Augustine, F. L., Barrett, H. H., Dereniak, E. L., Matherson, K. F., Perry, D. L., Venzon,

9. $Cd_{1-x}Zn_xTe$ Spectrometers for Gamma and X-Ray Applications

J. E., Woolfenden, J. M., and Young, E. T. (1994). *IEEE Conf. Rec. Nucl. Sci. Symp. Med. Imag. Conf.* **2**, 1381.
Bell, S. L., and Sen, S. (1985). *J. Vac. Sci. Technol.* **A3**, 112.
Bruder, M., Schwarz, H. H., Schmitt, R., Maier, H., and Mottler, M. D. (1990). *J. Cryst. Growth* **101**, 266.
Bruder, M., Figgemeier, H., Schmitt, R., and Maier, H. (1993). *Mater. Sci. Eng.* **B16**, 40.
Bube, C. H. (1967). In *The Physics and Chemistry of II–VI Compounds,* ed. M. Aven and J. S. Prener, 659. John Wiley & Sons, New York.
Burshtein, Z., Jayatirtha, H. N., Burger, A., Butler, J. F., Apotovsky, B., and Doty, F. P. (1993). *Appl. Phys. Lett.* **63**, 102.
Butler, J. F., Lingren, C. L., and Doty, F P. (1992). *IEEE Trans. Nucl. Phys.* **39**, 605.
Butler, J. F., Apotovsky, B., Niemela, A., and Sipila, H. (1993a). In *X-Ray Detector Physics and Applications,* Proceedings of the SPIE, **2009**, 121. SPIE, Bellingham, WA.
Butler, J. F., Doty, F. P., Apotovsky, B. Lajzerowicz, L., and Verger, L. (1993b). *Materials Science and Engineering* **B16**, 291.
Butler, J. F., Doty, F. P., Apotovsky, B., Friesenhahn, S. J., and Lingren, C. (1993c). In *Semiconductors for Room-Temperature Radiation Detector Applications* **302**, 497, ed. R. B. James, T. E. Schlesinger, P. Siffert, and L. Franks. Materials Research Society, Pittsburgh.
Butler, J. F., Friesenhahn, S.A., Lingren, C., Apotovsky, B., Simchon, N., Doty, F. R., Ashburn, W. L., and Dillon W. (1994). *IEEE Conf. Rec. Nucl. Sci. Symp. Med. Imag. Conf.* **1**, 565.
Cheuvart, P., El-Hanani, U., Schneider, D., and Triboulet, R. (1990). *J. Cryst. Growth* **101**, 270.
Dabrowski, A. J., Szymczyk, W. M., Kusmiss, J. H., Drummond, W., and Ames, L. (1983). *Nucl. Instr. Methods* **213**, 89.
Doty, F. P. (1994). EMIS Data Reviews. In Properties of Narrow Gap Cd-Based Compounds, ed. P. Capper. Series No. 10, p. 540. INSPEC, London.
Doty, F. P., Butler, J. F., Schetzina, J. F., and Bowers, K. A. (1992). *J. Vac. Sci. Technol.* **B10**, 1418.
Doty, F. P., Friesenhahn, S. J., Butler, J. F., and Hink, P. L. (1993). *Proc. SPIE* **1945**, 145.
Doty, F. P., Barber, H. B., Augustine, F. L., Butler, J. F., Apotovsky, B. A., Young, E. T., and Hamilton, W. (1994). *Nucl. Instr. and Meth* **A353**, 356.
Egarievwe, S. U., Salary, L., Chen, K. T., Burger, A., and James, R. B. (1994). In *Gamma-Ray Detector Physics and Applications* **2305**, 167. ed. E. Aprile. SPIE, Bellingham, WA.
Eisen, Y., and Horovitz, Y. (1994). *Nucl. Instr. and Meth.* **A353**, 60.
Eissler, E., Parnham, K., Glick, B., and Johnson, C. (1994). Unpublished data.
Fitzpatrick, B. J. (1988). *J. Cryst. Growth* **86**, 106.
Gaalema, S. (1985). *IEEE Trans. Nucl. Sci.* **NS-32**, 417.
Glick, B., Parnham, K., Eissler, E., Kramer, F., Sherbin, J., and Johnson, C. (1993). Unpublished data.
Glick, B., Eissler, E., Parnham, K., and Cameron, S. (1994). Presentation at the 1994 Symposium on Radiation Measurements and Applications.
Goorsky, M., Yoon, H., Schieber, M., and James, R. B. (1995). Unpublished data.
Hage-Ali, M., Stuck, R., Scharager, C., and Siffert, P. (1979). *IEEE Trans. Nucl. Sci.* **NS-26**, 281.
Hamilton, W. J., Rhiger, D. R., Sen, S., Kalisher, M. H., James, K., Ried, C. P., Gerrish, V., and Baccash, C. O. (1994). Proc. of IEEE Nucl. Sci. Sym. and Medical Imaging Conf., *IEEE Trans. Nucl. Sci.* **41**, 989.
Hazlett, T., Cole, H., Squillante, M. R., Entine, G., Sugars, G., Fecych, W., and Tench, O. (1986). *IEEE Trans. Nucl. Sci.* **NS-33**, 332.
Hecht, K. (1932). *Zeits. Phys.* **77**, 235.
Hess, R., DeAntonis, P., Morton, E. J., and Gilboy, W. B. (1994). *Nucl. Instru. and Meth.* **A353**, 76.
Hodgkinson, J. A., Howes, J. H., and Totterdell, D. H. J. (1979). *Nucl. Instr. and Meth.* **164**, 469.
James, R. B., Waymire, D., Burger, A., Egarievwe, S., Salary, L., Shah, K., and Lund, J. (1995). Unpublished.
Johnson, C. J., Eissler, E. E., Cameron, S E., Kong, Y., Fan, S., Jovanovic, S., and Lynn, K. G. (1993). In *Semiconductors for Room-Temperature Radiation Detector Applications* **302**, 463, ed. R. B. James, T. E. Schlesinger P. Siffert, and L. Franks. Materials Research Society, Pittsburgh.
Johnson, S. M., Sen, S., Konkel, W. H., and Kalisher, M. H. (1990). In *Physics and Chemistry of*

Mercury Cadmium Telluride and Novel IR Detector Materials, ed. D. G. Seiler, 1897. AIP, New York.
Johnson, S. M., Rhigar, D. R., Rosbeck, J. P., Peterson, J. M., Taylor, S. M., and Boyd, M. E. (1992). *J. Vac. Sci. Technol.* **B10**, 1499.
Jones, L. T., and Woollam, P. B. (1975). *Nucl. Instr. and Meth.* **124**, 591.
Kimura, H., and Komiya, H. (1973). *J. Cryst. Growth* **20**, 283.
Kikuma, I., Kikuchi, A., Yageta, M., Sekine, M., and Furukoshi, J. (1979). *J. Cryst. Growth* **98**, 302.
Knoll, G. F., and McGregor, D. S. (1993). In *Semiconductors for Room-Temperature Radiation Detector Applications* **302**, 3, ed. R. B. James, T. E. Schlesinger, P. Siffert, and L. Franks. Materials Research Society, Pittsburgh.
Kotani, T., and Tatsumi, M. (1987). Japanese patent 87-17 5969.
Lorenz, M., van Ryn, J., Merk, H., Markert, M., and Eisert, W. G. (1994). *Nucl. Instru. and Meth.* **A353**, 448.
Lund, J., Olsen, R., Schieber, M., and James, R. B. (1994a). Unpublished.
Lund, J., James, R. B., McGregor, D. S., and Olsen, R. (1994b). Unpublished.
Malm, H. L., and Martini, M. (1974). *IEEE Trans. Nucl. Sci.* **NS-21**, 322.
Meyer, B. K., Hofmann, D. M., Stadler, W., Salk, M., Eiche, C., and Benz, K. W. (1993). In *Semiconductors for Room-Temperature Radiation Detector Applications* **302**, 189, ed. R. B. James, T. E. Schlesinger, P. Siffert, and L. Franks. Materials Research Society, Pittsburgh.
Musa, A., and Ponpon, J. P. (1983). *Nucl. Instru. Meth.* **216**, 259.
Niemela, A., and Sipila, H. (1994). In Proc. of IEEE Nucl. Sci. Sym. and Medical Imaging Conf., *IEEE Trans. Nucl. Sci.* **41**, 1054.
Ohmori, H., Iwase, Y., and Ohno, R. (1993). *Mat. Sci. and Eng.* **B16**, 283.
Olsen, R., James, R. B., Antolak, A., and Wang, C. (1994a). In *Proceedings for 1994 International Nuclear Materials Management Meeting,* Vol. 23, p. 589. Institute of Nuclear Materials Management, Northbrook, Illinois.
Olsen, R., Lund, J., McGregor, D., James, R. B., Eissler, E., Parnham, K., and Johnson, C. (1994b). Unpublished data.
Parsons, A., Stahle, C. M., Lisse, C. M., Babu, S., Gehrels, N., Teegarden, B. J., and Shu, P. (1994). In *Gamma-Ray Detector Physics and Applications* **2305**, 121. ed. E. Aprile. SPIE, Bellingham, WA.
Polichar, R. M., Schirato, R. C., and Reed, J. H. (1992). In *X-Ray Detector Physics and Applications,* Proc. of the SPIE, **1736**, 43. SPIE, Bellingham, WA.
Polichar, R. M., Schirato, R. C., and Reed, J. H. (1994). *Nucl. Instr. and Meth.* **A353**, 349.
Raiskin, E., and Butler, J. F. (1988). *IEEE Trans. on Nucl. Science* **NS-35**, 82.
Richter, M., and Siffert, P. (1992). *Nucl. Instr. and Meth.* **A323**, 529.
Ruhter, W. D., and Gunnick, R. (1994). *Nucl. Instr. and Meth.* **A353**, 716.
Ryan, F. J., Shin, S. H., Edwall, D. D., Pasko, J. G., Khoshnevisan, M., Westmark, C. I., and Fuller, C. (1985). *Appl. Phys. Lett.* **46**, 274.
Sakai, E. (1982). *Nucl. Instru. Meth.* **196**, 121.
Sato, T., Sato, K., Ishida, S., Kiri, M., Hirooka, M., Yamada, M., and Kanamori, H. (1995). *IEEE Trans. Nucl. Sci.,* in press.
Schieber, M., Roth, M., James, R. B., Yao, W., and Goorsky, M. (1995). *J. Cryst. Growth* **146**, 15.
Schirato, R. C., Polichar, R. M., Reed, J. H., and Smith, S. T. (1993). Proc. SPIE **2009**, 48.
Schirato, R. C., Polichar, R. M., and Reed, J. H. (1994). In *X-Ray and UV Detectors,* ed. R. B. Hoover and Mark W. Tate, Vol. 2278, p. 47. SPIE, Bellingham, WA.
Schlesinger, T. E., Van Scyoc, J., Toney, J., David, D., and James, R. B. (1995). Unpublished.
Sen, S., and Stannard, J. E. (1993). In *Semiconductors for Room-Temperature Radiation Detector Applications* **302**, 391, ed. R. B. James, T. E. Schlesinger, P. Siffert, and L. Franks. Materials Research Society, Pittsburgh.
Sen, S., Konkel, W. H., Tighe, S. J., Bland, L. G., Sharma, S. R., and Taylor, R. E. (1988). *J. Cryst. Growth* **86**, 111.
Sher, H., and Montroll, E. W. (1975). *Phys. Rev.* **B12**, 2455.
Shin, S. H. (1985). *IEEE Trans. Nucl. Sci.* **NS-32**, 487.

Siffert, P. (1978). *Nucl. Instru. Meth.* **216**, 259.
Siffert, P., Berger, J., Scharager, C., Cornet, A., Stuck, R., Bell, R. O., Serreze, H. B., and Wald, F. V. (1976). *IEEE Trans. Nucl. Science* **NS-23**, 159.
Siffert, P., Hage-Ali, M., Stuck, R., and Cornet, A. (1977). *Rev. de Phys. Appl.* **12**, 335.
Soria, E., and James, R. B. (1995). Unpublished data.
Squillante, M., Olschner, F., and Lund, J (1995). Unpublished data.
Tiedje, T., and Rose, A. (1980). *Solid State Commun.* **37**, 49.
Triboulet, P., Neu, G., and Fotouhi, B. (1983). *J. Cryst. Growth* **65**, 262.
Trivedi, S. B., and Rosemeier, R. B. (1939). In *Optoelectronic Materials and Devices, Packaging, and Interconnects 2*, **154**, 94. SPIE, Bellingham, WA.
Vydyanath, H. R., Ellsworth, J., Kennedy, J. J., Dean, B., Johnson, C. J., Neugebauer, G. T., Sepich, J., and Liao, P. K. (1992). *J. Vac. Sci. Tech.* **10**, 1476.
Whited, R. C., and Schieber, M. M. (1979). *Nucl. Instru. Meth.* **162**, 113.
Woldseth, R. (1973). *X-Ray Energy Spectrometry*. KEVEX, Burlingame, CA.
Zanio, K. (1978). In *Semiconductors and Semimetals*, **13**, 103, ed. R. K. Willardson and A. C. Beer. Academic Press, New York.
Zanio, K., Krajenbrink, F., and Montano H. (1974). *IEEE Trans. Nucl. Sci.* **NS-21**, 315.

Chapter 10

Gallium Arsenide Radiation Detectors and Spectrometers

D. S. McGregor
SANDIA NATIONAL LABORATORIES
LIVERMORE, CALIFORNIA

J. E. Kammeraad
LAWRENCE LIVERMORE NATIONAL LABORATORY
LIVERMORE, CALIFORNIA

I. INTRODUCTION	386
II. BASIC PROPERTIES OF GaAs	391
1. Band Structure, Effective Mass, Density of States, and Intrinsic Carrier Concentration	391
2. Mobility and Velocity	394
3. Charge Carrier Lifetimes	395
4. Ionization Energy and the Fano Factor	399
5. Techniques of Material Growth	400
6. Compensation in Bulk GaAs	401
III. GENERAL DETECTOR OPERATION	403
IV. EPITAXIAL GaAs DETECTORS	407
1. Detector Configurations	408
2. Detector Performance	409
3. Discussion	413
V. BULK GaAs DETECTORS OPERATED IN QUANTUM PULSE MODE	414
1. Detector Configurations	414
2. I–V Characteristics	416
3. C–V Characteristics	418
4. Active Region Measurements	420
5. Radiation Measurements	425
6. Proposed Models for Observed Behavior	427
7. Discussion	431
VI. BULK GaAs PHOTOCONDUCTIVE DETECTORS OPERATED IN CURRENT MODE	432
VII. SUMMARY	437
References	437

List of Symbols

a	Unit cell dimension	HL1	Trap label for Cr deep acceptor
A	Detector contact area		
A^*	Effective Richardson constant	I_e	Induced current from electron motion
As_{Ga}	Arsenic substitutional on gallium site	I_h	Induced current from hole motion
B	Radiative recombination probability	I_{tot}	Total induced current
		J	Current density
CCE	Charge collection efficiency	J_{st}	Saturation current density
C_A	Detector active region capacitance	k	Boltzmann's constant
		k	Wave vector
C_S	Detector substrate capacitance	LEC	Liquid encapsulated Czochralski
C_T	Detector total parallel plate capacitance	LFO	Low frequency oscillation
		LPE	Liquid phase epitaxy
C_e	Electron capture coefficient	L_A	Detector active region length
C_h	Hole capture coefficient	L_S	Detector substrate region length
CEF	Carrier extraction factor		
χ	Electron affinity	λ_e^*	Electron mean free drift length
e_n	Electron emission rate		
E	Band energy	λ_h^*	Hole mean free drift length
E	Energy of interacting quantum of radiation	MBE	Molecular beam epitaxy
		m_0	Electron rest mass
E_g	Bandgap energy	m_c^*	Carrier effective mass
ϵ_s	Dielectric constant	m_e^*	Electron effective mass
ϵ	Average ionization energy	m_h^*	Hole effective mass
ϵ_α	Average ionization energy for alpha particles	m_Γ^*	Electron effective mass near the Γ conduction band edge
ϵ_β	Average ionization energy for beta particles	m_L^*	Electron effective mass in the L conduction band
EL2	Main deep donor level in undoped GaAs	m_X^*	Electron effective mass in the X conduction band
ESI	Epitaxial substrate interface	m_{lh}^*	Light hole effective mass
F	Fano factor	m_{hh}^*	Heavy hole effective mass
FWHM	Full-width at half-maximum	m_l^*	Longitudinal effective mass
ϕ	Schottky barrier height	m_t^*	Transverse effective mass
G_{ch}	Charge gain	μ_e	Electron drift mobility
Ga_{As}	Gallium substitutional on arsenic site	μ_h	Hole drift mobility
		n	Free electron concentration
h	Planck's constant	n_0	Equilibrium electron hole concentration
HB	Horizontal Bridgman		

10. GALLIUM ARSENIDE RADIATION DETECTORS AND SPECTROMETERS

Symbol	Description	Symbol	Description
n_i	Intrinsic carrier concentration	ϱ_e	Electron carrier extraction factor
n_m	Majority carrier concentration	ϱ_h	Hole carrier extraction factor
n_r	Refractive index	q	Unit electronic charge
N	Atom density	Q^*	Total induced charge
N	Diode ideality factor	$\langle Q^* \rangle$	Average total induced charge
N_b	Majority impurity concentration	$\langle (Q^*)^2 \rangle$	Average square of the total induced charge
N_a	Shallow acceptor concentration	Q_{in}	Total incident charge on detector
N_d	Shallow donor concentration	SI	Semiinsulating
N_{AA}	Deep acceptor concentration	σ_e	Electron capture cross section
N_{AA}^-	Ionized deep acceptor concentration	σ_h	Hole capture cross section
N_{DD}	Deep donor concentration	σ_{Q^*}	Standard deviation of induced charge
N_{DD}^+	Ionized deep donor concentration	σ_{q^*}	Fractional standard deviation of induced charge
N_C	Density of states in the conduction band	T	absolute temperature
N_V	Density of states in the valence band	t_e	Electron transit time
		t_h	Hole transit time
N_T	Total recombination center or trap density	τ	Minority carrier lifetime
		τ_e^*	Electron mean free drift time
N_0	Total number of excited electron–hole pairs	τ_h^*	Hole mean free drift time
		V_{bi}	Contact or junction built-in potential
N_e	Number of excited electrons		
N_h	Number of excited holes	V_B	Bias voltage
ω	Modulation frequency	V_R	Reverse bias voltage
p	Free hole concentration	VGF	Vertical gradient freeze
p_0	Equilibrium free hole concentration	VZM	Vertical zone melt
		v_e	Electron velocity
R_r	Radiative recombination rate	v_h	Hole velocity
R_{srh}	Shockley–Read–Hall recombination rate	x_e	Transit distance to move an electron from its point of origin to the active region boundary
R_S	Detector substrate resistance		
R_A	Detector active region resistance	x_h	Transit distance to move a hole from its point of origin to the active region boundary
ρ	Material electrical resistivity		
ρ_i	Material intrinsic resistivity		

W	Detector active region width for planar structure	Z	Detector impedance
		$\bar{Z}$	Average atomic number

I. Introduction

The general requirement for room temperature operation of a semiconducting material as a nuclear detector and spectrometer is a relatively large bandgap energy so that thermal generation of charge carriers is kept to a minimum. Conversely, the requirement for a high resolution gamma ray spectrometer is a small bandgap energy so that a large number of electron–hole pairs are created for an absorbed quantum of ionizing radiation. Therefore, a compromise is necessary if a semiconducting material is to be considered for a room temperature operated radiation spectrometer. There is indication that the optimum bandgap energy for such a spectrometer should be near 1.4 to 1.5 eV (Swierkowski and Armantrout, 1975; Armantrout *et al.*, 1977). The material under consideration should also have a relatively high average atomic number if used in gamma ray spectroscopy to increase the gamma ray interaction probability. High charge carrier mobilities and long charge carrier lifetimes are also needed to ensure efficient charge carrier extraction and minimal effects from position dependent charge collection (Day, Dearnaley, and Palms, 1967; Knoll and McGregor, 1993). Gallium arsenide (GaAs) is a semiconductor with several material characteristics that are well balanced for such a detector (Sakai, 1982; Cuzin, 1987).

GaAs is a III–V compound semiconductor that has received attention as a possible room temperature operated radiation detector since the early 1960s (Harding *et al.*, 1960; Mayer, 1962; Barraud, 1963b). GaAs radiation detectors have been studied by a multitude of different research groups for over three decades (see Table I), and GaAs detectors have been designed to operate in individual quantum pulse mode and current mode. "Individual quantum pulse mode" refers to the identification of individual quanta of radiation (Knoll, 1989; Knoll and McGregor, 1993), whereas "current mode" refers to the measurement of a short intense burst of radiation for timing information (Kammeraad *et al.*, 1992). Early researchers listed GaAs as having the best compromise in material characteristics for a room temperature gamma ray spectrometer (Dearnaley and Northrop, 1964), and GaAs was the first compound semiconductor to demonstrate high resolution at room temperature for gamma rays (Eberhardt, Ryan, and Tavendale, 1970). The average density and the average atomic number for GaAs are approximately the same as those for Ge, presently the most commonly used semiconductor for high resolution gamma ray spectroscopy. Hence, the gamma ray interaction probabilities per unit path length for Ge and GaAs are approximately the same. This chapter addresses the present status of GaAs as a radiation detector. Some discussion of the material characteristics are provided with ample references for further information.

TABLE I
GaAs Detectors

Material	Year	Comments	Reference
SI bulk GaAs	1960	Conductivity counters	(Harding et al., 1960)
	1962	Conductivity counters	(Mayer, 1962)
	1963	Conductivity counters, Schottky diodes, nonuniform electric fields	(Barraud, 1963b)
	1964		(Dearnaley and Northrop, 1964)
	1967		(Akutagawa et al., 1967)
	1971	Fe and Cr doped material, energy resolution >20%	(Afanas'ev, 1971)
	1972		(Kobayashi et al., 1972)
	1985	p-i-n diodes, noisy, high leakage currents	(McGregor et al., 1985)
	1990	Schottky diodes, 5% FWHM for α, radiation hardness tests	(Bertin et al., 1990)
	1991	p-i-n strip detectors, low CCE	(Buttar et al., 1991)
	1991	Schottky diodes	(Sumner et al., 1991)
	1992	Schottky diodes, 2.5% FWHM for α, 33% FWHM for 122 keV γ, nonuniform electric fields	(McGregor et al., 1992a)
	1992	Schottky diodes	(Beaumont et al., 1992)
	1992	Schottky diodes	(McGregor et al., 1992b,c)
	1992	Schottky diodes, 18% FWHM for 122 keV γ	(Benz et al., 1992)
	1992	Schottky diodes	(Sumner et al., 1992)
	1992	Schottky diodes, radiation hardness tests	(Karpinski et al., 1992)
	1992	Schottky diodes, leakage current tests	(Grant and Sumner, 1992)
	1992	Conductivity counter arrays	(Spooner et al., 1992)
	1992	p-i-n strip detectors	(Ashman et al., 1992)
	1993	Schottky diodes, radiation hardness tests	(Beaumont et al., 1993a,b)
	1993	Nonuniform electric fields	(Beaumont et al., 1993c)
	1993	Schottky diodes, p-i-n diodes, nonuniform electric fields, 13% FWHM for 122 keV γ	(Berwick et al., 1993a,b)

continues

TABLE I
CONTINUED

Material	Year	Comments	Reference
	1994	Schottky diodes	(Bencivelli et al., 1994)
	1994	Schottky diodes, nonuniform electric fields, electric field models	(Beaumont et al., 1994a,b)
	1994	electric field calculations	(McGregor et al., 1994a,b)
	1994		(Kubicki et al., 1994)
	1994	p-i-n diodes	(Dogru et al., 1994)
n-type bulk GaAs			
	1966	Schottky diodes, 7.7% FWHM for α	(Kobayashi and Takayanagi, 1966)
	1971	Avalanche detectors, noisy, poor α resolution	(Kobayashi and Takayanagi, 1971)
	1991	Conductivity counters	(Spooner et al., 1991)
	1992	Conductivity counter arrays	(Spooner et al., 1992)
LPE GaAs			
	1970	Schottky diodes, 60–80 μm thick, 2–3% FWHM for 122 keV γ	(Eberhardt et al., 1970)
	1971	Schottky diodes, 60–80 μm thick, 2–3% FWHM for 122 keV γ	(Eberhardt et al., 1971a,b)
	1971	≈50 μm thick	(Hesse and Gramann, 1971)
	1972	Schottky diodes, 0.43% FWHM for α, 12.3% FWHM for 122 keV γ	(Kobayashi and Sugita, 1972)
	1972	Schottky diodes	(Tavendale and Lawson, 1972)
	1972	Schottky diodes, 70 μm thick, 0.43% FWHM for α, 3.8% FWHM for 122 keV γ	(Kobayashi et al., 1972)
	1972	Schottky diodes, 30–120 μm thick, 0.36% FWHM for α, 5% FWHM for 60 keV γ	(Gibbons and Howes, 1972)

1972	Schottky diodes	(Hesse et al., 1972)
1973	Schottky diodes, biomedical probes	(Kobayashi et al., 1973)
1976	Schottky diodes, Cr doped LPE, 3.1% FWHM for 122 keV γ, double sided structures	(Kobayashi et al., 1976)
1992	Schottky diodes, 200 μm thick, 4.5% FWHM for 60 keV γ	(Alexiev and Butcher, 1992)
1993	π–ν junction detectors	(Chmill et al., 1993)
1994	π–ν junction detectors	(Chmill et al., 1994)
1994	Schottky diodes, 130 μm thick	(Sumner et al., 1994)

VPE GaAs

1971	very thin ($\approx 10 \mu$m)	(Hesse and Gramann, 1971)
1972		(Kobayashi et al., 1972)
1972	very thin ($\approx 8 \mu$m), 4.4% FWHM for 60 keV γ	(Hesse et al., 1972)
1993	π–ν junction detectors	(Chmill et al., 1993)
1994	π–ν junction detectors	(Chmill et al., 1994)

Neutron damaged

1986	photoconductors, resolving time < 100 psec	(Wagner et al., 1986)
1986	photoconductors, resolving time < 40 psec	(Wang et al., 1986)
1989	photoconductors, resolving time < 30 psec	(Wang et al., 1989)
1989	photoconductors, resolving time < 60 psec	(Friant et al., 1989)
1991	photoconductors	(Wang et al., 1991)
1992	photoconductors	(Kammeraad et al., 1992)
1992	photoconductors	(Moy et al., 1992)
1993	photoconductors, 140 psec resolving time	(Brullot et al., 1993)

TABLE II
Properties of GaAs at 300 K

Property	Symbol	Value	Reference
Crystal structure		Zincblende	(Blakemore, 1982; Sze, 1981)
Unit cell dimension (Å)	a	5.6533	(Sze, 1981; Tiwari, 1992)
Atom density (cm^{-3})	N	4.42×10^{22}	(Sze, 1981)
Density (g/cm^3)		5.317	(Blakemore, 1982)
Average atomic number	$\bar{Z}$	32 (31 Ga, 33 As)	
Dielectric constant	ϵ_s	12.85	(Blakemore, 1982)
		13.1	(Sze, 1981)
Effective density of states (cm^{-3})			
Conduction band	N_C	4.7×10^{17}	(Sze, 1981)
		4.21×10^{17}	(Blakemore, 1982)
Valence band	N_V	7.0×10^{18}	(Sze, 1981)
		9.51×10^{18}	(Blakemore, 1982)
Electron effective mass			
Γ_6 valley	m^*_Γ	$0.067\ m_0$	(Sze, 1981; Tiwari, 1992)
		$0.0632\ m_0$	(Blakemore, 1982)
L_6 valley	m^*_L	$0.55\ m_0$	(Blakemore, 1982)
X_6 valley	m^*_X	$0.85\ m_0$	(Blakemore, 1982)
Hole effective mass			
Heavy holes	m^*_{hh}	$0.5\ m_0$	(Blakemore, 1982)
		$0.45\ m_0$	(Sze, 1981)
Light holes	m^*_{lh}	$0.088\ m_0$	(Blakemore, 1982)
		$0.082\ m_0$	(Sze, 1981)
Drift mobility (cm^2/V-s)			
Electron (Γ_6 valley)	μ_e	>8000	(Blakemore, 1982; Sze, 1981; Tiwari, 1992)
Hole	μ_h	400	(Sze, 1981; Tiwari, 1992)
Energy gap (eV)	E_g		
Γ_6 valley		1.424	(Sell et al., 1974)
L_6 valley		1.707	(Aspnes, 1976)
X_6 valley		1.899	(Aspnes, 1976)
Intrinsic carrier concentration (cm^{-3})	n_i	1.79×10^6	(Sze, 1981)
		2.25×10^6	(Blakemore, 1982)
Intrinsic resistivity (Ω-cm)	ρ_i	$\approx 10^8$	(Sze, 1981)
Electron affinity (eV)	χ	4.07	(Sze, 1981)
Intrinsic Fermi energy level (eV)	$E_{Fi} - E_V$	0.752	(Blakemore, 1982)
Refractive index in visible range	n_r	3.347	(Tiwari, 1992)
Minority carrier lifetimes (seconds)	τ		
High purity LPE GaAs		4×10^{-5}	(Ryan and Eberhardt, 1972)
Bulk GaAs (low doped)		$10^{-9} - 10^{-8}$	(Hwang, 1972; Ritter et al., 1987; Dudenkova and Nikitin, 196)
Bulk GaAs (high doped $> 5 \times 10^{18}$/cm^3)		$< 10^{-9}$	(Hwang, 1972)
Neutron irradiated GaAs		$2 \times 10^{-11} - 10^{-10}$	(Wagner et al., 1986; Wang et al., 1989; Friant et al., 1989)
Fano factor (at 125 K)	F	0.18	(Eberhardt et al., 1971)
Average ionization energy (eV/e-h pair)			
Alpha particles	ϵ_α	4.3 eV	(Eberhardt et al., 1971; Kobayashi et al., 1972b)
Electrons	ϵ_β	4.6 eV	(Wittry and Kyser, 1965; Kobayashi et al., 1972)

II. Basic Properties of GaAs

The intrinsic and extrinsic properties of GaAs have been the topics of numerous books and journal articles and an abundance of literature on the material can be readily found. Therefore, only the major properties of GaAs that affect its performance as a radiation detector are discussed in this section. The interested reader will find a review article by Blakemore (1982) most helpful in providing additional information concerning the intrinsic properties of GaAs. Also available are several books devoted to (or having sections devoted to) GaAs materials, characterization, and device physics (Sze, 1981; DiLorenzo and Khandelwal, 1982; Willardson and Beer, 1984; Howes and Morgan, 1985; Shur, 1987; Look, 1989; Williams, 1990; Tiwari, 1992). Table II in the text lists several of the basic material properties of GaAs.

1. BAND STRUCTURE, EFFECTIVE MASS, DENSITY OF STATES, AND INTRINSIC CARRIER CONCENTRATION

GaAs single crystals form in the zincblende (sphalerite) lattice with unit cell dimension (a) equal to 5.653 Å. The average atomic number for stoichiometric GaAs is 32 (Ga 31, As 33) and the material density is 5.3174 ± 0.0026 g/cm^3 (Madelung, 1964). The variation of energy with reduced wave vector for the upper valence bands and lower conduction bands is shown in Fig. 1 (Blakemore, 1982).

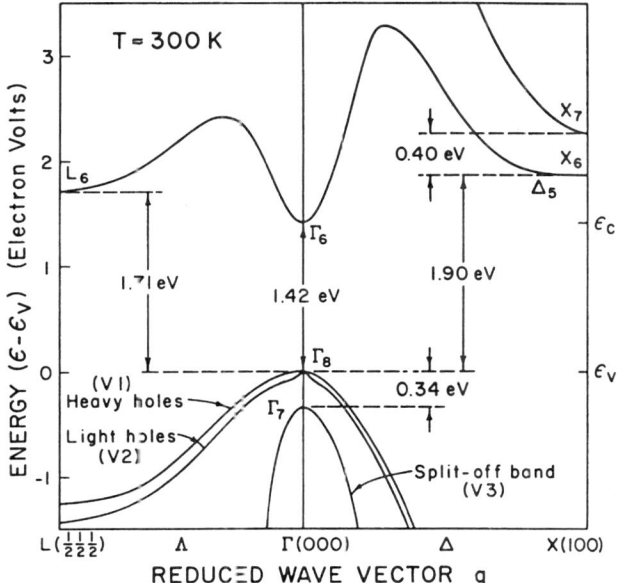

FIG. 1. Energy variation as a function of the reduced wave vector near the valence band maxima and conduction band minima. (Reprinted with permission from Blakemore, 1982 and the American Institute of Physics.)

The smallest energy gap between the valence band and the conduction band is denoted as the bandgap (E_g). It is evident from Fig. 1 that GaAs has a direct bandgap located at the Brillouin zone center (Γ (000)). The bandgap energy as a function of temperature is well approximated by (Thurmond, 1975)

$$E_g(T) = 1.519 - \frac{5.405 \times 10^{-4} T^2}{(T + 204)} \text{ eV} \tag{1}$$

where T is the absolute temperature. From Eq. (1), the room temperature (300 K) bandgap energy (E_g) is 1.422 eV. Two other conduction band minima occur at the L_6 ($\frac{1}{2}\frac{1}{2}\frac{1}{2}$) valley and at the X_6(100) valley. At 300 K, the conduction band minima at L_6 and X_6 appear at 1.71 eV and 1.90 eV above the valence band maximum, respectively.

In an intrinsic semiconductor, charge carriers are produced by excitation of electrons from the valence band into the conduction band. Electrons can move freely in the conduction band, much like electrons in vacuum; however, the electrons are acted upon by the periodic potential of the crystal. As a result, the effective mass of an electron is different from that of an electron in free space. The effective mass concept allows for the conduction electrons and holes in a semiconductor to be treated as classical charged particles. The carrier effective mass defined by the band curvature is

$$m_c^* = \left(\frac{h}{2\pi}\right)^2 \left(\frac{d^2 E}{dk^2}\right)^{-1} \tag{2}$$

where h is Planck's constant, k represents the wave vector, and E is the band energy. Near the Γ_6 band edge, the electron effective mass at 300 K is $0.063 m_0$, where m_0 is the mass of a free electron (Blakemore, 1982). Although the direct Γ_6 conduction band deviates from a parabolic potential, the effective density of states for nondegenerate material is generally represented by

$$N_C = 2\left(\frac{2\pi m_\Gamma^* kT}{h^2}\right)^{3/2} \tag{3}$$

where k is Boltzmann's constant, h is Planck's constant, and m_Γ^* is the effective mass of an electron near the direct conduction band minimum. Blakemore (1982) includes a correction factor for the conduction band density of states, hence Eq. (3) becomes

$$N_C = 2\left(\frac{2\pi m_\Gamma^* kT}{h^2}\right)^{3/2} (1 - 1.93 \times 10^{-4} T - 4.19 \times 10^{-8} T^2), \tag{4}$$

$(100 < T < 1200 \text{ K})$.

Electrons in the upper indirect conduction band valleys have longitudinal and transverse effective mass components denoted m_l^* and m_t^*, respectively. For the L_6 ($\frac{1}{2}\frac{1}{2}\frac{1}{2}$) valley, the low temperature electron effective masses have been estimated as $m_l^* \approx 1.9m_0$ and $m_t^* \approx 0.075m_0$ (Aspnes and Studna, 1973; Aspnes, 1976). The overall low temperature electron effective mass associated with the four ellipsoid L_6 valley is estimated as (Aspnes and Studna, 1973; Aspnes, 1976)

$$m_L^* = [16 m_l^* (m_t^*)^2]^{1/3} \approx 0.56 m_0. \tag{5}$$

The electron effective mass in the L_6 valley decreases only slightly at room temperature (300 K) to $m_L^* \approx 0.55 m_0$ (Blakemore, 1982). For the $X_6(100)$ valley, the electron effective masses have been estimated as $m_l^* \approx 1.9 m_0$ and $m_t^* \approx 0.19 m_0$ (Aspnes, 1976, 1977). Assuming a three ellipsoid model, the electron effective mass is estimated to be

$$m_X^* = [9 m_l^* (m_t^*)^2]^{1/3} \approx 0.85 m_0. \tag{6}$$

Three valence bands are apparent and all have their maxima at the Brillouin zone center. The heavy hole band (V_1 at Γ_8) and the light hole band (V_2 at Γ_8) are degenerate at the zone center. The lower splitoff valence band (V_3) is separated by 0.34 eV from the heavy hole and light hole bands at the zone center, and it is largely ignored for most calculations since hole occupancy is negligible at room temperature. Blakemore (1982) reports the effective mass at 300 K for light holes as $m_{lh}^* = 0.088 m_0$ and for heavy holes as $m_{hh}^* = 0.50 m_0$. The effective density of available states in the valence band is found through the addition of the two degenerate upper valence band state densities. Hence,

$$N_V = 2\left(\frac{2\pi kT}{h^2}\right)^{3/2} [(m_{lh}^*)^{3/2} + (m_{hh}^*)^{3/2}] = 2\left(\frac{2\pi kT m_h^*}{h^2}\right)^{3/2} \tag{7}$$

where m_{lh}^* is the light hole effective mass, m_{hh}^* is the heavy hole effective mass, and m_h^* is the density of states hole effective mass represented by

$$m_h^* = [(m_{lh}^*)^{3/2} + (m_{hh}^*)^{3/2}]^{2/3}. \tag{8}$$

Assuming a negligible population of holes in the splitoff band, the density of states effective mass as calculated from Eq. (8) is

$$m_h^* = [(0.088 m_0)^{3/2} + (0.50 m_0)^{3/2}]^{2/3} = 0.524 m_0. \tag{9}$$

For nondegenerate material used for nuclear detectors, the concentration of free electrons in the conduction band is

$$n_0 = N_C \exp\left(\frac{E_F - E_C}{kT}\right) \tag{10}$$

and the concentration of free holes in the valence band is

$$p_0 = N_V \exp\left(\frac{E_V - E_F}{kT}\right). \tag{11}$$

The intrinsic carrier concentration can be found with

$$n_i = \sqrt{n_0 p_0} = \sqrt{N_C N_V} e^{-E_g/2kT} \tag{12}$$

which clearly demonstrates the relation between the bandgap energy and the intrinsic free carrier concentration. The effective density of states in the conduction and valence bands also have an effect on n_i; however, mainly the bandgap energy predetermines whether or not the intrinsic free carrier concentration is suitably low for room temperature operation.

At a reported room temperature bandgap energy of 1.424 eV (Sell, Casey, and Wecht, 1974), GaAs has a significantly larger bandgap energy than Si (1.125 eV) and Ge (0.663 eV) (Thurmond, 1975; Sze, 1981). Thus the room temperature thermal leakage current is greatly reduced for GaAs compared to Si and Ge, and GaAs detectors do not need to be cooled during operation (as do Ge detectors). Blakemore (1982) reports the effective density of states in the conduction band at 300 K to be 4.21 × 10^{17}/cm^3 and the effective density of states in the valence band to be 9.51 × 10^{18}/cm^3. With a room temperature bandgap energy of 1.424 eV (Sell et al., 1974), Eq. (12) yields the intrinsic carrier concentration for GaAs as 2.4 × 10^6/cm^3, which is similar to the values reported by Blakemore (1982) and Sze (1981). The resulting intrinsic material resistivity is greater than 10^8 Ω-cm.

2. MOBILITY AND VELOCITY

Application of an electric field allows for electrons in the lower Γ_6 valley to gain energy and transfer to the upper L_6 valley (Ridley and Watkins, 1961; Hilsum, 1962), resulting in effective mass changes as well as mobility changes. Electrons in the direct Γ_6 valley can have high mobilities (above 8000 cm^2/V-sec at room temperature), resulting in high velocities at low electric fields. Application of higher electric fields results in the scattering of electrons into the indirect L_6 valley, in which the electron velocity is reduced due to their increase in effective mass and decrease in mobility. The transferred electron effect gives rise to a condition in which the average group velocity of the electrons decreases as the population of electrons in the L_6 valley increases to a considerable fraction of the overall free electron population. At certain applied bias voltages, the transferred electron effect results in the appearance of oscillations (known as the "Gunn effect") due to the instability inherent with negative differential resistance (Gunn, 1963). For stoichiometric GaAs, the electric field at which the average electron

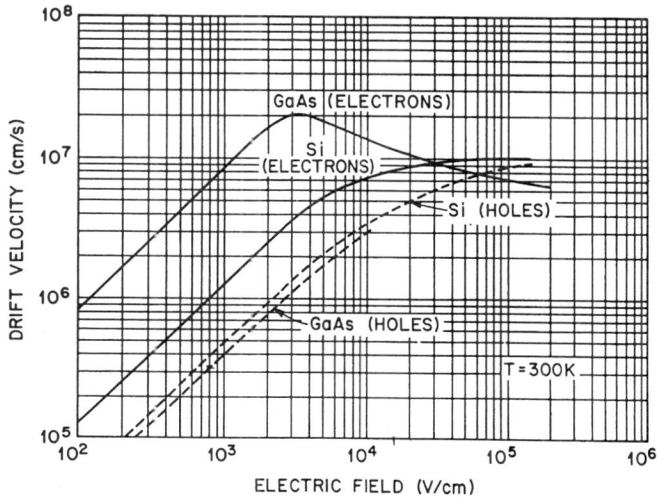

FIG. 2. Electron and hole drift velocities as a function of electric field for GaAs and Si. (From Sze, 1985, © 1985. Reprinted with permission from John Wiley & Sons, Inc.)

velocity begins to noticeably decrease is near 3×10^3 V/cm. Figure 2 shows the electron and hole velocities as a function of electric field. It should be noticed that under such conditions where the transferred electron effect is present, the use of the direct conduction band mobility is valid for only low electric field conditions (below 3×10^3 V/cm for GaAs).

Electrons in high quality material may reach maximum velocities near 2×10^7 cm/sec at electric fields near 3×10^3 V/cm. However, the low hole mobility (400 cm^2/V-sec) creates a condition such that the hole velocity ($\approx 10^6$ cm/sec at an electric field of 3×10^3 V/cm) is much less than the average electron velocity. Increasing the electric field will increase the average hole velocity at the expense of reducing the average electron velocity. To achieve similar electron and hole velocities, the electric field must be increased well above 10^4 V/cm such that both the average electron and hole velocities approach saturation, which is generally the preferred case for GaAs radiation spectrometers operated in pulse mode. Unfortunately, under such conditions the advantage of high electron mobility and velocity is compromised.

3. CHARGE CARRIER LIFETIMES

The average time that a free charge carrier is available for conduction is a major factor in determining the expected resolution for a specific detector geometry and size (Day *et al.*, 1967; Knoll and McGregor, 1993). GaAs, having a direct bandgap, ultimately has shorter theoretical recombination lifetimes than the carrier lifetimes expected for indirect bandgap semiconductors such as Ge and Si. However, the observed lifetimes for GaAs are generally much shorter than the theoretical radiative recombination times predicted by Hall (1959); hence, other mechanisms are assisting in the recombination process.

Hall (1959) proposed that the radiative recombination time for minority carriers in a semiconductor can be described by

$$\tau_c = \frac{1}{Bn_m} \tag{13}$$

where B is the radiative recombination probability and n_m is the majority carrier concentration. The net radiative recombination rate per cubic cm is (Hall, 1959)

$$R_r = B(np - n_i^2) \tag{14}$$

where n is the electron concentration, p is the hole concentration, and n_i is the intrinsic carrier concentration. For a direct bandgap semiconductor, the value of B is given by (Hall, 1959)

$$B = 0.58 \times 10^{-12} n_r \left(\frac{m_0}{m_e^* + m_h^*}\right)^{3/2}$$
$$\left(1 + \frac{m_0}{m_e^*} + \frac{m_0}{m_h^*}\right) \left(\frac{300}{T}\right)^{3/2} E_g^2 \; \frac{cm^3}{sec} \tag{15}$$

where n_r is the refractive index, m_e^* is the electron effective mass, m_h^* is the hole effective mass, T is the absolute temperature, and E_g is the bandgap energy in eV. The recombination probability described by Eq. (15) is similar to that derived by Varshni (1967). With the following values of electron effective mass in the Γ_6 valley $m_e^* = m_l^* = 0.063 m_0$, hole effective mass $m_h^* = 0.50 m_0$, index of refraction $n_r = 3.347$, and bandgap energy $E_g = 1.42$ eV, Eq. (15) indicates that the radiative recombination probability at 300 K is

$$B = 1.75 \times 10^{-10} \; \frac{cm^3}{sec}. \tag{16}$$

For very pure material with a relatively low trap density, the carrier lifetimes should be relatively long. For example, n-type material with a doping concentration of $n = 10^{14}/cm^3$ should have a minority carrier radiative recombination time of $\tau_h = 57.2$ μsec.

Studies on high purity epitaxial GaAs indicate that relatively long hole lifetimes have been observed ($\tau_h \approx 40$ μsec) (Eberhardt, Ryan, and Tavendale, 1971b; Ryan and Eberhardt, 1972). Generally, much shorter lifetimes are quoted for GaAs which are usually only a few nanoseconds (Mayburg, 1961; Ritter, Weiser, and Zeldov, 1987). As suggested by Ryan and Eberhardt (1972), the short carrier lifetimes generally observed in GaAs are most likely due to the presence of a high density of trapping and recombination centers. In relatively low doped GaAs material with a high point defect density (or compensated semiinsulating GaAs material), trap assisted recombination is expected to be the dominant process rather than direct radiative recombination.

Studies of highly doped n-type GaAs material demonstrated a reduction in hole lifetime at doping concentrations approaching $10^{18}/cm^3$ (Hwang, 1969, 1971, 1972; Casey, Miller, and Pinkas, 1973; Casey and Stern, 1976). The hole lifetimes were nearly independent of the doping concentration for concentrations below $8 \times 10^{17}/cm^3$, implying that mechanisms other than radiative recombination are dominant (Hwang, 1972). As the doping concentration approaches $10^{18}/cm^3$, the decreasing radiative recombination lifetimes become significant and reduce the overall lifetimes. Hwang (1972) reports a minimum in the radiative lifetime at a doping concentration near $1.5 \times 10^{18}/cm^3$ in which the radiative lifetime no longer decreases with increasing doping concentration. However, the overall hole lifetime continues to decrease with increasing doping concentration. A rapid decrease in hole lifetime is observed for doping concentrations above $3 \times 10^{18}/cm^3$, which is attributed to the increased influence of other nonradiative recombination processes (Hwang, 1971, 1972; Casey et al., 1973; Casey and Stern, 1976).

Semiinsulating GaAs material subjected to severe neutron damage demonstrates shorter carrier lifetimes than unirradiated material with reported lifetimes reduced from a few nanoseconds to values below 100 psec (Wagner, Bradley, and Hammond, 1986; Wang et al., 1989; Friant et al., 1989). Such behavior provides evidence that carrier lifetimes are largely influenced by the density of trapping and recombination centers. Shockley–Read–Hall recombination describes the trap assisted net recombination rate for a single set of traps as (Hall, 1952; Shockley and Read, 1952)

$$R_{srh} = \frac{np - n_i^2}{\tau_h(n_1 + n) + \tau_e(p_1 + p)} \tag{17}$$

with

$$p_1 = n_i \exp\left(\frac{E_{Fi} - E_T}{kT}\right) \tag{18}$$

$$n_1 = n_i \exp\left(\frac{E_T - E_{Fi}}{kT}\right) \tag{19}$$

where τ_e is the electron minority carrier lifetime, τ_h is the hole minority carrier lifetime, E_{Fi} is the intrinsic Fermi energy level, and E_T is the recombination center energy level. The minority carrier lifetimes are defined as

$$\tau_{e,h} = \frac{1}{C_{e,h}N_T} = \frac{1}{\sigma_{e,h}v_{e,h}N_T} \qquad (20)$$

where N_T is the recombination center concentration, $C_{e,h}$ is the carrier (electrons or holes) capture coefficient, $\sigma_{e,h}$ is the carrier capture cross section, and $v_{e,h}$ is the carrier thermal velocity. If it is assumed that the main recombination center energy is equal to the intrinsic Fermi energy ($E_T = E_{Fi}$) and the corresponding electron and hole lifetimes are equal ($\tau_e = \tau_h$), then Eq. (17) reduces to

$$R_{\text{srh}} = \frac{np - n_i^2}{\tau(2n_i + n + p)} \qquad (21)$$

Equations (14) and (21) are equal to zero at equilibrium. Charge injected into the material will increase the recombination terms in Eqs. (14) and (21), and the higher rate between radiative recombination and Shockley–Read–Hall recombination will be the dominant process.

Recombination during the detector plasma time can become a source of charge carrier loss and statistical fluctuations in the signal pulse height. It has been suggested that recombination during the plasma time for a direct bandgap semiconductor will result predominantly in radiative recombination; hence, photons emitted from recombining electrons and holes will subsequently be reabsorbed to excite electron–hole pairs (Chmill *et al.*, 1994). This view should be approached with some caution since it will be true only if band to band radiative recombination is dominant over other recombination processes. Such a case may exist for very pure material with a very low density of traps and recombination centers. However, for material with a high density of defect centers it is more probable that recombination will take place through recombination centers rather than direct band to band radiative transitions (Blakemore, 1987), resulting in energy losses from phonons or the emission of subbandgap photons with a low probability of being absorbed in the detector. Lax (1960) suggests that trap and impurity assisted recombination will occur predominantly through excited states of a deep center via phonon emission rather than radiative capture. Henry and Lang (1977) suggest that multiphonon emission through deep traps is the predominant recombination process in GaAs. Regardless, charge carrier lifetimes and plasma recombination will most likely be influenced by trap assisted recombination rather than direct band to band radiative transitions. Measurements of minority carrier lifetimes in bulk GaAs generally report values between 250 psec and 10 nsec (Mayburg, 1961; Cusano, 1964; Dudenkova and Nikitin, 1967; Hwang, 1969, 1972; Goldstein, 1971; Acket and Scheer, 1971; Casey *et al.*, 1973; Casey and Stern, 1976; Nelson and Sobers, 1978; Haegel *et al.*, 1987; Ritter *et al.*, 1987), indicating that trap assisted recombination is most likely the dominant process rather than direct radiative recombination.

Usually carrier lifetimes are measured near equilibrium conditions and do not necessarily represent the carrier lifetimes in a reverse biased radiation detector. In the detector active region (or depletion region), recombination is not expected to be a major contributor to carrier loss since both carriers must be available to complete the recombination process (Sah, Noyce, and Shockley, 1957). After the electrons and holes have separated from the initial plasma, carrier loss during transport through the active region will most likely be due to trapping (Bertolini and Coche, 1968). The carrier mean free drift time (τ^*) is the expected lifetime of a carrier as a result of the combined effects of plasma time recombination and trapping (Bertolini and Coche, 1968). Reverse biased GaAs detectors fabricated from compensated semiinsulating (SI) material demonstrated long charge carrier collection tails (over 40 nsec) at low bias voltages, most likely a result of low hole velocities, carrier trapping, and detrapping (McGregor et al., 1992a). At higher voltages, the charge collection tail decreased to 9 nsec (indicating increased hole velocities and reduced effects from carrier trapping and detrapping), which corresponds closely to the expected carrier extraction time. It is postulated that the carrier mean free drift time may be greater than 10 nsec in SI GaAs. However, since the theoretical extraction time and the period over which 90% of the charge was collected were nearly the same, it is difficult to distinguish signal reduction due to trapping from reduction due to extraction of carriers from the detector.

4. Ionization Energy and the Fano Factor

The general expression for intrinsic detector resolution at FWHM is

$$R_{\text{FWHM}} = 2.355\sqrt{\epsilon F E} \qquad (22)$$

where E is the energy of the quantum of radiation, ϵ is the average energy required to excite an electron–hole (e–h) pair, and F is the Fano factor (Fano, 1947). High resolution is achieved provided that the values of ϵ and F are relatively small. The average ionization energy (ϵ) for GaAs has been reported to be between 4.2–4.7 eV/e–h pair (Wittry and Kyser, 1965; Sakai, 1982). Electron probe studies performed by Wittry and Kyser (1965) indicate that the average ionization energy for electron–hole pair formation should be less than 4.6 eV. Studies with epitaxial GaAs material indicate that the average ionization energy for alpha particles (ϵ_α) is between 4.2 and 4.5 eV (Eberhardt, Ryan, and Tavendale, 1970, 1971a,b; Goldstein, 1971), reportedly being 4.27 eV at 300 K and 4.5 eV at 85 K (Eberhardt et al., 1971a,b). Kobayashi et al. (1972) measured ϵ_α to be 4.32 eV when the injected charge was compared to a standard charge and 4.35 eV when the detector response was compared to a Si reference detector. The relationship between bandgap energy and ϵ_α proposed by Kobayashi et al. (1972) is

$$\epsilon_\alpha = 2.53 E_g + 0.74 \text{ eV} \qquad (23)$$

and the value of ϵ_α indicated by Eberhardt *et al.* (1971b) is

$$\epsilon_\alpha = 2.7E_g + 0.43 \text{ eV}. \tag{24}$$

The average ionization energy measured with conversion electrons (ϵ_β) was reported as 4.57 eV (Kobayashi *et al.*, 1972), matching closely the results of Wittry and Kyser (1965). Kobayashi *et al.* (1972) suggest that the difference in ϵ_α and ϵ_β may be due to charge collection inhomogeneity in the material. However, there is indication that the ionization energy may depend on the nature of the incident radiation (Knoll, 1989), and slight differences in ionization energy for Si detectors have been measured as well (Pehl *et al.*, 1968; Ryan, 1973; Langley, 1973).

The Fano factor was first quoted as having an upper limit of 0.24 ± 0.04 at 130 K for 60 keV and 122 keV gamma rays (Eberhardt *et al.*, 1970). Later results indicated that the value of the Fano factor was ≤0.18 ± 0.04 at 125 K (Eberhardt *et al.*, 1971a,b). Figure 3 shows a comparison of the theoretical energy resolution at FWHM calculated with Eq. (22) for Ge and GaAs. The comparison indicates that GaAs detectors should perform as high resolution gamma ray detectors.

5. Techniques of Material Growth

Liquid phase epitaxy (LPE) has been employed successfully for the production of room temperature GaAs radiation detectors. LPE is the oldest method of growing epitaxial films on GaAs substrates. The fundamental method consists of bringing a bulk GaAs substrate into contact with a melt of high purity GaAs (Bolger

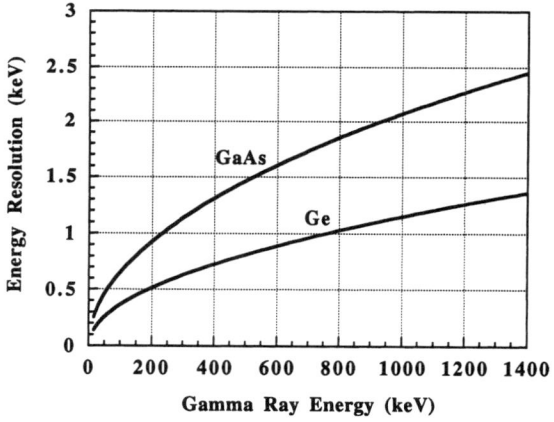

FIG. 3. Comparison of theoretical gamma ray energy resolution for Ge and GaAs detectors. The Fano factors and ionization energies used were $F = 0.08$, $\epsilon = 2.98$ eV for Ge (Knoll, 1989) and $F = 0.18$, $\epsilon = 4.3$ eV for GaAs. (Reprinted with permission from Eberhardt *et al.*, 1971b and Elsevier Science.)

et al., 1966; Hicks and Manley, 1969; Vilms and Garrett, 1972; Doi, Asano, and Migitaka, 1976). The substrate and melt are cooled, thus allowing for the formation of a thin film of high purity GaAs material. Segregation of impurities into the molten GaAs region during the cooling stage helps to produce high purity material. In vapor phase epitaxy (VPE) Ga and As atoms in the vapor phase are brought into contact with a GaAs substrate. The GaAs thin films are produced using the $AsCl_3$–Ga–H_2 reaction. Briefly described, hydrogen gas is forced through the liquid $AsCl_3$ into a reaction furnace holding a boat of Ga source material (Bolger *et al.*, 1966; Williams, 1990). The Ga source becomes saturated with As and forms a layer of GaAs. The GaAs substrates are loaded into the reactor and vapor etched. The substrate is maintained at a lower temperature than the source material and hydrogen is passed through the $AsCl_3$ and over the source material. Ga and As in the vapor phase are carried to the GaAs substrate where they react to form a film of high purity GaAs.

Several methods of bulk GaAs crystal growth have been developed to produce doped and undoped semiinsulating substrates. Overviews of some bulk growth techniques are given by AuCoin and Savage (1990) and Williams (1990). Bulk growth provides a method of producing large crystals of GaAs for use in the VLSI (very large scale integration) industry. Bulk growth methods generally consist of attaching a single crystal seed to a reservoir of molten GaAs. Methods for bulk growth include variations of horizontal Bridgman (HB) (Parsey *et al.*, 1981; Martin *et al.*, 1982; Lie *et al*, 1991), liquid encapsulated Czochralski (LEC) (Holmes *et al.*, 1982; Miyazawa *et al.*, 1984; Thomas *et al.*, 1984, 1988; Terashima *et al.*, 1984; Kashiwa *et al.*, 1990; Ware *et al.*, 1992), vertical Bridgman (Kremer *et al.*, 1990; Breivik *et al.*, 1992), vertical gradient freeze (VGF) (Gault, Monberg, and Clemans, 1986; Clemans *et al.*, 1989), and vertical zone melt (VZM) (Swiggard, 1989; Henry *et al.*, 1991). The methods differ in cooling techniques, but all use a seed crystal to order the GaAs ingot into a single crystal. For instance, LEC crystals are pulled vertically from molten GaAs located in a pyrolytic BN crucible. A seed crystal is dipped into the GaAs melt through the encapsulant and a single crystal is formed as the boule is pulled from the crucible. The vertical zone melt method employs the passage of a GaAs charge through a high temperature "spike" zone. A preliminary pass is used to conform the GaAs charge to the shape of the BN crucible. A second pass begins at the seed crystal to create a single crystal of bulk GaAs.

6. COMPENSATION IN BULK GAAS

Semiinsulating GaAs is realized through a careful balance between deep and shallow dopant levels and can be produced by the bulk crystal growth processes previously mentioned. A strong dependence of resistivity on stoichiometry is shown in Fig. 4, in which it is seen that high resistivity is achieved with an As atom fraction greater than 0.48 in the melt composition (Holmes *et al.*, 1982).

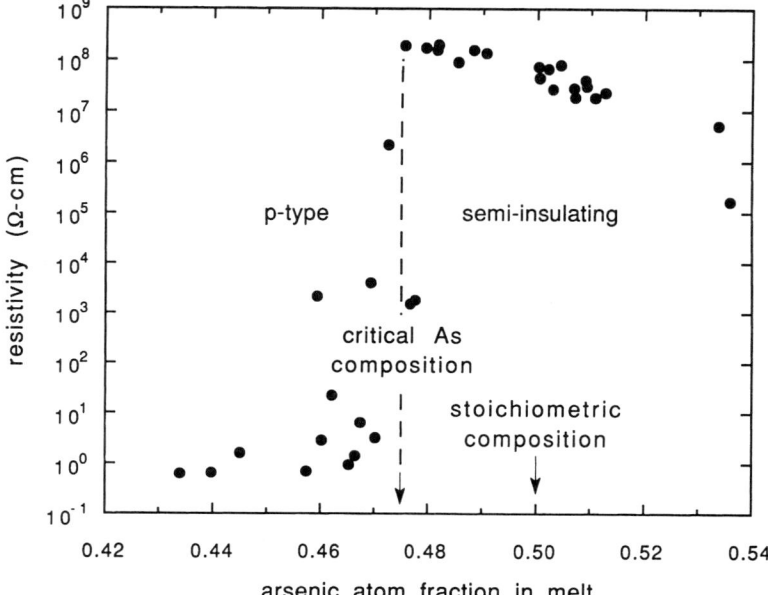

FIG. 4. Resistivity of LEC GaAs as a function of the arsenic fraction in the melt composition. (Reprinted with permission from Holmes et al., 1982 and the American Institute of Physics.)

Impurities such as C and O can be introduced during the growth process (Ware et al., 1992), while other impurities (Si, Zn, Pb, Fe, Sb, and S) can be introduced with the Ga and As starting materials. Carbon is the dominant shallow acceptor level in unintentionally doped bulk GaAs. Shallow acceptors can be compensated by the native defect deep donor EL2 to render the material semiinsulating.

The EL2 defect remains unidentified, however it is generally accepted that the defect is associated with the As antisite (As_{Ga}) (Thomas et al., 1984; Bourgoin, von Bardeleben, and Stievenard, 1988; Martin and Makram-Ebeid, 1992). There are models that attribute the EL2 defect to combinations of other intrinsic defects with the arsenic antisite (Bourgoin et al., 1988); however, there appears to be growing evidence that the EL2 center is the singular As antisite defect (Baraff, 1992; Schubert, 1993). The EL2 center is a double donor, with an energy level appearing approximately 0.75 eV below the conduction band edge and another level appearing approximately 0.25 eV lower. The defect is neutral when filled and positively charged when empty.

The deep location in the bandgap causes the EL2 centers to be only partially ionized at room temperature. Residual carbon acceptor impurities are almost completely ionized and will render the material p-type. For the case of EL2 deep donor compensation of carbon acceptors (or other shallow acceptors), the conditions for semiinsulating behavior in undoped LEC GaAs are

$$N_a > N_d \qquad (25)$$

$$N_{DD} > N_a - N_d \qquad (26)$$

where N_{DD} is the deep donor concentration, N_a is the shallow acceptor concentration, and N_d is the shallow donor concentration. Although some growth techniques allow for considerable reduction in carbon contamination, it is common to allow a low concentration of carbon to remain to meet the criteria in Eq. (25) and (26). Additionally, SI GaAs is grown slightly As rich to produce the proper concentration of deep donor EL2 centers throughout the crystal. The resulting GaAs has typical free carrier concentrations of $10^7/cm^3$ and resistivities greater than 10^7 Ω-cm. The density of EL2 centers is typically quoted between $7 \times 10^{15}/cm^3$ and $2 \times 10^{16}/cm^3$, depending on the growth process. Residual carbon acceptor concentrations generally range from $10^{14}/cm^3$ to $2 \times 10^{15}/cm^3$.

An alternative method for making SI GaAs is to intentially dope Si contaminated material with Cr Si acts as a shallow donor in GaAs and appears in high concentration in HB grown material due to contamination from the SiO_2 boat. Cr is a deep acceptor (HL1) in GaAs and acts to compensate the shallow donor impurities. The nature of the recombination mechanism in Cr-doped GaAs was investigated by Papastamatiou and Papaioannou (1990). They found that the carrier lifetimes are strongly affected by the degree of compensation between the EL2 donor levels and HL1 levels. The compensation is sensitive to the concentrations of shallow levels introduced by contamination.

III. General Detector Operation

Many GaAs detectors operate in individual quantum pulse mode in which information pertaining to the initial energy of the interacting quantum of radiation is preserved. A semiconductor detector coupled to a charge sensitive preamplifier can be used to identify distinct voltage output pulses from radiation quanta (energetic particle or photon) provided that the system electronic and thermal noise are sufficiently low. The amplitude of the output pulse is proportional to the charge induced by carrier motion in the detector, which often gives a relative measurement of the energy deposited by the interacting radiation quantum. The individual pulses can be accumulated, stored, and displayed as a pulse height spectrum on a multichannel analyzer. The pulse height spectrum provides a relative measurement of the spectrum of energies absorbed in the detector active volume. For this reason individual quantum pulse mode operation is generally employed for radiation spectroscopy measurements.

Many GaAs detectors are operated in current mode, in which a burst of energy is measured from multiple interacting quanta of radiation. If the lifetimes of excited charged carriers are relatively short, the induced current observed will decay rapidly as the carriers recombine or become trapped. Under such circumstances, the induced current does not yield information on the energy spectrum of the absorbed radiation, but can be used to measure the transient characteristics of the interacting radiation.

In either case, individual quantum pulse mode or current mode, the electrical signal produced from a semiconductor detector is due to the motion of charge carriers excited by the interacting radiation in the detector volume. The derivation of the signal pulse formation can be addressed through conservation of energy arguments (Knoll, 1989). The basic derivation assumes that no appreciable external current flows during the charge carrier extraction time and the pulse process shaping time is reasonably greater than the carrier extraction time. The energy required to move the carriers across the device active region must come at the expense of the initial electrostatic energy stored across the detector capacitance. The voltage across the detector continuously drops as the charge carriers transit across the active region. The product of the signal voltage and the detector capacitance is referred to as the "induced charge" produced by the carrier drift. Pulse formation terminates when the carriers move into a region of very small or zero electric field. Charge carriers that drift to the ends of the active region are commonly referred to as "collected." The maximum signal is produced if all carriers produced by a radiation interaction drift to the ends of the detector active region, which is commonly referred to as "complete charge collection." However, the term can be misleading since the output signal from a semiconductor detector begins immediately with the formation of charge carriers within the active volume. It is not necessary for charge carriers to be transported to the ends of the detector active region for charge to appear at the terminals of the device. The drift of the charge over a short distance from the initial point of origin will contribute to the induced charge.

The solution to the induced current produced by the motion of point charges in an uniform electric field has been derived by Shockley (1938), Ramo (1939), and Jen (1941), and is often used to describe the pulse formation in a semiconductor detector. In the case of a uniform field, the current and the induced charge formed from charge carrier transit across the active region in a planar detector can be described by

$$I = \frac{dQ^*}{dt} = \frac{qN_0}{W}\left(\frac{dx_e}{dt} + \frac{dx_h}{dt}\right) = \frac{qN_0}{W}(v_e + v_h) \quad (27)$$

where N_0 is the initial number of electron–hole pairs excited from the interacting radiation, q is the unit electronic charge, W is the width of the detector active region, v represents the charge carrier velocity, and the e and h subscripts represent electrons and holes, respectively. It should be noted that for p-i-n diodes, fully depleted detectors, and photoconductors with only ohmic contacts, the active region generally represents the full width of the detector volume. However, there are cases in which the electric field does not fully extend across the detector volume, in which case the active region refers to the portion of the detector in which an electric field is present (or where the applied voltage appears). If charge carriers become trapped or recombine before they exit the detector active region, Eq. (27) becomes

$$I(t) = \frac{dQ^*}{dt} = \frac{qN_0}{W}(v_e \exp[-t/\tau_e^*] + v_h \exp[-t/\tau_h^*]) \qquad (28)$$

where τ^* represents the carrier mean free drift time. In the ideal case, where charge is produced at a confined point in a detector (as might be expected from photoelectric interactions), the total induced charge for a reverse biased diode with blocking contacts is found by integrating Eq. (28) over the corresponding carrier extraction times, t_e and t_h; hence, the induced charge is

$$Q^* = \int_0^{t_e} I_e(t)\,dt + \int_0^{t_h} I_h(t)\,dt \qquad (29)$$
$$= \frac{qN_0}{W}\{v_e \tau_e^*(1 - \exp[-t_e/\tau_e^*]) + v_h \tau_h^*(1 - \exp[-t_h/\tau_h^*])\}$$

where $I_e(t)$ and $I_h(t)$ refer to the induced current formed by electrons and holes, respectively. Often the product of the carrier mobility and the carrier mean free transit time (or lifetime) is used to describe the expected carrier transport across a detector active region. However, as discussed earlier, the velocity characteristics for electrons in GaAs are determined by the distribution of electrons in the direct Γ_6 valley and the indirect L_6 valley. The mobility quoted for electrons located in the Γ_6 valley describes the velocity characteristics only for low field conditions (less than 3×10^3 V/cm), hence the $\mu\tau^*$ product becomes less meaningful at the higher electric fields generally employed for GaAs spectrometers. An alternative dimensionless index for charge collection is the carrier extraction factor (CEF), which takes into account the carrier velocity, carrier mean free drift time, and active region width (Knoll and McGregor, 1993). Defining the carrier extraction factor for any carrier (electrons or holes) as

$$\varrho_{e,h} = \frac{x_{e,h} \tau_{e,h}^*}{W t_{e,h}} = \frac{v_{e,h} \tau_{e,h}^*}{W} = \frac{\lambda_{e,h}^*}{W} \qquad (30)$$

where $x_{e,h}$ represents the transit distance required to move a carrier from its point of origin to the active region boundary, $t_{e,h}$ is the carrier transit time, and $\lambda_{e,h}^*$ is the carrier mean free drift length; the solution to the induced charge is

$$Q^* = qN_0\{\varrho_e(1 - \exp[-x_e/\varrho_e W]) + \varrho_h(1 - \exp[-x_h/\varrho_h W])\} \qquad (31)$$

The mean value and standard deviation of Eq. (31) can give a measure of the expected gamma ray resolution of a detector, provided that the electron and hole CEFs are known (Knoll and McGregor, 1993). Defining the percent energy resolution as

$$100\sigma_{q*} = 100\frac{\sqrt{\langle(Q*)^2\rangle - \langle Q*\rangle^2}}{\langle Q*\rangle} = 100\frac{\sigma_{Q*}}{\langle Q*\rangle} \quad (32)$$

where σ_{Q*} is the standard deviation of the induced charge and σ_{q*} is the fractional standard deviation of the induced charge, the expected energy resolution as a function of the CEF values can be found (Fig. 5). Energy resolution better than 0.2% requires that the CEF values for both charge carriers be greater than 50. If a group of charge carriers has a high CEF value (electrons, for instance) and the other does not, the resolution can suffer dramatically as indicated in Fig. 5. The CEFs can be increased by increasing the carrier velocity, increasing the carrier mean free drift times, or decreasing the detector active region width. Since the carrier velocities are saturation limited for any particular semiconductor, a practical gamma ray spectrometer requires that the carrier mean free drift times (or lifetimes) be relatively long. However, the values for $\tau*$ are generally determined by the density of

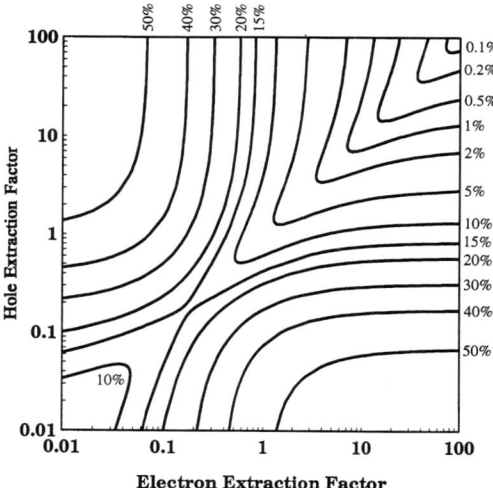

FIG. 5. The detector percent energy resolution, when defined as $100\sigma_{q*}$, is shown as a function of the carrier extraction factors (CEFs). Best energy resolution is accomplished with large CEFs (above 50). Although resolution improves for very low CEFs, noise contributions are expected to degrade the overall detector performance. Note that, for a Gaussian peak, the conventional definition for resolution is FWHM/$\langle Q*\rangle$, or 2.35 times the values shown. (Reprinted from Knoll and McGregor, 1993 with permission.)

trapping and recombination centers in the detecting material, which presents a materials problem that is not always easily solved. As a result, the CEF values are generally increased by producing appropriately thin detectors.

Photoconductive detectors used in current mode operation must be considered a bit differently. The basic configuration for a photoconductive detector is a block of semiconducting material with two ohmic contacts at opposite ends. The device operates with a voltage bias applied across the detector bulk and the free carrier population (or resistivity) defines the current flowing through the detector. High resistivity materials are generally used to ensure that leakage currents that can contribute to the detector noise are relatively low. A pulse of ionizing radiation excites electron–hole pairs in the detector and increases the free carrier population, hence increasing the measured current. The transient current response is observed for high speed current mode photoconductors. In the ideal case, charge carriers removed from the device at one contact are replaced by injected carriers at the opposite contact. The process decays as a function of the carrier mean free drift time (τ^*) until the current is restored back to its initial equilibrium value. As a result, high speed photoconducting detectors require very short mean free drift times (or lifetimes), a case quite opposite of that required for a spectrometer. Photoconductive gain is realized when the carrier transit time across the device active region is small in comparison to the carrier mean free drift time ($\tau^*_{e,h} \gg t_{e,h}$). Many GaAs photoconductive detectors are designed such that the carrier mean free drift time is significantly less than the carrier extraction time for the initially excited electron–hole pairs ($t_{e,h} \gg \tau^*_{e,h}$). In either case, the total induced charge as a result of carrier motion is found to be

$$Q^* = \int_0^\infty [I_e(t) + I_h(t)]dt = qN_0(\varrho_e + \varrho_h). \tag{33}$$

IV. Epitaxial GaAs Detectors

Radiation detectors fabricated from epitaxial GaAs were the first compound semiconductor gamma ray spectrometers to demonstrate high resolution at room temperature (Eberhardt *et al.*, 1970, 1971a,b; Kobayashi *et al.*, 1972; Kobayashi and Sugita, 1972; Hesse and Gramann, 1971; Gibbons and Howes, 1972; Hesse, Gramann, and Höppner, 1972). The GaAs material was grown by either liquid phase epitaxy (Eberhardt *et al.*, 1970, 1971a,b; Kobayashi *et al.*, 1972b; Kobayashi and Sugita, 1972; Hesse and Gramann, 1971; Gibbons and Howes, 1972; Hesse *et al.*, 1972) or vapor phase epitaxy (Kobayashi *et al.*, 1972; Hesse and Gramann, 1971; Hesse *et al.*, 1972), which restricted the active thicknesses of the detectors to 200 μm or less. Presently, high purity epitaxial GaAs Schottky barrier detectors demonstrate better energy resolution for gamma rays than detectors fabricated from bulk GaAs crystals. The configurations, electrical properties, and spectroscopic performance of epitaxial GaAs detectors are summarized in this section.

1. DETECTOR CONFIGURATIONS

Epitaxial based GaAs detectors consisted primarily of a blocking Schottky contact on the sensitive surface with either an alloyed ohmic contact or Schottky contact on the back surface. The detector designs reported from 1970 to present are fundamentally very similar, although some differences can be noted. This subsection describes the device configurations for epitaxial GaAs detectors reported by several different groups.

Eberhardt et al. (1970, 1971a,b) fabricated detectors from high purity LPE layers on n-type $\langle 100 \rangle$ oriented substrates. The epitaxial material was supplied and grown by Hicks as described in the literature (Hicks and Manley, 1969). The initial LPE layers ranged from 65 to 95 μm thick, but ranged from only 60 to 80 μm thick after detector processing. The electrical properties of the LPE layers were reported to be 33 Ω-cm resistivity, 6×10^{13} carriers per cm^3 net carrier concentration, and 8200 cm^2/V-sec electron mobility at 300 K. The basic device structures consisted of 3×3 mm die with full surface area Ga–In eutectic ohmic contacts alloyed to the n-type substrates (Eberhardt et al., 1971b). The alloying process was performed in air at 475 K for a period of 5 min. Au Schottky contacts 1.5 or 2 mm in diameter were evaporated onto the high purity LPE layers. The die were then clamped into Al holders with Au foils pressed against the Schottky contacts and In foils pressed against the Ga–In contacts.

Detectors reported by Gibbons and Howes (1972) had GaAs epitaxial layers (also grown by Hicks; see Hicks and Manley, 1969) ranging from 30 μm to 120 μm thick and free carrier concentrations ranging from 5×10^{12}/cm^3 to 10^{14}/cm^3. Samples approximately 6 mm in diameter were cut from the epitaxial wafers with a fine air abrasive jet. Au or Pd Schottky contacts 2.5 or 3 mm in diameter were evaporated onto the epitaxial surfaces. The substrates were attached to aluminum holders with colloidal Ag paste and thin Au wire was used to connect to the Schottky barrier with Ag cement or In pads. Hesse and Gramann (1971) and Hesse et al. (1972) reported on detectors fabricated from LPE and VPE material. Measurements indicated free carrier concentrations ranging from approximately 10^{13}/cm^3 to 10^{15}/cm^3 in the LPE layers. Schottky contacts 300 to 500 μm in diameter were evaporated from Al, Au, or Pd.

Epitaxial and bulk GaAs detectors were reported by Kobayashi et al. (1972). Of two LPE wafers reported, the free carrier concentration was 9.3×10^{13}/cm^3 for one wafer and 2×10^{14}/cm^3 for the other. The epitaxial thickness for both wafers was approximately 70 μm. The detectors were fabricated from 5×5 mm square die sliced from the epitaxial wafers. Metal contacts were fabricated by evaporating Au on the epitaxial surface and Al on the substrate surface through a 2.5 mm diameter mask. Kobayashi et al. (1976) also reported on GaAs detectors fabricated from Fe doped LPE layers. The transition metal Fe acts as a deep acceptor in GaAs and compensates residual shallow donor impurities (such as Si) (Milnes, 1973). The free carrier concentration for the Fe doped wafers investigated ranged from $2 \times 10^{10}/cm^3$ to $10^{14}/cm^3$. The substrates had epitaxial layers grown on both sides of a n-type substrate; however, one side was completely lapped off in many cases so that only one epitaxial layer was used for the detector, and contact could be made directly to the substrate. Die sizes for the single epitaxial layer detectors were 5×5 mm. Au evaporation was used for the Schottky contact, and Al evaporation was used for contact to the n-type substrate. Double epitaxial detectors are also described (Kobayashi et al., 1976). The detector die sizes were 4×7 mm rectangles and an area 3×4 mm was lapped off of one side of the samples to make contact to the n-type substrate. Au contacts were used as Schottky barriers on the epitaxial layers, and Al was used for contact to the n-type substrates.

Epitaxial detectors reported by Alexiev and Butcher (1992) were fabricated from LPE layers ranging up to 220 μm thick grown on n^+ GaAs substrates. The LPE layers had free carrier concentrations ranging from $2 \times 10^{13}/cm^3$ to $2 \times 10^{14}/cm^3$. The detectors were fabricated from 5 mm square sections cleaved from the LPE wafers. Similar to Kobayashi et al. (1972), Al was used for metal contact to the n^+ substrate and Au Schottky barriers 2 mm in diameter were evaporated onto the LPE surfaces.

2. DETECTOR PERFORMANCE

Of the detectors fabricated by Eberhardt et al. (1970, 1971a,b) only five satisfactory detectors resulted from three wafers processed. The satisfactory detectors demonstrated good energy resolution for low energy gamma rays ranging from 2–3% FWHM for 122 keV gamma rays at room temperature (see Fig. 6). The best energy resolution was observed at 130 K with 1.07% FWHM for 60 keV gamma rays from ^{241}Am. Resolution for gamma rays was observed to broaden with increasing temperature, demonstrating 4.2% FWHM energy resolution at 295 K for 60 keV gamma rays, 16.3% FWHM resolution at 323 K, and 33.6% FWHM resolution at 373 K. Conversion electron peaks at 62 keV and 84 keV were observed from ^{109}Cd with resolution better than 1.3 keV FWHM. Longer range conversion electrons from ^{137}Cs (624 keV and 656 keV) and ^{207}Bi (482 keV, 975 keV, and 1.048 MeV) did not show full absorption peaks since the epitaxial thicknesses were less than the particle ranges. Full energy peaks observed for 5.5 MeV alpha particles from ^{241}Am demonstrated energy resolution of 0.4% FWHM at 295 K and 0.18% FWHM for 8.785 MeV alpha particles from ^{212}Po at 210 K.

In some cases, the epitaxial–substrate interface (ESI) was reported to degrade the performance of the epitaxial GaAs detectors (Eberhardt et al., 1971b; Tavendale and Lawson, 1972). Such anomalous interfacial layers generally ranged

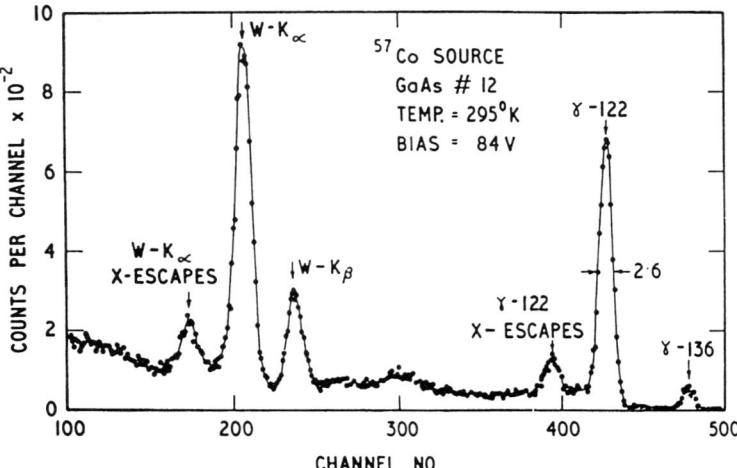

FIG. 6. Room temperature differential pulse height spectrum from a LPE GaAs detector exposed to ^{57}Co gamma rays. Also apparent are fluorescent x-rays from the tungsten backing on the source (energies in keV). (Reprinted with permission from Eberhardt et al., 1971b and Elsevier Science.)

0.2–2.0 μm thick and were composed primarily of a net acceptor concentration $(N_a - N_d)$ between $10^{14}/cm^3$ and $10^{16}/cm^3$ (Blocker, Cox, and Hasty, 1970), although Tavendale and Lawson (1972) reported on layers 7 μm thick. A characteristic of all detectors reported in the literature (Eberhardt et al., 1971b) was the monotonic increase in pulse height with increasing reverse bias for alpha particles and gamma rays. Multiple peaking, a phenomenon where monoenergetic gamma rays or alpha particles give rise to two or more separate peaks in the pulse height spectrum, was observed with alpha particle irradiation as the bias voltage approached full depletion. Beyond full depletion, the multiple peaks collapsed back into a single peak and the observed pulse heights saturated. The signal pulses all showed a slow component in the rise time, when dual peaking was observed, which disappeared at increased voltages. The satellite peaks and the slow tail components also disappeared when illuminated with infrared radiation (Tavendale and Lawson, 1972).

The cause of the multiple peaking and the slow rise time components were hypothesized as being consequences of uneven depletion into the high resistivity p-type anomalous interface layer between the substrate and the epitaxial region of the detectors (Eberhardt et al., 1971b; Tavendale and Lawson, 1972). However, Kobayashi et al. (1972) reported the appearance of multiple peaking from bulk GaAs detectors that did not have ESI layers, and multiple peaks were observed primarily from detectors with poor ohmic contacts. Multiple peaking is a phenomenon also observed with radiation damaged Si, in which case the introduction of traps promoted the appearance of satellite peaks that also disappeared with increasing bias (Dearnaley, 1963; Dearnaley and Northrop, 1964; George and

Gunnersen, 1964; Liu and Coleman, 1971). Liu and Coleman (1971) noted that slow rising pulses were observed to accompany the appearance of satellite peaks in severely damaged Si detectors. The appearance of satellite peaks is most likely related to distinct differences in the induced charge measured from the active region in the detector. This may in fact come about from nonuniformities in the electric field distribution across the detectors or from irregularities in carrier transport through regions of higher trap density.

The disappearance of multiple peaking under infrared illumination (Tavendale and Lawson 1972) was attributed to the deionization (emptying) of deep acceptor levels in the ESI region. Under such circumstances, deep acceptor compensation of donor levels would decrease, thus increasing the conductivity and reducing the capacitive effects of the ESI region. The observed disappearance in the slow tail component under infrared illumination could also be due to enhanced detrapping of charge carriers from trapping centers. Since the introduction of trapping centers has been correlated to the appearance of slow rising pulse components in Si detectors, it stands to reason that trapping centers in GaAs detectors may also contribute to the phenonmenon. Slow tail components in bulk GaAs detectors (without ESI layers) have been observed (McGregor et al., 1992a), and the slow tail component decreased dramatically with increased bias voltage. Infrared illumination can excite trapped charge carriers from trapping centers, hence reducing the detrapping time (and slow pulse component) and reducing the effect of charge loss on resolution.

Gibbons and Howes (1972) reported room temperature spectroscopic results with typical energy resolution of 0.36% FWHM for 5.5 MeV alpha particles from ^{241}Am. The best room temperature energy resolution for 5.5 MeV alpha particles was 0.28% FWHM. Room temperature measurements of 60 keV gamma rays from ^{241}Am demonstrated a typical energy resolution of 5% FWHM, decreasing to a best energy resolution of 3.7% FWHM at 255 K. The detectors showed a wide variation in reverse leakage current, which was attributed primarily to surface leakage rather than bulk generation. Detectors fabricated from the same sample demonstrated similar I–V characteristics, whereas detectors fabricated from different samples would be markedly different. Low leakage currents were generally observed from material with high surface uniformity.

LPE GaAs detectors reported by Kobayashi et al. (1972) and Kobayashi and Sugita (1972) were 70 μm thick and demonstrated room temperature energy resolution of 3.8% FWHM for 122 keV gamma rays from ^{57}Co and 0.43% FWHM for 5.5 MeV alpha particles from ^{241}Am (Fig. 7). Other LPE detectors from a separate sample demonstrated resolution of 1.45% FWHM for 5.5 MeV alpha particles. Conversion electrons from ^{57}Co (115 keV and 129 keV) were observed with energy resolution of 7.3% FWHM. Multiple peaking effects were not observed from the epitaxial detectors. Radiation damage degradation was observed from alpha particles after a dose of 10^{10} α/cm^2, however, the degradation was only a few percent of the initial noise width, and recovery of the noise width was observed after room temperature annealing.

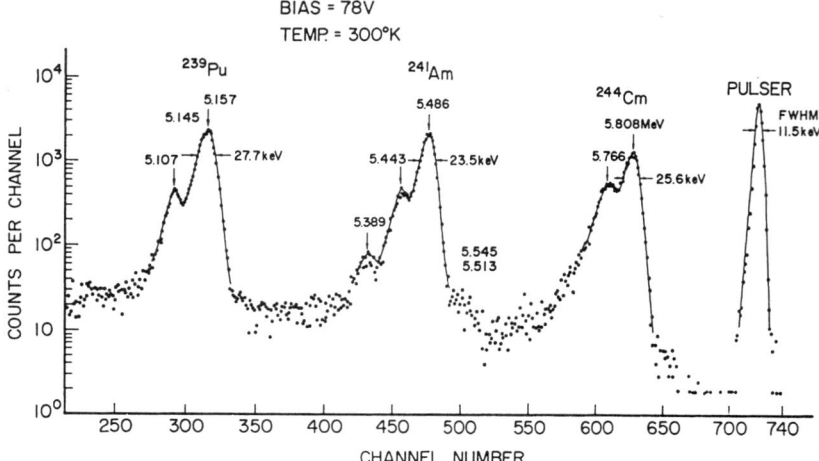

FIG. 7. Room temperature differential pulse height spectra from a LPE GaAs detector exposed to alpha particles. (Reprinted with permission from Kobayashi and Sugita, 1972 and Elsevier Science.)

Kobayashi et al. (1973) reported on the performance of these LPE GaAs detectors as biomedical probes. Gamma rays from ^{99m}Tc (140 keV) and ^{133}Xe (81 keV) were resolved at room temperature. The counting efficiency for beta particles (from ^{133}Xe) was reported to be good, but the gamma ray counting efficiency (for ^{99m}Tc) was small due to the relatively small active detector volumes. However, the LPE GaAs detectors performed with superior counting efficiency to Si probes under similar conditions. Fe-doped epitaxial GaAs detectors were later reported (Kobayashi et al., 1976). Spectroscopic measurements yielded good resolution with 5% FWHM energy resolution for 60 keV gamma rays and 3.1% FWHM energy resolution for 122 keV gamma rays. Gamma rays from ^{235}U (185 keV) were also observed, but with decreased efficiency. Attempts were made by Kobayashi et al. (1976) to improve the gamma ray detection efficiency by fabricating double sided epitaxial layer detectors, hence increasing the overall active volume. The counting rate was reported to increase by 20% when the detectors were operated in common mode as opposed to one detector operated in single mode.

Vapor phase epitaxial detectors were studied by Kobayashi et al. (1972); however, the detectors apparently did not perform as gamma ray spectrometers. VPE GaAs detectors reported by Hesse and Gramann (1971) and Hesse et al. (1972) were thin (≈ 8 μm) by comparison to LPE detectors. Energy resolution was reported to be 4.4% FWHM for 60 keV gamma rays for one such VPE detector. Due to the poor efficiency of their thin active layers, higher energy gamma rays were difficult to detect.

Results reported by Alexiev and Butcher (1992) demonstrate similar results for LPE GaAs detectors, indicating that the state of the art in this area has not changed a great deal in the intervening years. The reported detectors varied in reverse leakage current and only those detectors with leakage currents of 3 nA or less at full depletion showed good spectral resolution. The lower leakage current detectors demonstrated 4.5% FWHM energy resolution at room temperature and 5% FWHM energy resolution at 313 K for ^{241}Am 60 keV gamma rays. The leakage currents were observed to decrease from approximately 10^{-9} A at room temperature to 10^{-13} A at 220 K, indicating that leakage currents were due primarily to thermal processes. Alexiev and Butcher (1992) suggest that the reverse leakage currents were predominantly due to surface leakage. Since thermal mechanisms were responsible for the observed leakage currents, it is also possible that thermionic emission over the reverse biased Schottky barriers contributed to the leakage currents at room temperature.

3. DISCUSSION

Thin epitaxial GaAs detectors demonstrated the feasibility of using GaAs as a high resolution gamma ray spectrometer at room temperature. The most recent report on LPE GaAs gamma ray detectors demonstrated good energy resolution at room temperature for 60 keV and 122 keV gamma rays for detectors as thick as 200 μm (Alexiev and Butcher, 1992). Charge collection efficiency has been quoted to exceed 95% for high purity LPE layer GaAs detectors (Eberhardt et al., 1970; Kobayashi et al., 1972).

In all cases, the VPE and LPE material was limited to thicknesses near 200 μm or less and proved difficult to reproduce. Material and detector characteristics differed significantly from various samples, demonstrating difficulties in material quality control. The Compton scatter cross section for GaAs becomes significant by comparison to the photoelectric absorption cross section at energies approaching 140 keV. Hence, thin epitaxial GaAs detectors are inefficient as high energy gamma ray detectors, and utilization is restricted to spectroscopic measurements of charged particles, x-rays, and low energy gamma rays. Nevertheless, GaAs epitaxial gamma ray spectrometers may be useful under conditions where room temperature operation is of more concern than detector efficiency.

Detectors fabricated from LPE and VPE GaAs are inefficient, expensive, difficult to reproduce, and generally offer no satisfactory alternative to Ge or Si based spectrometers. Although it is possible that advances in material growth technology may eventually allow for the realization of much thicker epitaxial GaAs detectors, presently it appears that epitaxial GaAs detectors will be restricted to thicknesses on the order of only a few hundred microns. The obvious solution is to fabricate GaAs detectors from bulk grown material rather than epitaxial material. Unfortunately, bulk GaAs has historically been inferior in quality when compared to epitaxial material. Due to material problems, research concerning GaAs as a possible detector material was largely discontinued during the late 1970s and early 1980s. Recently, interest has renewed concerning bulk GaAs as a possible detector, and this is the topic of the next section.

V. Bulk GaAs Detectors Operated in Quantum Pulse Mode

Bulk GaAs was first investigated as a radiation detector in the early 1960s (Harding et al., 1960; Mayer, 1962; Barraud, 1963b; Dearnaley and Northrop, 1964) and bulk GaAs particle spectrometers were first reported in the mid 1960s (Kobayashi and Takayanagi, 1966). Bulk GaAs has been investigated as a potential gamma ray spectrometer (Sumner et al., 1991; McGregor et al., 1992a; Benz et al., 1992; Berwick et al., 1993a), as a minimum ionizing particle detector (Bertin et al., 1990; Beaumont et al., 1992, 1993b, 1994b), and as a detector for dark matter searches (Spooner et al., 1991, 1992). The configurations, electrical properties, and performance of bulk GaAs detectors operated in individual quantum pulse mode are summarized in this section.

1. Detector Configurations

Early bulk GaAs detector configurations consisted primarily of high resistivity conductivity counters with ohmic contacts at either end of a sample (Harding et al., 1960; Mayer, 1962). Detectors reported by Harding et al. (1960) fabricated from bulk GaAs with resistivity of 10^6 Ω-cm had dimensions of $0.1 \times 0.2 \times 1.0$ cm. Early Schottky contact based detectors were described by Barraud (1963b) and Kobayashi and Takayanagi (1966). The detectors described by Barraud (1963b) were fabricated from high resistivity bulk GaAs ($>10^7$ Ω-cm), prepared by the controlled introduction of oxygen[1] during the growth process, with alloyed Ga-In ohmic contacts and Al Schottky contacts. Detectors described by Kobayashi and Takayanagi (1966) were fabricated from n-type bulk GaAs with resistivity of 173 Ω-cm and electron mobility of 4400 cm^2/V-sec. The detectors were approximately 500 μm thick with alloyed Sn ohmic contacts on one side and 4 mm diameter Au Schottky contacts (100 Å thick) evaporated on the opposite side. Kobayashi and Takayanagi (1971) also report on later detectors fabricated from bulk GaAs, many of which were fabricated in the same manner.

Schottky contact diodes have been the most commonly utilized detector configurations over recent years. GaAs characteristically has a high density of surface states that "pin" the Fermi level at the metal–semiconductor interface (Bardeen,

---

[1] Oxygen was at one time thought to be related to the main electron trap ($E_c - 0.75$ eV) in SI bulk GaAs responsible for the compensation process. Although oxygen is correlated to the presence of deep levels in GaAs (Skowronski, 1992), Schubert (1993) points out that the deep level states produced by oxygen in GaAs are still not well understood. Huber et al. (1979) demonstrated that the main electron trap, labeled EL2 by Martin, Mitonneau, and Mircea (1977), was not related to the presence of oxygen in GaAs. Martin et al. (1982) postulated that the presence of oxygen during the bulk growth process acted as a getter for Si contamination. Hence, it has been suggested that the compensation mechanism in oxygen controlled HB bulk GaAs actually consists of EL2 compensation of shallow levels, which are primarily acceptors due to the reduction in the shallow donor concentration (Martin and Makram-Ebeid, 1992).

1947; Rideout, 1975). The pinning of the Fermi level forces a condition in which the Schottky barrier height is a function of the filled surface state density rather than the metal work function. The effect makes alteration of the barrier height difficult, but allows for reasonable reproducibility of Schottky contact barriers.

Bertin et al. (1990) and Beaumont et al. (1992) fabricated bulk GaAs detectors from undoped LEC semiinsulating (SI) wafers with resistivity ranging from 7.5×10^7 Ω-cm to 1.2×10^8 Ω-cm. The diodes ranged from 125 μm to 500 μm thick and consisted of alloyed Au/Ge/Ni ohmic contacts on one side and Ti/Pt/Au Schottky contacts on the opposite side. Schottky contact diodes described elsewhere (McGregor et al., 1992a,c) were fabricated from undoped SI LEC GaAs wafers. The wafers were diced into 5.08×5.08 mm squares and polished to thicknesses ranging from 40 μm to 760 μm thick. Si_3N_4 was plasma deposited to provide passivation and protect the surfaces during processing. Circular regions 4.32 mm in diameter were etched to the bare GaAs and Au, Ge, and Ni layers were evaporated into the regions to produce an ohmic contact. Annealing was performed at 410°C in N_2 for 1 min. Proton implantation was performed around a protected region on the opposite side to help reduce surface leakage. Regions 4.32 mm in diameter were then etched to the bare and undamaged GaAs, and the die were bonded into BN collars and connector rings with high resistivity potting epoxy. Ti/Au layers 500 Å thick were evaporated to form Schottky contacts on the front sides, and Au was evaporated over the ohmic contact sides to make electrical contact to the connector rings. The final device cross section is shown in Fig. 8.

Detector configurations described by Buttar et al. (1991), Ashman et al. (1992), and Berwick et al. (1993a) are p-i-n detectors. Under reverse bias, the p and n regions serve as blocking contacts to injected current while providing the means to help facilitate efficient carrier extraction. The detectors described by Ashman et al. (1992) are p-i-n strip detectors, fabricated on $\langle 100 \rangle$ oriented 400 μm thick undoped SI LEC GaAs substrates. Si doped layers (10^{18}/cm^3) 5000 Å thick were grown by molecular beam epitaxy (MBE) on one side of the SI LEC GaAs wafers to produce an n^+ layer. InGe/Au strips were evaporated onto the n^+ layers and annealed at 420°C to form ohmic contact. Grooves were etched between the InGe/

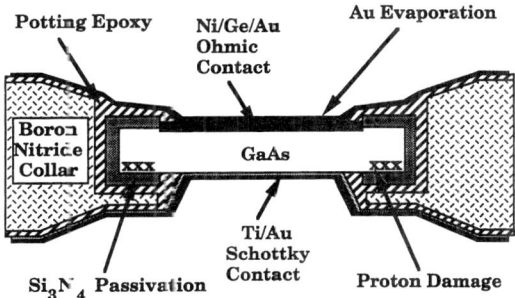

FIG. 8. Schottky barrier detector cross section (from McGregor et al., 1992a)

Au strips down to the SI GaAs substrate to isolate the readout strips. The wafer was then diced into sections and some of the sections thinned to 200 and 300 μm thick. A Au/Zn/Au layer was then evaporated onto the back surfaces of the die and annealed at 420° C to form a p^+ contact to the SI substrate. GaAs p-i-n detectors reported by Berwick et al. (1993a) were fabricated from multiwafer annealed SI LEC bulk GaAs material (Oda et al., 1992). Successive layers of Au, AuGe, Au, and Ni were evaporated on a 1 × 1 cm area bulk GaAs substrate. The layered contact was annealed at 420° C for 2 min to form an n^+ contact. The opposite side had four contact pads 3 mm in diameter evaporated from consecutive Au, Zn, and Au layers. The Au/Zn/Au contacts were annealed at 500° C for 2 min to form p^+ contacts.

2. I–V CHARACTERISTICS

Early bulk conductivity counters with injecting contacts at both ends of a block of GaAs often demonstrated undesirable low frequency oscillations (LFOs) under bias (Mayer, 1962; Barraud, 1963a,b; Dearnaley and Northrop, 1964). Such oscillations have been observed by a number of different groups and have been attributed to the presence of deep levels in the material (Northrop, Thornton, and Trezise, 1964; Holonyak and Bevacqua, 1963; Kaminska et al., 1982; Johnson, 1989; Derhacobian and Haegel, 1991). The GaAs devices that demonstrated LFOs were primarily n-i-n structures, although p-i-n structures have demonstrated LFOs as well (Derhacobian and Haegel, 1991). Kaminska et al. (1982) and Johnson (1989) attribute the LFOs observed in undoped SI LEC GaAs to field enhanced electron capture by EL2 deep donor antisite defect centers.

The forward current density due to thermionic emission for a Schottky diode is generally represented by (Sze, 1981)

$$J(V) = J_{st}\left(\exp\left[\frac{qV}{NkT}\right] - 1\right) \quad (34)$$

where

$$J_{st} = A^* T^2 \exp\left[\frac{-q\phi}{kT}\right] \quad (35)$$

and V is the applied voltage, q is the charge of an electron, k is Boltzmann's constant, T is the absolute temperature, A^* is the effective Richardson constant, ϕ is the Schottky barrier height, and N is the ideality factor of the current behavior. Pure thermionic emission is demonstrated for diodes with ideality factors equal to unity. Schottky diode detectors fabricated from undoped SI LEC GaAs material characteristically demonstrate rectifying behavior; however, the forward currents as a function of voltage are slow rising compared to ideal conditions (Bertin *et al.*, 1990; Sumner *et al.*, 1991; McGregor *et al.*, 1992a,b; Benz *et al.*, 1992; Karpinski *et al.*, 1992; Ashman *et al.*, 1992). The ideality factors reported elsewhere ranged from 5 to 40 depending on the detector thickness (McGregor *et al.*, 1992a), indicating that mechanisms other than simple thermionic emission were responsible for the observed forward currents. The turn-on voltage (trap filled limit voltage) for forward bias current increased with detector thickness (Fig. 9a), in which case thin detectors (40 μm) demonstrated much lower turn-on voltages than relatively thicker detectors ($\approx$250 μm) (McGregor *et al.*, 1992a). The slow rising I–V relationship and thickness dependent turn-on voltages are characteristics typically found in space charge limited, trap filled insulating materials (Lampert, 1956;

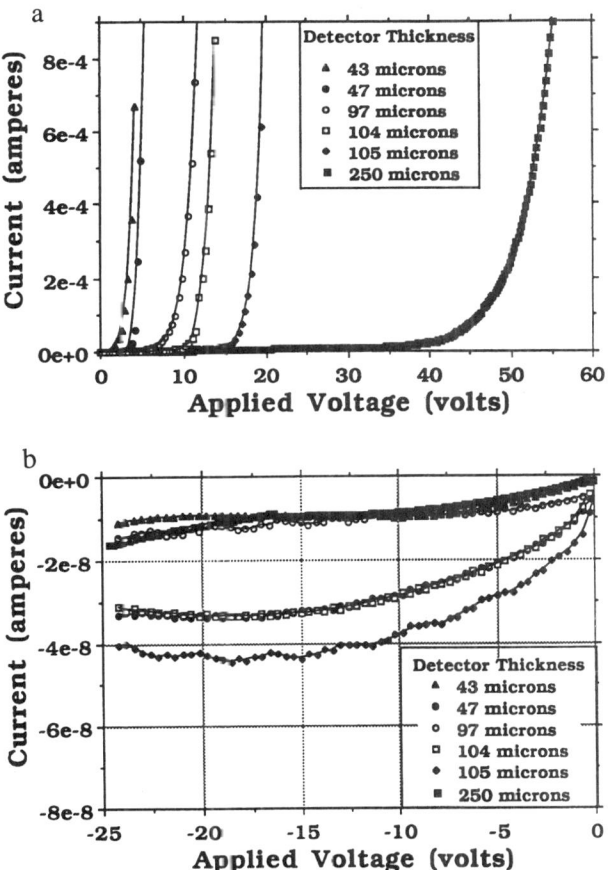

FIG. 9. Characteristic I–V curves for several thicknesses of Schottky diode detectors fabricated from SI LEC bulk GaAs showing the (a) forward currents and (b) the reverse leakage currents (from McGregor *et al.*, 1992a).

Wright, 1959; Lampert and Mark, 1970). Similar results were reported for bulk GaAs p-i-n diodes (Dogru et al., 1994), in which thickness dependent forward bias turn-on voltages were also observed.

Reported reverse leakage current densities range from 2.5 nA/mm^2 to over 100 nA/mm^2 (Bertin et al., 1990; Sumner et al., 1991; McGregor et al., 1992a; Grant and Sumner, 1992). Surface currents can contribute to the observed leakage currents (McGregor, White, and Weichold, 1985), but can be reduced by employing proper surface preparation techniques and surface passivation (Pearton, Haller, and Elliot, 1984; McGregor et al., 1992a; Karpinski et al., 1992). Reverse bias breakdown voltage has been observed to increase with detector width, generally requiring between 1 to 2 V per micron of detector thickness. Poor surface preparation can result in lower breakdown voltages as well as higher leakage currents. The reverse bias leakage currents do not appear to have thickness dependent characteristics (Fig. 9b), indicating that reverse leakage is controlled primarily by mechanisms other than bulk effects. Low temperature studies by Sumner et al. (1991) demonstrated a large reduction in leakage current with decreasing temperature, hence demonstrating that thermal effects were principal in contributing to reverse leakage currents. Although bulk generation can contribute to thermal leakage currents, it was concluded that thermionic emission over the Schottky barrier was the primary cause of reverse bias leakage current.

3. C–V CHARACTERISTICS

The modulated C–V measurement is an often employed method to correlate capacitance with the detector active region width as a function of reverse bias (Yang, 1978; Look, 1989). The method requires the application of a small signal voltage across a detector under bias. The capacitance is defined as

$$C = \left|\frac{dQ}{dV}\right| = \frac{A\epsilon_s}{W} \qquad (36)$$

where A is the contact area. In a modulated C–V measurement, it is often assumed that the detector active region capacitance represents the only significant element in the circuit and the out of phase current measured on a capacitance bridge yields the true capacitance as a function of active region depth. Under such an assumption, the depletion width can be derived from the small signal capacitance represented by

$$I = \frac{V}{Z} = j\omega V_R C \qquad (37)$$

where A is the contact area, Z is the detector impedance, and ω is the modulation frequency (Look, 1989). C–V measurements performed by Kobayashi and Takayanagi (1966) on early bulk GaAs detectors ($\rho = 173$ Ω-cm) indicated that the active regions with their devices extended 60 μm with a bias of 40 V. The diodes broke down at 80 V, indicating that the device active regions extended no further than 80 μm at breakdown voltage.

In reality, diodes have a significant resistance between the rectifying contact and the ohmic contact, and modulated C–V measurements on very high resistivity materials can often lead to erroneous conclusions regarding the active region width. The high resistivity of SI LEC GaAs creates a condition in which C–V measurements at frequencies above 1 kHz demonstrate no change in capacitance, a result that can be misinterpreted as full extension of the active region across the diodes. Ignoring contributions to the impedance from the contacts, a Schottky diode detector has resistance (R_A) in parallel with capacitance (C_A) in the active region and resistance (R_S) in parallel with capacitance (C_S) in the undepleted region or substrate (McGregor et al., 1992a). The impedance across the diode is described by

$$Z = \left[\frac{R_A}{1 + j\omega R_A C_A} + \frac{R_S}{1 + j\omega R_S C_S} \right] \tag{38}$$

In the limit that the resistivity of the substrate becomes comparably high as the active region, the impedance reduces to

$$Z \approx \frac{\rho(L_A + L_S)}{A(j\omega\rho\epsilon_s + 1)} \tag{39}$$

where ρ is the resistivity and L_A and L_S are the lengths of the active region and substrate region, respectively. The current is approximated by

$$I \approx V_R \left(\frac{1}{(R_D + R_S)} + j\omega C_T \right) \tag{40}$$

where C_T is the series combination of C_A and C_S and is equivalent to the parallel plate capacitance between the Schottky contact and the ohmic contact. Sufficiently high frequencies will also reduce the capacitance described in Eq. 38 to the parallel plate capacitance C_T. In either case, the diode appears to have a constant capacitance (C_T) between the rectifying and ohmic contact (as experimentally observed).

Generally, the resistivity is higher in the active region than the substrate because free carriers are being removed as they are generated. If the modulation frequency is lowered substantially, charge carriers in the substrate will have ample time to short the substrate capacitance. The small signal circuit appears to consist of only the active region capacitance C_A in series with the substrate resistance R_S. A change in capacitance may be observable; however, the high series resistance creates a situation in which the real and imaginary current terms are difficult to separate, hence a reliable measurement of the active region width becomes difficult (Wiley and Miller, 1975. Wiley, 1978).

It is often assumed that a change in bias voltage creates an immediate change in depletion thickness. Such an assumption is valid as long as the carrier emission rate is greater than the measurement modulation frequency ($e_n \gg \omega$) (Look, 1989). For most practical purposes, such a case exists for shallow dopant impurities. However, the GaAs described in most of this section is semiinsulating material that has been compensated with deep levels (in particular, EL2). The deep level EL2 has a room temperature emission rate of approximately 0.1/sec for electric fields below 10^5 V/cm (Makram-Ebeid, 1981; Look, 1989), indicating that the modulation frequency must be substantially lower than 1 kHz to avoid erroneous conclusions concerning the active region thickness. Capacitance changes with reverse bias have been demonstrated by Beaumont et al. (1993a,c) at low frequencies (below 1 kHz). The observed changes in capacitance with bias voltage may be due to both of the effects previously discussed. Presently, C–V measurements on SI GaAs do not render active region width information with acceptable confidence, and other methods have been employed to study the actual detector active region widths as a function of bias voltage (Barraud, 1963b; McGregor et al., 1992a, 1994a,b; Beaumont et al., 1993c; Berwick et al., 1993a,b).

4. ACTIVE REGION MEASUREMENTS

Undoped SI LEC GaAs material is high in resistivity and has a very low free charge carrier density, hence it was commonly assumed that Schottky contact detectors fabricated from SI LEC GaAs had active regions that extended completely across the devices with little or no applied voltage. C–V measurements performed at frequencies above 1 kHz seemed to support the idea of full depletion, since no apparent change in capacitance was observed with increasing reverse bias. As previously discussed, C–V measurements on high resistivity materials can lead to erroneous conclusions. Charged particle pulse height analysis (McGregor et al., 1992a,b; 1994a,b), optical probing (Barraud, 1963a,b), electron beam probing (Beaumont et al., 1993c), infrared transmission (Berwick et al., 1993a,b), and direct probing on cleaved samples (Berwick et al., 1993a,b) have demonstrated that substantial voltage is required to extend the active regions completely across the detectors.

a. Pulse Height Analysis

Monoenergetic protons (from an accelerator, for instance) and alpha particles have discrete penetration ranges in a medium. The straggle for alpha particles and protons is generally only a few percent of the total range. Energy deposition from protons and alpha particles in a medium can be described by the Bragg ionization curve (Evans, 1982), in which electron–hole pair excitation is greatest near the end of the charged particle range. The effect is manipulated by noting that only those electron–hole pairs created within the detector active region will contribute to the induced charge as they are swept from the detector; hence, the induced charge (or pulse height) can give a relative measure of the active region width as a function of the applied voltage.

If a detector is irradiated from the front (Schottky contact), the expected result under ideal conditions is an increase in pulse height with applied reverse bias until the active region extends beyond the particle range. Any further increase in voltage should demonstrate relatively little change in pulse height, since all of the particle energy is being deposited within the active region. If the detector is irradiated from the back contact (ohmic contact), no pulses will be seen until the active region reaches the end of the particle range. As the active region encounters the particle range a dramatic increase in pulse height will be observed, since the highest density of charge is created near the end of the charged particle path length. The pulse height will continue to increase until the active region extends completely across the detector; hence, all of the charge is created within the detector active region. Trapping, recombination, and high substrate series resistance will introduce difficulties with charge collection (Dodge *et al.*, 1964, 1965; Stuck *et al.*, 1972); however, the general pulse height response should retain its characteristic form. The measurement technique is described in detail elsewhere (McGregor *et al.*, 1992c, 1994b).

Reverse biased Schottky contact detectors fabricated from compensated bulk GaAs (Barraud, 1963b; McGregor *et al.*, 1992a, 1994a; Beaumont *et al.*, 1992, 1993a) irradiated from the front and back have shown definite dependencies on their active region widths with applied voltage. In many cases, alpha particle pulses were not observed when the detectors were irradiated at the anode (Barraud, 1963b; Bertin *et al.*, 1990; Buttar *et al.*, 1991), indicating that the detector active regions did not extend completely across the devices. Although undoped SI LEC GaAs material has low equilibrium free carrier concentrations ($\approx 10^7/cm^3$), the active regions do not extend across the detectors until appreciable voltage is applied. Figure 10 shows the pulse height responses for 45 μm and 100 μm Schottky contact bulk SI GaAs detectors to 3.18, 5.5, and 8.78 MeV alpha particles with approximate ranges in GaAs of 10, 20, and 40 μm, respectively. The pulse height responses show that the detector active regions clearly require considerable applied voltage before extending completely across the detectors. The point at which the pulse height abruptly increases under back side (ohmic contact) irradiation gives an indication of the active region depth. The active region widths demonstrated a near linear dependence on voltage, requiring approximately 1 V per micron of active region.

b. Optical Measurements

Optical probe measurements using the photoelectric effect were used to profile the electric field across reverse biased Schottky contact detectors (Barraud,

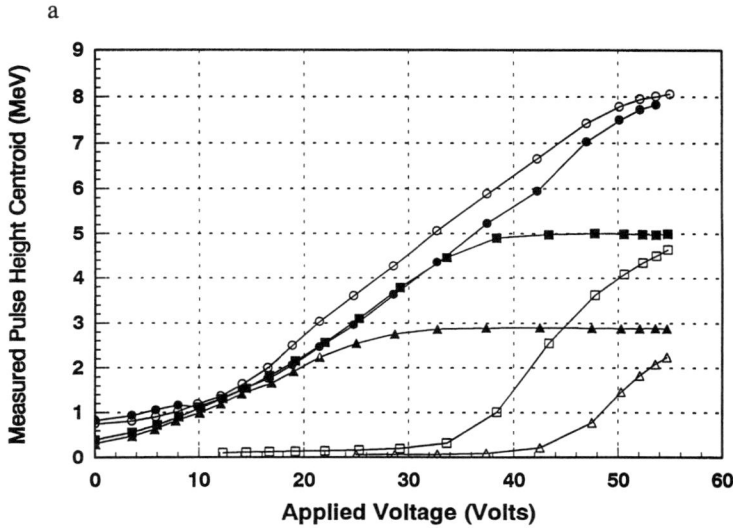

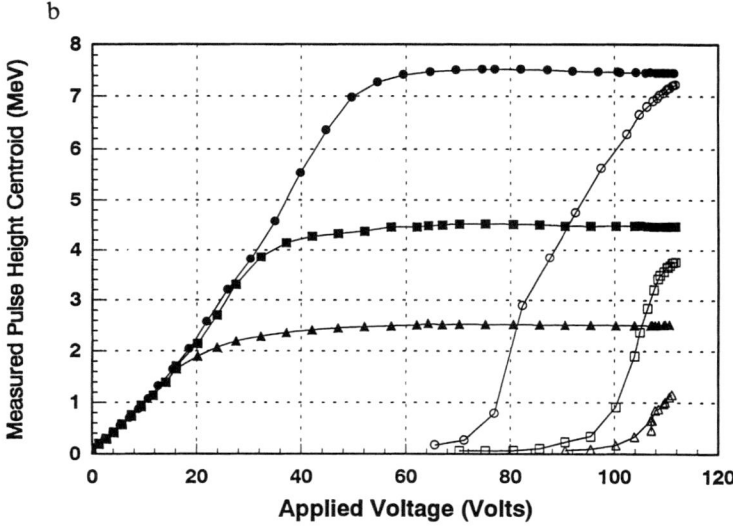

FIG. 10. Pulse height as a function of bias voltage for (a) a 45 μm Schottky contact bulk SI GaAs detector and (b) a 100 μm Schottky contact bulk SI GaAs detector. The detectors were irradiated in vacuum from the front contact (Schottky contact) and back contact (ohmic contact) with 3.18, 5.5, and 8.78 MeV alpha particles. (-○-, back/8.78 MeV; -□-, back/5.5 MeV; -△-, back/3.1 MeV; -●-, front/8.78 MeV; -■-, front/5.5 MeV; -▲-, front/3.1 MeV.) Reprinted with permission from McGregor *et al.*, 1994a and Elsevier Science.

1963b). The detectors were fabricated from high resistivity bulk GaAs material compensated by the controlled introduction of oxygen.[2] Barraud (1963b) observed that the active region width did not follow a $\sqrt{V}$ dependence and reported a dependence closer to $V^{0.65}$. The electric field was split into two regions: a high field region and a low field region. The high field region was fairly constant and extended into the detector bulk, followed by an abrupt drop in field strength to very low values. Barraud (1963b) observed that increasing the reverse bias extended the active region further into the detector, but did not appreciably increase the field strength. The production of a flat electric field region was attributed to partial compensation of positive charge centers by trapped electrons in the high field region. Barraud (1963b) reported an active region length of 1 mm at a reverse bias of 800 V.

Through the use of infrared transmission measurements, the change in optical absorption edge with applied electric field can be manipulated to yield active region width information. The room temperature optical absorption coefficient for GaAs changes abruptly near the bandgap energy from 8000/cm at 1.42 eV down to 10/cm at 1.37 eV (Blakemore, 1982). The optical threshold energy can be altered through the Franz–Keldysh effect (Franz, 1958; Keldysh, 1958) in which the optical threshold energy is reduced by the application of an electric field. The effect is small and has been measured to be only a 0.2 meV shift for an electric field of 5000 V/cm (Moss, 1961).

Since the absorption edge is abrupt in GaAs, a change in the location of the absorption edge will have a dramatic effect on the optical absorption of photons with energies near the bandgap energy. Berwick *et al.* (1993a,b) observed the absorption of 900 nm light in bulk GaAs *p-i-n* and Schottky diode detectors as a function of reverse bias voltage, in which separate regions of optical absorption were clearly evident. The nonuniform optical absorption indicates that the internal electric field is also nonuniform with distinctively different regions in electric field magnitude.

c. Electron Beam Measurements

Schottky barrier bulk GaAs detectors were cleaved and probed along the side with a 10 keV electron beam from a scanning electron microscope (Beaumont *et al.*, 1993c). The electron beam was scanned over the area between the detector contacts, and the detector output signal was observed as function of the detector bias voltage and electron beam location. Beaumont *et al.* (1993c) reported that a signal was observed only when the beam was scanned within a restricted active

---

[2] Oxygen introduction during crystal growth should effectively remove Si shallow donor contaminants introduced by quartz ampoules and boats (Skowronski, 1992; Martin and Makram-Ebeid, 1992), and it is suggested that oxygen controlled HB grown GaAs is actually EL2 compensated material (Martin *et al.*, 1982). Hence, it is possible that the bulk GaAs material described by Barraud (1963b) is similar to present day EL2 compensated material. However, the matter is uncertain due to the lack of documentation in the manuscript.

region near the reverse biased Schottky contact, and the width of the active region increased with increasing bias voltage. The reported data indicates a non-$\sqrt{V}$ dependence on the active region width. Additionally, voltage contrast measurements with the SEM indicated the presence of two distinctively different electric field regions between the detector contacts.

d. Direct Probe Measurements

Direct probing of a semiconductor surface has been used to measure the voltage as a function of position (Northrop et al., 1964; Panousis, Krambeck, and Johnson, 1969; Berwick et al., 1993a,b). The electric field at the surface can be calculated from the measured voltage profile. The surface electric field is assumed to give a representative measurement of the electric field in the bulk. Measurements performed by Berwick et al. (1993a,b) on cleaved detectors show that the electric field in reverse biased diodes fabricated from undoped SI LEC GaAs do not extend across the detectors until appreciable voltage is applied.

The detectors were cleaved along a $\langle 110 \rangle$ direction to reduce surface leakage. The surface electric field increases in width and magnitude with increasing reverse bias (Fig. 11). A fairly flat electric field region extends from the p^+ contact into the bulk (indicating a region of low space charge) followed by an abrupt drop in electric field strength (indicating a region of high space charge) (Berwick et al., 1993a,b). A bias voltage greater than 700 V was required to extend the electric field completely across a 500 μm device. The active region demonstrates a nearly

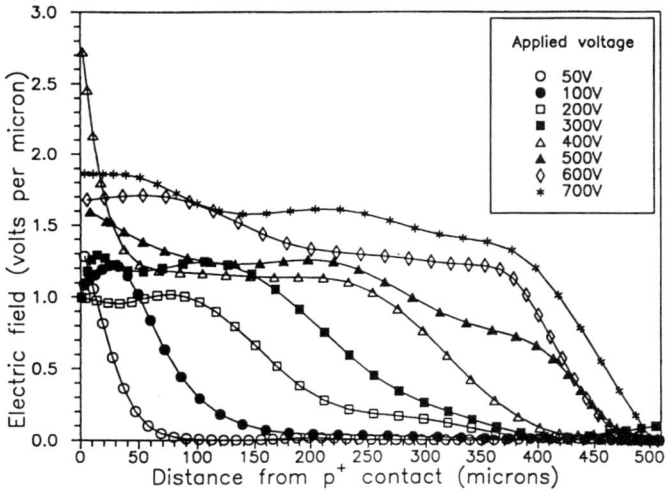

FIG. 11. The electric field within a GaAs detector for applied voltages ranging from 50 V to 700 V obtained from surface voltage measurements. (Reprinted with permission from Berwick et al., 1993b and Institute of Physics Publishing.)

linear dependence on applied voltage, requiring between 1.3 to 2 V per micron of active region (Buttar, private communication, 1994). The electric field profiles reported by Berwick et al. (1993a,b) are similar to the electric field profiles reported earlier by Barraud (1963b).

5. RADIATION MEASUREMENTS

Schottky contact detectors fabricated from n-type bulk GaAs by Kobayashi and Takayanagi (1966) demonstrated energy resolution of 7.7% FWHM for 5.15 MeV alpha particles. Avalanche GaAs detectors studied by Kobayashi and Takayanagi (1971) suffered relatively poor energy resolution. The avalanche multiplication factor increased to a maximum of 2 as the bias was raised from 70 to 90 V, however the spectral resolution broadened significantly. Afanas'ev (1971) reported on detectors fabricated from high resistivity Cr doped bulk GaAs. Charged particle energy resolution was poor for the detectors—demonstrating only 20% energy resolution for both ^{233}U alpha particles and ^{137}Cs beta particles.

Schottky contact detectors fabricated from undoped SI LEC GaAs have reported energy resolution for 5.5 MeV alpha particles ranging from 2% to 5% FWHM (Bertin et al., 1990; McGregor et al., 1992a). Charge collection efficiency for alpha particle and proton irradiation generally decreased as the total detector volume increased, ranging from over 90% charge collection for 40 μm detectors (McGregor et al., 1992a) to, in some cases, a maximum of less than 50% for much thicker detectors (Beaumont et al., 1993a). Charge collection efficiency for electrons has been reported to be superior to holes for thin ($\approx$100 μm) detectors (McGregor et al., 1992a,b,c, 1994a); however, there are reports of higher hole collection efficiency than electron collection efficiency for thicker ($\approx$400 μm) detectors (Beaumont et al., 1993a; Berwick et al., 1993a). Alpha particle resolution was usually reported to be best when the detectors were irradiated from the front contact (Schottky contact) as opposed to the back contact (ohmic contact), indicating that electrons were being collected with less statistical fluctuation than holes (Beaumont et al., 1992; McGregor et al., 1994a).

Schottky contact detectors only 45 μm thick were biased such that the electric field extended completely across the devices (as indicated by pulse height analysis) and irradiated from the front and back with ^{228}Th alpha particles (McGregor et al., 1994a). ^{228}Th emits several discrete energies of alpha particles ranging from 5.3 MeV to 8.78 MeV. The range of the lowest energy alpha particles (5.3 MeV) extends less than 20 μm in GaAs and the range of the highest energy alpha particles (8.78 MeV) extends 40 μm in GaAs (Ziegler and Biersack, 1990). If a 45 μm detector is irradiated from the front, the Bragg ionization distribution indicates that the low energy particles excite most of the charge carriers in the first half of the detector and the high energy particles excite most of the charge carriers in the last half of the detector. As a result, the induced charge will be electron dominant for the low energy alpha particles and hole dominant for the high energy particles. The opposite case is true if the detector is irradiated from the back.

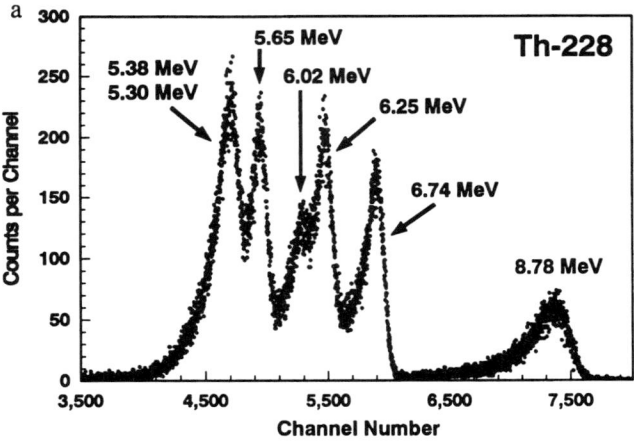

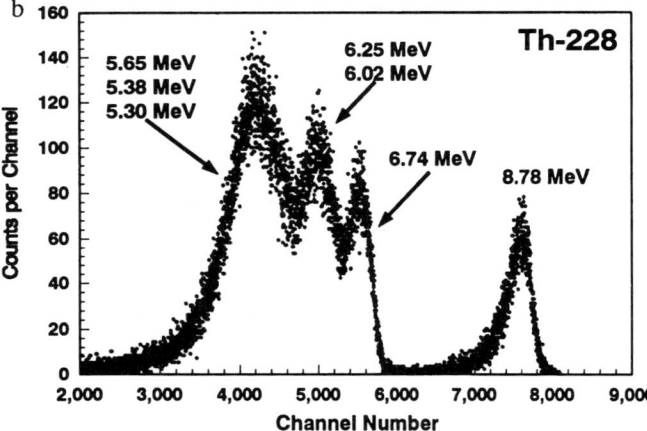

FIG. 12. Room temperature differential pulse height spectrum of alpha particles from a ^{228}Th source taken with a 45 μm SI bulk GaAs detector when irradiated from (a) the front (Schottky contact) and (b) the back (ohmic contact). (Reprinted with permission from McGregor *et al.*, 1994a and Elsevier Science.)

Figure 12 shows front and back irradiation spectra from a 45 μm detector under fully depleted and identical conditions. In both cases, the resolution of the electron dominant signal is better than the resolution of the hole dominant signal, indicating that electrons are collected from the detector with less statistical fluctuation than the holes. It is possible that the apparent increase in hole collection efficiency for thicker detectors is due to hole detrapping.

Gamma ray spectroscopic results from bulk GaAs detectors have been unimpressive at best and do not compare well to the results achieved with LPE GaAs detectors. Bulk GaAs detectors irradiated with low energy gamma rays (22–122 keV) demonstrated room temperature energy resolutions between approxi-

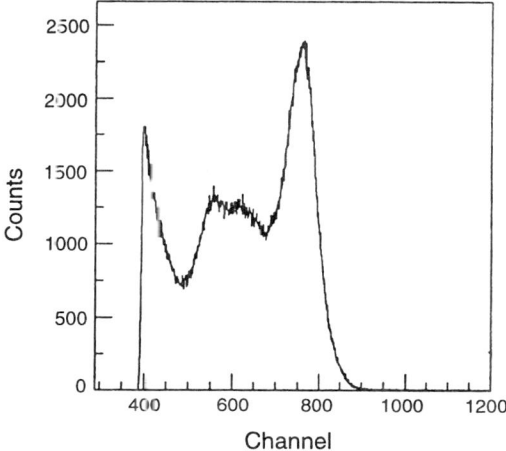

FIG. 13. Differential pulse height spectrum from a SI bulk GaAs p-i-n diode exposed to ^{57}Co 122 keV gamma rays. Quoted energy resolution is 13% at FWHM. (From Berwick et al., 1993a with permission.)

mately 20% to 35% FWHM (McGregor et al., 1992a,b; Benz et al., 1992; Sumner et al., 1992). The poor energy resolution is attributed primarily to poor charge carrier collection and position dependent charge collection. Pulses from higher energy gamma rays (662 keV) have been observed as only a continuum of counts with no apparent full energy peaks. Thicker detectors ($>>750$ μm) demonstrated gamma ray continua for 60 keV, 122 keV, and Compton scattered 663 keV gamma rays with no indication of full energy peaks (McGregor et al., 1992a), again indicating severe charge collection difficulties. The endpoint channels of the gamma ray continua demonstrated a linear relationship with gamma ray energy, indicating a linear production of electron–hole pairs with energy. Berwick et al. (1993a) reported results on 400 μm thick p-i-n devices fabricated from multiwafer annealed (Oda et al., 1992) bulk GaAs material. Quoted energy resolution was 13% FWHM for ^{57}Co 122 keV gamma rays (Fig. 13).

6. PROPOSED MODELS FOR OBSERVED BEHAVIOR

Bulk GaAs material with dominant shallow dopants had active region widths that appeared to follow a generally accepted $\sqrt{V}$ dependence (Kobayashi and Takayanagi, 1966) in which the active region or depletion region is described by

$$W(V) = \sqrt{\frac{2\epsilon_s(V_{bi} - V_R)}{qN_b}} \qquad (41)$$

where V_{bi} is the built-in junction potential, N_b is the dominant dopant concentration, and V_R is the reverse bias potential. Equation (41) is derived with the assumption that the dopants are uniformly distributed and fully ionized (as expected with shallow dopants). Although the free carrier concentration for undoped SI LEC GaAs is very low ($\approx 10^7/cm^3$), the reported bulk GaAs devices behaved more as extrinsic devices rather than intrinsic devices. However, the active region widths for SI GaAs material commonly demonstrated dependencies quite different than described by Eq. (41) (Barraud, 1963b; McGregor et al., 1992a; Beaumont et al., 1993c; Berwick et al., 1993a; McGregor et al., 1994a).

Models have been proposed to account for the discrepancy that take into account the effect of deep compensation centers (McGregor et al., 1992a, 1994a,b; Beaumont et al., 1993a). Undoped SI LEC GaAs is heavily compensated with the arsenic antisite deep donor labeled EL2 (Martin et al., 1977, 1980; Bourgoin et al., 1988). The material is semiinsulating due to a careful concentration balance between the native defect EL2 and residual shallow carbon acceptor impurities and possibly other native acceptor defects. If the capture cross section of a deep level is assumed to be constant and the reverse leakage current is assumed negligible, the solution to the electric field in a reverse biased diode demonstrates distinctively different regions (Ryvkin et al., 1968; Milnes, 1973; McGregor et al., 1992a, 1994a). Under equilibrium conditions, only a fraction of the compensating deep donors are ionized in the detector bulk and the Fermi energy level is pinned near the middle of the energy gap. The model indicates that the Schottky contact potential will cause full ionization of the deep donors in the region near the contact, hence creating a finite region of high positive space charge. Application of increasing reverse voltage extends the region of complete deep level ionization further into the detector bulk. Consequently, the nonuniform ionization distribution promotes the appearance of a two zone electric field distribution across the detectors (Ryvkin et al., 1968; McGregor et al., 1992a, 1994a). The electric field decreases rapidly near the Schottky contact with a slope approximately equal to qN_{DD}/ϵ_s, and the high field region width would continue to demonstrate a $\sqrt{V}$ dependence. Further from the contact, the deep donors are no longer fully ionized and the electric field decreases less rapidly. The model demonstrates the possible perturbation of an electric field in the presence of deep donors; however, it fails to account for experimentally observed results.

The conventional assumption for a reverse biased Schottky junction is that all deep donor levels below the n quasi Fermi energy level are filled (neutral) and all above are empty (positively charged). It is also generally assumed that the reverse leakage current is negligible, hence the quasi-Fermi levels would be practically constant through the device active region. However, the relationship is derived for near equilibrium conditions, and it becomes clear that the devices are not at equilibrium at the high bias voltages usually applied during operation. Rhoderick and Williams (1988) suggest that, although the actual electron quasi-Fermi level in the detector active region rises considerably as it approaches the metal contact (Crowell and Beguwala, 1971), the relation between the carrier capture rate and emission rate at a deep center determines the probability of a trap being occupied. Barraud (1963b) attributes the elongation of the active region in SI GaAs detectors to compensation of positively charged centers by the trapping of electrons in high electric fields. Panousis *et al.* (1969) suggest that electric field enhanced changes in the ratio between the electron and hole emission rates can result in a significant change in the overall ionized deep acceptor concentration in Au doped Si. The effect described by Panousis *et al.* (1969) can promote the appearance of a quasi-neutral region across the bulk of reverse biased Au doped Si diodes with n^+ contacts. It has been suggested that a similar condition may arise in bulk GaAs due to electric field enhanced changes in the ratio between the electron capture and electron emission rates at deep donor centers (McGregor *et al.*, 1994a,b).

The steady state ionized deep donor concentration can be approximated by (Look, 1989)

$$N_{DD}^+ \approx \frac{N_{DD}}{1 + n\sigma_e v_e/e_n} \tag{42}$$

where n is the free electron concentration, v_e is the electron thermal velocity, σ_e is the deep donor electron capture cross section, and e_n is the emission rate. There is evidence that the electron capture cross section for EL2 increases with increasing electric field (Prinz and Bobylev, 1980; Prinz and Rechkunov, 1983), and it is suggested that the effect is the possible cause of LFOs observed in SI GaAs (Kaminska *et al.*, 1982; Johnson, 1989). Models of injected charge through a forward biased ohmic contact indicate that injected electrons can populate the deep donor centers and produce a neutralization effect (Adams *et al.*, 1993). It is suggested that substantial leakage current through a reverse biased GaAs Schottky detector may create a condition at high electric fields in which the compensating deep donor centers (EL2) are driven toward neutrality by field enhanced electron capture, hence promoting the appearance of a quasi-neutral region (McGregor *et al.*, 1994a,b).

The emission rate (e_n) is assumed to be constant at electric fields below 10^5 V/cm (Makram-Ebeid, 1981), and the capture rate ($n\sigma_e v_e$) is assumed to increase with electric field due to increased velocity, increased leakage current, and field enhanced increases in the capture cross section. The actual EL2 capture cross section as a function of electric field is not known, and the field enhanced capture model uses an arbitrary function to describe the capture effect (McGregor *et al.*, 1994a,b). The field enhanced capture model indicates that two predominant electric field regions should appear as a result of the difference in space charge in the low and high field regions. At high electric fields, the electron capture rate exceeds the electron emission rate, which reduces the number of ionized deep donors to a concentration near the shallow acceptor concentration. The result is a region of low net positive space charge. At low electric fields, the lower electron capture cross section decreases the electron capture rate such that more deep donors become ionized, hence increasing the positive space charge. A fairly flat electric field should appear near the Schottky contact due to the low space charge density and an abrupt decrease in electric field should appear further in the bulk due to the high concentration of ionized deep donors. The model also demonstrates a practically linear dependence on the active region width with applied voltage. Hence, the field enhanced capture model presents a plausible explanation for the observed alpha particle pulse height curves (McGregor et al., 1992a,b, 1994a,b) and the measured two zone electric field profiles (Barraud, 1963b; Berwick et al., 1993a,b).

However it should be pointed out that the field enhanced capture cross section model utilizes an arbitrary "filling function" to describe the capture characteristics of the deep donor centers, and the model is simply a hypothesis that has not yet been proven. The simple field enhanced model also assumes that the EL2 deep donor centers become completely ionized under reverse bias near the end of the detector active region, which translates into a region of high charge density nearly equivalent to the net charge density of donor and acceptor centers. As measured by Berwick et al. (1993a,b) this does not appear to be the case. Simple calculations from Fig. 11 indicate that the charge density in the lower field region ranges between approximately 10^{12} and $10^{13}/cm^3$. The low net ionization density measured under steady state conditions represents a relatively small change from the charge balanced neutral equilibrium conditions. The infrared measurement technique employed by Berwick et al. (1993b) to measure the change in EL2 concentration is generally insensitive to such small changes, which provides a simple explanation for the lack of any observed increase in EL2 deep donor ionization as reported.

A model proposed by Kubicki et al. (1994) provides a possible explanation for the low net charge density in the detector active region. The model is based on earlier models reported by Crowell and Beguwala (1971), Rhoderick (1972), and Ma et al. (1988). The model includes the effect of reverse bias current on the quasi-Fermi level in the active region, which is calculated to effectively pin near the deep donor states in the active region. Ma et al. (1988) suggest that such an effect may account for observed reductions in EL2 DLTS signals with decreasing Schottky barrier height. The position of the quasi-Fermi level as related to the conduction band is calculated within the Fermi–Dirac distribution function, and the location of the quasi-Fermi level was found by simultaneously solving Poisson's equation and the electron continuity equation. Hole current was assumed negligible and was not addressed. The solution to the system of equations indicated that a very low space charge density would be present in the detector active region ($\approx 4.3 \times 10^{12}/cm^3$) with an abrupt increase in space charge density near the end of the active region. The electric field is highest near the Schottky contact and decreases linearly with depth into the detector. The model indicates that the active region of the detector, the region in which the electric field appreciably influences the charge carrier motion, should extend 300 μm with an application of 400 V reverse bias. The model also indicates that the EL2 deep donor centers would not be fully ionized in the detector bulk primarily due to the effect of leakage currents. The active region, while holding the bias at 400 V, is calculated to increase with increasing leakage current density, ranging from 185 μm at a leakage current of 8 nA/mm^2 to 315 μm at a leakage current of 25 nA/mm^2. The model also indicates that the active region width will increase as the total deep donor level density decreases. However, the model does not take into account changes in carrier capture and emission rates with electric field. The model also does not demonstrate the appearance of a two zone electric field distribution as reported by Barraud (1963b) and Berwick (1993a,b), where a fairly flat electric field region extends from the Schottky contact followed by a decreasing lower field region.

7. Discussion

Most of the recent work performed on bulk GaAs detectors report results from undoped SI LEC GaAs material. The high resistivity and relatively low concentration of impurities make the material attractive as a candidate for radiation spectrometers. Unfortunately, two major problems arise that must be resolved before SI bulk GaAs becomes a viable candidate for a room temperature gamma ray spectrometer.

The first major problem is the electric field distortion presently attributed to the high concentration of native deep donors in the material. Although portions of the electric field may be fairly constant as reported by Berwick et al. (1993a,b), the required voltage for full extension of the active region across a few mm thickness is comparable to that required for HPGe detectors of several cm in thickness. The efficiency of GaAs gamma ray detectors of such small size can not compete with present day HPGe detectors.

The second major problem is the fact that carrier mean free drift times (or lifetimes) are relatively short. The short carrier lifetimes are attributed to the high concentrations of various electron and hole traps in the bandgap (Martin et al., 1977; Mitonneau, Martin, and Mircea, 1977; Elliott et al., 1982; Figielski, 1984; Bourgoin et al., 1988; Look, 1989). The high density of traps and recombination centers create a situation in which the carrier extraction factors (ϱ_e and ϱ_h) are ultimately reduced to levels inconsistent with that necessary for high resolution gamma ray spectroscopy. Even if the electric fields could be extended across several cm of thickness the carrier extraction factors would be too small for useful spectrometers. Assuming saturated carrier velocities, a 1 mm thick detector with electron and hole lifetimes of 5 nsec would have CEF values approximately equal to 0.5, corresponding to an expected resolution (from Fig. 5) of 10.4% when defined as $100\sigma_{q*}/\langle Q*\rangle$. The differential pulse height spectrum is typically skewed in the presence of severe carrier trapping; however, if the spectrum were in fact Gaussian the conventional definition of resolution would be approximately 25% at FWHM. If the carrier mean free drift times are uneven so that the value of τ_e^* is 10 nsec and the value of τ_h^* is 5 nsec, then the resolution would broaden to 12.5% or, in the case of a Gaussian spectrum, 30% at FWHM. A GaAs detector with resolution (defined as $\sigma_{q*}/\langle Q*\rangle$) within 1% would require increasing the electron and hole mean free drift times greater than 100 nsec for a 1 mm thick detector and greater than 1 μsec for a 1 cm thick detector.

It appears that the problems observed with bulk GaAs are related to intrinsic defects and impurities. The realization of viable room temperature bulk GaAs gamma ray detectors depends on the production of bulk material with significantly reduced trap and impurity concentrations. Theory suggests that the carrier lifetimes can be significantly increased with the reduction of trap and impurity concentrations (Hall, 1952, 1959; Shockley and Read, 1952). This appears to have been demonstrated with high purity LPE GaAs material (Ryan and Eberhardt, 1972), and detectors fabricated from high purity LPE GaAs have demonstrated good energy resolution at room temperature (Eberhardt *et al.*, 1970, 1971a,b; Kobayashi *et al.*, 1972b; Hesse and Gramann, 1971; Gibbons and Howes, 1972; Hesse *et al.*, 1972; Alexiev and Butcher, 1992). The GaAs described in these previous works was of the highest purity available and was not compensated with deep level impurities or native defects. It seems only reasonable that efforts toward the production of bulk GaAs spectrometers in the future should be directed toward the improvement of the bulk material itself.

Bulk GaAs has been demonstrated as an alternative alpha particle, beta particle, and conversion electron spectrometer (Bertin *et al.*, 1990; McGregor *et al.*, 1992a; Benz *et al.*, 1992); however, the detectors show no practical advantage over commercially available Si based detectors. Gamma rays at energies of 60 keV and 122 keV have been measured at room temperature with energy resolution ranging from 13% to 35% FWHM (McGregor *et al.*, 1992a; Benz *et al.*, 1992; Sumner *et al.*, 1992; Berwick *et al.*, 1993a). However, room temperature operation at the expense of such poor resolution hardly competes with cooled HPGe detectors or other more developed room temperature gamma ray detection materials. Practical bulk GaAs gamma ray spectrometers will become a reality when improved crystal growth processes are implemented to produce high purity material with significantly reduced impurity and native defect trap concentrations.

VI. Bulk GaAs Photoconductive Detectors Operated in Current Mode

Detectors are used in current mode to measure signals from transient events that cause many particles to interact in the detector in a relatively short time. The detector output, the convolution of the detector impulse response with the rate of interactions, provides a good measure of the transient event if the impulse response is sufficiently fast. Detectors made of bulk GaAs can be used in the current mode to detect photons and charged particles over a wide range of energies. Many applications also exist for the detection of visible light, but will not be addressed in this report. These detectors, often called "photoconductive detectors," effectively operate as variable resistors. In a typical application, ohmic contacts are fabricated on two opposing faces of a cylindrical or parallelepiped crystal cut from bulk LEC GaAs material. An external bias is applied to the detector, and one measures either the induced current from the detector or the voltage drop across the load resistor placed in series with the detector. These detectors have been used to measure transient signals in various room temperature applications including Compton spectrometers, recoil proton detectors, x-ray detectors, and gamma ray detectors (Wagner et al., 1986; Wang et al., 1986, 1989; Friant et al., 1989; Kammeraad et al., 1992; Moy et al., 1992). Incident radiation generates charge carriers in the material, thereby increasing its effective conductivity. A change in conductivity proportional to the rate of illumination is desired for ease of use in the current mode. However, it is interesting to note that a linear response is obtained only if the concentration of charge carriers (n) is less than the concentration of traps (N_T). Kittel (1971) shows that if $n \gg N_T$, then the conductivity has a square root dependence on the illumination rate. Thus a nonlinear response could be obtained for incident pulses that are sufficiently long that charge carrier traps become substantially filled. Wagner et al (1986) have observed that long pulses incident upon GaAs and InP can cause a superlinear response when electron traps become sufficiently filled. Thus charge carrier traps are essential to detectors operating in the current mode. The issue becomes one of determining how to obtain the appropriate type of traps and the required concentration. At present this is more of an art than a science.

Methods used to introduce deep level traps in GaAs include Cr doping, and irradiation of either Cr doped or undoped GaAs with electrons, gamma rays, heavy ions, and neutrons. Of these, neutron irradition is the most effective for producing traps that significantly shorten the carrier lifetime and hence improve the time response. This method was first used for detectors in current mode by Wagner et al. (1986), who irradiated GaAs with neutrons from a reactor. The GaAs crystals were wrapped in indium foil and enclosed in a boron compound to filter out low energy neutrons that would cause transmutation doping in the material. The higher energy neutrons are believed to introduce traps or recombination centers by causing structural damage in the GaAs crystal. The formation of deep electron traps in undoped GaAs has been studied by Martin et al. (1984). Wang et al. (1989) used 14 MeV neutrons to generate structural defects in Cr-doped and undoped GaAs, and measured the amplitude and pulse width of the detector response as a function of neutron fluence (Fig. 14). The charge carrier lifetime and mobility were found to decrease monotonically with increasing neutron fluence. For fluences of about 5×10^{15} n/cm^2, the FWHM of the response was measured to be less than 30 psec. This time resolution is obtained at the expense of sensitivity in which the carrier mean free drift lengths (λ_e^* and λ_h^*) are significantly reduced. The electron and hole CEF values will be small, in which case the induced current should be practically position independent of the radiation interaction location (Knoll and McGregor, 1993). The gain–bandwidth tradeoff is apparent even for relatively low levels of neutron irradiation. An example is shown in Fig. 15, which compares the signals from two Cr-doped GaAs detectors without and with 500 sec of neutron irradiation (roughly a total of 6×10^{14} n/cm^2) in the Omega West reactor at Los Alamos National Laboratory. The long tail evident on the signal from the undamaged detector is significantly reduced at the expense of amplitude.

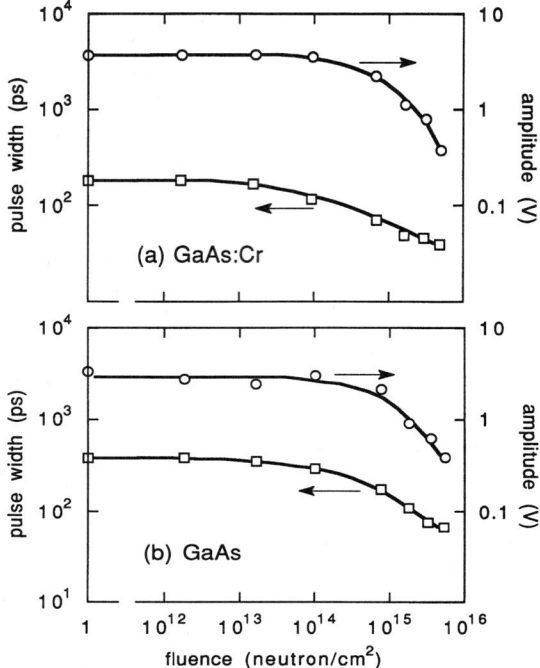

FIG. 14. Pulse width and amplitude of the response from (a) a Cr-doped and (b) an undoped GaAs detector as a function of neutron irradiation fluence. (Reprinted with permission from Wang *et al.*, 1989 and the American Institute of Physics.)

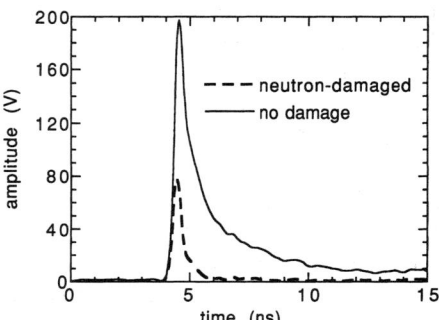

FIG. 15. Comparison of the response of two similar GaAs photoconductive detectors, where one detector was irradiated for 500 sec in a reactor. The detectors were made of SI bulk GaAs that was heavily doped with chromium. Detector dimensions were 5 × 16 × 7 mm. Ohmic contacts were applied to the 5 × 16 mm faces. The detectors were biased to 7 kV and were excited by 30 psec pulses (0.4 pC per pulse) of 8 MeV electrons. The electron pulses were collimated to 3 mm and struck the detector at the center of the 5 × 16 mm face (parallel to the applied electric field). (Reprinted with permission from Kammeraad *et al.*, 1992 and the Society of Photo-Optical Instrumentation Engineers.)

10. GALLIUM ARSENIDE RADIATION DETECTORS AND SPECTROMETERS

The response of a photoconductive detector to an impulse of penetrating radiation is easy to derive. The total current induced in the circuit is the sum of the contributions from electron current and hole current

$$I_{\text{tot}}(t) = \frac{qN_e(t)v_e(t)}{W} + \frac{qN_h(t)v_h(t)}{W} = I_e(t) + I_h(t) \qquad (43)$$

where N is the number of generated carriers, W is the detector width between the electrodes, v is the carrier velocity, and the e and h subscripts represent electrons and holes, respectively. The carrier population as a function of time decreases exponentially. We consider the situation in which the carrier lifetimes are much less than the carrier extraction times, as is the case for detectors with considerable neutron damage. For low electric fields in which μ_e and μ_h describe a linear relationship between carrier velocities and the applied electric field, the total current can be represented by

$$I_{\text{tot}}(t) = \frac{qN_0 V_B}{W^2} (\mu_e \exp[-t/\tau_e] + \mu_h \exp[-t/\tau_h]) \qquad (44)$$

where N_0 is the total number of electron–hole pairs excited and V_B is the applied bias voltage. At higher fields, the carrier mobilities no longer describe the velocity characteristics and the induced current is more appropriately described by Eq. (28). The total charge (Q^*) induced in the circuit, as represented by Eq. (33), is proportional to the electron and hole carrier extraction factors. The CEF values can be affected by the concentration of charge carrier traps and recombination centers. Charge carriers can scatter from either type of defect and the carrier mobilities are reduced, hence reducing the average carrier drift velocity. The carrier lifetimes can be reduced because the defects act as traps, recombination centers, or both.

The previously described response is appropriate to both uniform and nonuniform illumination of the detector provided that incident radiation penetrates the detector. Consider the situation in which a highly collimated, energetic beam is incident upon a neutron damaged detector that is large by comparison to the beam diameter. The response is observed to be independent of the beam position in accordance with Eq. (31) (Kammeraad et al., 1992). This is the case even if the beam strikes the detector sufficiently far from the electrodes so that none of the generated carriers is likely to drift to an electrode (very small CEFs). For example, taking $\mu_e = 1400$ cm^2/V-sec and $\tau_e = 250$ psec for neutron damaged GaAs (Wang et al., 1989) and using typical values of $V_B = 2500$ V and $W = 5$ mm, the drift time from the center to an electrode is 35 nsec, and the probability for drifting to the electrode is extremely small ($\exp[-35/0.25]$ and $\rho_e = 0.0035$). Thus, the response can not be explained simply by the number of charges swept to the electrodes. Models assuming that charge carriers must reach the electrodes before a charge can be measured would predict that the response is exponentially dependent (in the present case) on position; it would also predict a signal delay corresponding to the drift time. No such delay is observed and the displacement current is propagated instantaneously in the material.

If the incident radiation does not penetrate the detector, the response could be position dependent. One possible reason is that, for very shallow penetrations, the effect of the material surface could be significant. Also, in some geometries the electric field may vary with position, causing the drift velocities and carrier lifetimes (which can also depend on the electric field) to be position dependent. Furthermore, if the material is not homogeneous, the carrier lifetimes could be position dependent, regardless of whether the incident radiation is deeply penetrating or not. Position dependent responses have been measured for visible laser light incident upon GaAs photoconductive detectors in a stripline geometry (Moy *et al.*, 1992).

The charge gain of the detector for incident particles (which differs from photoconductive gain) is defined as

$$G_{ch} = \frac{Q^*}{Q_{in}} \tag{45}$$

where Q_{in} represents the total charge of the incident pulse of radiation on the detector and Q^* represents the total charge induced by the excited electron–hole pairs. For penetrating electrons, typical gains are on the order of 10^3; i.e., the sensitivity is on the order of 10^{-16} C/e^-. For protons that stop in the detector, sensitivities of 10^{-13} C/p^+ have been obtained (Wang, Flatley, and Pocha, 1991). For gamma rays, sensitivities of 10^{-8}–10^{-9} C/rad have been obtained (Friant *et al.*, 1989). The gain depends upon the choice of bias voltage since the carrier velocities depend upon the electric field strength. However, the gain cannot be increased indefinitely by increasing the bias voltage, since the carrier velocities are saturation limited. The time response also has a dependence on the applied electric field, although rather weakly. Typical electric fields for these devices are 500 to 100 V/cm with leakage currents of 10^{-7} to 10^{-6} A at room temperature. The choice of bias is also important to the dynamic range of the detector, in which a higher bias generally allows for a larger dynamic range. Nonlinearity of the peak output voltage can typically be expected for signals exceeding about 10% of the bias voltage, a condition that arises simply because of the effective resistance of the GaAs in the "voltage divider" circuit. Linear responses up to 5 A peak instantaneous current are typically reported. Furthermore, these detectors can perform reproducibly quite far into the nonlinear regime, where meaningful results can be "unfolded" if impulse responses have been measured.

A final consideration in the use of GaAs photoconductive detectors is the areal density of the incident beam. If the incident beam is highly focused, one must consider the density of the generated charge carriers rather than just the total amount. A detector may respond linearly to energy deposited uniformly throughout its volume, but may respond nonlinearly if the same energy is deposited in a smaller portion of the volume. This is because the electric field that arises from the separating plasma of electrons and holes can become comparable to the applied field. In addition, nonlinearities can be introduced if the local density of charge carriers exceeds that of the charge carrier traps, as described earlier in this section.

VII. Summary

GaAs has shown promise as a room temperature operated radiation detector and spectrometer. Schottky barrier detectors fabricated from LPE GaAs have demonstrated good energy resolution for low energy gamma rays and x-rays indicating that GaAs in pure form is a viable candidate for room temperature gamma ray spectroscopy. Epitaxial detectors are restricted to thin films generally no thicker than 200 μm, hence their application is restricted to low energy gamma rays. Thicker spectrometers fabricated from bulk GaAs have shown good energy resolution at room temperature for charged particles; however, gamma ray resolution suffers due to short carrier lifetimes and poor charge carrier extraction from the devices. The realization of thick bulk GaAs high resolution spectrometers requires advances in bulk material growth processes so that impurities and native defects are significantly reduced. Crystals cut from bulk Cr-doped and undoped GaAs have been used successfully in the current mode to detect high speed transient responses from charged particles and gamma rays. The desired carrier lifetimes are short (opposite of the spectrometer case), which can be achieved through severe neutron damaging of the bulk GaAs crystal. The detectors are relatively rugged and time resolution down to 30 psec has been observed at the expense of detector gain. GaAs offers a reasonable compromise between the fundamental material characteristics generally desired for room temperature operated detectors; and with advances in material production, GaAs detectors may become viable alternative detectors for radiation detection and spectroscopy.

Acknowledgments

The authors express their gratitude to R. J. Anderson, C. M. Buttar, R. B. James, G. F. Knoll, D. C. Lock, L. S. Pan, M. Schieber, T. E. Schlesinger, and S. P. Swierkowski for their helpful suggestions and for critically reviewing the presented subject matter.

References

Acket, G. A., and Scheer, J. J. (1971). *Solid State Electron.* **14,** 167.
Adams, J. C., Capps, C. D., Falk, R. A., and Ferrier, S. G. (1993). *Appl. Phys. Lett.* **63,** 633.

Afanas'ev, V. F. (1971). *Sov. Phys. Semicond.* **4**, 1171.
Akutagawa, W., Zanio, K., and Mayer, J. W. (1967). *Nucl. Instr. and Meth.* **55**, 383.
Alexiev, D., and Butcher, K. S. A. (1992). *Nucl. Instr. and Meth.* **A317**, 111.
Armantrout, G. A., Swierkowski, S. P., Sherohman, J. W., and Yee, J. H. (1977). *IEEE Trans. Nucl. Sci.* **NS-24**, 121.
Ashman, J. G., Booth, C. N., Buttar, C. M., Combley, F. H., Dogru, M., Grey, R., Hill, G., Hou, Y., and Houston, P. (1992). In *GaAs Detectors and Electronics for High Energy Physics*, ed. C. del Papa, P. G. Pelfer, and K. Smith, p. 222. World Scientific, Singapore.
Aspnes, D. E. (1976). *Phys. Rev. B* **14**, 5331.
Aspnes, D. E. (1977). In *Gallium Arsenide and Related Compounds*, No. 33b, p. 110. Institute of Physics, London.
Aspnes, D. E., and Studna, A. A. (1973). *Phys. Rev. B* **7**, 4605.
AuCoin, T. R., and Savage, R. O. (1990). In *Gallium Arsenide Technology*, ed. D. K. Ferry. p. 47 H. W. Sams, Indianapolis.
Baraff, G. A. (1992). In *Deep Centers in Semiconductors*, 2nd ed., ed. S. T. Pantelides, p. 547. Gordon and Breach, Philadelphia.
Bardeen, J. (1947). *Phys. Rev.* **71**, 717.
Barraud, A. (1963a). *Comptes Rendus* **256**, 3632.
Barraud, A. (1963b). *Comptes Rendus* **257**, 1263.
Beaumont, S. P., et al. (1992). *Nucl. Instr. and Meth.* **A322**, 472.
Beaumont, S. P., et al. (1993a). *Nucl. Instr. and Meth.* **A326**, 313.
Beaumont, S. P., et al. (1993b). *IEEE Trans. Nucl. Sci.* **NS-40**, 1225.
Beaumont, S. P., et al. (1993c). *Nucl. Phys. B,* Suppl. 32, 296.
Beaumont, S. P., et al. (1994a). *Nucl. Instr. and Meth.* **A342**, 83.
Beaumont, S. P., et al. (1994b). *Nucl. Instr. and Meth.* **A348**, 514.
Bencivelli, W., Bertin, R., Bertolucci, E., Bottigli, U., D'Auria, S., Del Papa, C., Fantacci, M. E., Randaccio, P., Rosso, V., and Stefanini, A. (1994). *Nucl. Instr. and Meth.* **A338**, 549.
Benz, K. W., Irsigler, R., Ludwig, J., Rosenzweig, J., Runge, K., Schäfer, F., Schneider, J., and Webel, M. (1992). *Nucl. Instr. and Meth.* **A322**, 493.
Bertin, R., D'Auria, S., del Papa, C., Fiori, F., Lisowski, B., O'Shea, V., Pelfer, P. G., Smith, K., and Zichichi, A. (1990). *Nucl. Instr. and Meth.* **A294**, 211.
Bertolini, G., and Coche, A. (eds.). (1968). *Semiconductor Radiation Detectors*. North-Holland Press, Amsterdam.
Berwick, K., Brozel, M. R., Buttar, C. M., Cowperthwaite, M., and Hou, Y. (1993a). In MRS Proc., *Semiconductors for Room-Temperature Radiation Detector Applications*, ed. R. B. James, T. E. Schlesinger, P. Siffert, and L. Franks, **302**, p. 363. MRS, Pittsburgh.
Berwick, K., Brozel, M. R., Buttar, C. M., Cowperthwaite, M., and Hou, Y. (1993b). In Inst. Phys. Conf. Ser., *Defect Recognition and Image Processing in Semiconductors and Devices*, No. 135, p. 305.
Blakemore, J. S. (1982). *J. Appl. Phys.* **53**, R123.
Blakemore, J. S. (1987). *Semiconductor Statistics*. Dover Books, New York.
Blocker, T. G., Cox, R. H., and Hasty, T. E. (1970). *Solid-State Communic.* **8**, 1313.
Bolger, D. E., Franks, J., Gordon, J., and Whitaker, J. (1966). *Proc. Internat. Symp. on GaAs*, p. 16.
Bourgoin, J. C., von Bardeleben, H. J., and Stievenard, D. (1988). *J. Appl. Phys.* **64**, R65.
Breivik, L., Brozel, M. R., Stirland, D. J., and Tüzemen, S. (1992). *Semicond. Sci. Technol.* **7**, A269.
Brullot, B., Galli, R., Lecat, X., Rubbelynck, C., and Pochet, T. (1993). In MRS Proc., *Semiconductors for Room-Temperature Radiation Detector Applications*, ed. R. B. James, T. E. Schlesinger, P. Siffert, and L. Franks, **302**, p. 369. MRS, Pittsburgh.
Buttar, C. M., Combley, F. H., Dawson, I., Dogru, M., Harrison, M., Hill, G., Hou, Y., and Houston, P. (1991). *Nucl. Instr. and Meth.* **A310**, 208
Casey, H. C., Jr., and Stern, F. (1976). *J. Appl. Phys.* **47**, 631.
Casey, H. C., Jr., Miller, B. I., and Pinkas, E. (1973). *J. Appl. Phys.* **44**, 1281.
Chmill, V. B., Chuntonov, A. V., Sergeev, V. A., Smol, A. V., Tsyupa, Yu.P., Vorobiev, A. P., Gor-

dienko, A. I., Khludkov, S. S., Koretskaya, O. B., Potapov, A. I., and Tolbanov, O. P. (1993). *Nucl. Instr. and Meth.* A**326**, 310.
Chmill, V. B., Chuntonov, A. V., Vorobiev, A. P., Khludkov, S. S., Koretski, A. V., Potapov, A. I., and Tolbanov, O. P. (1994). *Nucl. Instr. and Meth.* A**340**, 328.
Clemans, J. E., Ejim, T. I., Gault, W. A., and Monberg, E. M. (1989). *AT&T Tech. J.* **68**, 29.
Crowell, C. R., and Beguwala, M. (1971). *Solid State Elec.* **14**, 1149.
Cusano, D. A. (1964). *Solid State Communic.* **2**, 353.
Cuzin, M. (1987). *Nucl. Instr. and Meth.* A**253**, 407.
Day, R. B., Dearnaley, G., and Palms, J. M. (1967). *IEEE Trans. Nucl. Sci.* **NS-14**, 487.
Dearnaley, G. (1963). *IEEE Trans. Nucl. Sci.* **NS-10**, 106.
Dearnaley, G., and Northrop, D. C. (1964). *Semiconductor Counters for Nuclear Radiations.* E&FN Spon, London.
Derhacobian, N., and Haegel, N. M. (1991). *Phys. Rev.* B **44**, 12754.
DiLorenzo, J. V., and Khandelwal, D. D. (1982). *GaAs FET Principles and Technology.* Artech House, Dedham, MA.
Dodge, W. R., Domen, S. R. Hirshfeld, A. T., and Hoppes, D. D. (1964). *IEEE Trans. Nucl. Sci.* **NS-11**, 238.
Dodge, W. R., Domen, S. R. Hirshfeld, A. T., and Hoppes, D. D. (1965). *IEEE Trans. Nucl. Sci.* **NS-12**, 295.
Dogru, M., et al. (1994). *Nucl. Instr. and Meth.* A**348**, 510.
Doi, A., Asano, T., and Migitaka, M. (1976). *J. Appl. Phys.* **47**, 1589.
Dudenkova, A. V., and Nikitin, V. V. (1967). *Sov. Phys. Solid State* **8**, 2432.
Eberhardt, J. E., Ryan, R. D. and Tavendale, A. J. (1970). *Appl. Phys. Lett.* **17**, 427.
Eberhardt, J. E., Ryan, R. D., and Tavendale, A. J. (1971a). Proceedings of the International Symposium on Cadmium Telluride, Strasbourg, paper no. 29.
Eberhardt, J. E., Ryan, R. D. and Tavendale, A. J. (1971b). *Nucl. Instr. and Meth.* **94**, 463.
Evans, R. D. (1982). *The Atomic Nucleus.* Krieger, Malabar, Florida.
Elliott, K. R., Holmes, D. E., Chen, R. T., and Kirkpatrick, C. G. (1982). *Appl. Phys. Lett.* **40**, 898.
Fano, U. (1947). *Phys. Rev.* **72**, 26.
Figielski, T. (1984). *Appl. Phys.* A **35**, 255.
Friant, A., Saliou, C., Galli, R., and Barday, S. (1989). *Nucl. Instr. and Meth.* A**283**, 318.
Franz, W. (1958). *Z. Naturforsch* A **13**, 484.
Gault, W. A., Monberg, E. M., and Clemans, J. E. (1986). *J. Crystal Growth* **74**, 491.
George, G. G., and Gunnersen, E. M. (1964). *Nucl. Instr. and Meth.* **25**, 253.
Gibbons, P. E., and Howes, J. H. (1972). *IEEE Trans. Nucl. Sci.* **NS-19**, 353.
Goldstein, B. (1971). *J. Appl. Phys.* **42**, 2570.
Grant, S. M., and Sumner, T J. (1992). In *GaAs Detectors and Electronics for High Energy Physics*, ed. C. del Papa, P. G. Pelfer, and K. Smith, p. 101. World Scientific, Singapore.
Gunn, J. B. (1963). *Solid State Communic.* **1**, 88.
Haegel, N. M., Winnacker, A., Leo, K., Rühle, W. W., and Gisdakis, S. (1987). *J. Appl. Phys.* **62**, 2946.
Hall, R. N. (1952). *Phys. Rev.* **87**, 387.
Hall, R. N. (1959). *Proc. IEE* **106B** (Suppl. 17), 923.
Harding, W. R., Hilsum, C. Moncaster, M. E., Northrop, D. C., and Simpson, O. (1960). *Nature* **187**(4735), 405.
Henry, C. H., and Lang, D. V. (1977). *Phys. Rev.* B **15**, 989.
Henry, R. L., Nordquist, P. E. R., Gorman, R. J., and Qadri, S. B. (1991). *J. Crys. Growth* **109**, 228.
Hesse, K., and Gramann, W. (1971). Proceedings of the International Symposium on Cadmium Telluride, Strasbourg, paper no. 30.
Hesse, K., Gramann, W., and Höppner, D. (1972). *Nucl. Instr. and Meth.* **101**, 39.
Hicks, H. G. B., and Manley, D. F. (1969). *Solid State Commun.* **7**, 1463.
Hilsum, C. (1962). *Proc. IRE* **50**, 185.
Holonyak, N., Jr., and Bevacqua, S. F. (1963). *Appl. Phys. Lett.* **2**, 71.
Holmes, D. E., Chen, R. T., Elliott, K. R. and Kirkpatrick, C. G. (1982). *Appl. Phys. Lett.* **40**, 46.

Howes, M. J., and Morgan, D. V. (eds.). (1985). *Gallium Arsenide Materials, Devices and Circuits.* John Wiley & Sons, New York.
Huber, A. M., Linh, N. T., Valladon, M., Debrun, J. L., Martn, G. M., Mitonneau, A., and Mircea, A. (1979). *J. Appl. Phys.* **50,** 4022.
Hwang, C. J. (1969). *J. Appl. Phys.* **40,** 3731.
Hwang, C. J. (1971). *J. Appl. Phys.* **42,** 4408.
Hwang, C. J. (1972). *Phys. Rev. B* **6,** 1355.
Jen, C. K. (1941). *Proc. IRE* **29,** 345.
Johnson, D. A. (1989). Thesis Arizona State University.
Kaminska, M., Parsey, J. M., Lagowski, J., and Gatos, H. C. (1982). *Appl. Phys. Lett.* **41,** 989.
Kammeraad, J. E., Sale, K. E., Wang, C. L., and Baltrusaitis, R. M. (1992). *Proc. SPIE* **1734,** 242.
Karpinski, W., Kubicki, T., Lübelsmeyer, K., Toporowsky, M., Wallraff, W., Heime, K., and Wüller, R. (1992). *Nucl. Instr. and Meth.* **A323,** 635.
Kashiwa, M., Otoki, Y., Seki, M., Taharasako, S., and Okubo, S. (1990). *Hitachi Cable Rev.* **9,** 55.
Keldysh, L. V. (1958). *Sov. Phys. JETP* **34,** 788.
Kittel, C. (1971). *Introduction to Solid State Physics,* 4th ed. John Wiley & Sons, New York.
Knoll, G. F. (1989). *Radiation Detection and Measurement,* 2nd ed. John Wiley & Sons, New York.
Knoll, G. F., and McGregor, D. S. (1993). In MRS Proc., *Semiconductors for Room Temperature Radiation Detector Applications,* ed. R. B. James, T. E. Schlesinger, P. Siffert, and L. Franks, **302,** 3. MRS, Pittsburgh.
Kobayashi, T., and Sugita, T. (1972). *Nucl. Instr. and Meth.* **98,** 179.
Kobayashi, T., and Takayanagi, S. (1966). *Nucl. Instr. and Meth.* **44,** 145.
Kobayashi, T., and Takayanagi, S. (1971). *Nucl. Instr. and Meth.* **95,** 365.
Kobayashi, T., Sugita, T., Koyama, M., and Takayanagi, S. (1972). *IEEE Trans. Nucl. Sci.* **NS-19,** 324.
Kobayashi, T., Sugita, T., Takayanagi, S., Iio, M., and Sasaki, Y. (1973). *IEEE Trans. Nucl. Sci.* **NS-20,** 310.
Kobayashi, T., Kuru, I., Hojo, A., and Sugita, T. (1976). *IEEE Trans. Nucl. Sci.* **NS-23,** 97.
Kremer, R. E., Francomano, D., Beckhart, G. H., and Burke, K. M. (1990). *J. Mater. Res.* **5,** 1468.
Kubicki, T., Lübelsmeyer, K., Ortmanns, J., Pandoulas, D., Syben, O., Toporowsky, M., and Xiao, W. J. (1994). *Nucl. Instr. and Meth.* **A345,** 468.
Lampert, M. A. (1956). *Phys. Rev.* **103,** 1648.
Lampert, M. A., and Mark, P. (1970). *Current Injection in Solids.* Academic Press, New York.
Langley, R. A. (1973). *Nucl. Instr. and Meth.* **113,** 109.
Lax, M. (1960). *Phys. Rev.* **119,** 1502.
Lie, K. H., Hsu, J. T., Guo, Y. D., and Chen, T. P. (1991). *J. Crys. Growth* **109,** 205.
Liu, Y. M., and Coleman, J. A. (1971). *IEEE Trans. Nucl. Sci.* **NS-18,** 192.
Look, D. C. (1989). *Electrical Characterization of GaAs Materials and Devices.* John Wiley & Sons, New York.
Ma, Q. Y., Schmidt, M. T., Wu, X., Evans, H. L., and Yang, E. S. (1988). *J. Appl. Phys.* **64,** 2469.
Madelung, O. (1964). *Physics of III–V Compounds.* John Wiley & Sons, New York.
Makram-Ebeid, S. (1981). In MRS Proc., *Defects in Semiconductors,* ed. J. Narayan and T. Y. Tan, **2,** p. 495. North-Holland Press, New York.
Martin, G. M., and Makram-Ebeid, S. (1992). In *Deep Centers in Semiconductors,* 2nd ed., ed. S. T. Pantelides, p. 457. Gordon and Breach, Philadelphia.
Martin, G. M., Mitonneau, A., and Mircea, A. (1977). *Electron. Lett.* **13,** 191.
Martin, G. M., Farges, J. P., Jacob, G., Hallais, J. P., and Poiblaud, G. (1980). *J. Appl. Phys.* **51,** 2840.
Martin, G. M., Jacob, G., Hallais, J. P., Grainger, F., Roberts, J. A., Clegg, B., Blood, P., and Poiblaud, G. (1982). *J. Phys. C: Solid State Phys.* **15,** 1841.
Martin, G. M., Estéve, E., Langlade, P., and Makram-Ebeid, S. (1984). *J. Appl. Phys.* **56,** 2655.
Mayburg, S. (1961). *Solid-State Elec.* **2,** 195.
Mayer, J. W. (1962). *Nucleonics* **20,** 60.

McGregor, D. S., White, R. D., and Weichold, M. H. (1985). *TEES Tech. Rept. Ser.,* Texas A&M University, p. 80.
McGregor, D. S., Knoll, G. F., Eisen, Y. and Brake, R. (1992a). *IEEE Trans. Nucl. Sci.* **NS-39,** 1226.
McGregor, D. S., Knoll, G. F., Eisen, Y. and Brake, R. (1992b). *Nucl. Instr. and Meth.* **A322,** 487.
McGregor, D. S., Knoll, G. F., Eisen, Y., and Brake, R. (1992c). In *GaAs Detectors and Electronics for High Energy Physics,* ed. C. del Papa, P. G. Pelfer, and K. Smith, p. 30. World Scientific, Singapore.
McGregor, D. S., Rojeski, R. A., Knoll, G. F., Terry, F. L., Jr., East, J., and Eisen, Y. (1994a). *Nucl. Instr. and Meth.* **A343,** 527.
McGregor, D. S., Rojeski, R. A., Knoll, G. F., Terry, F. L., Jr., East, J., and Eisen, Y. (1994b). *J. Appl. Phys.* **75,** 7910; McGregor, D. S., Rojeski, R. A., Knoll, G. F., Terry, F. L., Jr., East, J., and Eisen, Y. (1995). *J. Appl. Phys.* **77,** 1331. (Errata)
Milnes, A. G. (1973). *Deep Impurities in Semiconductors.* John Wiley & Sons, New York.
Mitonneau, A., Martin, G. M., and Mircea, A. (1977). *Electron. Lett.* **13,** 666.
Miyazawa, S., Honda, T., Ishii, Y., and Ishida, S. (1984). *Appl. Phys. Lett.* **44,** 410.
Moss, T. S. (1961). *J. Appl. Phys.* **32**(Suppl. 10), 2136.
Moy, K. J., Wang, C. L., Flatley, J. E., Pocha, M. D., Davis, B. A., and Wagner, R. S. (1992). *Proc. SPIE* **1734,** 152.
Nelson, R. J., and Sobers, R. G. (1978). *J. Appl. Phys.* **49,** 6103.
Northrop, D. C., Thornton, P. R., and Trezise, K. E. (1964). *Solid-State Electron.* **7,** 17.
Oda, O., Yamamoto, H., Seiwa, M., Kano, G., Inoue, T., Mori, M., Shimakura, H., and Oyake, M. (1992). *Semicond. Sci. Techno.* **7,** A215.
Panousis, P. T., Krambeck, R. H., and Johnson, W. C. (1969). *Appl. Phys. Lett.* **15,** 79.
Papastamatiou, M. J., and Papaioannou, G. J. (1990). *J. Appl. Phys.* **68,** 1094.
Parsey, J. M., Jr., Nanishi, Y. Lagowski, J., and Gatos, H. C. (1981). *J. Electrochem. Soc.* **128,** 936.
Pearton, S. J., Haller, E. E., and Elliot, A. G. (1984). *Appl. Phys. Lett.* **44,** 684.
Pehl, R. H., Goulding, F. S., Landis, D. A., and Lenzlinger, M. (1968). *Nucl. Instr. and Meth.* **59,** 45.
Prinz, V. Ya., and Bobylev, B. A. (1980) *Sov. Phys. Semicond.* **14,** 1097.
Prinz, V. Ya., and Rechkunov, S. N. (1983). *Physica Status Solidi (b)* **118,** 159.
Ramo, S. (1939). *Proc. IRE* **27,** 584.
Rhoderick, E. H. (1972). *J. Phys. D* **5,** 1920.
Rhoderick, E. H., and Williams, R. H. (1988). *Metal-Semiconductor Contacts.* Clarendon Press, Oxford.
Rideout, V. L. (1975). *Solid State Elec.* **18,** 541.
Ridley, B. K., and Watkins, T. B. (1961) *Proc. Phys. Soc. London* **78,** 293.
Ritter, D., Weiser, K., and Zeldov, E. (1987). *J. Appl. Phys.* **62,** 4563.
Ryan, R. D. (1973). *IEEE Trans. Nucl. Sci.* **NS-20,** 473.
Ryan, R. D., and Eberhardt, J. E. (1972). *Solid-State Elec.* **15,** 865.
Ryvkin, S. M., Makovsky, L. L., Strokan, N. B., Subashieva, V. P., and Khusainov, A. Kh. (1968). *IEEE Trans. Nucl. Sci.* **NS-15,** 226.
Sah, C. T., Noyce, R. N., and Shockley, W. (1957). *Proc. IRE* **45,** 1228.
Sakai, E. (1982). *Nucl. Instr. and Meth.* **196,** 121.
Schubert, E. F. (1993). *Doping in III–V Semiconductors.* Cambridge University Press, Cambridge.
Sell, D. D., Casey, H. C., Jr., and Wecht, K. W. (1974). *J. Appl. Phys.* **45,** 2650.
Shockley, W. (1938). *J. Appl. Phys.* **9,** 635.
Shockley, W., and Read, W. T., Jr. (1952). *Phys. Rev.* **87,** 835.
Shur, M. (1987). *GaAs Devices and Circuits.* Plenum Press, New York.
Skowronski, M. (1992). In *Deep Centers in Semiconductors,* 2nd ed., ed. S. T. Pantelides, p. 379. Gordon and Breach, Philadelphia.
Spooner, N. J. C., Bewick, A., Holmes, S. N., Phillips, C. C., Quenby, J. J., Stradling, R. A., Sumner, T. J., Thomas, R. H., and Wang, P. D. (1991). *Nucl. Instr. and Meth.* **A310,** 227.
Spooner, N. J. C., Bewick, A., Quenby, J. J., Smith, P. F., and Lewin, J. D. (1992). In *GaAs Detectors*

and Electronics for High Energy Physics, ed. C. del Papa, P. G. Pelfer, and K. Smith, p. 156. World Scientific, Singapore.
Stuck, R., Ponpon, J. P., Siffert, P., and Ricaud, C. (1972). *IEEE Trans. Nucl. Sci.* **NS-19,** 270.
Sumner, T. J., Grant, S. M., Bewick, A., Li, J. P., Spooner, N. J. C., Smith, K., and Beaumont, S. P. (1991). *Proc. SPIE* **1549,** 256.
Sumner, T. J., Grant, S. M., Bewick, A., Li, J. P., Smith, K., and Beaumont, S. P. (1992). *Nucl. Instr. and Meth.* **A322,** 514.
Sumner, T. J., Grant, S. M., Alexiev, D., and Butcher, K. S. A. (1994). *Nucl. Instr. and Meth.* **A348,** 518.
Swierkowski, S. P., and Armantrout, G. A. (1975). *IEEE Trans. Nucl. Sci.* **NS-22,** 205.
Swiggard, E. M. (1989). *J. Crys. Growth* **94,** 556.
Sze, S. M. (1981). *Physics of Semiconductor Devices,* 2nd ed. John Wiley & Sons, New York.
Sze, S. M. (1985). *Semiconductor Devices, Physics and Technology.* John Wiley & Sons, New York.
Tavendale, A. J., and Lawson, E. M. (1972). *IEEE Trans. Nucl. Sci.* **NS-19,** 318.
Terashima, K., Katsumata, T., Orito, F., and Fukuda, T. (1984). *Jap. J. Appl. Phys.* **23,** L302.
Thomas, R. N., Hobgood, H. M., Eldridge, G. W., Barrett, D. L., Braggins, T. T., Ta, L. B., and Wang, S. K. (1984). In *Semiconductors and Semimetals,* ed. R. K. Willardson and A. C. Beer, **20,** p. 1. Academic Press, Orlando, FL.
Thomas, R. N., McGuigan, S., Eldridge, G. W., and Barrett, D. L. (1988). *Proc. IEEE* **76,** 778.
Thurmond, C. D. (1975). *J. Electrochem. Soc.* **122,** 1133.
Tiwari, S. (1992). *Compound Semiconductor Device Physics.* Academic Press, Boston.
Varshni, Y. P. (1967). *Phys. Stat. Sol. (b)* **19,** 459.
Vilms, J., and Garrett, J. P. (1972). *Solid State Electron.* **15,** 443.
Wagner, R. S., Bradley, J. M., and Hammond, R. B. (1986). *IEEE Trans. Nucl. Sci.* **NS-33,** 250.
Wang, C. L., Eckels, J. D., Morgan, W. V., Pocha, M. D., Slaughter, D. R., Davis, B. A., Kania, D. R., and Wagner, R. S. (1986). *Rev. Sci. Instrum.* **57,** 2182.
Wang, C. L., Pocha, M. D., Morse, J. D., Singh, M. S., and Davis, B. A. (1989). *Appl. Phys. Lett.* **54,** 1451.
Wang, C. L., Flatley, J. E., and Pocha, M. D. (1991). *Energy and Technology Review,* Lawrence Livermore Laboratory (Sept.–Oct.), 15.
Ware, R. M., Doering, P. J., Freidenreich, B., Koegl, R. T., and Collins, T. (1992). *Semicond. Sci. Technol.* **7,** A224.
Wiley, J. D. (1978). *IEEE Trans. Electron Dev.* **ED-25,** 1317.
Wiley, J. D., and Miller, G. L. (1975). *IEEE Trans. Electron Dev.* **ED-22,** 265.
Willardson, R. K., and Beer, A. C. (eds.). (1984). *Semiconductors and Semimetals,* **20.** Academic Press, Orlando, FL.
Williams, R. (1990). *Modern GaAs Processing Methods.* Artech House, Norwood, MA.
Wittry, D. B., and Kyser, D. F. (1965). *J. Appl. Phys.* **36,** 1387.
Wright, G. T. (1959). Proc. IEE, paper No. 2928 E, 915.
Yang, E. (1978). *Fundamentals of Semiconductor Devices.* McGraw-Hill, New York.
Ziegler, J. F., and Biersack, J. P. (1990). TRIM-90, Version 90.05.

CHAPTER 11

Lead Iodide Crystals and Detectors

J. C. Lund
ADVANCED MATERIALS RESEARCH DEPARTMENT
SANDIA NATIONAL LABORATORY
LIVERMORE, CALIFORNIA

F. Olschner
RADIATION MONITORING DEVICES, INC.
WATERTOWN, MASSACHUSETTS

A. Burger
NASA CENTER FOR PHOTONIC MATERIALS AND DEVICES
PHYSICS DEPARTMENT
FISK UNIVERSITY
NASHVILLE, TENNESSEE

I. Introduction	444
II. Physical Properties	444
1. *Crystal Structure and Lattice Properties*	444
2. *Semiconducting Properties*	445
III. Preparation of Lead Iodide Crystals	445
1. *Phase Behavior*	446
2. *Purification*	446
3. *Crystal Growth*	448
4. *Summary of Crystal Growth Preparation*	451
IV. Radiation Detector Fabrication and Implementation	451
1. *Detector Fabrication*	451
2. *Electronic Readout*	453
3. *Radiation Testing*	456
V. Potential Applications of Lead Iodide	459
1. *X-Ray Spectrometers*	459
2. *Gamma Ray Detectors*	459
3. *Flux Detectors*	462
VI. Conclusion	463
1. *Summary*	463
2. *Future Research Directions*	463
References	463

I. Introduction

Lead iodide is a high atomic number semiconducting material with a bandgap of 2.3 eV and thus a good candidate material for use as radiation detectors. Because of its attractive physical properties, lead iodide has been under sporadic investigation as a detector material since the 1970s (Roth and Willig, 1971; Manfredotti et al., 1977). In this chapter, we review the properties of lead iodide and radiation detectors fabricated from it. Our discussion begins with a brief review of the physical properties of lead iodide crystals followed by a description of methods to produce crystals of this material for detector use. Next, we discuss the electrical properties of crystals that have been produced in the laboratory, followed by a discussion of the properties of detectors fabricated from these crystals. Finally, we briefly outline the potential applications of lead iodide detectors and speculate on the future of this promising material.

II. Physical Properties

Many of the studies on lead iodide single crystals report extreme difficulties in determining the physical properties due to formidable problems in sample preparation. Therefore, some of the data in the literature should be considered in the context of not only different purity levels but also possible damage induced during sample preparation.

1. Crystal Structure and Lattice Properties

Lead iodide has a layer structure similar to CdI_2, including the well-known presence of polytypes caused by the ease of dislocation generation in this family of crystals.

The lead iodide single crystal basically consists of molecular layers Pb–I–Pb, where the Pb ions are octahedrally coordinated (Palosz, 1983). The 2H polymorph has a hexagonal unit cell that belongs to the point group P_{3mi} (Minagawa, 1975). The bonding between the layers is of the molecular (van der Waals) type. The interlayer van der Waals bonding is very weak and, as a result, bulk lead iodide crystals are delicate. Lead iodide crystals cleave very easily perpendicular to the c-axis and have a "flaky" nature.

The reported unit cell values at room temperature are $a = 4.557$ Å and $c = 6.979$ Å (Sirdesmuckh and Deshpande, 1972). Vapor grown lead iodide crystals have been reported to have the 12H structure, occasionally being mixed with 4H polytypes (Minagawa, 1975). Minagawa also reports the 12H polytype being related to the processing at high temperature, while the low temperature modification was reported to be 2H (Hanoka and Vand, 1968) and more recently during melt growth by the zone refining technique (Lund et al., 1989). The effect of impurities on polytypism was clearly observed only in gel growth (Hanoka and

Vand, 1968). From annealing studies it was found that the 2H–12R transformation is a reversible one (Minagawa, 1975).

As a result of the crystalline layer structure, some anisotropy has been shown for several lattice properties: the thermal expansion coefficients $1/a(da/dT)$ and $1/c(dc/dT)$ have been measured and values of $4.0 \times 10^{-5} \ K^{-1}$ and $3.6 \times 10^{-5} \ K^{-1}$, respectively, have been reported (Sears, Klein, and Morrison, 1979); the same study produced, at room temperature, force constants ratios of $[k_1/k_0]^{\text{shear}} = 0.031$ and $[k_1/k_0]^{\text{compression}} = 0.14$, where the subscripts 1 and 0 refer to inter- and intralayer, respectively.

The lattice vibration properties of lead iodide have also been intensively studied using infrared reflectance (Lucovsky et al., 1976), inelastic neutron scattering (Dorner, Ghosh, and Harbeke, 1976), and Raman spectroscopy (Grisel and Schmid, 1976).

2. SEMICONDUCTING PROPERTIES

Lead iodide is a wide bandgap semiconductor. At room temperature a direct bandgap of 2.58 eV was determined from optical measurements (Blossey, 1971). The band structure was studied using the empirical tight binding approximation (Doni et al., 1972) and by the empirical pseudopotential method (Schluter and Schluter, 1974). The electronic structure was investigated using optical methods (Harbeke and Tosatti, 1972) and by photoemission (Matsukawa and Ishii, 1976). In the region of the $n = 1$ exciton transition; the doublet nature of the free exciton luminescence band was explained (Belyi et al., 1977) by a longitudinal–transverse splitting of the photoexciton states.

Electric measurements (Olschner et al., 1989) of melt grown lead iodide single crystals have revealed a large similarity between lead iodide and mercuric iodide properties; namely, an electric resistivity greater than $10^{13} \ \Omega$ cm and a mean energy required to produce an electron–hole pair $\epsilon = 4.9$ eV. This energy was found (Lund et al., 1989) to deviate from the "main sequence" of semiconductors, which follows Klein's relationship (Klein, 1967). A report on room temperature measurements of drift mobilities (parallel to the c-axis) is available (Minder, Ottaviani, and Canali, 1976). That study produced values of 8 and 2 cm^2 V^{-1} sec^{-1} for the room temperature mobilities of electrons and holes, respectively.

Very few data on defect levels are available. Thermally stimulated currents measurements have identified two energy levels induced by heat treatments at 0.46 and 0.56 eV, which have been attributed (Batlog et al., 1979) to the Pb$^+$ centers and to the anionic vacancies, respectively.

III. Preparation of Lead Iodide Crystals

The preparation of detector grade lead iodide crystals consists of two steps: the purification of starting material and the actual growth of crystals. Before we consider the details of the processes used to produce lead iodide crystals we will first examine the phase behavior of the lead iodide system as these properties will largely determine suitable crystal growth techniques.

1. PHASE BEHAVIOR

As a piece of lead iodide is heated from room temperature it may undergo two fates, depending on its purity. If the piece is impure it may undergo a series of solid–solid phase transitions between various polytypic phases before reaching the melting point. However, if the piece of lead iodide is of high purity (the situation encountered in the preparation of detector material) it will simply remain in a single solid phase until its melting point at 408°C.

The various polytypic phases of lead iodide have been extensively studied (Chaudhary and Trigunayat, 1983; Palosz, 1983; Rao and Srivistava, 1978); and although these phases are of great interest from a crystallographic point of view, their existence can be safely ignored by an investigator interested in detectors. It has been established (Rao and Srivastava, 1978; Chand and Trigunayat, 1977) that reasonably pure lead iodide is all of one polytype (2H).

It is worth noting that the lack of any solid–solid phase transitions in pure lead iodide sharply distinguishes it from the related material mercuric iodide. Because of a destructive solid–solid phase transition between the orthorhombic and tetragonal phases at 130°C, it is not possible to grow detector grade mercuric iodide material directly from the melt. By contrast, the lack of any solid–solid phase transition in pure lead iodide crystals allows the growth of crystals directly from the melt, instead of using the vapor phase methods used with mercuric iodide.

Of great importance to the processing of detector grade lead iodide crystals is the vapor pressure of lead iodide as a function of temperature. A graph of this is shown in Fig. 1. Examination of Fig. 1 indicates that lead iodide has a negligible vapor pressure at room temperature but an appreciable pressure at its melting point of 408°C. The low vapor pressure at room temperature (compared with mercuric iodide) prevents lead iodide detectors from degrading under laboratory conditions even without encapsulation. On the other hand, the relatively high vapor pressure at the melting point necessitates certain precautions when purifying and growing crystals by melt growth techniques.

2. PURIFICATION

The biggest difference between lead iodide detectors recently fabricated (Lund et al., 1988, 1989, 1992) and those produced in initial efforts (Roth and Willig, 1971; Manfredotti et al., 1977) is in the degree of purity of the crystals. It has been found that detectors fabricated from extensively purified materials exhibit significantly improved performance compared to those from unpurified materials. The principal technique used to purify the lead iodide has been liquid zone refin-

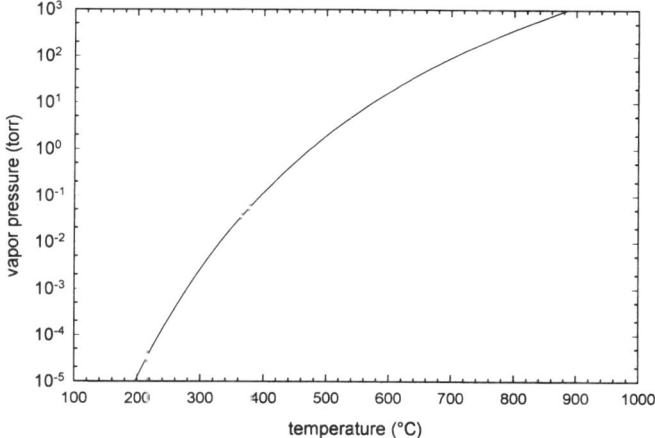

FIG. 1. Vapor pressure versus temperature of lead iodide.

ing. Although other methods are possible (such as the vapor phase zone refining methods used in mercuric iodide development), no mention of techniques other than liquid zone refining appears in the literature.

a. Zone Purification

Zone purification was developed by Pfann and has been a widely used purification method for materials that melt congruently. The details of zone refining have been described in detail elsewhere (Pfann, 1978; Sloan and McGhie, 1988), and we will only review the basis of the technique here for clarity. Zone refining relies on the differential solubility of an impurity in the liquid and the solid as its physical basis. The technique is implemented by melting a portion of the material, a *zone,* which is then slowly passed through the solid starting material. Frequently, this process is repeated many times (multiple *passes*) to increase the efficacy of the process. The result of zone refining is a charge that is much purer at one end than at the other (assuming the segregation coefficients of all of the impurities are less than 1).

An important parameter in zone refining is the *segregation coefficient:* the ratio of the solubility of the impurity in the solid to the solubility in the melt. Usually, the segregation coefficient of an impurity is less than unity (i.e., the impurity is more soluble in the melt than in the solid). Obviously, it is desirable for the impurities in the lead iodide starting material to have segregation coefficients significantly different than unity, otherwise a large number of zone passes are required to purify the material. The number of zone refining passes is also important in determining the purity of the resultant material. The number of passes may be increased by using more than one heater and by repeating the process many times.

Two types of conventional zone refining can be used for lead iodide purification: horizontal and vertical. Horizontal zone refining is the easiest to implement; however, it is ineffective when volatile impurities are present and has other problems due to the vapor pressure of lead iodide. Because lead iodide has an appreciable vapor pressure at its melting point, significant vapor transport of lead iodide from the liquid zone may occur during horizontal refining. To combat this problem it is necessary to pass the zones through the feed material at fairly rapid rates (≈ 10 cm hr^{-1}) to prevent the loss of the zone by vapor transport. At rapid zone speeds the *effective* segregation coefficient is reduced and more passes are required to attain the same purification levels than could be attained at slow speeds (<1 cm hr^{-1}). Thus, counterintuitively, it may require much longer to purify material at fast zone speeds than at slow ones.

Vertical zone refining is, in principle, the most effective method for purifying lead iodide; however, it is difficult to implement in practice. The difficulties with vertical zone refining arise because of the difference in density between liquid and solid lead iodide. Because liquid lead iodide is lower in density than the solid, any expansion of a confined liquid zone will cause large pressures to be created in the zone refining ampoule, resulting in ampoule breakage. The problem is minimized for a "lifting" configuration (zone melting originates at top of feed material) versus a "dropping" configuration but even under lifting conditions zone size fluctuations may occur because of temperature variations, resulting in ampoule breakage. For this reason recent work in the purification of lead iodide for detector use has usually employed the horizontal zone refining method.

Despite the success of horizontal zone refining in lead iodide, many obvious improvements could be made to the purification procedure applied to lead iodide starting material. These improvements include successful application of the vertical zone refining method and perhaps the introduction of a new purification method that relies on a different physical process. For instance, vapor phase refining as implemented with mercuric iodide would appear to be an attractive refining procedure.

3. Crystal Growth

The lack of any significant solid–solid phase transitions in lead iodide allows it to be grown from the melt using conventional crystal growth techniques. The primary method used to grow lead iodide crystals to date has been the Bridgman method. The Czochralski method and vapor phase techniques are also attractive options for lead iodide crystal growth, although to date these methods have not produced large crystals. In addition to the melt growth techniques, extensive work has been performed on the growth of lead iodide from gels. The details of these crystal growth techniques follow.

a. Bridgman Method

The Bridgman crystal growth method is perhaps the simplest melt based technique and it has been used extensively for the growth of lead iodide crystals for

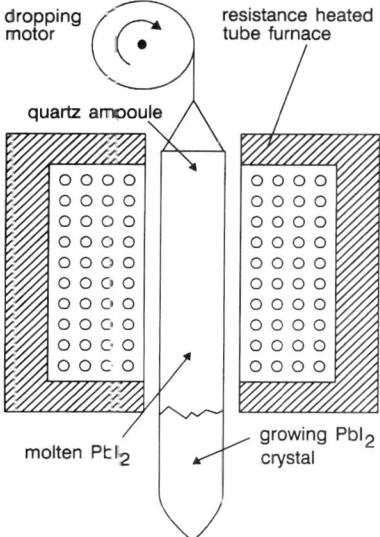

FIG. 2. The growth of lead iodide crystals by the Bridgman method. A quartz crucible containing molten lead iodide is slowly dropped from a tube furnace. As the molten lead iodide encounters the sharp negative temperature gradient at the bottom of the furnace, it precipitates into a solid crystal.

detector applications. The primary advantage of the Bridgman method is its simplicity and ease of implementation. A diagram illustrating the apparatus used for Bridgman growth is shown in Fig. 2. The primary disadvantage of the Bridgman method is that the growing lead iodide crystal remains in contact with the growth ampoule. Contact with the quartz crucible can lead to the introduction of impurities, and if the growing crystal sticks to the surface of the crucible, large thermal strains may be introduced in the crystal as it cools. Despite these problems, fairly large (>5 cm^3) oriented single crystals of lead iodide have been grown by this method (Zhang et al., 1992).

Detector grade lead iodide crystals have also been grown by the horizontal Bridgman technique. The apparatus used to implement this method is very similar to that used for horizontal refining. Unfortunately, horizontal Bridgman methods share the disadvantages of the horizontal refining methods: vapor transport forces a rapid growth speed.

b. Czochralski Method

An attractive technique for the growth of lead iodide crystals is the Czochralski method, commonly used for the growth of silicon and GaAs crystals. However, attempts to grow lead iodide crystals by this method have not been successful to date due to the physical properties of lead iodide. The difficulties in Czochralski lead iodide growth originate from vapor transport of the melt and the low thermal conductivity of solid lead iodide. In a Czochralski system the energy released by freezing the melt must be removed by the growing crystal. In a system such as silicon this is easily accomplished because silicon melts at a high temperature (1415°C), where radiative heat loss is dominant; in addition, solid silicon is a good thermal conductor. By contrast, lead iodide melts at 408°C, where radiative processes are less significant, and solid lead iodide is a poor thermal conductor. Thus, the only way to grow lead iodide crystals by the Czochralski method would be to pull the crystal very slowly from the melt. However, as the growth speed is reduced, the problem of vapor transport from the melt is aggravated. One solution to these problems would be to use the liquid encapsulated Czochralski method (LEC), which is commonly used for the growth of GaAs and InP. To utilize LEC with lead iodide, however, it would be necessary to develop a suitable encapsulant. The encapsulant commonly used for LEC work, boric oxide, softens at too high a temperature for use in lead iodide growth.

The great advantage of the Czochralski method is that lead iodide crystal can grow from the melt without any constraints or thermal distortions due to contact with the crucible. Although the Czochralski method has not been successfully applied to lead iodide crystal growth it remains a promising technique and, with some research, could become a preferred method for growing large volume detector grade crystals.

c. Vapor Phase Growth

Another potential method for growing detector grade lead iodide crystals is the vapor phase method. Although reports of vapor phase platelet growth appear in the literature (Manfredotti *et al.,* 1977), it appears little work has been done in growing lead iodide crystals using this technique. The methods used for the production of large mercuric iodide crystals using this method would seem to be directly applicable to lead iodide growth. Vapor phase growth has the advantage over some melt growth methods that no contact with the crucible is required, minimizing thermal stresses on the growing crystal.

d. Gel Growth

One method of growing lead iodide crystals that deserves mention, if only for historical purposes, is the growth of lead iodide crystals from aqueous silicate gels. The gel method (Heinisch, 1970) has been extensively used for the growth of lead iodide crystals for crystallographic studies (Patel and Rao, 1981; Chand and Trigunayat, 1977). However, this method holds little promise for use as technique for producing detector grade crystals. In particular, the gel method has two large disadvantages: it relies on growth from an aqueous system, which tends to introduce impurities; and the crystals produced by the gel technique grow very slowly to a size that is inadequate for detector fabrication.

4. SUMMARY OF CRYSTAL GROWTH PREPARATION

The preparation of detector grade lead iodide crystals consists of two distinct steps: the purification of starting materials and the growth of crystals. Purification using the horizontal zone refining method has been successfully used to produce detector starting material; however, the vertical zone refining method is probably a superior method (but requires more care in its implementation). Another possible method for purifying lead iodide is to use vapor phase distillation or sublimation methods, but no work has been reported with this approach.

Most lead iodide detectors built to date have been fabricated from crystals grown by the Bridgman method. However, the Bridgman method may introduce strain induced defects into the growing crystal because of contact made between the growth crucible and the growing crystal. Other techniques such as Czochralski and vapor phase methods show promise and may eliminate the deficiencies of the Bridgman method, but these methods have not yet been successfully applied.

IV. Radiation Detector Fabrication and Implementation

1. DETECTOR FABRICATION

Lead iodide x-ray and gamma ray detectors are simple devices: they consist of carefully prepared slabs of lead iodide (typically 50 μm to 1 mm in thickness), with electrical contacts placed on opposite sides. A potential applied between the contacts produces an electric drift field between them, sweeping out mobile electric charges produced during ionizing events, such as an x-ray interaction. This current is ultimately read by a suitable low noise charge–sensitive preamplifier, and the event is recorded. Lead iodide detectors are very similar to mercuric iodide detectors, which at this time are much more widely used. Because of the similarities in the physical properties of lead iodide to mercuric iodide, one would naturally expect that the fabrication methods of lead iodide detectors would also be similar to that of mercuric iodide detectors. There are some subtle differences between these two materials, however, which require some modifications in fabrication methods. This subsection will focus on the differences between lead iodide and mercuric iodide detector fabrication techniques.

Typically, it is desired to cut slices of lead iodide 50 μm to 1 mm in thickness for an x-ray spectrometer; a number of tradeoffs must be considered when choosing the detector thickness for a particular application. Although the majority of these considerations are concerned with maximizing the detector performance, the practicality of producing crystal slices of the desired thickness and area is also a factor. The optimal detector thickness depends on tradeoffs among several considerations. These considerations include the detector stopping efficiency, active detection area, electronic noise (which is a function of the detector capacitance and leakage current), and the practicality of building the detector (detectors <30 μm in thickness are difficult to make).

Assuming that the lead iodide crystal sample has been grown in a boule or some large solid charge, methods must be used to extract a wafer section from this charge. In mercuric iodide detector fabrication technology, string–etchant saws are presently used to make the appropriate cuts in the crystal. In general, the advantages of string–etchant saws are significant. They do not mechanically abrade the crystal or impart large forces to the crystal faces. In addition, these saws can be made to cut in precise and straight lines, when carefully applied. Mercuric iodide is easily etched by a water solution of potassium iodide (KI), which is used to coat the saw string and is itself unaffected by the solution.

Attempts at applying this sawing technology to lead iodide were initially foiled by a combination of the material's lower solubility in KI solution and the formation of a white precipitate on the lead iodide surface. After performing a survey of some likely lead iodide etchants, we have found that the best etchant for lead iodide is a saturated solution of sodium iodide in water. We have made some initial experiments, successfully cutting through lead iodide boules using a string saw with this etchant at room temperature.

Because lead iodide has a highly layered crystal structure, it easily cleaves along one particular orientation (the c-axis plane). This is analogous to mercuric iodide, which exhibits similar cleavage properties. Mechanical cleavage of lead iodide crystals was the preferred method used to cut wafers from the lead iodide boule until recently. Small wafers can be mechanically cleaved in a manner that has traditionally required only a scalpel and a steady hand. Depending on the steadiness or luck of the technician, very thin sections of the boule may be sliced off that contain subsections of appropriate thickness, uniformity, and area. A practical lower limit to these thicknesses, however, is approximately 50–100 μm and is dependent on the desired detector area. Large area, thin wafers are the most difficult to fabricate reproducibly and with good uniformity. This problem is worsened by the fact that the cleavage planes in lead iodide boules are often visibly warped; thus making a mechanical cleaving device difficult to implement.

It has been recently discovered that excellent thin detectors can be separated from the boule by a "tape extraction" method. In this method, tape (or some other adhesive surface) is applied to the face of the lead iodide boule and then removed, thereby delaminating a thin lead iodide layer. The lead iodide layer is then removed from the tape for subsequent processing. The lead iodide layers created in this fashion are approximately 30 to 50 μm thick. Some of our best results with low energy x-ray detectors have been obtained with detectors fabricated with this method.

Contacts can be made to lead iodide using a number of technologies, the easiest of which is a colloidal conductive paint, such as Eccocoat. Evaporated metal, such as gold and palladium also work well. Because lead iodide has a lower room temperature vapor pressure than mercuric iodide, vacuum deposition techniques are easily applicable to lead iodide. Generally speaking, these techniques are difficult to employ with mercuric iodide due to its volatility and the fact that it will contaminate vacuum systems. It should be mentioned that it is very important, in the

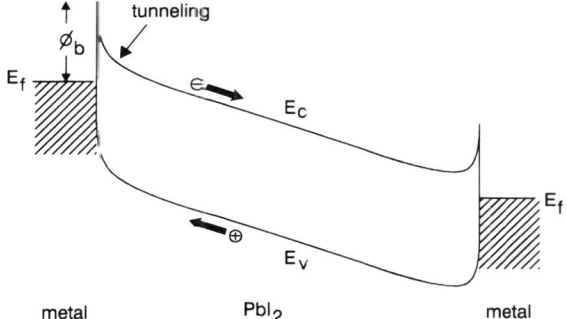

FIG. 3. The energy bands within a lead iodide crystal with metallic contacts. With a perfect lead iodide crystal one would expect a large potential barrier, ϕ_b, to prevent injection of electrons and holes from the metal into the crystal. However, the presence of a large density of ionized energy states (especially at the semiconductor surface) causes substantial "band bending," which enables current injection to occur via quantum mechanical tunneling. If these energy states were not present, the flat band condition would exist and an effective Schottky barrier contact would be maintained.

interest of minimizing the electronic noise, to fabricate low resistance contacts to the lead iodide detectors. Carbon "painted" contacts (e.g., Eccocoat) can often give problems in this regard, due to problems in their application. If a carbon contact is badly applied, resistances of several hundred ohms can be introduced in series to the detector, which produces much thermal noise. Silver paint, being more conductive, gives less problem in this regard; and we have used it successfully.

Because of lead iodide's large bandgap, virtually all contacting materials should theoretically exhibit a relatively large potential barrier (>1 eV) to its surface, creating "Schottky barrier" type contacts. In practice, however, it is observed that ohmic (or "injecting") contacts are often produced. This is due to the large number of defect or impurity related energy states in the lead iodide bandgap. Ionization of these states produces a thin space charge region under the contact that lowers the band potential of the bulk, whereby electrons can be injected via quantum mechanical tunneling. This is diagrammed in Fig. 3. For this reason, all contacting materials have been found to be ohmic on lead iodide surfaces, and lead iodide detectors are therefore considered to be *photoconductive* detectors.

2. ELECTRONIC READOUT

The electronic readout of a lead iodide spectroscopic x-ray detector system takes on the general form assumed by mercuric iodide detection systems and is diagrammed in Fig. 4. Because the leakage current passing through lead iodide detectors is relatively small (approximately tens of picoamperes at most), dc coupled charge sensitive preamplifiers can be used to read the detectors. Pulse

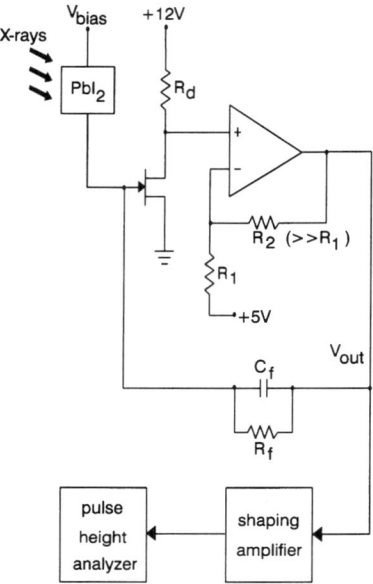

FIG. 4. The electronics used to read out a lead iodide detector in the pulse height spectrometer mode. The most critical part of the circuit is the low noise, charge sensitive preamplifier whose individual components are explicitly shown. The feedback resistor, R_f, is often replaced in practice by more sophisticated reset techniques such as optical reset or drain reset.

shaping amplifiers are subsequently used to read the preamplifier output, which are in turn read by pulse height analyzers, creating a histogram of events comprising the pulse height spectrum.

Because the amount of charge ionized by an interacting x-ray is relatively small (in lead iodide, a 5 keV x-ray produces 1000 mobile electron–hole pairs), the preamplifier must operate with a minimum of electronic noise, or else the acquired histogram of pulse events (energy spectra) will be significantly broadened by this noise. Worse still, the noise level under some conditions may be so severe that the entire signal is obscured by the noise and totally lost. For these reasons, the design of the preamplifier systems that read lead iodide x-ray detectors must be made to operate with a minimum of noise.

The components of electronic noise from a semiconductor detector system are classified according to how they are modeled as circuit elements with respect to the detector (in parallel or in series) and according to their frequency spectra (white or $1/f$) (Radeka, 1988). *Series thermal noise* is due to properties of first stage transistor in the preamplifier and proportional to the total capacitance at the preamplifier input (which is the sum of the detector capacitance, the gate capacitance of the input JFET, and the stray coupling capacitance). *Parallel noise* sources include the detector leakage current noise (*shot noise*) and the thermal noise of any existing feedback resistor in the preamplifier. *Series 1/f noise* is proportional to the input capacitance (as are all series noise sources) and affected by the properties of the detector contacts. Finally, there is *parallel 1/f noise,* which is affected by the presence of lossy dielectrics in the vicinity of the preamplifier input. Thorough analysis of the various noise sources generally encountered with semiconductor detectors has been presented in the literature (Radeka, 1988; Goulding, 1977; Goulding and Landis, 1982).

To prevent stray capacitance from significantly increasing the electronic noise, the lead iodide detector must be located very close to the "front end" FET, preferably within a centimeter or two. This eliminates the possibility of using the lead iodide detector remotely from the preamplifier, or attached by a cable/connector system, and requires the detector and preamplifier to share the same housing. Care must be taken to ensure that the housing is insulated from mechanical vibration, as the system can be quite microphonic. In addition, the detector and preamplifier should occupy a very dry atmosphere, as humidity can cause excess leakage current and noise. We have found that dry nitrogen is a bad atmosphere for detectors, because nitrogen (ionized by incoming radiation) can drift within the detector housing and contribute a large unwanted signal. Dry air, however, works fine in this application.

Charge sensitive preamplifiers are traditionally used to read semiconductor spectrometer detectors because the amplifier gain does not depend on the input capacitance. The preamplifiers act as integrating transimpedance amplifiers, providing an output that is the time integral of the current pulse (= charge). Resetting the preamplifiers can be done via several different methods and represents the principal difference between the various charge sensitive preamplifier systems available today.

The easiest and most common method of pulse resetting used in charge sensitive preamplifiers is to provide a bleed resistor in parallel to the feedback capacitor. The preamplifier will then reset with a time constant of $R_f C_f$, where R_f and C_f are the feedback resistor and capacitor. Typically, $C_f \approx 0.5$ pF and $R_f \approx 5$ GΩ, which then results in a reset time constant of a couple milliseconds. The problem with providing the parallel resistor reset is that the resistor introduces a significant thermal noise source to system. More important, the resistors are usually a significant source of $1/f$ noise; some resistors are much worse than others, however; so the preamplifier designer will find that it is worth his or her time to search for a source of "good" resistors.

Other preamplifier designs use different methods of reset. Optical reset is the most common alternate method of reset, which illuminates the "front end FET" with light, typically from a red LED. This can be performed either continuously or in a pulsed mode and generally results in better noise performance than the resistive feedback preamplifier circuit. This design works well when properly implemented; however, this method has a number of subtleties that must be considered when designing the detector housing. We have used *continuous drain feedback* preamplifiers with lead iodide detectors and achieved our best results with this design. A number of researchers have reported on drain feedback preamplifier circuits; we have used a modification of that described by Bertuccio, Rehak, and Xi, 1993).

3. Radiation Testing

Testing lead iodide x-ray detectors as high resolution energy spectrometers requires an electronics setup similar to that shown in Fig. 4, which was described previously. Radiation testing of lead iodide detectors is generally performed using the same techniques used in mercuric iodide detector radiation testing, which has been the subject of many articles over the last two decades. In this subsection, we will describe a few of the observations made while using lead iodide detectors as x-ray spectrometers.

When initializing the detector, bias must be applied slowly (as with mercuric iodide detectors) or the detector will undergo breakdown. Breakdown is usually a reversible effect, identified by a rapid rise in leakage current and noise. After an optimal bias voltage is reached, the detector output will typically be characterized by a random "bursting" phenomenon during the subsequent settling period in which the lead iodide detector becomes electrically stabilized. This period typically lasts a duration of roughly 1 to 10 min. The choice of an optimal bias voltage requires a tradeoff between the desire to maximize the bias voltage to obtain good charge collection efficiency and avoiding this breakdown effect.

Figures 5 and 6 show good pulse height spectra obtained from lead iodide de-

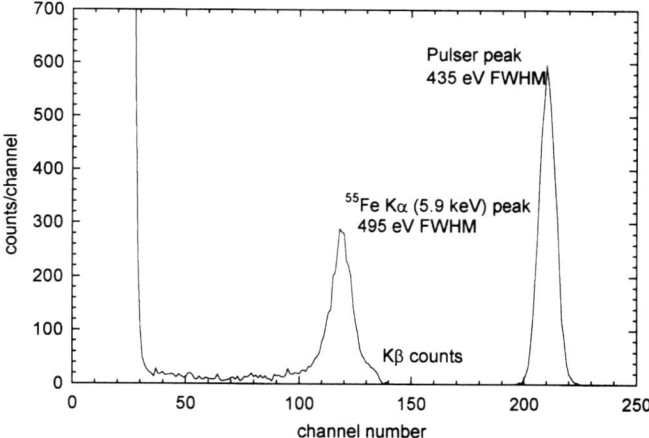

FIG. 5. Pulse height spectrum obtained by irradiating a lead iodide detector with an ^{55}Fe isotopic x-ray source. The 5.9 keV peak is observed to have an energy resolution of 495 keV FWHM. The electronic noise is 38 electrons rms, which is 435 eV FWHM of lead iodide noise (using the ionization rate of 4.9 eV/charge pair). The detector is 0.2 mm^2 area, <100 μm thick, operating at a bias of 15 V, and 6 μsec integration time.

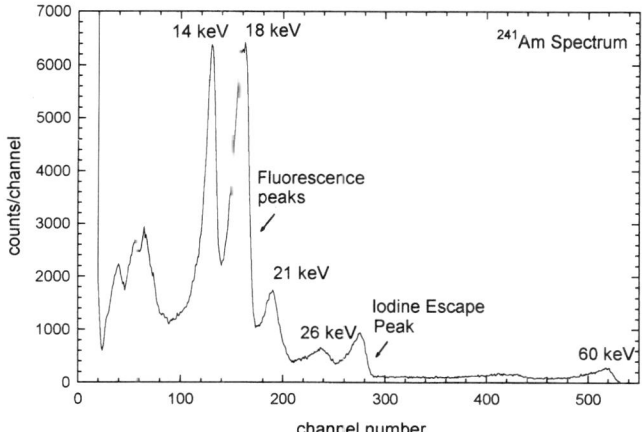

FIG. 6. Pulse height spectrum obtained by irradiating a lead iodide detector with a ^{241}Am isotopic photon source. Visible are the x-ray fluorescence peaks, the iodine escape peak, and the 60 keV gamma ray emission. This detector is 1 mm² in area, <100 μm thick, at a bias of 10 V, and 6 μsec integration time.

tectors irradiated with ^{55}Fe and ^{241}Am, respectively. As is true with mercuric iodide x-ray detectors, it is important to irradiate the negatively biased contact in a lead iodide x-ray spectrometer detector because of the greater charge collection efficiency at this region. As this figure demonstrates, the energy resolution of lead iodide detectors at low energies is strongly influenced by electronic noise, thus the identification of the noise sources is important to improve detector performance.

The individual components of the total noise in a lead iodide detector system—*series thermal, parallel thermal (+shot)*, and *1/f noise*—may be resolved by determining the amplitude of the noise as a function of amplifier integration time. This is performed by injecting a test pulser into the detector preamplifier system and measuring the broadening of the pulser amplitude with a pulse height analyzer system for different values of the linear amplifier's time constant. The results of an experiment of this kind are shown in Fig. 7. As Fig. 7 illustrates, the dominant noise sources in typical lead iodide detector systems are the series thermal and 1/f noise components. Because of the very high resistivity of lead iodide, small area detectors (<1 cm²) do not produce appreciable amounts of shot noise. The series thermal noise may be minimized by selecting preamplifier input FETs with good g_m/C_{gate} values (Radeka, 1988) and by minimizing the stray capacitance and the capacitance of the detector itself. In a detection system, "$1/f$" noise can be of the "series" or "parallel" type; parallel $1/f$ noise is the easiest to eliminate, as it arises from the presence of lossy dielectrics in the vicinity of the preamplifier input. Minimizing this noise component generally requires repackaging the FET in a low loss header (such as Teflon), mounting the lead iodide detector itself on a low loss material (fused silica, alumina or Teflon), and avoiding the use of resistive

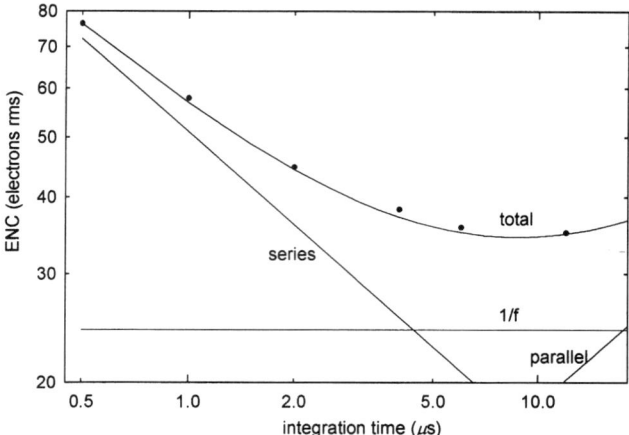

FIG. 7. Measured noise of a lead iodide detector as a function of linear amplifier integration time. This was a small lead iodide detector (0.2 mm² area, <100 μm thick), operating at a bias of 15 V. A least squares fit of the data to the noise model produced the estimates of the individual noise components shown.

feedback designs in the preamplifier (see Subsection IV.2). Series $1/f$ noise depends on the nature of the electrical contact made to the detector and therefore on the detector material properties and the contacting processes. We have found that lead iodide detectors fabricated using the etchant–string saw or the tape extraction method have lower series $1/f$ noise than detectors razor cleaved from the boule (Shah et al., 1994).

As has been the problem during the development of other compound semiconductor x-ray detector materials, electric polarization in lead iodide devices is sometimes observed to degrade detector performance. An operator of lead iodide spectrometer detectors will usually recognize the effects of polarization by noting the continual degradation of the charge collection efficiency during a period after bias is applied. Typically, this time period might last from seconds to hours, depending on many of the operating parameters.

Polarization is the result of the build-up of trapped or fixed charges in a semiconductor. This effect reduces the electric field in many regions of the device, resulting in lower charge collection efficiency. In addition to reducing the charge collection efficiency of the detector, polarization also is observed to reduce the detector leakage current. In fact, the reduction of measured detector leakage current with time (under constant temperature) can serve as an indicator of polarization. The more general subject of the "dielectric relaxation properties of insulators" has been reviewed decades ago in a number of articles (Gupta and Van Overstaeten, 1975, 1976; Lampert, 1956; Simmons, 1968, 1972; Simmons and Taylor, 1972b).

The most desirable method to use in the quest to reduce polarization in lead iodide detectors would be to attempt to reduce the density of mid-bandgap energy states in the detector crystals, especially near the contacted surfaces where they tend to be problematic. For this reason, careful surface preparation of the crystals, including the contacting materials themselves, are very important. It is also important to minimize the density of trapping states within the bulk of the lead iodide used to fabricate the detectors.

V. Potential Applications of Lead Iodide

Because of its high atomic number, lead iodide is an attractive material for use in radiation detector applications. At the present time, the electron and hole transport properties of available lead iodide crystals determine the range of applications. Because the drift length of electrons in currently available lead iodide is only about 500 μm (for typical electric fields of 5×10^3 V cm^{-1}) the maximum thickness of detectors with good energy resolution is also only about 500 μm. The thickness restriction limits the utility of lead iodide in spectrometer applications to photons of lower than 30 keV (usually x-rays). However, lead iodide might also be useful for higher energy photons if energy resolution is not required (e.g., gamma ray counters).

1. X-Ray Spectrometers

The most likely near term application of lead iodide detectors is in room temperature x-ray spectroscopy. Small area detectors have been fabricated with good energy resolution (<500 eV FWHM at 5.9 keV) at room temperature and above. Devices of this type would be useful in industrial and laboratory x-ray fluorescence equipment, where use of a conventional x-ray spectrometer (liquid nitrogen cooled Si(Li) detector or proportional gas detector) is impractical.

A particularly attractive feature of lead iodide x-ray spectrometers is their ability to operate at elevated temperatures. Good energy resolution has been reported with lead iodide detectors up to approximately 100°C. This high temperature performance would allow their use in certain industrial applications where no other suitable detector exists.

The quality of lead iodide detector performance stands to improve the most from improvements in the technology of cutting and etching the crystal surfaces. We foresee that the recent discovery of a good etchant (saturated sodium iodide solution in water) will probably result in higher performance detectors in the near future.

2. Gamma Ray Detectors

Because of the very high atomic number of its constituent atoms lead iodide has a very high gamma ray stopping power and is an attractive material for constructing gamma ray detectors. The attenuation coefficient for high energy pho-

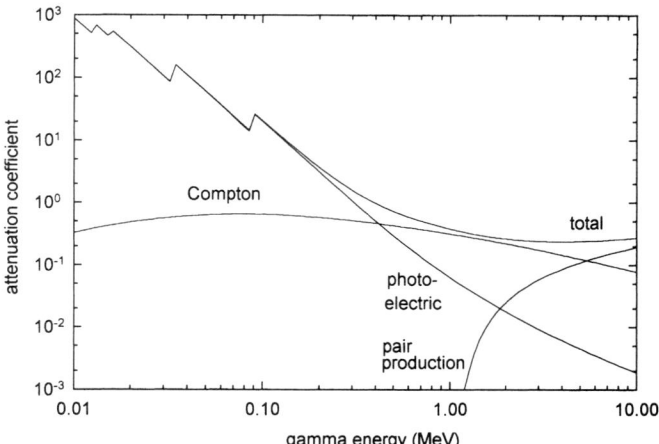

FIG. 8. Gamma ray attenuation coefficient versus photon energy for lead iodide. Cross sections were obtained using the computer code XCOM (Berger and Hubbell, 1987) and a mass density of 6.16 g cm^{-3} was assumed in the calculations.

tons is shown in Fig. 8. In principle two types of gamma ray detectors could be constructed from lead iodide: spectrometers, where energy resolution is of great importance; and simple counters, which require little or no pulse height energy solution. We discuss both of these devices in this subsection.

Unfortunately, it is not possible to exploit the excellent gamma ray stopping power of lead iodide in a spectrometer at the present time. Lead iodide detectors of the thickness required in gamma ray studies (>1 cm), would have poor energy resolution due to charge collection problems. It is worth examining these charge collection problems in some detail, however, to estimate by how much lead iodide crystals must be improved before they can be used effectively in gamma ray spectrometers. The charge collection in a lead iodide detector can be examined quantitatively by considering a model of charge collection in the lead iodide detector based on the Hecht (Hecht, 1932; Akutagawa and Zanio, 1969) model. This model computes the charge collection efficiency η as a function of position within the detector and may be written

$$\eta(x) = \frac{\lambda_e}{d}\left[1 - e^{(x-d)/\lambda_e}\right] + \frac{\lambda_h}{d}\left[1 - e^{-x/\lambda_h}\right] \quad (1)$$

where d is the thickness of the detector and λ is the mean drift length of the electron (λ_e) or hole (λ_h) in the semiconductor. The drift length for a device with a constant electric field may be expressed as

$$\lambda = \frac{\mu \tau V}{d} \quad (2)$$

where μ is the mobility, τ is the trapping lifetime, and V is the potential applied

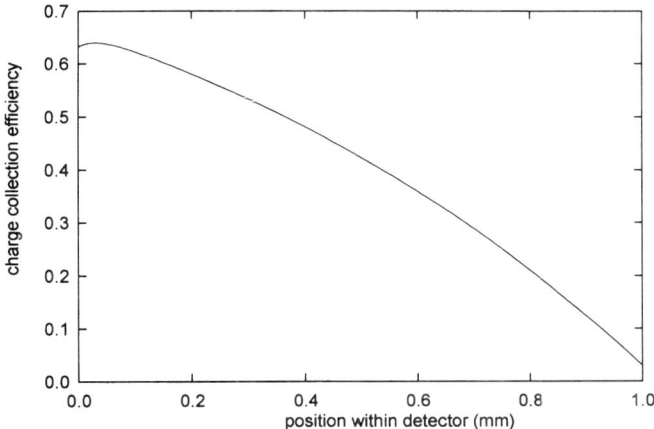

FIG. 9. Charge collection efficiency as a function of position through a lead iodide detector. The detector was assumed to be 1.0 mm thick and operated at a bias of 1000 V, values of $(\mu\tau)_e = 1 \times 10^{-5}$ cm^2 V^{-1} and $(\mu\tau)_h = 3 \times 10^{-7}$ cm^2 V^{-1} were also assumed in the calculations.

across the thickness of the detector. A plot of Eq. (1) is shown in Fig. 9 for a 1.0 cm thick lead iodide detector operated at a bias of 1000 V and using experimentally determined values of $(\mu\tau)_e = 1 \times 10^{-5}$ cm^2 V^{-1} and $(\mu\tau)_h = 3 \times 10^{-7}$ cm^2 V^{-1}. Examination of Fig. 9 indicates that η is not constant throughout the thickness of the detector. If such a detector were used as a spectrometer of high energy gamma rays (which interact throughout the thickness of the detector), distortion of the pulse height spectra would result (see Fig. 10). This distortion would result because the pulse height would depend on the position of gamma ray interaction in the detector. The only way to remove the charge collection distortion and make lead iodide detectors useful for gamma ray spectroscopy would be to increase the $\mu\tau$ product of electrons and holes in the detectors. Using Eqs. (1) and (2) we can compute how much of an improvement in energy resolution would result if we improved $(\mu\tau)_e$ and $(\mu\tau)_h$ in a lead iodide crystal, and these results are shown in Fig. 10. As the figure indicates, the mobility lifetime product of electrons and holes in lead iodide crystals must be improved by at least two orders of magnitude over present crystals for lead iodide to become an effective gamma ray spectrometer material.

For gamma ray counters, where it is not necessary to have good pulse height energy resolution (only the total number of counts above some threshold is required) lead iodide may be a good choice. The electrical properties of existing lead iodide crystals should be suitable for the construction of efficient counters. Computing the overall efficiency of a lead iodide counter as a function of gamma ray energy is a difficult computation problem requiring Monte Carlo techniques for its solution (because the transport of Compton scattered photons must be included). However, we can assume that a counter with a low energy threshold of 100 keV should be effective as a counter for energetic gammas (>500 keV). Un-

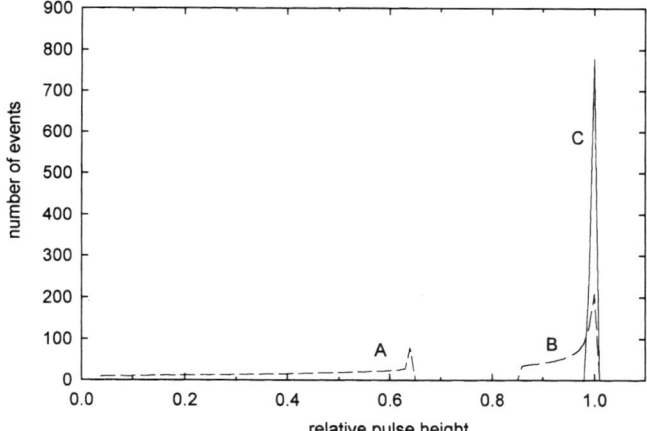

FIG. 10. Noise free pulse height spectra that would be obtained by irradiating three hypothetical lead iodide detectors with the same number (1000) of gamma rays from a monoenergetic deeply penetrating source. Curve A assumes electron and hole transport properties like those of currently available lead iodide. Curve B assumes two orders of magnitude increase in the transport properties (i.e., $(\mu\tau)_e = 1 \times 10^{-4}$ cm^2 V^{-1} and $(\mu\tau)_h = 3 \times 10^{-6}$ cm^2 V^{-1}). Curve C assumes $(\mu\tau)_e = 1 \times 10^{-3}$ cm^2 V^{-1} and $(\mu\tau)_h = 3 \times 10^{-5}$ cm^2 V^{-1}. These calculations indicate the improvements in electron and hole transport properties required (approximately two orders of magnitude over existing material) to make high performance gamma ray spectrometers from lead iodide.

der this assumption we estimate that counters as thick as 5 mm could be constructed (operated at 1000 V bias) before charge collection problems begin to seriously reduce the efficiency of the device.

3. Flux Detectors

Another potential application of lead iodide detectors is for detectors for measuring high level x-ray and gamma ray fluxes. Semiconductor detectors can be operated in a lowered frequency response mode that does not measure individual particles (like counters and pulse height spectrometers), but instead produces an output current proportional to the product of the continuous flux and the energy of the incident radiation. Such detectors might be useful in medical and industrial tomography. The use of such devices with lead iodide is, to date, completely unexplored. The high atomic number and high resistivity would indicate that lead iodide crystals might be promising in this application. Conversely, the poor electron lifetimes that adversely affect lead iodide gamma ray spectrometers may also be a problem with flux detectors. The short lifetimes would reduce the signal amplitude produced from the detector and may induce space charge polarization of the detector at high flux rates.

VI. Conclusion

1. Summary

Lead iodide is an interesting and potentially useful material for the fabrication of radiation detectors. In general, lead iodide crystals and detectors are very similar to those produced by the more well-known detector material mercuric iodide but lead iodide is currently in a more primitive state of development because fewer researchers have investigated it. Like its cousin mercuric iodide, the biggest flaw with presently available lead iodide crystals is the transport properties of electrons and holes in the crystals (mobility lifetime product). On the one hand, one might expect development of lead iodide to proceed rapidly compared to mercuric iodide because it is an easier material from which to grow large crystals. On the other hand, lead iodide appears to be even more fragile than mercuric iodide and consequently the development of detector fabrication methods may be more difficult. In general though, lead iodide must be considered a strong candidate for room temperature x-ray detectors as good performance detectors have been demonstrated after a relatively short development effort.

2. Future Research Directions

The last decade saw a substantial improvement in the development of lead iodide detectors. Most of this improvement can be attributed to the use of zone refining methods to improve the purity of starting material used for detector fabrication. There is some evidence that the mobility–lifetime product (and, hence, the energy resolution) of present lead iodide detectors is dominated by structural imperfections in the crystals and no longer by extrinsic chemical impurities. To reduce the structural defects present in lead iodide detector crystals will require research into improved crystal growth techniques that minimize thermal stresses imparted to the crystal, the development of improved device fabrication procedures that minimize damage imparted to the crystal by the device manufacturing process, and research into the effects of annealing to remove any defects that remain in the completed detector. If these methods can successfully increase the mobility–lifetime product of electrons and holes in lead iodide, there is a good probability that it will become one of the leading detector materials in the future.

References

Agrawal, W. K. (1971). *Phys. Lett.* **A34,** 82.
Agrawal, W. K. (1981). *J. Cryst. Growth* **53,** 574.
Akutagawa, W., and Zanio, K. (1969). *J. Appl. Phys.* **40,** 3838.
Batlog, I., Piticu, I., Constantinescu, M., Ghita G., and Ghita, L. (1979). *Phys. Stat. Sol. (a)* **52,** 103.
Belyi, N. M., Blonskii, I. V., Gorban, I. S., Gubanov, V. A., Lysenko, V. G., Sushkevich, T. N., and Timofeev, V. B. (1977). *Sol. Phys. Solid State* **19**(8), 1302.

Berger, M. J., and Hubbell, J. H. (1987). *XCOM: Photon Cross Sections on a Personal Computer.* Center for Radiation Research, National Bureau of Standards, Gaithersburg, MD.
Bertuccio, G., Rehak, P., and Xi, D. (1993). *Nucl. Inst. and Meth.* **A326**, 71.
Blossey, D. F. (1971). *Phys. Rev.* **B3**, 1382.
Chand, M., and Trigunayat, G. C. (1977). *J. Crystal Growth* **39**, 299.
Chaudhary, S. K., and Trigunayat, G. C. (1983). *J. Crystal Growth* **62**, 398.
Doni, E., Grosso, G., and Sparvieri, G. (1972). *Solid State Commun.* **11**, 493.
Dorner, B., Ghosh, R. E., and Harbeke, G. (1976). *Phys. Stat. Sol.* **73**.
Goulding, F. S. (1977). *Nucl. Inst. and Meth.* **142**, 213.
Goulding, F. S., and Landis, D. A. (1982) *IEEE Trans. Nuc. Sci.* **NS-29**, 1125.
Grisel, A., and Schmid, P. (1976). *Phys. Sta. Sol. (b)* **73**, 587.
Gupta, H. M., and Van Overstraeten, R. J. (1975). *J. Appl. Phys.* **46**, 2675.
Gupta, H. M., and Van Overstraeten, R. J. (1976). *J. Appl. Phys.* **47**, 1003.
Hanoka, J. I., and Vand, V. (1968). *J. Appl. Phys.* **39**, 5288.
Harbeke, G., and Tosatti, E. (1972). *Phys. Rev. Lett.* **28**, 1567.
Hecht, K. (1932). *Z. Physik* **77**, 235.
Heinisch, H. K. (1970). *Crystal Growth in Gels.* Pennsylvania State University Press, University Park.
Klein, C. A. (1967). *J. Appl. Phys.* **4**, 2029.
Lampert, M. A. (1956). *Phys. Rev.* **103**, 1648.
Lucovsky, G., and White, R. M. (1977). *Nuovo Cimiento* **B38**, 290.
Lucovsky, G., White, R. M., Liang, W. Y., Zallen, R., and Schmid, P. H. (1976). *Solid State Commun.* **18**, 811.
Lund, J. C., Shah, K. S., Squillante, M. R., and Sinclair, F. (1988). *IEEE Trans. Nuc. Sci.* **NS-35**, 89.
Lund, J. C., Shah, K. S., Squillante, M. R., Moy, L. P., Sinclair, F., and Entine, G. (1989). *Nuc. Inst. and Method.* **A283**, 299.
Lund, J. C., Shah, K. S., Olschner, F., Zhang, J., Moy, L. P., Medrick, S., and Squillante, M. R. (1992). *Nucl. Inst. and Meth.* **A322**, 464.
Manfredotti, C., Murri, R., Quirini, A., and Vasanelli, L. (1977). *IEEE Trans. Nuc. Sci.* **NS-24**, 126.
Matsukawa, T., and Ishii, T. (1976). *J. Phys. Soc. Japan* **41**, 1285.
Minagawa, T. (1975). *Acta Cryst.* **A31**, 823.
Minder, R., Ottaviani, G., and Canali, C. (1976). *J. Phys. Chem. Solids* **37**, 417.
Olschner, F., Lund, J. C., Shah, K. S., and Squillante, M. R. (1989). *ICFA Instrum. Bull.* No. 7, 9.
Palosz, B. (1983). *Phys. Stat. Sol. (a)* **80**, 11.
Patel, A. R., and Rao, A. V. (1981). *Indian J. of Pure and Appl. Phys.* **19**, 685.
Pfann, W. G. (1978). *Zone Melting.* Robert E. Kreiger Publishing, Huntington, NY.
Radeka, V. (1988). *Ann. Rev. Nucl. Part. Sci.* **38**, 217.
Rao, M., and Srivastava, O. N. (1978). *J. Phys. D: Appl. Phys.* **11**, 919.
Roth, S., and Willig, W. R. (1971). *Appl. Phys. Lett.* **18**, 328.
Schluter, I. C., and Schluter, M. (1974). *Phys. Rev.* **B9**, 1652.
Sears, W. M., Klein, M. L., and Morrison, J. A. (1979). *Phys. Rev.* **B19**, 2305.
Shah, K. S., Lund, J. C., Olschner, F., Bennett, P., Zhang, J., Moy, L. P., and Squillante, M. R. (1994). *Nuc. Inst. and Method.* Phys. Res. **A353**, 85.
Simmons, J. G. (1968). *Phys. Rev.* **166**, 912.
Simmons, J. G. (1972). *Phys. Rev. B* **5**, 553.
Simmons, J. G., and Taylor, G. W. (1972a). *Phys. Rev. B* **6**, 4793.
Simmons, J. G., and Taylor, G. W. (1972b). *Phys. Rev. B* **6**, 4804.
Sirdesmuckh, D. B., and Deshpande, V. T. (1972). *Current Sci.* **41**, 210.
Sloan, J. S., and McGhie, A. R. (1988). *Techniques of Melt Crystallization.* John Wiley & Sons, New York.
Zhang, J., Shah, K. S., Lund, J. C., Olschner, F., Moy, L. P., Daley, K., and Squillante, M. R. (1992). *Nucl. Inst. and Meth.* **A322**, 499.

CHAPTER 12

Other Materials: Status and Prospects

Michael R. Squillante and Kanai S. Shah

RADIATION MONITORING DEVICES, INC.
WATERTOWN, MASSACHUSETTS

I. INTRODUCTION 465
II. DETECTOR MATERIALS 467
   1. *Overview* 467
   2. *Fundamentals of Crystal and Device Preparation* 468
   3. *III–V Materials* 469
   4. *II–VI Semiconductors* 473
   5. *Thallium Bromide* 475
   6. *Amorphous Silicon* 477
   7. *Ternary Materials* 478
   8. *Other, Less Studied, Crystalline Materials* 481
   9. *Other, Less Studied, Thin Film Materials* 482
III. CURRENT STATUS AND PROSPECTS 484
   1. *Comparison of Material Properties* 484
   2. *Future Directions* 486
   3. *Summary and Conclusions* 487
   References 487

I. Introduction

In many instances, the ability of scientists and engineers to make nuclear radiation measurements is limited only by the properties of available radiation detectors. Because of this, research into new semiconductor materials for radiation detection has been, and continues to be, a very active field. The chapters preceding this one have discussed the status of those materials that are mature and available commercially as well as those that have the highest near term potential to be useful for practical measurements.

This chapter reviews recent research on "new" materials, i.e., those materials that have promise based on known or anticipated physical and electronic properties but have not yet been proven to be suitable for use as detector materials. Data reported in the literature that are relevant to nuclear radiation detection applications are tabulated and compared and an assessment of potential future progress is made. Several excellent review articles have been published over the past few years (Squillante *et al.*, 1993c; Lund *et al.*, 1992; Bencivelli *et al.*, 1991; Olschner *et al.*, 1989c; Cuzin, 1987; Sakai, 1982; Armantrout *et al.*, 1977) that contain additional data on many of the materials presented here and on several materials that are not included because no significant progress has been made on them for several years.

Semiconductor nuclear detectors have been used for decades but are experiencing a dramatic increase in interest at the present time. This increase is due partly to the availability of better quality materials and advances in semiconductor processing technology that make possible the fabrication of new and better device structures. However, one of the most important factors that has stimulated interest in semiconductor detectors is the availability of powerful and relatively inexpensive computers and electronic circuitry for powering and reading out semiconductor devices. In particular, these advances have influenced the current direction of research on imaging devices.

All of these advances have had a direct effect on the desire to identify and develop new materials to solve problems that cannot be solved using the three "traditional" semiconductor detector materials: silicon, germanium, and cadmium telluride. For example, for years it was possible to only dream about a gamma ray imaging detector system composed of a large room temperature detector with very high stopping power. Producing such a system was, for all practical purposes, impossible: even if suitable material were available, the cost of the required preamplifiers, amplifiers, interface electronics and the mainframe computer was completely prohibitive. Since there was no practical way to use a large detector array, there was little incentive for doing the materials science research on the new semiconductor materials needed to build it. Now, however, computers are extremely inexpensive, hybrid preamps are available at very low cost on single- and multiamp hybrid chips, and extremely cost effective interface cards are available and easy to use. A practical gamma ray imaging system could easily be built today that would certainly be a commercial success, if only better semiconductor detector materials were available. This is just one example of how advances in other technologies have fueled the interest in new semiconductor materials and spurred the recent research on them.

Historically, the driving force for new materials was to identify the "ideal" material, which had the performance characteristics of germanium but was easy to make and operated at room temperature (Mayer, 1968). More recently, expectations have become more realistic and the drive to find and develop new materials usually results from the need to solve a specific technical problem relating to applications with special needs that cannot be satisfactorily met with traditional detector materials. Typical requirements include ruggedness, better energy resolution, higher stopping power, smaller size, and high temperature operation.

The search for better performance usually does not mean looking for a totally new material. Shortcomings in one material often lead to research in another very similar material. An example of this process is the history of research on PbI_2 at Radiation Monitoring Devices, Inc., Watertown, Massachusetts. For 20 years,

HgI$_2$ has been the leading contender as the next practical detector material and recently has been used in several commercial products. But, as discussed in previous chapters, in addition to its strengths as a detector material, HgI$_2$ has deficiencies, some of which have been solved (Squillante, Shah, and Moy, 1990; Shah, Squillante, and Entine, 1990a; Iwanczyk et al., 1989) but others have yet to be overcome, and the desire to circumvent these problems led directly to research on the very similar material, PbI$_2$. Research on PbI$_2$ has shown that this material does indeed solve many of the problems associated with HgI$_2$: it can be grown from the melt and it is chemically and electrically very stable (Lund et al., 1988a, 1989). It does not, however, solve all the problems, since, like HgI$_2$, it has a layered crystal structure and is soft and fragile. These characteristics led to research on another promising heavy metal halide, TlBr, which was chosen partly because it has a different crystal structure than HgI$_2$ and PbI$_2$. TlBr is discussed later in this chapter.

II. Detector Materials

1. OVERVIEW

Most of the semiconductor materials investigated for room temperature radiation detectors are commercially available, have been used in other applications, and are "new" only with respect to their use as detectors. This distinguishes them from mature materials for radiation detectors that have been sold commercially for years, as well as those that have been extensively studied and whose properties are well understood.

Of the many semiconductor materials available, only three have been regularly used for commercial radiation detectors—Si, Ge, and CdTe—and only Si and CdTe are used at room temperature. Silicon is the semiconductor material most often used for low energy x-rays and ionizing particles; germanium is the material of choice when spectral quality is a prime concern; and for applications where room temperature operation, small size, and high sensitivity is required, CdTe is often used. Other materials such as HgI$_2$ and GaAs have been extensively studied for many years, thus significant breakthroughs are not expected in these materials. There is one possible exception, however, as there appears to be great promise in a relatively new Si device structure, the silicon drift diode (Avset et al., 1990; Gatti, Rehak, and Walten, 1984; Kemmer et al., 1987; Hall, 1988). Although no other materials have been studied as extensively as Si, Ge, CdTe, and HgI$_2$, a considerable level of interest has recently developed in four materials: GaAs, TlBr, amorphous Si (a-Si), and PbI$_2$. PbI$_2$ is discussed in an earlier chapter and will not be discussed here other than to include it in Table I. GaAs, TlBr and a-Si are discussed in the following sections.

In addition to single crystal detectors, several materials have been examined for use in radiographic imaging in polycrystalline or amorphous thin film form. For example, Se and CdS have been used in the photoconductive mode as xeroradiographic films, and thin films of a number of materials including CdS, CdSe and GaAs have been used to fabricate arrays of diodes or transistors for digital imaging.

2. Fundamentals of Crystal and Device Preparation

A variety of methods are used to grow semiconductor materials for nuclear sensor applications (Sloan and McGhie, 1988). Si, Ge, GaAs, and InP are typically grown by the Czochralski method, although the float zone technique is also popular for Si growth (Pfann, 1978). The Bridgman crystal growth method is also used, especially in the early stages of development in a new material, because it is a relatively straightforward technique that can be implemented without a large capital investment. Vertical zone and horizontal zone melt growth techniques are used on a variety of materials including CdTe and PbI_2. Solution growth techniques can also yield good results; the best CdTe available at this time is grown by a vertical solution zone technique, the traveling heater method (THM) (Wald and Entine, 1978; Siffert, 1978). THM has also been used for other II–VI materials.

Vapor growth of single crystals for some semiconductor materials has been investigated, but only HgI_2 crystals are regularly grown from the vapor phase (Schieber et al., 1978; Lamonds, 1983; Faile et al., 1980; Squillante et al., 1983). HgI_2 must be grown by vapor phase growth because of a solid–solid phase transition at 127°C that destroys the quality of any crystals grown from the melt. In general, when it is possible, growth from the melt or solution is preferred because vapor growth processes are generally much slower than liquid techniques. Thin films, however, are readily grown epitaxially on selected substrates from the vapor phase or the liquid phase. Most thin films of the materials of interest for nuclear detectors are presently grown by one of several modifications of the chemical vapor deposition technique (CVD). Since the films are thin, the relatively slow growth rates from the vapor are not a factor, and CVD processes allow tight control over growth parameters and film stoichiometry.

A major part of any development effort on new materials is research to identify appropriate device fabrication procedures including finding etching procedures and workable electrode structures. In general, two generic types of devices are fabricated on the crystalline materials during the early stages of developing new materials: photoconductors and photodiodes. These devices have a relatively simple configuration: parallel planar electrodes are vacuum evaporated, plated or painted onto both surfaces of cut or cleaved wafers that have been cleaned, polished and etched. The electrodes are selected from materials that form ohmic contacts for photoconductors and Schottky barrier contacts for diodes. As a material matures and its properties are better understood, more sophisticated electrode structures, diffused junctions and specialized surface treatments are often used to modify and improve the performance. Thin semiconductor films can also be tested by fabricating simple ohmic and diode structures, but for imaging sensors, device fabrication makes use of existing photolithographic technology to build arrays of sophisticated multi-layer diodes, photodiodes and transistors.

3. III–V MATERIALS

a. Gallium Arsenide

GaAs, a III–V semiconductor, is presently one of the most technologically important electronic materials. It is useful for the fabrication of very fast electronic devices because of the very high mobility (μ) and short lifetime (τ) (see Table I). Most GaAs is produced in single crystal form using the Czochralski crystal growth method. Large ingots are sliced into wafers that are used in device processing. Thin films of GaAs grown by CVD on various substrates can also be used for the fabrication of devices.

For many years, GaAs has seemed to be an obvious candidate for use in radiation detectors because of its very high electron and hole mobilities and the very advanced state of GaAs manufacturing. Good results were reported in the early 1970s for GaAs films (Eberhardt, Ryan, and Tavendale, 1970, 1971; Kobayashi et al., 1972, 1976). These films were 60 to 80 μm thick and simple surface barrier devices operated at room temperature exhibited FWHM noise of 2.9 keV, as measured by a pulser and 2.95 keV FWHM resolution for ^{57}Co 122 keV photons. These results show that GaAs inherently possesses the necessary properties for the production of high quality nuclear detectors.

Unfortunately, although GaAs has been studied by many researchers for nuclear detection, devices made on single crystal GaAs wafers have yet to show significant progress. Only particle detectors and photon detectors with relatively poor energy resolution have been obtained because of the extremely short charge carrier lifetimes and relatively low resistivity. These two features, which make GaAs attractive for high speed electronic devices, work together in such a way that the noise in the devices caused by leakage current increases faster than the signal as the bias voltage is increased in order to improve charge collection efficiency.

The potential still exists for GaAs to become a practical material for nuclear detector fabrication despite the problems (Benz et al., 1992; Eiche et al., 1993; McGregor et al., 1992). This is because most of the past research and development on GaAs has been specifically directed at the production of very fast devices, such as high speed transistors, that require short lifetimes and low resistivities. These characteristics are obtained intentionally by impurity doping and other treatments. With appropriate materials science research, it should be possible to prepare material with longer charge carrier lifetime.

Since the bandgap of GaAs is 1.5 eV, it is possible to grow GaAs crystals with room temperature resistivities much higher than those normally available from commercial sources. Researchers at the Lawrence Livermore National Laboratory

TABLE I

PROPERTIES OF SEMICONDUCTOR MATERIALS AT 25°C

Material	Atomic Number	Density g/cm^3	Band-gap eV	Melting Point °C	Knoop Hardness	Crystal Structure	Ionicity	Dielectric Constant
Ge	32	5.33	0.67	958	692	Cubic	0	16
Si	14	2.33	1.12	1412	1150	Cubic	0	11.7
CdTe	48, 52	6.2	1.44	1092	45	Hexagonal	0.61	11
CdZnTe	48, 30, 52	≈ 6	1.5–2.2	1092–1295				
CdSe	48, 34	5.81	1.73	>1350		Hexagonal	0.6	10.6
CdZnSe	48, 30, 34	≈ 5.5	1.7–2.7	1239–1520				
HgI$_2$	80, 53	6.4	2.13	250 (127†)	<10	Tetragonal	0.67	8.8
TlBrI	81, 35, 53	7.5	2.2–2.8	405–480	40	Cubic		
GaAs	31, 33	5.32	1.43	1238	750	Cubic	0.23	12.8
InI	49, 53	5.31	2.01	351	27	Orthorhombic	0.8	26
GaSe	31, 34	4.55	2.03	960		Hexagonal	0.53	8
diamond	6	3.51	5.4	4027	10^4	Cubic	0	5.5
TlBr	81, 35	7.56	2.68	480	12	Cubic	0.81	29.8
PbI$_2$	82, 53	6.2	2.32	402	<10	Hexagonal	0.8	
InP	49, 15	4.78	1.35	1057	535	Cubic	0.38	12.5
ZnTe	30, 52	5.72	2.26	1295		Cubic	0.62	9.7
HgBrI	80, 35, 53	6.2	2.4–3.4	229–259	14	Orthorhombic		
a-Si	14	2.3	1.8				0	11.7
a-Se	34	4.3	2.3				0	6.6
BP	5, 15	2.9	2	dl400	4700	Cubic	0.01	11
GaP	31, 15	4.13	2.24	1750		Cubic		
CdS	48, 16	4.82	2.5	1477		Hexagonal	0.58	11.6
SiC	14, 6	3.2	2.2			Cubic		
AlSb	13, 51	4.26	1.62			Cubic		
PbO	82, 8	9.8	1.9	886				
BiI$_3$	83, 53	5.78	1.73	408		Hexagonal		
ZnSe	30, 34	5.42	2.58			Cubic		8.1

Note: Materials are listed in order of decreasing $\mu\tau(e)$ at room temperature.
*Estimated for 20% Zn.
**Estimated.
†Solid/solid phase transition.

(LLNL), Livermore, California, have prepared high resistivity (>10^9 Ω-cm), chromium doped GaAs (Wang *et al.*, 1989, 1991). Photoconductor detectors made from this material are being tested to replace silicon *p-i-n* diodes, which are used in very high radiation environments. The GaAs detector is 50 times faster (≈100 psec FWHM compared to ≈5 nsec for Si). With respect to radiation hardness, it is estimated that the GaAs detectors are 10,000 times harder to 14 MeV neutrons than silicon diodes. Thus, in spite of the limitations of GaAs, there has been a recent renewal of interest in thin film GaAs detectors for very high flux applications (Garconnet *et al.*, 1992).

TABLE I (*Continued*)

E_{pair} eV	Resistivity (25°C) Ω-cm	Electron Mobility cm²/V sec	Electron Lifetime sec.	Hole Mobility cm²/V · sec	Hole Lifetime sec.	$\mu\tau(e)$ Product cm²/V	$\mu\tau(h)$ Product cm²/V
2.96	50	3900	$>10^{-3}$	1900	1×10^{-3}	>1	>1
3.62	up to 10^4	1400	$>10^{-3}$	480	2×10^{-3}	>1	≈ 1
4.43	10^9	1100	3×10^{-6}	100	2×10^{-6}	3.3×10^{-3}	2×10^{-4}
5.0*	10^{11}	1350	10^{-8}	120	5×10^{-8}	1×10^{-3}	6×10^{-6}
5.5**	10^8	720	10^{-8}	75	10^{-6}	7.2×10^{-4} $\approx 10^{-4}$	7.5×10^{-5}
4.2	10^{13} 10^{10}	100	10^{-8}	4	10^{-5}	10^{-4} 9×10^{-5}	4×10^{-5}
4.2	10^7 10^{11}	8000	10^{-8}	400	10^{-7}	8×10^{-5} 7×10^{-5}	4×10^{-6}
4.5		75	5×10^{-7}	45	2×10^{-7}	3.5×10^{-5}	9×10^{-5}
13.25		2000	10^{-8}	1600	$<10^{-8}$	2×10^{-5}	$<1.6 \times 10^{-5}$
6.5	10^{12}	6	2.5×10^{-6}			1.6×10^{-5}	1.5×10^{-6}
4.9	10^{12}	8	10^{-8}	2		8×10^{-6}	
4.2	10^7	4600	1.5×10^{-9}	150	$<10^{-7}$	4.8×10^{-6}	$<1.5 \times 10^{-5}$
7.0**	10^{10} 5×10^{13}	340	4×10^{-9}	100	7×10^{-7}	1.4×10^{-6} 1×10^{-6}	7×10^{-5} $<1 \times 10^{-7}$
4	10^{12}	1	6.8×10^{-9}	.005	4×10^{-6}	6.8×10^{-8}	2×10^{-8}
7	10^{12}	.005	10^{-8}	.14	10^{-6}	5×10^{-9}	1.4×10^{-7}
6.5**	1	10	10^{-8}				
7.0**		120		120			
7.8**		300		50			
9.0**		400(α)					
5.05	$<10^4$	300		400			
6.47							
5.5**	10^{12}						
8.0**		100					

Researchers in high energy physics are also interested in GaAs for use in high flux, high energy accelerator facilities, and Schottky diode and *p-i-n* diode structures are being studied for this application (Bertin *et al.*, 1990; McGregor *et al.*, 1991; Beaumont *et al.*, 1992; Sumner *et al.*, 1992). Frequently detector systems for high energy physics experiments are huge and require the use of very large areas of detector coverage. In addition, high data rates and spatial information are often needed in this field. Thus, GaAs is attractive because of the existence of a mature, large scale fabrication technology, the fast response, and high radiation hardness, properties that are more important than the limited energy resolution.

b. Indium Phosphide

Another III–V material, indium phosphide (InP), is very similar to GaAs in its electronic properties, and for the same basic reason as GaAs, InP has the potential for making good nuclear detectors (Hammond, 1981). Like GaAs, InP is being developed as a material for use in high speed devices where it has the potential faster response than GaAs.

In addition, indium has a relatively high Z of 49, so that InP would have a stopping power similar to that of CdTe for gamma rays, and it has the unique property of having a high cross section for neutrinos (Raghaven, 1976). Lund *et al.* (1988b), Olschner *et al.* (1989b) and Suzuki, Fukuda, and Nagashima (1989) studied InP neutrino detectors. In this application, a proposed cubic meter of InP detectors would offer the same overall sensitivity as the $^{37}Cl \rightarrow {}^{37}Ar$ neutrino detector in operation at the Homestake mine in South Dakota, which uses approximately 400 m^3 of tetrachloroethylene and would also provide low energy and timing data not possible using the tetrachloroethylene detector (Rowley, Cleveland, and Davis, 1984). Because of this property, indium containing detector materials like InP have the potential to allow physicists to finally solve the long standing "neutrino problem" of solar physics (Bahcall, 1981). Lund *et al.* (1988b) tested InP crystals doped with Fe, Zn, and Cu. Wafers of these materials were used to fabricate gamma ray and particle detectors up to 1 cm^3 in volume. The best value of mobility-lifetime product ($\mu\tau$) was obtained for the Fe doped crystals: $\mu_e\tau_e = 5 \times 10^{-6}$ cm^2/V, $\mu_h\tau_h \leq 10^{-7}$ cm^2/V. The energy to create one charge pair was measured to be 4.2 eV.

c. Boron Phosphide

A third potential III–V detector material is boron phosphide, BP. BP is a low Z material and would have relatively poor stopping power for x-rays and gamma rays. However, ^{10}B, which makes up almost 20% of naturally occurring boron, has one of the highest cross sections for thermal neutrons and thus provides the potential for a solid state neutron detector that would have higher sensitivity than any existing detector (Kumashiro and Okada, 1985; Kumashiro *et al.*, 1988). BP has a bandgap of 2.0 eV, which makes it a candidate for room temperature operation. Unfortunately, like GaAs and InP, the charge carrier lifetimes are extremely short.

The BP thermally decomposes at 1400° C and cannot be grown from the melt. Layers of BP have been grown by CVD on various substrates, a technology that could ultimately be useful for fabricating arrays for neutron imaging detectors. One severe difficulty encountered with BP is that it has very small crystal lattice constants, and no suitable substrates have lattice constants that match well enough to grow high quality films. This factor has, thus far, limited the performance of BP detectors. In spite of this problem, Lund *et al.* (1990) fabricated detectors that

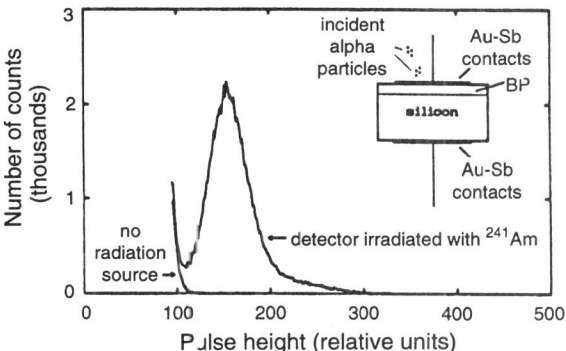

FIG. 1. ^{241}Am alpha particle spectrum obtained using a BP-on-silicon device (from Lund et al., 1990).

could detect neutrons and alpha particles. Fig. 1 shows an ^{241}Am alpha particle spectrum from a BP-on-Si device.

4. II–VI Semiconductors

Although CdTe is by far the most developed II–VI detector material, the other II–VI materials have also been studied. In principle, many of them should have properties that surpass CdTe, but none has achieved its full potential.

a. Cadmium Sulfide

CdS has been studied for many decades in numerous electronic devices. It has been studied in crystalline form and in thin film form for nuclear radiation detection. In crystalline form, high concentrations of deep trap states have always limited its usefulness in this application. The trap states in CdS thin films, however, make it a good photoconductive material and thus it has received attention for fabricating large area imaging devices that operate in photoconductive mode.

Thin amorphous and polycrystalline layers of semiconductors have been used in radiation imaging in combination with several readout techniques (Stanton, 1979). One technique examined intensively for several years is xeroradiography (Wolfe et al., 1987). The detector systems are relatively complex mechanically and, like film, require postexposure image processing. Xeroradiography is an extension of the xerographic process (Pfister, 1979) in which x-rays, instead of light, are used to excite the photoconductor material. The x-rays form a latent image on a large polycrystalline or amorphous plate of semiconductor, and then the image is transferred to paper in the same way as a photocopy. It was first discovered at the Battelle Memorial Institute in 1944 that the surface charge on a photoconductor could be depleted by x-rays (DeWerd, 1982).

The important parameters for this approach are essentially the same as for crystalline detectors. In particular, the charge carrier mobility, lifetime, and resistivity must achieve certain values for the approach to succeed. Typical properties are film thickness ≈ 30 μm, resistivity 10^{11} to 10^{12} Ω-cm to limit dark currents that will decrease the low level sensitivity of the device, and $\mu\tau$ product $> 5 \times 10^{-6}$ cm^2/V. As with optical xerography, amorphous selenium (a-Se) layers met these requirements and were initially used as the semiconductor in xeroradiography (Schein, 1988). However, CdS offers similar performance and much higher stopping power. Mehendru and Eisra (1986) successfully fabricated x-ray sensitive layers for radiographic use using CdS.

b. Cadmium Selenide

Roth and Burger (1986) and Roth et al. (1987) studied the properties of CdSe crystals grown by the temperature gradient solvent zone (TGSZ) method (Burger and Roth, 1984a). Crystals with resistivities of 10^7 to 10^8 Ω-cm were obtained. Devices with Aquadag carbon electrodes were fabricated on Br–MeOH etched slices of 0.2 to 1 mm thick. These devices were tested as x-ray detectors and impressive results were obtained. The noise appears to be limited by the leakage current, and the bandgap of 1.73 eV suggests that it should be possible to grow material with resistivity higher than CdTe ($\approx 10^9$ Ω-cm). which should improve the performance in the future.

c. Zinc Telluride

Because of its similarity to CdTe, its wider bandgap, and its high atomic number, ZnTe has seemed to be a good candidate for use in nuclear detectors for many years. The material properties of ZnTe have been studied for numerous applications (Ribeiro and Pautrat, 1973; Larson and Stevenson, 1973). Various elemental and alloy electrode materials have been investigated for fabricating nuclear detectors (Tupenevich and Kononenko, 1978; Hajghassem, Brown, and Luqman, 1987; Luqman, Brown, and Hajghassem, 1985). However, no one has yet been able to grow ZnTe crystals with sufficiently good charge carrier properties for nuclear detectors. Although the bandgap is nearly 2.3 eV, good quality, high resistivity material has proven somewhat elusive. High resistivity ZnTe crystals (10^8 to 10^{10} Ω-cm) have been grown by the traveling heater method, but the crystals were heavily doped with thallium and suffered from high levels of deep traps and had very poor $\mu\tau$ product of $<1 \times 10^{-5}$ cm^2/V (Saulnier, Squillante, and Entine, 1984). Although these detectors provided no spectroscopy information, they operated as gamma ray counters up to 150°C. Recently, Aurora Technology has grown ZnTe by high pressure Bridgman with a resistivity of 10^8 Ω-cm which increase to 4×10^9 Ω-cm after annealing (Butler, private communications, 1993). A photopeak at 60 keV for ^{241}Am was observed, and perhaps, further research into ZnTe may finally result in useful performance.

5. THALLIUM BROMIDE

Thallium bromide (TlBr) is emerging as a promising material for the development of room temperature, low noise, high resolution x-ray and gamma ray spectrometers. The interest in TlBr is due to the its high average atomic number (Z_{Tl} = 81, Z_{Br} = 35), high density (7.5 g/cm^3), and wide semiconducting bandgap (2.7 eV). The photon stopping power of TlBr is greater than any of the other semiconductors discussed and is approximately that of bismuth germanate.

Thallium bromide was first investigated as a radiation detector material by Hofstadter and others (Hofstadter, 1949; Rahman and Hofstadter, 1984; Rahman et al., 1987). The performance of these early detectors, however, was limited by the purity of the crystals used in their fabrication.

TlBr has a CsCl type cubic crystal structure; it melts congruently at 480° C; and it does not exhibit any solid–solid phase transition below its melting point. Hence good quality crystals of TlBr can be grown directly from the melt. In fact, large area (>3 cm diameter) crystals of TlBr are commercially grown (for infrared window application) using the Bridgman process.

Recently, it was found that considerable improvement in the TlBr detector performance, especially its resistivity and its charge transport properties, could be obtained by zone purifying commercially available TlBr prior to crystal growth (Shah et al., 1989, 1990b; Olschner et al., 1990, 1992). Radiation detector grade, single crystals of TlBr (up to 5 cm long and 1.4 cm in diameter) have also been grown by Bridgman methods. Single crystal TlBr boules are cut into wafers (about 1 mm thick) using a wire saw. These wafers are lapped, polished, and etched using a 5% bromine in methanol solution. Paint-on style carbon contacts are then applied to these crystals. Other contacting schemes such as evaporated Au and Pd films have also been tested with similar results in detector performance.

Zone refining TlBr starting material has led to considerable improvements in the electrical properties of TlBr crystals. The mobility–lifetime product of electrons and holes are 1.3×10^{-5} cm^2/V and 1.5×10^{-6} cm^2/V, respectively (Shah et al., 1990b), as compared to early values reported by Hofstadter and Rahman for the first TlBr detectors ($\mu_e \tau_e \approx 1 \times 10^{-8}$ cm^2/V) (Rahman and Hofstadter, 1984; Rahman et al., 1987). The zone refined crystals also have high electrical resistivity (1×10^{11} Ω-cm). Further zone refining of the feed material failed to improve the results, and it is likely that chemical impurities are no longer limiting the resistivity and $\mu_e \tau_e$ product, and other effects such as crystal defects are responsible for the relatively low $\mu \tau$ product. While harder than HgI$_2$ and PbI$_2$, TlBr

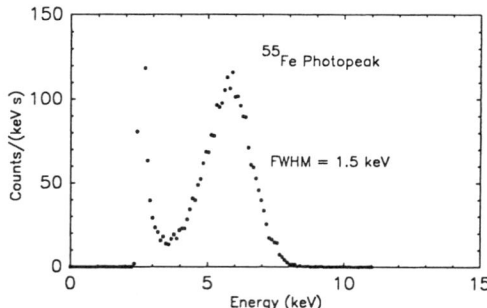

FIG. 2. ⁵⁵Fe spectrum (5.9 keV x-rays) obtained using a TlBr detector. (Reproduced with permission from Shah *et al.*, 1990b and Elsevier Science Publishers.)

is not a very hard material (Knoop hardness of 12) and can be damaged during the fabrication processes.

Figure 2 shows an ^{55}Fe spectrum (5.9 keV x-rays) recorded with a TlBr detector (1.2 mm² area, 100 μm thick) and the resolution of the 5.9 keV photopeak was estimated to be 1.5 keV (FWHM). Figure 3 shows an ^{241}Am spectrum measured with a TlBr detector. While the lower energy peaks corresponding to 14, 18, 21, and 25 keV are well resolved, the resolution of the 60 keV peak is degraded due to "hole tailing" effect. This resolution could be somewhat improved by increasing the detector bias, as shown in the inset to Fig. 3.

While these small detectors exhibit adequate energy resolution, most commercial applications would require substantially larger detector sizes. Fortunately, TlBr crystals are not very difficult to grow, and the Bridgman crystal growth process is well suited to scaling up the crystal volume. However, to maintain the energy resolution at these larger sizes, further improvements in the charge transport parameters as well as the resistivity of TlBr crystals are needed.

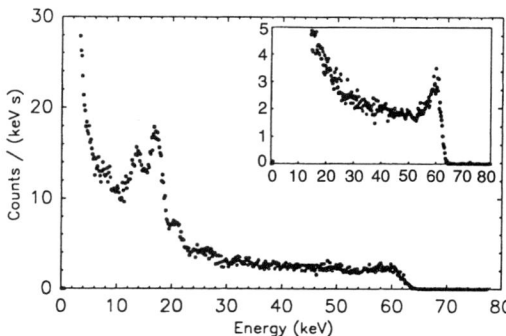

FIG. 3. ^{241}Am spectrum obtained using a TlBr detector. (Reproduced with permission from Shah *et al.*, 1990b and Elsevier Science Publishers.)

6. AMORPHOUS SILICON

Amorphous silicon appears to be an attractive material for making two dimensional, position sensitive x-ray and particle detectors. Amorphous silicon (a-Si) has been under intense investigation for over a decade for use in low cost photovoltaic solar cells and more recently for use in electronic devices, displays, and imaging optical sensors. A large technological base for producing a-Si films and fabricating devices and device arrays has grown out of solar cell processing and the silicon wafer processing industry, and because of this a-Si offers the potential for very low cost production of large area detector arrays. Although it is made primarily of silicon and many of the processes for fabricating devices are derived from those for single crystal silicon, a-Si differs significantly from silicon in its electronic properties. For example, the bandgap is 1.7 eV and the values of electron and hole mobilities and lifetimes are orders of magnitude less than silicon, which limit its usefulness, at present, to very thin devices that can be suitable for visible light, charged particle, and low energy x-ray detection.

Layers of a-Si are typically deposited on metal or ceramic substrates using plasma enhanced or glow discharge chemical vapor deposition. In these processes, silane gas (SiH_4) decomposes and amorphous silicon layers containing up to 20% hydrogen grow. Substrates with areas of hundreds of square centimeters are common. Devices are fabricated using silicon wafer stepper technology. Typically, p-i-n photodiodes have been investigated on a-Si for imaging (Street, 1992; Perez-Mendez, 1991; Kaplan et al., 1986; Naruse and Hatayama, 1987; Antonuk et al., 1990; Hamel et al., 1991; Cho et al., 1992; Qureshi et al., 1989).

The optical imaging sensors have the potential to be used for gamma ray imaging when coupled to a scintillator (Manfredotti et al., 1992; Equer, 1992). Thin film transistors (TFTs) for amplifying the photodiode signal for readout purposes have been fabricated directly on the a-Si surface (Fugieda et al., 1991), and very high quality images have been reported using a-Si p-i-n–TFT arrays coupled to standard Kodak Lanex Regular medical film screen as shown in Fig. 4.

Several researchers are extending this sensor technology to include imaging detectors that detect x-rays or particles directly (Fujieda et al., 1990). The photon stopping power of silicon is low, and since the useful thickness of the layers is limited by the electronic properties of the a-Si to less than 200 microns, these devices are useful for only very low x-ray energies. However, to improve the stopping power, thin layers of a-Si have been used in conjunction with intervening layers of metals to increase x-ray and gamma ray stopping power (Naruse and Hatayama, 1989). The Si detects the photoelectron emitted from the metal layers. A variety of device configurations are possible, and both Schottky diodes and p-i-n diodes can be fabricated. Furthermore, by choosing the appropriate materials, the metal layers can be used as the device electrodes. While this significantly enhances the stopping power, most of the energy information from the photons is lost. In addition, neutron converters have been used in conjunction with a-Si to make position sensitive neutron detectors (Mireshghi et al., 1992). Another ap-

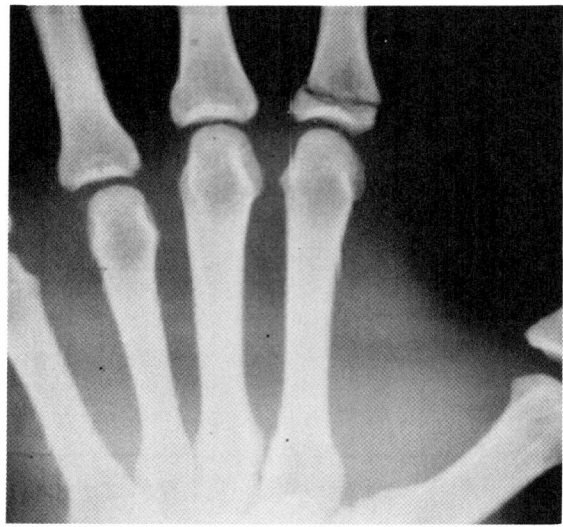

FIG. 4. X-ray image taken using an a-Si imaging array coupled to standard medical film screen.

proach to improve the sensitivity of a-Si arrays to x-rays has been to couple them to vapor grown CsI(Tl) layers (Jing *et al.,* 1993). The CsI(Tl) layers are patterned in such a way that each a-Si pixel is coupled to an individual CsI(Tl) scintillator. The scintillator thickness is adjusted to obtain good x-ray stopping efficiency, and the optical photons emitted by the scintillator are then detected by a-Si sensors.

7. Ternary Materials

Markakis (1988) indirectly spurred interest in ternary materials when he coupled a 1 inch diameter CsI(Tl) scintillator crystal to a 1 inch HgI_2 detector with a transparent liquid electrode and obtained the best spectrum ever reported for any photodetector coupled to a scintillator. He obtained 5% FWHM for the 662 keV photopeak from ^{137}Cs. The high optical quantum efficiency of the HgI_2 for the light emitted by the scintillator allowed it to surpass photomultiplier tubes even though the PMTs have high gain and low noise.

Markakis's achievement demonstrated that high resistivity semiconductors can compete with other optical detectors for use with scintillator crystals for the detection of gamma rays. His results have spurred an interest in other researchers to investigate this further. The light output of a scintillator is peaked at a specific wavelength, which gives it its characteristic color; thus, to achieve high quantum efficiency, a photodetector must have an appropriate spectral response. Only a small number of binary semiconductors have bandgaps of the correct energy to be sensitive to the visible light emitted by common scintillators. In addition, most semiconductors have a narrow sensitivity range: they are transparent to wavelengths with energies less than the bandgap and the sensitivity drops off quickly at shorter wavelengths. This limited range is due to the absorption coefficient for direct bandgap materials, which is very high and increases steeply for shorter wavelengths. Thus, the higher the photon energy is, the closer to the surface it is absorbed. All semiconductors have high concentrations of defects near the surface caused by surface oxides, dangling bonds, and mechanical damage resulting from device fabrication, which cause the charge carrier lifetimes to be very short. These defects result in poor charge collection for the electrons generated by photons that stop near the surface.

Therefore, it is necessary to maximize the overlap in the spectral response of the detector with the spectral emission of the scintillator. One way to do this is to combine binary compounds to make ternary semiconducting compounds with tailored bandgaps. Such ternary materials will have a bandgap between the two binaries that varies as a function of the composition.

a. $TlBr_xI_{1-x}$

$TlBr_xI_{1-x}$ is a ternary semiconductor and is based on a solid solution of TlI in TlBr (Zhang *et al.*, 1993). The interest in $TlBr_xI_{1-x}$ is primarily due to the possibility of varying the bandgap of the ternary material from 2.3 eV ($x = 0.3$) to 2.7 eV ($x = 1$) (Ikedo *et al.*, 1986), which can be used to create material with a tuned optical response. This property is particularly useful in the development of photodetectors with optical response tuned to match the emission spectrum of the common inorganic scintillators such as BGO, CsI(Tl), or CsI(Na).

$TlBr_xI_{1-x}$ photodetectors fabricated from zone refined and Bridgman grown crystals have recently been evaluated as scintillation spectrometers and have successfully detected 5.5 MeV α-particles (^{241}Am source) with 20% resolution (FWHM) at room temperature (Shah *et al.*, 1994). The $TlBr_xI_{1-x}$ photodetectors do not match the performance of established optical detectors such as photomultiplier tubes or silicon photodiodes at present, and the principal factor limiting their performance is the relatively low electrical resistivity (about 10^{10} Ω-cm). This leads to high leakage current, and thereby high shot noise, in the $TlBr_xI_{1-x}$ scintillation spectrometers. Improvements in $TlBr_xI_{1-x}$ crystal purity and quality as well as the development of blocking contacts on $TlBr_xI_{1-x}$ crystals are needed for $TlBr_xI_{1-x}$ photodetectors to become commercially useful.

b. $HgBr_xI_{2-x}$ *Detectors*

$HgBr_xI_{2-x}$ is another ternary semiconductor formed by mixing $HgBr_2$ and HgI_2, and its bandgap can be varied from 2.1 eV to 3.4 eV. Also, unlike HgI_2, for most of the composition range ($x > 0.4$), $HgBr_xI_{2-x}$ has no solid–solid phase change, and hence high quality crystals can be grown directly from the melt. Recently $HgBr_xI_{2-x}$ has been investigated as a nuclear and optical detector material (Zhou, et al., 1993). Due to its high vapor pressure, the zone refining technique is not easily applicable to purification of $HgBr_xI_{2-x}$, and multiple step sublimation, which is commonly used in purifying HgI_2 (Lamonds, 1983), is used to purify $HgBr_xI_{2-x}$. The crystals of $HgBr_xI_{2-x}$ were grown by Bridgman process and evaluated as photodetectors in scintillation spectrometers. $HgBr_xI_{2-x}$ photodetectors coupled to CsI(Na) scintillator successfully detected 5.5 MeV α-particles (^{241}Am source) with 20% resolution (FWHM) at room temperature. Unlike $TlBr_xI_{1-x}$, the resistivity of $HgBr_xI_{2-x}$ crystals is very high ($>10^{13}$ Ω-cm). However, the charge transport properties in $HgBr_xI_{2-x}$ are poor ($\mu\tau_e \approx 10^{-6}$ cm^2/V) and limit the performance of the $HgBr_xI_{2-x}$ detectors. This limitation may be due to material purity and better purification could lead to improved detector performance.

c. Ternary II–VI Materials

Ternary materials have also been investigated as detectors to detect x-rays directly. In most cases, the ternaries have not achieved the charge carrier properties of either of the binary materials from which they are made. However, there are numerous reasons to study ternaries, such as the presence of a destructive phase transition in HgI_2 that is not present in $HgBr_xI_{2-x}$ with $x > 0.4$. Ternaries tend to be harder than binary compounds, and it can often be easier to grow large single crystals of ternaries. Also, it may be useful to have a slightly higher bandgap than a binary, which can be achieved with the addition of another component. Usually, however, a tradeoff in performance parameters is encountered in this approach, and the ternary material is not able to achieve the same level of performance as the binary.

Roth et al. (1986, 1987; Burger and Roth, 1984b; Burger, Roth, and Schieber, 1985) investigated $Cd_xZn_{1-x}Se$ by adding ZnSe ($E_{gap} = 2.67$ eV) to CdSe ($E_{gap} = 1.73$ eV) to increase the bandgap. They obtained detector grade material with $x = 0.7$, which had a bandgap of 2.14 eV. The detectors were able to detect ^{55}Fe 5.9 keV x-rays with about 30% FWHM energy resolution. These researchers obtained better results with the binary CdSe (as discussed earlier) and concluded that surface states, not bandgap was dominating the leakage current in the devices. Butler, Lingren, and Doty (1992a,b) have used Zn to increase the bandgap of CdTe and prepared $Cd_xZn_{1-x}Te$ (CZT) with 4 to 20% Zn. High pressure Bridgman process has been used to grow large volume CZT crystals. CZT detectors with 1 cm^2 area and 2.5 cm thick have been fabricated from such crystals and have been operated with 5% resolution (FWHM) for 662 keV gamma rays (^{137}Cs source) (Glick et al., 1994). Meyer et al. (1993) has prepared another ternary material $CdTe_{0.9}Se_{0.1}$. The resistivity obtained was 7×10^9 Ω-cm, but significant hole-tailing was observed due to reduced $\mu\tau$ product for holes.

8. OTHER, LESS STUDIED, CRYSTALLINE MATERIALS

A large number of semiconductor materials have been considered and studied over the years, including Bi_2S_3 (Wald, Bullitt, and Bell, 1975), AlSb (Yee, Swierkowski, and Sherohman, 1977), and many others (Armantrout, 1977; Sakai, 1982), but most have shown no real promise for detector applications and there are no published reports of continued research on them, although A. Witt and P. Becla at the Massachusetts Institute of Technology are investigating the growth of AlSb crystals at the time of this writing (Becla, private communication, 1993). Bridgman growth has resulted in small grain polycrystalline ingots, but large grains were obtained by Czochralski growth. The material has an indirect bandgap of 1.6 eV. Initial results are encouraging and very exciting, with the hole mobility values of several hundred cm^2/Vs. Resistivities to date were low, $<10^4$, but this is attributed to impurities, and the material is moisture sensitive. Both n- and p-type material can be grown.

a. Diamond

The possibility of fabricating diamond detectors has been enticing researchers for decades (Kuzolov, 1975, 1977). Diamond is attractive because it is an element and because it is extremely hard and rugged, has a wide bandgap (6.6 eV), and very high resistivity. This material offers the same benefits described previously for GaAs, very fast response and very high radiation hardness, and thus may find use in high energy physics experimentation. Unfortunately, it is a very expensive material and hard to grow in high quality crystal form. Recently, there has been an interest in making devices using thin films of diamond grown by CVD (Beetz et al., 1991; Kagan et al., 1993).

b. Gallium Selenide

GaSe was investigated by Manfredotti et al. (1974, 1975) in the early 1970s as a nuclear detector material and alpha particle detectors were obtained. Sakai et al. (1988) and Nakatani et al. (1989) have investigated this material further and reported on the fabrication of devices on GaSe platelets that were cleaved from Bridgman grown ingots. A variety of metallic electrodes in symmetric and nonsymmetrical configurations were tested, as was the use of a guard ring structure to reduce leakage currents. Alpha particle detectors with about 5% FWHM at 5.5 MeV (^{241}Am) were obtained.

c. Indium Iodide

InI is a wide bandgap semiconductor ($E_{gap} = 2.01$ eV) with a base centered orthorhombic crystal structure (Ohno et al., 1984). Recently, InI has shown con-

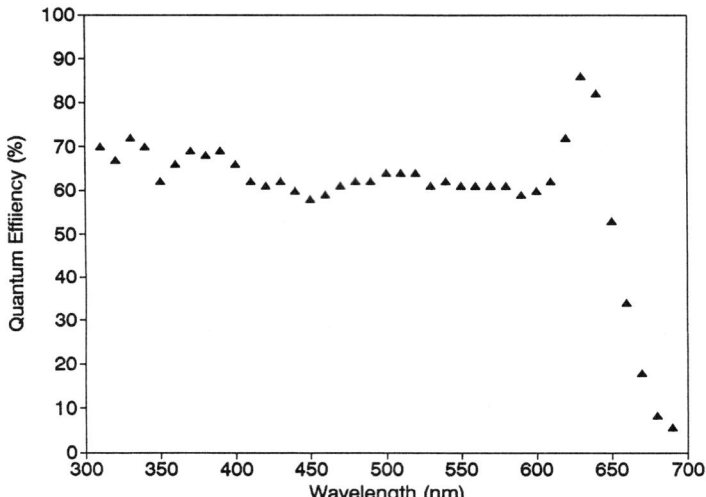

FIG. 5. Spectral response of photodetectors fabricated from an InI crystal.

siderable promise as a nuclear and optical detector material (Shah *et al.,* 1992b; Squillante *et al.,* 1993d). Single crystals of InI were grown by the Bridgman process. These crystals showed high electrical resistivity (10^{11} Ω-cm), good charge transport properties ($\mu_h \tau_h = 7 \times 10^{-5}$ cm²/V, $\mu_e \tau_e = 5 \times 10^{-6}$ cm²/V), and Knoop hardness of 27. Photodetectors fabricated from the InI crystals showed high quantum efficiency (>60%) in the 300 nm to 600 nm wavelength region (Fig. 5) and the photoresponse was found to be uniform over the active detector area. InI photodetectors coupled to CsI(Tl) scintillator were successfully operated as spectrometers at room temperature, detecting 5.5 MeV α-particles (^{241}Am) with 11.7% resolution (FWHM), as well as 662 keV gamma rays (^{137}Cs) with 17.3% resolution, as shown in Fig. 6. InI detectors were capable of operating as counters up to 150°C. Based on these early results, InI has the potential of being an important nuclear detector material. Further investigation of material purification, crystal growth and detector fabrication is warranted to explore the full potential of this material.

9. OTHER, LESS STUDIED, THIN FILM MATERIALS

Numerous other materials have been studied that show promise for certain applications but that have not achieved that promise. The recent experience with PbI$_2$ shows that, despite earlier discouraging results, any of these materials may have great future potential. CdSe has been studied for decades for use in the fabrication of TFTs (Wilson and Gutierrez, 1965; de Beats *et al.,* 1990; Van Calster, 1985). Other materials that have been studied for thin film diodes and TFTs in-

12. OTHER MATERIALS: STATUS AND PROSPECTS

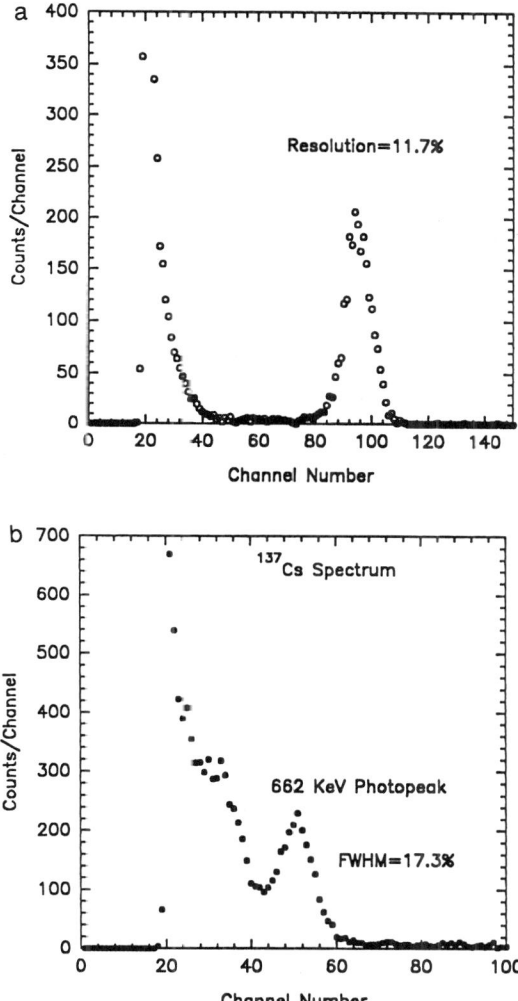

FIG. 6. Spectrum taken with an InI photodetector coupled to CsI(Tl): (a) is an ^{241}Am alpha spectrum and (b) is a ^{137}Cs, 662 keV gamma ray spectrum (from Shah *et al.,* 1992).

clude CdSe, CdS, Te, InAs, PbS, InSb, PbTe, HgSe, and SnO_2 (Maissel and Glang, 1970).

a. TlBr Films

In addition to the development of nuclear detectors from single crystals of TlBr, vacuum sublimed TlBr films have also been studied for use in radiation detectors

(Olschner et al., 1992). These films can be produced in large areas at relatively low cost and hence are promising for some applications such as medical imaging.

Polycrystalline TlBr films (100 to 150 μm thick, 10 cm² area) have been deposited on metal coated glass substrates by resistively heating a charge of zone purified TlBr in a quartz crucible kept under high vacuum (10^{-6} Torr) (Olschner et al., 1992). Such TlBr films have shown electrical resistivity of about 10^{10} Ω-cm and have been successfully operated as single photon counters detecting ^{137}Cs (662 keV), ^{57}Co (122 keV), and ^{241}Am (60 keV) at room temperature. The films have also shown relatively good $\mu\tau$ product (2.8×10^{-6} cm²/V) for electrons.

The electrical resistivity and the charge transport properties of TlBr single crystals are better than those for the sublimed films, as would be expected, and hence TlBr crystals would be the choice for applications where energy resolution is of prime importance. However, for imaging applications such as xeroradiography, TlBr films are potentially promising.

III. Current Status and Prospects

1. Comparison of Material Properties

While the "new" materials have yet to approach the overall performance of the mature materials, there is good reason to expect some breakthroughs in the near future, because these materials are being studied more intensively than they have been for many years. Also, there are already certain applications where some of these materials can perform in ways the standard materials cannot. For example, PbI_2 and InI can operate above 100°C, and TlBr has the highest stopping power of the semiconducting materials that have been examined.

Table I summarizes the properties of the semiconductor materials. The values in the table are the best reported values, taken from the articles cited in each of the sections pertaining to a material. The three most important parameters for obtaining good signal to noise ratios in most applications are the photon stopping power, the mobility–lifetime product, and the resistivity. These are the primary parameters limiting the quantum efficiency, the electronic charge collection and the noise. For good performance, both the mobility and the lifetime must be good. Silicon has by far the best values of $\mu\tau$ product.

Figure 7 shows the linear attenuation coefficients for some of the materials discussed. The photon stopping power increases strongly as a function of the atomic number. This dependence is obviously important for charged particle as well as photons. It is even very important for low energy x-rays that do not penetrate deeply into any material, but since they are stopped closer to the surface of high Z materials, the charges generated are easier to collect and thus detectors made from high Z materials like HgI_2 and PbI_2 have very good energy resolution at low energies.

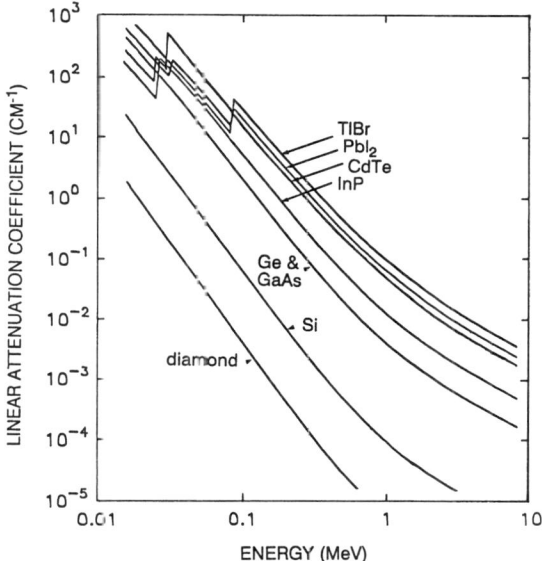

FIG. 7. Linear attenuation coefficient of several nuclear detector materials.

It is interesting to note the values of energy to create a charge pair. Klein (1968) identified a relationship for E_{pair} that approximated a straight line with value of roughly $3 \times E_{gap}$ per charge pair for semiconductor detectors. For years, HgI_2 seemed to be an isolated exception to this rule with E_{pair} about $2 \times E_{gap}$. In 1988 Lund et al. (1988a and Shah et al., 1989) showed that at least one other line appears to exist for semiconductors as shown in Figure 8.

No dramatic correlation stands out in the table between any of the listed parameters and the E_{pair} line a material falls on. The materials on the newer line

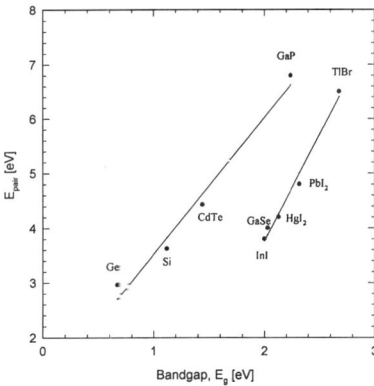

FIG. 8. Bandgap, E_g, versus E_{pair} relationship for various detector materials.

produce a higher signal due to the lower energy needed to produce a charge pair, and this factor would be a useful guide to researchers seeking new materials if a correlation between the materials and the line they are on could be identified. One parameter, the ionicity (Keating, 1966), appears to have at least a weak relationship. Materials on the new line tend to have higher ionicity.

2. FUTURE DIRECTIONS

a. *Thin Film Imaging Detector*

Clearly the area generating the most excitement (based on the number of publications in the field) is that of thin film imaging detector arrays. Advances in thin film imaging sensor technology have been impressive and, given the very high level of interest, continued progress is all but assured. When developed, these imaging detectors will be useful for low energy x-ray applications such as x-ray diffraction and for imaging higher energy x-rays and gamma rays in medical and industrial radiography when coupled to scintillators and intensifying screens.

b. *Potential New Materials*

Research on related semiconductor materials in other fields will also have an impact on future developments in nuclear detector materials. Many of these areas are far better funded than nuclear detector materials research, thus quicker progress can be made, as was the case for the recent progress in $Cd_xZn_{1-x}Te$, which was developed as a subsatrate material for infrared detector fabrication (Bruder *et al.*, 1990). For example, research into materials for use as detectors for extreme ultraviolet is currently a very active area (Davis *et al.*, 1988). One approach is to use semiconductors with very wide bandgaps, such as diamond (Marchywka *et al.*, 1991; Gildenblat *et al.*, 1990). As is the case with nuclear detectors, researchers are considering potential materials by extrapolating from well understood materials. Extending from the existing knowledge base, GaAs and related compounds, GaAsP, GaAlP, and GaP are under study (Krumrey *et al.*, 1988). Another III–V material under consideration is BN (Ahmad and Lichtman, 1989). UV detector research is a very active area that will probably enjoy a considerable amount of funding over the next decade, and significant progress is to be expected. It is likely that some of these materials will function well as low energy x-ray or particle materials and should play a part in future developments in this area.

In another field, Tanaka *et al.* (1985) reported the growth of large single crystals of YB_{66}. This material is being tested for use as a material for fabricating soft x-ray monochromators (Wang *et al.*, 1991). It is a semiconductor, but its electronic properties have not yet been studied. This material deserves consideration for neutron detectors because of the high percentage of boron.

3. Summary and Conclusions

The interest in developing new detector materials has increased partly because of developments in other fields that affect the potential utility of any new materials developed. The prospects for the successful development of improved detector materials is higher than it has been in over a decade, especially in the area of position sensitive imaging arrays of detectors. The next few years should see developments that far exceed those of the past few decades.

References

Ahmad, N., and Lichtman, D. (1989). *Sensors and Actuators* **18,** 397.
Albee, A. L., and Economou, T. E. (1988). *IEEE Trans. Nucl. Sci.* **NS-35,** 356.
Antonuk, L. E., Yorkston, J., Boudry, J., Longo, M. J., Jimenez, J., and Street, R. A. (1990). *IEEE Trans. Nucl. Sci.* **37,** 165.
Armantrout, G. A., Swierkowski, S. P., Sherohman, J. W., and Yee, J. H. (1977). *IEEE Trans. Nucl. Sci.* **NS-24**(Feb.), 121.
Avset, B. S., Ellison, J. A., Evensen, L., Hansen, T.-E., Roe, S., and Wheadon, R. (1990). *Nucl. Inst. and Meth.* **A288,** 131.
Bahcall, J. N. (1981). Proc. Int. Conf. Neutrino Physics and Astrophysics, Hawaii, 1.
Beaumont, S., Bertin, R., Bibi, F., Booth, C. N., Buttar, C., Carraresi, C., Cindolo, F., Colocci, M., Combley, F. H., Dalgi, F., D'Auria, S., del Papa, C., de Maria, C., Dogru, M., Edwards, M., Fiori, F., Foster, F., Francescato, A., Hou, Y., Houston, P., Jones, O. B., Lynch, J. G., Lisowski, B., Matheson, J., Nava, F., Nuti, M., O'Shea, V., Ottaviani, P., Pischedda, M., Pelfer, P. G., Raine, C., Santana, J., Saunders, I., Seller, P. H., Shankar, K., Sharp, P. H., Skillicorn, J. O., Sloan, T., Smith, K. M., Tartoni, N., ten Have, I., Turnbull, R. M., Vanni, U., Vinattieri, A., and Zichichi, A. (1992). *Nucl. Inst. and Meth.* **A322,** 472.
Beetz, C. P., Lincoln, B., Winn, D. R., Segall, K., Vegas, M., and Wall, D. (1991). *IEEE Trans. Nucl. Sci.* **NS-38,** 107.
Bencivelli, W., Bertolucci, E., Bottigli, U., Del Guerra, A., Messineo, A., Nelson, W. R., Randaccio, P., Rosso, V., Russo, P., and Stefanini, A. (1991). *Nucl. Inst. and Meth.* **A310,** 210.
Benz, K. W., Irsigler, R., Ludwig, J., Rosenzweig, K., Runge, K., Schäfer, F., Schneider, J., and Webel, M. (1992). *Nucl. Inst. and Meth.* **A322,** 493.
Bertin, R., D'Auria, S., Del Papa, C., Fiori, F., Lisowski, B., O'Shea, V., Pelfer, P. G., Smith, K., and Zichichi, A. (1990). *Nucl. Inst. and Meth.* **A294,** 211.
Bruder, M., Schwarz, H.-J., Schnitt, R., and Maier, H. (1990). *J. Cryst. Growth* **101,** 266.
Burger, A., and Roth, M. (1984a). *J. Cryst. Growth* **67,** 507.
Burger, A., and Roth, M. (1984b). *J. Cryst. Growth* **70,** 386.
Burger, A., Roth, M., and Schieber, M. (1985). *IEEE Trans. Nucl. Sci.* **NS-32,** 556.
Butler, J. F., Lingren, C., and Doty, F. P. (1992a). *IEEE Trans. Nucl. Sci.* **NS-39,** 605.
Butler, J. F., Doty, F. P., and Lingren, C. (1992b). *Proc. SPIE* **1734,** 131.
Cho, G., Qureshi, S., Drewery, J. S., Jing, T., Kaplan, S. N., Lee, H., Mireshghi, A., Perez-Mendez, V., and Wildermuth, D. (1992). *IEEE Trans. Nucl. Sci.* **NS-39,** 641.
Cuzin, M. (1987). *Nucl. Inst. and Meth.* **A253,** 407.
Davis, R. F., Sitar, Z., Williams, B. E., Kong, H. S., Kim, H. J., Palmour, J. W., Edmond, J. A., Ryu, J., Glass, J. T., and Carter, C. H., Jr. (1988). *Materials Sci. and Engineering* **B1,** 77.
De Beats, J., VanFleteren, J., DeRyke, L., Doutreloigne, J., Van Calster, A., and De Visschere, P. (1990). *IEEE Trans. Elect. Dev.* **ED-37,** 636.

DeWerd, L. A. (1982). In *Handbook of Medical Physics*, ed. R. G. Waggener. CRC Press, Cleveland.
Eberhardt, J. E., Ryan, R. D., and Tavendale, A. J. (1970). *App. Phys. Lett.* **17,** 427.
Eberhardt, J. E., Ryan, R. D., and Travendale, A. J. (1971). *Nucl. Inst. and Meth.* **94,** 463.
Eiche, C., Fiederle, M., Weese, J., Maier, D., Ludwig, J., and Benz, K. W. (1993). *Semiconductors for Room Temperature Radiation Detector Applications,* ed. R. B. James, P. Siffert, T. E. Schlesinger, and L. Franks. Materials Research Society, Pittsburgh, PA.
Equer, B. (1992). *Nucl. Inst. and Meth.* **A322,** 457.
Faile, S. P., Dabrowski, G. C., Huth, G. C., and Iwanczyck, J. S. (1980). *J. Cryst. Growth* **50,** 752.
Fraint, A., and Mellet, J. (1991). Presented at the *7th International Workshop on Room Temperature Semiconductor X- and Gamma Ray Detectors and Associated Electronics,* Ravello, Italy, Sept. 1991.
Fujieda, I., Cho, G., Conti, M., Drewery, J., Kaplan, S. N., Perez-Mendez, V., Qureshi, S., and Street, R. A. (1990). *IEEE Trans. Nucl. Sci.* **NS-37,** 124.
Fujieda, I., Nelson, S., Street, R. A., and Weisfield, R. L. (1991). *Mat. Res. Soc. Symp.* **219,** 537.
Fujieda, I., Nelson, S., Street, R. A., and Weisfield, R. L. (1992). *IEEE Trans. Nucl. Sci.* **NS-39,** 1056.
Garconnet, J.-P., Bourgade, J.-L., Nail, M., Schirmann, D., and Cuzin, M. (1992). Presented at the APS, 9th Topical Conference on High Temperature Plasma Diagnostics, Santa Fe, NM.
Gatti, E., Rehak, P., and Walten, J. T. (1984). *Nucl. Inst. and Meth.* **226,** 129.
Gildenblat, G. S., Grot, S. A., Hatfield, C. W., Badzian, A. R., and Badzian, T. (1990). *IEEE Elect. Devices Lett.* **11,** 371.
Glick, B., Eissler, E., Parnham, K., and Cameron, S. (1994). Symposium on Radiation Measurements and Applications, Ann Arbor, MI.
Hajghassem, H. S., Brown, W. D., and Luqman, M. M. (1987). *Microelectron Reliab.* **27,** 677.
Hall, G. (1988). *Nucl. Inst. and Meth.* **A273,** 559.
Hamel, L. A., Dubeau, J., Pochet, T., and Equer, B. (1991). *IEEE Trans. Nucl. Sci.* **NS-38**(Apr.), 251.
Hammond, R. B. (1981). *Electron Devices Meeting Technical Digest.*
Hofstadter, R. (1949). *Nucleonics* (Apr.), 2.
Ikedo, M., Watari, M., Tateisha, F., and Ishiwatari, H. (1986). *Appl. Phys.* **60,** 3035.
Iwanczyk, J. S., Wang, Y. J., Bradley, J. B., Conley, J. M., Albee, A. L., and Economou, T. E. (1989). *IEEE Trans. Nucl. Sci.* **NS-36,** 841.
Jing, T., Goodman, C. A., Cho, G., Drewery, J., Hong, W. S., Lee, H., Kaplan, S. N., Mireshghi, A., Perez-Mendez, V., and Wildermuth, D. (1994). *IEEE Trans. Nucl. Sci.,* **41,** 903..
Kagan, H., Gan, K. K., Kass, R., Malchow, R., Morrow, F., Palmer, W., White, C., Zhao, S., Pan, L., Han, S., Kania, D., Lee, M., Kim, S., Sannes, F., Schnetzer, S., Stone, R., Thomson, G., Sugimoto, Y., Fry, A., Kanda, S., Olsen, S., and Franklin, N. (1993). *Semiconductors for Room Temperature Radiation Detector Applications,* ed. R. B. James, P. Siffert, T. E. Schlesinger, and L. Franks. Materials Research Society, Pittsburgh, PA. **302,** 257.
Kaplan, S. N., Morel, J., Mulera, T. A., Perez-Mendez, V., and Churmacher, G. (1986). *IEEE Trans. Nucl. Sci.* **NS-33,** 351.
Keating, B. N. (1966). *J. Phys. Rev.* **145,** 637.
Kemmer, J., Lutz, G., Belau, E., Prechtel, U., and Welser, W. (1987). *Nucl. Inst. and Meth.* **A253,** 378.
Klein, C. A. (1968). *IEEE Trans. Nucl. Sci.* **NS-15,** 307.
Kobayashi, T., Sugita, T., Koyanna, M., and Takayanagi, S. (1972). *IEEE Trans. Nucl. Sci.* **NS-19,** 324.
Kobayashi, T., Kuru, I., Hojo, A., and Sugita, T. (1976). *IEEE Trans. Nucl. Sci.* **NS-23,** 97.
Kozolov, S. F., Stuck, R., Hage-Ali, M., and Siefert, P. (1975). *IEEE Trans. Nucl. Sci.* **NS-22,** 160.
Kozolov, S. F., Konolova, E. A., Kuznetsow, Y. A., and Salikov, Y. A. (1977). *IEEE Trans. Nucl. Sci.* **NS-24,** 235.
Krumrey, M., Tegeler, E., Barth, J., Krisch, M., Schafers, F., and Wolf, R. (1988). *Applied Optics* **27**(Oct.), 4336.
Kumashiro, Y., and Okada, Y. (1985). *Appl. Phys. Lett.* **47,** 64.

Kumashiro, Y., Kudo, K., Matsumoto, K., Okado, Y., and Koshiro, T. (1988). *J. Less-Common Metals* **143,** 71.
Lamonds, H. A. (1983). *Nucl. Inst. and Meth.* **213,** 5.
Larson, T. L., and Stevenson, D. A. (1973). *J. Appl. Phys.* **44,** 843.
Lund, J. C., Shah, K. S., Squillante, M. R., and Sinclair, F. (1988a). *IEEE Trans. Nucl. Sci.* **NS-35,** 89.
Lund, J. C., Olschner, F., Sinclair, F., and Squillante, M. R. (1988b). *Nucl. Inst. and Meth.* **A272,** 885.
Lund, J. C., Shah, K. S., Squillante, M. R., Moy, L. P., Sinclair, F., and Entine, G. (1989). *Nucl. Inst. and Meth.* **A283,** 299.
Lund, J. C., Olschner, F., Ahmed, F., and Shah, K. S. (1990). In *Diamond, Boron Nitride, Silicon Carbide and Related Wide Bandgap Semiconductors,* ed. J. T. Glass, R. F. Messier, and N. Fujimori. Proceedings of the MRS **162.**
Lund, J. C., Olschner, F., Shah, K. S., and Squillante, M. R. (1992). *Proc. SPIE* **1734,** 140.
Luqman, M. M., Brown, W. D., and Hajghassem, H. S. (1985). *J. Electronic Materials* **16,** 123.
Maissel, L. I., and Glang, R. (eds.). (1970). *Handbook of Thin Film Technology,* Chapter 20. McGraw-Hill, New York.
Manfredotti, C., Murri, R., and Vasanelli, L. (1974). *Nucl. Inst. and Meth.* **115,** 349.
Manfredotti, C., Murri, R., Quirini, A., and Vasanelli, L. (1975). *Nucl. Inst. and Meth.* **131,** 457.
Manfredotti, C., Faccio, F., Fizzotti, F., and Marchisio, R. (1992). *Nucl. Inst. and Methods* **A322,** 483.
Marchywka, M., Hochedez, J. F., Geis, M. W., Socker, D. G., Moses, D., and Goldberg, R. T. (1991). *Applied Optics* **30**(Dec.), 5011.
Markakis, J. M. (1988). *IEEE Trans. Nucl. Sci.* **NS-35,** 356.
Mayer, J. W. (1968). In *Semiconductor Detectors,* ed. G. Bertolini and A. Coche. Wiley Interscience, New York.
McGregor, D. S., Knoll, G. F., Eisen, Y., and Brake, R. (1991). *IEEE Trans. Nucl. Sci.* **NS-38,** 90.
McGregor, D. S., Knoll, G. F., Eisen, Y., and Brake, R. (1992). *Nucl. Inst. and Meth.* **A322,** 487.
Mehendru, P. C., and Eisra, S. C. K. (1985). *J. of Tech.* **24,** 576.
Meyer, B. K., Hofman, D. M., Stadtler, W., Salk, M., Eiche, C., and Benz, K. W. (1993). *Semiconductors for Room Temperature Radiation Detector Applications,* ed. R. B. James, P. Siffert, T. E. Schlesinger, and L. Franks. Materials Research Society, Pittsburgh, PA.
Mireshgi, A., Cho, G., Drewery, J., Jing, T., Kaplan, S. N., Perez-Mendez, V., and Wildermuth, D. (1992). *IEEE Trans. Nucl. Sci.* **NS-39,** 635.
Nakatani, H., Sakai, E., Tatsuyama, T., and Takeda, F., *et al.* (1989). *Nucl. Inst. and Meth.* **A283,** 303.
Naruse, Y., and Hatayama, T. (1987). Proc. 4th Int. Conf. Solid State Sensors and Actuators, Tokyo, 262.
Naruse, Y., and Hatayama, T. (1989). *IEEE Trans. Nucl. Sci.* **NS-36**(Apr.), 1347.
Ohno, N., Yoshida, M., Nakamura, H., Nakahara, J., and Kobayashi, K. (1984). *J. Phys. Soc. Jpn* **53,** 1548.
Olschner, F., Toledo-Quinones, M., Shah, K. S., and Squillante, M. R. (1990). *IEEE Trans. Nucl. Sci.* **NS-37,** 1162.
Olschner, F., Lund, J. C., Squillante, M. R., and Kelly, D. L. (1989b). *IEEE Trans. Nucl. Sci.* **NS-36,** 210.
Olschner, F., Lund, J. C., Shah, K. S., and Squillante, M. R. (1989c). *ICFA Instrum. Bull.* No. 7, 9.
Olschner, F., Shah, K. S., Lund, J. C., Zhang, J., Daley, K., Medrick, S., and Squillante, M. R. (1992). *Nucl. Inst. and Meth.* **A322,** 504.
Perez-Mendez, V. (1991). In *Physics and Applications of Amorphous and Microcrystalline Semiconductor Devices,* ed. J. Kanicki. Artech House, Norwood, MA.
Pfann, W. G. (1978). *Zone Melting.* R. E. Krieger Publishing, Huntington, NY.
Pfister, G. (1979). *Contemp. Phys.* **20,** 449.
Qureshi, S., Perez-Mendez, V., Kaplan, S. N., Fujieda, I., Cho, G., and Street, R. A. (1989). *IEEE Trans. Nucl. Sci.* **NS-36,** 194.
Raghavan, R. S. (1976). *Phys. Rev. Lett.* **37,** 259.
Rahman, L. U., and Hofstadter, R. (1984). *Phys. Rev.* **B29,** 3500.

Rahman, L. U., Fisher, W. A., Hofstadter, R., and Shen, J. (1987). *Nucl. Inst. and Meth.* **A261,** 427.
Ribeiro, C. A., and Pautrat, J. L. (1973). *Solid State Comm.* **13,** 589.
Roth, M., and Burger, A. (1986). *IEEE Trans. Nucl. Sci.* **NS-33,** 407.
Roth, M., Burger, A., Nissenbaum, J., and Schieber, M. (1987). *IEEE Trans. Nucl. Sci.* **NS-34,** 465.
Rowley, J. K., Cleveland, B. T., and Davis, R., Jr. (1984). *Proc. Conf. Solar Neutrinos and Neutrino Astronomy,* 1. AIP Press, New York.
Sakai, E. (1982). *Nucl. Inst. and Meth.* **196,** 121.
Sakai, E., Nakatani, H., Tatsuyama, C., and Takeda, F. (1988). *IEEE Trans. Nucl. Sci.* **NS-35,** 85.
Saulnier, K., Squillante, M. R., and Entine, G. (1984). *Solid State Nuclear Sensor for High Temperature Down Hole Logging,* Final Report, DOE Contract No. DE-AC02-83ER80058.
Schein, L. B. (1988). *Electrophotography and Development Physics.* Springer-Verlag, Berlin.
Schieber, M., Beinglass, I., Dishon, G., Holzer, A., and Yaron, G. (1978). *IEEE Trans. Nucl. Sci.* **NS-25,** 71.
Shah, K. S., Lund, J. C., Olschner, F., Moy, L., and Squillante, M. R. (1989). *IEEE Trans. Nucl. Sci.* **NS-36,** 199.
Shah, K. S., Squillante, M. R., and Entine, G. (1990a). *IEEE Trans. Nucl. Sci.* **NS-37,** 152.
Shah, K. S., Olschner, F., Moy, L., and Squillante, M. R. (1990b). *Nucl. Inst. and Meth.* **A299,** 57.
Shah, K. S., Moy, L., Zhang, J., Medrick, S., Olschner, F., and Squillante, M. R. (1992). *Proc. SPIE* **1734,** 161.
Shah, K. S., Lund, J. C., Olschner, F., Zhang, J., Moy, L. P., and Squillante, M. R. (1994). *IEEE Trans. Nucl. Sci.* **NS-41,** 2715.
Siffert, P. (1978). *Nucl. Inst. and Meth.* **150,** 1.
Sloan, G. J., and McGhie, A. R. (1988). *Techniques in Chemistry,* **19.** Wiley Interscience, New York.
Squillante, M. R., Lis, S., Hazlett, T., and Entine, G. (1983). *Mat. Res. Soc.* **16,** 191.
Squillante, M. R., Shah, K. S., and Moy, L. (1990). *Nucl. Inst. and Meth.* **A288,** 79.
Squillante, M. R., Zhou, C., Moy, L. P., Zhang, J., and Shah, K. (1993a). *IEEE Trans. Nucl. Sci.* **NS-40** (Aug.).
Squillante, M. R., Cole, H., Waer, P., and Entine, G. (1993b). *Semiconductors for Room Temperature Radiation Detector Applications,* ed. R. B. James, P. Siffert, T. E. Schlesinger, and L. Franks. Materials Research Society, Pittsburgh, PA. **302,** 507.
Squillante, M. R., Zhang, J., Zhou, C., Bennett, P., and Moy, L. (1993c). *Semiconductors for Room Temperature Radiation Detector Applications,* ed. R. B. James, P. Siffert, T. E. Schlesinger, and L. Franks. **302,** 319.
Squillante, M. R., Zhou, C., Zhang, J., Moy, L., and Shah, K. S. (1993d). *IEEE Trans. Nucl. Sci.* **NS-40,** 364.
Stanton, L. (1979). In *The Physics of Medical Imaging: Recording System Measurements and Techniques,* ed. A. G. Haus. AIP Press, New York.
Street, R. A. (1992). *MRS Bulletin* (Nov.), 70.
Sumner, T. J., Grant, S. M., Bewick, A., Li, J. P., Smith, K., and Beaumont, S. P. (1992). *Nucl. Inst. and Meth.* **A322,** 514.
Suzuki, Y., Fukuda, Y., and Nagashima, Y. (1989). *Nucl. Inst. and Meth.* **A275,** 142.
Tanaka, T., Otari, S., and Ishizawa, Y. (1985). *J. Cryst. Growth* **73,** 31.
Tupenevich, P. A., and Kononenko, V. K. (1978). *J. Applied Spectroscopy* **28,** 592.
Van Calster, A. (1985). *Thin Solid Films* **126,** 219.
Wald, F. V., and Entine, G. (1978). *Nucl. Inst. and Meth.* **150,** 13.
Wald, F. V., Bullit, J., and Bell, R. O. (1975). *IEEE Trans. Nucl. Sci.* **NS-22,** 246.
Wang, C. L., Pocha, M. D., Morse, J. D., Singh, M. S., and Davis, B. A. (1989). *Appl. Phys. Lett.* **54,** 1451.
Wang, C. L., Flatley, J. E., and Pocha, M. D. (1991). *Energy and Technology Review* (Sept.–Oct.), 20.
Weckler, G. P. (1992). Seventh International Workshop on Room Temperature Semiconductor X- and Gamma Ray Detectors and Associated Electronics, Ravello, Italy.
Wilson, H. L., and Gutierrez, W. A. (1965). *J. Electrochem. Soc.* **112,** 85.

Wolfe, J. N., Buck, K. A., Salane, M., and Parekh, N. J. (1987). *Radiology,* 305.
Yee, J. T., Swierkowski, S. P., and Sherohman, J. W. (1977). *IEEE Trans. Nucl. Sci.* **NS-24,** 1962.
Zhang, J., Cirignano, L., Daley, K., and Squillante, M. R. (1993). *Semiconductors for Room Temperature Radiation Detector Applications,* ed. R. B. James, P. Siffert, T. E., Schlesinger, and L. Franks. Materials Research Society, Pittsburgh, PA. **302,** 329.
Zhou, C., Squillante, M. R., Moy, L., and Bennett, P. (1993). *Semiconductors for Room Temperature Radiation Detector Applications,* ed. R. B. James, P. Siffert, T. E. Schlesinger, and L. Franks. **302,** 357.

CHAPTER 13

Characterization and Quantification of Detector Performance

Vernon M. Gerrish*

EG&G ENERGY MEASUREMENTS, INC.
SANTA BARBARA OPERATIONS
GOLETA, CALIFORNIA

I. INTRODUCTION	493
II. X-RAY AND GAMMA RAY SPECTROSCOPY	496
1. *Interaction of X-Rays and Gamma Rays with Matter*	496
2. *Detector Response Function*	498
3. *Charge Collection*	499
4. *Electronics*	502
5. *Detector Performance*	503
6. *Polarization*	510
7. *Radiation Damage Resistance*	513
III. ELECTRONIC CHARACTERIZATION	513
1. *Bulk Measurements*	513
2. *Contacts*	520
IV. CORRELATION OF MATERIAL PROPERTIES WITH DETECTOR PERFORMANCE	524
V. CONCLUDING REMARKS	527
References	527

I. Introduction

Interest in the use of wide bandgap semiconductor materials as room temperature radiation detectors has grown rapidly in recent years, due, in part, to the improved performance and increasing availability of these devices. Wide bandgap solid state detectors offer high resolution, compact size, low power requirements, operation over a wide temperature range, insensitivity to magnetic fields, and in certain materials, good radiation damage resistance. In many applications room temperature semiconductor detectors offer an attractive alternative to sodium iodide (NaI) scintillator–photomultiplier tube (PMT) combinations (which are bulky and have limited energy resolution), germanium (Ge) detectors (which must be cooled), or silicon (Si) detectors (which have very low gamma ray peak efficiencies). Wide bandgap solid state detectors have been used in x-ray fluorescence

---

*Present Address: Constellation Technology Corporation, St. Petersburg, Florida.

(Warburton et al., 1986), medical imaging (Barber et al., 1991; Entine et al., 1989), industrial and environmental monitoring of radiation (Entine et al., 1989; Patt et al., 1989; Droms et al., 1976; Friant et al., 1989), geological exploration (Entine et al., 1989), astrophysics (Ricker, Vallerga, and Wood, 1983; Bradley et al., 1989; Economou and Iwanczyk, 1989), and high energy physics research (Patt et al., 1989; Becchetti et al., 1983). Two high Z compound semiconductors, mercuric iodide (HgI_2) and cadmium telluride (CdTe), have attracted the most attention with regard to room temperature detector applications. In view of the important role of electronic characterization techniques in the development of these materials, the emphasis throughout this chapter will be on the characterization of HgI_2 and CdTe detector performance and important material properties using electronic methods.

The ability to obtain high purity material and grow large single crystals containing a relatively low number of defects and the development of suitable fabrication techniques are necessary prerequisites for the production of good quality detectors. In addition, properties desirable for solid state detectors include high detection efficiency, low leakage current, good charge collection, high charge carrier mobilities, radiation damage resistance, and stable response. Wide bandgap semiconductor detector materials tend to exhibit low leakage currents, which reduces the associated current noise and permits room temperature operation. For gamma ray energies greater than about 50 keV the energy resolution of room temperature semiconductor detectors is often limited more by incomplete charge collection than by electronic noise (Knoll, 1989; Debertin and Helmer, 1988). Charge trapping due to impurities or lattice defects limits the thickness of spectrometers which can be fabricated and still exhibit good charge collection. High charge carrier mobilities would be an advantage due to the reduction in the charge collection times. However, wide bandgap materials tend to have low charge carrier mobilities due to polar lattice scattering. In wide bandgap materials mobility–lifetime products often limit detector thicknesses to a few millimeters or less. A list of properties of some room temperature detector materials is given in Table I.

Considerable effort has gone into controlling the purity (Skinner et al., 1989; Burger et al., 1991), stoichiometry (Zanio, 1978; Burger et al., 1990), and growth (Zha, Piechotka, and Kaldis, 1991; Raiskin and Butler, 1988) of HgI_2 and CdTe crystals in order to improve the charge transport properties, and hence the energy resolution, of large volume nuclear detectors. Single crystals of HgI_2 up to 1 kg are grown by the vapor transport method and exhibit good uniformity. Recent advances in producing high purity HgI_2 by chemical synthesis (Skinner et al., 1989) have resulted in yields of 60–80% for spectrometer grade detectors of 1–4 mm thickness and 3–6 cm^2 area (Gerrish and van den Berg, 1990; Gerrish, 1992a). In the past, production of HgI_2 detectors was restricted to small scale research programs but recent efforts at commercial production should lead to greater availability. For CdTe, doping with chlorine or indium is often used to compensate the residual impurities and native defects that inevitably occur in compound semiconductors. Progress in the area of contact deposition has over-

TABLE I

PROPERTIES OF SOME ROOM TEMPERATURE NUCLEAR DETECTOR MATERIALS

Material	Average Z	Density (g/cm³)	Bandgap (eV)	Mobility (cm²/V-s) e: electron h: hole	Resistivity (Ωcm)	$\mu \cdot \tau$ Product (cm²/V) e: electron h: hole	Best % FWHM Energy Resolution at Room Temperature
HgI$_2$	62	6.4	2.2	e: 100 h: 4	10^{14}	e: 1×10^{-3} h: 8×10^{-5}	5.9 keV: 5.9% 60 keV: 2.7% 662 keV: 1.0%
CdTe	50	6.2	1.5	e: 1050 h: 100	10^9	e: 8×10^{-4} h: 2×10^{-4}	5.9 keV: 19% 60 keV: 2.9% 662 keV: 1.2%
Cd$_{1-x}$Zn$_x$Te	41–50	5.7–6.2	1.6–2.4	e: 1120 h: 200	10^{11}	e: 1×10^{-5} h: 1×10^{-6}	30 keV: 8.4% 122 keV: 7%
GaAs	32	5.3	1.4	e: 8500 h: 400	10^8	—	60 keV: 37% 122 keV: 33%
CdSe	41	5.8	1.7	e: 720 h: 75	10^{12}	e: 2×10^{-5} h: 1×10^{-6}	60 keV: 14%
PbI$_2$	67	6.2	2.3	e: 8 h: 2	10^{12}	e: 8×10^{-6} h: 9×10^{-7}	5.9 keV: 16%
TlBr	58	7.5	2.7	e: 7.7 h: —	10^{12}	e: 3×10^{-6} h: 2×10^{-6}	60 keV: 13%
CdS	32	4.9	2.4	e: 240 h: 50	10^{13}	—	122 keV: 6.6%

Data from Bertin et al. (1990); Burger et al. 1983, 1985; Dabrowski et al. 1978; Doty et al. 1992; Eichinger and Kallmann 1974; Hess et al. 1972; Kusmiss et al. 1983; Lund et al. 1989; Olschner et al. 1990; Sakai 1982; Shah et al. 1989.

come the polarization problems associated with chlorine doped CdTe. In addition, large (10 kg) undoped high resistivity CdTe crystals have been successfully grown by the high pressure Bridgeman method (Raiskin and Butler, 1988). This method has also been used to produce detector quality crystals of Cd$_{1-x}$Zn$_x$Te. These developments have contributed to the increased commercial availability in recent years of good quality CdTe detectors of 0.5–2.0 mm thickness and up to 150 mm² area. Cd$_{1-x}$Zn$_x$Te detectors of similar dimensions are also produced commercially.

Semiconductor nuclear detectors operate by directly converting radiation into an electrical signal. An alternative approach is the use of a scintillator material, which emits visible light when exposed to x-rays or gamma rays. A PMT or photodiode can then be used to convert the scintillation light to an electrical signal. Due to its high sensitivity to visible light, HgI$_2$ has been used successfully to detect scintillation light. The high quantum efficiency and low leakage current of HgI$_2$ has allowed the fabrication of photodiodes with up to 1.5 in. diameter contacts exhibiting room temperature energy resolution exceeding that of NaI/PMT devices (Markakis, 1988). In this application, electron–hole pairs are generated at the HgI_2 surface by the highly absorbed scintillation light, and since only the electrons traverse the detector thickness, hole trapping problems are avoided.

II. X-Ray and Gamma Ray Spectroscopy

Room temperature semiconductor detectors have evolved during the past two decades from simple counters to high resolution spectrometers used most often for measuring x-rays and gamma rays. Steady progress in controlling material properties has increased yields of spectrometer grade detectors and led to the introduction of devices utilizing these materials into new areas. The routine characterization and quantification of detector performance plays a key role in material and device development. In particular, x-ray and gamma ray spectroscopy with semiconductor detectors allows the determination of important device performance characteristics such as energy resolution, efficiency, stability, and radiation damage resistance. Before discussing aspects of detector characterization that are unique to room temperature semiconductor materials, the physical principles underlying radiation measurement with semiconductor detectors will be reviewed briefly.

1. Interaction of X-Rays and Gamma Rays with Matter

X-rays and gamma rays interact with matter mainly by three basic processes: photoelectric absorption, Compton scattering, and pair production. In photoelectric absorption, which predominates at energies below a few hundred keV, the incident photon is absorbed by an atomic electron (usually an innermost or K shell electron), which then generates electron–hole pairs by collisions within the lattice. Since all of the incident photon energy is deposited in the detector, measurement of the photon energy is possible. Compton scattered photons, however, deposit only a fraction of their energy in the detector and so do not, in general, contribute to the spectral peak. This produces the "Compton shelf," which is observed in gamma ray spectra as a continuum of counts to the low energy side of the full energy peak. However, the Compton scattered photon may interact further with the detector, either by additional Compton scattering or by the photoelectric process, so that multiple interaction events may be an important contribution to the full energy peak. At gamma ray energies greater than about a hundred keV thick detector geometries offer higher "peak to Compton ratios" due to the greater probability of these multiple interaction events occurring within the detector volume. For energies greater than twice the rest mass energy of the electron (1.022 MeV) pair production may occur and will actually be the dominant type of interaction at energies above several MeV. In this type of interaction the energy of the incident photon is converted to an electron–positron pair. The positron eventually annihilates with an electron producing two 511 keV photons. Unless these photons are reabsorbed in the detector, the pair production process will not contribute to the full energy peak. The photoelectric cross section increases much more rapidly with increasing atomic number than does either the Compton cross section or the pair production cross section so that high Z materials offer the capability for high peak detection efficiencies and high peak to Compton ratios from comparatively small detector volumes.

The absorption of photons as they pass through an absorbing medium follows an exponential law:

$$I = I_0 e^{-\mu x} \tag{1}$$

where I_0 is the incident photo flux in cm^{-2} sec^{-1}, I is the flux at a depth x (cm) in the absorber, and μ is the linear attenuation coefficient in cm^{-1}. The linear attenuation coefficient is the sum of the contributions due to the individual interactions:

$$\mu = \mu_{PE} + \mu_C + \mu_{PP} \tag{2}$$

where PE = photoelectric, C = Compton, PP = pair production. These quantities depend on the energy of the incident photon and the atomic number and density of the absorbing medium. A comparison of photoelectric, Compton, and pair production linear attenuation coefficients for Si, Ge, HgI$_2$, and CdTe is shown in Fig. 1.

With regard to radiation measurements, the important feature of all of these interactions is that electron–hole pairs are generated in a detector and can be de-

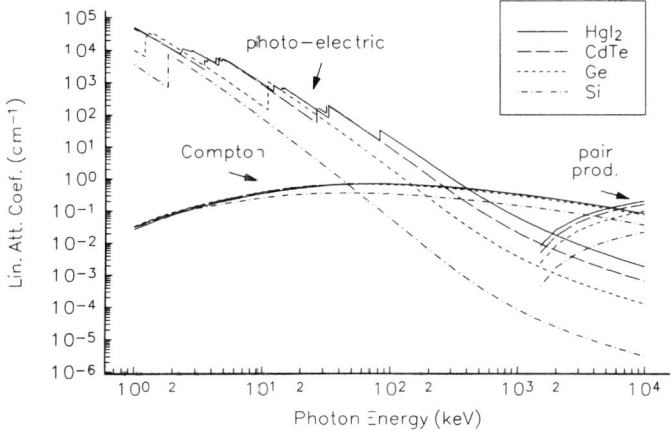

FIG. 1. Photoelectric, Compton scattering, and pair production linear attenuation coefficients for (—), HgI$_2$; (– –), CdTe; (---), Ge; and (– · –), Si.

tected by an external circuit. For x-rays and gamma rays the number of electron–hole pairs is found experimentally to be proportional to the energy of the incident photon. The proportionality constant, the energy to create an electron–hole pair, depends on the semiconductor material but is on the order of a few eV. The number of electron–hole pairs generated by an interaction places a statistical limit on the energy resolution of the detector. In scintillation spectrometers the conversion of radiation energy to scintillation photons is inefficient, resulting in a smaller number of charge carriers. Therefore, statistical fluctuations in the number of charge carriers imposes a limit on the energy resolution achievable with this type of spectrometer. For semiconductor detectors in which photons interact directly with the detector, the number of charge carriers generated is much larger so that the statistical limit on the energy resolution is greatly reduced. For room temperature semiconductor detectors other factors, such as charge trapping and electronic noise, limit the energy resolution.

2. Detector Response Function

The response of real semiconductor detectors includes features related to the types of photon interactions described previously as well as additional characteristics due to multiple interactions, escape of secondary x-rays or gamma rays, or incomplete charge collection. Photoelectric interactions produce the full energy peak, or photopeak, characteristic of a monoenergetic x-ray or gamma ray source. If charge trapping or recombination is present, the photopeak will tend to be broader and will usually have an asymmetrical shape as counts are shifted to lower energy. Compton scattering results in counts appearing in a continuum of energies to the low energy side of the photopeak up to some maximum channel somewhat below the photopeak. This maximum energy is determined by the scattering of the incident photon at a 180° angle (the Klein–Nishina formula) and is referred to as the "Compton edge." As mentioned previously, multiple photon interactions may contribute to the photopeak so that the ratio of counts in the photopeak to the counts in the Compton shelf may be considerably higher than the corresponding ratio of photon cross sections. The probability of multiple interactions increases with increasing detector size and increasing atomic number so that large volume, high Z detectors exhibit high peak detection efficiencies.

The energy resolution capability of a nuclear detector is usually defined in terms of the spectral peak width or full-width at half-maximum (FWHM) in eV, keV, or as a percentage of the peak energy. Also, the peak to Compton ratio is often quoted as indication of the performance of a gamma ray spectrometer. This is the ratio of the maximum counts/channel in the photopeak to the average counts/channel in the Compton shelf (just below the Compton edge). A detector that exhibits high energy resolution and a high "photofraction" (the ratio of counts in the photopeak to the total counts) would have a high peak to Compton ratio. Another figure of merit sometimes quoted for gamma ray spectrometers is the peak to valley ratio. This is the ratio of the maximum counts/channel in the photopeak region to the minimum counts/channel in the valley between the Compton edge and the photopeak. This number indicates the capability of a detector to distinguish the photopeak from the Compton continuum.

A feature that tends to be more prominent in the spectral response of high Z detectors is the presence of "x-ray escape peaks," which are one or more secondary peaks occurring to the low energy side of the main photopeak. These are due to events in which a photoelectric interaction takes place at an atom near the surface of the detector. In the photoelectric interaction an inner shell electron is liberated from the atom and, subsequently, an x-ray is produced by deexcitation of the atom. Usually the x-ray is reabsorbed, but if it escapes from the detector then the total energy deposited in the detector is lowered by an amount equal to the binding energy of the inner electron. This process produces an x-ray escape peak at a characteristic energy which is about 10–90 keV lower than the photopeak energy in most high Z materials In compound semiconductors an x-ray escape peak occurs for each of the constituent elements. Several factors tend to cause x-ray escape peaks to be more prominent in detectors composed of high Z materials (as compared to Si or Ge). These include the higher energy of the characteristic x-rays, the greater surface to volume ratio for typical detector dimensions, and the higher probability of stopping incident photons sufficiently near the surface. Often, however, the energy resolution of high Z semiconductor detectors may not be sufficient to fully resolve all of the x-ray escape peaks from the main photopeak.

For gamma rays of sufficient energy to cause appreciable pair production interactions, additional escape peaks may occur. This is because one, or both, of the 511 keV annihilation photons may escape from the detector resulting in secondary peaks at 511 keV ("single escape") and 1022 keV ("double escape") below the main photopeak, respectively. Alternatively, the annihilation photons may Compton scatter adding a continuum of counts in the energy range between the double escape peak and the photopeak.

Compton scattering of gamma rays from materials surrounding the detector often leads to the presence of a "backscatter peak" in the spectral response at about 200 keV. Also, x-rays may be emitted by material surrounding the detector due to photoelectric interactions that cause characteristic x-ray escape events in the materials. Other processes that may lead to additional complicating features in the detector response function include secondary electron escape, bremmstrahlung escape, secondary radiations produced by the gamma ray source, and "sum peaks" due to coincidence events. A fuller discussion of these topics is available in Knoll, 1989.

3. CHARGE COLLECTION

When an electric field is applied to a semiconductor that is exposed to ionizing radiation, the electron–hole pairs generated separate and drift toward opposite electrodes. In a solid state detector this induces a signal whose amplitude is proportional to the energy of the incident photon. For accurate determination of the energy of the interacting photon, it is desirable to collect all of the electron–hole pairs so that each interaction contributes to the full energy peak.

Charge trapping or recombination may prevent full charge collection and limit the energy resolution of semiconductor detectors. Typically, hole trapping limits the performance of thick spectrometers when detecting highly penetrating gamma radiation, as both electrons and holes must be collected throughout the entire active volume of the detector. Under many experimental conditions, the pulse shape in a semiconductor detector can be described by the Hecht relation (Hecht, 1932). For an incident photon interacting at a distance, x, below the cathode of a planar detector of thickness, L, having a uniform electric field, V/L, in the absence of detrapping effects, the collected charge as a function of time due to the electrons is given by

$$Q_e(t) = \begin{bmatrix} Q_0 \frac{T_e}{t_e}\left(1 - e^{-\frac{t}{T_e}}\right) & \text{if } t < \left(1 - \frac{x}{L}\right)t_e \\ Q_0 \frac{T_e}{t_e}\left(1 - e^{-\left(1-\frac{x}{L}\right)\frac{t_e}{T_e}}\right) & \text{if } t \geq \left(1 - \frac{x}{L}\right)t_e \end{bmatrix} \quad (3)$$

where Q_0 is the total electron charge (in coulombs) generated by the photon interaction, τ_e is the electron lifetime in seconds, and x and L are in cm. The electron transit time for an interaction very close to the cathode, t_e, is given by

$$t_e = \frac{L^2}{\mu_e V} \quad (4)$$

where V is the applied bias in volts, and μ_e is electron mobility in cm^2/V/sec. Similarly, the hole contribution is

$$Q_h(t) = \begin{bmatrix} Q_0 \frac{T_h}{t_h}\left(1 - e^{-\frac{t}{T_h}}\right) & \text{if } t < \frac{x}{L}t_h \\ Q_0 \frac{T_h}{t_h}\left(1 - e^{-\frac{xt_h}{LT_h}}\right) & \text{if } t \geq \frac{x}{L}t_h \end{bmatrix} \quad (5)$$

where the subscript h indicates the corresponding parameters for holes, and the hole transit time for an interaction very close to the anode, t_h, is

$$t_h = \frac{L^2}{\mu_h V}. \quad (6)$$

The total collected charge, $Q(t)$, is

$$Q(t) = Q_e(t) + Q_h(t). \quad (7)$$

13. Characterization and Quantification of Detector Performance

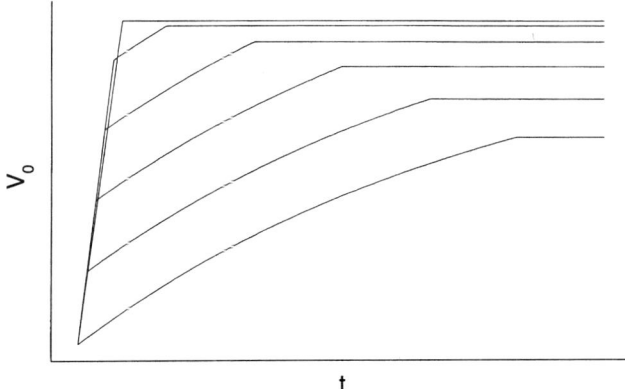

Fig. 2. Charge pulse shapes for various photon interaction depths in a planar detector base on the Hecht relation.

The charge pulse shapes for different photon interaction depths are shown in Fig. 2. Values for electron and hole (trapping time)/(transit time) ratios were used, which might be typical for a detector in which hole trapping dominates. The fast rising portion of the pulse is due to the electron component while the slower rising portion is due to the holes, which often have a mobility about an order of magnitude lower than the electrons in many of the high Z materials. The slight curvature in the hole component is due to charge trapping and results in a smaller pulse amplitude for pulses containing a larger hole component. The charge pulse amplitude, Q_{max}, is the total collected charge, $Q(t)$, from Eq. (7) evaluated at the time at which both the electrons and holes are collected (the electron or hole transit time, whichever is greater)

$$Q_{max} = Q_0 \frac{\mu_e \tau_e V}{L^2}\left(1 - e^{-\frac{(L-x)L}{\mu_e \tau_e V}}\right) + Q_0 \frac{\mu_h \tau_h V}{L^2}\left(1 - e^{-\frac{xL}{\mu_h \tau_h V}}\right). \quad (8)$$

The charge collection efficiency for a photon interaction depth, x, is

$$\eta(x) = \frac{Q_{max}}{Q_0}. \quad (9)$$

Charge trapping causes variations in the charge collection efficiency for different photon interaction depths, which results in spectral peak broadening, asymmetrical peak shapes, and possibly reduced detection efficiency. The charge collection efficiency can be improved by simply raising the bias voltage, but in practice, the maximum operating voltage is limited by leakage current and breakdown. In cases where one type of charge carrier suffers significant trapping but

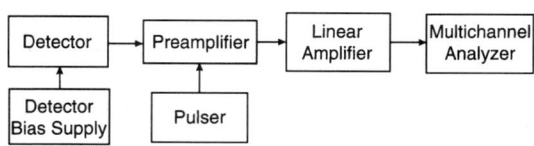

FIG. 3. Diagram of electronics for a basic spectrometer system.

the other carrier exhibits good transport, good detection efficiency can still be obtained for low energy radiation, which is absorbed near the appropriate detector surface.

4. ELECTRONICS

A block diagram of a basic spectrometer system is shown in Fig. 3. The preamplifier is generally an ac coupled or dc coupled, charge sensitive configuration. A low noise design is particularly important for very low energy measurements, and the preamplifier should be located as close to the detector as possible to minimize stray capacitance and microphonic noise. A pulser is often used to verify the stability of the system and determine the electronic noise contribution to the spectral peak broadening. When the detector leakage current is sufficiently low a pulsed optical feedback preamplifier design (Iwanczyk, 1989) is sometimes used to eliminate the noise contribution of the feedback resistor. The postamplifier is typically a Gaussian shaping amplifier that improves the signal to noise ratio by limiting the bandwidth.

Because of the relatively low mobilities in most high Z materials the charge collection times of room temperature semiconductor detectors can be several microseconds or longer, so that long shaping time constants may be required to minimize the spectral peak broadening due to "ballistic deficit" effects. Shaping times of $10-20$ μsec are commonly used with HgI_2 detectors of a few millimeters thickness. Long shaping times may also be favored by signal to noise considerations. If the detector leakage current is low, the preamplifier feedback resistor value may be made high so that the associated Johnson (thermal) noise contribution is reduced. Under these circumstances long shaping times are preferred. Detectors that have a higher leakage current may exhibit optimal energy resolution at shorter shaping times.

Other pulse filtering electronics are sometimes used with semiconductor detectors and offer certain advantages. Triangular shaping (Radeka and Karlovac, 1967) postamplifiers are available and offer a modest improvement in signal to noise, which may be particularly important for low energy applications. Gated integrators (Radeka, 1972) are also available and are useful for obtaining improved performance at high count rates. This is possible because shorter amplifier shaping times can be used that reduce pulse pile-up. Without gated integration the shorter shaping times would produce ballistic deficit. This results in attenuation of long rise time pulses that degrade signal to noise and may also cause spectral peak broadening due to pulse rise time variations. In view of the long charge collection times associated with many room temperature semiconductor detector devices, gated integration is useful for minimizing the system "dead time" in high count rate applications to acceptable levels. Finally, charge trapping correction techniques (Gerrish, Williams, and Beyerle, 1987; Richter and Siffert, 1992) have been developed to enhance the energy resolution of semiconductor detectors using electronic pulse filtering methods. These techniques correct for pulse amplitude variations due to bulk hole trapping, and so may give very significant improvement in gamma ray energy resolution but do not improve low energy (x-ray) performance. In general, for a particular application the use of more sophisticated electronics to improve the overall system performance must be sufficient to justify the increase in system complexity.

5. DETECTOR PERFORMANCE

As x-ray and gamma ray spectrometers, HgI_2 and CdTe exhibit good energy resolution and high detection efficiency. At present, detectors can be fabricated at thicknesses that are 100% efficient for x-rays in the energy range 5–30 keV and exhibit good stability. At higher energies, HgI_2 detectors of 1–4 mm thickness and 3–6 cm^2 area can be fabricated that routinely exhibit energy resolution of 2–4 keV at 60 keV and 10–30 keV at 662 keV (Gerrish and van den Berg, 1990; Gerrish 1992a). Examples of HgI_2 ^{241}Am and ^{137}Cs spectra are shown in Fig. 4 (from EG&G Energy Measurements). CdTe detectors are commercially available at thicknesses up to 2 mm and 150 mm^2 area. CdTe detectors of a few mm^2 area exhibit energy resolution as high as 4–6 keV at 60 keV and 18–30 keV at 662 keV (Dabrowski et al., 1978; Siffert, 1983; Hazlett et al., 1986). Figure 5 shows examples of CdTe spectra (from Radiation Monitoring Devices, Inc.). The high atomic numbers of these materials provide higher photoelectric cross sections for gamma rays than germanium or sodium iodide, resulting in higher peak efficiencies for equivalent thicknesses. This also allows the fabrication of much thinner detectors having equivalent detection efficiencies. Figure 6 shows a comparison of a ^{137}Cs spectrum from a 0.4 in. thick, 1.5 × 1.5 in. area HgI_2 detector, and a 2 in. diameter × 3 in. NaI scintillator coupled to a photomultiplier tube. Both detectors have roughly equivalent resolution and efficiency but the HgI_2 detector offers a considerable reduction in size and weight.

At photon energies below 50 keV the energy resolution of a spectrometer system may be limited mainly by electronic noise (Knoll, 1989). For room temperature semiconductor spectrometers, the thermal noise contribution due to the detector leakage current may be important. In this regard, the high resistivity of HgI_2, about 10^{14} Ω-cm, is an advantage in x-ray spectroscopy. Typical leakage currents of 10–100 picoamps permit room temperature energy resolution as high as

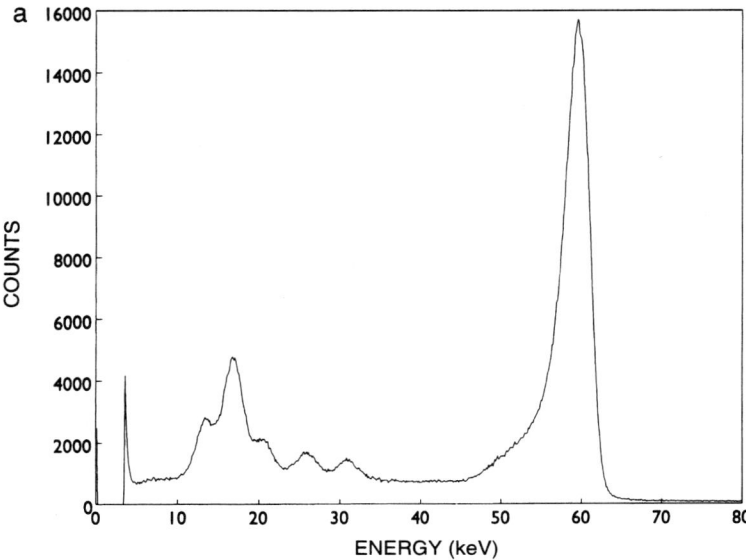
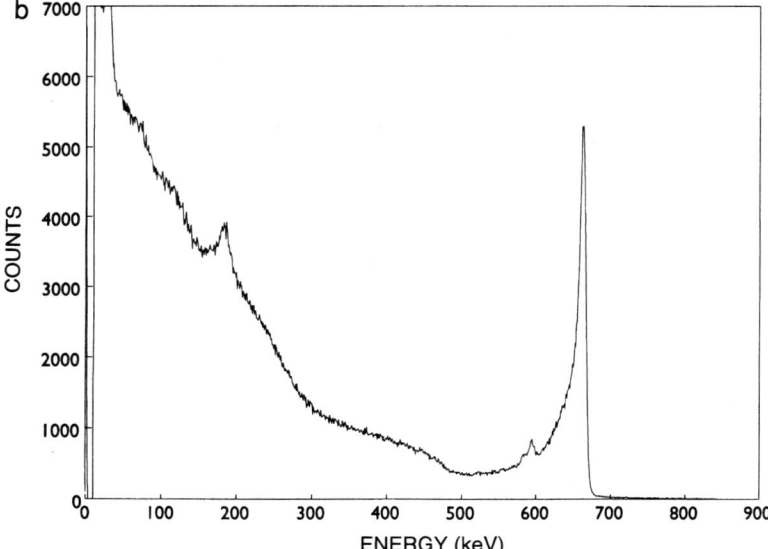

FIG. 4. (a) ^{241}Am spectrum from a 1.8 mm × 4.41 cm^2 HgI$_2$ detector at 2000 V bias. Resolution at 60 keV is 6.4% FWHM. (b) ^{137}Cs spectrum from a 2.2 mm × 3.24 cm^2 HgI$_2$ detector at 2400 V bias. Resolution at 662 keV is 1.7% FWHM.

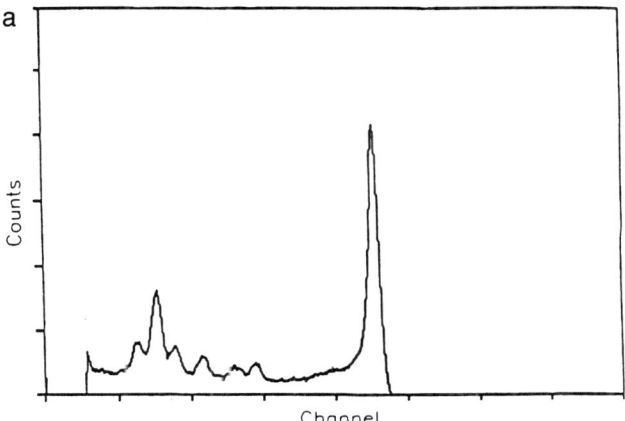

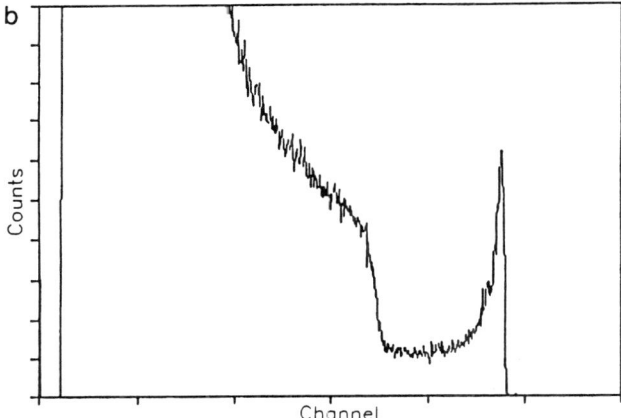

FIG. 5. (a) ^{241}Am spectrum from a 2 mm × 4 mm^2 CdTe detector at 60 V bias. Resolution at 60 keV is 6.7% FWHM. (b) ^{137}Cs spectrum from a 2 mm × 4 mm^2 CdTe detector at 60 V bias. Resolution at 662 keV is 2.7% FWHM. (Reprinted with permission of Radiation Monitoring Devices, Inc.)

350 eV at 6 keV for 0.5 mm × 2 mm^2 detectors (Iwanczyk, 1989). Figure 7 shows an ^{55}Fe spectrum acquired with a HgI$_2$ detector of these dimensions (from EG & G Energy Measurements, Inc.). The lower resistivity of CdTe (around 10^9 Ω-cm) limits its performance as a spectrometer at very low energies. One approach to increasing the resistivity of CdTe is the replacement of some fraction of the Cd with Zn. Resistivities in excess of 10^{11} Ω-cm are attained in Cd$_{1-x}$Zn$_x$Te detectors using a Zn fraction of 20%. Commercially available 1.5 mm × 25 mm^2 Cd$_{1-x}$Zn$_x$Te detectors achieve room-temperature energy resolution as high as

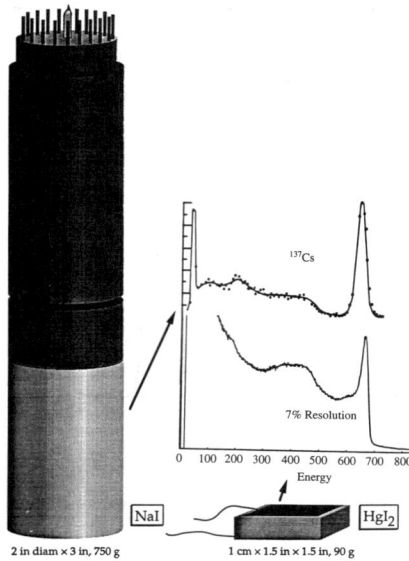

FIG. 6. Comparison of a 2 in. diameter × 3 in. NaI/PMT combination with a 1 cm × 2.25 cm² HgI$_2$ detector. The detectors have roughly equal efficiency and energy resolution for 662 keV gamma rays.

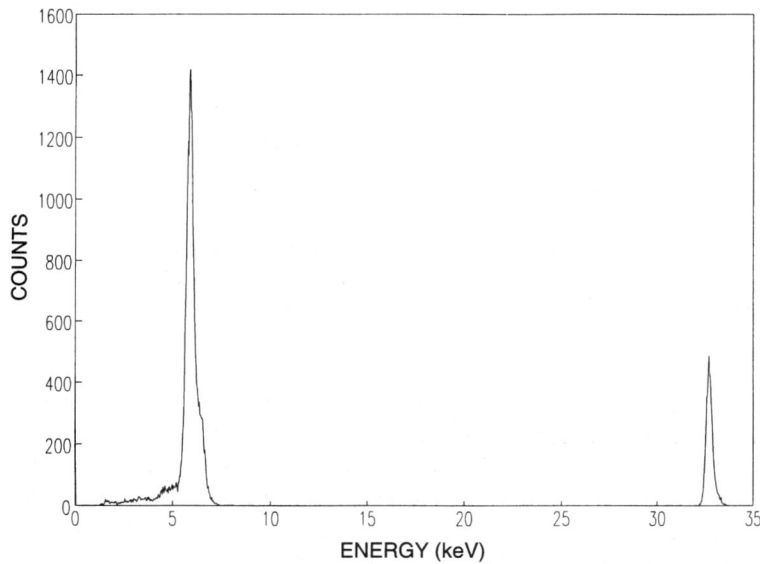

FIG. 7. ^{55}Fe spectrum from a 0.5 mm × 3.14 mm² HgI$_2$ detector at 1000 V bias. Resolution at 5.9 keV is 420 eV FWHM (7.1% FHWM).

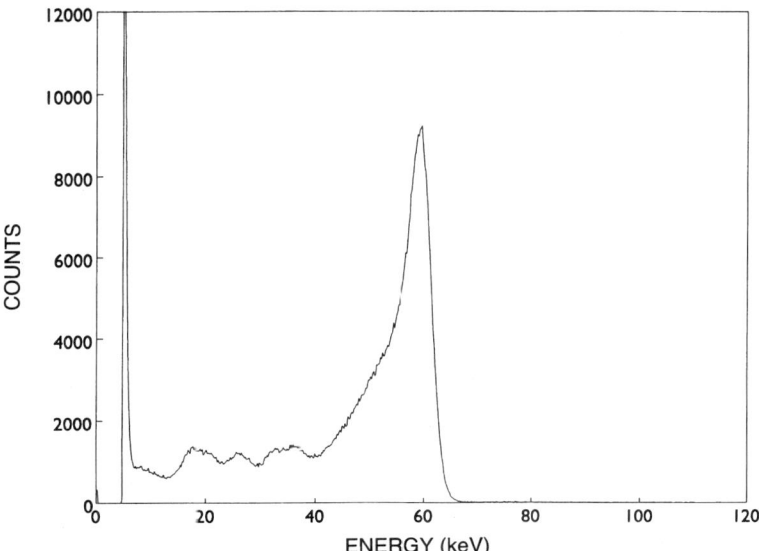

FIG. 8. ^{241}Am spectrum from a 1.5 mm × 100 mm² $Cd_{1-x}Zn_xTe$ detector at 160 V bias. Resolution at 60 keV is 11% FWHM.

2.5 keV at 30 keV (Doty *et al.*, 1992). Figure 8 shows a plot of an ^{241}Am spectrum obtained from a 1.5 mm × 100 mm² $Cd_{1-x}Zn_xTe$ detector (from Aurora Technologies Corporation).

Detector capacitance also contributes to the system noise so that the detector area and thickness may to a large extent determine the energy resolution at low energy. The capacitance and associated noise is greater for larger area detectors but to some extent this can be compensated for by increasing the detector thickness. For shallow penetrating radiation such as x-rays only one type of charge carrier (usually electrons) is required to traverse the entire thickness so that relatively thick detectors can be fabricated and still retain good charge collection properties.

This principle is also used in HgI_2 photodetectors. Gamma ray spectrometers using HgI_2 photodetectors have achieved the highest reported energy resolution for room temperature gamma ray scintillation spectroscopy: 5.0% FWHM at 662 keV obtained with 1 in. diameter × 2 mm thick HgI_2 photodetector coupled with a 1 in. diameter × 1 in. thick Cs(Tl) scintillator. An example is shown in Fig. 9. Cesium iodide scintillators are used with HgI_2 because of the better spectral match than NaI. Solid state HgI_2 photodetectors offer significant advantages over conventional PMTs. They are compact, exhibit higher quantum efficiency, are insensitive to magnetic fields, operate at low bias voltages, and do not require precise bias voltage regulation. Because only electron charge collection is re-

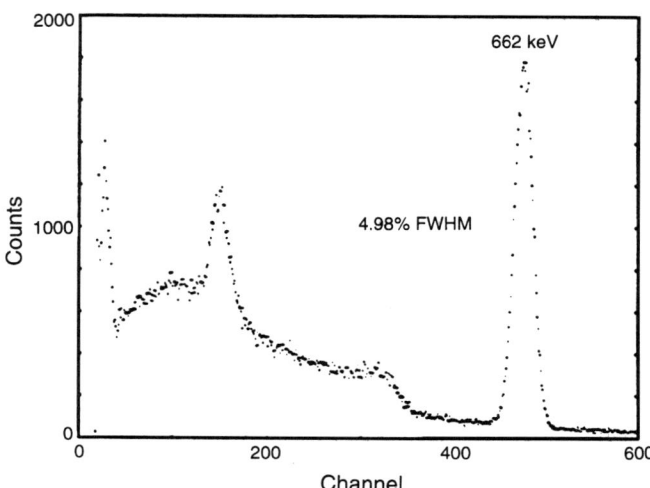

FIG. 9. ^{137}Cs spectrum from a 1 in. diameter × 2 mm thick HgI$_2$ photodetector coupled to a 1 in. diameter × 1 in. CsI(Tl) scintillator. Room temperature energy resolution at 662 keV is 4.98% FWHM.

quired, full charge collection is obtained at approximately 600 V for 2 mm thick HgI$_2$ photodetectors.

Semiconductor detectors are relatively insensitive to bias voltage fluctuations compared to NaI/PMT combinations. The energy resolution of a semiconductor detector may degrade if the bias voltage is lowered much below the optimum operating voltage (typically about 1 kV/mm for HgI$_2$ and 30–100 V/mm for CdTe) but spectral peak positions often remain relatively stable over a wide range of bias voltages, providing that the electric field is high enough to fully deplete the detector. (Under this condition the detector is said to be operating in the *saturation region*). Also, many wide bandgap room temperature semiconductor detectors may be operated over a fairly wide temperature range without significant performance degradation, particularly within the range of typical ambient operating temperatures (-10 to $60°C$). However, some loss of resolution at low energy is to be expected when operating at elevated temperatures, due to the increase in thermally generated leakage currents. The temperature dependence of the resolution at 5.9 keV (^{55}Fe source) obtained with a 0.5 mm × 12.6 mm^2 thick HgI$_2$ detector is shown in Figure 10 along with a second order polynomial fit of the data. In this experiment the detector and electronics were cooled together so that the improved resolution at lower temperature may be due to both a reduction in electronic noise and processes associated with the detector (such as changes in charge mobility and leakage current). In this sample an optimum energy resolution was observed at about 0°C. (Thermal stresses due to packaging may account for the degradation below 0°C.)

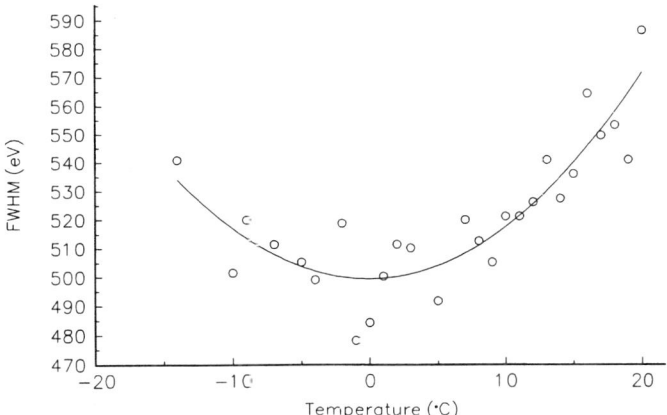

FIG. 10. Temperature dependence of the energy resolution of a 0.5 mm × 12.6 mm² HgI₂ x-ray detector at 5.9 keV (⁵⁵Fe source) along with least squares second order polynomial fit.

For HgI$_2$ detectors of 2 mm thickness, only about 8–30% of incident gamma rays in the range 300–1000 keV are stopped. Detectors up to 1 cm or more in thickness have been used to gain increased detection efficiency but hole trapping tends to compromise energy resolution (Beyerle et al., 1983). The best gamma ray energy resolution is obtained with HgI$_2$ detectors less than 4 mm in thickness. A similar situation exists with CdTe and limits detector thicknesses to about 2–4 mm. Typically, hole trapping is evident as an asymmetrical peak shape—a "tail" on the low energy side of the photopeak, evidence of which can be seen in Figs. 4 and 5. Progress in controlling impurities and stoichiometry in both HgI$_2$ and CdTe in recent years has considerably reduced the severity of the hole trapping problem (Skinner et al., 1989; Burger et al., 1990, 1991). Additionally, stacked configurations of thinner detectors have been used to obtain high efficiency without sacrificing energy resolution (McKee et al., 1988).

An alternative approach to improving the energy resolution of thick detectors is the use of pulse processing techniques that correct for charge trapping effects electronically (Beyerle, Gerrish, and Hull, 1986; Gerrish et al., 1987; Richter and Siffert, 1992). These methods typically use analog or digital filtering techniques to measure the photon interaction depth or pulse rise time. This information may be used to discriminate against long rise time pulses, which exhibit more pronounced charge trapping. Greatly improved energy resolution can be achieved using this approach but at the cost of lower detection efficiency. However, methods have also been developed to correct for variations in pulse amplitude due to trapping without rejecting pulses. Using hole trapping compensation techniques, energy resolution in the range of 1–2% FWHM at 662 keV is achieved using HgI$_2$ detectors of 1–4 mm thickness and 3–6 cm² area without loss of efficiency. Ex-

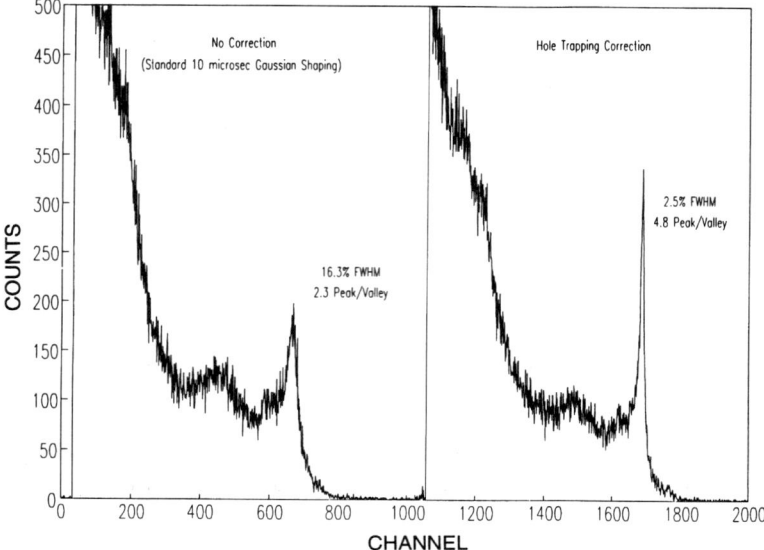

FIG. 11. ^{137}Cs spectra from a 2.8 mm × 6.4 cm² area HgI$_2$ detector at 2000 V bias. Left: Standard Gaussian shaping amplifier with 10 μsec time constant. Energy resolution at 662 keV: 16.3% FWHM. Peak to valley ratio: 2.3:1. Right: Hole trapping correction based on rise time compensation. Energy resolution at 662 keV: 2.5% FWHM. Peak to valley ratio: 4.8:1.

amples of ^{137}Cs spectra from a HgI$_2$ detector with and without electronic hole trapping compensation are shown in Fig. 11.

6. Polarization

Polarization in HgI$_2$ and CdTe detectors is an area in which considerable progress has been made, in both understanding its causes and developing methods to eliminate it. Polarization is observed in the spectral response curve as a gradual change in the energy resolution, count rate, or peak position over time. In HgI$_2$ polarization is known to be due, in part, to a gradual degradation of the material when exposed to air (Scott, 1975). Encapsulation methods have been developed for HgI$_2$ detectors that have greatly improved their long term stability (Squillante, Shah, and Moy, 1990). Encapsulated HgI$_2$ gamma ray detectors have been operated continuously for periods of more than three years with no signs of significant degradation (Gerrish, 1992b). Another type of polarization associated with HgI$_2$ x-ray detectors has been described as a "dead layer" effect, which causes a gradual reduction in the x-ray peak counts or peak position over a period of several hours after application of bias voltage. This has been interpreted either as evidence for fabrication induced damage (Holzer and Schieber, 1980), space charge formation (Mohammed-Brahim, Friant, and Mellet, 1984), or electrodrift and diffusion of impurities (Lanyi, Dikant, and Ruzicka, 1978). In any case, improvements in fabrication methods and material purification appear to have eliminated this problem for the thin HgI_2 detectors (<1 mm) used as x-ray detectors (Iwanczyk, 1989). It has been observed that thicker detectors (2–5 mm) are more likely to show this type of polarization which suggests that bulk properties play a role (K. James et al., 1992).

An interesting type of polarization sometimes observed in HgI_2 spectrometers manifests initially as a pronounced "tail" on the high energy side of the gamma ray peak. If the detector bias is maintained for several days, this feature gradually disappears from the spectrum as these counts shift down to the photopeak. An example of this behavior is shown in Fig. 12. This figure shows a sequence of ^{137}Cs spectra acquired over a 19-day period with a 2.1 mm × 645 mm² HgI_2 detector at 2000 V bias. (Individual spectra are offset.) In addition, it has been observed that the high energy tailing is *decreased* when the bias voltage is raised. However, the final peak position is independent of bias voltage over a wide voltage range. Furthermore, the hole lifetime is found to increase during the period in which the counts shift down to the photopeak. These observations, along with

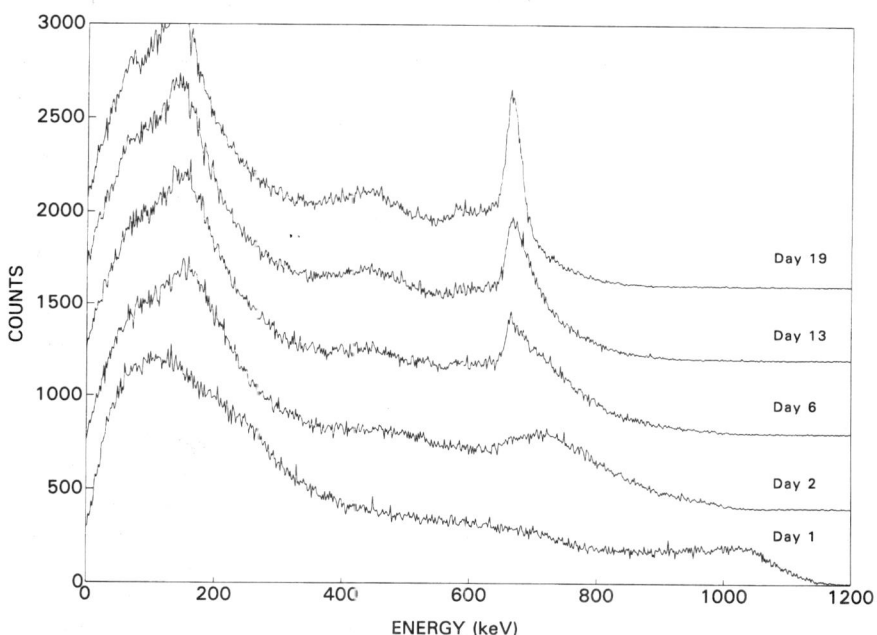

FIG. 12. ^{137}Cs Spectra from a 2.1 mm × 6.4 cm² HgI_2 detector at 2000 V bias. Spectra were acquired over a 19-day period and show polarization effects associated with Auger recombination process.

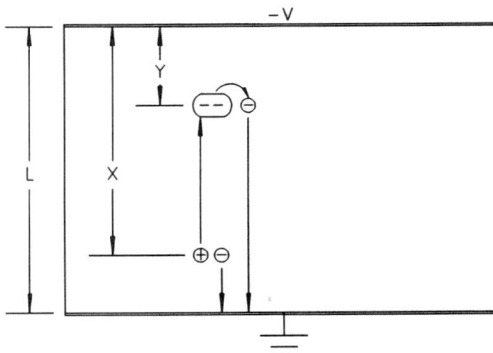

FIG. 13. Diagram of charge multiplication due to Auger recombination in HgI$_2$ detectors.

measurements of the mean energy to generate an electron–hole pair in HgI$_2$ detectors and light conditioning experiments, suggest that the appearance of gamma ray counts at energies well above the full energy peak is due to charge multiplication associated with an Auger recombination process (Gerrish, 1992a). In this process a charged center containing two or more trapped electrons may act as a recombination site for a hole. The energy released by recombination is transferred to a remaining trapped electron, which is then ejected into the conduction band. Figure 13 illustrates the charge multiplication process: an electron–hole pair is generated at a distance, X, from the negative electrode. Due to the electric field the electron and hole separate and move to corresponding electrodes. After traveling a distance, $X - Y$, the hole recombines at an Auger center at Y. An electron is released, which then moves to the positive electrode. Depending on where the Auger recombination occurs in the detector, charge multiplication of between 1 and 2 is allowed by this mechanism. This behavior is obviously undesirable from the standpoint of stable detector performance, and fortunately, recent improvements in material processing have considerably reduced the severity of this problem (Gerrish, 1992b). The high energy tailing can also be eliminated by a brief illumination of the positive electrode with above bandgap energy light.

In CdTe detectors, polarization occurs as a gradual degradation in detector performance over a period of several hours after applying bias voltage. This is a result of electron trapping, which causes an accumulation of space charge and a reduction of the depletion region thickness. This phenomenon is generally associated with the chlorine doping used to produce high resistivity material (Zanio, 1978). However, advances in contact deposition, material purification, doping methods, and crystal growth appear to have eliminated the polarization problems in CdTe detectors (Roth, 1989). The development of suitable electrical contacts to control the charge injection levels in chlorine doped CdTe detectors is one approach that has been successful. Another is the development of the high pressure Bridgeman method of growing undoped, purified CdTe crystals. Also, the replacement of a fraction of the Cd in CdTe detectors with Zn results in high resistivity, nonpolarizing material (Butler, Lingren, and Doty, 1991). At present, however, the energy resolution of $Cd_{1-x}Zn_xTe$ detectors for high energy gamma ray energies (>100 keV) is inferior to that obtained from high resistivity CdTe detectors.

7. Radiation Damage Resistance

For certain detector applications in areas such as astrophysics and accelerator experiments, radiation damage resistance is an important consideration. It has recently been demonstrated that HgI_2 detectors exhibit no degradation in performance when exposed to a fluence of 10^8 1.5 GeV protons/cm^2 (Patt et al., 1990). This fluence is estimated to be equivalent to the cosmic radiation dose expected during 1 year in space. (In the same experiment degradation in high purity Ge detectors was observed at a fluence of less than 4×10^7 protons/cm^2.) Resistance to fast neutron radiation damage at fluences up to 10^{15} neutrons/cm^2 has also been reported with HgI_2 detectors (Becchetti et al., 1983). The high radiation damage resistance of HgI_2 may be due to several reasons. The high atomic weights of Hg and I (as compared to Ge or Si) means that greater kinetic energy is required to displace these atoms from their lattice sites. Also, a self-compensation mechanism may occur in HgI_2 that would tend to neutralize radiation induced defects. Finally, the mean free path of electrons in HgI_2 is usually large compared to the detector thickness so that an appreciable reduction in the electron lifetime can be sustained before significant spectral effects occur. This would be relevant to the x-ray or alpha particle response because of the shallow penetration of these radiations. In any case, it appears that HgI_2 detectors may be suitable for operation in a high radiation environment.

III. Electronic Characterization

A large number of electronic characterization techniques have been developed to try to understand the relationship between fundamental material parameters and detector performance (Schroder, 1990; Sze, 1981). Resistivity, charge carrier mobilities, trapping parameters, defect densities, and contact properties can be determined using electronic methods. At present, an extensive literature exists on the electronic characterization of semiconductor materials, and so only a brief overview will be presented of the techniques most widely used with wide bandgap nuclear detector materials.

1. Bulk Measurements

In relating material parameters to detector performance it is useful to apply characterization methods that provide information about the bulk properties of a material independent of surface properties or effects caused by detector fabrication. In this regard, transient charge techniques are extremely useful in the determination of the bulk electronic transport properties of high resistivity materials. Charge carrier mobilities and lifetimes and parameters associated with the capture and release of charges from trapping centers can be obtained with these methods. Generally, transient charge techniques involve the analysis of a waveform produced by irradiation with x-rays, alpha particles, or pulsed ultraviolet or visible light.

a. Transient Charge Techniques

A commonly used transient charge measurement is the time of flight (TOF) method for the determination of charge carrier mobilities (Schroder, 1990). In this technique the drift velocity, or transit time, of charge carriers generated at one of the contacts is measured. Electron–hole pairs are produced by irradiating a contacted surface with shallow penetrating radiation such as alpha particles or pulsed laser light. The charge transferred by the resulting current pulse is measured using a charge sensitive or current sensitive preamplifier, and the output voltage is displayed on an oscilloscope from which the pulse rise time can be read. When the electrons (holes) traverse the entire detector thickness, L, the electron (hole) mobility, μ, is related to the pulse rise time, t_r, by

$$\mu = \frac{L^2}{Vt_r} \qquad (10)$$

where V is the applied bias voltage and t_r is in seconds. The rise time is often measured over a range of bias voltages or temperatures.

As a means of generating charge carriers for TOF measurements, pulsed laser light offers certain advantages to using an alpha source. Plasma effects associated with the high density of electron–hole pairs generated along the track of an incident alpha particle can effect the pulse rise time and shape (Knoll, 1989). Laser light, in contrast, can illuminate the entire contact area so that the electron–hole density is greatly reduced for pulses of comparable amplitude. Illumination of the full contact area also tends to average any variations in transport properties over the detector volume. Finally, the amplitude of pulses generated by a laser can be easily controlled by the use of optical filters. Disadvantages of pulsed lasers are the more elaborate experimental setup required and the necessity of a semitransparent contact.

Using the TOF method, electron and hole mobilities have been measured in HgI_2 (Kusmiss *et al.*, 1983; Kurtz *et al.*, 1987; Roth *et al.*, 1987) and CdTe (Canali *et al.*, 1971a) and, near room temperature, are found to decrease with increasing temperature. This is consistent with the view that phonon scattering is the primary mechanism that controls the mobilities (except at very low temperatures, where lattice vibrations are reduced and scattering at defects or impurities may dominate). It should be mentioned that the Hall effect method for determining mobilities is usually not applicable to high resistivity materials due to the very low current densities. However, Hall mobilities have been measured with low resistivity CdTe and values in agreement with the drift mobilities were obtained (Zanio, 1978).

In addition to the transit time, t_r, the trapping time (mean free drift time), τ, of the charge carriers can be measured by performing a suitable analysis of the waveform shape (Kusmiss et al., 1983; Roth et al., 1987). For single carrier transport in which detrapping of charge carriers can be ignored and a uniform electric field distribution is assumed, the pulse shape is given by the Hecht relation (Eq. (7) with $x = 0$ for electron transport, or $x = 1$ for hole transport):

$$Q(t) = \begin{bmatrix} Q_0 \dfrac{\tau}{t_r}\left(1 - e^{-\frac{t}{\tau}}\right) & (0 < t \le t_r) \\ Q_0 \dfrac{\tau}{t_r}\left(1 - e^{-\frac{t_r}{\tau}}\right) & (t_r < t) \end{bmatrix} . \qquad (11)$$

Figure 14 shows a plot of a hole charge pulse created in a HgI_2 detector by an alpha particle along with a computer generated fit of the Hecht relation. A hole

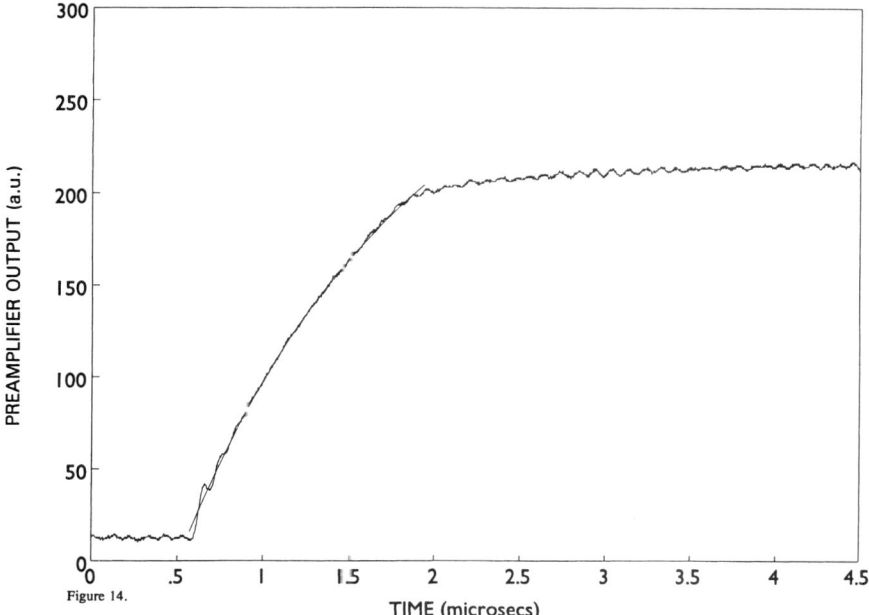

Figure 14.

FIG. 14. A 6 Mev alpha particle pulse (hole collection) from a charge sensitive preamplifier connected to a 0.6 mm × 20 mm² HgI_2 detector. Also shown is a fit based on the Hecht relation from which a hole mobility of 2.7 cm²/V · sec, and a hole trapping time of 1.3 μsec, were obtained.

mobility of 2.7 cm^2/V · s and a hole trapping time of 1.3 μsec was obtained by fitting Eq. (11). When detrapping of charge carriers is present, the waveform analysis is more complex, but the problem has been analyzed and the results applied to high resistivity CdTe to obtain detrapping times and thermal activation energies of traps (Zanio, Akutagawa, and Kikuchi, 1968). Data from TOF measurements of the hole mobility of high resistivity In doped CdTe have been interpreted in terms of a trap controlled mobility, i.e., the drift of holes is controlled by the successive trapping and detrapping events (Ottaviani et al., 1973). The concentrations and thermal activation energies of the traps were estimated and evidence for electric field induced lowering of the trap barrier (Poole–Frenkel effect) was observed.

If a sufficiently high density of charge carriers is injected at the contact (by a pulse of intense laser light, for example) so that the near-surface electric field distribution is greatly disturbed, space charge limited (SCL) transients may be observed. This situation has been analyzed by Many and Rakavy (1962) and the theory applied to HgI$_2$ (Cho et al., 1975) and CdTe (Canali, Ottaviani, and Martini, 1971b) to obtain mobilities and trapping times.

Frequently, mobility–lifetime products for electrons and holes are measured rather than separately determining the charge mobilities and lifetimes. For a given detector geometry and field strength, the mobility–lifetime products determine the charge collection efficiency and are often quoted when evaluating nuclear detector materials. For an applied electric field, E, the mobility–lifetime product, $\mu\tau$, determines the mean free drift length, λ, of charge carriers in the material

$$\lambda = \mu\tau E. \quad (12)$$

A commonly used method to determine the mobility–lifetime product is to measure the charge collection efficiency of injected charge carriers as a function of the applied bias voltage (Mayer, 1967). The charge collection efficiency is given by Eq. (9) with $x = 0$, or $x = 1$, for electron, or hole, collection, respectively. The electric field at which the charge collection efficiency, η, is equal to 1/2, E_{hmc}, is

$$E_{\text{hmc}} = \frac{0.63L}{\mu\tau}. \quad (13)$$

The x-ray, alpha particle, or pulsed laser peak channel can be measured as a function of the bias voltage applied to the detector. The field at which the peak channel attains half its saturation value can be determined and application of Eq. (13) gives the mobility–lifetime product. A drawback of this method is that surface related effects may also limit the collection efficiency of the injected charge carriers. This tends to reduce the measured values of the mobility–lifetime products and may lead to a dependence of these values on detector thickness (as has been observed in HgI$_2$) (Levi, Schieber, and Burshtein, 1983).

b. Detector Response Mapping

Defects in crystal structure occurring during crystal growth or caused by detector fabrication may play an important role in limiting detector performance. A method that has been used to determine the importance of crystallinity variations is the response mapping of detectors exposed to a collimated radiation source. The response of large area HgI_2 detectors to x-rays and gamma rays has been mapped with a computer controlled X–Y positioning system and compared with crystallinity and microhardness variations (K. James *et al.*, 1992). A correlation was observed between optically visible defects, microhardness, and variations in detector performance. Distinct regions of varying optical clarity were identified indicative of variations in extended defect density within the crystal. The hazy regions exhibit well-defined growth stria composed of voids or semitransparent precipitates. It was found that crystal regions having a *lower* density of optically visible defects are associated with a "dead layer," which reduces the amplitude of the x-ray and low energy gamma ray response.

As an example, Figure 15a shows a drawing of a 3 mm thick slice cut from the center of a HgI_2 crystal (perpendicular to the c-axis) based upon visual inspection of the visible light transmittance of the sample. Regions of varying optical clarity were observed and are indicated in the drawing as "clear" and "hazy." These patterns are associated with the crystal growth directions. Figure 15b shows a map (1 mm² pixel size) of the response of the same sample to 60 keV gamma rays from a collimated ^{241}Am source. (The mean penetration depth of 60 keV gamma rays in HgI_2 is about 0.25 mm, or about 8% of the sample thickness.) The 60 keV peak position is plotted in Fig. 15b so that lighter regions of the plot correspond to higher peak positions and the difference in peak position between light and dark regions is about 30–50%. Two regions of reduced charge collection (lower peak position) are observable in the plot that correlate with the two optically clear regions of the sample. The clear regions are also harder and more brittle than the hazy regions. An interpretation of this data is that gettering of point defects to extended defects in the hazy regions may reduce their concentration in the surrounding lattice so that the hazy regions exhibit better electron transport. However, there is evidence that the extended defects may also be associated with increased hole trapping, which tends to degrade the high energy gamma ray response in the hazy regions (K. James *et al.*, 1992).

c. Thermally Stimulated Current and Current–Voltage Measurements

The measurement of thermally stimulated currents (TSC) has been used to characterize traps in HgI_2 (Whited and van den Berg, 1977; Tadjine *et al.*, 1983; Gelbart *et al.*, 1977; Schlesinger *et al.*, 1992) and CdTe (Zanio, 1978; Mancini *et al.*, 1977; Scharager *et al.*, 1975). This technique provides information on trap centers such as thermal activation energies, trap concentrations, and capture cross sec-

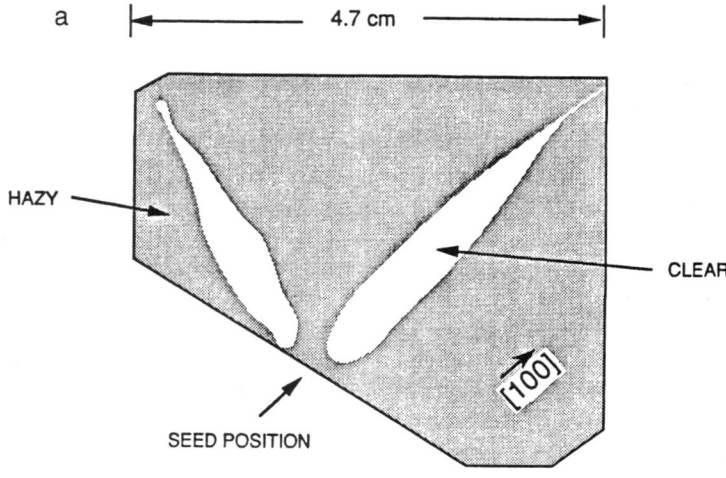

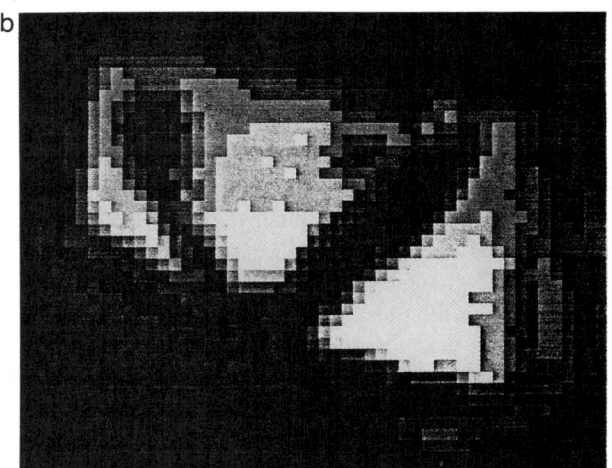

FIG. 15. (a) Sketch of the optical clarity of a 3-mm-thick cross-sectional slice from HgI_2 crystal. "Clear" and "hazy" regions are identified. (b) ^{241}Am peak position respose map of cross-sectional slice. Light regions indicate higher response, dark regions are lower response.

tions. With proper experimental conditions, electron and hole traps can be distinguished. The experimental procedure requires that a sample be placed under bias and cooled to a low temperature, typically either 4 K or 77 K. The sample temperature is then raised while the leakage current and heating rate are recorded. Thermal detrapping of carriers from defects causes peaks in the leakage current at certain temperatures corresponding to specific trapping centers. If the sample has a contact that is at least semitransparent to visible light, the contact may be illuminated with above bandgap energy light so that either electrons or holes traverse the bulk. When this is done at low temperature, electron or hole traps can be preferentially filled and the two types of traps can be distinguished.

The density of traps associated with a particular TSC peak can be estimated by integrating the area under the peak since this corresponds to the total charge released by the traps. Densities of electronically active traps as low as 10^9 cm^{-3} can easily be detected in high resistivity materials. This corresponds to about 1 trap per 10^{12} atoms, which makes this an extremely sensitive technique. However, the calculation of thermal activation energies and capture cross sections from TSC spectra is complicated by the necessity of making certain assumptions about the relative rates of detrapping, retrapping, and recombination of charge carriers in the material. In spite of these difficulties, a number of researchers have analyzed TSC peaks in HgI$_2$ and CdTe and numerous levels have been observed in both materials. Some studies have attempted to correlate trap levels obtained from TSC peaks with native defects, impurities, or deviations from stoichiometry in HgI$_2$ and CdTe nuclear detectors. TSC spectra have been measured in Hg and I doped HgI$_2$ samples and observed TSC peaks were attributed to I or Hg vacancies or Hg interstitials (Whited and van den Berg, 1977). The authors report that Hg doping reduces the hole mobility–lifetime product by two orders of magnitude and a TSC peak observed at 220°C is assigned to a Hg interstitial. The effect of various metal contacts on the TSC spectra of HgI$_2$ samples has been studied (Schlesinger *et al.*, 1992). The data shows that the trap type and concentration is a function of the contacting metal. (Also, a good summary is presented of TSC results on HgI$_2$ obtained from the literature.) TSC has been used to study the effect of anealing gold contacts on CdTe in various atmospheres (Mergui *et al.*, 1992a,b). Comparison of TSC spectra with current–voltage, mobility–lifetime, and polarization measurements suggest that anealing in a hydrogen atmosphere may improve CdTe nuclear detector performance.

The use of TSC to characterize deep level defects in semiconductors has largely been replaced by the technique of deep level transient spectroscopy (DLTS). The technique is similar to TSC, except that DLTS is done using a pulsed excitation of the sample (usually pulsed voltage or illumination) and the resulting capacitance or current transient is measured over some temperature range. DLTS data is much simpler to interpret than TSC data, but the experimental equipment used with DLTS is more elaborate. The development of automated data acquisition systems for DLTS has aided the extensive investigation of defects in materials such as Si and GaAs, but to date, DLTS techniques have not been widely used with room temperature nuclear detector materials. However, the DLTS technique of photoinduced current transient spectroscopy (PICTS) is well suited to the study of defects in wide bandgap materials and has been applied to CdTe (Isett and Raychaudhuri, 1984; Tomitori *et al.*, 1985; Hage-Ali and Siffert, 1992), CdS (Yoshie and Kamihara, 1983), GaAs (Zhang *et al.*, 1987), and InP (Singh and Anderson, 1988). In one study on CdTe (Hage-Ali and Siffert, 1992) PICTS data is presented that indicates that hydrogen ion implantation followed by annealing at 150°C markedly decreases the concentration of defect levels associated with cadmium vacancies while increasing the bulk resistivity by more than two orders of magnitude.

When injecting (ohmic) contacts can be applied to a semiinsulating (wide bandgap) material, current–voltage measurements (I–V) can be used to determine the total trap density. This technique utilizes Lampert's theory of space charge limited currents for an insulator with traps (Lampert and Mark, 1970) and has been applied to CdSe (Mancini *et al.*, 1977), CdS (Lampert and Mark, 1970), and GaAs (Lampert and Mark, 1970). For example, trap densities on the order of 10^{10} cm^{-3} have been measured in CdSe with this method. Characterization of semiconductor materials using current–voltage measurements is generally easier than performing nuclear measurements or chemical analysis; and where injecting contacts can readily be fabricated, this technique can be used to determine overall impurity or doping levels.

2. Contacts

The ability to form suitable conductive contacts is critical to the fabrication of practical devices. In general, at least one of the detector contacts should form a blocking, or rectifying, Schottky barrier to create a depletion region and minimize the leakage current due to charge injection. This results in a diode structure. A typical HgI_2 or CdTe detector is fabricated with identical contacts on either side so that a symmetrical back to back diode structure is formed. The detectors can then be biased in either direction. (However, for nonpenetrating radiation, irradiation of the cathode is preferred because of the better charge transport of electrons as compared to holes.) The contact must be chemically and mechanically stable and should not create charge trapping or recombination in the near-surface region. Also, care must taken to minimize edge leakage currents. Palladium or colloidal carbon are commonly used as contacts on HgI_2. To obtain transparent contacts needed for HgI_2 photodetectors, indium tin oxide (ITO), electrolyte, or conducting polymer contacts have been applied successfully. For CdTe detectors, gold, aluminum, or platinum contacts are commonly used.

A simplified model of a metal-semiconductor interface is shown in Fig. 16 for zero bias. The conduction band, valence band, Fermi level, and bandgap energies are labeled E_C, E_V, E_F, and E_g, respectively. Some band bending occurs in the semiconductor due to the presence of space charge. The metal–semiconductor interface gives rise to a potential barrier so that an energy, $q\phi_B$, is required to raise the electron from the metal Fermi level to the semiconductor conduction band. The energy difference between the conduction band and the Fermi level, $q\phi_{BN}$, is actually slightly larger but an electron entering the semiconductor from the metal sees a smaller potential barrier due to image forces (Henisch, 1984). This is known as the "Schottky effect." The height of the Schottky barrier depends on

13. CHARACTERIZATION AND QUANTIFICATION OF DETECTOR PERFORMANCE   521

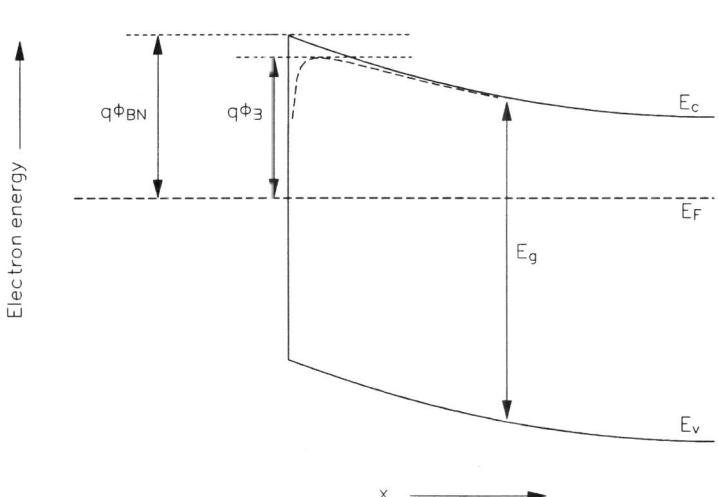

FIG. 16. Diagram of metal–semiconductor interface at zero bias. Lowering of potential barrier height due to image force (Schottky effect) is shown.

the electric field in this region and is reduced slightly when a reverse bias voltage is applied. For blocking contacts, lowering the Schottky barrier height accounts for the increase in the leakage current as the applied bias voltage is raised. The dominant charge injection process for reverse-biased Schottky barriers on intrinsic or moderately doped semiconductors is thermionic emission of electrons from the metal. For a detailed discussion of semiconductor contacts, see Henisch (1984) and Sze (1981).

a. Schottky Barrier Height Measurements

Blocking contacts on intrinsic or compensated wide bandgap materials may be obtained by depositing a metal or other conductive material that forms a high Schottky barrier (>0.5 eV). When the resulting junction is reverse biased, the dominant charge injection mechanism is thermionic emission from the contact. The expression for thermionic emission is

$$J = A^* T^2 e^{-\frac{q\phi_B}{kT}} \tag{14}$$

where J is the current density in C/cm^2/s, $A^* = 4\pi q k^2 m^*/h^3$ is the effective Richardson constant (approximately 120 A/cm^2/K^2), m^* is the electron effective mass, q is the electron charge, h is Planck's constant, ϕ_B is the Schottky barrier height in volts, k is Boltzmann constant, and T is absolute temperature. The barrier height depends on the electric field at the contact as

$$\phi_B = \phi_{B0} - \sqrt{\frac{qE_c}{4\pi\epsilon}} \quad (15)$$

where ϕ_{B0} is the barrier height at zero bias, E_c is the electric field at the cathode, and ϵ is the high frequency dielectric constant in $C^2/J/cm$. In the absence of band bending near the contact (i.e., "flat band" conditions), E_c is equal to the bulk electric field that is just the applied bias voltage divided by the detector thickness, V/L. Therefore, at voltages sufficient for full depletion, a plot of $\log(I)$ as a function of $V^{1/2}$ should give a straight line whose intersection with the voltage axis will give the barrier height. This method has been applied to HgI_2 detectors and values for the barrier height near 1.1 eV were obtained with Pd and carbon contacts (Mellet and Friant, 1989). By analyzing the slope of the $\log(I)$–$V^{1/2}$ curve additional barrier parameters were obtained including the ratio of the cathode electric field to the bulk electric field, which is found to be considerably greater than unity. This ratio has been interpreted as a measure of the blocking properties of HgI_2 contacts and may be an indication of the surface quality. Figure 17 shows a plot of I–V data from a 1.6 mm $\times$ 6.4 cm^2 HgI_2 detector with Pd contacts along with a least squares fit, which gives a barrier height of 1.08 eV. Using this method, measurements have been made for various metal, conducting oxide, electrolyte, and conducting polymer contacts on HgI_2 (Gerrish, 1991), and in each case the

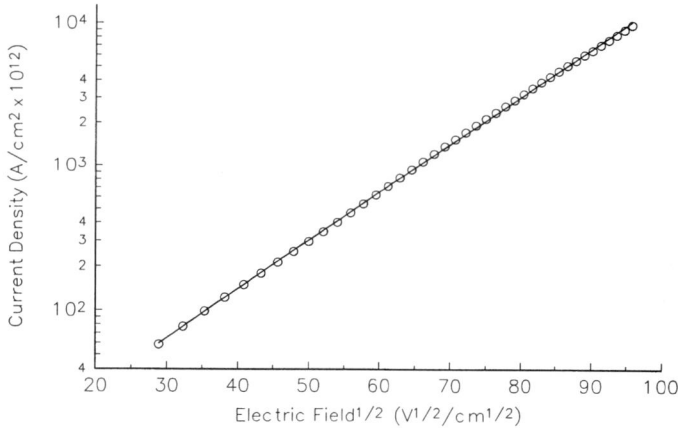

FIG. 17. I–V curve from a 1.6 $\times$ 7.8 mm^2 HgI_2 detector (Pd contact) along with least squares linear fit from which a barrier height of 1.08 eV is obtained.

barrier heights were near 1.1 eV. These data suggest that there is a pinning of the Fermi level near mid-bandgap in HgI_2, probably due to the existence of surface states formed at the HgI_2 surface prior to contact deposition. In contrast, the Schottky barrier heights of metal contacts on CdTe exhibit a weak dependence on the metal work function which allows some control over the degree of band bending and charge injection (Zanio, 1978).

The Schottky barrier height may also be measured by the photoelectric method. In this method photons having an energy greater than the barrier height energy, but less than the bandgap energy, are incident on the contacted surface of the semiconductor and inject electrons into the bulk. A bias voltage is applied and the photocurrent is measured as a function of the photon wavelength. The dependence of the photocurrent–incident photon, I_{ph}, on the photon energy, $h\nu$, is given, to good approximation, by (Schroder, 1990)

$$I_{ph} = C(h\nu - \phi_B)^2 \qquad (16)$$

where C is independent of the photon energy, and $h\nu$ and ϕ_B are in eV. A plot of $I_{ph}^{1/2}$ as a function of the photon energy gives a straight line whose intercept with the energy axis determines the barrier height. This method has been applied to HgI_2 and confirms the values obtained from I–V curve measurements (R. James, 1991). Figure 18 shows a plot of data from a Pd-contacted HgI_2 detector along with a least squares fit from which a barrier height of 1.11 eV is obtained.

I–V curves have also been used to evaluate contacts on CdTe detectors. In one study (Mergui et al., 1992a) it was found that chemically deposited gold contacts

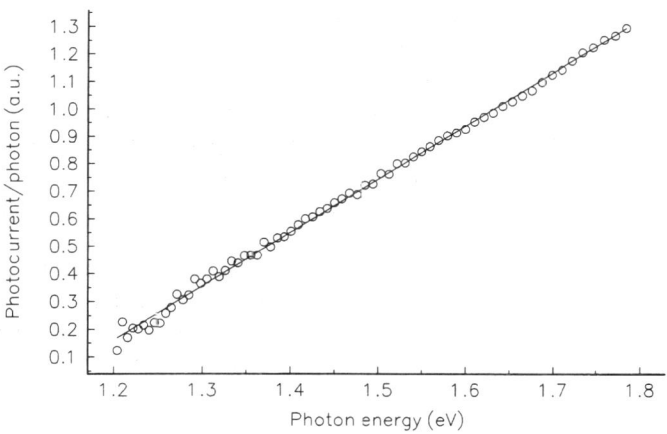

FIG. 18. Photoresponse curve (subbandgap energy photons) of a 2.3 mm × 6.2 cm² HgI_2 detector (semitransparent Pd contact) along with a least squares linear fit from which a barrier height of 1.11 eV is obtained.

exhibited lower leakage current and more ohmic behaviour than either evaporated or electrolytically applied gold. In another study (Verger *et al.*, 1992) chemically deposited palladium contacts on chlorine doped CdTe were also found to produce lower leakage current and more ohmic behavior than either sputtered or evaporated gold.

b. Surface Recombination

For characterizing devices used to detect nonpenetrating radiation, the surface recombination velocity is often a parameter of interest. Low temperature photoluminescence is the most common method of measuring surface recombination in semiconductor materials but electronic methods are also used. For example, the dependence of the charge collection efficiency on the mean penetration depth of x-rays and low energy gamma rays of various energies is one method of measuring "dead layer" thicknesses. In HgI_2, surface recombination velocities of electrons and holes have been determined by measuring the charge collection efficiency as a function of applied bias voltage and separating out the effect of bulk trapping (Levi *et al.*, 1983). Values on the order of 10^4 cm/sec were obtained for both electrons and holes. By comparing the data with bulk trapping parameters, the conclusion was drawn that electron collection in HgI_2 is limited mainly by surface recombination, while hole collection is bulk limited. This result was confirmed by a more recent study (Roth *et al.*, 1987).

IV. Correlation of Material Properties with Detector Performance

In addition to electronic methods, a number of optical, chemical, and physical analytical techniques have been used to obtain important information about the material properties of wide bandgap semiconductors that may not be obtainable by electronic techniques alone. These include measurements of crystallinity and stoichiometry, mapping surface topography, and impurity analysis. Chemical and physical characterization techniques in conjunction with electronic and optical methods can, in principle, allow the identification and quantitative analysis of electrically active impurities and lattice defects. Correlation of these measurements with detector response data can then determine which material properties are important to the performance of nuclear detectors. Although research on HgI_2 and CdTe in this area is more limited, those studies that have been done provide valuable information about the role of material properties in determining detector performance.

A number of studies have been done to correlate extended defects in HgI_2 with detector performance using defect etching. Selective etching is used to characterize imperfections in the crystal lattice such as dislocations or growth stria. The density and geometry of etch pits produced by defect etching may be used to gauge the structural homogeneity of the sample and allow a determination of the type of crystalline defect. In one study (Randtke and Ortale, 1977), etch pitch density variations in the range 10^3–10^6 cm^{-2} were measured on several cross-sectional slices from HgI_2 crystals and correlated with variations in x-ray and low energy gamma ray response. It was found that regions of the samples with etch pit counts of 10^3–10^4 cm^{-2} showed acceptable detector performance while higher concentrations were correlated with degraded detector response. Recent detector mapping experiments (K. James et al., 1992) indicate that regions of HgI_2 having lower densities of optically visible defects exhibit longer hole lifetimes and better high energy gamma ray resolution. However, these same regions tend to exhibit a "dead layer" at the surface that degrades the low energy response. In another study (Milstein, James, and Georgeson, 1989), results from defect etching, combined with cathodoluminescence and microhardness characterization indicate that regions of HgI_2 crystals containing linear etch pit patterns are more brittle and tend to exhibit poorer performance than regions containing randomly oriented defects.

Some studies have looked for correlations between HgI_2 or CdTe detector performance and data acquired with photoluminescence measurements. Photoluminescence spectroscopy is a nondestructive characterization tool, most often used for the identification of shallow level, and sometimes deep level, impurities that produce radiative recombination. The technique is quite sensitive but absolute defect concentrations are difficult to obtain. In photoluminescence measurements a semiconductor surface is illuminated with high intensity above bandgap energy light and the intensity of the luminescence due to radiative recombination is measured as a function of the luminescence wavelength. Peaks in the photoluminescence spectrum are generally associated with the recombination of excitons (bound states of electron–hole pairs) or with band to defect level or band to band transitions. For the peaks to be resolved the sample must be cooled, usually to liquid He temperature. Correlations are observed between features in the photoluminescence spectra of HgI_2 samples and gamma ray detector performance (Bao et al., 1992), although at present, the defects responsible have not been positively identified. Also, low temperature photoluminescence has been used to study the defects created by applying various metal contacts on HgI_2 (R. James et al., 1992). The rapid diffusion of copper and silver in HgI_2 at room temperature has also been observed (R. James et al., 1990). Photoluminescence peaks in uncompensated CdTe have been correlated with gamma ray performance and mass spectroscopy measurements (Triboulet et al., 1974). Improved gamma ray performance was obtained using highly purified CdTe grown by the traveling heater method (THM), and these results correlated well with impurity related photoluminescence features. In another study of chlorine doped CdTe crystals grown by radio frequency (RF-THM) and ordinary traveling heater methods (OR-THM) (Shoji, Onabi, and Hiratate, 1992), it was found that the RF-THM method produced a more uniform chlorine concentration and better gamma ray detector performance than the OR-THM method.

Deviations from stoichiometry are thought to have an important effect on the electronic properties of compound semiconductors. However, chemical methods of measuring stoichiometry are generally not very sensitive and usually do not indicate how the elemental components are incorporated in the lattice structure. An alternative method that has been used to determine stoichiometry in HgI_2 is the technique of differential scanning calorimetry (DSC) (Burger et al., 1989; Nicolau, Dupuy, and Kabsch, 1989). In DSC a sample is gradually heated while the heat absorbed by the sample is measured. A heat absorption peak occurs at a particular temperature when the material undergoes a phase transition such as melting or a change from one crystal structure to another. In one study it was shown that DSC could be used to detect excess Hg or Hg_2I_2 in HgI_2 (Burger et al., 1989). Elevated levels of Hg_2I_2 were observed in powdered samples, and this was interpreted as evidence that iodine loss occurs at the HgI_2 surface. Doping experiments combined with electronic measurements have established that excess Hg in HgI_2 is associated with increased hole trapping and poorer gamma ray detector performance (Whited and van den Berg, 1977).

Chemical analysis techniques such as mass spectroscopy, total carbon analysis, and ion chromatography have been used to analyze bulk impurities in HgI_2 (Steinberg et al., 1989; Piechotka and Kaldis, 1986; Muheim, Kobayashi, and Kaldis, 1983; Hermon, Roth, and Schieber, 1992; Cross, 1992). To date, only limited work has been done with regard to correlating specific impurities with detector performance, although the measurements have provided guidelines for improving material processing (Skinner et al., 1989). Presently, such efforts are underway (Cross, 1992), due, in part, to the observation that detector performance is strongly correlated with the starting material from which the HgI_2 crystals are grown (Gerrish and van den Berg, 1990; Gerrish, 1992b). Also, a recent study has correlated increased total organic carbon concentrations in HgI_2 with deterioration in hole lifetimes (Hermon et al., 1992). HgI_2 surfaces have been studied using Rutherford backscattering spectrometry and Auger electron spectroscopy (Felter et al., 1989). Evidence for loss of iodine at HgI_2 surfaces in vacuum was observed, along with significant interdiffusion of Pd and HgI_2 in Pd-contacted detectors.

X-ray topography, and x-ray and gamma ray rocking curves are related techniques that are used to determine structural imperfections in crystals (Schroder, 1990). In x-ray topography a sample is exposed to monochromatic x-rays and the diffracted x-rays are detected on photographic film. Because of the high attenuation of x-rays in high Z materials, reflection topography is generally used as opposed to transmission topography. In x-ray rocking curve characterization, the x-ray beam is highly collimated and the sample is slowly rotated or "rocked" about an axis normal to the diffraction plane while the intensity of the scattered x-rays are recorded as a function of angle. These techniques have been used to characterize HgI_2 single crystals and cross-sectional slices and the results compared with detector performance data (Schieber et al., 1989; Keller, 1991). In one study a good correlation was observed between the width of both x-ray and gamma ray rocking curves, x-ray topographs, and detector performance (Schieber et al., 1989).

V. Concluding Remarks

Room temperature nuclear detectors are becoming important in a wide range of scientific and industrial applications where the combination of high resolution, compact size, and room temperature operation offers an important advantage over conventional nuclear detector technology. However, the production of high quality solid state spectrometers based on high Z compound semiconductors presents unique challenges compared with conventional semiconductor materials such as Si or Ge. The need to obtain high purity, stoichiometric crystals has motivated the development of new material processing methods and improved crystal growth procedures. While major advances have been made, it is expected that additional improvements in detector performance and production yields will be achieved through the further development and refinement of these, and newer, methods. The characterization techniques described in this chapter undoubtedly will continue to play a vital role in this work.

ACKNOWLEDGMENTS

This document has been authorized by a contractor of the U.S. government under Contract No. DE-AC08-88NV10617. Accordingly, the U.S. government retains a nonexclusive, royalty-free license to publish or reproduce the published form of this contribution, or allow others to do so, for U.S. government purposes.

REFERENCES

Bao, X. J., Schlesinger, T. E., James, R. B., Harvey, S. J., Cheng, A. Y., Gerrish, V., and Ortale, C. (1992). *Nucl. Instr. and Meth.* **A317**, 194.

Barber, H. B., Barrett, H. H., Hickernell, T. S., Kwo, D. P., Woolfenden, J. M., Entine, G., and Baccash, C. (1991). *Med. Pys.* **18**(2), 373.

Becchetti, F. D., Raymond, R. S., Ristinen, R. A., Schnepple, W. F., and Ortale, C. (1983). *Nucl. Instr. and Meth.* **213**, 127.

Bertin, R., D'Auria, S., del Papa, C., Fiori, F., Lisowski, B., O'Shea, V., Pelfer, P. G., Smith, K., and Zichichi, A. (1990). *Nucl. Instr. and Meth.* **A294**, 211.

Beyerle, A., Hull, K., Markakis, J., Lopez, B., and Szymczyk, W. M. (1983). *Nucl. Instr. and Meth.* **213**, 107.

Beyerle, A., Gerrish, V., and Hull, K. (1986). *Nucl. Instr. and Meth.* **A242**, 443.

Bradley, J. G., Conley, J. M., Albee, A. L., Iwanczyk, J. S., Dabrowski, A. J., and Warburton, W. K. (1989). *Nucl. Instr. and Meth.* **A283**, 348.

Burger, A., Shilo, I., and Schieber, M. (1983). *IEEE Trans. Nucl. Sci.* **NS-30**(1), 368.

Burger, A., Roth, M., and Schieber, M. (1985). *IEEE Trans. Nucl. Sci.* **NS-32**(1), 556.

Burger, A., Morgan, S., Jiang, H., Silberman, E., Schieber, M., van den Berg, L., Keller, L., and Wagner, C. N. J. (1989). *Nucl. Instr. and Meth.* **A283**, 130.

Burger, A., Morgan, S., He, C., and Silberman, E. (1990). *J. Cryst. Growth* **99**, 988.

Burger, A., Henderson, D. O., Morgan, S. H., and Silberman, E. (1991). *J. Crys. Growth* **109**, 304.

Butler, J. F., Lingren, C. L., and Doty, F. P. (1991). *Conference Record,* IEEE Nucl. Sci. Symp. and Medical Instr. Conference, Santa Fe NM, 120.

Canali, C., Martini, M., Ottaviani, G., Quaranta, A. A., and Zanio, K. R. (1971a). *Nucl. Instr. and Meth.* **96,** 561.
Canali, C., Ottaviani, G., and Martini, M. (1971b). *Appl. Phys. Lett.* **19**(3), 51.
Cho, Z. H., Watt, M. K., Slapa, M., Tove, P. A., Schieber, M., Davies, T., Schnepple, W., Randtke, P., Carlston, R., and Sarid, D. (1975). *IEEE Trans. Nucl. Sci.* **NS-22**(1).
Cross, E. (1992). EG&G Energy Measurements, Inc., quarterly progress report (April).
Dabrowski, A. J., Iwanczyk, J., Szymczyk, W. M., Kokoshinegg, P., Steltzhammer, J., and Triboulet, R. (1978). *Nucl. Instr. and Meth.* **150,** 25.
Debertin, K., and Helmer, R. G. (1988). *Gamma- and X-Ray Spectrometry with Semiconductor Detectors.* North-Holland Press,
Doty, F. P., Butler, J. F., Schetzina, J. F., and Bowers, K. A. (1992). *J. Vac. Sci. Technol.* **B 10**(4), 1418.
Droms, C. R., Langdon, W. R., Robison, A. G., and Entine, G. (1976). *IEEE Trans. Nucl. Sci.* **NS-23**(1), 498.
Economou, T., and Iwanczyk, J. (1989). *Nucl. Instr. and Meth.* **A283,** 352.
Eichinger, P., and Kallmann, H. (1974). *Appl. Phys. Lett.* **25**(11), 676.
Entine, G., Waer, P., Tiernan, T., and Squillante, M. R. (1989). *Nucl. Instr. and Meth.* **A283,** 282.
Felter, T. E., Stulen, R. H., Schnepple, W. F., Ortale, C., and van den Berg, L. (1989). *Nucl. Instr. and Meth.* **A283,** 195.
Friant, A., Mellet, J., Barrandon, G., and Csakvary, E. (1989). *Nucl. Instr. and Meth.* **A283,** 227.
Gelbart, U., Yacoby, Y., Beinglass, I., and Holzer, A. (1977). *IEEE Trans. Nucl. Sci.* **NS-24**(1), 135.
Gerrish, V. (1991). EG&G Energy Measurements, Inc., quarterly progress report (January).
Gerrish, V. (1992a). *Nucl. Instr. and Meth.* **A322,** 402.
Gerrish, V. (1992b). EG&G Energy Measurements quarterly progress report (April).
Gerrish, V., and van den Berg, L. (1990). *Nucl. Instr. and Meth.* **A299,** 41.
Gerrish, V., Williams, D., and Beyerle, A. (1987). *IEEE Trans. Nucl. Sci.* **NS-34**(1), 85.
Hage-Ali, M., and Siffert, P. (1992). *Nucl. Instr. and Meth.* **A322,** 313.
Hazlett, T., Cole, H., Squillante, M. R., Entine, G., Sugars, G., Fecych, W., and Tench, O. (1986). *IEEE Trans. Nucl. Sci.* **NS-33**(1), 332.
Hecht, K. (1932). *Z. Physik* (Berlin), 235.
Henisch, H. K. (1984). *Semiconductor Contacts.* Clarendon Press, Oxford.
Hermon, H., Roth, M., and Schieber, M. (1992). *Nucl. Instr. and Meth.* **A322,** 442.
Hess, K., Gramann, W., and Höppner, D. (1972). *Nucl. Instr. and Meth.* **101,** 39.
Holzer, A., and Schieber, M. (1980). *IEEE Trans. Nucl. Sci.* **NS-27**(1), 266.
Isett, L. C., and Raychaudhuri, P. K. (1984). *J. Appl. Phys.* **55**(1), 3605.
Iwanczyk, J. S. (1989). *Nucl. Instr. and Meth.* **A283,** 208.
James, K., Gerrish, V., Cross, E., Markakis, J., Nason, D., Marschall, J., Milstein, F., and Burger, A. (1992). *Nucl. Instr. and Meth.* **A322,** 390.
James, R. B. (1991). *Bull. Am. Phys. Soc.* **36,** 4247.
James, R. B., Bao, X. J., Schlesinger, T. E., Ortale, C., and van den Berg, L. (1990). *Bull. Am. Phys. Soc.* **35,** 707.
James, R. B., Bao, X. J., Schlesinger, T. E., Cheng, A. Y., Baccash, C., and van den Berg, L. (1992). *Nucl. Instr. and Meth.* **A322,** 435.
Keller, L. (1991). Unpublished report.
Knoll, G. (1989). *Radiation Detection and Measurement,* 2nd ed. John Wiley & Sons, New York.
Kusmiss, J. H., Iwanczyk, J. S., Barton, J. B., Dabrowski, A. J., and Seibt, W. (1983). Proceedings of the IEEE Conference, Amman, Jordan, 185.
Kurtz, S. R., Hughes, R. C., Ortale, C., and Schnepple, W. F. (1987). *J. Appl. Phys.* **62**(10), 4308.
Lampert, M. A., and Mark, P. (1970). *Current Injection in Solids.* Academic Press, New York.
Lanyi, S., Dikant, J., and Ruzicka, M. (1978). *Acta Phys. Slov.* **28**(3), 210.
Levi, A., Schieber, M. M., and Burshtein, Z. (1983). *J. Appl. Phys.* **54**(5), 2472.
Lund, J. C., Shah, K. S., Squillante, M. R., Moy, L. P., Sinclair, F., and Entine, G. (1989). *Nucl. Instr. and Meth.* **A283,** 299.

Mancini, A. M., Manfredotti, C., de Blasi, C., Micocci, G., and Tepore, A. (1977). *Revue de Physique Appliquee* **12,** 255.
Many, A., and Rakavy, G. (1962). *Phys. Rev.* **126**(6), 1980.
Markakis, J. (1988). *Nucl. Instr. and Meth.* **A263,** 499.
Mayer, J. W. (1967). *J. Appl. Phys.* **38**(1), 296.
McKee, B. T. A., Goetz, T., Hazlett, T., and Forkert, L. (1988). *Nucl. Instr. and Meth.* **A272,** 825.
Mellet, J., and Friant, A. (1989). *Nucl. Instr. and Meth.* **A283,** 199.
Mergui, S., Hage-Ali, M., Koebel, J. M, and Siffert, P. (1992a). *Nucl. Instr. and Meth.* **A322,** 375.
Mergui, S., Hage-Ali, M., Koebel, J. M, and Siffert, P. (1992b). *Nucl. Instr. and Meth.* **A322,** 381.
Milstein, F., James, T. W., and Georgeson, G. (1989). *Nucl. Instr. and Meth.* **A285,** 500.
Mohammed-Brahim, T., Friant, A., and Mellet, J. (1984). *IEEE Trans. Nucl. Sci.* **NS-32**(1), 581.
Muheim, J. T., Kobayashi, T., and Kaldis, E. (1983). *Nucl. Instr. and Meth.* **213,** 39.
Nicolau, Y. F., Dupuy, M., and Kabsch, Z. (1989). *Nucl. Instr. and Meth.* **A283,** 149.
Olschner, F., Toledo-Quinones, M., Shah, K. S., and Lund, J. C. (1990). *IEEE Trans. Nucl. Sci.* **NS-37**(3), 1162.
Ottaviani, G., Canali, C., Jacoboni, C., and Quaranta, A. A. (1973). *J. Appl. Phys.* **44**(1), 360.
Patt, P. E., Beyerle, A. G., Dolin, R. C., and Ortale, C. (1989). *Nucl. Instr. and Meth.* **A283,** 215.
Patt, B. E., Dolin, R. C., DeVore, T. M., and Markakis, J. M. (1990). *Nucl. Instr. and Meth.* **A299,** 176.
Piechotka, M., and Kaldis, E. (1986). *J. Crys. Growth* **79,** 469.
Radeka, V. (1972). *Nucl. Instr. and Meth.* **99,** 525.
Radeka, V., and Karlovac, N. (1967). *Nucl. Instr. and Meth.* **52,** 86.
Raiskin, E., and Butler, J. F. (1988). *IEEE Trans. Nucl. Sci.* **NS-35**(1), 81.
Randtke, P. T., and Ortale, C. (1977). *IEEE Trans. Nucl. Sci.* **NS-24**(1), 129.
Richter, M., and Siffert, P. (1992). *Nucl. Instr. and Meth.* **A322,** 529.
Ricker, G. R., Vallerga, J. V., and Wood, D. R. (1983). *Nucl. Instr. and Meth.* **213,** 133.
Roth, M. (1989). *Nucl. Instr. and Meth.* **A283,** 291.
Roth, M., Burger, A., Nisserbaum, J., and Schieber, M. (1987). *IEEE Trans. Nucl. Sci.* **NS-34**(1), 465.
Sakai, E. (1982). *Nucl. Instr. and Meth.* **196,** 121.
Scharager, C., Muller, J. C., Stuck, R., and Siffert, P. (1975). *Phys. Stat. Sol.(a)* **31,** 247.
Schieber, M., Ortale, C., van den Berg, L., Schnepple, W., Keller, L., Wagner, C. N. J., Yelon, W., Ross, F., Georgeson, G., and Milstein, F. (1989). *Nucl. Instr. and Meth.* **A283,** 172.
Schlesinger, T. E., Bao, X. J., James, R. B., Cheng, A. Y., Ortale, C., and van den Berg, L. (1992). *Nucl. Instr. and Meth.* **A322,** 414.
Schroder, D. K. (1990). *Semiconductor Material and Device Characterization.* John Wiley & Sons, New York.
Scott, R. (1975). *Appl. Phys. Lett.* **27**(2), 99.
Shah, K. S., Lund, J. C., Olschner, F., Moy, L., and Squillante, M. R. (1989). *IEEE Trans. Nucl. Sci.* **NS-36**(1), 199.
Shoji, T., Onabe, H., and Hiratate, Y. (1992). *Nucl. Instr. and Meth.* **A322,** 324.
Siffert, P. (1983). *Mat. Res. Soc. Symp. Proc.* **16,** 87.
Singh, A., and Anderson, W A. (1988). *J. Appl. Phys.* **64**(8), 3999.
Skinner, N. L., Ortale, C., Schieber, M. M., and van den Berg, L. (1989). *Nucl. Instr. and Meth.* **A283,** 119.
Squillante, M. R., Shah, K. S., and Moy, L. (1990). *Nucl. Instr. and Meth.* **A288,** 79.
Steinberg, S., Kaplan, I., Schieber, M., Ortale, C., Skinner, N., and van den Berg, L. (1989). *Nucl. Instr. and Meth.* **A283,** 123.
Sze, S. M. (1981). *Physics of Semiconductor Devices,* 2nd ed. John Wiley and Sons, New York.
Tadjine, A., Gosselin, D., Koebel, J. M., and Siffert, P. (1983). *Mat. Res. Soc. Symp. Proc.* **16,** 217.
Tomitori, M., Kuriki, M., Ishii, S., Fuyuki, S., and Hayakawa, S. (1985). *Jap. J. Appl. Phys.* **24**(5), L329.
Triboulet, R., Marfaing, Y., Cornet, A., and Siffert, P. (1974). *J. Appl. Phys.* **45**(6), 2759.
Verger, L., Cuzin, M., Gaude, G., Glasser, F., Mathy, F., Rustique, J., and Schaub, B. (1992). *Nucl. Instr. and Meth.* **A322,** 357.

Warburton, W. K., Iwanczyk, J. S., Dabrowski, A. J., Hedman, B., Penner-Hahn, J. E., Roe, A. L., Hodgson, K. O., and Beyerle, A. (1986). *Nucl. Instr. and Meth.* **A246,** 558.
Whited, R. C., and van den Berg, L. (1977). *IEEE Trans. Nucl. Sci.* **NS-24**(1), 165.
Yoshie, O., and Kamihara, M. (1983). *Jap. J. Appl. Phys.* **22,** 621.
Zanio, K. (1978). *Semiconductors and Semimetals,* **13,** *Cadmium Telluride.* Academic Press, New York.
Zanio, K., Akutagawa, W., and Kikuchi, R. (1968). *J. Appl. Phys.* **39**(6), 2818.
Zha, M., Piechotka, M., and Kaldis, E. (1991). *J. Cryst. Growth* **115,** 43.
Zhang, H., Aoyagi, Y., Iwai, S., and Namba, S. (1987). *Appl. Phys. Lett.* **50**(6), 341.

CHAPTER 14

Electronics for X-Ray and Gamma Ray Spectrometers

Jan S. Iwanczyk and Bradley E. Patt

XSIRIUS, INC.
CAMARILLO, CALIFORNIA

I. Introduction 531
II. Electronic Noise Limited Systems 534
   1. The Charge Sensitive Preamplifier and Sources of Electronic Noise 536
   2. Configurations for Low Noise Preamplifiers 543
   3. Low Noise FET Structures 547
III. Statistical Noise Limited Systems 548
IV. Trapping Noise Limited Systems 549
   1. Specialized Electronics Accommodating Long Shaping Times 552
   2. Single Carrier Techniques 555
   3. Charge Deficit Correction 557
V. Miniaturized Electronics and Multielement Systems 558
   References 559

I. Introduction

Low noise, high resolution spectrometers based upon Si(Li) and Ge detectors have been used in many laboratory applications. Their wide use in portable field instruments, industrial applications, space applications, and even analytical laboratory instrumentation has, however, been limited because of the difficulties associated with achieving the appropriate level of cooling for these detectors in a manner that is amenable to field, industrial, space, and laboratory applications and because of the cost and labor involved in their maintenance and operation. The development of detectors and electronics based upon compound wide bandgap and high Z semiconductors is driven by the potential of these materials to operate as spectrometer grade nuclear radiation detectors at or near room temperature, thus overcoming the necessity for cryogenic cooling, the associated apparatus, and the maintenance labor.

The required properties for room temperature detectors are discussed elsewhere in this book. The best candidates are HgI_2, CdTe, GaAs, and $Cd_{1-x}Zn_xTe$.

The operation of a solid state detector is analogous to that of an ionization chamber in which charge generated by the absorption of radiation can be collected to form a signal by which radiation is detected. Moreover, spectrometry applications require the capability of measuring a signal that is proportional to the energy of the incident radiation. The classical approach to spectrometer design has been to integrate the induced current arising form the movement of electrons and holes in the detector material under the influence of an applied electric field. The general strategy for developing optimum electronics for use with these detectors is to develop a low noise, charge sensitive preamplifier to "collect" all of the charge resulting from ionization in the detector. The signal thus produced is then shaped using a suitable amplifier designed to give near optimum filtering for the best signal to noise ratio (SNR). However, the SNR may be compromised to provide a signal compatible with the digital to analog conversion circuitry used to determine the signal amplitude. The details of the amplifier design are based upon the shape of the pulses from the preamplifier. A typical spectroscopy system is presented in Fig. 1.

This general strategy is in fact an oversimplification of the problem. To achieve optimum performance from the electronics, they must be tailored to the specific electrical properties of the detector within the energy range that the spectrometer will be applied, as well as to the specific application. The factors limiting performance will be shown to be very much related to the spectral region of interest for the detector. Under certain conditions it will be better not to attempt to collect all of the charge produced by ionization of the incoming radiation. A strategy for obtaining the best performance from a given detector that depends upon the specific energy range of application for the detector will be outlined.

For many spectroscopy applications, the "energy resolution" is one of the most important parameters. The energy resolution is a direct measure of the capability of the spectrometer to resolve photopeaks that are close in energy. For example, in x-ray fluorescence applications this determines the capability among other things to differentiate between proximal elements in the periodic table. The energy resolution is typically expressed as the full-width at half-maximum (FWHM) of a peak in the pulse height spectrum. The FWHM is usually given in hundreds of

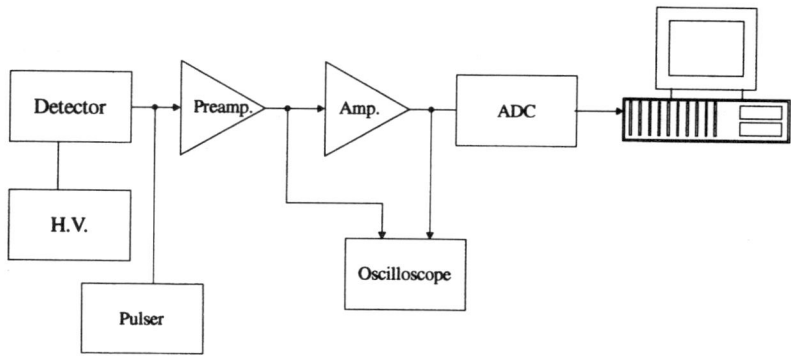

FIG. 1. Block diagram of typical laboratory spectroscopy system.

electron volts (eV) for x-ray spectrometers and in percent of peak position for gamma ray spectrometers. Many factors influence the energy resolution. The broadening of the spectrum can be represented by the quadrature sum of the contributions due to each of these factors (this applies to all uncorrelated terms). In a properly set up spectroscopy system the predominant factors are (Dabrowski et al., 1983)

1. The electronic noise due to the input stage of the preamplifier including the capacitance and leakage current of the detector itself,
2. The broadening of the linewidth due to the stochastic nature of the generation of charge,
3. The fluctuations due to trapping of charge carriers in the detector.

Other factors influencing the energy resolution include the escape of electrons produced near the surface during the charge generation process, pulse pile-up effects (especially in high count rate applications), edge effects due to both the finite size of the detector and the geometry of the electrodes, microphonics, and time dependent polarization in the detector. These effects can to some extent be minimized by the selection of detector, proper detector fabrication techniques, and correct packaging of the detector and electronics.

The standard deviation of the distribution resulting from the three predominant factors just identified relates to the noise linewidth. It is characterized by the variance of each of these processes in the following way

$$\sigma(E) = [\sigma^2(E_e) + \sigma^2(E_s) + \sigma^2(E_t)]^{1/2} \qquad (1)$$

where

$\sigma^2(E_e)$ is the variance in the collected charge due to the electronic noise in the spectrometer,

$\sigma^2(E_s)$ is the variance of the collected charge due to the statistical nature of the generation of charge carriers in the detector, and

$\sigma^2(E_t)$ is the variance of the collected charge due to the probability of trapping of charge carriers.

The variation of these components with energy is shown in Fig. 2. The electronic noise component is independent of energy. It is symmetrical and due to several noise sources arising from the detector, the components and materials used in the input stage of the preamplifier, the placement of the components, and specific parameters of the shaping amplifier. The statistical noise is proportional to the square root of the energy of the incident radiation. This term also produces a symmetrical broadening of the photopeak. The spread due to charge loss associated with trapping is more complex than the other two terms. In the case of low energy photons, interactions occur in a shallow region near the entrance electrode and the charge collection is dominated by a single carrier. Biasing is chosen to

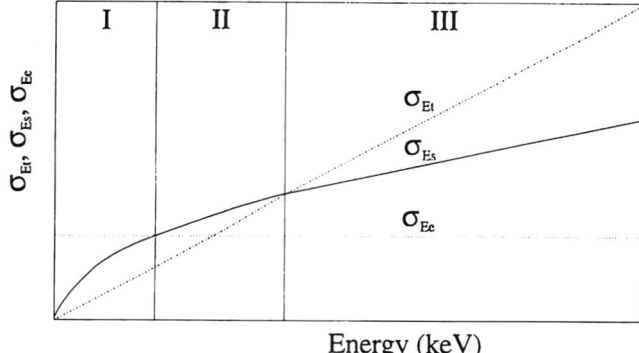

FIG. 2. Contributions of various noise sources.

favor the more mobile carrier. In this case, surface recombination/trapping, detrapping and the spread in drift lengths caused by inhomogeneities in the detector material play a major role in the trapping component. The particular shape of the photopeak will be affected by this spread and will be energy dependent. At higher energies interactions occur throughout the detector volume and the geometrical aspect of trapping plays a major role. The resulting shape of the photopeak is dependent on the energy of the incident radiation.

Because of the energy dependent characteristics of the various components of peak broadening as shown in Fig. 2, a different source of resolution broadening dominates in each of the three energy ranges: I, low energy region; II, intermediate energy region; and III, high energy region. In region I, at the low energies, the resolution is limited by the electronic noise. Whereas the electronic noise remains constant (it is independent of energy), the statistical term is proportional to the square root of the energy; and at some point in the intermediate energy region, region II, the resolution becomes limited by the statistical broadening. For high energies (region III), the resolution is always limited by charge trapping.

Based on these very different factors influencing the spectral response of compound semiconductor detectors in the low, intermediate, and high energy regions, the goal in this chapter is to discuss the dominant sources of noise in each of the energy regions and lay out a strategy for enhancing the spectral response of detectors whose applications are in each of these energy regions. Specific ways to develop electronics optimized to minimize the detrimental effect of the dominant spectral broadening in each of the different energy application regions will be identified.

II. Electronic Noise Limited Systems

In the low energy x-ray region (<20 keV) the x-ray penetration is very shallow (within 100 μm of the surface) for the high Z semiconductor materials (e.g., HgI_2

and CdTe). If the detectors are made thick enough that the x-ray penetration depth is small compared with the detector thickness, the system will be virtually reduced to single carrier charge collection. Because the mobility–lifetime product ($\mu\tau$) for electrons is much better than for holes, the electrons exhibit significantly less trapping; and it thus makes sense that the detector should be biased so that the charge collected is mainly due to electrons (entrance electrode negatively biased). In this case, the contribution to the energy linewidth due to incomplete charge collection can be relatively low (Dabrowski and Huth, 1978). In addition, the statistical broadening in this energy range is less than 215 eV for photon energies up to 20 keV (using $F = 0.1$ in Eq. (26) (Iwanczyk et al., 1984)).

The result of these factors coupled with low detector capacitance and the exceptionally low leakage currents exhibited by compound semiconductors even when operated near room temperature forms the basis for x-ray spectroscopy without the need for cryogenic cooling. The factor limiting the resolution of x-ray spectrometers has and continues to be associated with the electronic noise. In this regard much of the work in x-ray spectrometer development has been directed toward the development of ultralow noise electronics and, in particular, the first amplification stage, with the result that these systems can operate at or near room temperature and provide performance comparable to cryogenically cooled systems. Figure 3 shows a Mn K_α spectrum with resolution of 198 eV (FWHM) obtained with a 5 mm^2 HgI$_2$ detector operated at about 0° C and with the input field effect transistor (FET) cooled to about $-40°$ C.

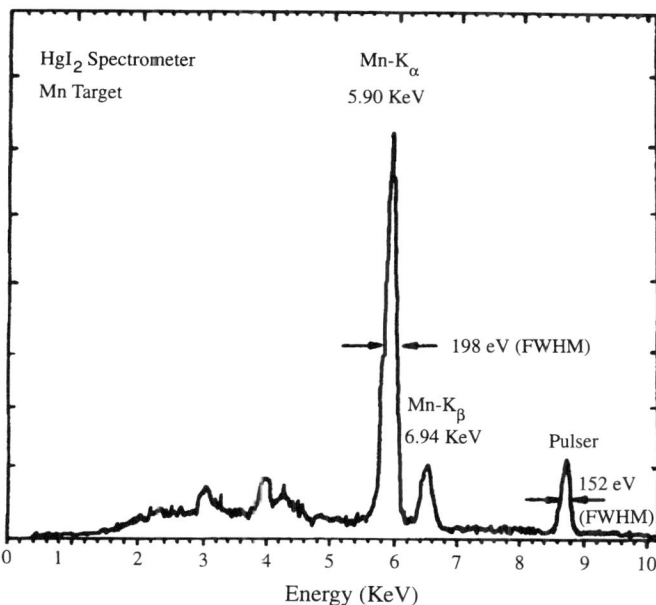

FIG. 3. Mn K_α spectrum obtained with a HgI$_2$ detector (from Iwanczyk et al., 1989).

In the next subsection we will discuss the configuration of the charge sensitive preamplifier, the origin and impact of the main sources of noise, and what can be done to reduce the impact of these noise sources.

1. THE CHARGE SENSITIVE PREAMPLIFIER AND SOURCES OF ELECTRONIC NOISE

The typical configurations, ac coupled and direct coupled, of the charge sensitive preamplifier are shown in Figs. 4a,b, respectively. The detector is biased by a high voltage supply that is conditioned externally by passive filtering. The preamplifier consists of a sensing capacitor C_f that produces the input charge to output voltage conversion and a feedback resistor R_f that is used in the direct coupled circuit to stabilize the dc operation of the preamplifier and provide a pathway by which the detector and FET leakage current and the current generated by the x-rays may be drained. Last, the preamplifier has a substantial open loop gain $A(f)$ associated with it.

In Fig. 5 we show the signal equivalent circuit for the detector–preamplifier configuration in Fig. 4b. The parallel shunt resistance R'_p is due mainly to the detector resistance, and the capacitance C'_p is due to the detector capacitance (0.5–3 pF for x-ray detectors), the input FET capacitance (3–5 pF for 2N4416 FETs), and the stray capacitance at the input. The output signal V_O due to the charge Q_{in}

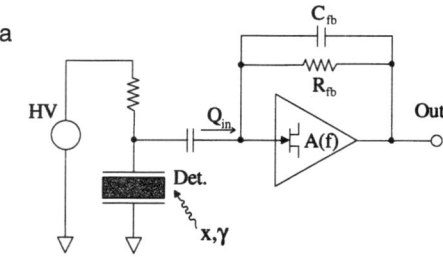

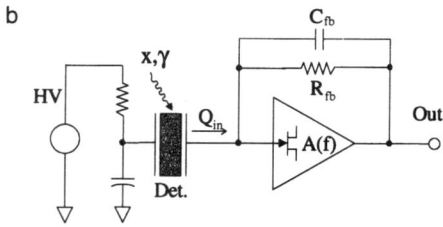

FIG. 4. Schematic of charge sensitive preamplifier in the (a) ac coupled and (b) direct coupled configurations.

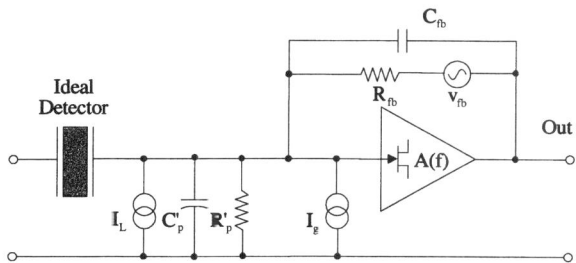

FIG. 5. Schematic diagram of preamplifier signal equivalent circuit.

deposited in the detector is given by

$$V_O \approx -Q_{in}/C_f \qquad (2)$$

where $Q_{in} \approx eE/w$, e is the charge on an electron ($e = 1.6 \times 10^{-19}$ coulombs), E is the energy of the incident radiation, and w is the mean energy required for creation of an electron–hole pair (in eV). Also shown in Figure 5 is the detector leakage current I_L and the FET gate leakage current I_g.

The electronic noise related to the preamplifier is due to four main sources: (1) series "white" noise, (2) parallel "white" noise, (3) excess "$1/f$" noise, and (4) generation–recombination noise caused by traps present in the gate depletion region of the input field effect transistor.

Each of these noise sources can be considered as a generalized noise source characterized by its power spectral density $\mathbf{S}_j(f)$ (Llacer, 1975). The noise sources are filtered by the preamplifier and amplifier network which have a combined transfer function $H_j(f)$, resulting in an output power spectral density

$$\mathbf{S}_{O_j}(f) = \mathbf{S}_j(f)|H_j(f)|^2. \qquad (3)$$

The noise power at the output due to the jth noise source is the jth output power spectral density integrated over the entire spectrum

$$W_j = \int_{-\infty}^{\infty} S_{O_j}(f)df \qquad (4)$$

which by substitution of (3) into (4) and transformation to the time domain yields an expression for the output noise power in terms of the autocorrelation of the noise source

$$W_j = \int_{-\infty}^{\infty} R_j(\tau) \int_{-\infty}^{\infty} h_j(t)h_j(t+\tau)dtd\tau \qquad (5)$$

where $R_j(\tau)$ is the autocorrelation of the jth noise source, and $h_j(t)$ is the impulse response (inverse Fourier transform of the transfer function $H_j(f)$) for the jth noise source, which is dependent on where the noise source occurs in the network.

Relationships for each of these noise sources in terms of the noise power as given in (5) will be developed, each of which will have the form

$$W_j = G \langle N_j^2 \rangle \qquad (6)$$

where G_j is a function of the characteristics of the jth noise source and the network describing it's input–output response, and $\langle N_j^2 \rangle$ is a coefficient dependant on the pulse shaping used in the amplifier.

Figure 6 shows the schematic representation of the preamplifier model incorporating the fundamental noise sources. In Fig. 6 we model the detector as an ideal noiseless, zero capacitance, and infinite resistance source producing the charge Q_{in}. The noise due to the detector leakage current is modeled as the parallel current source i_L. The detector capacitance is represented by the shunt capacitance C_d. The detector resistance is modeled by the shunt resistor R_p in series with the parallel resistance noise source ν_p. There is a term due to the detector series resistance R_{sd} and corresponding series noise voltage source ν_{sd}, and similar term A_{sd} representing the detector series $1/f$ resistance.

The parallel $1/f$ noise is modeled by the Norton equivalent current source in parallel with a frequency dependent factor modeled by the frequency dependent resistance $A_{1/f}$.

The preamplifier contributes a series noise term ν_s due to the FET channel resistance R_s and series $1/f$ noise, which is represented by the term A_s. The noise due to the FET gate leakage is modeled as the parallel current source i_g. The FET input capacitance is represented by the parallel capacitance C_g. Last, there is a voltage noise source ν_{fb} due to the feedback resistor R_{fb}. With the noise sources removed from the preamplifier in this way, the preamplifier itself is modeled as a noiseless gain $A(f)$.

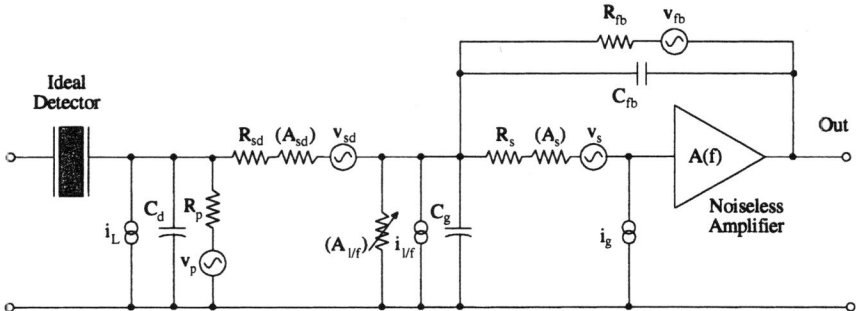

FIG. 6. Noise circuit for the charge sensitive preamplifier.

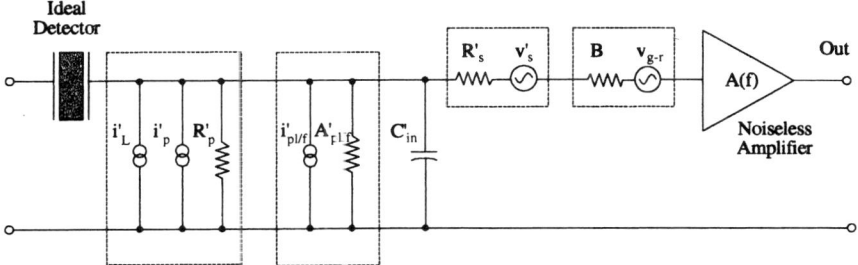

FIG. 7. Noise equivalent circuit for the charge sensitive preamplifier.

a. Series "White" Noise

Series "white" noise is due to the detector series resistance and fluctuations in currents flowing in the channel of the input FET of the preamplifier, which produce short (delta) current pulses in the output of the preamplifier. Hence, this type of noise has been referred to as "shot noise" or "delta noise." In Fig. 7 the detector series noise and the FET series noise are combined by defining a single series noise resistance R'_s (Llacer, 1975)

$$R'_s = R_s + \frac{R_{sd}C_d^2}{C_{in}^2} \tag{7}$$

where C_{in} is the total input capacitance. The noise source v'_s associated with the combined resistance R'_s has power spectral density (Johnson, 1928; Nyquist, 1928)

$$\frac{v_s^2}{\delta f} = 4kTR'_s. \tag{8}$$

Solving for the noise power in (5) we can represent the noise power due to the series noise sources (Gillespie, 1953; Llacer, 1975) as

$$W_s = 2kTR'_s C_{in}^2 \langle N_s^2 \rangle \tag{9}$$

where $\langle N_s^2 \rangle$ is a coefficient proportional to $1/\tau$, where τ is the peaking time of the postamplifier and the particular pulse shaping network used. Values of $\langle N_s^2 \rangle$ for single RC–CR, seventh order Gaussian, and triangular shaping are (Goulding and Landis, 1982)

$$\langle N_s^2 \rangle = \begin{cases} 1.87/\tau & \text{RC–CR (semi-Gaussian)} \\ 2.53/\tau & \text{Seventh order Gaussian} \\ 2/\tau & \text{Triangle.} \end{cases} \tag{10}$$

From (9) and (10) we see that, for a given type of pulse shaping and for a given peaking time, to maximize the signal to parallel noise ratio, we must minimize the parallel capacitance C_{in} and the series resistance R_s. Because the FET series resistance R_s is proportional to $1/g_m$ ($R_s = 0.67/g_m$ according to Van der Ziels equation, where g_m is the FET transconductance), the selected FET should in addition to low capacitance also have large g_m. To reduce stray capacitance, the input stage should be designed so that the FET is external to the preamplifier and moreover in as close proximity to the detector as possible.

Cooling the FET can significantly reduce the series noise. This improvement comes about because the FET equivalent noise voltage is directly dependent on temperature and also the transconductance of the FET increases with decreasing temperature. Cooling may be easily achieved by using a miniature thermoelectric cooler with significant improvement of the noise; however, it is not possible to obtain the optimum operating temperature using thermoelectric cooling. For example, the optimal temperature for the 2N4416 FET is 140–180 K, which requires cryogenic cooling.

b. Parallel "White" Noise

Parallel "white" noise is associated with fluctuations in the current flowing in the detector and the FET input circuit (gate source) (Goulding, 1972), which look like delta currents that are integrated by the input capacitance to produce voltage steps at the preamplifier input. Hence, this type of noise is referred to as "step noise." The main sources of this noise are the fluctuations in the detector leakage current and input FET gate current, and the generation of parallel noise resistance.

i. Leakage Current Noise. Because the detector parallel resistance R_p and the reactance due to the detector $1/2\pi f C_d$ are much greater than the detector series resistance R_{sd} at frequencies within the passband of typical systems, I_L and I_g are essentially in parallel and can be combined into the single parallel current I'_L as shown in Fig. 7

$$I'_L = I_L + I_g. \tag{11}$$

The power spectral density of the fluctuations in current I'_L is (Schottky, 1918)

$$\frac{i'^2_L}{\delta f} = 2qI'_L \tag{12}$$

from which we can represent the noise power due to the series noise source as (Goulding and Landis, 1982; Llacer, 1975)

$$W_{I_p} = qI'_L \langle N_p^2 \rangle \tag{13}$$

14. ELECTRONICS FOR X-RAY AND GAMMA RAY SPECTROMETERS

where $\langle N_p^2 \rangle$ is a coefficient proportional to the peaking time, τ, and dependent on the particular pulse shaping used. Values of $\langle N_p^2 \rangle$ for single RC–CR, seventh order Gaussian, and triangular shaping are (Goulding, 1972)

$$\langle N_p^2 \rangle = \begin{cases} 1.87\tau & \text{RC–CR (semi-Gaussian)} \\ 0.67\tau & \text{Seventh order Gaussian} \\ 0.67\tau & \text{Triangle.} \end{cases} \quad (14)$$

Cooling the detector and FET can help to reduce I_L and I_g and the associated leakage noise.

ii. Parallel Resistance Noise. The parallel resistors R_p and R_{fb} both generate noise terms represented by the voltage noise sources v_p and v_{fb}. Each of these sources has associated with it a power spectral density given by $v^2/\delta f = 4kTR$, which can be converted to a Norton equivalent current source $i = v/R$ with power spectral density of $i^2/\delta f = 4kT/R$. If we form a parallel combination of the current noise sources associated with R_p and R_{fb}, the spectral power density is

$$\frac{i_p'^2}{\delta f} = \frac{4kT}{R_p'} \quad (15)$$

where R_p' is the parallel combination of R_p and R_{fb} as shown in Fig. 7.

The noise power due to R_p' can be expressed as

$$W_{R_p'} = \frac{2kT}{R_p'} \langle N_p^2 \rangle \quad (16)$$

where $\langle N_p^2 \rangle$ is the coefficient defined in (14).

The parallel noise terms due to the detector and FET leakage currents (13) and the parallel resistance (16) can be combined, and the resulting term for the noise power due to parallel "white" noise sources is

$$W_p = \left[\frac{2kT}{R_p'} + qI_L' \right] \langle N_p^2 \rangle. \quad (17)$$

c. Excess "1/f" Noise

Excess $1/f$-type noise arises from the detector, feedback resistor, FET, and other components and materials used in the detector interface and the electronic circuits.

Selection of materials in contact (or close proximity) with the input such as dielectrics used for the detector substrates, and for encapsulating the FET is critical, as these materials exhibit dielectric "losses" (Radeka, 1968) that appear in parallel with the input and contribute excess parallel $1/f$ noise. These dielectrics may be considered passive frequency dependent conductances $G(f)$ that generate a thermal noise and appear as noise current sources with the power spectral density

$$\frac{i_\epsilon^2}{\delta f} = 4kTG(f) \tag{18}$$

which gives rise to a noise power term due to the dielectric

$$W_\epsilon = \frac{kTG_0}{2\pi^2 f_0} \langle N_\epsilon^2 \rangle \tag{19}$$

where G_0 is the conductance measured at frequency f_0 and $\langle N_\epsilon^2 \rangle$ is a coefficient that is dependent on the transfer function of the network. PTFE is one of the materials of choice for minimizing $1/f$ noise associated with dielectrics. To reduce $1/f$ noise associated with the as-bought encapsulated FETs (Goulding, 1977) used in the preamplifier input, it has become a standard practice in low noise preamplifier implementation to use unencapsulated (or specially reencapsulated) FETs.

Excess noise also results from the feedback resistor. This is due to the construction of the resistor, its method of fabrication, the particular materials used, and nonuniformity of the resistance film. Excess noise varies by up to an order of magnitude for the same value resistors made by different manufacturers.

To be consistent with the nomenclature defined in (6) the parameter $A_{1/f}$ is defined to represent all of the $1/f$ noise sources in the circuit, and $\langle N_{1/f}^2 \rangle$ to represent the associated coefficient in the noise power term for the $1/f$ noise

$$W_{1/f} = A_{1/f} \langle N_{1/f}^2 \rangle \tag{20}$$

The $1/f$ noise has no dependency on the peaking time, but depends strongly on the type of filter used in the shaping amplifier. Cusp, transversal, and time variant filters have all been suggested to improve the $1/f$ characteristics of the shaping amplifier, however each has its limitations in implementation.

d. Generation–Recombination Noise

A major factor in the selection of FETs for noise limited electronics is the variability in the generation–recombination noise of the FET, which is caused by carrier traps in the gate depletion layer (Sah, 1964). The noise power is specified by (Goulding, 1977)

$$W_{g-r} = BC_{in}^2 \langle N_{g-r}^2 \rangle \tag{21}$$

where $\langle N_{g-r}^2 \rangle$ is a function of the peaking time τ and the trapping time τ_t, but is independent of the particular shaping network used, B is a constant, and C_{in} is the total input capacitance.

e. Electronic Noise Linewidth

The noise linewidth can be presented as a function of the noise power terms (9), (17), (20), and (21) derived for each of the sources described. The noise power is normalized with respect to w, the mean energy required to create an electron–hole pair (in eV) for the particular detector material and given in terms of the FWHM by (Llacer, 1975)

$$\Delta E_n = \frac{2.355 w}{e} \left\{ \sum_j W_j \right\}^{1/2} \tag{22}$$

where ΔE_n is the linewidth (FWHM) due to the electronic noise (in eV), and e is the electron charge. Thus

$$\Delta E_n = 2.355 \frac{w}{e} \left\{ \left[qI'_L + \frac{2kT}{R'_p} \right] \langle N_p^2 \rangle + 2kTR'_s C_{in}^2 \langle N_s^2 \rangle \right. \\ \left. + A_{1/f} \langle N_{1/f}^2 \rangle + BC_{in}^2 \langle N_{g-r}^2 \rangle \right\}^{1/2}. \tag{23}$$

As has been shown, the $1/f$ noise term is a flat function of the shaping time, the series noise is proportional to the inverse of the shaping time and thus it is dominant at very short shaping times, whereas the parallel noise term is proportional to the shaping time and thus is dominant at very long shaping times. Because these terms add in quadrature, the total noise minimum occurs at the particular shaping time for which the series and parallel noise terms are equal. The effect of the $1/f$ term is then to increase the total noise value in the region of the noise minimum.

2. Configurations for Low Noise Preamplifiers

The resistor feedback preamplifier was used for the discussion of the noise sources in the preamplification stage of electronic noise limited spectrometer systems.

The feedback resistor is used in this circuit to stabilize the dc operation of the amplifier and provide a pathway by which the leakage current of the detector and FET may be drained (see Fig. 4). The aim of the alternative configurations is to provide a mechanism for discharging C_f while eliminating the noise associated with the feedback resistor.

Drain feedback, "injection electrode" feedback, dc light coupled feedback, and pulsed light feedback (PLF) preamplifiers have been investigated as replacements for the conventional resistor feedback preamplifier. Schematic diagrams of the various circuit configurations are shown in Fig. 8. Selection of a particular configuration will depend on the specific application and the premium placed on ob-

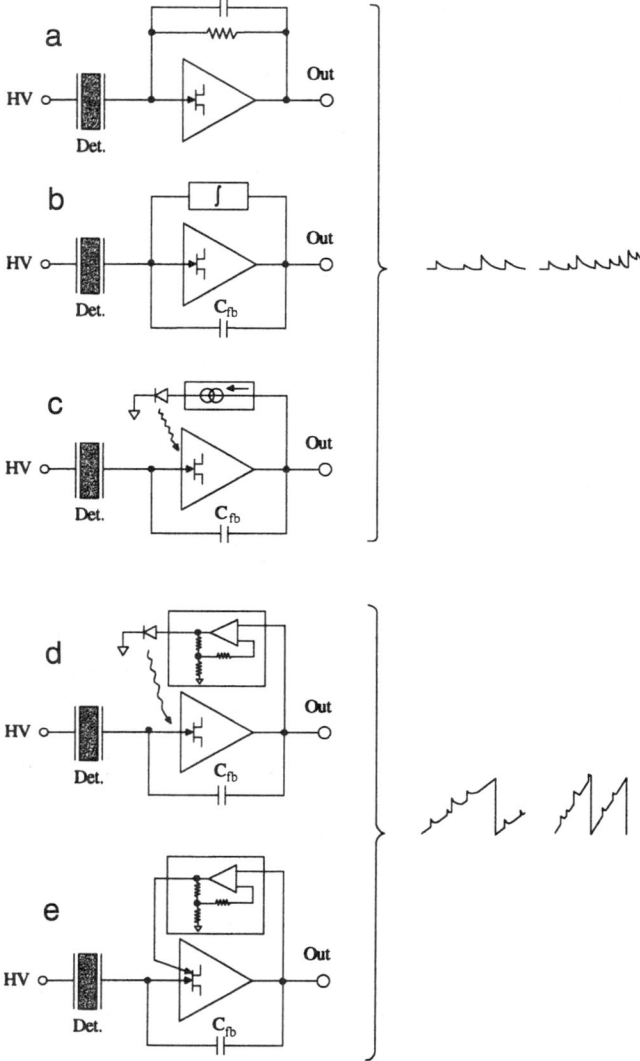

FIG. 8. Block diagram schematics of preamplifier circuits: (a) resistor feedback; (b) drain feedback; (c) dc light feedback; (d) pulse light feedback; (e) "injection" feedback. The shape of the output of each preamplifier is shown at the right for both low and high count rates.

jectives such as electronic noise, throughput, crosstalk between neighboring channels, or simplicity of the design and implementation.

Of the configurations shown in Fig. 8 the pulsed light feedback preamplifier has received the most attention because (1) no additional steady state current is flow-

14. ELECTRONICS FOR X-RAY AND GAMMA RAY SPECTROMETERS

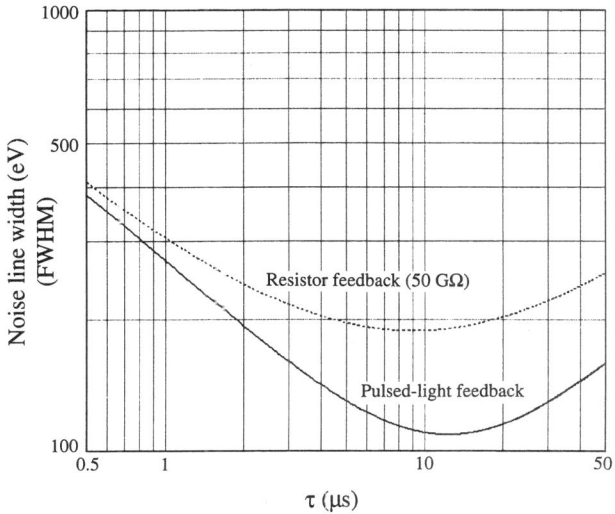

FIG. 9. Calculated noise linewidths (FWHM) versus peaking time for HgI_2 with resistor feedback and pulsed light feedback preamplifiers.

ing in the FET, (2) the amplifier may be gated off during the light pulse thus avoiding any distortions in the output signal, and (3) it provides the capability for high count rates. The calculated noise linewidths (FWHM) versus peaking time are plotted in Fig. 9. Equation (23) was used with $I_D = 1$ pA, $I_g = 0.2$ pA, $R_p = 10^{15}$ Ω, $R_{fb} = 50$ GΩ, $g_m = 5$ mS, $R_s = 0.67/g_m$, $R_{sd} = 10$ Ω, $C_d = 0.75$ pF, $C_{in} = 3$ pF, and $w = 4.2$ (HgI_2). A value of 140 eV (FWHM) was used for the excess $1/f$ noise due to the feedback resistor, the generation–recombination noise was ignored, and triangular shaping was assumed. Minimum noise linewidths of 190 eV for the resistor feedback, and 110 eV for the pulsed light feedback preamplifiers were obtained at the respective optimum shaping times. Using the PLF technique for HgI_2 detectors electronic noise better than 160 eV has been obtained. The difference between the theoretical and practically obtained result being due to excess noise in the other components which was not accounted for in the theoretical calculation.

The schematic diagram of the PLF circuit is shown in Fig. 10. The early designs of PLF preamplifiers used a separate photodiode (Goulding, Walton, and Malone, 1969) in the input stage, which was optically coupled to a light-emitting diode (LED). It was immediately recognized that this added stray capacitance and additional leakage. The input stage of the modern PLF (Iwanczyk *et al.*, 1981, 1987) consists of a low noise, high transconductance FET such as the 4416 or the Interfet SNJL01, which is removed from its can and refabricated using mechanical materials such as PTFE or ceramics with low dielectric constants to reduce stray capacitance, provide low leakage current, and low moisture absorption. The LED is

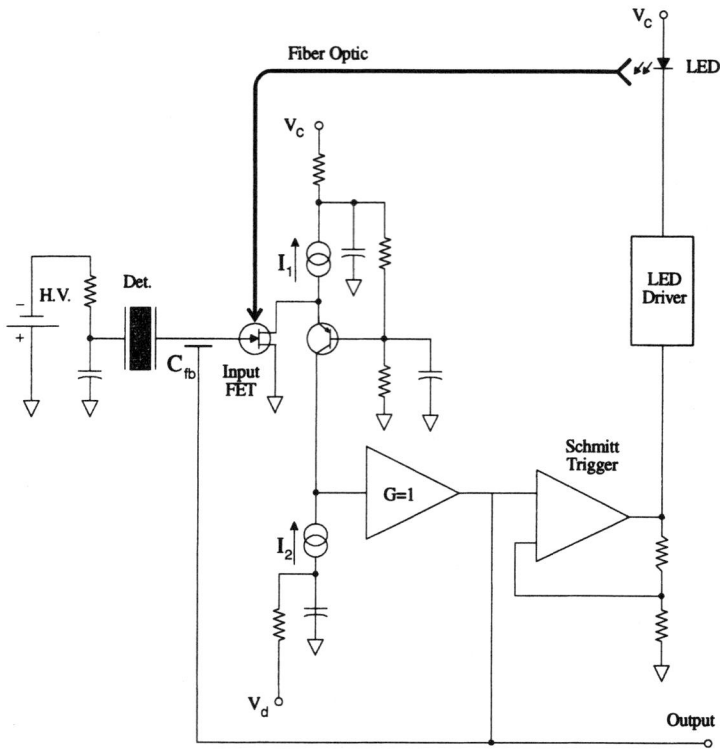

FIG. 10. Circuit diagram of the pulsed light feedback preamplifier.

optically coupled directly to the FET. The use of decanned FETs rather than the IC chips themselves is preferred because the FETs may be preselected prior to fabrication of the low noise part. The LED/FET structure can be assembled to form a modular component that can be tested separately from the rest of the preamplifier. To optimize the design of the input stage for multielement detector array systems, the LEDs may be mounted externally with an interface to the FETs via optical fibers.

The feedback capacitor C_{fb} is usually constructed in the vicinity of the input FET by bringing a wire near the FET gate. Feedback capacitance of about 0.2 pF is typically obtained in this way. In the PLF preamplifier the leakage current charges the input capacitance (essentially C_{fb}), and the output follows a constant ramp as shown in Fig. 8. Once a predetermined value set by the Schmitt trigger resistor divider is reached, the LED driver is triggered. A light pulse (*reset pulse*) is generated that shines on the input FET, and C_{fb} is discharged. This promptly brings the output back to its starting value, and the whole process starts over again. An inhibit pulse can be generated to gate off the further amplification stages during the reset pulse. Several commercial preamplifiers include the inhibit pulse generating circuitry. The preamplifier output is coupled to a standard Gaussian or triangular shaping amplifier.

The reset rate of the PLF preamplifier may be determined from the difference between the detector leakage current and the gate leakage current. The reset rate is

$$\Delta t \approx \frac{C_f \Delta V}{\Delta i} \qquad (24)$$

where C_f is the feedback capacitance, ΔV is the voltage swing of the ramp, and Δi is the difference between the detector leakage current and the FET gate leakage current.

For a detector leakage current of about 1 pA and FET gate leakage of about 0.2 pA, it can be determined that a 2 V reset swing is realized if the reset rate is about 2 Hz when C_{fb} is 0.25 pF

The current sources I_1 and I_2 in Fig. 10 greatly improve the noise performance of the preamplifier by precise regulation of the current in the first stage of the preamplifier. The pulses generated by ionization events in the detector cause a small ΔI to flow in the first stage. The input impedance of the current source I_2 is extremely large and a relatively large voltage pulse is produced at the input to the push–pull output stage.

3. Low Noise FET Structures

The standard FET (such as the 2N4416) used in low noise preamplifiers is a three electrode (source–gate–drain) structure. The FET is typically biased by setting the drain current (less than I_{DSS}) and defining the drain potential. This establishes a nonzero gate bias voltage. On the other hand, four terminal FET structures (Kandiah and White, 1981; Howes, Deighton, and Smith, 1984) with two separate gates can be operated, with the top gate and the bottom gate (substrate) biased independently. Thus the input gate may be operated at the optimal bias (for low noise) while the substrate gate is adjusted to define the drain current. This gives the lowest noise configuration. Additionally, because of the different diffusion processes forming the two gate junctions, the top gate has a lower input capacitance than the substrate gate. Thus the tetrode FET provides higher g_m/C_{in} (transconductance/input capacitance). The noise voltage is typically higher (R_s in (7) is slightly higher), but because of the lower capacitance the product $R'_s C_{in}$ in the series noise term (9) is lower (by up to 20%). Further improvement has been achieved through the development of coaxial FET configurations. The coaxial structure may provide further reduction of C_{in} (Howes, private communication).

The "Pentafet" (Nashashibi and White, 1990) combines a low noise, four element FET structure with an integrated restore mechanism that makes *electronic reset* possible by using the extra electrode to short the source–gate junction. This can potentially reduce excess noise from the dielectrics used for packaging the light feedback interconnect.

III. Statistical Noise Limited Systems

Whereas the electronic noise is constant with energy, the statistical contribution is proportional to $\sqrt{E}$. There is therefore an energy above which statistical broadening will dominate over electronic noise.

The broadening of the linewidth due to the stochastic nature of the generation of charge is described in terms of the Fano factor (Fano, 1946, 1947) and is usually determined empirically. The variance in the number of generated pairs is modeled by Poisson statistics and is given by (Restelli and Rota, 1968)

$$\sigma_s^2 = \overline{(N - \overline{N})^2} = F\overline{N} \tag{25}$$

where N is the number of generated pairs, $\overline{N}$ is the mean number of generated pairs that is proportional to the energy of the incident radiation, and F is the Fano factor, which is a scalar that is less than unity, that is introduced to account for deviation from pure Poisson statistics.

Historically, the published experimental values published for the Fano factor have decreased with improved materials and electronics development. The relationship given in (25) can be rewritten in terms of the contribution to the pulse-width as measured in terms of eV (FWHM) as

$$\Delta E_S = 2.355w\sqrt{\sigma_s^2} = 2.355\sqrt{FEw} \tag{26}$$

where ΔE_s is the energy resolution (FWHM) measured in eV, F is the Fano factor (material dependent), E is the energy of the incident radiation in eV, and w is the mean energy to create an electron–hole pair (material dependent).

Thus, for example, at 100 KeV in HgI_2 we can expect that the statistical noise linewidth will be about 483 eV (FWHM). In reality the statistical noise is rarely the limiting factor because of the tremendous amount of hole trapping in compound semiconductors. A plot of the noise contributions (similar to Fig. 2) using typical values of parameters for HgI_2 detectors ($\Delta E_n = 180$ eV, $F = 0.1$, $w = 4.2$ eV/ehp) is shown in Fig. 11. The contribution due to trapping was found from measured values of the total noise linewidth (250 eV at 5900 eV, 880 eV at 60 keV, 1370 at 88 keV, 1940 at 122 keV, and 2% at 662 keV). Clearly, in the case of HgI_2, there is no range of energies in which the statistical contribution dominates the noise. Moreover, above about 10–20 keV, trapping noise always begins to dominate. For this reason, no further discussion is needed on this topic. The effect of trapping and the strategy for designing electronics for trapping noise limited detector systems will be discussed in the next section.

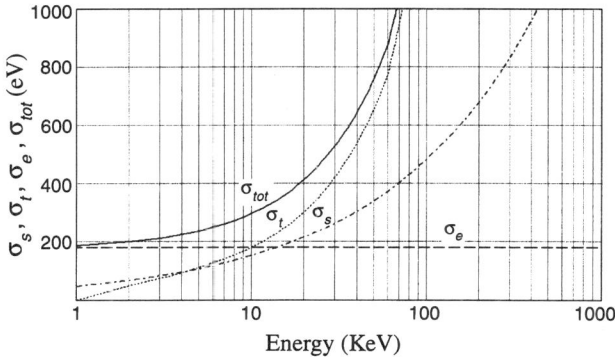

FIG. 11. Noise contributions using typical HgI_2 parameters.

IV. Trapping Noise Limited Systems

The dependence of the spectral performance of a detector on charge trapping is very complex. At low energies the degree of material inhomogeneity determines the level of the trapping effect. This is because, at low energies, the radiation interactions occur within a very narrow region near the entrance electrode, and it is simply the spread in drift lengths $\lambda = \mu\tau E$ caused by variations in electric field E, carrier mobility μ, and trapping time τ, associated with volume nonuniformities produces a random charge loss component (Dabrowski, 1982). The energy spectrum resulting from single charge carrier collection when the electric field in the detector is uniform and detrapping is negligible is governed by the Hecht relation (Mayer, 1968). The energy resolution under these conditions is proportional to energy.

For higher energies interactions occur throughout the volume of the detector, and the geometrical aspect of trapping plays a dominant role. This is particularly evident in the high Z wide bandgap materials (e.g., HgI_2, CdTe) most of which show very large disparities in the mobility–lifetime products ($\mu\tau$) and consequently exhibit large differences in the mean drift lengths for electrons and holes, resulting in a geometrical trapping effect dependent on the depth of interaction. The geometrical effect is thus proportional to the incident radiation energy (Akutagawa and Zanio, 1969; Zanio, 1970; Bell, 1971) and is the dominant source of peak broadening at high energies.

Thus in the high energy range spectroscopy is always limited by degraded charge collection efficiency, due mostly to poor hole mobility and short hole trapping times. A 1 cm thick HgI_2 detector at the maximum practical bias of 10,000 V will have hole transit times $T_h = L/\mu_h E > 25$ μsec ($L = 1$ cm, $\mu_h = 4$ cm^2/V-sec, $E = 10,000$ V/cm) exceeding the corresponding trapping time (2–20 μsec).

Thus the hole collection is very poor, and the resulting energy spectra exhibit peak broadening with large tails toward low energies.

This effect is exaggerated in practical application for gamma ray spectroscopy because thicker detectors are needed in this energy range to ensure the required absorption efficiency. Moreover, even if trapping were not such a great problem, extremely long shaping times would be needed to collect the holes for events corresponding to the maximum transit times. The use of sufficiently long shaping times for these long transit time events would impose severe limitations on throughput. The result of these factors is that in very thick detectors the hole contribution to the total collected charge is typically very low.

On the other hand, the value of $\mu\tau$ for electrons is much larger than it is for holes in the compound semiconductors (Armantrout et al., 1977; Bube, 1978). The electron transit times $T_e = L/(\mu_e E) \sim 1$ μsec ($L = 1$ cm, $\mu_e = 100$ cm^2/V-sec, $E = 10,000$ V/cm) are much shorter that the corresponding electron trapping time (10 μsec), allowing for quite good collection efficiency for electrons from even thick sections of the material. In the case of HgI$_2$ good electron collection is obtained for detectors up to 1 cm thick.

The classical method of gamma ray spectrometry involves integration over time of the induced current arising from the movement of both the electrons and the holes. The shortcoming of these classical methods is that to obtain good spectral performance adequate collection of both carriers is required, and this is simply not possible in most of the compound semiconductors. In contrast, by implementing special electronic circuits that take advantage of the large disparity between the $\mu\tau$ values for the electrons and holes, it is possible to substantially improve the spectral performance of the detector system. Of course, quite different approaches are needed than in the electronic noise limited systems. To see how these special approaches are applied it is necessary to first look at the nature of the pulses produced.

The instantaneous current induced in a planar detector is composed of components due to electrons and holes,

$$i(t) = i_e(t) + i_h(t) \tag{27}$$

with

$$i_e(t) = \begin{cases} I_e e^{-t/\tau_e} & 0 < t \leq t_e \\ 0 & \text{Else} \end{cases} \tag{28}$$

$$i_h(t) = \begin{cases} I_h e^{-t/\tau_h} & 0 < t \leq t_h \\ 0 & \text{Else} \end{cases} \tag{29}$$

where $i_e(t)$ is the electron component of the current, $i_h(t)$ is the hole component of the current, I_e is the amplitude of the electron component, I_h is the amplitude

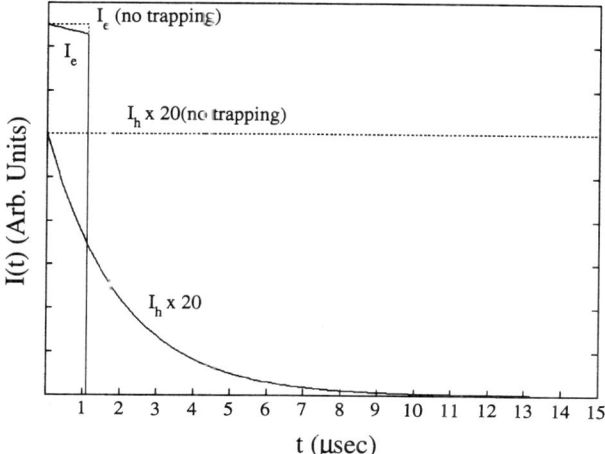

FIG. 12. Electron and hole instantaneous currents in a HgI$_2$ detector. The dashed curves represent currents in the absence of trapping, and solid curves are with trapping.

of the hole component, t_e is the electron transit time, t_h is the hole transit time, τ_e is the mean electron trapping time, and τ_h is the mean hole trapping time. As an example, the electron and hole components of the current (with and without trapping) induced by an event interacting at a depth of 5 mm in a 1 cm thick HgI$_2$ detector are shown in Fig. 12. The following values were used for the detector parameters: $\mu_e = 85$ cm^{2}V^{-1}sec^{-1}, $\mu_h = 3$ cm^{2}V^{-1}sec^{-1}, $V = 5000$ V, $\tau_e = 4 \times 10^{-5}$ sec, and $\tau_h = 2 \times 10^{-6}$ sec. The electron and hole transit times are $t_e = (D - d)D/[\mu_e V]$, and $t_h = dD/[\mu_h V]$, where d is the depth of interaction (as shown in Fig. 13) for the particular photon event, D is the detector thickness, μ_e and μ_h are the electron and hole mobilities, and V is the detector bias.

The charge pulses corresponding to the preamplifier output signals result from integration of the current pulses. The electron trapping can be neglected, as it has negligible influence on the attainable energy resolution. In this case the expres-

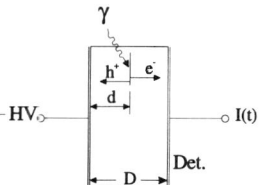

FIG. 13. Definition of terms used to describe a gamma ray interaction at depth d in a detector of thickness D.

sions for the electron and hole charge pulses are given by

$$q_e(t) = \begin{cases} 0 & t \leq 0 \\ Q_e t/t_e & 0 \leq t < t_e \\ Q_e & t > t_e \end{cases} \quad (30)$$

$$q_h(t) = \begin{cases} 0 & t \leq 0 \\ Q_h \dfrac{\tau_h}{t_h}(1 - e^{-t/\tau_h}) & 0 \leq t < t_h \\ Q_h \dfrac{\tau_h}{t_h}(1 - e^{-t_h/\tau_h}) & t > t_h \end{cases} \quad (31)$$

where $q_e(t)$ and $q_h(t)$ are the electron and hole charge–pulse components, and Q_e and Q_h are the amplitudes of the electron and hole components. The total charge is given by the sum of the electron and hole components

$$q(t) = q_e(t) + q_h(t). \quad (32)$$

1. SPECIALIZED ELECTRONICS ACCOMMODATING LONG SHAPING TIMES

Several techniques for resolution enhancement based upon adjusting the shaping time constant to mitigate ballistic deficit have been reported. It is important to note that ballistic deficit results from incomplete signal processing and not from loss of charge due to trapping in the detector. Loss of energy resolution occurs due to ballistic deficit when conventional Gaussian shaping is used in the postam-

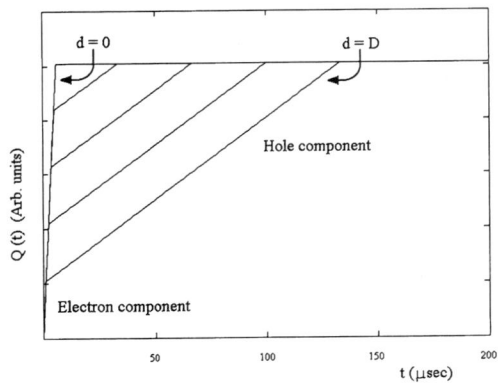

FIG. 14. Charge pulses following gamma ray interactions at different depths in the detector. The fast slope corresponds to the electron component, and the slower slope, to the hole component.

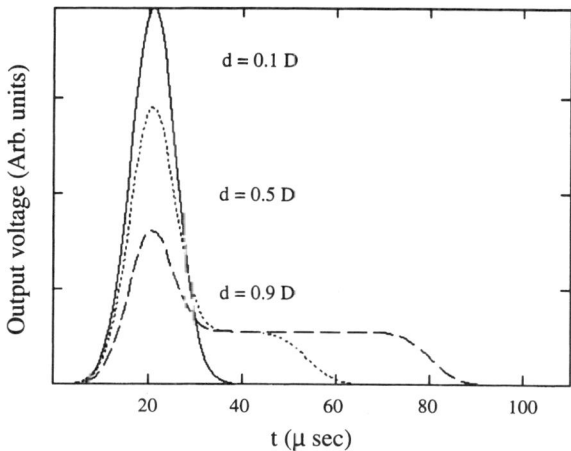

FIG. 15. Pulse shapes at the output of a Gaussian shaping amplifier with a short shaping time constant ($\sigma = 5$ μsec, $t_{h\,Max} = 70$ μsec) for an HgI_2 detector.

plifier stage. Interactions at various depths in the detector produce the pulse shapes shown in Fig. 14, which are derived from (30) and (31). Ballistic deficit then occurs when the filter time constant is smaller than the maximum charge transit time. The result is incomplete charge collection for events that occur deep within the detector that have long transit times. Simulated pulse shapes obtained by applying the pulses in Fig. 12 to the input of a Gaussian shaping amplifier with $\sigma < t_h$ are shown in Fig. 15. The ratio of the pulse amplitudes is 1 to 0.76 for pulses near the entrance electrode versus half of the detector thickness.

a. Long Shaping Time Constants

By using conventional Gaussian shaping with a shaping time constant that is long compared with the maximum carrier transit time, it is possible to eliminate ballistic deficit effects and collect all the carriers that are not trapped. The simulated pulse shapes obtained for Gaussian shaping with $\sigma = 2t_h$ shown in Fig. 16 illustrate this effect clearly. On the other hand, this technique can not be practically implemented because of the severe pulse pile-up that would occur even at moderate input count rates. Additionally, using very long shaping times results in low frequency noise problems and does not provide the optimum signal to noise ratio.

b. Stacked Detector Arrangements

Because ballistic and trapping deficits are negligible in thin detector sections, it is possible to conceive of stacked detector geometries composed of a stack of

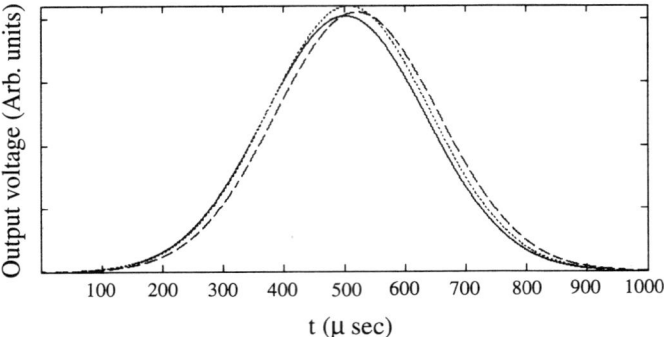

FIG. 16. Pulse shapes at the output of a Gaussian shaping amplifier with a long shaping time constant ($\sigma = 140$ μsec, $t_{h\,Max} = 70$ μsec) for an HgI_2 detector.

detector sections where each section is thin enough to eliminate ballistic and trapping deficits, yet the total thickness of the stack is sufficient to ensure good absorption efficiency. Stacks connected in both series and parallel configurations are possible; however, each has its disadvantages. Series connected stacks require extremely high voltage supplies, whereas parallel connection of detectors will result in very large input capacitance. Nonetheless, this arrangement is a serious consideration especially for high energy (>1 MeV) gamma ray spectroscopy.

c. Gated Integrators

At least two types of time variant filters are referred to in the literature as "gated integrators." Each of these two very different processing schemes is used for a different purpose. One scheme reduces the dead time and pile-up effects associated with a fixed long shaping time by gating off the integration on the tail end of the Gaussian and rapidly discharging the signal following readout. The implications for improved throughput are obvious. This type of system is especially useful for high count rate applications employing "prompt restore" preamplifiers, which generate reset pulses after each signal pulse (Kandiah, Smith, and White, 1975).

The second reference to the gated integrator involves a system that integrates the output of a conventional Gaussian shaping amplifier to reduce the effect of ballistic deficit. This scheme works because the fall time of the original pulses depends on the depth of interaction as shown in Fig. 15. If these pulses are integrated as shown in Fig. 17, the resulting pulse heights are more uniform (Konrad, 1968; Radeka, 1968, 1972) than the original charge pulses. The pulse height is determined by sampling the output of the integrator at some time $t > t_{h\,Max}$. The integration time is varied based on the pulse rise times.

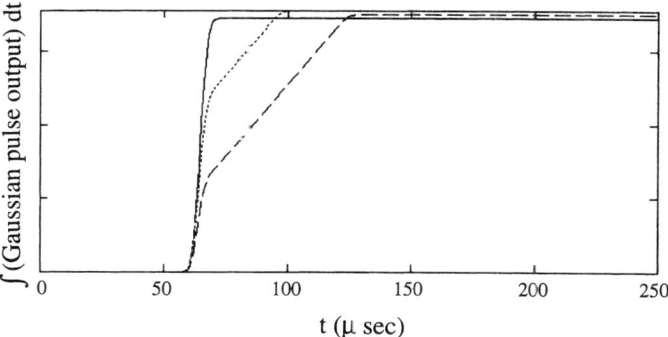

FIG. 17. Pulse shapes following integration of the pulses at the output of a Gaussian shaping amplifier.

d. *Adaptive Filters*

An interesting approach is that of an adaptive filter, in which a longer shaping time is used for interactions characterized by longer transit times (Desi, private communication). Thus, the transit time is measured on a pulse by pulse basis and the appropriate shaping time ($\tau \gg t_h$) is used. The merit of this technique is that the maximum count rate for full charge collection (in the absence of trapping) is possible because the pulse widths are only as long as they need to be.

2. SINGLE CARRIER TECHNIQUES

a. *Rise Time Discrimination*

One technique tried in the past for HgI_2 and CdTe detectors is based upon the selection of events in which the interactions occur near the negative electrode and the signal is dominated by the electron carriers, thus avoiding ballistic deficit and trapping deficit. Rise time discrimination has been used in the implementation of these systems (Jones and Woollam, 1975; Whited, Schieber, and Randtke, 1976). In practice only a fraction of the detector volume is used, and the efficiency is very poor because most of the events are rejected.

b. *Current Pulse Measurement*

The electron component can be determined by measuring the amplitude of the ionization current pulse before charge collection is completed rather than measuring the total amount of charge produced (Szymczyk *et al.*, 1983; Beyerle *et al.*,

1983). Assume that the carrier mobilities μ_e and μ_h and the electric field intensity E are independent of the positions of the drifting carriers and, furthermore, that the electron trapping is negligible, then for fixed measurement times $t = t_m$, where t_m is less than the time of electron collection. The instantaneous current flowing in the detector may be described by

$$I(t \le t_m) = I_e + I_h \tag{33}$$

Thus the current measured is constant and proportional to the energy of the incident gamma ray. This particular high energy spectral enhancement method is well suited to most of the compound semiconductors because of the relatively low electron mobilities exhibited at room temperature (compared with cryogenic detectors). Thus the mean time before trapping of the electrons is greater than the measurement time and reasonable values of t_m can be used. Because only part of the total charge is being collected, the electronic noise must be kept low to make use of this technique at lower gamma ray energies.

The circuit diagram for one of the approaches used to implement the constant current pulse concept is shown in Fig. 18. A standard resistor feedback, charge sensitive preamplifier (as described in Subsection II.1) is used to obtain the charge pulse Q. This is followed by a differentiator that produces the current pulse I. When this approach is used, the time constants for differentiation and integration in the semi-Gaussian shaping amplifier that follows should be less than the electron transit time to preserve proportionality to the current pulse rather than the charge pulse. The shorter is the shaping time chosen, the better is the approximation to the current pulse; but on the other hand, for shaping times shorter than the optimum, the noise is increased due to the wider bandwidth, and the signal is simultaneously reduced. Experimental results obtained at 662 keV for a 1 cm thick HgI_2 detector using both long (10 μsec) and short (0.5 μsec) shaping times (Iwanczyk, 1983) showed improved peak efficiency and higher peak to valley ratio with slightly degraded resolution (17%). This technique is limited by the uniformity of the electric field both spatially and with depth in the detector. This limitation can, however, be overcome by the technique described in the following subsection.

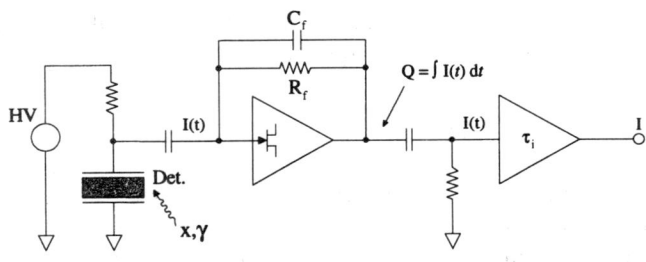

FIG. 18. Circuit for measuring constant current pulses.

c. Current Pulse Plus Depth of Interaction Measurement

A scheme based upon the preceding technique improves upon the basic concept for single carrier current spectroscopy by including the measurement of the interaction depth, d, via timing methods (Warburton and Iwanczyk, 1987). The system then extrapolates the total charge based upon the electron pulse and the depth and makes corrections for field nonuniformity. Computer simulations showed that for 1–2 cm thick HgI_2 detectors 1% resolution at 662 keV with 6–10% photopeak efficiency is in principle possible.

The major setback of this system is the complexity of the circuitry needed to adequately determine d. Jitter in the timing measurement is the major contribution to the spectral broadening. Because the timing error is a function of d, being worst when $d \approx D$, the technique can be enhanced by selectively rejecting events that occur at the near-surface regions of the detector. On the other hand, rejection of a large number of pulses degrades the counting efficiency. Mapping of $E(d)$ can be done for each detector to determine an a priori look-up table. This data can then be included on a pulse by pulse basis using the look-up table as part of the processing to improve the energy resolution. Real time implementation of the technique is possible.

d. Electron-Only Induced Signals

A somewhat different single-carrier technique involves the use of novel detector configurations for enhancing electron collection and suppressing hole collection. There are particular electronics requirements for each specific configuration like semi-spherical detectors (Iwanczyk, 1983) and drift-chambers.

3. CHARGE DEFICIT CORRECTION

a. Adaptive Gain

An interesting technique is the use of adaptive gain filters to compensate for charge loss. One adaptive filter approach attempts to minimize the geometrical effect of trapping by using an adaptive gain system, in which the gain is varied on a pulse-by-pulse basis as function of the charge transit time and, therefore, the depth of interaction to compensate for trapping and ballistic deficit (Warburton and Iwanczyk, 1987; Kurtz, 1978).

b. Electron Plus Total Charge Measurement

A hole trapping correction can be made by measuring the total charge due to both the electrons and the holes and the separate electron component. The total charge is measured by using a long shaping time, whereas the electron component

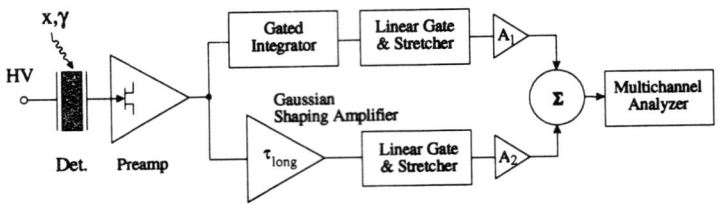

FIG. 19. Block diagram of electronics for "linear approximation" hole deficit correction.

is measured by using a short shaping time. The relative contribution of the electron component thus determined can be used to estimate the interaction depth, d. The trapping deficit due to an interaction at d can then be overcome by an empirical correction that essentially scales the data based upon the estimated depth. This can be accomplished by postprocessing the data on a computer (Finger et al., 1984). Energy resolution of 2.6% FWHM was achieved when this approach was used for a 310 μm detector that gave 8.3% FWHM using standard techniques.

The trapping correction has also been implemented in real time using a linear combination of the full charge and the electron component signals (Gerrish, Williams, and Beyerle, 1987) as shown in the "linear approximation" circuit of Fig. 19. Resolution was improved using this technique from 7.0% FWHM for a 2.3 mm HgI_2 detector at 3800 V bias without the correction to 3.1% FWHM with the correction. The peak to valley ratio improved from 2.5:1 without the correction circuit to 5.2:1 with the linear approximation circuit. A disadvantage of this system is that the technique is most useful for fairly thin detectors in which the hole trapping is not so severe to start with.

An advantage of the computer processing for hole deficit correction is that the correction is not limited to linear correction as shown in Fig. 19. On the other hand, this does not preclude one from using non-computer based electronics to implement more sophisticated corrections than the linear correction. Exponential corrections have also been applied by simply using a modified pole–zero circuit to achieve the required scaling of the pulses (Gerrish et al., 1987).

V. Miniaturized Electronics and Multielement Systems

Historically, little effort has been invested in the miniaturization of x-ray and gamma ray spectrometer electronics. The lack of motivation to miniaturize the electronics is understandable given the size and weight of the cooling systems and associated vacuum cryostat necessary for operation of Si[Li] and Ge detectors. These detector systems require cooling to liquid Nitrogen (LN_2) temperatures. Typically this has been achieved by cryogenic cooling. The cryogenic systems are large and bulky and do not lend themselves well to field use. More recently efforts have also been directed towards Peltier cooling for Si(Li) detectors. The large temperature differential (90° C) demanded of the Peltier cooled systems requires multistage coolers and the use of vacuum pumps. On the other hand, spectrometers based on high Z compound semiconductors perform optimally at or near room temperature and require no bulky cooling apparatus.

The relaxation of the cooling requirements for these types of detectors opens up a whole new area of novel spectrometer systems for portable field instruments, industrial applications, space applications, analytical laboratory instrumentation, and high density multidetector systems. The development of miniaturized and computerized amplification and processing electronics for single element and multielement systems is seen as one of the major areas for future development, which encompasses many important applications including (1) handheld instrumentation systems where the additional bulk and weight of the LN_2 are a drawback, (2) implanted detector applications where the need for miniaturization of the detector and the electronics is critical, (3) remote site operations where the minimal size and weight as well as computer control are highly desirable to enable transportation to remote sites and remote field operation of the equipment, (4) industrial quality control lines where small size, low cost, simplicity, and ruggedness are required, (5) laboratory applications where room temperature design vastly simplifies the design and construction of systems, ultimately allowing for lower overall cost, (6) large scale x-ray energy dispersive array detector systems where the absence of cryogenic requirements simplifies the overall design, allows higher packing densities, and greatly lowers the cost of such systems, (7) portable x-ray and gamma ray cameras for remote terrestrial and space imaging, and (8) miniature multielement and imaging systems for medical applications.

REFERENCES

Akutagawa, W., and Zanio, K. (1969). *J. Appl. Phys.* **40,** 3838.
Armantrout, G. A., Swierkowski, S. P., Sherohman, J. W., and Yee, J. H. (1977). *IEEE Trans. Nucl. Sci.* **NS-24,** 121.
Bell, R. O. (1971). *Nucl. Instr. & Meth.* **93,** 341.
Beyerle, A., Hull, K., Markakis, J., Lopez, B., and Szymczyk, W. M. (1983). *Nucl. Instrum. Methods* **213,** 107–113.
Bube, R. H. (1978). *Photoconductivity of Solids,* p. 268. Robert E. Krieger, Huntington, NY.
Dabrowski, A. J. (1982). *Advances in X-Ray Analysis,* 25, 1–21.
Dabrowski, A. J., and Huth, G. C. (1978). *IEEE Trans. Nucl. Sci.* **NS-25**(1), 205.
Dabrowski, A. J., Szymczyk, W. M., Iwanczyk, J. S., Kusmiss, J. H., Drummond, W., and Ames, L. (1983). *Nucl. Instrum. Methods* **213,** 89–94.
Fano, U. (1946). *Phys. Rev.* **70,** 44.
Fano, U. (1947). *Phys. Rev.* **72,** 26.
Finger, M., Prince, T. A., Padgett, L., Prickett, S., and Schnepple, W. (1984). *IEEE Trans. Nucl. Sci.* **NS-31,** 384.
Gerrish, V. M., Williams, D. J., and Beyerle, A. G. (1987). *IEEE Trans. Nucl. Sci.* **NS-34,** 85.
Gillespie, A. B. (1953). *Signal Noise and Resolution in Nuclear Counter Amplifiers,* pp. 24–29. Pergamon Press, London.
Goulding, F. S. (1972). *Nucl. Instrum. Methods* **100,** 493–504.

Goulding, F. S. (1977). *Nucl. Instrum. Methods* **142**, 213–223.
Goulding, F. S., and Landis, D. A. (1982). *IEEE Trans. Nucl. Sci.* **NS-29**(3), 1125.
Goulding, F. S., Walton, J., and Malone, D. F. (1969). *Nucl. Instrum. Methods* **71**, 273–279.
Howes, J. H., Deighton, M. O., and Smith, A. J. (1984). *IEEE Trans. Nucl. Sci.* **NS-31**(1), 470.
Iwanczyk, J. S. (1983). Jordan International Electrical and Electronic Engineering Conference, Amman, 357–363.
Iwanczyk, J. S., Dabrowski, A. J., Huth, G. C., Del Duca, A., and Schnepple, W. F. (1981). *IEEE Trans. Nucl. Sci.* **NS-28**(1), 579.
Iwanczyk, J. S., Dabrowski, A. J., Huth, G. C., and Drummond, W. (1984). *Advances in X-Ray Analysis* **27**, 405–414.
Iwanczyk, J. S., Dabrowski, A. J., Dancy, B. W., Patt, B. E., DeVore, T. M., Del Duca, A., Ortale, C., Schnepple, W. F., Barksdale, J. E., and Thompson, T. J. (1987). *IEEE Trans. Nucl. Sci.* **NS-34** (1), 124.
Iwanczyk, J. S., Wang, Y. J., Bradley, J. G., Conley, J. M., Albee, A. L., and Economou, T. E. (1989). *IEEE Trans. Nucl. Sci.* **NS-36**(1), 841.
Johnson, J. B. (1928). *Phys. Rev.* **32**(July), 97.
Jones, L. T., and Woollam, J. M. (1975). *Nucl. Instrum. Methods* **124**, 591–595.
Kandiah, K., and White, G. (1981). *IEEE Trans. Nucl. Sci.* **NS-28**(1), 613.
Kandiah, K., Smith, A. J., and White, G. (1975). *IEEE Trans. Nucl. Sci.* **NS-22**, 2058.
Konrad, M. (1968). *IEEE Trans. Nucl. Sci.* **NS-15**, 268.
Kurtz, R. (1978). *Nucl. Instr. and Meth.* **150**, 91.
Llacer, J. (1975). Second ISPRA Nuclear Electronics Symposium, Stresa, Italy, 20–23 and Lawrence Berkeley Laboratory Report # LBL-3671.
Mayer, J. W. (1968). In *Semiconductor Detectors*, ed. G. Bertolini and A. Coche. North-Holland Press.
Nashashibi, T., and White, G. (1990). *IEEE Trans. Nucl. Sci.* **NS-37**(2), 452–456.
Nyquist, H. (1928). *Phys. Rev.* **32**(July), 110.
Radeka, V. (1968a). Int. Symp. on Nuclear Electronics, Versailles, France.
Radeka, V. (1968b). *IEEE Trans. Nucl. Sci.* **NS-15**, 455.
Radeka, V. (1972). *Nucl. Instr. and Meth.* **99**, 525.
Restelli, G., and Rota, A. (1968). In *Semiconductor Detectors*, ed. G. Bertolini and A. Coche. North Holland Press, pp. 86–87.
Sah, C. T. (1964). *Proc. IEEE* **52**, 795.
Schottky, W. (1918). *Ann. Physik.* **32**, 541.
Szymczyk, W. M., Dabrowski, A. J., Iwanczyk, J. S., Kusmiss, J. H., Huth, G. C., Hull, K., Beyerle, A., and Markakis, J. (1983). *Nucl. Instrum. Methods* **213**, 115–122.
Warburton, W. K., and Iwanczyk, J. S. (1987). *Nucl. Instr. and Meth. in Phys. Res.* **A254**, 123–128.
Whited, R. C., Shieber, M. M., and Randtke, P. T. (1976). *J. Appl. Phys.* **47**, 2230.
Zanio, K. (1970). *Nucl. Instr. & Meth.* **83**, 288.

CHAPTER 15

Summary and Remaining Issues for Room Temperature Radiation Spectrometers

Michael Schieber

SANDIA NATIONAL LABORATORIES
LIVERMORE, CALIFORNIA
(AND HEBREW UNIVERSITY OF JERUSALEM)

R. B. James

ADVANCED MATERIALS RESEARCH DEPARTMENT
SANDIA NATIONAL LABORATORIES
LIVERMORE, CALIFORNIA

T. E. Schlesinger

DEPARTMENT OF ELECTRICAL AND COMPUTER ENGINEERING
CARNEGIE MELLON UNIVERSITY
PITTSBURGH, PENNSYLVANIA

I. INTRODUCTION 561
II. MATERIALS REQUIREMENTS 562
III. ISSUES IN HgI$_2$ DETECTOR TECHNOLOGY 563
   1. *Precursors and Starting Materials* 564
   2. *Purification and Crystal Growth* 566
   3. *Device Fabrication* 569
   4. *Nuclear Spectroscopic Results* 574
IV. MATERIALS ISSUES IN CdTe AND CdZnTe 575
   1. *Purification, Precursors, and Growth of CdTe and CdZnTe* . . 575
   2. *Device Fabrication* 578
   3. *Nuclear Spectroscopic Data* 579
V. UNRESOLVED PROBLEMS AND CONCLUSIONS 580
   References 581

I. Introduction

An intense research and development effort in wide bandgap (E_g) semiconductors has been motivated by the need for lightweight, handheld, portable nuclear radiation spectrometer systems capable of operating at room temperature. A wide range of semiconductor materials are at various stages of development for this application, and many of these have been reviewed in this volume. Among these materials, HgI$_2$ and CdTe, with E_g (T = 300 K) of 2.13 and 1.5 eV, respectively, are in the most advanced state of development. In this chapter, we discuss the primary issues that still have to be addressed to improve HgI$_2$ and CdTe detector technology. By considering these materials, we will elucidate the issues facing all semiconductors that have thus far demonstrated potential as room temperature nuclear spectrometers.

This chapter will review (1) the materials requirements, (2) the state of the art, particularly in HgI$_2$ and CdTe, (3) comparison with the technology of other materials such as Cd$_{1-x}$Zn$_x$Te with x = 0.10 to 0.20, and (4) areas of future research.

II. Materials Requirements

The most important criteria upon which one bases the selection of a semiconductor material to serve as a spectrometer have been reviewed in this volume and elsewhere (Schieber, Roth, and Schnepple, 1993). These criteria include the energy gap, E_g; average atomic number, Z; charge transport properties (mobility, μ, lifetime, τ, or their products, $\mu_e\tau_e$ or $\mu_h\tau_h$, for electrons and holes respectively); the electrical resistivity, ρ; and the homogeneity of the material in terms of purity, stoichiometry, and absence of structural defects. A large E_g (>1.4 eV) will prevent thermal generation of carriers and thus ensure a low dark current. However, if E_g is greater than about 3.0 eV the energy required for electron–hole pair formation will be increased to the point of decreasing the detector efficiency. Large average Z is important since this will enhance the interaction of the incident photons with the material. As discussed in Chapter 1, the attenuation coefficient is proportional to Z, Z^2, or Z^5 for Compton, pair production, and photoelectric effects, respectively. A large value of Z will diminish the thickness L of the detector required for reasonable detector efficiency. For example, to absorb 20% of the photons at an energy of 500 keV with a Z of about 60, a thickness L of about 0.3 cm is required. Highly uniform material is critical, since this allows for the fabrication of large volume detectors that provide a large interaction volume and hence high efficiency. High resistivity is generally required (>10^7 Ωcm) so that the material can sustain a large electric field (>10^4 V/cm) and maintain a low dark current. Finally, the mobility–lifetime product should be high to allow for charge transport across the detector, as discussed in Chapter 1, with this product at least 10^{-4} cm^2/V for the slower carrier type, which is usually the holes. If this latter requirement can be met it would lead to a drift length for holes, λ, of up to 1 cm. For good charge collection one ideally would require that $\lambda \sim 10L$ (Knoll and McGregor, 1993). Table I summarizes requirements that can reasonably be achieved for this application in 1994.

In any given material two parameters of Table I, the $\mu_h\tau_h$ product and L, can be significantly increased through an improvement in the understanding of the

15. SUMMARY AND REMAINING ISSUES

TABLE I

MATERIAL REQUIREMENTS FOR ROOM TEMPERATURE
NUCLEAR SPECTROMETERS

Z	E_g (eV)	$\mu_h \tau_h$ (cm²/V)	E (V/cm)	L (cm)
>40	1.4–3.0 eV	$>10^{-4}$	$>10^4$	>1

materials science problems associated with the growth and fabrication of detectors. A high $\mu_h \tau_h$ product requires close to perfect stoichiometry and crystallinity, absence of defects, and in particular the absence of active impurities that cause trapping and decrease both electron and holes lifetimes. A large L requires homogeneous material properties through the bulk of the detector. HgI_2, CdTe, and $Cd_{0.8}Zn_{0.2}Te$ (Schieber, Hermon, and Roth, 1993; Butler et al., 1993) make up the list of the best materials for room temperature radiation spectrometers. The most extensively studied of these materials, for nuclear radiation detector applications, is HgI_2 and CdTe. In the following we consider the technological steps required to produce HgI_2 detectors and examine the types of defects each step introduces in the material. We then summarize the remaining problems facing the HgI_2 detectors. We next consider the analogous issues for CdTe and other candidate materials and also present the technological problems associated with these materials.

III. Issues in HgI_2 Detector Technology

The sequence of steps associated with the fabrication of a nuclear radiation spectrometer is shown in Fig. 1. By considering this sequence in HgI_2, we can analyze each step and identify the defects introduced into the crystal. The primary defects are crystalline imperfections, impurities, and departures from stoichiometry.

A	B	C	D	E
Precursors	Synthesis of Starting Materials	Purification Vapor-Solid Melt-Solid Solution-Solid	Crystal Growth	Cutting and Polishing
F	**G**	**H**	**I**	**J**
Electrode Deposition	Encapsulation	Attachment to a Rigid Precontacted Substrate	Contact to FET and First Stage of Amplification	Detector System

FIG. 1. Technological steps to produce radiation detectors.

1. PRECURSORS AND STARTING MATERIALS

Commercially available reagent grade HgI_2 is produced worldwide in small batches. With this as a precursor the crystal grower can never be completely certain that consistently stoichiometric material with the same impurity content is always received. In fact, measurements show that a large variation in impurity content is typically observed from one batch to another (Soria and James, 1994). On the other hand, $HgCl_2$ has many commercial applications and is produced in larger quantities. By using $HgCl_2$ one can expect a more uniform supply for the HgI_2 precursor. KI is also produced in large quantities and synthesis of HgI_2 by way of the reaction, $HgCl_2 + KI = HgI_2 + 2KCl$, leads to more uniform starting material (Lamonds, 1983; Skinner et al., 1989; van den Berg, 1989). The primary defects at this stage of the process are impurities (Muhein, Kobayashi, and Kaldis, 1983; Steinberg et al., 1989; Cross, Mroz, and Olivares, 1993), which can be as follows:

1. Anions. These particularly include halogens but also NO_3^- and SO_4^{2-}, which may stem from previous reactions of the precursors with acids.

2. Hydrocarbons. These stem from the iodine, a molecular crystal, which originates from decomposed seaweed. The hydrocarbons are intercalates in the crystal structure of iodine. Figure 2 shows the relative distribution of alkane hydrocarbons as a function of the carbon number for starting nonpurified and for a large vapor grown HgI_2 crystal. The C_{16} is the hydrocarbon most frequently distributed in the starting material (Fig. 2a), while C_{26} is the hydrocarbon most frequently distributed in HgI_2 crystals (Fig. 2b) (Steinberg et al., 1989). Thus, although purification diminishes the amount of hydrocarbons, the handling of the material during purification and crystal growth introduces new hydrocarbons previously not present. Careful analysis of hydrocarbons in the starting material and finished crystals using gas chromatography has been reported; however, to perform the analysis, the HgI_2 material was dissolved in hexane. This allows for the analysis of the hydrocarbons with $C > C_6$ but precludes the analysis of C_1 to C_6 hydrocarbons. The total oxidized carbon, which includes all hydrocarbons and graphitic carbon, has been analyzed routinely, however, the results do not show any clear correlation with detector properties (Steinberg et al., 1989). The deterioration of τ_e and τ_h has recently been demonstrated by doping experiments with many kinds of hydrocarbons, such as aliphatic aromatic and oxyhydrocarbons (Schieber et al., 1993). Only the latter two proved to be very harmful to the detector performance.

3. Metallic impurities. The most common is Fe, which has its origins in the iron vessels used to produce the precursors. Other impurities such as Cd, Cu, or Ag may have accompanied the natural minerals and ores from which Hg or I precursors are refined. Additionally, impurities such as SiO_2, Al_2O_3, B_2O_2, or CaO may be incorporated from the glass reactor vessels. Figure 3 shows a typical trace analysis of HgI_2 using plasma coupled ion atomic emission spectroscopy.

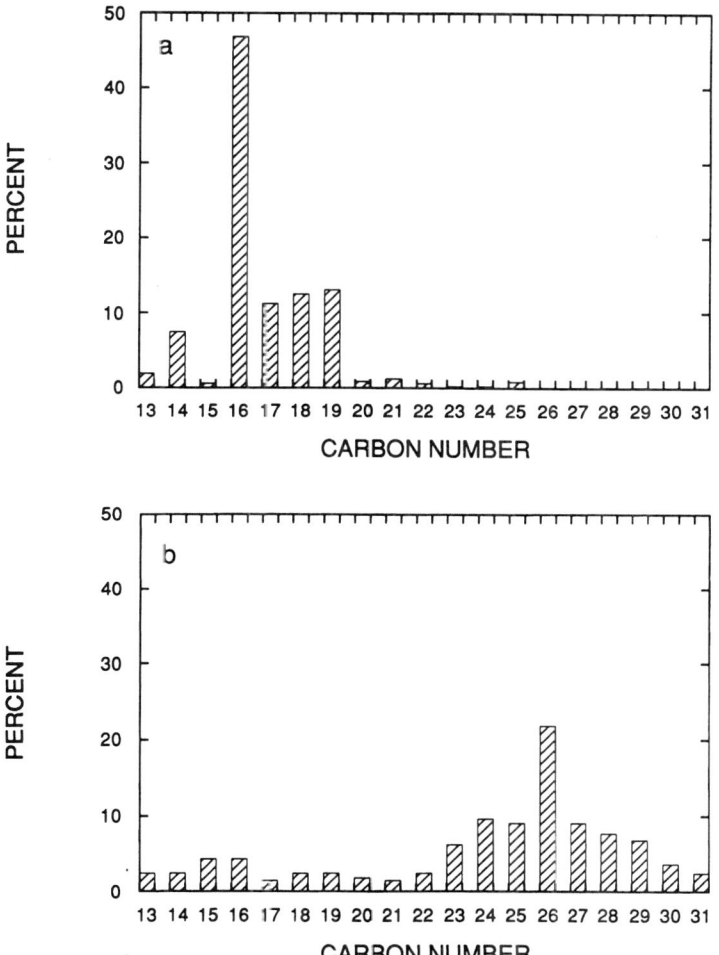

FIG. 2. Distribution of alkane hydrocarbon impurities as a function of carbon number in HgI_2 material: (a) starting material and (b) a single crystal. (Reprinted with permission from Steinberg et al., 1989 and Elsevier Science.)

One can see some of the typical impurities in the pbb range (Soria and James, 1994). Although there is some literature (Steinberg et al., 1989; Cross et al., 1993) on the chemical analysis of HgI_2, these analyses are often performed on nonpurified materials, where many impurities are present in the high ppm range (10^{-6}). In purified material or single crystals, most impurities and, particularly the metal impurities, are in the pbb (10^{-9}) range. Such analyses are extremely difficult to perform and these analysis techniques for HgI_2 are currently being developed (Soria and James, 1994). Perfecting the methods of impurity analysis in the 10^{-9}

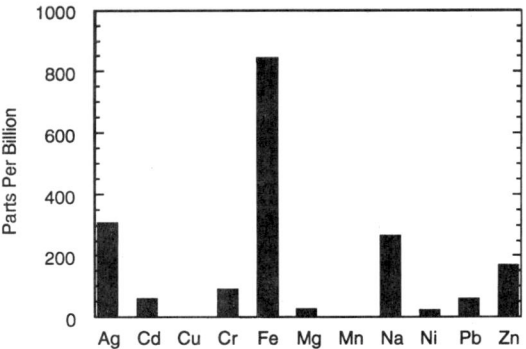

FIG. 3. Trace metal analysis of HgI_2 material using Ion-coupled plasma atomic emission spectroscopy (Soria and James, 1994).

range is one of the remaining problems to be solved in the processing technology associated with the fabrication of these nuclear spectrometers. This information, in turn, must be used to recommend alternative purification and crystal growth techniques, as well as improved handling practices.

2. Purification and Crystal Growth

The main objective of both purification and the final crystal growth processes is indeed to eliminate any impurities (Lamonds, 1983; van den Berg, 1989; Faile et al., 1980; Schieber et al., 1983). This is accomplished by a combination of steps that involve (1) solid–vapor–solid transport (sublimation) in both closed and continuously pumped open systems, (2) solid–melt–solid transport by solidification or zone refining, and (3) dissolution purification in the synthesis stage. In this last technique the precursors are dissolved and recrystallized in high purity solvents, which dissolve part of the impurities and these can then be separated by filtration. The polyethylene vapor transport process used to grow HgI_2 platelets involves sublimation of HgI_2 in the molten phase (Faile et al., 1980; Schieber et al., 1983; Burger, Roth, and Schieber, 1982; Hermon, Roth, and Schieber, 1992a). It is assumed that the mixed melt HgI_2/polyethylene dissolves organic molecules such as hydrocarbons that are left behind as a residue while the sublimed HgI_2 has a higher purity. The high temperature of $\sim 140°C$ used for sublimation and $\sim 260°C$ for melting or zone refining during the purification stages partially decompose the hydrocarbons into graphite, which is often seen as a black residue.

Crystal growth is the next and perhaps most critical step in the fabrication process. The growth temperature is limited to $\sim 110°C$ due to the solid state phase transformation at $\sim 130°C$ of red tetragonal α-HgI_2 to the yellow orthorhombic β-HgI_2 that, upon cooling, transforms destructively back to the red phase. Crystal growth methods have been reviewed by Schieber and co-workers (1983). Growth techniques include both solution growth from DMSO (Nicolau, 1980) and the more successful vapor growth method, from which large crystals weighing up to 3 kg have been grown (van den Berg, 1993). Typical large sized crystals vary between 20–2000 gram in weight and are adequate for the fabrication of gamma ray detectors. The first large crystals were grown by Scholz (1974). This method was later modified and improved by Schieber, Schnepple, and van den Berg, (1976) and further improved by van den Berg, who also grew more perfect crystals under microgravity conditions in space (van den Berg, 1993). Originally this method involved periodic reversal of the temperature gradient between the growth and source zone to allow reevaporation of imperfect layers on secondary nucleated particles. However, improved temperature control of the two zones has allowed for continuous growth without the gradient reversal. This method is still referred to, however, as the temperature oscillation method (TOM). A representation of the TOM growth apparatus was shown in this text in the chapter on growth of HgI_2. One disadvantages of the TOM is the use of ampoules sealed in vacuum, which can be used for only one growth.

A more recent development includes a two-part ampoule as shown in Fig. 4, which can be used for several growths. In Fig. 4 one can also see a movable three dimensional cold finger. This is used to select the best quality nucleus desired for further growth while allowing for reevaporation of all other undesired nuclei (Hermon, Roth, and Schieber, 1992b). In any of these variations, the crystal grows in the sealed ampoule on a glass substrate to which it adheres and from which it must

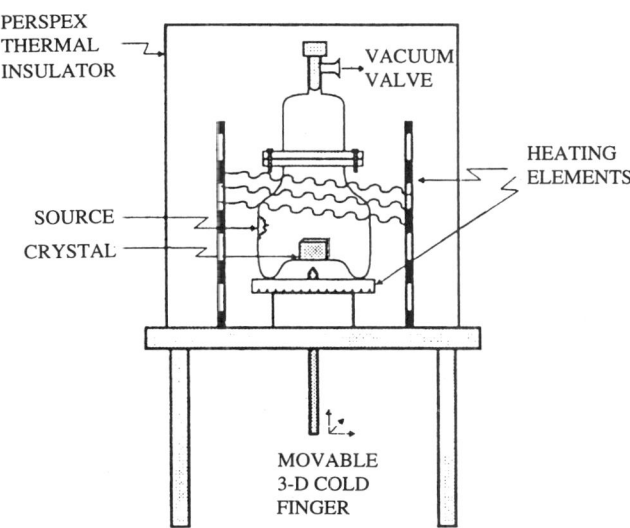

FIG. 4. An improved two part ampoule used for vapor growth of HgI_2. (Reprinted with permission from Hermon *et al.*, 1992a and Elsevier Science.)

be removed after opening the ampoule. The TOM allows for the growth of very large crystals and is currently used to produce large HgI_2 crystals at a number of sites, including EG&G/EM in Santa Barbara (van den Berg, 1993), XSIRIUS, Inc. in Santa Barbara (Skinner *et al.*, 1989), the Hebrew University of Jerusalem (Schieber *et al.*, 1993), and the Federal Institute of Technology in Zurich (Piechotka and Kaldis, 1992).

Another vapor growth method produces platelets, and this has the advantage of dispensing with the need to slice and polish the large crystals grown by TOM. This technique uses polyethylene mixed with nonpurified or purified HgI_2, which are melted together in a hot zone and then evaporated and sublimed in a temperature gradient (Faile *et al.*, 1980; Schieber *et al.*, 1983; Burger *et al.*, 1982). Many yellow platelets are formed in the temperature range above 130°C that disintegrate during the destructive phase transformation upon cooling. Below 130°C many red platelets are sublimed, some of which can produce thin high quality x-ray detectors. Typical sizes of platelets are $\sim 6 \times 4 \times 0.1$ mm^3 with some platelets as thick as ~ 0.5 mm.

An alternative approach for producing smaller sizes of crystals for x-ray spectrometers that also do not require slicing but may be cleaved or polished to reduce the thickness of the crystals has been developed by Omaly and coworkers (1983). This method is essentially similar to the platelet growth method; however, in this technique, the crystals are forced to grow in the center of a tube by a temperature gradient while the impurities with a higher vapor pressure than HgI_2 are forced to condense in the cold sink. Typical crystal sizes are $5 \times 5 \times 2$ mm^3. The disadvantage of both platelet and cold sink growth methods is that each crystal is grown at a different temperature in the gradient, and therefore the concentration of impurities or deviation from stoichiometry varies from crystal to crystal. Thus, the detector quality varies with the actual growth temperature of each individual platelet. In addition thicker platelets may include low angle grain boundaries, which affect detector quality. The yield of high quality uniform crystals is therefore small. Nevertheless, the advantages of not slicing and polishing the large TOM crystal is extremely important, and spectrometers produced from platelets grown by a modified version of this growth method have already been commercialized by TN Technologies, Inc.

One of the most important remaining issues in crystal growth technology is growth striations. These striations stem from periodic temperature fluctuations, which themselves most likely stem from the environment outside the furnace. The relatively low growth temperature ~ 110°C of HgI_2 from the vapor phase makes it more sensitive to small temperature fluctuations in the surrounding room temperature of ~ 25° compared to a crystal grown from the melt at ~ 1000°C. Therefore it is critical to employ temperature controllers that can stabilize the growth and source temperature to better than 10^{-2}°C and at the same time protect the growth environment from temperature fluctuations larger than 10^{-1}°C. This requirement for precise temperature control and stability was recognized early in the development of the growth technologies. However, even the most well-stabilized room environments have not been adequate to prevent the temperature fluctuations that cause growth striations. The striations were observable in early crystals grown in 1975 by optical microscopy (Schieber et al., 1977). Even in more recently grown crystals, growth striations are observed by x-ray topography on as grown (110) faces (Keller, Wang, and Cheng, 1993). Thus, these striations continue to exist despite the significant improvements in temperature control of both the growth process and the environment. It should be noted that crystalline perfection and homogeneity are two of the most critical parameters determining the detector quality (Schieber et al., 1989). It may be inferred that the better crystallinity of the vapor phase crystals of HgI_2 grown in space stems from the much more accurate temperature controllers specially developed for the microgravity space crystal growth experiment. Thus, better temperature controllers and the elimination of growth striations remain open problems.

3. DEVICE FABRICATION

Let us now return to Fig. 1, which shows the technological sequence associated with the fabrication of a nuclear spectrometer. After growth the crystal must be removed from the ampoule. In the case of large vapor grown crystals, they are cut with a string saw, consisting of a thread immersed in a 10% KI water solution, into blocks of the desired cross section. These blocks are then sliced into the final wafers with a thickness approaching that of the final spectrometer ~1 to 5 mm. Each wafer is then polished on a rotating metal disk covered with a synthetic polishing cloth. Each of these operations of cutting and polishing is performed with a 5–10% KI solution followed by a rinse in deionized water.

A number of technical difficulties must be overcome in these processing steps. To produce detector slices with the crystallographic c-axis perpendicular to the plane of the wafer, the crystal must be oriented. Often this is accomplished without precise x-ray diffraction or optical method determination of the c-axis. This results in misorientations of as much as 5° as shown in Fig. 5 (Goorsky and Schie-

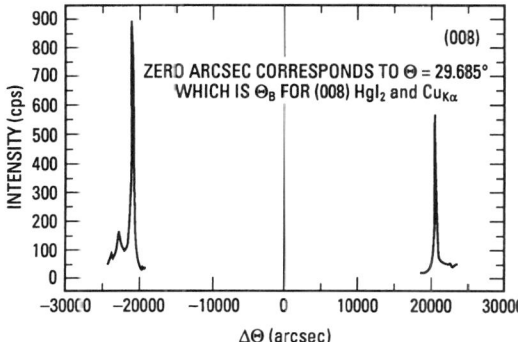

FIG. 5. X-ray diffraction scan of (008) peak of HgI_2 detector plate showing a misfit of 5.75° (Goorsky and Schieber, 1994).

ber, 1994). Figure 5 shows a double crystal x-ray diffraction measurement of a bad detector in which a 180° rotation was performed on the (008) plane and a misalignment of 5.75° can be seen from the asymmetry of the peaks. This misalignment suggests the need of a quantitative rather than "eyeball" orientation of the crystal before cutting. The other fabrication steps mentioned previously can also introduce crystal defects. For example, the removal of the crystal from the pedestal of the growth ampoule will introduce defects depending on the crystalline orientation of the basal plane of the grown crystal relative to the pedestal. This is due to the high structural anisotrophy of α-HgI_2 ($c_0/a_0 = 2.85$) and the very strong directional dependence of its thermal expansion coefficients α (Keller et al., 1993). If a crystal is grown with the c-axis, [001], vertical to the pedestal, it will rest on its a–b plane. In this case $\alpha_{11} = 11 \pm 4$ ($10^{-6}/°C$), is of the same order of magnitude as $\alpha = 3.3 \times 10^{-6}/°C$ of the Pyrex glass pedestal. However, if the crystal is grown with the c-axis parallel to the pedestal, resting on its a–c plane, there is factor of 5 difference between this α_{33} and α for the Pyrex glass. This suggest defects or strain are present in the crystal region close to the growth pedestal, and this has been documented by an increased broadening in rocking curves (Schieber et al., 1989). Mercuric iodide, in its a–b plane, appears to be a better match to evaporated Pd, $\alpha = 11 \times 10^{-6}/°C$, which is important as a contact metal. Wire sawing and etching using KI can also produce serious problems. It has been demonstrated that etching with KI solution changes the PL spectrum, whereas etching, for example, with acetone does not alter this spectrum (David et al., 1993). Furthermore, studies have shown that KI is an efficient agent in introducing unwanted impurities such as Cu or Ag (James et al., 1993; Van Scyoc et al., 1993), which have serious consequences on the performance of HgI_2 spectrometers even at levels below 1 ppm. Therefore, the identification of alternative high purity etching agents that do not diffuse or transport impurities into the detector crystal is an additional issue to be addressed.

Cleavage or string sawing still remains a controversial problem, despite evidence of an advantage of string sawing in terms of crystalline perfection (Yelon et al., 1981; Nissenbaum et al., 1983). It is clear that cutting HgI_2 crystals with a diamond saw or with a diamond impregnated wire saw heats the surface and neighboring bulk to above the temperature of the red to yellow phase transformation. Solution sawing, which does not heat the surface, does, however, produce a rough and uneven surface. This rough surface must subsequently be polished, which leads to an excessive loss of material. The ideal etchant and technique for obtaining smooth HgI_2 surfaces with low surface recombination rates has yet to be determined.

One recent investigation (Keller et al., 1993) seems to indicate that as-grown, macro-defects such as large subgrains or crystal bending extend through millimeters of material and cannot be removed by polishing or deep etching. Strongly oriented bands, which most likely are a result of changes in the rate of crystal growth, are always present in as-grown as well as sawn surfaces. Selective etching appears to occur along these bands, and it is assumed that stoichiometric variations make such chemical attack possible. String sawing strains the edges of samples but does not introduce damage to the interior. The cleaving of c-planes introduces a considerable amount of strain and can lead to macro-defects if the process is forced. Polishing appears to introduce specific, local defects but also helps to restore the sample surface to crystalline uniformity. Etching with clean solutions does little, except to clean and prepare the immediate surface for contact deposition. The combined sawing, polishing, and etching process generally leads to homogeneous surfaces with limited macro-defects but care must be taken with crystal orientation, deformation by handling, and the use of extra clean chemicals. However, to date few publications address the depth of damage due to processing.

After slicing, polishing, and etching it is assumed that all the crystalline damage is removed. At this point the metal contacts are deposited on the HgI_2 wafer. The deposition of metal electrodes is usually done by thermal evaporation or sputtering. Metals, primarily Pd, have been used successfully, although in the early 1970s simply painting a colloidal solution of carbon "aquadag" with a painter's brush onto the wafer was also employed successfully. Thin platinum or palladium wires are then attached with the colloidal carbon solution. The detector is encapsulated in paralene, a polymer that protects the detector from further evaporation of iodine, if exposed to heat or vacuum conditions. The detector is then attached to an alumina or Teflon substrate, the wires are connected to the first preamplification stage, and the spectrometer is encased in a metal enclosure with a beryllium window, through which the radiation impinges on the detector.

Alternative contacts to Pd have been studied theoretically (Cheng, 1993) and experimentally by PL and other methods (James *et al.*, 1989, 1990; Bao *et al.*, 1991, 1990a,b). These studies included a study of transparent electrodes that allow the use of HgI_2 as a photodetector in conjunction with scintillator materials (Iwanczyk *et al.*, 1983; Markakis and Cheng, 1989). The particular contacts studied included Pd, C, In, and Sn, which are the constituents of the transparent contact indium-tin oxide (ITO) as well as Cu, Ag, Au, and Pt. Among all these electrodes, Cu was found to be devastating to the detectors due to Cu diffusion. Severe interactions were also observed for Ag deposited on HgI_2. It appears that the free iodine that results from the photochemical decomposition of HgI_2 is extremely reactive. Thermodynamic data predict that many metals or semiconductors will react with this free iodine or decompose HgI_2. For example, Al reacts according to $2Al + 3HgI_2 = 2AlI_3 + 3Hg$ (Cheng, 1993). PL measurements indicate that even Pd, a noble metal, reacts with I_2 probably producing intermediary layers of PdI_2.

We now consider the issue of stoichiometry both in the as-grown and the processed crystal. Figure 6 shows the phase diagram of the $Hg-I_2$ binary system where only Hg_2I_2 is present in addition to HgI_2. Figure 7 shows the extent to which HgI_2 can be doped with Hg_2I_2 and I_2 to produce nonstoichiometric solid solutions with compositions of $HgI_2-Hg_2I_2$ or HgI_2-I_2 (Schieber *et al.*, 1993; Hermon, 1993; Hermon *et al.*, 1993). These studies, in which analyzed samples of HgI_2 had been doped with either Hg or I_2, heated, and quenched from various

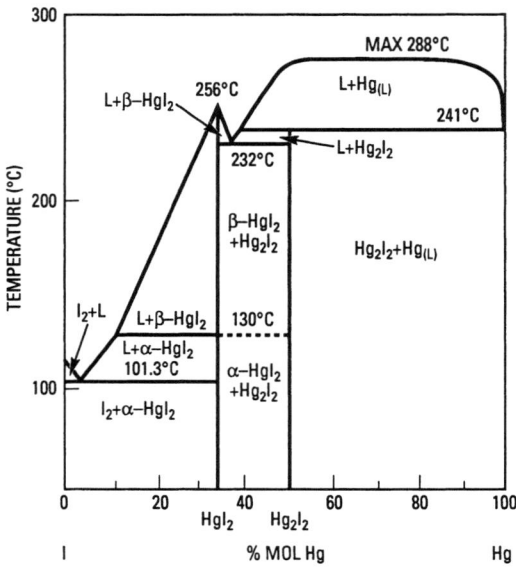

FIG. 6. Phase diagram of Hg–I. The off-stoichiometric phases of HgI$_2$ are Hg$_2$I$_2$ and I$_2$ (Hermon, 1993).

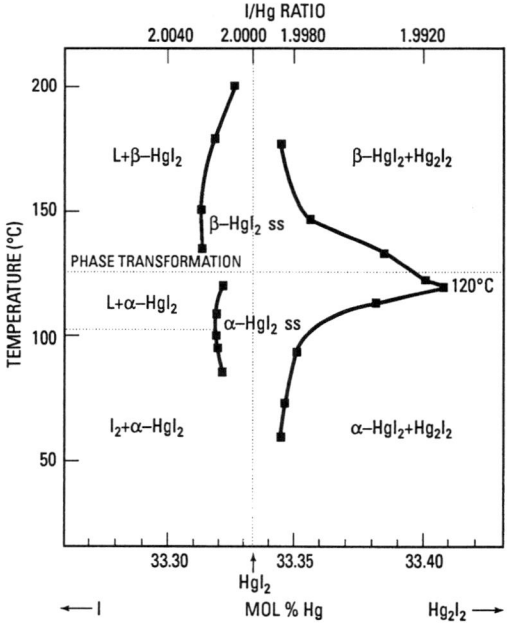

FIG. 7. Phase diagram of the near stoichiometric region of the HgI$_2$ compound, showing amounts of Hg$_2$I$_2$ and I$_2$ found by chemical analysis in HgI$_2$ samples doped, quenched from the high temperature to room temperature, and immediately analyzed (Hermon, 1993).

temperatures, showed that Hg_2I_2 can dissolve relatively large amounts of both Hg_2I_2 and I_2, forming nonstoichiometric solid solutions. HgI_2 is composed of two high vapor pressure elements. However, I_2 has the higher vapor pressure and therefore has a tendency to evaporate according to the equation $2HgI_2 = Hg_2I_2 + I_2$. The Hg_2I_2 dissolves in the HgI_2 forming Hg-rich HgI_2 and the free I_2 diffuses to the outer surface before leaving the crystal. This causes deviations from stoichiometry and associated native defects. Fortunately, Hg_2I_2 has a somewhat higher vapor pressure than HgI_2, and therefore any HgI_2 freshly crystallized from the vapor phase is very close to the ideal stoichiometry. However, crystallographic sites previously occupied by Hg^+ or excess I_2 cause defects such as vacancies, interstitials, and complex molecules. The molecules are formed by association with remaining impurities such as metals, anions, and hydrocarbons; and they may be active traps in detectors.

Departure from stoichiometry has been studied using low temperature photoluminescence spectroscopy. Merz and co-workers (1983) first studied the open tube sublimation process used to purify raw materials for vapor phase TOM crystal growth. The effect of sublimation on the 77 K photoluminescence spectra of HgI_2 powders has been discussed previously in this text (see the chapter on optical properties of HgI_2). Also discussed previously is the spectra obtained from single crystals grown from materials doped with either iodine or mercury. From these studies it was concluded that the band 2 to band 1 ratio in the photoluminescence spectra is an indicator of the relative concentration of iodine in the material, and the subsequent studies on the single crystal material appeared to agree with this hypothesis.

The effect of stoichiometry was further studied by Bao and co-workers (1992) and has also been discussed previously in this volume. The results of these experiments were found to be consistent with the results of the previous study (Merz et al., 1983) in that a low band 1 intensity relative to band 2 is an indication of iodine deficiency. The origin of the changes in the band 1 region appears to be changes in the P3/P2 ratio (these peaks dominate the band 1 region of the spectrum) with a large P3/P2 ratio observed in stoichiometric (non-iodine deficient) material. The gradual loss of iodine due to evaporation during storage in air and in vacuum has been studied by performing aging experiments and observing changes in the low temperature PL (James et al., 1990; Bao et al., 1991, 1990a,b). Aging in air causes the P3 peak to decrease and gradually disappear while P2 becomes dominant. This is exactly the reverse process from that seen in the etching study. After about 20 days, the spectrum is predominantly a featureless broad strong peak slightly shifted away from P2 to P3. From the gradual change of the spectra, it appears that this broad peak derives from P2. From a large number of measurements, it was found that typically the effect of aging can be observed in about 24 hr for a freshly etched sample stored in a desiccator. The effect of vacuum exposure on the near-bandgap photoluminescence spectra of HgI_2 is very similar to that of chemical etching. Unlike aging in air, where there seems to be a preferential loss of iodine, the vacuum exposure removes the surface layer congruently. This is probably due to the high rate of removal of the HgI_2 molecules under vacuum. The effect of vacuum exposure is thus similar to chemical etching rather than an aging effect. The removal of iodine from the outer faces of HgI_2 has also been studied by recent ellipsometric studies (Yao and John, 1993; Schieber et al., 1994). These studies reveal an aging effect pointing to departure from stoichiometry by depletion of iodine from a previously iodine rich surface. Prior to aging and immediately after storage, the surface was very smooth, while after aging it became quite rough with a surface roughness on the order of 1.5 μm after 545 hr.

One can summarize these results by concluding that the departure from stoichiometry presents a more serious problem for device fabrication than for crystal growth. In both cases, the problem can be minimized if prior to crystal growth the starting material is purified again by open tube sublimation, and if during fabrication the detector is etched adequately, contacts deposited on freshly etched surfaces and immediately encapsulated. Nevertheless, in terms of the long term stability of spectrometers, free iodine continues to be a problem. This free iodine produced by the decomposition of HgI_2 is extremely chemically active and can explain why even an inert contact material such as Pd can react with HgI_2 over long periods of times. This problem is even more serious for HgI_2 photodiodes if very thin (semitransparent) layers of Pd are used and the alternative transparent ITO contacts have indium or tin rich regions. Thus the decomposition of HgI_2, the presence of free I_2, and the need for alternative (nonreactive) contacts is a continuing issue for this material.

Finally, any of the other stages of device fabrication and handling may produce material problems. It has been observed by us that the very best detectors can be destroyed by device specialists who were not aware of the relative softness of the HgI_2 material and the ease of cleavage of the (001) planes. It should also be noted that the more pure is the material, the softer it becomes and the more susceptible to plastic deformation, thereby presenting serious handling problems during the process of device fabrication.

4. Nuclear Spectroscopic Results

Perhaps one of the best results obtained with HgI_2 detectors has been reported by Gerrish (1993) and was presented in the chapter on detector characterization of this volume. However, even this detector was only 1.7 mm thick. On the positive side it was also reported that, in 1992, about one third of all detectors tested at EG&G/EM had a resolution less than 5% (at 662 keV), were about 1 to 4 mm thick, and with the best $\mu_h \tau_h$ product of $\sim 1.4 \times 10^{-5}$ cm^2/V. However, most of the gamma ray detectors will exhibit some degree of polarization or change in the spectral response after application of bias. The change is oftentimes beneficial; i.e., it improves the detector nuclear response as was shown previously in this volume (Gerrish, 1992). While it is beyond the scope of this chapter to discuss polarization effects, one point should be made: there is a correlation between purification, crystal growth, and defects produced during detector fabrication and the detector quality. Each of the steps introduces its own specific set of defects. Understanding these defects will allow for improvements in the technology to the level desired and to an increased development of HgI_2 x-ray and gamma ray field spectrometers.

IV. Materials Issues in CdTe and CdZnTe

CdTe and CdZnTe have the advantage over HgI_2 in that they can be grown from the melt, which allows for generally much larger growth rates than those obtained from vapor phase growth. However, these materials also suffer from the fact that they decompose thermally by evaporation of Cd leaving the CdTe melt rich in Te. The Te then precipitates around grain boundaries in the crystal. The vapor pressure of Cd at the melting point of CdTe (1092°C) is 0.65 atm. whereas for Te it is only 5.0×10^{-3} atm. (Zanio, 1978). This is the origin of the problems associated with deviation from stoichiometry encountered with CdTe, and these problems are often more severe than for HgI_2. Solid solutions of ZnTe and CdTe increase the melting point of the solid solution $Cd_xZn_{1-x}Te$ (Zabdyr, 1984) compared to CdTe, since the melting point of ZnTe is about 1290°C. The vapor pressure due to thermal dissociation is smaller for (Cd, Zn)Te than for pure CdTe, thus leaving fewer (Cd, Zn) vacancies than in CdTe. With E_g for ZnTe of ~2.3 eV the bandgap of (Cd, Zn)Te is greater than that of CdTe, and this is another advantage of (Cd, Zn)Te alloys compared to CdTe. In this chapter we shall review briefly the technology of CdTe; that is, (1) precursors, purification, and crystal growth methods and potential defects associated with these processes as compared to HgI_2, (2) device fabrication, and (3) nuclear spectroscopic response. We shall treat CdTe and (Cd, Zn)Te together although the majority of the numerical data is more readily available for CdTe. Finally, (4) we shall compare the problems of HgI_2 with those of CdTe and $Cd_xZn_{1-x}Te$. The latter material with $0.05 > x < 0.20$ is an emerging new material for room temperature nuclear spectrometers and published data are growing rapidly on its properties for detector applications.

1. Purification, Precursors, and Growth of CdTe and CdZnTe

The precursor elements, Cd, Zn, and Te are commercially available today at a quoted purity better than $6N$. Therefore, the purification of starting materials is less of a problem for Cd and for Te, since these elements have been thoroughly studied and developed for the technology of infrared detectors based on (Hg, Cd)Te. Still the purity of commercially available Te is insufficient for nuclear spectrometers and may require further purification, which is usually accomplished by zone refining. The purification of Zn may also be at a less developed stage than the purification of Cd. However, when compared with HgI_2, the starting materials for CdTe are of much higher quality. Nevertheless, in view of the progress achieved for HgI_2 detectors with purer starting materials, continued improvement in starting materials would also benefit CdTe and CdZnTe radiation detectors, and further development of even higher purity precursors remains an open problem.

Crystals, such as (Cd, Zn)Te, grown from the melt have the advantage that the melt mixes the elements, which presynthesizes the precursor material. However, due to the difference in vapor pressure between Cd and Te, it is advantageous to presynthesize the precursors. This can be done in a closed sealed ampoule under continuous stirring or in a rocking furnace to avoid breaking the ampoule due to overpressure of the elements and the heat of reaction. Growth methods for CdTe have been studied over the years, and most techniques involve growth from the melt. Some sublimation methods from the vapor phase have also been used (Zanio, 1978; Zabdyr, 1984; Ebina, Saito, and Takahashi, 1973; Reno and Jones, 1992; Grasza et al., 1992). We shall review here melt growth methods that were applied to the growth of crystals used as nuclear spectrometers and detectors. It should be noted that for many years, mainly in the 1980s, CdTe was the preferred substrate crystal used for thin film infrared detectors, and therefore hundreds of millions of dollars of research were invested in its growth. The requirements placed on bulk material for application in nuclear spectrometers are much more stringent than the requirements placed on material that serves as a substrate for the growth of thin film devices. Thus, it appears that the primary benefit to the nuclear radiation detector community is the emergence of CdZnTe, also used as a substrate, as one of the leading candidate materials for room temperature nuclear spectrometers. This same observation can be made regarding GaAs substrate material when used for nuclear spectrometers.

The primary problem for CdTe melt growth is the thermal dissociation pressure at the melting point. Cd evaporates at a much higher rate than Te (Greenberg, Guskov, and Cazarev, 1992) leaving the crystal nonstoichiometric and causing the Te to precipitate in the crystal. Therefore, to overcome this problem, semiinsulating CdTe is doped with chlorine to $\sim 10^{18}/cm^3$, which compensates the Cd vacancies, though the actual defect concentration is much lower $\sim 10^6/cm^3$ (Hage-Ali and Siffert, 1992). The crystal growth methods of pure CdTe, CdTe doped with chlorine, and CdZnTe from the melt include (1) the vertical Bridgman (VB) method, (2) the traveling heating method (THM), usually with a liquid Te zone, (3) the horizontal Bridgman (HB) method, and (4) the high pressure Bridgman (HPB) method. Occasionally during growth using VB or HB, a slight overpressure of Cd is used to reduce losses of Cd. In growth by VB, HB or THM, chlorine doping is typically used.

a. Vertical Bridgman Growth

A typical vertical Bridgman technique (without Cd overpressure) used to grow CdTe and $Cd_{0.9}Zn_{0.1}Te$ and crystals doped with chlorine has been described by

Meyer and co-workers (1993). The materials are synthesized from elemental Cd and Te, 7N purity, and Zn or Se 5N purity, and the crystals doped by adding $CdCl_2$ or $TeCl_4$ up to concentrations of 2×10^{19} cm^{-3}. Doping by Cl gas can also be done in a specially designed dosage apparatus. The weight of the corresponding Cl gas quantities are condensed at LN_2 temperatures into the synthesis ampoule. Both doping techniques assume complete incorporation of the dopant. The synthesis and the crystal growth are performed in the same sealed quartz glass ampoule. After a homogenization time of 6 hr the ampoule is moved with a pulling rate of 1 mm/hr (24 mm/day). Afterwards the ampoule is cooled down with a cooling rate of 20°C/hr (Rudolph and Muhlberg, 1993).

b. Traveling Heater Method

A schematic representation of the THM furnace used for CdTe growth has been shown in the chapter on growth of CdTe in this volume. The furnace may use a radio frequency (RF) zone heater or can also have resistance zone heaters (Shoji, Onabe, and Hiratate, 1992; Ohmori, Iwase, and Ohno, 1993). It appears that the RF THM produces more uniform crystals. Chlorine doping is also used in the THM crystal growth method. The amount of chlorine added is 0.6–0.7% relative to the Te weight in CdTe. The feed crystals are usually grown by the Bridgman method. The hot zone temperature is ~750°C (much lower than 1054°C, which is the melting point of CdTe) and a typical growth rate is ~4 mm/day. Resistivity is higher than 10^9 Ω-cm, and the chlorine concentration in the final crystal is ~10^{17} cm^{-3}. THM crystals show good nuclear spectroscopic performance (Iwase et al., 1992), however major problems are the small diameter crystals and the high detector prices.

c. Horizontal Bridgman

Originally known as the gradient freeze method (Matveev et al., 1969), it was developed further by using a stationary crucible and furnace and electronically displacing the temperature gradient (Cheuvart et al., 1990). To correct the stoichiometry, a special Cd vapor feeding source adds Cd to the melt. This method is also used for the growth of (Cd, Zn)Te. The crystal grows freely and unconstrained as with closed crucibles used for ordinary VB or HB methods. Also, in the case of HB, it is necessary to presynthesize CdTe or (Cd, Zn)Te. The reaction occurs near 800°C, at the melting point of Te, but at this temperature the maximum loss of Cd takes place. Therefore, synthesis must be done by feeding Cd vapor to the molten Te, which reacts slowly to produce CdTe. Such low pressure feeding of Cd may not be enough to produce stoichiometric high resistivity detector grade CdTe or CdZnTe crystals from which large thickness detectors can be fabricated.

d. High Pressure Bridgman

The high pressure Bridgman method results in improved electrical charge transport properties compared to crystals grown by THM, and it has the advantage of allowing for large uniform crystals as obtained by Bridgman techniques. A high pressure of ~100 atm. is used. Preparation and properties of (Cd, Zn)Te (CZT) crystals grown by the HPB method along with the metallurgical and electrical properties obtained have been described elsewhere (Raiskin and Butler, 1988; Butler *et al.*, 1993a,b; Johnson *et al.*, 1993). A stoichiometric composition and a high electrical resistivity is achieved intrinsically and without the introduction of external dopants such as Cl. A schematic representation of a HPB furnace is shown in Fig. 8. The crystals of CdTe have intrinsic resistivities of $>10^9$ Ωcm while the crystals of $Cd_{0.8}Zn_{0.2}Te$ (CZT) have resistivities of $>10^{11}$ Ωcm. If ultrahigh purity starting materials are used, HPB crystals may be the best spectroscopic grade detectors prepared so far, particularly when large volume detectors are desired. It should also be noted that CZT detectors have shown the greatest rate of progress as a function of time (1989–1994) relative to CdTe (1960–1994) or HgI_2 (1973–1994) and show great promise in the near future. However, it should also be noted that much of the rapid progress made in this material may be a consequence of the lessons learned in these other material systems.

2. DEVICE FABRICATION

A boule of CdTe or CZT may contain several large crystal grains. They can be diamond sawn using an inside diameter saw, and they are etched with Br–methanol (5% Br_2) then chemically or mechanically polished and etched to remove the

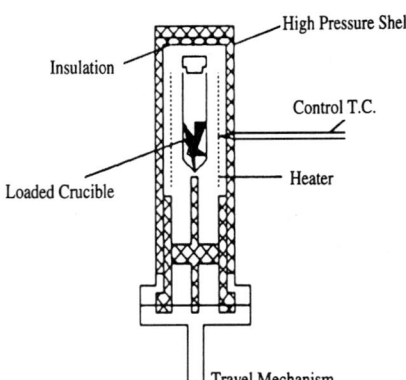

FIG. 8. Schematic diagram of high pressure Bridgman furnace used to grow CdTe and $Cd_{1-x}Zn_xTe$ crystals. (Reprinted with permission from Butler *et al.*, 1993b and Materials Research Society.)

15. SUMMARY AND REMAINING ISSUES

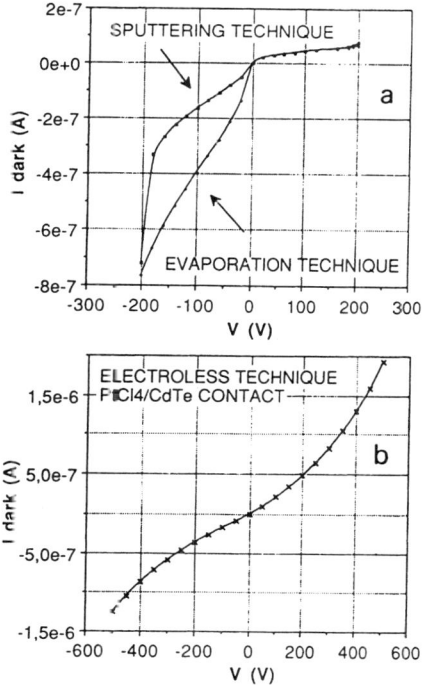

FIG. 9. Current voltage characteristics at 300 K for (a) sputtered or evaporated Au ohmic contacts on CdTe:Cl and (b) electroless $PtCl_4$ on CdTe:Cl. (Reproduced with permission from Verger et al., 1992 and Elsevier Science.)

sawing damage, which may be as much as 100–200 μm deep. Gold electrodes are then applied electrolessly (Musa and Ponpon, 1983) from a AuCl solution for ohmic contacts and then thoroughly cleaned from the chemical electrode treatment. The advantage of chemical electroless deposition of Pt ohmic electrodes over sputtered or evaporated Au electrodes is shown in Fig. 9, where the dark current vs. voltage shows a symmetrical behavior for positive and negative biases (Verger et al., 1992). Work in developing sputtered contacts is continuing also. An alternative contact is a p-i-n structure combined with Peltier cooling. This approach was employed by Khusainov and reported to achieve better spectral resolution (Khusainov, 1992). However, cooling is generally to be avoided for compact, portable, or unattended operation of nuclear detectors and spectrometers.

3. NUCLEAR SPECTROSCOPIC DATA

While the number of high resolution spectra reported at 662 keV is small, it is possible to obtain commercial CdTe detectors that provide resolutions of better than 5% from Eurorad-Strasbourg, France, and RMD, Inc. The newer detectors of

HPB–CdZnTe are very much improved in their homogeneity and are more promising for the manufacturer of radiation detectors with much greater thicknesses and, hence, greater stopping power. The resolutions are expected to be better than 5% for ^{137}Cs, which is the best quoted resolution for scintillator detectors operating at room temperature.

V. Unresolved Problems and Conclusions

The comparison between desired material properties and those obtained for efficient nuclear radiation spectrometers operating at room temperature can be summarized as follows. The primary limitation common to all large bandgap nuclear radiation detectors is homogeneity over a large detector volume. The thicknesses typically seen today are ~4 mm for HgI_2, ~1.5 mm for CdTe, and may soon reach a value of 10 mm for (Cd, Zn)Te. In other promising detector materials such as PbI_2, thicknesses of only ~0.3 mm are reported at best. It appears that the reasons limiting detector materials to small thickness are different in the different material systems. HgI_2 appears to be limited by photochemical decomposition process leading to nonstoichiometric material. The shortcomings of CdTe technology also stem from the departure from stoichiometry. In CdTe the thermal dissociation pressure of Cd at the melting point is very high. The evaporation of Cd during melt growth of CdTe cannot be corrected even by crystal growth under partial pressures of Cd of ~1 atm. The deviation from stoichiometry results in Cd vacancies and complex defects. Thus to achieve semiinsulating CdTe having a desired high resistivity of ~10^9 Ωcm, Cl doping is necessary; and this in turn causes other inhomogeneity problems due to photoionization of the defect complexes under the influence of applied electric field. Apparently this problem has been reduced in CdTe by adding Zn and growing (Cd, Zn)Te. While Zn does have a lower vapor pressure than Cd, the solution of this problem is achieved by using a high pressure of ~100 atm. during crystal growth, which reduces the partial evaporation of Cd or Zn. It is very possible that GaAs suffers from a similar set of problems as CdTe. The vapor pressure of As over GaAs during melt growth is 2 atm. (Wenzl et al., 1991). The material as grown is not stoichiometric, and to produce high resistivity, it is necessary to dope with Cr or grow under As rich conditions to produce EL2. It is very possible that crystal growth at high pressures of ~100 atm. may also improve the quality of GaAs for these applications. All other materials considered in the last decade, such as PbI_2, TlBr, CdSe, or other wide bandgap compounds, have never been studied in as much detail. For example, PbI_2 is one of the more promising materials, and it has not yet been studied for departure from stoichiometry, presence of oxyiodide, impurity content, and so on.

It therefore appears that in any candidate material investigations of crystalline perfection, impurity analysis, and departure from stoichiometry after each of the technological steps shown in Fig. 1 must be undertaken to achieve the desired levels of device performance. Characterization methods discussed in this volume for HgI_2 and CdTe detectors can be used for other materials. In-depth research to identify and understand the defects and their correlation with detector fabrication will lead to better radiation detectors. It is only through systematic programs of this sort that any material will ultimately achieve its full potential in this application, both in individual device performance, device yield, and long term stability.

REFERENCES

Bao, X. J., Schlesinger, T. E., James, R. B., Stulen, R. H., Ortale, C., and van den Berg, L. (1990a). *J. Appl. Phys.* **67**, 7265.
Bao, X. J., Schlesinger, T. E., James, R. B., Cheng, A. Y., and Ortale, C. (1990b). *Mat. Res. Soc. Symp. Proc.* **163**, 1027.
Bao, X. J., Schlesinger, T. E., James, R. B., Gentry, G. L., Cheng, A. Y., and Ortale, C. (1991). *J. Appl. Phys.* **69**, 4247.
Bao, X. J., James, R. B., Fung, C. Y., Schlesinger, T. E., Cheng, A. Y., Ortale, C., and Van den Berg, L. (1992). *X-Ray Detector Physics and Applications*, ed. R. B. Hoover, SPIE **1736**, 60.
Burger, A., Roth, M., and Schieber, M. (1982). *J. Cryst. Growth* **55**, 526.
Butler, J. F., Doty, F. B., and Apotovsky, B. (1993a). *CdTe and related Cd rich alloys*, ed. R. Triboulet, W. R. Wilcox, and O. Oda, p. 291. Mater. Res. Proc., Strasbourg.
Butler, J. F., Doty, F. P., Apotovsky, B., Friesenhahn, S. J., and Lingren, C. (1993b). *Semiconductors for Room Temperature Radiation Detector Applications,* ed. R. B. James, T. E. Schlesinger, P. Siffert, and L. Franks, **302**, p. 497. Material Research Soc., San Francisco.
Cheng, A. Y. (1993). *Semiconductors for Room Temperature Radiation Detector Applications,* ed. R. B. James, T. E. Schlesinger, P. Siffert, and L. Franks, **302**, p. 141. Material Research Soc., San Francisco.
Cheuvart, P., El-Hanani, U., Schneider, D., and Triboulet, R. (1990). *J. Cryst. Growth* **101**, 270.
Cross, E., Mroz, E., and Olivares, J. A. (1993). *Semiconductors for Room Temperature Radiation Detector Applications,* ed. R. B. James, T. E. Schlesinger, P. Siffert, and L. Franks, **302**, p. 61. Material Research Soc., San Francisco.
David, D. C., Van Scyoc, J. M., Khudatyan, M., James, R. B., Anderson, R. J., and Schlesinger, T. E. (1993). *Semiconductors for Room Temperature Radiation Detector Applications,* ed. R. B. James, T. E. Schlesinger, P. Siffert, and L. Franks, **302**, p. 147. Material Research Soc., San Francisco.
Ebina, A., Saito, K., and Takahashi, T. (1973). *J. Appl. Phys.* **44**, 3659.
Faile, P., Dabrowsky, A. J., Huth, G. C., and Iwanczyk, J. J. (1980). *J. Crys. Growth* **50**, 752.
Gerrish, V. M. (1992). *Nucl. Instr. and Meth.* **A322**, 402.
Gerrish, V. M. (1993). *Semiconductors for Room Temperature Radiation Detector Applications,* ed. R. B. James, T. E. Schlesinger, P. Siffert, and L. Franks, **302**, p. 129. Material Research Soc., San Francisco.
Goorsky, M., and Schieber M. (1994) Unpublished.
Grasza, K., Zuzga-Grasza U., Szadkowski, A., and Grodzicka, E. (1992). *J. Cryst. Growth* **123**, 519.
Greenberg, J. H., Guskov, V. N., and Cazarev, V. B. (1992). *Mater. Res. Bull.* **27**, 997.
Hage-Ali, M., and Siffert, P. (1992). *Nucl. Instr. and Meth.* **A322**, 312.
Hermon, H. (1993). Thesis Hebrew University of Jerusalem (in Hebrew).
Hermon, H., Roth, M., and Schieber, M. (1992a). *Nucl. Inst. Meth.* **A322**, 432.
Hermon, H., Roth, M., and Schieber, M. (1992b). *Nucl. Inst. Meth.* **A322**, 442.
Hermon, H., Roth, M., Schieber, M., and Shamir, J. (1993). *Bull. Mater. Soc.* **28**, 229.

Iwanczyk, J. S., Barton, J. B., Dabrowski, A. J., Kusmiss, J. H., Szymczyk, W. M., Markakis, J., Schnepple, W. F., and Lynn, R. (1983). *Nucl. Instr. Methods* **213**, 123.

Iwase, Y., Funaki, M., Onosuka, A., and Ohmori, M. (1992). *Nucl. Inst. Met.* **A322**, 628.

James, R. B., Bao, X. J., Schlesinger, T. E., Markakis, J., Cheng, J. M., and Ortale, C. (1989). *J. Appl. Phys.* **66**, 2578.

James, R. B., Bao, X. J., Schlesinger, T. E., Ortale, C., and Cheng, A. Y. (1990). *J. Appl. Phys.* **67**, 2571.

James, R. B., Bao, X. J., Schlesinger, T. E., Cheng, A. Y., and Gerrish, V. M. (1993). *Semiconductors for Room Temperature Radiation Detector Applications,* ed. R. B. James, T. E. Schlesinger, P. Siffert, and L. Franks, **302**, p. 103. Material Research Soc., San Francisco.

Johnson, C. J., Eissler, E. E., Cameron, S. E., Keng, Y., Fan, S., Jovanovic, S., and Lynn, K. G. (1993). *Semiconductors for Room Temperature Radiation Detector Applications,* ed. R. B. James, T. E. Schlesinger, P. Siffert, and L. Franks, **302**, p. 463. Material Research Soc., San Francisco.

Keller, L., Wang, E. X., and Cheng, A. Y. (1993). *Semiconductors for Room Temperature Radiation Detector Applications,* ed. R. B. James, T. E. Schlesinger, P. Siffert, and L. Franks, **302**, p. 183. Material Research Soc., San Francisco.

Khusainov, A. (1992). *Nucl. Inst. and Meth.* **A322**, 335.

Knoll, G., and McGregor, D. S. (1993). *Semiconductors for Room Temperature Radiation Detector Applications,* ed. R. B. James, T. E. Schlesinger, P. Siffert, and L. Franks, **302**, p. 3. Material Research Soc., San Francisco.

Lamonds, H. A. (1983). *Nucl. Inst. and Meth.* **213**, 5.

Markakis, J. M., and Cheng, A. Y. (1989). *Nucl. Instr. Meth. Phys. Res.* **A283**, 236.

Matveev, O. A., Prokofev, S. V., and Rud, Y. V. (1969). *Izv. Akad. Nauk USSR, Neorgan. Mater.* **5**, 1175.

Merz, J. C., Wu, Z. L., van den Berg, L., and Schnepple, W. F. (1983). *Nucl. Instr. and Meth.* **213**, 51.

Meyer, B. K., Hofmann, D. M., Stadler, W., Salk, M., Eiche, C., and Benz, W. B. (1993). *Semiconductors for Room Temperature Radiation Detector Applications,* ed. R. B. James, T. E. Schlesinger, P. Siffert, and L. Franks, **302**, p. 189. Material Research Soc., San Francisco.

Muhein, J. T., Kobayashi, T., and Kaldis, E. (1983). *Nucl. Inst. and Meth.* **213**, 39.

Musa, A., and Ponpon, J. P. (1983). *Nucl. Inst. and Meth.* **216**, 259.

Nicolau, I. F. (1980). *J. Cryst. Growth* **48**, 45.

Nissenbaum, J., Shilo, I., Burger, A., Levi, A., Schieber, M., Keller, L., and Wagner, C. N. J. (1983). *Nucl. Inst. and Meth.* **213**, 27.

Ohmori, J. M., Iwase, Y., and Ohno, R. (1993). *CdTe and related Cd rich alloys,* ed. R. Triboulet, W. R. Wilcox, and O. Oda, p. 383. Mater., Res. Soc., Strasbourg.

Omaly, J., Robert, M., Brisson, P., and Cadoret, R. (1983). *Nucl. Inst. and Meth.* **213**, 19.

Piechotka, M., and Kaldis, E. (1992). *Nucl. Inst. and Meth.* **A322**, 387.

Raiskin, E., and Butler, J. F. (1988). *IEEE Trans. on Nucl. Sci.* **NS-35**, 82.

Reno, J. L., and Jones, E. D. (1992). *Phys. Rev. B* **45**, 1449.

Rudolph, P., and Muhlberg, M. (1993). *CdTe and related Cd rich alloys,* ed. R. Triboulet, W. R. Wilcox, and O. Oda, p. 8. Mater., Res. Soc., Strasbourg.

Schieber, M., Schnepple, W. F., and van den Berg, L. (1976). *J. Cryst. Growth* **33**, 125.

Schieber, M., Beinglass, I., Dishon, G., and Holzer, A. (1977). *Current Topics in Materials Science,* ed. E. Kaldis and H. J. Scheel, **2**, p. 280. North-Holland Press, Amsterdam.

Schieber, M., Roth, M., and Schnepple, W. F. (1983). *J. Cryst. Growth* **65**, 353.

Schieber, M., Ortale, C., van den Berg, L., Schnepple, W. F., Keller, L., Wagner, C. N. J., Yelon, W., Ross, F., Georgson, G., and Milstein, F. (1989). *Nucl. Inst. and Meth.* **A283**, ITL.

Schieber, M., Hermon, H., and Roth, M. (1993). *Semiconductors for Room Temperature Radiation Detector Applications,* ed. R. B. James, T. E. Schlesinger, P. Siffert, and L. Franks, **302**, p. 347. Material Research Soc., San Francisco.

Schieber, M., Roth, M., Yao, H., DeVries, M., James, R., and Goorsky, M. (1995). *J. Cryst. Growth* **146**, 15.

Scholz, H. (1974). *ACTA Electronica* **17,** 69.
Shoji, T., Onabe, H., and Hiratate, Y. (1992). *Nucl. Inst. and Meth.* **A322,** 324.
Skinner, N. L., Ortale, C., Schieber, M M., and van den Berg, L. (1989). *Nucl. Instr. Meth. Phys. Res.* **A283,** 119.
Soria, E., and James, R. B (1994). Unpublished.
Steinberg, S., Kaplan, I., Schieber, M., Ortale, C., and Skinner, N. (1989). *Nucl. Inst. and Meth.* **A283,** 123.
Van den Berg, L. (1989). *Nucl. Inst. and Meth.* **A283,** 123.
Van den Berg, L. (1993). *Semiconductors for Room Temperature Radiation Detector Applications,* ed. R. B. James, T. E. Schlesinger, P. Siffert, and L. Franks, **302,** p. 73. Material Research Soc., San Francisco.
Van Scyoc, J. M., Schlesinger, T. E., James, R. B., Cheng, A. Y., Ortale, C., and van den Berg, L. (1993). *Semiconductors for Room Temperature Radiation Detector Applications,* ed. R. B. James, T. E. Schlesinger, P. Siffert, and L. Franks, **302,** p. 115. Material Research Soc., San Francisco.
Verger, L., Cuzin, M., Gaude, G., Glasser, F., Mathy, F., Rustirue, J., and Schaub, B. (1992). *Nucl. Inst. and Meth.* **A322** 357.
Wenzl, H., Dahlen, A., Fatah, A., Petersen, S., Mila, K., and Henkel, D. (1991). *J. Cryst. Growth* **109,** 191.
Yao, H., and John, B. (1993). *Semiconductors for Room Temperature Radiation Detector Applications,* ed. R. B. James, T. E. Schlesinger, P. Siffert, and L. Franks, **302,** p. 341. Material Research Soc., San Francisco.
Yelon, W. B., Alkire, R. W., Schieber, M. M., van den Berg, L., Rasmussen, S. E., Christensen, H., and Schneider, J. R. (1981). *J. Appl. Phys.* **52,** 4604.
Zabdyr, L. A. (1984). *J. Electrochem. Soc.* **131,** 2157.
Zanio, K. (1978). *Semiconductors and Semi-Metals Series: Cadmium Telluride.* Academic Press, New York.

Index

1/f noise, 454

A

As antisite, 402
Auger recombination, 163, 512
α particles, 280
α-HgI_2, 87
absorption, 176, 187, 497
absorption bands, 192
absorption coefficient, 176, 183, 185, 190, 193, 345
absorption cross section, 413
absorption depth, 6
absorption efficiencies, 42
absorption spectra, 200
accelerator, 471
acceptor, 33
acoustic vibrations, 194
activation energy, 135, 136, 370
active region, 420
adaptive filter, 555
adaptive gain filters, 557
aging, 210, 368
alloy, 341
alpha particles, 2
amorphous germanium, 44
amorphous layers, 473
amorphous silicon, 467, 477
amplifying FET, 71
ampoule, 92, 567
anisotropy, 131
annealing, 67, 276, 295
applications, 1, 326, 459, 559
archeological artifacts, 12
areal density, 436
array geometry, 356
arrays, 375
art, 12
astronomy, 18
astrophysics, 18, 328
atomic absorption, 261
atomic force microscopy, 99
atomic number, 4, 9, 24
attenuation coefficient, 5, 10, 298, 459, 484, 497

B

Br-methanol, 269
Bragg ionization, 425
Bridgman, 234, 294, 337, 401, 448, 576
β-HgI_2, 87
backscatter peak, 499
ballistic deficit, 502, 552
band structure, 170
bandgap, 7, 172, 177, 292, 340, 369, 392, 481
barrier, 114, 278
barrier height, 155, 522
barriers, 363
bending modes, 191
beta particles, 2
bias voltage, 507
binding energy, 35, 196
blocking contact, 116, 521
blood flow, 16
boron nitride, 238
boron phosphide, 472
breakdown, 456
breakdown voltage, 296
broadening mechanisms, 351
bulk material, 432

C

C-V measurements, 418
CCD arrays, 11
$Cd_{1-x}Zn_xTe$, 336
CdTe, 219
Compton ratio, 50
Compton edge, 498

586 INDEX

Compton scattering, 3, 4, 24, 496
Czochralski, 26, 225, 449
cadmium selenide, 474
cadmium sulfide, 473
cadmium telluride, 1
cadmium zinc telluride, 1
capacitance, 37, 283, 296, 418, 455, 457, 507, 535, 538
capacitance transient, 35
capacitance-voltage, 66
capacitor, 317
capture cross section, 309, 398
capture probability, 134
capture rate, 429
capture time, 56
carrier collection, 113
carrier collection efficiencies, 357
carrier concentration, 394
carrier diffusion, 160
carrier drift length, 7
carrier drift mobilities, 32
carrier drift velocity, 123
carrier extraction factors, 406
carrier lifetime, 150, 280, 343, 345, 386
carrier mobilities, 129, 386, 514
carrier mobility, 159
carrier transport, 32
carrier traps, 113
carrier-phonon scattering, 176
cathodoluminescence, 197, 281
cerebral blood flow, 329
charge collection, 51, 60, 275, 280, 343, 350, 494, 500, 508
charge collection efficiency, 120, 125, 157, 306, 460, 516
charge collection times, 502
charge density, 430
charge gain, 436
charge injection, 363
charge multiplication, 512
charge pulse, 3, 551
charge sensitive preamplifiers, 453
charge trapping, 500
charged particle, 23
charged radiations, 2
chemical analysis, 327, 526
chemical composition, 18
chemical etching, 138
chemical potential, 222
chloride, 271
cleaning technique, 136
cleavage, 452

cleaving, 100
coal mining, 13
coaxial detector, 28, 40, 361
collected charge, 118
collection efficiency, 9, 425, 303
color, 103
compensate, 6
compensated material, 282
compensation, 6, 246, 254, 294
complete charge collection, 404
composition, 220
computer processing, 558
conductance, 542
conduction band, 172, 250
conduction bands, 175
conductivity, 32, 433
conductivity counters, 416
congruent sublimation, 87, 224
constant current pulse, 556
constant photocurrent method, 282
contact materials, 143
contacts, 210, 270, 274, 306, 408, 520, 571
contaminants, 349
convective flow, 94
correction, 557
coulomb interaction, 4
counters, 461
counting efficiency, 412
crystal geometries, 40
crystal perfection, 101
crystal structure, 86
crystalline purity, 198
cubic crystal, 475
current, 435, 550
current density, 416
current mode, 386
current source, 538
current-voltage curves, 341, 353
cyclotron resonance, 131, 148

D

dark conductivity, 114
dark current, 10, 114, 117, 133
dark resistivity, 113
dead layer, 73, 159, 510, 525
dead layers, 77, 158
dead time, 503
decomposition, 580
deep donor, 428
deep level transient spectroscopy, 35, 519
deep levels, 37, 133, 283

deep trap, 196
defect levels, 248
defects, 225, 247, 274
degradation, 68
delta noise, 539
density of states, 392
depletion, 24, 159
depletion layer, 298
depletion region, 427
depletion width, 313
deposition, 273
detection efficiency, 412, 503
detection limit, 261
detection sensitivity, 78
detector arrays, 25
detector assemblies, 71
detector module, 375
detrapping, 160, 162, 324
detrapping times, 134
diamond, 481
dicing, 350
dielectric constant, 185, 193
dielectric relaxation time, 118
dielectrics, 541
differential scanning calorimetry, 103
differentiator, 556
diffuse, 225
diffusion, 212, 226, 241, 273, 525
digital filtering, 509
diode, 24
dislocation densities, 27
dislocations, 36, 38, 54, 102
disordered region, 66
dissolution, 273
distortion, 461
distribution coefficient, 28 31, 92
donor, 33, 226
dopants, 213
doping, 206, 274, 282
dosimetry, 16, 328
double crystal, 99
double crystal x-ray diffraction, 570
drift lengths, 61, 120
drift times, 431
drift velocity, 8, 159, 514

E

EL2, 402
edge leakage, 520
effective mass, 131, 132, 392
effective mobility, 118

efficiency, 304, 307
elastic scattering, 3
electric drift field, 451
electric field, 424, 429
electrical contacts, 154, 512
electrical properties, 113
electroabsorption, 184
electrode, 211, 468
electrodes, 10, 579
electroding, 350
electrodrift, 191
electroless contacts, 271, 295
electroless deposition, 271
electron beam, 280, 423
electron mobility, 9, 118, 128
electron spin resonance, 282
electron trapping, 52
electron-hole pairs, 4
electron-positron pair, 4
electronic noise, 47, 537
electronic readout, 453
electronics, 456, 532
electrons, 2
elemental analysis, 18
ellipsometry, 179, 267
emission coefficient, 313
emission probability, 134, 145
emission rate, 55, 429
emission spectra, 200
emission time, 59
encapsulation, 231, 276, 510
energy linearity, 69
energy resolution, 427, 532
energy spectroscopy, 39
enthalpy, 251
entropy, 251
environmental remediation, 17
epitaxial GaAs, 396
epitaxial layers, 409
equilibrium, 222
equilibrium constant, 224
equivalent circuit, 536
escape peaks, 10, 79, 499
etch, 38
etch pits, 38, 268. 346, 524
etchant, 209, 452
etched surface, 295
etching, 207, 268, 524
evaporated contacts, 73, 270
exchange splitting, 189
exciton, 196
exciton absorption, 184

exciton transition, 188, 200, 445
exothermicity, 92
explosives, 12
extended defects, 54, 524
extraction times, 405
extraordinary spectrum, 182, 188

F

Fano factor, 399, 548
FET, 72, 457, 540, 545, 547
Franz-Keldysh effect, 190, 423
Frenkel defects, 226
fabrication, 70, 207, 350, 520, 563, 569, 578
fast neutrons, 52
feedback, 543
feedback capacitor, 546
field effect transistor, 535
filtering, 532
fluctuations, 533, 539
fluorescence, 263
flux, 462
forced flux, 97
four point probe, 278
four-point-contact, 300
free exciton, 342
front contact, 72
full ionization, 428
furnaces, 236, 241

G

Gallium arsenide, 386
Gaussian shaping, 552
gain-bandwidth, 433
gallium arsenide, 469
gallium selenide, 481
gas filled detectors, 7
gating, 554
gel technique, 450
geological samples, 12
geometrical effect, 549
glow discharge, 477
grain boundaries, 38, 575
grains, 268
graphite, 231
growth, 25, 96, 232, 234, 316, 337, 446, 448, 566
growth morphology, 105
growth rates, 106, 236, 240
growth speed, 244
growth techniques, 468

H

Hall coefficient, 32
Hall effect, 228, 301
Hall mobilities, 515
Hall mobility, 126, 130
Hecht relation, 304, 515
HgI_2, 112
HPGe, 25
heavy hole, 176
high pressure, 238, 338, 578
hole mobility, 9, 128
hole tailing, 373, 476
hole trapping, 31, 38, 52
homogeneity, 225, 525, 580
horizontal zone refining, 448
hydrocarbons, 91, 191
hydroxides, 29

I

I-V characteristics, 117
I-V relationship, 417
II-VI materials, 473
III-V materials, 472
International Microgravity Laboratory, 95
IR microscopy, 266
ideality factors, 417
illumination, 435
imaging, 14, 375, 477
imaging system, 466
impedance, 419
implant, 44
impulse response, 432
impurities, 28, 192, 230, 349
impurity, 24, 260, 564
impurity concentration, 32
inclusions, 31, 348
incomplete charge collection, 535
indirect bandgap, 184
indium iodide, 481
indium phosphide, 472
induced charge, 404, 407
induced current, 532
industrial applications, 327, 330
industries, 13
infrared transmission, 192
ingot, 279
inhomogeneities, 33, 534
injecting contact, 116
injection electrode, 543
instability, 260

INDEX 589

integrator, 324, 554
interaction depths, 501
interstitials, 252
intrinsic efficiency, 6
iodine doping, 138
iodine partial pressure, 103
ion chromatography, 265
ion implantation, 276
ionic bonding, 221
ionicity, 486
ionization chamber, 531
ionization energy, 399
ionized acceptor, 68
isothermal currents, 144

J

JFET, 454
Johnson noise, 502
junction, 298
junction diode, 42

K

K escape peaks, 79
Knoop hardness, 482

L

L escape peaks, 79
Laplace's equations, 243
lapping, 315
large volume, 38, 356
layered structure, 86, 444
lead iodide, 443
leakage current, 24, 113, 275, 296, 354, 411, 469, 503, 536
lifetimes, 8, 126, 150, 157, 469, 472
light pulse, 545
line broadening, 74
linewidth broadening, 533
liquid encapsulated Czochralski, 450
liquid phase epitaxy, 401
liquidus, 220
lithium, 30
lithium diffused, 43
local density approximation, 172
low cost, 477
low frequency oscillations, 416
luminescence, 196

M

M-π-n structures, 320
M-S-M structures, 275
magnetic field, 189
magnetic resonance, 282
majority carrier, 53
mass action, 247, 249
mass action law, 224
materials, 467
medical applications, 328
melt temperature, 236
melting, 205
melting point, 221
mercuric bromo-iodide, 479
mercuric iodide, 1, 86
mercurous iodide, 87
mercury doping, 138
metal-semiconductor interface, 520
microgravity, 94
microphonic noise, 46, 502
mineral exploration, 12
miniaturization, 558
minority carrier, 53
microscopy, 183
mobilities, 131, 157
mobility, 8, 280, 316, 343, 394, 469
mobility-lifetime products, 121, 125, 343, 516, 549
$\mu\tau$ product, 9, 254, 275, 279, 303, 405, 461, 474
$\mu\tau$E, 8
multiple peaking, 410

N

Norton equivalent, 541
n-i-p structures, 295, 320
narcotics, 12
native defects, 225
neutral impurities, 132
neutrino, 472
neutron activation, 263
neutron damage, 64
neutron imaging, 472
neutrons, 2
noise, 24, 310, 339, 350, 366, 454, 457, 469, 494, 533
noise linewidth, 533, 543
noise power, 542
noise threshold, 372
noise width, 411
nonuniform trap concentrations, 63

nuclear activation, 261
nuclear physics, 25
nuclear power, 16
nuclear waste, 16
nucleation, 236

O

ohmic contacts, 7, 32, 260, 268, 414
operating voltage, 501, 508
optical absorption, 423
optical clarity, 517
optical microscopy, 98
optical phonons, 201
optimum bandgap, 386
ordinary spectrum, 181, 188
orthorhombic, 481, 566
orthorhombic phase, 87
overpressure, 237
oxidation, 28
oxide, 270
oxides, 29

P

Peltier refrigerators, 364
Pentafet, 547
Poisson statistics, 548
Poole-Frenkel effect, 114, 516
p-i-n detectors, 415
p-i-n devices, 427
p-i-n diodes, 362, 404, 477
pair generation energy, 305
pair production, 3, 24, 496
parallel noise, 454
partial compensation, 423
partial pressure, 222
peak broadening, 534
peak shift, 324
peak width, 498
peak-to-background ratio, 158, 366
peak-to-valley ratio, 214, 373
peaking time, 540
personal safety, 17
phase diagram, 220, 571
phase transformation, 566
phase transitions, 446
phonon dispersion, 194
phonon replica, 201
phonons, 194
photo-electric absorption, 3
photo-Hall, 129

photo-Hall effect, 121
photocapacitance, 148, 154
photocapacitance spectroscopy, 145
photoconductive detectors, 432
photoconductive mode, 473
photoconductive process, 8
photoconductivity, 148
photoconductors, 468, 470
photocurrent, 148, 150
photodetectors, 478, 479
photodiodes, 468, 495, 545
photoelectric absorption, 24, 496
photoelectric absorption coefficient, 126
photoelectric effect, 421
photoelectric interactions, 291
photoelectric method, 523
photoelectrons, 160
photoemission, 148, 170
photofraction, 498
photogenerated carriers, 195
photoinduced current transient, 246
photoinduced current transient spectroscopy, 145, 283
photoionization, 190
photoluminescence, 194, 196, 198, 202, 246, 281, 342, 525, 573
photoluminescence spectroscopy, 143
photomagnetoelectric effect, 121
photon counters, 293
photopeak, 10, 498
photopeak escape, 302
photoquenching, 153
photoresponse, 148, 155
photothermal ionization spectroscopy, 33
physics, 18
planar detectors, 361
platelet growth, 98
platelets, 97
point defects, 55, 56
polariton, 200
polarization, 148, 160, 175, 180, 187, 238, 253, 295, 314, 368, 458, 510
polarization effect, 9
polaron, 174
polished surfaces, 270
polishing, 267, 315
pollution monitoring, 12
polyethylene, 97
polytypes, 444
portable spectrometers, 355
postprocessing, 557
power spectral density, 537

INDEX 591

preamp, 357
preamplifier, 455, 458, 502, 532, 536
precipitates, 239, 245, 348, 575,
precursor, 564, 575
probes, 15
proton, 23
proton damage, 64
proton implantation, 415
protons, 2
pseudopotential, 172, 445
pulse shaping, 538
pulse amplitude, 8, 501
pulse filtering, 502
pulse height spectra, 351, 421
pulse mode, 386
pulse processing, 509
pulse resetting, 455
pulse shape discrimination, 322
pulse shapes, 501
pulse shaping, 39, 47, 541
pulsed laser, 514
pulsed light, 544
pulsed optical feedback, 365
pulser, 366, 457
purification, 28, 229, 244, 338, 564, 566
purity, 91, 209, 446, 494

Q

quality parameter, 319
quantum efficiencies, 7
quantum pulse mode, 403
quasi-Fermi level, 430

R

Raman, 193
RC networks, 360
radiation damage, 55, 64, 513
radiation hardness, 471
radiative recombination, 194, 202, 396
radioactive materials, 11
radiography, 14, 486
reaction kinetics, 274
reactivity, 89
reactor fuel, 327
recoil nucleus, 4
recombination, 195, 393
recombination centers, 9
recombination lifetimes, 395
rectifying contacts, 7
reflectance spectroscopy thermometry, 104

reflection, 178
reflectivity, 188
refractive index, 190, 193
refractive indices, 179
reset, 455
resistivity, 9, 245, 277, 293, 300, 340, 354, 407
resolution, 24, 33, 47, 309, 319, 399
resolution degradation, 65, 67
response mapping, 517
restralen region, 191
reverse leakage current, 418
rise times, 324
rocking curve, 100, 346

S

SCF current, 124
SCL current, 124
Schockley-Read-Hall recombination, 397
Schottky barrier, 73, 295, 407, 453
Schottky barrier contact, 8, 43
Schottky contact, 408, 414
Schottky defects, 226
Schottky effect, 114, 520
Spacelab 3, 95
safeguard, 327
satellite peaks, 410
satellites, 18
saturation region, 508
scanner, 375
scanning electron microscopy, 99
scatter cross section, 413
scattering, 514
scintillation, 18, 495
scintillator, 478, 503
secondary ion mass spectrometry, 265
security, 11
seeding, 102
segregation coefficient, 229, 261, 263, 447
self-compensation, 513
semiinsulating material, 122, 128
series resistance, 538
shallow levels, 132
shaping time, 366, 502, 550, 556
sheet resistance, 267
short circuit current, 151
shot noise, 454, 539
signal amplitude, 313
signal-to-noise ratio, 532
slicing, 350
slip, 95
solid solutions, 573, 575

solidus, 227, 250
solidus line, 294
solubility, 86, 227, 233
solution depletion, 239
solution growth, 98
solvent extraction, 91
solvents, 239
space, 328
space charge, 453
space charge free current, 122
space charge limited current, 122
space charge limited transients, 516
space crystal, 96
space program, 94
spark mass spectrography, 260
spectral resolution, 47
spectral response, 574
spectrometer, 562
spherical detectors, 319
spin, 176
sputtering, 155
stability, 9, 311
stacked detector, 553
staining, 268
statistical noise, 548
step noise, 540
stoichiometry, 88, 107, 197, 207, 221, 269, 338, 401, 526, 573, 580
stopping efficiency, 451
stopping power, 24, 460, 474, 477
stray capacitance, 502
stresses, 90
striations, 568
string-etchant saws, 452
stumatite, 244
subgrains, 570
sublimation, 88, 91, 205, 224, 480, 574
substrate, 338, 576
sum peaks, 499
supercooling, 242
surface, 208, 424
surface aging, 185
surface barrier, 270
surface behavior, 72
surface charge, 33
surface currents, 418
surface leakage, 411
surface morphology, 99
surface passivation, 44
surface preparation, 306
surface properties, 266
surface recombination, 125, 149, 156, 524
surface state density, 415
surface states, 33, 154
surface temperature, 104
synchrotron, 102
synthesis, 231, 245, 564
system noise, 368

T

TSC, 133
TSC current, 137
TSC spectra, 138
tail, 511
tail factor, 75
tailing, 253, 310
thallium bromo-iodide, 479
tape extraction, 452
temperature, 152, 370
temperature coefficient, 178
temperature control, 567
temperature dependence, 204
temperature distribution, 242
temperature fluctuations, 568
temperature oscillation method, 93
temperature response, 370
temperature-composition, 220
ternary materials, 478
testing, 456
tetragonal, 87, 170, 566
tetrode FET, 547
thallium bromide, 467, 475
thermal conductivity, 236, 449
thermal decomposition, 87
thermal diffusivity, 89
thermal dissociation, 576
thermal evaporation, 155
thermal expansion coefficients, 90
thermal gradients, 27
thermal noise, 454
thermal quenching, 152
thermal strain, 101
thermal velocity, 56, 309
thermally stimulated currents, 133, 445, 517
thermionic emission, 521
thermoelectric cooling, 321
thermostimulated current, 246, 283
thin film, 482
tight binding, 445
time of flight, 280, 514
time resolution, 433
time response, 433
tomography, 14
toxicity, 92
transferred electron, 394

transient charge techniques, 514
transit time, 8, 10, 50, 119, 121, 293, 405, 514
transmission, 203
transmission electron microscopy, 99
transmission images, 376
transport, 113, 463
trap concentration, 134
trap density, 143
trap levels, 133
trapped carriers, 154
trapped charge, 59
trapped electron, 162
trapping, 54, 160, 549
trapping center density, 293
trapping compensation, 509
trapping time, 119, 134, 135
traps, 141
traveling heater, 240
traveling heater method, 294, 577
treaty verification, 12
triaxial, 102
triaxial x-ray diffraction, 347
tumor, 15, 329
tuned optical response, 479
twin planes, 38
twins, 268, 348
two dimensional array, 376

U

uncharged radiations, 2
uniaxial stress, 188
unit cell, 86

V

van der Pauw, 32, 278, 300
vacancies, 252, 294
valence band, 392

valence bands, 175
van der Waals bonding, 444
vapor growth, 568
vapor phase, 222, 450
vapor pressure, 88, 223, 446
vapor transport, 95, 103, 449
vertical gradient freeze, 401
vertical stacking, 358
vertical zone melt, 401
vertical zone refining, 448
vibrational frequencies, 192
vibrational modes, 191
volatility, 30

W

waveform analysis, 516
white noise, 540
work function, 320

X

x-ray fluorescence, 12, 532
x-ray imager, 11
x-ray topography, 526
xeroradiography, 473

Y

yields, 496

Z

zero crossing, 323
zinc telluride, 474
zincblende, 391
zone center, 201
zone melting, 233
zone purification, 447
zone refining, 28, 91, 444, 475

Contents of Volumes in This Series

Volume 1 Physics of III–V Compounds

C. Hilsum, Some Key Features of III–V Compounds
Franco Bassani, Methods of Band Calculations Applicable to III–V Compounds
E. O. Kane, The k-p Method
V. L. Bonch-Bruevich, Effect of Heavy Doping on the Semiconductor Band Structure
Donald Long, Energy Band Structures of Mixed Crystals of III–V Compounds
Laura M. Roth and Petros N. Argyres, Magnetic Quantum Effects
S. M. Puri and T. H. Geballe, Thermomagnetic Effects in the Quantum Region
W. M. Becker, Band Characteristics near Principal Minima from Magnetoresistance
E. H. Putley, Freeze-Out Effects, Hot Electron Effects, and Submillimeter Photoconductivity in InSb
H. Weiss, Magnetoresistance
Betsy Ancker-Johnson, Plasma in Semiconductors and Semimetals

Volume 2 Physics of III–V Compounds

M. G. Holland, Thermal Conductivity
S. I. Novkova, Thermal Expansion
U. Piesbergen, Heat Capacity and Debye Temperatures
G. Giesecke, Lattice Constants
J. R. Drabble, Elastic Properties
A. U. Mac Rae and G. W. Gobeli, Low Energy Electron Diffraction Studies
Robert Lee Mieher, Nuclear Magnetic Resonance
Bernard Goldstein, Electron Paramagnetic Resonance
T. S. Moss, Photoconduction in III–V Compounds
E. Antončik and J. Tauc, Quantum Efficiency of the Internal Photoelectric Effect in InSb
G. W. Gobeli and F. G. Allen, Photoelectric Threshold and Work Function
P. S. Pershan, Nonlinear Optics in III–V Compounds
M. Gershenzon, Radiative Recombination in the III–V Compounds
Frank Stern, Stimulated Emission in Semiconductors

Volume 3 Optical of Properties III–V Compounds

Marvin Hass, Lattice Reflection
William G. Spitzer, Multiphonon Lattice Absorption
D. L. Stierwalt and R. F. Potter, Emittance Studies
H. R. Philipp and H. Ehrenveich, Ultraviolet Optical Properties
Manuel Cardona, Optical Absorption above the Fundamental Edge
Earnest J. Johnson, Absorption near the Fundamental Edge
John O. Dimmock, Introduction to the Theory of Exciton States in Semiconductors
B. Lax and J. G. Mavroides, Interband Magnetooptical Effects

H. Y. Fan, Effects of Free Carries on Optical Properties
Edward D. Palik and George B. Wright, Free-Carrier Magnetooptical Effects
Richard H. Bube, Photoelectronic Analysis
B. O. Seraphin and H. E. Bennett, Optical Constants

Volume 4 Physics of III–V Compounds

N. A. Goryunova, A. S. Borschevskii, and D. N. Tretiakov, Hardness
N. N. Sirota, Heats of Formation and Temperatures and Heats of Fusion of Compounds $A^{III}B^{V}$
Don L. Kendall, Diffusion
A. G. Chynoweth, Charge Multiplication Phenomena
Robert W. Keyes, The Effects of Hydrostatic Pressure on the Properties of III–V Semiconductors
L. W. Aukerman, Radiation Effects
N. A. Goryunova, F. P. Kesamanly, and D. N. Nasledov, Phenomena in Solid Solutions
R. T. Bate, Electrical Properties of Nonuniform Crystals

Volume 5 Infrared Detectors

Henry Levinstein, Characterization of Infrared Detectors
Paul W. Kruse, Indium Antimonide Photoconductive and Photoelectromagnetic Detectors
M. B. Prince, Narrowband Self-Filtering Detectors
Ivars Melngalis and T. C. Harman, Single-Crystal Lead-Tin Chalcogenides
Donald Long and Joseph L. Schmidt, Mercury-Cadmium Telluride and Closely Related Alloys
E. H. Putley, The Pyroelectric Detector
Norman B. Stevens, Radiation Thermopiles
R. J. Keyes and T. M. Quist, Low Level Coherent and Incoherent Detection in the Infrared
M. C. Teich, Coherent Detection in the Infrared
F. R. Arams, E. W. Sard, B. J. Peyton, and F. P. Pace, Infrared Heterodyne Detection with Gigahertz IF Response
H. S. Sommers, Jr., Macrowave-Based Photoconductive Detector
Robert Sehr and Rainer Zuleeg, Imaging and Display

Volume 6 Injection Phenomena

Murray A. Lampert and Ronald B. Schilling, Current Injection in Solids: The Regional Approximation Method
Richard Williams, Injection by Internal Photoemission
Allen M. Barnett, Current Filament Formation
R. Baron and J. W. Mayer, Double Injection in Semiconductors
W. Ruppel, The Photoconductor-Metal Contact

Volume 7 Application and Devices
PART A

John A. Copeland and Stephen Knight, Applications Utilizing Bulk Negative Resistance
F. A. Padovani, The Voltage-Current Characteristics of Metal-Semiconductor Contacts
P. L. Hower, W. W. Hooper, B. R. Cairns, R. D. Fairman, and D. A. Tremere, The GaAs Field-Effect Transistor
Marvin H. White, MOS Transistors

G. R. Antell, Gallium Arsenide Transistors
T. L. Tansley, Heterojunction Properties

PART B

T. Misawa, IMPATT Diodes
H. C. Okean, Tunnel Diodes
Robert B. Campbell and Hung-Chi Chang, Silicon Carbide Junction Devices
R. E. Enstrom, H. Kressel, and L. Krassner, High-Temperature Power Rectifiers of $GaAs_{1-x}P_x$

Volume 8 Transport and Optical Phenomena

Richard J. Stirn, Band Structure and Galvanomagnetic Effects in III–V Compounds with Indirect Band Gaps
Roland W. Ure, Jr., Thermoelectric Effects in III–V Compounds
Herbert Piller, Faraday Rotation
H. Barry Bebb and E. W. Williams, Photoluminescence I: Theory
E. W. Williams and H. Barry Bebb, Photoluminescence II: Gallium Arsenide

Volume 9 Modulation Techniques

B. O. Seraphin, Electroreflectance
R. L. Aggarwal, Modulated Interband Magnetooptics
Daniel F. Blossey and Paul Handler, Electroabsorption
Bruno Batz, Thermal and Wavelength Modulation Spectroscopy
Ivar Balslev, Piezopptical Effects
D. E. Aspnes and N. Bottka, Electric-Field Effects on the Dielectric Function of Semiconductors and Insulators

Volume 10 Transport Phenomena

R. L. Rhode, Low-Field Electron Transport
J. D. Wiley, Mobility of Holes in III–V Compounds
C. M. Wolfe and G. E. Stillman, Apparent Mobility Enhancement in Inhomogeneous Crystals
Robert L. Petersen, The Magnetophonon Effect

Volume 11 Solar Cells

Harold J. Hovel, Introduction; Carrier Collection, Spectral Response, and Photocurrent; Solar Cell Electrical Characteristics; Efficiency; Thickness; Other Solar Cell Devices; Radiation Effects; Temperature and Intensity; Solar Cell Technology

Volume 12 Infrared Detectors (II)

W. L. Eiseman, J. D. Merriam, and R. F. Potter, Operational Characteristics of Infrared Photodetectors
Peter R. Bratt, Impurity Germanium and Silicon Infrared Detectors
E. H. Putley, InSb Submillimeter Photoconductive Detectors
G. E. Stillman, C. M. Wolfe, and J. O. Dimmock, Far-Infrared Photoconductivity in High Purity GaAs

Volume 13 Cadmium Telluride

Kenneth Zanio, Materials Preparation; Physics; Defects; Applications

Volume 14 Lasers, Junctions, Transport

N. Holonyak, Jr. and M. H. Lee, Photopumped III–V Semiconductor Lasers
Henry Kressel and Jerome K. Butler, Heterojunction Laser Diodes
A. Van der Ziel, Space-Charge-Limited Solid-State Diodes
Peter J. Price, Monte Carlo Calculation of Electron Transport in Solids

Volume 15 Contacts, Junctions, Emitters

B. L. Sharma, Ohmic Contacts to III–V Compound Semiconductors
Allen Nussbaum, The Theory of Semiconducting Junctions
John S. Escher, NEA Semiconductor Photoemitters

Volume 16 Defects, (HgCd)Se, (HgCd)Te

Henry Kressel, The Effect of Cyrstal Defects on Optoelectronic Devices
C. R. Whitsett, J. G. Broerman, and C. J. Summers, Crystal Growth and Properties of $Hg_{1-x}Cd_xSe$ alloys
M. H. Weiler, Magnetooptical Properties of $Hg_{1-x}Cd_xTe$ Alloys
Paul W. Kruse and John G. Ready, Nonlinear Optical Effects in $Hg_{1-x}Cd_xTe$

Volume 17 CW Processing of Silicon and Other Semiconductors

James F. Gibbons, Beam Processing of Silicon
Arto Lietoila, Richard B. Gold, James F. Gibbons, and Lee A. Christel, Temperature Distributions and Solid Phase Reaction Rates Produced by Scanning CW Beams
Arto Leitoila and James F. Gibbons, Applications of CW Beam Processing to Ion Implanted Crystalline Silicon
N. M. Johnson, Electronic Defects in CW Transient Thermal Processed Silicon
K. F. Lee, T. J. Stultz, and James F. Gibbons, Beam Recrystallized Polycrystalline Silicon: Properties, Applications, and Techniques
T. Shibata, A. Wakita, T. W. Sigmon, and James F. Gibbons, Metal-Silicon Reactions and Silicide
Yves I. Nissim and James F. Gibbons, CW Beam Processing of Gallium Arsenide

Volume 18 Mercury Cadmium Telluride

Paul W. Kruse, The Emergence of $(Hg_{1-x}Cd_x)Te$ as a Modern Infrared Sensitive Material
H. E. Hirsch, S. C. Liang, and A. G. White, Preparation of High-Purity Cadmium, Mercury, and Tellurium
W. F. H. Micklethwaite, The Crystal Growth of Cadmium Mercury Telluride
Paul E. Petersen, Auger Recombination in Mercury Cadmium Telluride
R. M. Broudy and V. J. Mazurczyck, (HgCd)Te Photoconductive Detectors

M. B. Reine, A. K. Soad, and T. J. Tredwell, Photovoltaic Infrared Detectors
M. A. Kinch, Metal-Insulator-Semiconductor Infrared Detectors

Volume 19 Deep Levels, GaAs, Alloys, Photochemistry

G. F. Neumark and K. Kosai, Deep Levels in Wide Band-Gap III–V Semiconductors
David C. Look, The Electrical and Photoelectronic Properties of Semi-Insulating GaAs
R. F. Brebrick, Ching–Hua Su, and Pok-Kai Liao, Associated Solution Model for Ga–In–Sb and Hg–Cd–Te
Yu. Ya. Gurevich and Yu. V. Pleskon, Photoelectrochemistry of Semiconductors

Volume 20 Semi-Insulating GaAs

R. N. Thomas, H. M. Hobgood, G. W. Eldridge, D. L. Barrett, T. T. Braggins, L. B. Ta, and S. K. Wang, High-Purity LEC Growth and Direct Implantation of GaAs for Monolithic Microwave Circuits
C. A. Stolte, Ion Implantation and Materials for GaAs Integrated Circuits
C. G. Kirkpatrick, R. T. Chen, D. E. Holmes, P. M. Asbeck, K. R. Elliott, R. D. Fairman, and J. R. Oliver, LEC GaAs for Integrated Circuit Applications
J. S. Blakemore and S. Rahimi, Models for Mid-Gap Centers in Gallium Arsenide

Volume 21 Hydrogenated Amorphous Silicon
Part A

Jacques I. Pankove, Introduction
Masataka Hirose, Glow Discharge; Chemical Vapor Deposition
Yoshiyuki Uchida, dc Glow Discharge
T. D. Moustakas, Sputtering
Isao Yamada, Ionized-Cluster Beam Deposition
Bruce A. Scott, Homogeneous Chemical Vapor Deposition
Frank J. Kampas, Chemical Reactions in Plasma Deposition
Paul A. Longeway, Plasma Kinetics
Herbert A. Weakliem, Diagnostics of Silane Glow Discharges Using Probes and Mass Spectroscopy
Lester Gluttman, Relation between the Atomic and the Electronic Structures
A. Chenevas-Paule, Experiment Determination of Structure
S. Minomura, Pressure Effects on the Local Atomic Structure
David Adler, Defects and Density of Localized States

Part B

Jacques I. Pankove, Introduction
G. D. Cody, The Optical Absorption Edge of a-Si: H
Nabil M. Amer and Warren B. Jackson, Optical Properties of Defect States in a-Si: H
P. J. Zanzucchi, The Vibrational Spectra of a-Si: H
Yoshihiro Hamakawa, Electroreflectance and Electroabsorption
Jeffrey S. Lannin, Raman Scattering of Amorphous Si, Ge, and Their Alloys
R. A. Street, Luminescence in a-Si: H
Richard S. Crandall, Photoconductivity
J. Tauc, Time-Resolved Spectroscopy of Electronic Relaxation Processes
P. E. Vanier, IR-Induced Quenching and Enhancement of Photoconductivity and Photoluminescence
H. Schade, Irradiation-Induced Metastable Effects
L. Ley, Photoelectron Emission Studies

Part C

Jacques I. Pankove, Introduction
J. David Cohen, Density of States from Junction Measurements in Hydrogenated Amorphous Silicon
P. C. Taylor, Magnetic Resonance Measurements in a-Si: H
K. Morigaki, Optically Detected Magnetic Resonance
J. Dresner, Carrier Mobility in a-Si: H
T. Tiedje, information about band-Tail States from Time-of-Flight Experiments
Arnold R. Moore, Diffusion Length in Undoped a-Si: H
W. Beyer and J. Overhof, Doping Effects in a-Si: H
H. Fritzche, Electronic Properties of Surfaces in a-Si: H
C. R. Wronski, The Staebler-Wronski Effect
R. J. Nemanich, Schottky Barriers on a-Si: H
B. Abeles and T. Tiedje, Amorphous Semiconductor Superlattices

Part D

Jacques I. Pankove, Introduction
D. E. Carlson, Solar Cells
G. A. Swartz, Closed-Form Solution of I–V Characteristic for a-Si: H Solar Cells
Isamu Shimizu, Electrophotography
Sachio Ishioka, Image Pickup Tubes
P. G. LeComber and W. E. Spear, The Development of the a-Si: H Field-Effect Transistor and Its Possible Applications
D. G. Ast, a-Si: H FET-Addressed LCD Panel
S. Kaneko, Solid-State Image Sensor
Masakiyo Matsumura, Charge-Coupled Devices
M. A. Bosch, Optical Recording
A. D'Amico and G. Fortunato, Ambient Sensors
Hiroshi Kukimoto, Amorphous Light-Emitting Devices
Robert J. Phelan, Jr., Fast Detectors and Modulators
Jacques I. Pankove, Hybrid Structures
P. G. LeComber, A. E. Owen, W. E. Spear, J. Hajto, and W. K. Choi, Electronic Switching in Amorphous Silicon Junction Devices

Volume 22 Lightwave Communications Technology
Part A

Kazuo Nakajima, The Liquid-Phase Epitaxial Growth of IngaAsp
W. T. Tsang, Molecular Beam Epitaxy for III–V Compound Semiconductors
G. B. Stringfellow, Organometallic Vapor-Phase Epitaxial Growth of III–V Semiconductors
G. Beuchet, Halide and Chloride Transport Vapor-Phase Deposition of InGaAsP and GaAs
Manijeh Razeghi, Low-Pressure Metallo-Organic Chemical Vapor Deposition of $Ga_xIn_{1-x}AsP_{1-y}$ Alloys
P. M. Petroff, Defects in III–V Compound Semiconductors

Part B

J. P. van der Ziel, Mode Locking of Semiconductor Lasers
Kam Y. Lau and Ammon Yariv, High-Frequency Current Modulation of Semiconductor Injection Lasers
Charles H. Henry, Spectral Properties of Semiconductor Lasers
Yasuharu Suematsu, Katsumi Kishino, Shigehisa Arai, and Fumio Koyama, Dynamic Single-Mode Semiconductor Lasers with a Distributed Reflector
W. T. Tsang, The Cleaved-Coupled-Cavity (C^3) Laser

Part C

R. J. Nelson and N. K. Dutta, Review of InGaAsP InP laser Structures and Comparison of Their Performance
N. Chinone and M. Nakamura, Mode-Stabilized Semiconductor Lasers for 0.7–0.8- and 1.1–1.6-μm Regions
Yoshiji Horikoshi, Semiconductor Lasers with Wavelengths Exceeding 2 μm
B. A. Dean and M. Dixon, The Functional Reliability of Semiconductor Lasers as Optical Transmitters
R. H. Saul, T. P. Lee, and C. A. Burus, Light-Emitting Device Design
C. L. Zipfel, Light-Emitting Diode-Reliability
Tien Pei Lee and Tingye Li, LED-Based Multimode Lightwave Systems
Kinichiro Ogawa, Semiconductor Noise-Mode Partition Noise

Part D

Federico Capasso, The Physics of Avalanche Photodiodes
T. P. Pearsall and M. A. Pollack, Compound Semiconductor Photodiodes
Takao Kaneda, Silicon and Germanium Avalanche Photodiodes
S. R. Forrest, Sensitivity of Avalanche Photodetector Receivers for High-Bit-Rate Long-Wavelength Optical Communication Systems
J. C. Campbell, Phototransistors for Lightwave Communications

Part E

Shyh Wang, Principles and Characteristics of Integratable Active and Passive Optical Devices
Shlomo Margalit and Amnon Yariv, Integrated Electronic and Photonic Devices
Takaoki Mukai, Yoshihisa Yamamoto, and Tatsuya Kimura, Optical Amplification by Semiconductor Lasers

Volume 23 Pulsed Laser Processing of Semiconductors

R. F. Wood, C. W. White, and R. T. Young, Laser Processing of Semiconductors: An Overview
C. W. White, Segregation, Solute Trapping, and Supersaturated Alloys
G. E. Jellison, Jr., Optical and Electrical Properties of Pulsed Laser-Annealed Silicon
R. F. Wood and G. E. Jellison, Jr., Melting Model of Pulsed Laser Processing
R. F. Wood and F. W. Young, Jr., Nonequilibrium Solidification Following Pulsed Laser Melting
D. H. Lowndes and G. E. Jellison, Jr., Time-Resolved Measurements During Pulsed Laser Irradiation of Silicon
D. M. Zebner, Surface Studies of Pulsed Laser Irradiated Semiconductors
D. H. Lowndes, Pulsed Beam Processing of Gallium Arsenide
R. B. James, Pulsed CO_2 Laser Annealing of Semiconductors
R. T. Young and R. F. Wood, Applications of Pulsed Laser Processing

Volume 24 Applications of Multiquantum Wells, Selective Doping, and Superlattices

C. Weisbuch, Fundamental Properties of III–V Semiconductor Two-Dimensional Quantized Structures: The Basis for Optical and Electronic Device Applications
H. Morkoc and H. Unlu, Factors Affecting the Performance of (Al, Ga)As/GaAs and (Al, Ga)As/InGaAs Modulation-Doped Field-Effect Transistors: Microwave and Digital Applications
N. T. Linh, Two-Dimensional Electron Gas FETs: Microwave Applications
M. Abe et al., Ultra-High-Speed HEMT Integrated Circuits

D. S. Chemla, D. A. B. Miller, and P. W. Smith, Nonlinear Optical Properties of Multiple Quantum Well Structures for Optical Signal Processing
F. Capasso, Graded-Gap and Superlattice Devices by Band-Gap Engineering
W. T. Tsang, Quantum Confinement Heterostructure Semiconductor Lasers
G. C. Osbourn et al., Principles and Applications of Semiconductor Strained-Layer Superlattices

Volume 25   Diluted Magnetic Semiconductors

W. Giriat and J. K. Furdyna, Crystal Structure, Composition, and Materials Preparation of Diluted Magnetic Semiconductors
W. M. Becker, Band Structure and Optical Properties of Wide-Gap $A_{1-x}^{II}Mn_xB^{VI}$ Alloys at Zero Magnetic Field
Saul Oseroff and Pieter H. Keesom, Magnetic Properties: Macroscopic Studies
T. Giebultowicz and T. M. Holden, Neutron Scattering Studies of the Magnetic Structure and Dynamics of Diluted Magnetic Semiconductors
J. Kossut, Band Structure and Quantum Transport Phenomena in Narrow-Gap Diluted Magnetic Semiconductors
C. Riquaux, Magnetooptical Properties of Large-Gap Diluted Magnetic Semiconductors
J. A. Gaj, Magnetooptical Properties of Large-Gap Diluted Magnetic Semiconductors
J. Mycielski, Shallow Acceptors in Diluted Magnetic Semiconductors: Splitting, Boil-off, Giant Negative Magnetoresistance
A. K. Ramadas and R. Rodriquez, Raman Scattering in Diluted Magnetic Semiconductors
P. A. Wolff, Theory of Bound Magnetic Polarons in Semimagnetic Semiconductors

Volume 26   III–V Compound Semiconductors and Semiconductor Properties of Superionic Materials

Zou Yuanxi, III–V Compounds
H. V. Winston, A. T. Hunter, H. Kimura, and R. E. Lee, InAs-Alloyed GaAs Substrates for Direct Implantation
P. K. Bhattachary and S. Dhar, Deep Levels in III–V Compound Semiconductors Grown by MBE
Yu. Yu. Gurevich and A. K. Ivanov-Shits, Semiconductor Properties of Superionic Materials

Volume 27   High Conducting Quasi-One-Dimensional Organic Crystals

E. M. Conwell, Introduction to Highly Conducting Quasi-One-Dimensional Organic Crystals
I. A. Howard, A Reference Guide to the Conducting Quasi-One-Dimensional Organic Molecular Crystals
J. P. Pouquet, Structural Instabilities
E. M. Conwell, Transport Properties
C. S. Jacobsen, Optical Properties
J. C. Scott, Magnetic Properties
L. Zuppiroli, Irradiation Effects: Perfect Crystals and Real Crystals

Volume 28   Measurement of High-Speed Signals in Solid State Devices

J. Frey and D. Ioannou, Materials and Devices for High-Speed and Optoelectronic Applications
H. Schumacher and E. Strid, Electronic Wafer Probing Techniques
D. H. Auston, Picosecond Photoconductivity: High-Speed Measurements of Devices and Materials
J. A. Valdmanis, Electro-Optic Measurement Techniques for Picosecond Materials, Devices, and Integrated Circuits
J. M. Wiesenfeld and R. K. Jain, Direct Optical Probing of Integrated Circuits and High-Speed Devices

G. Plows, Electron-Beam Probing
A. M. Weiner and R. B. Marcus, Photoemissive Probing

Volume 29 Very High Speed Integrated Circuits: Gallium Arsenide LSI

M. Kuzuhara and T. Nazaki, Active Layer Formation by Ion Implantation
H. Hasimoto, Focused Ion Beam Implantation Technology
T. Nozaki and A. Higashizaka, Device Fabrication Process Technology
M. Ino and T. Takada, GaAs LSI Circuit Design
M. Hirayama, M. Ohmori, and K. Yamasaki, GaAs LSI Fabrication and Performance

Volume 30 Very High Speed Integrated Circuits: Heterostructure

H. Watanabe, T. Mizutani, and A. Usui, Fundamentals of Epitaxial Growth and Atomic Layer Epitaxy
S. Hiyamizu, Characteristics of Two-Dimensional Electron Gas in III–V Compound Heterostructures Grown by MBE
T. Nakanisi, Metalorganic Vapor Phase Epitaxy for High-Quality Active Layers
T. Nimura, High Electron Mobility Transistor and LSI Applications
T. Sugeta and T. Ishibashi, Hetero-Bipolar Transistor and LSI Application
H. Matsueda, T. Tanaka, and M. Nakamura, Optoelectronic Integrated Circuits

Volume 31 Indium Phosphide: Crystal Growth and Characterization

J. P. Farges, Growth of Discoloration-free InP
M. J. McCollum and G. E. Stillman, High Purity InP Grown by Hydride Vapor Phase Epitaxy
T. Inada and T. Fukuda, Direct Synthesis and Growth of Indium Phosphide by the Liquid Phosphorous Encapsulated Czochralski Method
O. Oda, K. Katagiri, K. Shinohara, S. Katsura, Y. Takahashi, K. Kainosho, K. Kohiro, and R. Hirano, InP Crystal Growth, Substrate Preparation and Evaluation
K. Tada, M. Tatsumi, M. Morioka, T. Araki, and T. Kawase, InP Substrates: Production and Quality Control
M. Razeghi, LP-MOCVD Growth, Characterization, and Application of InP Material
T. A. Kennedy and P. J. Lin-Chung, Stoichiometric Defects in InP

Volume 32 Strained-Layer Superlattices: Physics

T. P. Pearsall, Strained-Layer Superlattices
Fred H. Pollack, Effects of Homogeneous Strain on the Electronic and Vibrational Levels in Semiconductors
J. Y. Marzin, J. M. Gerard, P. Voisin nad J. A. Brum, Optical Studies of Strained III–V Heterolayers
R. People and S. A. Jackson, Structurally Induced States from Strain and Confinement
M. Jaros, Microscopic Phenomena in Ordered Superlattices

Volume 33 Strained-Layer Superlattices: Materials Science and Technology

R. Hull and J. C. Bean, Principles and Concepts of Strained-Layer Epitaxy
William J. Schaff, Paul J. Tasker, Mark C. Foisy, and Lester F. Eastman, Device Applications of Strained-Layer Epitaxy
S. T. Picraux, B. L. Doyle, and J. Y. Tsao, Structure and Characterization of Strained-Layer Superlattices

E. Kasper and F. Schaffler, Group IV Compounds
Dale L. Martin, Molecular Beam Epitaxy of IV–VI Compound Heterojunction
Robert L. Gunshor, Leslie A. Kolodziejski, Arto V. Nurmikko, and Nobuo Otsuka, Molecular Beam Epitaxy of II–VI Semiconductor Microstrutures

Volume 34 Hydrogen in Semiconductors

J. I. Pankove and N. M. Johnson, Introduction to Hydrogen in Semiconductors
C. H. Seager, Hydrogenation Methods
J. I. Pankove, Hydrogenation of Defects in Crystalline Silicon
J. W. Corbett, P. Deák, U. V. Desnica, and S. J. Pearton, Hydrogen Passivation of Damage Centers in Semiconductors
S. J. Pearton, Neutralization of Deep Levels in Silicon
J. I. Pankove, Neutralization of Shallow Acceptors in Silicon
N. M. Johnson, Neutralization of Donor Dopants and Formation of Hydrogen-Induced Defects in n-Type Silicon
M. Stavola and S. J. Pearton, Vibrational Spectroscopy of Hydrogen-Related Defects in Silicon
A. D. Marwick, Hydrogen in Semiconductors: Ion Beam Techniques
C. Herring and N. M. Johnson, Hydrogen Migration and Solubility in Silicon
E. E. Haller, Hydrogen-Related Phenomena in Crystalline Germanium
J. Kakalios, Hydrogen Diffusion in Amorphous Silicon
J. Chevalier, B. Clerjaud, and B. Pajot, Neutralization of Defects and Dopants in III–V Semiconductors
G. G. DeLeo and W. B. Fowler, Computational Studies of Hydrogen-Containing Complexes in Semiconductors
R. F. Kiefl and T. L. Estle, Muonium in Semiconductors
C. G. Van de Walle, Theory of Isolated Interstitial Hydrogen and Muonium in Crystalline Semiconductors

Volume 35 Nanostructured Systems

Mark Reed, Introduction
H. van Houten, C. W. J. Beenakker, and B. J. van Wees, Quantum Point Contacts
G. Timp, When Does a Wire Become an Electron Waveguide?
M. Büttiker, The Quantum Hall Effect in Open Conductors
W. Hansen, J. P. Kotthaus, and U. Merkt, Electrons in Laterally Periodic Nanostructures

Volume 36 The Spectroscopy of Semiconductors

D. Heiman, Spectroscopy of Semiconductors at Low Temperatures and High Magnetic Fields
Arto V. Nurmikko, Transient Spectroscopy by Ultrashort Laser Pulse Techniques
A. K. Ramdas and S. Rodriguez, Piezospectroscopy of Semiconductors
Orest J. Glembocki and Benjamin V. Shanabrook, Photoreflectance Spectroscopy of Microstructures
David G. Seiler, Christopher L. Littler, and Margaret H. Wiler, One- and Two-Photon Magneto-Optical Spectroscopy of InSb and $Hg_{1-x}Cd_xTe$

Volume 37 The Mechanical Properties of Semiconductors

A.-B. Chen, Arden Sher and W. T. Yost, Elastic Constants and Related Properties of Semiconductor Compounds and Their Alloys
David R. Clarke, Fracture of Silicon and Other Semiconductors

Hans Siethoff, The Plasticity of Elemental and Compound Semiconductors
Sivaraman Guruswamy, Katherine T. Faber and John P. Hirth, Mechanical Behavior of Compound Semiconductors
Subhanh Mahajan, Deformation Behavior of Compound Semiconductors
John P. Hirth, Injection of Dislocations into Strained Multilayer Structures
Don Kendall, Charles B. Fleddermann, and Kevin J. Malloy, Critical Technologies for the Micromachining of Silicon
Ikuo Matsuba and Kinji Mokuya, Processing and Semiconductor Thermoelastic Behavior

Volume 38 Imperfections in III/V Materials

Udo Scherz and Matthias Scheffler, Density-Functional Theory of sp-Bonded Defects in III/V Semiconductors
Maria Kaminska and Eicke R. Weber, EL2 Defect in GaAs
David C. Look, Defects Relevant for Compensation in Semi-Insulating GaAs
R. C. Newman, Local Vibrational Mode Spectroscopy of Defects in III/V Compounds
Andrzej M. Hennel, Transition Metals in III/V Compounds
Kevin J. Malloy and Ken Khachaturyan, DX and Related Defects in Semiconductors
V. Swaminathan and Andrew S. Jordan, Dislocations in III/V Compounds
Krzysztof W. Nauka, Deep Level Defects in the Epitaxial III/V Materials

Volume 39 Minority Carriers in III–V Semiconductors: Physics and Applications

Niloy K. Dutta, Radiative Transitions in GaAs and Other III–V Compounds
Richard K. Ahrenkiel, Minority-Carrier Lifetime in III–V Semiconductors
Tomofumi Furuta, High Field Minority Electron Transport in p-GaAs
Mark S. Lundstrom, Minority-Carrier Transport in III–V Semiconductors
Richard A. Abram, Effects of Heavy Doping and High Excitation on the Band Structure of GaAs
David Yevick and Witold Bardyszewski, An Introduction to Non-Equilibrium Many-Body Analyses of Optical Processes in III–V Semiconductors

Volume 40 Epitaxial Microstructures

E. F. Schubert, Delta-Doping of Semiconductors: Electronic, Optical, and Structural Properties of Materials and Devices
A. Gossard, M. Sundaram, and P. Hopkins, Wide Graded Potential Wells
P. Petroff, Direct Growth of Nanometer-Size Quantum Wire Superlattices
E. Kapon, Lateral Patterning of Quantum Well Heterostructures by Growth of Nonplanar Substrates
H. Temkin, D. Gershoni, and M. Parish, Optical Properties of Ga1-$_x$In$_x$As/InP Quantum Wells

Volume 41 High Speed Heterostructure Devices

F. Capasso, F. Beltram, S. Sen, A. Pahlevi, and A. Y. Cho, Quantum Electron Devices: Physics and Applications
P. Solomon, D. J. Frank, S. L. Wright, and F. Canora, GaAs-Gate Semiconductor–Insulator–Semiconductor FET
M. H. Hashemi and U. K. Mishra, Unipolar InP-Based Transistors
R. Kiehl, Complementary Heterostructure FET Integrated Circuits
T. Ishibashi, GaAs-Based and InP-Based Heterostructure Bipolar Transistors

H. C. Liu and T. C. L. G. Sollner, High-Frequency-Tunneling Devices
H. Ohnishi, T. More, M. Takatsu, K. Imamura, and N. Yokoyama, Resonant-Tunneling Hot-Electron Transistors and Circuits

Volume 42 Oxygen in Silicon

F. Shimura, Introduction to Oxygen in Silicon
W. Lin, The Incorporation of Oxygen into Silicon Crystals
T. J. Schaffner and D. K. Schroder, Characterization Techniques for Oxygen in Silicon
W. M. Bullis, Oxygen Concentration Measurement
S. M. Hu, Intrinsic Point Defects in Silicon
B. Pajot, Some Atomic Configurations of Oxygen
J. Michel and L. C. Kimerling, Electrical Properties of Oxygen in Silicon
R. C. Newman and R. Jones, Diffusion of Oxygen in Silicon
T. Y. Tan and W. J. Taylor, Mechanisms of Oxygen Precipitation: Some Quantitative Aspects
M. Schrems, Simulation of Oxygen Precipitation
K. Simino and I. Yonenaga, Oxygen Effect on Mechanical Properties
W. Bergholz, Grown-in and Process-Induced Effects
F. Shimura, Intrinsic/Internal Gettering
H. Tsuya, Oxygen Effect on Electronic Device Performance

Volume 43 Semiconductors for Room Temperature Nuclear Detector Applications

R. B. James and T. E. Schlesinger, Introduction and Overview
L. S. Darken and C. E. Cox, High-Purity Germanium Detectors
A. Burger, D. Nason, L. Van den Berg, and M. Schieber, Growth of Mercuric Iodide
X. J. Bao, T. E. Schlesinger, and R. B. James, Electrical Properties of Mercuric Iodide
X. J. Bao, R. B. James, and T. E. Schlesinger, Optical Properties of Red Mercuric Iodide
M. Hage-Ali and P. Siffert, Growth Methods of CdTe Nuclear Detector Materials
M. Hage-Ali and P. Siffert, Characterization of CdTe Nuclear Detector Materials
M. Hage-Ali and P. Siffert, CdTe Nuclear Detectors and Applications
R. B. James, T. E. Schlesinger, J. Lund, and M. Schieber, $Cd_{1-x}Zn_xTe$ Spectrometers for Gamma and X-Ray Applications
D. S. McGregor, J. E. Kammeraad, Gallium Arsenide Radiation Detectors and Spectrometers
J. C. Lund, F. Olschner, and A. Burger, Lead Iodide
M. R. Squillante, and K. S. Shah, Other Materials: Status and Prospects
V. M. Gerrish, Characterization and Quantification of Detector Performance
J. S. Iwanczyk and B. E. Patt, Electronics for X-Ray and Gamma Ray Spectrometers
M. Schieber, R. B. James, and T. E. Schlesinger, Summary and Remaining Issues for Room Temperature Radiation Spectrometers.

ISBN 0-12-752143-7